Handbook of Food Analytical Chemistry

HANDBOOK OF FOOD ANALYTICAL CHEMISTRY

PIGMENTS, COLORANTS, FLAVORS, TEXTURE, AND BIOACTIVE FOOD COMPONENTS

Edited by

Ronald E. Wrolstad
Terry E. Acree
Eric A. Decker
Michael H. Penner
David S. Reid
Steven J. Schwartz
Charles F. Shoemaker
Denise Smith
Peter Sporns

WILEY-INTERSCIENCE

A JOHN WILEY & SONS, INC., PUBLICATION

For general information on our other products and services please contact our Customer Care Department within the U.S. at 877-762-2974, outside the U.S. at 317-572-3993 or fax 317-572-4002.

Wiley also publishes its books in a variety of electronic formats. Some content that appears in print, however, may not be available in electronic format.

Library of Congress Cataloging-in-Publication Data:

Handbook of food analytical chemistry / edited By Ronald E. Wrolstad . . . [et al.].
    p. cm.
Includes bibliographical references and index.
ISBN 0-471-66378-6 Volume 1 (cloth), ISBN 0-471-71817-3 Volume 2 (cloth)
ISBN 0-471-72187-5 (set)

1. Food--Analysis--Handbooks, manuals, etc. I. Wrolstad, Ronald E.
TX545.H34 2005
664′.07--dc22

                                                                        2004013225

Printed in the United States of America

10 9 8 7 6 5 4 3 2 1

Contents

ix   **Preface**

xi   **Foreword to Current Protocols in Food Analytical Chemistry**

xii   **Contributors**

F   **PIGMENTS AND COLORANTS**   **1**

F1 **Anthocyanins / 5**

F1.1   Extraction, Isolation, and Purification of Anthocyanins / 7

F1.2   Characterization and Measurement of Anthocyanins by UV-Visible Spectroscopy / 19

F1.3   Separation and Characterization of Anthocyanins by HPLC / 33

F1.4   Characterization of Anthocyanins by NMR / 47

F2 **Carotenoids / 71**

F2.1   Extraction, Isolation, and Purification of Carotenoids / 73

F2.2   Detection and Measurement of Carotenoids by UV/VIS Spectrophotometry / 81

F2.3   Chromatographic Techniques for Carotenoid Separation / 91

F2.4   Mass Spectrometry of Carotenoids / 107

F3 **Miscellaneous Colorants / 121**

F3.1   Betalains / 123

F3.2   Spectrophotometric and Reflectance Measurements of Pigments of Cooked and Cured Meats / 131

F3.3   Measurement of Discoloration in Fresh Meat / 139

F4 **Chlorophylls / 153**

F4.1   Overview of Chlorophylls in Foods / 155

F4.2   Extraction of Photosynthetic Tissues: Chlorophylls and Carotenoids / 165

F4.3   Chlorophylls and Carotenoids: Measurement and Characterization by UV-VIS Spectroscopy / 171

F4.4   Chromatographic Separation of Chlorophylls / 179

F4.5   Mass Spectrometry of Chlorophylls / 191

F5 **Strategies for Measurement of Colors and Pigments / 201**

F5.1   Overview of Color Analysis / 203

G   **FLAVORS**   **217**

G1 **Smell Chemicals / 223**

G1.1   Direct Sampling / 225

G1.2   Isolation and Concentration of Aroma Compounds / 235

G1.3   Identification and Quantitation of Aroma Compounds / 245

       G1.4   Stereodifferentiation of Chiral Odorants Using
                High-Resolution Gas Chromatography / 257

       G1.5   Analysis of Citrus Oils / 277

       G1.6   Solid-Phase Microextraction for Flavor Analysis / 301

       G1.7   Simulation of Mouth Conditions for Flavor Analysis / 313

       G1.8   Gas Chromatography/Olfactometry / 329

G2   Acid Tastants / 341

       G2.1   Titratable Activity of Acid Tastants / 343

       G2.2   Liquid Chromatography of Nonvolatile Acids / 351

H   TEXTURE/RHEOLOGY   363

H1   Viscosity of Liquids, Solutions, and Fine Suspensions / 367

       H1.1   Overview of Viscosity and Its Characterization / 369

       H1.2   Measuring the Viscosity of Non-Newtonian Fluids / 375

       H1.3   Viscosity Determination of Pure Liquids, Solutions, and Serums Using Capillary Viscometry / 385

       H1.4   Measuring Consistency of Juices and Pastes / 391

H2   Compressive Measurements of Solids and Semi-Solids / 395

       H2.1   General Compressive Measurements / 397

       H2.2   Textural Measurements with Special Fixtures / 405

       H2.3   Texture Profile Analysis / 417

H3   Viscoelasticity of Suspensions and Gels / 425

       H3.1   Dynamic or Oscillatory Testing of Complex Fluids / 427

       H3.2   Measurement of Gel Rheology: Dynamic Tests / 439

       H3.3   Creep and Stress Relaxation: Step-Change Experiments / 449

I   Bioactive Food Components   457

       I1   Polyphenolics / 461

       I1.1   Determination of Total Phenolics / 463

       I1.2   Extraction and Isolation of Polyphenolics / 471

       I1.3   HPLC Separation of Polyphenolics / 483

       I1.4   Proanthocyanidins: Extraction, Purification, and Determination of Subunit Composition by HPLC / 499

       I1.5   Identification of Flavonol Glycosides Using MALDI-MS / 511

       I1.6   Analysis of Isoflavones in Soy Foods / 519

APPENDICES AND INDEXES

Appendices / 537

       A.1   Abbreviations and Useful Data / 539

         A   Abbreviations Used in This Manual / 539

A.2   Laboratory Stock Solutions, Equipment, and Guidelines / 543

    A   Common Buffers and Stock Solutions / 543

    B   Laboratory Safety / 551

    C   Standard Laboratory Equipment / 553

A.3   Commonly Used Techniques / 555

    A   Introduction to Mass Spectrometry for Food Chemistry / 555

Suppliers Appendix / 563

Index / 597

PREFACE

Accurate and state-of-the-art analysis of food composition is of interest and concern to a divergent clientele including research workers in academic, government and industrial settings, regulatory scientists, analysts in private commercial laboratories, and quality control professionals in small and large companies. Some methods are empirical, some commodity specific, and many have been widely accepted as standard methods for years. Others are at the cutting edge of new analytical methodology and are rapidly changing. A common denominator within this diverse group of methods is the desire for detailed descriptions of how to carry out analytical procedures. A frustration of many authors and readers of peer-reviewed journals is the brevity of most Materials and Methods sections. There is editorial pressure to minimize description of experimental details and eliminate advisory comments. When one needs to undertake an analytical procedure with which one is unfamiliar, it is prudent to communicate first-hand with one experienced with the methodology. This may require a personal visit to another laboratory and/or electronic or phone communication with someone who has expertise in the procedure. An objective of *the Handbook of Food Analytical Chemistry* is to provide exactly this kind of detailed information which personal contact would provide. Authors are instructed to present the kind of details and advisory comments they would give to a graduate student or technician who has competent laboratory skills and who has come to them to learn how to carry out an analytical procedure for which the author has expertise.

Some basic food analytical methods such as determination of °brix, pH, titratable acidity, total proteins and total lipids are basic to food analysis and grounded in procedures which have had wide-spread acceptance for a long time. Others such as analysis of cell-wall polysaccharides, analysis of aroma volatiles, and compressive measurement of solids and semi-solids, require use of advanced chemical and physical methods and sophisticated instrumentation. In organizing *the Handbook of Food Analytical Chemistry* we chose to categorize on a disciplinary rather than a commodity basis. Included are chapters on water, proteins, enzymes, lipids, carbohydrates, colors, flavors texture/rheology and bioactive food components. We have made an effort to select methods that are applicable to all commodities. However, it is impossible to address the unique and special criteria required for analysis of all commodities and all processed forms. There are several professional and trade organizations which focus on their specific commodities, e.g., cereals, wines, lipids, fisheries, and meats. Their methods manuals and professional journals should be consulted, particularly for specialized, commodity-specific analyses.

This two-volume handbook is derived from another John Wiley & Sons publication, *Current Protocol in Food Analytical Chemistry*. That manual was published from January 2001–December 2003 in loose-leaf and CD-Rom format. That design permitted addition of new and revised units on a quarterly basis. The two-year compilation of these units makes for a very complete reference on food analytical methods.

FOREWORD TO CURRENT PROTOCOLS IN FOOD ANALYTICAL CHEMISTRY

Accurate, precise, sensitive, and rapid analytical determinations are as essential in food science and technology as in chemistry, biochemistry, and other physical and biological sciences. In many cases, the same methodologies are used. How does one, especially a young scientist, select the best methods to use? A review of original publications in a given field indicates that some methods are cited repeatedly by many noted researchers and analysts, but with some modifications adapting them to the specific material analyzed. Official analytical methods have been adopted by some professional societies, such as the Official Methods of Analysis (Association of Official Analytical Chemists), Official Methods and Recommendation Practices (American Oil Chemists' Society), and Official Methods of Analysis (American Association of Cereal Chemists).

The objective of *Current Protocols in Food Analytical Chemistry* is to provide the type of detailed instructions and comments that an expert would pass on to a competent technician or graduate student who needs to learn and use an unfamiliar analytical procedure, but one that is routine in the lab of an expert or in the field.

What factors can be used to predetermine the quality and utility of a method? An analyst must consider the following questions: Do I need a proximate analytical method that will determine all the protein, or carbohydrate, or lipid, or nucleic acid in a biological material? Or do I need to determine one specific chemical compound among the thousands of compounds found in a food? Do I need to determine one or more physical properties of a food? How do I obtain a representative sample? What size sample should I collect? How do I store my samples until analysis? What is the precision (reproducibility) and accuracy of the method or what other compounds and conditions could interfere with the analysis? How do I determine whether the results are correct, as well as the precision and accuracy of a method? How do I know that my standard curves are correct? What blanks, controls and internal standards must be used? How do I convert instrumental values (such as absorbance) to molar concentrations? How many times should I repeat the analysis? And how do I report my results with appropriate standard deviation and to the correct number of significant digits? Is a rate of change method (i.e., velocity as in enzymatic assays) or a static method (independent of time) needed?

Current Protocols in Food Analytical Chemistry will provide answers to these questions. Analytical instrumentation has evolved very rapidly during the last 20 years as physicists, chemists, and engineers have invented highly sensitive spectrophotometers, polarometers, balances, etc. Chemical analyses can now be made using milligram, microgram, nanogram, or picogram amounts of materials within a few minutes, rather than previously when grams or kilograms of materials were required by multistep methods requiring hours or days of preparation and analysis. *Current Protocols in Food Analytical Chemistry* provides state-of-the-art methods to take advantage of the major advances in sensitivity, precision, and accuracy of current instrumentation.

How do chemical analyses of foods differ from analyses used in chemistry, biochemistry and biology? The same methods and techniques are often used; only the purpose of the analysis may differ. But foods are to be used by people. Therefore, methodology to determine safety (presence of dangerous microbes, pesticides, and toxicants), acceptability (flavor, odor, color, texture), and nutritional quality (essential vitamins, minerals, amino acids, and lipids) are essential analyses. *Current Protocols in Food Analytical Chemistry* is designed to meet all these requirements.

John Whitaker
Davis, California

CONTRIBUTORS

Terry E. Acree
Cornell University
Geneva, New York

Ozlem Akpinar
Oregon State University
Corvallis, Oregon

Øyvind M. Andersen
University of Bergen
Bergen, Norway

Tom Berkelman
Amersham Pharmacia Biotech
San Francisco, California

Hugues Brevard
Nestle Research Center
Lausanne, Switzerland

Zvonko Burkus
University of Alberta
Edmonton, Canada

Claus Buschmann
Universitaet Karlsruhe
Karlsruhe, Germany

Pavinee Chinachoti
University of Massachusetts
Amherst, Massachusetts

Mary G. Chisholm
Behrend College, The Pennsylvania
    State University
Erie, Pennsylvania

Daren Cornforth
Utah State University
Logan, Utah

John Coupland
Pennsylvania State University
University Park, Pennsylvania

Neal E. Craft
Craft Technologies
Wilson, North Carolina

Susan L. Cuppett
University of Nebraska-Lincoln
Lincoln, Nebraska

Kathryn D. Deibler
Cornell University
Geneva, New York

Robert W. Durst
Oregon State University
Corvallis, Oregon

Wayne Ellefson
Covance Laboratories
Madison, Wisconsin

Cameron Faustman
University of Connecticut
Storrs, Connecticut

Mario G. Feruzzi
Ohio State University
Columbus, Ohio

E. Allen Foegeding
North Carolina State University
Raleigh, North Carolina

Anthony J. Fontana
Decagon Devices
Pullman, Washington

Torgils Fossen
University of Bergen
Bergen, Norway

Eric Fournier
University of Alberta
Alberta, Canada

Jane E. Friedrich
Cargill Incorporated
Minneapolis, Minnesota

Suzanne Frison
University of Alberta
Edmonton, Alberta, Canada

Sean Gallagher
Motorola, Inc.
Tempe, Arizona

Fernando García-Carreño
Centro de Investigaciones Biológicas
    (CIBNOR)
La Paz, Mexico

Harold W. Gardner
National Center for Agricultural
    Utilization Research, ARS, USDA
Peoria, Illinois

Trevor Gentry
Cornell University
Ithaca, New York

M. Mónica Giusti
University of Maryland
College Park, Maryland

N. Guizani
College of Agriculture, Sultan
    Qaboos University
Muscat, Sultanate of Oman

Sandra Harper
The Wistar Institute
Philadelphia, Pennsylvania

R.W. Hartel
University of Wisconsin
Madison, Wisconsin

M.L. Herrera
University of Wisconsin
Madison, Wisconsin

R. Hoover
Memorial University of
    Newfoundland
St. John's, Canada

Montana Camara Hurtado
Universidad Complutense de Madrid
Madrid, Spain

Shinya Ikeda
Osaka City University
Osaka, Japan

James A. Kennedy
Oregon State University
Corvallis, Oregon

Dae-Ok Kim
Cornell University
Geneva, New York

Sasithorn Kongruang
Oregon State University
Corvallis, Oregon

Magnus M. Kristjansson
University of Iceland
Reykjavik, Iceland

Randall I. Krohn
Pierce Chemical
Rockford, Illinois

Peter H. Krygsman
Bruker Ltd.
Milton, Canada

Theodore P. Labuza
University of Minnesota
St. Paul, Minnesota

Duane K. Larick
North Carolina State University
Raleigh, North Carolina

Chang Y. Lee
Cornell University
Geneva, New York

P.P. Lewicki
Warsaw Agricultural University
 (SGGW)
Warsaw, Poland

Yong Li
Purdue University
West Lafayette, Indiana

Hartmut K. Lichtenthaler
Universitaet Karlsruhe
Karlsruhe, Germany

Kevin Loughrey
Gretag Macbeth
South Deerfield, Massachusetts

George Lunn
Baltimore, Maryland

D. Julian McClements
University of Massachusetts
Amherst, Massachusetts

Richard E. McDonald
Food and Drug Administration
College Park, Maryland

Laurence D. Melton
University of Auckland
Auckland, New Zealand

Christian Milo
Nestlé Research Center
Lausanne, Switzerland

Yoshi Mochizuki
University of California
Davis, California

Magdi M. Mossoba
Food and Drug Administration
College Park, Maryland

Shuryo Nakai
University of British Columbia
Vancouver, British Columbia,
Canada

M. Angeles Navarrete del Toro
Centro de Investigaciones Biológicas
 (CIBNOR)
La Paz, Mexico

Toshiaki Ohshima
Tokyo University of Fisheries
Tokyo, Japan

Roger H. Pain
Jozef Stefan Institute
Ljubljana, Slovenia

James D. Parker
North Carolina State University
Raleigh, North Carolina

Kirk L. Parkin
University of Wisconsin
Madison, Wisconsin

Ronald B. Pegg
University of Saskatchewan
Saskatoon, Canada

Michael H. Penner
Oregon State University
Corvallis, Oregon

Amy Phillips
University of Connecticut
Storrs, Connecticut

Oscar A. Pike
Brigham Young University
Provo, Utah

Praphan Pinsirodom
University of Wisconsin
Madison, Wisconsin

M. Shafiur Rahman
Sultan Qaboos University
Muscat, Sultanate of Oman

Barbara Rasco
Washington State University
Pullman, Washington

W. S. Ratnayake
Memorial University of
 Newfoundland
St. John's, Canada

Joe M. Regenstein
Cornell University
Ithaca, New York

David S. Reid
University of California at Davis
Davis, California

Jody Renner-Nantz
University of California
Davis, Cailfornia

Khee C. Rhee
Texas A&M University
College Station, Texas

Deborah Roberts
Nestle Research Center
Lausanne, Switzerland

Gustavo A. Rodriguez
Prodemex
Los Mochis, Mexico

Luis E. Rodriguez-Saona
University of Maryland and Joint
 Institute for Food Safety and
 Applied Nutrition
Washington, D. C.

Michael D.H. Rogers
University of Guelph
Guelph, Canada

Rennie P. Ruiz
Hunt-Wesson, Inc.
Fullerton, California

Shyam S. Sablani
Sultan Qaboos University
Muscat, Sultanate of Oman

Steven J. Schwartz
Ohio State University
Columbus, Ohio

R.K. Scopes
LaTrobe University
Bundoora, Australia

K. John Scott
Institute of Food Research
Colney, United Kingdom

Fereidoon Shahidi
Memorial University of
 Newfoundland
St. John's, Canada

Michael H. Simonian
Beckman Coulter
Fullerton, California

Bronwen G. Smith
University of Auckland
Auckland, New Zealand

Denise M. Smith
University of Idaho
Moscow, Idaho

David W. Speicher
The Wistar Institute
Philadelphia, Pennsylvania

Peter Sporns
University of Alberta
Edmonton, Alberta, Canada

Feral Temelli
University of Alberta
Edmonton, Canada

Marvin A. Tung
University of Guelph
Guelph, Canada

Richard B. van Breemen
University of Illinois at Chicago
Chicago, Illinois

Saskia van Ruth
University College Cork
Cork, Ireland

Thava Vasanthan
University of Alberta
Edmonton, Canada

Joachim H. von Elbe
University of Wisconsin
Madison, Wisconsin

John R. L. Walker
University of Canterbury
Christchurch, New Zealand

Andrew L. Waterhouse
University of California, Davis
Davis, California

Bruce A. Watkins
Purdue University
West Lafayette, Indiana

Jochen Weiss
University of Tennessee
Knoxville, Tennessee

Peter Whittingstall
ConAgra Grocery Products
Irvine, California

Alan Williams
Amersham Pharmacia Biotech
Piscataway, New Jersey

Ronald E. Wrolstad,
Oregon State University
Corvallis, Oregon

Yu Chu Zhang
Ohio State University
Columbus, Ohio

Shengying Zhou
The Minute Maid Company
Apopka, Florida

F   PIGMENTS AND COLORANTS

INTRODUCTION

F1   Anthocyanins

F1.1   Extraction, Isolation, and Purification of Anthocyanins

F1.2   Characterization and Measurement of Anthocyanins by UV-Visible Spectroscopy

F1.3   Separation and Characterization of Anthocyanins by HPLC

F1.4   Characterization of Anthocyanins by NMR

F2   Carotenoids

F2.1   Extraction, Isolation, and Purification of Carotenoids

F2.2   Detection and Measurement of Carotenoids by UV/VIS Spectrophotometry

F2.3   Chromatographic Techniques for Carotenoid Separation

F2.4   Mass Spectrometry of Carotenoids

F3   Miscellaneous Colorants

F3.1   Betalains

F3.2   Spectrophotometric and Reflectance Measurements of Pigments of Cooked and Cured Meats

F3.3   Measurement of Discoloration in Fresh Meat

F4 Chlorophylls

F4.1 Overview of Chlorophylls in Foods

F4.2 Extraction of Photosynthetic Tissues: Chlorophylls and Carotenoids

F4.3 Chlorophylls and Carotenoids: Measurement and Characterization by
UV-VIS Spectroscopy

F4.4 Chromatographic Separation of Chlorophylls

F4.5 Mass Spectrometry of Chlorophylls

F5 Strategies for Measurement of Colors and Pigments

F5.1 Overview of Color Analysis

SECTION F
Pigments and Colorants

INTRODUCTION

Color is one of the most important quality attributes for consumer acceptance of foods. The initial impression of the quality and acceptability of a food product is judged on the basis of its visual appearance. Thus, the naturally occurring food pigments, which are the primary components absorbing visible light energy, provide coloration and represent important quality constituents of foods. In foods, there exist predominately five major classes of naturally occurring pigments. Within the plant kingdom, the most abundant pigments are the lipid-soluble chlorophylls and carotenoids. These pigments are ubiquitous in nature because of their prominent role in photosynthesis and photoprotection, and are thus found in all higher plants. The carotenoid pigments are often masked by the more dominant green chlorophyll pigments except during autumn, when the yellow-orange color of carotenoids become evident. Within the flavonoid group of phenolic compounds, are the water-soluble anthocyanins. These pigments are one of the most broadly distributed pigment groups in the plant world, responsible for the blue, purple, red, and orange colors of many plants including most fruits and berries. Plants containing betalains, another group of water-soluble pigments, have both red and yellow colors. These pigments are mostly limited to foods containing beet root, beet products, or beet juice concentrates. In animal and meat products, it is the heme pigments that are responsible for the reddish colors. A variety of different heme pigments can result, depending on whether the product is fresh, cured, or cooked meat. Additionally, the synthetic pigments, FD&C colorants, cannot be ignored, because a number of these certified dyes are used in many food products to impart or enhance appearance.

Measurement of both natural and synthetic pigments in foods presents an analytical challenge to food chemists. The diversity and complexity of naturally occurring pigments and their derivatives, and potential formation of pigmented decomposition products that contribute to food color, complicate analytical measurements. Most naturally occurring pigments are labile and may degrade into other colored components, especially during thermal treatment of foods. Often, pigments are compartmentalized or localized within plant or animal tissues. Difficulties may be encountered in liberating and extracting the pigments for analysis. Despite these obstacles, many excellent methodologies and techniques have been developed specifically for extraction, separation, identification, and quantitative measurement of pigments and colorants.

This section provides useful methods that are applicable to the measurement of colorants in foods. Chapter F1 describes techniques for measurement of anthocyanins. In *UNIT F1.1*, methods for extracting, isolating, and purifying anthocyanins are given. Special attention is given to the extraction conditions (solvent, pH, temperature) because of the need to prevent decomposition of these labile pigments. *UNIT F1.2* covers characterization and quantitation of the anthocyanins by ultraviolet-visible spectrophotometry—the standard technique for quantifying almost all colorants when solubilized in solvents. In *UNIT F1.3* high-performance liquid chromatographic (HPLC) separations are described, which are most appropriate for qualitative and quantitative measurements of individual pigments present in complex mixtures. In *UNIT F1.4*, anthocyanins are analyzed by NMR spectroscopy. This unit describes sample preparation, gives detailed information about the applications of various NMR techniques, and provides useful NMR data for reference.

Pigments and Colorants

A similar approach is considered for measurement of the carotenoid pigments (Chapter F2). Since carotenoids are antioxidants, care must be taken to prevent oxidation during extraction, isolation, and purification. These protocols are described in *UNIT F2.1*. Ultraviolet-visible spectroscopy, described in *UNIT F2.2*, is the method of choice for quantitation of pure carotenoids dissolved in organic solvents. Comparison to authentic standards or literature spectra is then employed for identification purposes. HPLC chromatographic methods to separate, quantify, and identify individual carotenoids are today the most common technique for analysis of most carotenoid mixtures or extracts of foods or biological tissues. Since many carotenoid pigments possess biological activity as provitamin A nutrients, HPLC analysis is of critical importance to measure nutritional content of plant foods or products with added carotenoids, such as beta carotene. Several of these methods are provided in *UNIT F2.3*. More sophisticated techniques, such as mass spectroscopy (MS), are becoming routine in analytical laboratories. *UNIT F2.4* describes useful methods for carotenoid analysis using MS techniques as well as coupled LC-MS methods, which are very powerful and sensitive techniques.

Chapter F3 covers analytical protocols for miscellaneous colorants. In *UNIT F3.1*, protocols applicable to the betalains are described. *UNIT F3.2* contains protocols for the heme pigments in cured meat products, and *UNIT F3.3* contains protocols for heme pigments in fresh meat products. Chapter F4 covers chlorophyll analysis. Given the variety of chlorophylls that may be found in fresh and processed foods, the chapter begins with an overview describing these compounds (*UNIT F4.1*). *UNIT F4.2* contains protocols on appropriate methods for extraction, isolation, and purification of chlorophylls from foods and plant tissues, and *UNIT F4.3* discusses UV/VIS spectrophotometric techniques for characterization of chlorophylls and their derivatives. If the equipment is available, HPLC mehods of separation are the procedures of choice to analyze both polar and nonpolar chlorophyll derivatives. Practical methods are summarized in *UNIT F4.4*. HPLC-MS techniques, which are described in *UNIT F4.5*, assist in the positive identification of these pigments. Additional chapters are planned for protocols describing the determination of synthetic colorants in foods; an overview of color analysis of foods is presented in *UNIT F5.1*.

Steven J. Schwartz

Anthocyanins

F1.1   Extraction, Isolation, and Purification of Anthocyanins   **F1.1.1**
Basic Protocol 1: Acetone Extraction and Chloroform Partition of Anthocyanins F1.1.1
Alternate Protocol: Methanol Extraction of Anthocyanins   F1.1.3
Support Protocol: Sample Preparation for Anthocyanin Purification   F1.1.4
Basic Protocol 2: Anthocyanin Purification   F1.1.4
Commentary   F1.1.6

F1.2   Characterization and Measurement of Anthocyanins by UV-Visible
Spectroscopy   **F1.2.1**
Basic Protocol 1: Total Monomeric Anthocyanin by the pH-Differential Method F1.2.1
Basic Protocol 2: Indices for Pigment Degradation, Polymeric Color, and
    Browning   F1.2.7
Reagents and Solutions   F1.2.8
Commentary   F1.2.8

F1.3   Separation and Characterization of Anthocyanins by HPLC   **F1.3.1**
Basic Protocol 1: Sample Preparation of Anthocyanins and Their HPLC
    Separation on Silica C_{18} Columns   F1.3.2
Alternate Protocol: Sample Preparation of Acylated Anthocyanins and Their
    Separation on Polymeric C_{18} Columns   F1.3.3
Basic Protocol 2: Anthocyanidins: Preparation and HPLC   F1.3.6
Basic Protocol 3: Saponification of Acylated Anthocyanins and Their HPLC
    Separation   F1.3.8
Commentary   F1.3.9

F1.4   Characterization of Anthocyanins by NMR   **F1.4.1**
Basic Protocol: Recording NMR Spectra for Assignments
    of Proton and Carbon Signals of Anthocyanins   F1.4.1
Support Protocol 1: Preparation of an Anthocyanin NMR Sample   F1.4.10
Support Protocol 2: Recording of the 1-D ^{1}H NMR Spectrum
    and Initial Preparations for 2-D Experiments   F1.4.11
Commentary   F1.4.12

Contents

1

Extraction, Isolation, and Purification of Anthocyanins

Anthocyanins are the flavonoid compounds that produce plant colors ranging from orange and red to various shades of blue and purple. This unit describes methods for extraction, isolation, and purification of anthocyanin pigments from plant tissues. These methods are essential laboratory operations prior to subsequent experimental work involving separation, characterization, and quantitation of the pigments (UNITS F1.2 & F1.3). The polar character of the anthocyanin molecule allows for its solubility in many different solvents such as alcohols, acetone, dimethyl sulfoxide, and water. The choice of extraction method should maximize pigment recovery with a minimal amount of adjuncts and minimal degradation or alteration of the natural state. Basic Protocol 1 describes the extraction of anthocyanins with acetone and their partition with chloroform. This procedure permits concentration of anthocyanin pigments in the aqueous phase while removing lipids, chlorophylls, and other water-insoluble compounds. The Alternate Protocol describes the extraction of anthocyanins with acidified methanolic solutions. Methanol is the most commonly used solvent for anthocyanin extraction because its low boiling point allows for rapid concentration of the extracted material. However, the resultant extract contains low-polarity contaminants and further purification may be necessary. Basic Protocol 2 describes a simple, fast, and effective method for purification of anthocyanins from polyphenolic compounds, sugars, and organic acids using solid-phase adsorption. The Support Protocol describes a method for preparing a finely powdered sample using liquid nitrogen. This produces a uniform composite sample with a high surface area, which allows for efficient pigment extraction.

ACETONE EXTRACTION AND CHLOROFORM PARTITION OF ANTHOCYANINS

In this method, acetone extracts the anthocyanins from the plant material, and chloroform partitioning further isolates and partially purifies the pigments. The addition of chloroform results in phase separation between the aqueous portion (which contains the anthocyanin, phenolics, sugars, organic acids, and other water-soluble compounds) and the bulk phase (which contains the immiscible organic solvents, lipids, carotenoids, chlorophyll pigments, and other nonpolar compounds). This method has the advantage of producing an extract with no lipophilic contaminants. The absence of a concentration step minimizes the risk of acid-dependent pigment degradation.

Materials

    Powdered plant material (see Support Protocol), frozen
    Acetone
    70% (v/v) aqueous acetone *or* aqueous acidified acetone: 70% aqueous acetone
        with 0.01% HCl
    Chloroform
    Acidified water: 0.01% (v/v) HCl in deionized, distilled water
    Waring Blender with stainless steel container (Waring) *or* general-purpose
        homogenizer

    Whatman no. 1 filter paper
    Buchner funnel
    Separatory funnel
    500-ml boiling flask
    Rotary evaporator with vacuum pump or water aspirator, 40°C

Contributed by Luis E. Rodriguez-Saona and Ronald E. Wrolstad

CAUTION: Chloroform is a toxic irritant and mild carcinogen; take suitable precautions (*APPENDIX 2B*) and work in a fume hood.

1. Mix ~50 g powdered plant material (accurately weighed and recorded) 1:1 (w/v) with acetone using a Waring Blender with stainless steel container or a general-purpose homogenizer.

   For most materials, a 1:1 ratio of sample to solvent should be used. For materials rich in pectic substances, however, a higher proportion of acetone may be required, e.g., 1:1.4 or 1:2.

   For dried samples or materials with high sugar content, the sample needs to be dispersed in water before extracting with acetone. A general guildine is to suspend dried sample in the amount of water that would be present in the fresh tissue. If a blender is used, it should be explosion proof.

   Alternatively, the plant material and acetone can be mixed with a chemical-resistant stir-bar.

2. Separate the anthocyanin extract (filtrate) from insoluble plant material by filtering the slurry through a Whatman no. 1 filter paper by vacuum suction using a Buchner funnel.

3. Reextract plant material with 70% (v/v) aqueous acetone until a clear or faintly colored solution is obtained. If plant material has a pH ≥4, use aqueous acidified acetone. Pool filtrates and discard plant material.

   Three subsequent extractions should be sufficient.

   The use of acidified acetone ensures that the aqueous fraction will be at a low pH where the anthocyanins are more stable, and that the chloroform/acetone solvents will be in an acidic environment.

   In the authors' experience, the pigments of a wide variety of plant materials have been stable in the aqueous acetone extract. However, problems have been encountered with pigment degradation for certain potato cultivars with high polyphenol oxidase activity. This problem can be circumvented by putting the aqueous acetone extract (uncapped) in a boiling water bath for 5 min. After heating, the volume of acetone lost by evaporation should be replenished. This enzyme inactivation step has been found unnecessary for most materials.

4. Transfer filtrate to a separatory funnel, add 2 vol chloroform, and gently mix by turning funnel upside down a few times. Store sample overnight at 4°C or until a clear partition between the two phases is obtained.

   CAUTION: *Chloroform/acetone is explosive if it comes in contact with strong alkali. Waste solvent should be stored in a labeled container strictly used for disposal of these solvents, and not mixed with other materials. Disposal is conducted according to implemented safety guidelines of the institution. If there is a possibility of the waste turning alkaline, some hydrochloric acid should be added to ensure it remains acidic.*

   For small sample sizes, a 1:1 ratio of filtrate to chloroform can be placed in a screw-cap test tube, mixed, and centrifuged to separate the phases. Rapid and clean phase separation is produced with this method.

5. Transfer the aqueous phase (upper portion) to a 500-ml boiling flask. Remove residual acetone/chloroform in a rotary evaporator at 40°C under vacuum.

   The presence of anthocyanin pigments in the aqueous phase is evidenced by the pink to red, purple, or blue color of the solution.

The evaporating flask should be less than one-half full for efficient solvent removal. Generally, the flask volume should be four to five times that of the solvent volume. Prolonged evaporation time should be avoided to minimize pigment degradation. Evaporation should be complete in 5 to 10 min.

6. Make up remaining aqueous extract to a known volume (usually 100 ml) with acidified deionized distilled water. If the sample is to be analyzed within 2 days, store extract at 4°C. For longer periods (up to 1 year or even longer), store at −18°C. Avoid repeated freezing and thawing.

METHANOL EXTRACTION OF ANTHOCYANINS

ALTERNATE PROTOCOL

This is the classical method of extracting anthocyanins from plant materials. This procedure involves maceration or soaking of the plant material in methanol containing a small concentration of mineral acid (e.g., HCl). Methanol extraction is a rapid, easy, and efficient method for anthocyanin extraction. However, a crude aqueous extract with several contaminants is obtained, and methanol evaporation can result in hydrolysis of labile acyl linkages, which is aggravated by the presence of HCl.

Additional Materials (also see Basic Protocol 1)

    Acidified methanol: 0.01% (v/v) HCl in methanol

1. Homogenize 50 g powdered plant material (accurately weighed and recorded) in 2 vol (w/v) acidified methanol. Allow it to macerate 1 hr.

   One hour should be sufficient time for anthocyanin extraction because of the high surface area of the powdered material. Materials are often allowed to extract overnight under refrigerated conditions, particularly when materials have been ground directly in the blender with acidified methanol and not previously powdered.

2. Filter slurry through a Whatman no. 1 filter paper by vacuum suction using a Buchner funnel.

3. Reextract plant material with acidified methanol until a faint-colored extract is obtained. Pool filtrates and discard plant material.

   Three subsequent extractions should be sufficient.

4. Transfer filtrates to a boiling flask and evaporate methanol in a rotary evaporator at 40°C under vacuum.

   The evaporating flask should be less than one-half full for efficient solvent removal. Prolonged evaporation time should be avoided to minimize pigment degradation. A marked reduction in the rate of evaporation as well as an apparent increase in viscosity in indicative that the residual liquid is mostly water.

   If pigment isolate is to be analyzed in an aqueous system, the extract should not be taken to dryness. For some analytical applications, the sample should be taken to dryness and then dissolved in methanol or other appropriate solvent. In the latter concentration stages, azeotropic removal of water can be facilitated by addition of methanol.

5. Make up remaining aqueous extract to a known volume with acidified deionized distilled water, water, methanol, or other appropriate solvent. If the sample is to be analyzed within 2 days, store extract at 4°C. For longer periods (up to 1 year or even longer), store at −18°C. Avoid repeated freezing and thawing.

   Extract should be made up in acidified water or water if continuing with Basic Protocol 2.

SAMPLE PREPARATION FOR ANTHOCYANIN PURIFICATION

The plant material is frozen with liquid nitrogen and powdered using a Waring Blender or mill suitable for use under extremely low temperatures. The use of liquid nitrogen minimizes anthocyanin degradation by lowering the temperature and providing a nitrogen environment. The fine powder maximizes pigment recoveries due to its high surface area and favors disruption of cellular compartments.

Materials

Liquid nitrogen
Plant material

Waring Blender with stainless steel container (Waring), M 20 IKA-Universal Mill (IKA Works), or equivalent
Teflon or stainless steel container (or equivalent container able to withstand −210°C)
Freeze-resistant container, high-density polyethylene (HDPE) or equivalent

1. Pour liquid nitrogen into a dry stainless steel Waring Blender container or M 20 IKA-Universal Mill grinding chamber and allow to evaporate. If operating a blender, turn on for few seconds. Repeat this procedure until container is well chilled.

    CAUTION: *Wear suitable protective goggles and gloves when pouring liquid nitrogen.*

2. Freeze 50 g plant material with liquid nitrogen in a separate container of Teflon, stainless steel, or other material that can withstand temperatures of −210°C.

    For most plant materials, a representative sample size of 50 g should allow for a reliable estimate of the anthocyanin content; however, if the amount of material is limited, the method can be applied to a minimum sample of 1 g.

    Most berries and seeds can be frozen as is, but other plant materials of larger size should be reduced in particle size to ~2 cm³ with a stainless steel knife.

3. Blend frozen plant material using the chilled container (step 1).

4. Open blender or mill, add liquid nitrogen, allow to evaporate, and repeat grinding process until a fine powder is obtained.

    CAUTION: *The blender or mill lid should not be tightly closed until liquid nitrogen has evaporated completely to avoid explosive separation of the lid. Lids can be modified with a venting stainless steel tube or chimney to avoid this hazard.*

5. Transfer powdered material to a freeze-resistant container for analysis, or store it immediately at −20°C.

    Some powdered materials are hygroscopic, making accurate weighing (see Basic Protocol 1 or Alternate Protocol) difficult. For quantitative analyses, it may be advisable to weigh fresh material (before freezing), and quantitatively transfer powdered material using an appropriate solvent.

ANTHOCYANIN PURIFICATION

Purification of anthocyanin-containing extracts is often necessary, as the solvent systems commonly used for extraction are not specific for anthocyanins. Considerable amounts of accompanying materials may be extracted and concentrated in the colored extracts, which can influence the stability and/or analysis of these pigments (Jackman and Smith, 1996). Anthocyanin purification using solid-phase extraction (Figure F1.1.1) permits the removal of several interfering compounds present in the crude extracts. Mini-columns containing C_{18} chains bonded on silica retain hydrophobic organic compounds (e.g.,

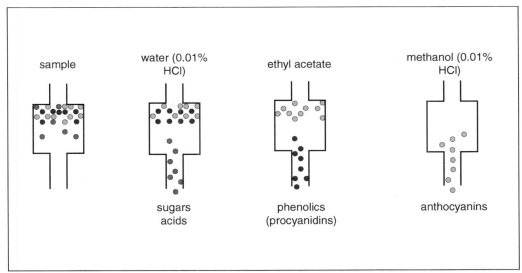

Figure F1.1.1 Solid-phase (C18) purification of anthocyanins. The sample components (represented by differentially shaded circles) are resolved by subsequent wash steps as indicated. The last wash, with acidified methanol, elutes anthocyanins. Acidified methanol and/or water should be used as solvents if electrospray mass spectrometry is to be carried out subsequently.

anthocyanins, phenolics), while allowing matrix interferences such as sugars and acids to pass through to waste. Washing the retained pigments with ethyl acetate will further remove phenolic compounds other than anthocyanins.

Materials

Methanol
Acidified water: 0.01% (v/v) HCl in deionized, distilled water
Aqueous anthocyanin extract (see Basic Protocol 1; see Alternate Protocol)
Ethyl acetate
Acidified methanol: 0.01% (v/v) HCl in methanol

C18 cartridge (with C_{18} sorbent bonded on silica: C_{18} Sep-Pak Cartidge (360 mg sorbent), Waters Chromotography; ODS-4 Octadecyl Silane (500 mg sorbent), Whatman; *or* equivalent)
50- to 100-ml boiling flask
Rotary evaporator with vacuum pump or water aspirator, 40°C
Freeze-resistant container (optional)

1. Condition a C18 cartridge by passing two column volumes methanol through the sorbent bed.

2. Pass three column volumes acidified deionized distilled water through cartridge to remove remaining methanol.

3. Force an aqueous anthocyanin extract through cartridge.

   The solid-phase extractant will become colored. Stop loading sample if excessive color is passing through the cartridge (see Critical Parameters and Troubleshooting, discussion of anthocyanin purification).

   The volume of extract applied to the cartridge will depend mainly on sample quantity, its anthocyanin content, and the amount of sorbent packing in the column. Usually a sample volume of 5 to 10 ml is used for a C18 solid-phase extraction cartridge containing 360 mg sorbent.

4. Wash cartridge with two column volumes acidified water to remove compounds not adsorbed (e.g., sugars, acids).

5. Wash cartridge with two column volumes ethyl acetate to remove polyphenolic compounds such as phenolic acids and flavonols.

> *This step is optional, depending on how critical it is to separate anthocyanins from other polyphenolics in subsequent analysis. The ethyl acetate fraction is enriched in polyphenolics such as flavonols, procyanidins, and cinnamates. If analysis of this fraction is desired, a cleaner isolate will be obtained if residual water is removed by passing a nitrogen gas stream through the cartridge for 2 to 3 min before applying ethyl acetate.*

6. Elute anthocyanin pigments with acidified methanol and collect in a 50- to 100-ml boiling flask.

7. Remove methanol in a rotary evaporator at 40°C under vacuum.

8. Redissolve pigments in acidified deionized distilled water or an appropriate HPLC mobile-phase solvent.

9. Store purified anthocyanin extract at 4°C if subsequent analysis will be performed within 24 hr. Store sample for longer periods at −15°C or lower (preferably at −70°C) in a freeze-resistant container to minimize pigment degradation.

COMMENTARY

Background Information

Anthocyanins are flavonoid compounds that are widely distributed in plants. They are responsible for blue, purple, violet, magenta, red, and orange plant coloration (Jackman and Smith, 1996). The range of colors associated with anthocyanins results from distinct and varied substitution of the parent $C_6C_3C_6$ (aglycone) nucleus (Figure F1.1.2), in addition to acylation patterns and various environmental influences (Jackman and Smith, 1992).

Extraction

The extraction of anthocyanins is the first step in determination of total as well as individual anthocyanins in any type of plant tissue (Fuleki and Francis, 1968). The choice of an extraction method is of great importance in the analysis of anthocyanins, and largely depends on the purpose of the extraction, the nature of the anthocyanins, and the source material. A good extraction procedure should maximize anthocyanin recovery with a minimal amount of adjuncts and minimal degradation or alteration of the natural (in vivo) state. Nevertheless, one can never be sure that the extracted pigment is exactly the one occurring in vivo (Brouillard and Dangles, 1994). Knowledge of the factors that influence anthocyanin structure and stability is vital and has been reviewed by Markakis (1982), Francis (1989), and Jackman and Smith (1996). It is also desirable that the extraction

procedure not be too complex, hazardous, time consuming, or costly.

In most fruits and vegetables the anthocyanin pigments are located in cells near the surface (Jackman et al., 1987). Extraction procedures have generally involved the use of acidic solvents, which denature the membranes of cell tissue and simultaneously dissolve pigments. The acid tends to stabilize anthocyanins, but it may also change the native form of the pigment in the tissue by breaking associations with metals, co-pigments, or other factors. Concentration procedures may also cause acid hydrolysis of labile acyl and sugar residues. Extraction with solvents containing hydrochloric acid may result in pigment degradation during concentration, and is one of the reasons why acylations with aliphatic acids had been overlooked in the past (Strack and Wray, 1989). To minimize the decomposition of pigments, the use of weaker organic acids (e.g., formic, acetic, citric, or tartaric acids) or small amounts (0.5% to 3%) of more volatile acids (e.g., trifluoroacetic acid) that can then be removed during pigment concentration has been proposed (Strack and Wray, 1994; Jackman and Smith, 1996). Alternatives that have been used successfully for extraction include 3% trifluoroacetic acid, 5% acetic acid, and 0.01% to 2.5% citric acid (Metivier et al., 1980; Odake et al., 1992; Baublis et al., 1994). Low hydrochloric acid concentrations (on the order of

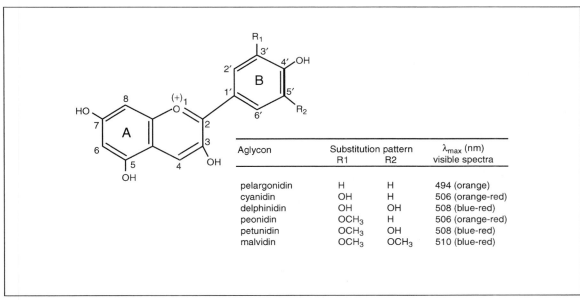

Figure F1.1.2 Structural and spectral characteristics of the major naturally occurring aglycons. The chemical structure indicates the two aromatic rings (**A**, **B**) as well as the R_1 and R_2 substitution sites.

Aglycon	Substitution pattern R1	R2	λ_{max} (nm) visible spectra
pelargonidin	H	H	494 (orange)
cyanidin	OH	H	506 (orange-red)
delphinidin	OH	OH	508 (blue-red)
peonidin	OCH_3	H	506 (orange-red)
petunidin	OCH_3	OH	508 (blue-red)
malvidin	OCH_3	OCH_3	510 (blue-red)

0.01% to 0.05%) and procedures in the absence of acid have also been suggested (Jackman et al., 1987; Strack and Wray, 1994). Procedures carried out in the presence of acid need to be performed with care to avoid acid-dependent pigment degradation.

A commonly used method is maceration of crushed or ground material for a few hours to overnight at 4°C in methanol containing small amounts of hydrochloric acid (<1%). The extracted material is usually too dilute for further analyses, and the extraction procedure is usually followed by evaporation of the methanol using vacuum and mild temperatures (30° to 40°C). The extraction of anthocyanins using ethanol acidified with citric acid (0.01%) instead of hydrochloric acid was reported by Main et al. (1978). Ethanol would be preferred for food use to avoid the toxicity of methanolic solutions. Citric acid is less corrosive than hydrochloric acid, chelates metals, maintains a low pH, and may have a protective effect during processing (Timberlake and Bridle, 1980). However, recoveries are not as high as those obtained with methanol, and concentration is more difficult because citric acid has a higher boiling point.

To obtain anthocyanins closer to their natural state, a number of researchers have performed the initial extraction using neutral solvents such as 60% methanol, *n*-butanol, cold acetone, acetone/methanol/water mixtures, or simply water (Jackman et al., 1987). Others have isolated anthocyanin pigments with mixtures of methanol/acetic acid/water (10:1:9, v/v/v; Takeda et al., 1986; Davies and Mazza, 1992), ethanol/acetic acid/water (12:1:24, v/v/v; Toki et al., 1991; 10:1:9, v/v/v; Hosokawa et al., 1995), and methanol/formic acid/water (50:5:45, v/v/v; Donner et al., 1997). Metivier et al. (1980) compared the efficiency of extraction with three different solvents—methanol, ethanol, and water—that were acidified with either hydrochloric acid or different organic acids, and found that methanol extraction was 20% more effective than ethanol and 73% more effective than water when used for anthocyanin recovery from grape pomace. They also reported that hydrochloric acid was most effective when used in combination with ethanol, whereas citric acid was more effective with methanol, and acetic acid was more effective with water.

Depending on the means of extraction, decreasing the ratio of extraction solvent to plant material could avoid the need for a concentration step (Jackman and Smith, 1996).

A second procedure uses acetone as the extracting solvent followed by partition with chloroform (Timberlake and Bridle, 1971; Wrolstad and Heatherbell, 1974; Abers and Wrolstad, 1979; Wrolstad et al., 1990). Timberlake and Bridle (1971) compared this extraction procedure with the usual acidified methanol extraction and concluded that the use of acetone with separation of the aqueous phase by addition of chloroform gave much cleaner and better-defined bands and enabled a better

Anthocyanins

assessment of anthocyanin composition. Recently, Wrolstad and Durst (1998) compared the acetone/chloroform method of extraction to the acidified methanol method on twenty samples (liquid and dry powder forms) derived from elderberries and cranberries that included juices, colorants, and nutraceutical preparations. Anthocyanin recoveries were comparable or up to 30% higher with the use of acetone. An advantage of this procedure is that the chloroform/acetone mixture will partition lipids, chlorophylls, and other water-insoluble materials from anthocyanins, yielding high recoveries of anthocyanins that require little concentration or further purification. The moisture content of the sample and the water from the secondary acetone extractions determine the final aqueous volume.

Purification of crude extracts

Frequently, the qualitative and quantitative analysis of anthocyanins is complicated by the presence of other compounds that may interfere with the measurements. The solvent systems generally used for extraction purposes are by no means specific for anthocyanins (Markakis, 1974; Jackman et al., 1987). Purification of anthocyanin-containing extracts is often necessary because considerable amounts of other compounds may also be extracted and concentrated. The variety and levels of other compounds will depend on the solvent and methodologies used. The presence of extraneous materials could influence the stability or analysis of anthocyanins. Therefore, the next step towards anthocyanin characterization is the prefractionation of those extracts.

When appreciable amounts of lipids, chlorophylls, or unwanted polyphenols are suspected to be present in anthocyanin-containing extracts, these materials may be removed by washing with petroleum ether, ethyl ether, diethyl ether, or ethyl acetate (Jackman and Smith, 1996).

Different resins have been used to clean up or prefractionate anthocyanins prior to isolation or characterization, including ion-exchange resins, polyamide powders, and gel materials. Chromatography on Dowex or Amberlite ion-exchange resins, as well as polyamide powders such as polyvinylpyrrolidone (PVP), have been used to isolate polar nonphenolic compounds from crude anthocyanin extracts. Column chromatography on Sephadex LH-20 can be used for fractionation of crude extracts and is also particularly useful for purification

of individual anthocyanins (Strack and Wray, 1989).

C18 cartridges are becoming popular because of their ease of use and high efficiency for fractionating anthocyanins. In an aqueous phase, anthocyanins are bound to the solid phase, whereas polar compounds such as acids and sugars can be washed away with acidified water. Using ethyl acetate to wash out polyphenolic compounds other than anthocyanins was suggested by Oszmianski and Lee (1990). Finally, a relatively pure anthocyanin extract can be removed from the column with slightly acidified methanol. Baldi et al. (1995) proposed a complex fractionation method for grape (*Vitis vinifera* L.) anthocyanin extracts that involved deposition of the sample on an Extrelut 20 cartridge to remove lipophilic and polyphenolic compounds with hexane and ethyl acetate, followed by adsorption of the sample onto a C18 cartridge to eliminate hydrophilic molecules (e.g., sugars, organic acids).

A key factor in the use of these purification techniques is the stability of anthocyanins under the conditions used as well as the ease of anthocyanin recovery from the column (Strack and Wray, 1989).

Critical Parameters and Troubleshooting

Sample preparation

Grinding the plant material in the presence of liquid nitrogen provides a uniform powdered composite sample; however, some powdered materials are very hygroscopic. With such materials, it is advisable to accurately weigh the plant material (before freezing) and quantitatively transfer the powdered material with an appropriate solvent.

Make sure that the container is completely dried before pouring the liquid nitrogen into the blender because any remaining water will freeze the blender bearings, and the blades will not revolve. If this happens, recover the material and rinse the blender with acetone and air dry. Chill the blender with small amounts of liquid nitrogen, allow to evaporate, and turn the blender on and off before adding frozen plant material.

When working with liquid nitrogen, do not tightly close the lid of the blender or mill container, so that the gas can escape.

Anthocyanin extraction

It is critical that the acetone/chloroform mixture be in an acid environment to avoid explosion hazards. The use of acidified acetone as extracting solvent will ensure a solution with low pH. In addition, the aqueous fraction will also have a low pH, which favors anthocyanin stability.

The authors have used ratios of acetone/chloroform varying from 1:1, 1:2, 1:2.4, to even 1:5. A greater proportion of chloroform reduces the amount of acetone in the aqueous phase and may eliminate the need for removal of acetone by rotary evaporation. This evaporative step, however, takes little time with no apparent anthocyanin degradation. Therefore, the authors favor using 1:1 or 1:2 ratios to avoid excessive use of the chloroform solvent.

Use a temperature of 30° to 40°C during rotary evaporation to avoid anthocyanin degradation. It is also important to monitor the process to ensure that the aqueous crude extract is not suctioned into the solvent reservoir. If this is going to happen, quickly release the vacuum and transfer the extract to a larger boiling flask and/or reduce the temperature of the water bath.

The crude anthocyanin extract contains a low level of hydrochloric acid (0.01%), which favors the formation of the stable flavylium ion. However, during concentration (and especially at dryness) the acid could cause the hydrolysis of labile acyl groups, co-pigments, or metal complexes that are part of the native form of the anthocyanin and are important for their stability. The risk of concentration of hydrochloric acid in the extract is limited with the acetone/chloroform procedure, as most of the extract contains water (high boiling point) with small levels of organic solvent. However, during methanol (low boiling point) extraction, the alcohol often evaporates to dryness with increased danger of anthocyanin decomposition.

An advantage of the acetone/chloroform method as compared to the methanol method is that the aqueous anthocyanin is not contaminated with lipophilic compounds. If the methanol method is used, the concentrated extract can be extracted with hexane, petroleum ether, or diethyl ether to remove unwanted lipophilic compounds after vacuum evaporation of the methanol (Fuleki and Francis, 1968).

The extracted anthocyanin-containing solution can be used for pigment quantitation by the pH-differential method (*UNIT F1.2*). Both the weight of the fresh material and the final volume of the extract should be accurately recorded when further analyses of pigment content will be carried out.

Anthocyanin purification

Anthocyanin purification steps are important for anthocyanin characterization. Removal of interfering compounds allows for more reliable HPLC separation, spectral information, mass spectra, and NMR spectra during the identification of anthocyanins in plant extracts.

Solid-phase extraction is used to adsorb the anthocyanin and other hydrophobic organic compounds and remove sugars and organic acids from the extract. It is important to avoid saturation of the C18 resin. Polyphenolic compounds along with anthocyanins have an affinity for the C18 column and will compete for binding sites. Use a 10- to 20-g sorbent cartridge placed on a filtration flask for a large extraction volume (50 to 250 ml) and intensely colored extracts (i.e., high levels of anthocyanins). Saturation of the C18 cartridge is clearly evidenced by the leaching of colored compounds (anthocyanins) from the resin. When this occurs, stop adding sample to the cartridge, wash the column with acidified water, and elute the anthocyanins before continuing the purification process.

Ethyl acetate is used on the C18-adsorbed anthocyanins to remove polyphenolic compounds. More efficient polyphenolic recovery will be accomplished if residual water is removed from the cartridge with a nitrogen gas stream for 2 to 3 min before application of ethyl acetate. After washing the column with ethyl acetate, water should not be added because some anthocyanins could be eluted and lost. Ethyl acetate removal of polyphenolics is particularly recommended if the anthocyanin fraction is to be subsequently analyzed by electrospray mass spectroscopy.

Anticipated Results

A homogeneous and fine powder should be obtained from the liquid nitrogen blending process. After extraction and purification of the plant material, a colored solution should be obtained. Depending on the nature of the predominant anthocyanin and its concentration, the coloration of the solution might be pink/red or purple. In the acetone/chloroform procedure, the organic solvent (lower phase) might show coloration due to the presence of chlorophyll (green) and/or carotenoids (yellow, red) in the plant material. The yield will vary depending on the type of materials used.

Time Considerations

The sample preparation should take 15 to 30 min, depending on the amount of material. The acetone extraction of the anthocyanin pigments usually takes 1 to 2 hr. After addition of chloroform, the solution should be stored overnight at 4°C to obtain well-defined phases. Alternatively, phase separation can be accelerated by centrifugation. This is particularly appropriate when working with small sample sizes. The sample matrix and the cleanness of the phase separation will influence the time for removal of residual solvents (acetone/chloroform) from the aqueous phase. It may be as short as 2 min or extended to 10 to 20 min. Methanol extraction is a faster procedure. Maceration of the nitrogen-powdered material for 1 hr is enough to extract the pigments. Therefore, if there is no need to remove lipophilic compounds from the concentrated extract, a total time of 2 to 4 hr is sufficient to extract the anthocyanins. Purification by solid-phase resin is a fast procedure and normally takes <1 hr.

Literature Cited

Abers, J.E. and Wrolstad, R.E. 1979. Causative factors of color deterioration in strawberry preserves during processing and storage. *J. Food Sci.* 44:75-78, 81.

Baldi, A., Romani, A., Mulinacci, N., Vincieri, F.F., and Casetta, B. 1995. HPLC/MS application to anthocyanins of *Vitis vinifera* L. *J. Agric. Food Chem.* 43:2104-2109.

Baublis, A., Spomer, A., and Berber-Jimenez, M.D. 1994. Anthocyanin pigments: Comparison of extract stability. *J. Food Sci.* 59:1219-1221, 1233.

Brouillard, R., and Dangles, O. 1994. Flavonoids and flower color. *In* The Flavonoids: Advances in Research Since 1986 (J.B. Harborne, ed.) pp. 565-588. Chapman & Hall, London.

Davies, A.J. and Mazza, G. 1992. Separation and characterization of anthocyanins of *Monarda fistulosa* by high-performance liquid chromatography. *J. Agric. Food Chem.* 40:1341-1345.

Donner, H., Gao, L., and Mazza, G. 1997. Separation and characterization of simple and malonylated anthocyanins in red onions, *Allium cepa* L. *Food Res. Int.* 30:637-643.

Francis, F.J. 1989. Food colorants: Anthocyanins. *Crit. Rev. Food Sci. Nutr.* 28:273-314.

Fuleki, T. and Francis, F.J. 1968. Quantitative methods for anthocyanins. 1. Extraction and determination of total anthocyanin in cranberry juice. *J. Food Sci.* 33:72-78.

Hosokawa, K., Fukunaga, Y., Fukushi, E., and Kawabata, J. 1995. Acylated anthocyanins from red *Hyacinthus orientalis*. *Phytochemistry* 39:1437-1441.

Jackman, R.L. and Smith, J.L. 1992. Anthocyanins and Betalains. *In* Natural Food Colorants (G.A.F. Hendry and J.D. Houghton, eds.) pp. 183-241. Blackie A&P, Great Britain.

Jackman, R.L. and Smith, J.L. 1996. Anthocyanins and Betalains. *In* Natural Food Colorants, 2nd ed. (G.A.F. Hendry and J.D. Houghton, eds.) pp. 244-309. Blackie and Son, Ltd., London.

Jackman, R.L., Yada, R.Y., and Tung, M.A. 1987. A Review: Separation and chemical properties of anthocyanins used for their qualitative and quantitative analysis. *J. Food Biochem.* 11:279-308.

Main, J.H., Clydesdale, F.M., and Francis, F.J. 1978. Spray drying anthocyanin concentrates for use as food colorants. *J. Food Sci.* 43:1693-1694, 1697.

Markakis, P. 1974. Anthocyanins and their stability in foods. *CRC Crit. Rev. Food Technol.* 4:437-456.

Markakis, P. 1982. Stability of anthocyanins in foods. *In* Anthocyanins as Food Colors, pp. 163-180. Academic Press, New York.

Metivier, R.P., Francis, F.J., and Clydesdale, F.M. 1980. Solvent extraction of anthocyanins from wine pomace. *J. Food Sci.* 45:1099-1100.

Odake, K., Terahara, N., Saito, N., Toki, K., and Honda, T. 1992. Chemical structures of two anthocyanins from purple sweet potato, *Ipomoea batatas*. *Phytochemistry* 31:2127-2130.

Oszmianski, J. and Lee, C.Y. 1990. Isolation and HPLC determination of phenolic compounds in red grapes. *Am. J. Enol. Vitic.* 41:202-206.

Strack, D. and Wray, V. 1989. Anthocyanins. *In* Methods in Plant Biochemistry, Vol. 1: Plant Phenolics (P.M. Dey and J.B. Harborne, eds.) pp. 325-359. Academic Press, San Diego.

Strack, D. and Wray, V. 1994. The anthocyanins. *In* The Flavonoids: Advances in Research Since 1986 (J.B. Harborne, ed.) pp. 1-19. Chapman & Hall, London.

Takeda, K., Harborne, J.B., and Self, R. 1986. Identification and distribution of malonated anthocyanins in plants of the compositae. *Phytochemistry* 25:1337-1342.

Timberlake, C.F. and Bridle, P. 1971. Anthocyanins in petals of *Chaenomeles speciosa*. *Phytochemistry* 10:2265-2267.

Timberlake, C.F. and Bridle, P. 1980. Anthocyanins. *In* Developments in Food Colours – 1 (J. Walford, ed.) pp. 115-149. Applied Science Publishers, London.

Toki, K., Yamamoto, N., Terahara, N., Saito, N., Honda, T., Inoue, H., and Mizutani, H. 1991. Pelargonidin 3-acetylglucoside in *Verbena* flowers. *Phytochemistry* 30:3828-3829.

Wrolstad, R.E. and Durst, R.W. 1998. Use of anthocyanin and polyphenolic analyses in authenticating fruit juices. *In* Proceedings of Fruit Authenticity Workshop, pp. 79-86. Montreal, Canada, September, 1999. EUROFINS Scientific, Nantes, France.

Wrolstad, R.E. and Heatherbell, D.A. 1974. Identification and distribution of flavonoids in Tamarillo fruit (*Cyphomandra betaceae* (Cav.) Sendt.). *J. Sci. Food Agric.* 25:1221-1228.

Wrolstad, R.E., Skrede, G., Lea, P., and Enersen, G. 1990. Influence of anthocyanin pigment stability in frozen strawberries. *J. Food Sci.* 55:1064-1065, 1072.

Key References

Oszmianski and Lee, 1990. See above.

A description of methods for isolating polyphenolics and anthocyanins from grapes by solid-phase extraction.

Strack and Wray, 1994. See above.

An excellent review of analytical methods for anthocyanin pigment analyses

Contributed by Luis E. Rodriguez-Saona
University of Maryland and Joint Institute
  for Food Safety and Applied Nutrition
Washington, D.C.

Ronald E. Wrolstad
Oregon State University
Corvallis, Oregon

Characterization and Measurement of Anthocyanins by UV-Visible Spectroscopy

Anthocyanin pigment content has a critical role in the color quality of many fresh and processed fruits and vegetables. Thus, accurate measurement of anthocyanins, along with their degradation indices, is very useful to food technologists and horticulturists in assessing the quality of raw and processed foods. Since many natural food colorants are anthocyanin derived (e.g., grape-skin extract, red-cabbage extract, purple-carrot extract), the same measurements can be used to assess the color quality of these food ingredients. In addition, there is intense interest in the anthocyanin content of foods and nutraceuticals because of possible health benefits such as reduction of coronary heart disease (Bridle and Timberlake, 1996), improved visual acuity (Timberlake and Henry, 1988), antioxidant activities (Takamura and Yamagami, 1994; Wang et al., 1997), and anticancer activities (Karaivanova et al., 1990; Kamei et al., 1995). Substantial quantitative and qualitative information can be obtained from the spectral characteristics of anthocyanins. The protocols described in this unit rely on the structural transformation of the anthocyanin chromophore as a function of pH, which can be measured using optical spectroscopy. The pH-differential method, a rapid and easy procedure for the quantitation of monomeric anthocyanins, is first described (see Basic Protocol 1). In addition, other auxiliary spectrophotometric techniques are used to measure the extent of anthocyanin polymerization and browning (see Basic Protocol 2).

TOTAL MONOMERIC ANTHOCYANIN BY THE pH-DIFFERENTIAL METHOD

Anthocyanin pigments undergo reversible structural transformations with a change in pH manifested by strikingly different absorbance spectra (Fig. F1.2.1). The colored oxonium form predominates at pH 1.0 and the colorless hemiketal form at pH 4.5 (Fig. F1.2.2). The pH-differential method is based on this reaction, and permits accurate and rapid measurement of the total anthocyanins, even in the presence of polymerized degraded pigments and other interfering compounds.

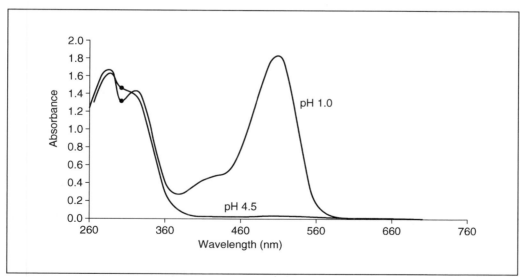

Figure F1.2.1 Spectral characteristics of purified radish anthocyanins (acylated pelargonidin-3-sophoroside-5-glucoside derivatives) in pH 1.0 and pH 4.5 buffers.

Contributed by M. Mónica Giusti and Ronald E. Wrolstad

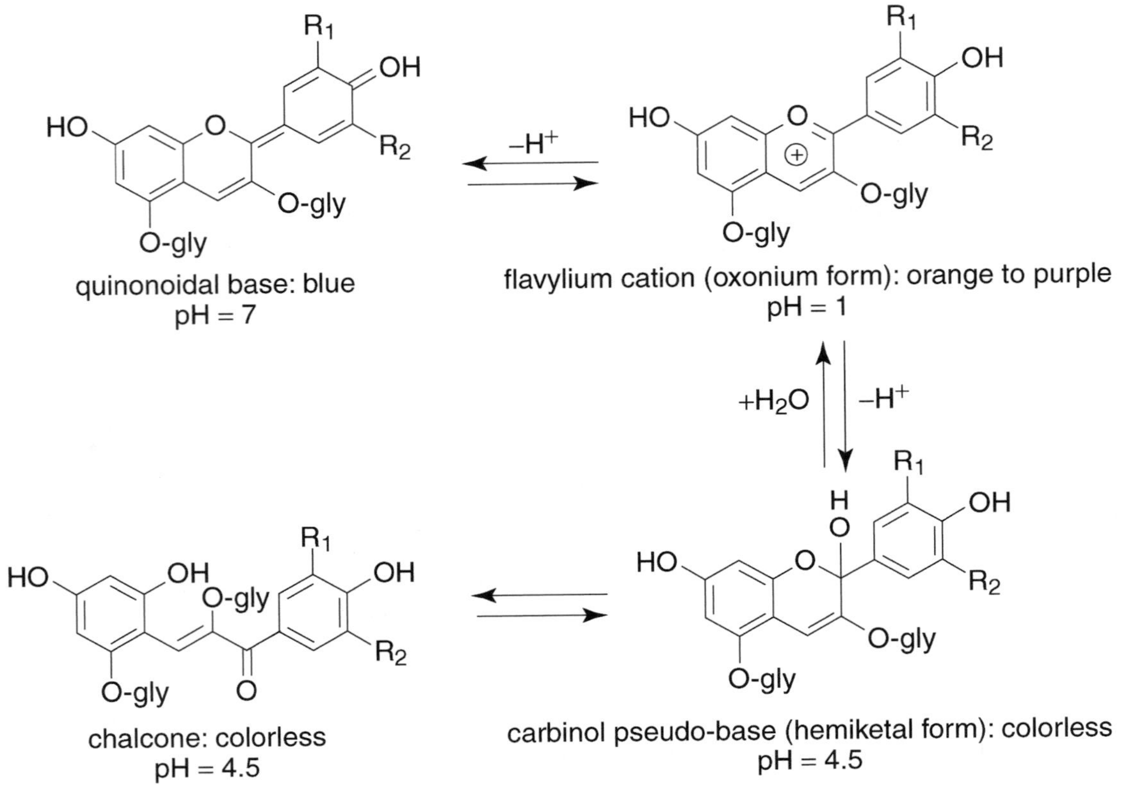

Figure F1.2.2 Predominant structural forms of anthocyanins present at different pH levels.

Materials

    0.025 M potassium chloride buffer, pH 1.0 (see recipe)
    0.4 M sodium acetate buffer, pH 4.5 (see recipe)

1. Turn on the spectrophotometer. Allow the instrument to warm up at least 30 min before taking measurements.

2. Determine the appropriate dilution factor for the sample by diluting with potassium chloride buffer, pH 1.0, until the absorbance of the sample at the $\lambda_{vis\text{-}max}$ (Table F1.2.1) is within the linear range of the spectrophotometer (i.e., for most spectrophotometers the absorbance should be less than 1.2). Divide the final volume of the sample by the initial volume to obtain the dilution factor (DF; for example see step 7).

   IMPORTANT NOTE*: In order to not exceed the buffer's capacity, the sample should not exceed 20% of the total volume.*

3. Zero the spectrophotometer with distilled water at all wavelengths that will be used ($\lambda_{vis\text{-}max}$ and 700 nm).

   Many spectrophotometers will allow for a rapid baseline correction to zero by using baseline adjust.

4. Prepare two dilutions of the sample, one with potassium chloride buffer, pH 1.0, and the other with sodium acetate buffer, pH 4.5, diluting each by the previously determined dilution factor (step 2). Let these dilutions equilibrate for 15 min.

Characterization and Measurement of Anthocyanins by UV-Visible Spectroscopy

F1.2.2

Table F1.2.1 Reported Molar Absorptivity of Anthocyanins

Anthocyanin[a]	Solvent system	$\lambda_{vis\text{-}max}$ (nm)	Molar absorptivity (ε)	Reference
Cyanidin (Cyd)				
Cyd	0.1% HCl in ethanol	510.5	24600	Schou, 1927
	0.1% HCl in ethanol	547	34700	Ribereau-Gayon, 1959
Cyd-3-ara	15:85 0.1 N HCl/ethanol	538	44400	Zapsalis and Francis, 1965
	15:85 0.1 N HCl/ethanol	535	44460	Fuleki and Francis, 1968a
Cyd-3,5-diglu	0.1 N HCl	520	30175	Niketic-Aleksic and Hrazdina, 1972
	Methanolic HCl	508.5	35000	Brouillard and El Hache Chahine, 1980
Cyd-3-gal	0.1% HCl in methanol	530	34300	Siegelman and Hendricks, 1958
	15:85 0.1 N HCl/ethanol	535	44900	Sakamura and Francis, 1961
	15:85 0.1 N HCl/ethanol	535	46200	Zapsalis and Francis, 1965
	15:85 0.1 N HCl/ethanol	535	46230	Fuleki and Francis, 1968a
	HCl in methanol	530	30200	Swain, 1965
Cyd-3-glu	Aqueous buffer, pH 1	510	26900	Jurd and Asen, 1966
	0.1 N HCl	520	25740	McClure, 1967
	1% HCl in methanol	530	34300	Siegelman and Hendricks, 1958
	10% ethanol, pH 1.5	512	18800	Heredia et al., 1998
Cyd-3-rut	Aqueous buffer, pH 0.9	510	7000	Figueiredo et al., 1996
	1% HCl	523	28840	Swain, 1965
Cyd-3-sam-5-glu	Aqueous buffer, pH 0.9	522	3600	Figueiredo et al., 1996
Cyd-3-sam-5-glu + sinapic + caffeic + malonic	Aqueous buffer, pH 0.9	538	21200	Figueiredo et al., 1996
Cyd-3-sam-5-glu + sinapic + ferulic	Aqueous buffer, pH 0.9	528	15100	Figueiredo et al., 1996
Cyd-3-sam-5-glu + sinapic + ferulic + malonic	Aqueous buffer, pH 0.9	538	20100	Figueiredo et al., 1996
Cyd-3-sam-5-glu + sinapic + p-coum + malonic	Aqueous buffer, pH 0.9	536	19000	Figueiredo et al., 1996
Cyd-3-soph-5-glu	Methanolic HCl	524	37150	Hrazdina et al., 1977
Cyd-3-soph-5-glu + malonic	Methanolic HCl	528	32360	Hrazdina et al., 1977
Cyd-3-soph-5-glu + sinapic	Methanolic HCl	528	37150	Hrazdina et al., 1977
Cyd-3-soph-5-glu + di-sinapic	Methanolic HCl	530	38020	Hrazdina et al., 1977
Cyd-3-soph-5-glu + ferulic	Methanolic HCl	528	32360	Hrazdina et al., 1977
Cyd-3-soph-5-glu + di-ferulic	Methanolic HCl	530	34670	Hrazdina et al., 1977
Cyd-3-soph-5-glu + p-coumaric	Methanolic HCl	526	38020	Hrazdina et al., 1977
Cyd-3-soph-5-glu + di-p-coumaric	Methanolic HCl	528	32360	Hrazdina et al., 1977
Delphinidin (Dpd)				
Dpd	0.1% HCl in ethanol	522.5	34700	Schou, 1927

continued

F1.2.3

Anthocyanin[a]	Solvent system	$\lambda_{vis\text{-}max}$ (nm)	Molar absorptivity (ϵ)	Reference
Dpd-3-glu	1% HCl in methanol	543	29000	Asen et al., 1959
	10% ethanol, pH 1.5	520	23700	Heredia et al., 1998
Malvidin (Mvd)				
Mvd	0.1% HCl in ethanol	520	37200	Schou, 1927
	0.1% HCl in ethanol	557	36200	Ribereau-Gayon, 1959
Mvd-3,5-diglu	0.1% HCl in ethanol	519	10700	Schou, 1927
	0.1% HCl in ethanol	545	10300	Ribereau-Gayon, 1959
	0.1 N HCl	520	37700	Niketic-Aleksic and Hrazdina, 1972
Mvd-3-glu	0.1% HCl in methanol	546	13900	Somers, 1966
	0.1% HCl in methanol	538	29500	Koeppen and Basson, 1966
	0.1 N HCl	520	28000	Niketic-Aleksic and Hrazdina, 1972
	Methanol, pH 1.0	535	36400	Metivier et al., 1980
	10% ethanol, pH 1.5	520	20200	Heredia et al., 1998
Mvd-3-glu + p-coum	0.1% HCl in methanol	536	30200	Koeppen and Basson, 1966
Pelargonidin (Pg)				
Pg	0.1% HCl in ethanol	504.5	17800	Schou, 1927
	0.025 M potassium chloride buffer, pH 1.0	505	18420	Giusti et al., 1999
	0.1% HCl in methanol	524	19780	Giusti et al., 1999
Pg-3,5-diglu	HCl in methanol	510	32360	Swain, 1965
Pg-3-(dicaffeoylglu)-soph-5-glu	Aqueous buffer, pH 0.8	512	28000	Dangles et al., 1993
Pg-3-glu	1% HCl in H$_2$O	496	27300	Jorgensen and Geissman, 1955
			36600	Wrolstad et al., 1970
	1% HCl	513	22390	Swain, 1965
	1% HCl in ethanol	516	31620	Swain, 1965
	0.025 M potassium chloride buffer, pH 1.0	496	15600	Giusti et al., 1999
	0.1% HCl in methanol	508	17330	Giusti et al., 1999
Pg-3-rut-5-glu + p-coumaric	0.025 M potassium chloride buffer, pH 1.0	504	32080	Giusti et al., 1999
	0.1% HCl in methanol	511	39591	Giusti et al., 1999
Pg-3-soph-5-glu	Aqueous buffer, pH 0.8	498	18000–20000	Dangles et al., 1993
	0.025 M potassium chloride buffer, pH 1.0	497	25370	Giusti et al., 1999
	0.1% HCl in methanol	506	30690	Giusti et al., 1999
Pg-3-soph-5-glu + ferulic	0.025 M potassium chloride buffer, pH 1.0	506	24140	Giusti et al., 1999
	0.1% HCl in methanol	507	29636	Giusti et al., 1999
Pg-3-soph-5-glu caffeoyl derivatives	Aqueous buffer, pH 0.8	498	18000-20000	Dangles et al., 1993
Pg-3-soph-5-glu + p-coumaric	0.025 M potassium chloride buffer, pH 1.0	506	28720	Giusti et al., 1999
	0.1% HCl in methanol	508	34889	Giusti et al., 1999

continued

F1.2.4

Table F1.2.1 Reported Molar Absorptivity of Anthocyanins, continued

Anthocyanin[a]	Solvent system	$\lambda_{vis\text{-}max}$ (nm)	Molar absorptivity (ε)	Reference
Pg-3-soph-5-glu + p-coumaric + malonic	0.025 M potassium chloride buffer, pH 1.0	508	33010	Giusti et al., 1999
	0.1% HCl in methanol	508	39785	Giusti et al., 1999
Pg-3-soph-5-glu + ferulic + malonic	0.025 M potassium chloride buffer, pH 1.0	508	31090	Giusti et al., 1999
	0.1% HCl in methanol	508	39384	Giusti et al., 1999
Peonidin (Pnd)				
Pnd	0.1% HCl in ethanol	511	37200	Schou, 1927
	15:85 0.1 N HCl/ethanol	532	40800	Sakamura and Francis, 1961
Pnd-3-ara	15:85 0.1 N HCl/ethanol	532	46100	Zapsalis and Francis, 1965
	15:85 0.1 N HCl/ethanol	532	46070	Fuleki and Francis, 1968a
Pnd-3,5-diglu	0.1 N HCl	520	36654	Niketic-Aleksic and Hrazdina, 1972
Pnd-3-gal	15:85 0.1 N HCl/ethanol	532	48400	Sakamura and Francis, 1961
	15:85 0.1 N HCl/ethanol	532	48400	Zapsalis and Francis, 1965
	15:85 0.1 N HCl/ethanol	531	48340	Fuleki and Francis, 1968a
Pnd-3-glu	0.1% HCl in methanol	536	11300	Somers, 1966
	10% ethanol, pH 1.5	512	14100	Heredia et al., 1998
Petunidin (Ptd)				
Ptd-3,5-diglu	0.1 N HCl	520	33040	Niketic-Aleksic and Hrazdina, 1972
	HCl in methanol	535	23440	Swain, 1965
Ptd-3-glu	0.1% HCl in methanol	546	12900	Somers, 1966
	10% ethanol, pH 1.5	520	18900	Heredia et al., 1998

[a]Abbreviations: ara: arabinoside; gal: galactoside; glu: glucoside; rut: rutinoside; sam: sambubioside; soph: sophoroside.

5. Measure the absorbance of each dilution at the $\lambda_{vis\text{-}max}$ and at 700 nm (to correct for haze), against a blank cell filled with distilled water.

All measurements should be made between 15 min and 1 hr after sample preparation, since longer standing times tend to increase observed readings.

Absorbance readings are made against water blanks, even if the samples are in buffer or bisulfite solutions, as buffer or bisulfite absorbance is nil at the measured wavelengths. The authors have compared the values obtained by using water as a blank as compared with buffer or bisulfite as blanks in different systems and have found no difference in the final values obtained for monomeric and/or polymeric anthocyanin content; on the other hand, reading the diluted samples against the corresponding buffer and/or bisulfite solution is more time-consuming and extends the procedure unnecessarily.

The samples to be measured should be clear and contain no haze or sediments; however, some colloidal materials may be suspended in the sample, causing scattering of light and a cloudy appearance (haze). This scattering of light needs to be corrected for by reading at a wavelength where no absorbance of the sample occurs, i.e., 700 nm.

Table F1.2.2 Molecular Weights of Anthocyanidins, Anthocyanins, and Acylating Groups Commonly Found in Nature[a]

Anthocyanidins	Pelargonidin	Cyanidin	Peonidin	Delphinidin	Petunidin	Malvidin
	271	287	301	303	317	331
Hex	180.2	180.2	180.2	180.2	180.2	180.2
Hex $-H_2O$[b]	162.2	162.2	162.2	162.2	162.2	162.2
Acd + 1 hex	433.2	449.2	463.2	465.2	479.2	493.2
Acd + 2 hex	595.4	611.4	625.4	627.4	641.4	655.4
Acd + 3 hex	757.6	773.6	787.6	789.6	803.6	817.6
Pent	150.0	150.0	150.0	150.0	150.0	150.0
Pent $-H_2O$[b]	132.0	132.0	132.0	132.0	132.0	132.0
Acd + 1 pent	403.0	419.0	433.0	435.0	449.0	463.0
Acd + 1 hex + 1 pent	565.2	581.2	595.2	597.2	611.2	625.2
Rhamnose	164.2	164.2	164.2	164.2	164.2	164.2
Rutinose	326.2	326.2	326.2	326.2	326.2	326.2
Rutinose $-H_2O$[b]	308.2	308.2	308.2	308.2	308.2	308.2
Acd + rutinose	579.2	595.2	609.2	611.2	625.2	639.2
Acd + rutinose + 1 hex	741.4	757.4	771.4	773.4	787.4	801.4
Acd + rutinose + 1 pent	711.2	727.2	741.2	743.2	757.2	771.2

Common acylating groups

		$-H_2O$[b]
p-Coumaric acid	164.2	146.2
Caffeic acid	180.2	162.2
Ferulic acid	194.2	176.2
Sinapic acid	224	206
Acetic acid	82	64
Propionic acid	96.1	78.1
Malonic acid	104.1	86.1
Succinic acid	118.1	100.1

[a]Abbreviations: hex: hexose; pent: pentose; acd: anthocyanidin.

[b]$-H_2O$ indicates a dehydrated sugar (water is lost upon forming a glycosidic bond).

6. Calculate the absorbance of the diluted sample (A) as follows:

$$A = (A_{\lambda \text{ vis-max}} - A_{700})_{\text{pH 1.0}} - (A_{\lambda \text{ vis-max}} - A_{700})_{\text{pH 4.5}}$$

7. Calculate the monomeric anthocyanin pigment concentration in the original sample using the following formula:

Monomeric anthocyanin pigment (mg/liter) = $(A \times \text{MW} \times \text{DF} \times 1000)/(\varepsilon \times 1)$

where MW is the molecular weight (Table F1.2.2), DF is the dilution factor (for example, if a 0.2 ml sample is diluted to 3 ml, DF = 15), and ε is the molar absorptivity (Table F1.2.1).

IMPORTANT NOTE: *The MW and ε used in this formula correspond to the predominant anthocyanin in the sample. Use the ε reported in the literature for the anthocyanin pigment in acidic aqueous solvent. If the ε of the major pigment is not available, or if the sample composition is unknown, calculate pigment content as cyanidin-3-glucoside, where MW = 449.2 and ε = 26,900 (see Background Information, discussion of Molar Absorptivity).*

The equation presented above assumes a pathlength of 1 cm.

Characterization
and Measurement
of Anthocyanins
by UV-Visible
Spectroscopy

F1.2.6

24

INDICES FOR PIGMENT DEGRADATION, POLYMERIC COLOR, AND BROWNING

Indices for anthocyanin degradation of an aqueous extract, juice, or wine can be derived from a few absorbance readings of a sample that has been treated with sodium bisulfite. Anthocyanin pigments will combine with bisulfite to form a colorless sulfonic acid adduct (Figure F1.2.3). Polymerized colored anthocyanin-tannin complexes are resistant to bleaching by bisulfite, whereas the bleaching reaction of monomeric anthocyanins will rapidly go to completion. The absorbance at 420 nm of the bisulfite-treated sample serves as an index for browning. Color density is defined as the sum of absorbances at the $\lambda_{vis\text{-}max}$ and at 420 nm. The ratio between polymerized color and color density is used to determine the percentage of the color that is contributed by polymerized material. The ratio between monomeric and total anthocyanin can be used to determine a degradation index.

Materials

Bisulfite solution (see recipe)
0.025 M potassium chloride buffer, pH 1.0 (see recipe)

1. Turn on the spectrophotometer and allow the instrument to warm up at least 30 min before taking measurements.

2. Determine the appropriate dilution factor for the sample by diluting with 0.025 M potassium chloride buffer, pH 1.0 until the absorbance of the sample at the $\lambda_{vis\text{-}max}$ is within the linear range of the spectrophotometer (i.e., for most spectrophotometers the absorbance should be less than 1.2). Divide the final volume of the sample by the initial volume to obtain the dilution factor (DF; for example see step 6).

3. Zero the spectrophotometer with distilled water at all wavelengths that will be used (420 nm, $\lambda_{vis\text{-}max}$, 700 nm).

   Many spectrophotometers will allow for a rapid baseline correction to zero by using baseline adjust.

4. Dilute the sample with distilled water using the dilution factor already determined (step 2). Transfer 2.8 ml of the diluted sample to each of two cuvettes. Add 0.2 ml of bisulfite solution to one and 0.2 ml distilled water to the other. Equilibrate for 15 min.

   It is critical that the pH not be adjusted to highly acidic conditions (e.g., pH 1) but rather be in the typical pH range of fruit juices and wines, or higher (e.g., pH 3). Highly acidic conditions will reverse the bisulfite addition reaction and render the measurement invalid.

5. Measure the absorbance of both samples at 420 nm, $\lambda_{vis\text{-}max}$, and 700 nm (to correct for haze), against a blank cell filled with distilled water.

   All measurements should be made between 15 min (see step 4) and 1 hr after sample preparation and bisulfite treatment. Longer standing times tend to increase observed readings.

   Absorbance readings are made against water blanks, even if the samples are in buffer or bisulfite solutions, as buffer or bisulfite absorbance is nil at the measured wavelengths. The authors have compared the values obtained by using water as a blank as compared with the use buffer or bisulfite as a blank in different systems and have found no difference in the final values obtained for monomeric and/or polymeric anthocyanin content; on the other hand, reading the samples against the corresponding buffer and/or bisulfite solution is more time-consuming and extends the procedure unnecessarily.

   The samples to be measured should be clear and contain no haze or sediments; however, some colloidal materials may be suspended in the sample, causing scattering of light and a cloudy appearance (haze). This scattering of light needs to be accounted for by reading at a wavelength where no absorbance of the sample occurs (i.e., 700 nm).

Anthocyanins

F1.2.7

6. Calculate the color density of the control sample (treated with water) as follows:

Color density = $[(A_{420\ nm} - A_{700nm}) + (A_{\lambda\ vis\text{-}max} - A_{700\ nm})] \times DF$

where DF is the dilution factor (for example, if 0.2 ml sample diluted to 3 ml, DF = 15)

7. Calculate the polymeric color of the bisulfite bleached sample as follows:

Polymeric color = $[(A_{420\ nm} - A_{700\ nm}) + (A_{\lambda\ vis\text{-}max} - A_{700\ nm})] \times DF$

8. Calculate the percent polymeric color using the formula:

Percent polymeric color = (polymeric color/color density) $\times$ 100

REAGENTS AND SOLUTIONS

Use deionized or distilled water in all recipes and protocol steps. For common stock solutions, see APPENDIX 2A; for suppliers, see SUPPLIERS APPENDIX.

Bisulfite solution

Dissolve 1 g of potassium metabisulfite ($K_2S_2O_5$) in 5 ml of distilled water.

This reagent must be prepared the same day as the readings; otherwise, it develops a yellow color that will contribute to the absorbance readings and interfere with the quantitation.

Potassium chloride buffer, 0.025 M, pH 1.0

Mix 1.86 g KCl and 980 ml of distilled water in a beaker. Measure the pH and adjust to 1.0 with concentrated HCl. Transfer to a 1 liter volumetric flask and fill to 1 liter with distilled water.

The solution should be stable at room temperature for a few months, but the pH should be checked and adjusted prior to use (see Critical Parameters).

Sodium acetate buffer, 0.4 M, pH 4.5

Mix 54.43 g $CH_3CO_2Na \cdot 3\ H_2O$ and ~960 ml distilled water in a beaker. Measure the pH and adjust to 4.5 with concentrated HCl. Transfer to a 1 liter volumetric flask and fill to 1 liter with distilled water.

The solution should be stable at room temperature for a few months, but the pH should be checked and adjusted prior to use (see Critical Parameters).

COMMENTARY

Background Information

Anthocyanin pigments are responsible for the attractive red to purple to blue colors of many fruits and vegetables. Anthocyanins are relatively unstable and often undergo degradative reactions during processing and storage. Measurement of total anthocyanin pigment content along with indices for the degradation of these pigments are very useful in assessing the color quality of these foods. Interest in the anthocyanin content of foods and nutraceutical preparations has intensified because of their possible health benefits. They may play a role in reduction of coronary heart disease (Bridle and Timberlake, 1996) and increased visual acuity (Timberlake and Henry, 1988), and also have antioxidant (Takamura and Yamagami,

1994; Wang et al., 1997) and anticancer properties (Karaivanova et al., 1990; Kamei et al., 1995). Anthocyanins have also found considerable potential in the food industry as safe and effective food colorants (Strack and Wray, 1994); interest in this application has increased in recent years. In 1980, the annual world production had been estimated as reaching 10,000 tons from grapes alone (Timberlake, 1980). Quantitative and qualitative anthocyanin composition are important factors in determining the feasibility of the use of new plant materials as anthocyanin-based colorant sources.

Frequently, it is desirable to express anthocyanin determinations in terms that can be compared with the results from different workers. The best way to express these results is in terms

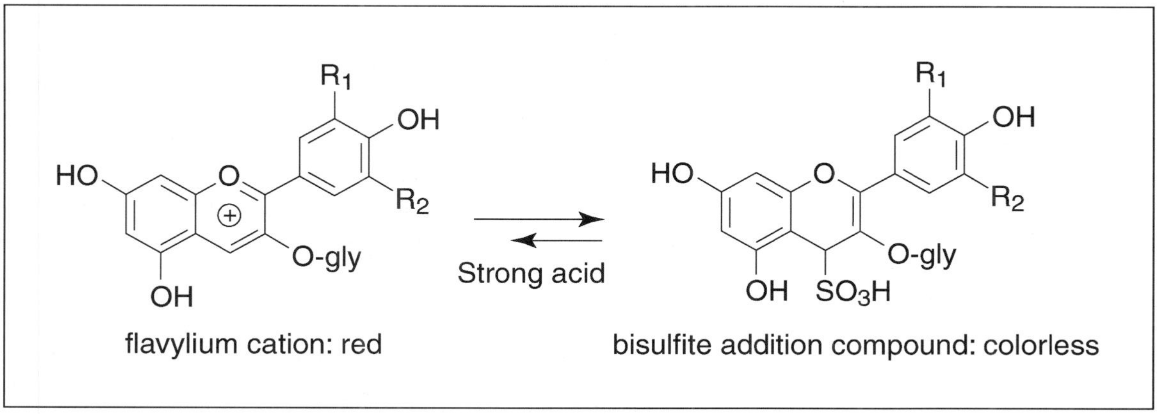

Figure F1.2.3   Formation of colorless anthocyanin-sulfonic acid adducts.

of absolute quantities of anthocyanins present (Fuleki and Francis, 1968a).

The total anthocyanin content in crude extracts containing other phenolic materials has been determined by measuring absorptivity of the solution at a single wavelength. This is possible because anthocyanins have a typical absorption band in the 490 to 550 nm region of the visible spectra (Figure F1.2.1). This band is far from the absorption bands of other phenolics, which have spectral maxima in the UV range (Fuleki and Francis, 1968a). In many instances, however, this simple method is inappropriate because of interference from anthocyanin degradation products or melanoidins from browning reactions (Fuleki and Francis, 1968b). In those cases, the approach has been to use differential and/or subtractive methods to quantify anthocyanins and their degradation products (Jackman and Smith, 1996).

The differential method (see Basic Protocol 1) measures the absorbance at two different pH values, and relies on the structural transformations of the anthocyanin chromophore as a function of pH (Fig. F1.2.1 and Fig. F1.2.2). This concept was first introduced by Sondheimer and Kertesz in 1948, who used pH values of 2.0 and 3.4 for analyses of strawberry jams (Francis, 1989). Since then, the use of other pH values has been proposed. Fuleki and Francis (1968b) used pH 1.0 and 4.5 buffers to measure anthocyanin content in cranberries, and modifications of this technique have been applied to a wide range of commodities (Wrolstad et al., 1982, 1995). The pH differential method has been described as fast and easy for the quantitation of monomeric anthocyanins (Wrolstad et al., 1995).

Subtractive methods (see Basic Protocol 2) are based on the use of bleaching agents that will decolor anthocyanins but not affect interfering materials. A measurement of the absorbance at the visible maximum is obtained, followed by bleaching and remeasuring to give a blank reading (Jackman et al., 1987). The two most used bleaching agents are sodium sulfite (Somers and Evans, 1974; Wrolstad et al., 1982) and hydrogen peroxide (Swain and Hillis, 1959).

By using both of these spectral procedures, accurate measurement of the total monomeric anthocyanin pigment content can be obtained, along with indices for polymeric color, color density, browning, and degradation. To determine total anthocyanin content, the absorbance at pH 1.0 and 4.5 is measured at the $\lambda_{vis-max}$ (Table F1.2.1) and at 700 nm, which allows for haze correction. The bisulfite bleaching reaction is utilized to generate the various degradation indices. While monomeric anthocyanins are readily bleached by bisulfite at product pH (Fig. F1.2.3), the polymeric anthocyanin-tannin and melanoidin pigments are resistant and will remain colored. Somers and Evans (1974) used this reaction in developing spectral methods for assessing the color quality of wines. The author's laboratory has found them useful for tracking color quality in a wide range of anthocyanin-containing foods (Wrolstad et al., 1982, 1995). Absorbance measurements are taken at the $\lambda_{vis-max}$ and at 420 nm on the bisulfite bleached and control samples. Color density is the sum of the absorbances at the $\lambda_{vis-max}$ and at 420 nm of the control sample, while polymeric color is the same measurement for the bisulfite treated sample. A measure of percent polymeric color is obtained as the ratio between these two indexes. The absorbance at 420 nm of the bisulfite-treated sample is an index for browning, as the accumulation of brownish

Anthocyanins

degradation products increases the absorption in the 400 to 440 nm range. The absorption of these compounds are in general not affected by the addition of a bisulfite solution.

Molar absorptivity

Regardless of the method used for anthocyanin quantitation, the determination of the amount present requires an absorptivity coefficient. Absorptivity coefficients have been reported as the absorption of a 1% solution measured through a 1-cm path at the $\lambda_{\text{vis-max}}$, or as a molar absorption coefficient. Absorptivity coefficients of some known anthocyanins have been reported by different researchers (Table F1.2.1). Through the years, there has been a lack of uniformity on the values of absorptivity reported, mainly due to the difficulties of preparing crystalline anthocyanin, free from impurities, in sufficient quantities to allow reliable weighing under optimal conditions (Fuleki and Francis, 1968a; Francis, 1982; Giusti et al., 1999). Other problems are that the anthocyanin mixtures may be very complicated, and not all absorptivity coefficients may be known. Even when they are known, it is necessary to first evaluate if the objective is the estimation of total anthocyanin content or the determination of individual pigments, and then to decide which absorption coefficient(s) to use. The absorptivity is dependent not only on the chemical structure of the pigment but also on the solvent used; preferably, the coefficient used should be one obtained in the same solvent system as the one used in the experiment. If the identity of the pigments is unknown, it has been suggested that it can be expressed as cyanidin-3-glucoside, since that is the most abundant anthocyanin in nature (Francis, 1989).

Spectral characteristics

Substantial information can be obtained from the spectral characteristics of anthocyanins (Fig. F1.2.1). Two distinctive bands of absorption, one in the UV-region (260 to 280 nm) and another in the visible region (490 to 550 nm) are shown by all anthocyanins. The different aglycons have different $\lambda_{\text{vis-max}}$, ranging from 520 nm for pelargonidin to 546 nm for delphinidin, and their monoglucosides exhibit their $\lambda_{\text{vis-max}}$ at about 10 to 15 nm lower (Strack and Wray, 1989). The shape of the spectrum may give information regarding the number and position of glycosidic substitutions and number of cinnamic acid acylations. The ratio between the absorbance at 440 nm and the absorbance at the $\lambda_{\text{vis-max}}$ is almost twice as much for anthocyanins with glycosidic substitutions in position 3 as compared to those with substitutions in positions 3 and 5 or position 5 only. The presence of glycosidic substitutions at other positions (e.g., 3,7-diglycosides) can be recognized because they exhibit a different spectral curve from those of anthocyanins with common substitution patterns. The presence of cinnamic acid acylation is revealed by the presence of a third absorption band in the 310 to 360 nm range (Figure F1.2.1), and the ratio of absorbance at 310 to 360 nm to the absorbance at the visible $\lambda_{\text{vis-max}}$ will give an estimation of the number of acylating groups (Harborne, 1967; Hong and Wrolstad, 1990). The solvent used for spectral determination will affect the position of the absorption bands, and therefore must be taken into consideration when comparing available data.

Critical Parameters and Troubleshooting

The pH of buffers should always be checked and adjusted prior to use. The use of buffers with lower or higher pH levels will result in under- or overestimations of the pigment content.

The accuracy of the results will be greatly affected by the accuracy of the volumetric measurements. Make sure that any volumetric flasks or pipets used for obtaining the appropriate dilutions are calibrated correctly.

For the methodologies described in this unit, all spectral measurements should be made between 15 min and 1 hr after the dilutions have been prepared. The observed readings tend to increase with time.

When working with several different samples, it may be acceptable to use one common approximate $\lambda_{\text{vis-max}}$ that is typical of all samples (i.e., 520 nm). The visible absorbance peak is broad, and measuring a few nanometers off $\lambda_{\text{vis-max}}$ will not significantly alter the estimated final values.

Serial dilutions are recommended to ensure accurate measurements of highly concentrated, high density, or dried samples. Perform a weight-by-volume dilution with distilled water to obtain a single-strength solution (e.g., usually around 10° Brix for fruit juices; UNIT H1.4), followed by a second dilution using 0.025 M potassium chloride buffer, pH 1.0. Both dilution factors must be considered when calculating monomeric anthocyanin content.

For example, 1 g of a 75° Brix juice concentrate was diluted to a final volume of 10 ml with distilled water (dilution factor = 10; assuming

Table F1.2.3 Anthocyanin Content of Some Common Fruits and Vegetables

Source	Pigment content (mg/100 g fresh weight)	Reference
Apples (Scugog)	10	Mazza and Miniati, 1993
Bilberries	300–320	Mazza and Miniati, 1993
Blackberries	83–326	Mazza and Miniati, 1993
Black currants	130–400	Timberlake, 1988
Blueberries	25–495	Mazza and Miniati, 1993
Red cabbage	25	Timberlake, 1988
Black chokeberries	560	Kraemer-Schafhalter et al., 1996
Cherries	4–450	Kraemer-Schafhalter et al., 1996
Cranberries	60–200	Timberlake, 1988
Elderberry	450	Kraemer-Schafhalter et al., 1996
Grapes	6–600	Mazza and Miniati, 1993
Kiwi	100	Kraemer-Schafhalter et al., 1996
Red onions	7–21	Mazza and Miniati, 1993
Plum	2–25	Timberlake, 1988
Red radishes	11–60	Giusti et al., 1988
Black raspberries	300–400	Timberlake, 1988
Red Raspberries	20–60	Mazza and Miniati, 1993
Strawberries	15–35	Timberlake, 1988
Tradescantia pallida (leaves)	120	Shi et al., 1992

a density of 1 g/ml for juice). Then, the appropriate dilution factor for the sample was determined by diluting 0.2 ml of the solution with 2.8 ml of 0.025 M potassium chloride buffer, pH 1.0 (dilution factor = 15). To calculate monomeric anthocyanin content, color density, or polymeric color, the dilution factor to use would be: $DF = (10 \times 15) = 150$.

The methodologies used to measure color density and polymeric color were developed for fruit juices, which naturally have an acidic pH. If the material to be measured has a pH in the neutral or alkaline range, the pH of the solution should be lowered with a weak acid. In these cases, the authors recommend the use of a 0.1 M citric acid buffer, pH 3.5, instead of distilled water to prepare the different dilutions.

Some potential interfering materials are other red pigments: FD&C Red No. 40, FD&C Red No. 3, cochineal, and beet powder (betalain pigments). The presence of alternative colorants may be suspected if the $\lambda_{vis-max}$ at pH 1.0 is high (550 nm, more typical of betalain pigments), or if a bright red coloration is found at pH 4.5 (potential presence of artificial dyes).

The presence of ethanol does not interfere with the assay at the levels typically encountered in wines (10% to 14%).

Highly acylated anthocyanins may not respond to pH changes the same way as anthocyanins with no or few acylating groups, and may not decolor as much as nonacylated or mono- or diacylated anthocyanins do at pH 4.5.

Anticipated Results

The anthocyanin content of different common fruits and vegetables is presented in Table F1.2.3. Anthocyanin-containing fruit or vegetable juices typically have pigment content ranging from 50 to 500 mg/liter. Anthocyanin-based natural colorants and nutraceuticals may have a much higher pigment concentration, on the order of a few grams/liter.

Fresh fruit or vegetable juices should have a low percentage of polymeric color (usually less than 10%), while processed samples and materials subjected to storage abuse will be much higher (30% or more). This is highly variable, dependent on the commodity, processing conditions, and storage history.

Always express anthocyanin pigment content in terms of the specific anthocyanin used for calculation, and specify molecular weight and ε utilized.

Time Considerations

Quantitation of anthocyanins can be achieved in <1 hr. It is necessary to wait for the spectrophotometer to warm up, and for the diluted samples to equilibrate at least 15 min. The absorbance readings take a few minutes.

Literature Cited

Asen, S., Stuart, N.W., and Siegelman, H.W. 1959. Effect of various concentrations of nitrogen, phosphorus and potassium on sepal color of *Hydrangea macrophylla*. *Am. Soc. Hort. Sci.* 73:495-502.

Bridle, P. and Timberlake, C.F. 1996. Anthocyanins as natural food colors-selected aspects. *Food Chem.* 58:103-109.

Brouillard, R. and El Hache Chahine, J.M. 1980. Chemistry of anthocyanin pigments. 6. Kinetic and thermodynamic study of hydrogen sulfite addition to cyanin. Formation of a highly stable Meisenheimer-type adduct derived from a 2-phenylbenzopyrylium salt. *J. Am. Chem. Soc.* 102:5375-5378.

Dangles, O., Saito, N., and Brouillard, R. 1993. Anthocyanin intramolecular copigment effect. *Phytochemistry* 34:119-124.

Figueiredo, P., Elhabiri, M., Saito, N., and Brouillard, R. 1996. Anthocyanin intramolecular interactions. A new mathematical approach to account for the remarkable colorant properties of the pigments extracted from *Matthiola incana*. *J. Am. Chem. Soc.* 118:4788-4793.

Francis, F.J. 1982. Analysis of anthocyanins. *In* Anthocyanins as Food Colors (P. Markakis, ed.) pp. 182-205. Academic Press, New York.

Francis, F.J. 1989. Food colorants: Anthocyanins. *Crit. Rev. Food Sci. Nutr.* 28:273-314.

Fuleki, T. and Francis, F.J. 1968a. Quantitative methods for anthocyanins. 1. Extraction and determination of total anthocyanin in cranberries. *J. Food Sci.* 33:72-78.

Fuleki, T. and Francis, F.J. 1968b. Quantitative methods for anthocyanins. 2. Determination of total anthocyanin and degradation index for cranberry juice. *J. Food Sci.* 33:78-82.

Giusti, M.M., Rodriguez-Saona, L.E., Baggett, J.R., Reed, G.L., Durst, R.W., and Wrolstad, R.E. 1998. Anthocyanin pigment composition of red radish cultivars as potential food colorants. *J. Food Sci.* 63:219-224.

Giusti, M.M., Rodriguez-Saona, L.E., and Wrolstad, R.E. 1999. Spectral characteristics, molar absorptivity and color of pelargonidin derivatives. *J. Agric. Food Chem.* 47:4631-4637.

Harborne, J.B. 1967. Comparative Biochemistry of the Flavonoids. Academic Press, London.

Heredia, F.J., Francia-Aricha, E.M., Rivas-Gonzalo, J.C., Vicario, I.M., and Santos-Buelga, C. 1998. Chromatic characterization of anthocyanins from red grapes. I. pH effect. *Food Chem.* 63:491-498.

Hong, V. and Wrolstad, R.E. 1990. Use of HPLC separation/photodiode array detection for characterization of anthocyanins. *J. Agric. Food Chem.* 38:708-715.

Hrazdina, G., Iredale, H., and Mattick, L.R. 1977. Anthocyanin composition of *Brassica oleracea* cv. Red Danish. *Phytochemistry* 16:297-301.

Jackman, R.L. and Smith, J.L. 1996. Anthocyanins and betalains. *In* Natural Food Colorants, 2nd ed. (G.A.F.Hendry and J.D. Houghton, eds.) Chpt. 8. Blackie & Son, Glasgow, Scotland.

Jackman, R.L., Yada, R.Y., and Tung, M.A. 1987. A review: Separation and chemical properties of anthocyanins used for their qualitative and quantitative analysis. *J. Food Biochem.* 11:279-308.

Jorgensen, E.C. and Geissman, T.A. 1955. The chemistry of flower pigmentation in *Antirrhinum majus* color genotypes. III. Relative anthocyanin and aurone concentrations. *Biochem. Biophys.* 55:389-402.

Jurd, L. and Asen, S. 1966. The formation of metal and "co-pigment" complexes of cyanidin 3-glucoside. *Phytochemistry* 5:1263-1271.

Kamei, H., Kojima, T., Hasegawa, M., Koide, T., Umeda, T., Yukawa, T., and Terabe, K. 1995. Suppression of tumor cell growth by anthocyanins in vitro. *Cancer Invest.* 13:590-594

Karaivanova, M., Drenska, D., and Ovcharov, R. 1990. A modification of the toxic effects of platinum complexes with anthocyans. *Eksp. Med. Morfol.* 29:19-24.

Koeppen, B.H. and Basson, D.S. 1966. The anthocyanin pigments of Barlinka grapes. *Phytochemistry* 5:183-187.

Kraemer-Schafhalter, A., Fuchs, H., Strigl, A., Silhar, S., Kovac, M., and Pfannhauser, W. 1996. Process consideration for anthocyanin extraction from Black Chokeberry (*Aronia meloncarpa* ELL). *In* Proceedings of the Second International Symposium on Natural Colorants, INF/COL II (P.C. Hereld, ed.), pp. 153-160. S.I.C. Publishing Col., Hamden, Ct.

Mazza, G. and Miniati, E. 1993. Introduction. *In* Anthocyanins in fruits, vegetables, and grains. (G. Mazza and E. Miniati, eds). CRC Press, Boca Raton, Fla.

McClure, J.W. 1967. Photocontrol of *Spirodela intermedia* flavonoids. *Plant Phys.* 43:193-200.

Metivier, R.P., Francis, F.J., and Clydesdale, F.M. 1980. Solvent extraction of anthocyanins from wine pomace. *J. Food Sci.* 45:1099-1100.

Niketic-Aleksic, G. and Hrazdina, G. 1972. Quantitative analysis of the anthocyanin content in grape juices and wines. *Lebensm. Wiss. U. Technol.* 5:163-165.

Ribereau-Gayon, P. 1959. Recherches sur les anthocyannes des vegetaux. Application au genre Vitis. Doctoral dissertation, p.118. University of Bordeaux. Librarie Generale de l'Ensignement, Paris.

Sakamura, S. and Francis, F.J. 1961. The anthocyanins of the American cranberry. *J. Food Sci.* 26:318-321.

Schou, S.A. 1927. Light absorption of several anthocyanins. *Helv. Chim. Acta.* 10:907-915.

Siegelman, H.W. and Hendricks, S.B., 1958. Photocontrol of alcohol, aldehyde and anthocyanin production in apple skin. *Plant Phys.* 33:409-413.

Somers, T.C. 1966. Grape phenolics: the anthocyanins of *Vitis vinifera*, var. Shiraz. *J. Sci. Food Agric.* 17:215-219.

Somers, T.C. and Evans, M.E. 1974. Wine quality: Correlations with colour density and anthocyanin equilibria in a group of young red wines. *J. Sci. Food Agric.* 25:1369-1379.

Strack, D. and Wray, V. 1989. Anthocyanins. *In* Methods in Plant Biochemistry, Vol. I, Plant Phenolics (P.M. Dey and J.B. Harborne. eds.). Academic Press, San Diego.

Strack, D. and Wray, V. 1994. The anthocyanins. *In* The Flavonoids: Advances in Research Since 1986. (J.B. Harborne, ed.). Chapman and Hall.

Swain, T. 1965. Analytical methods for flavonoids. *In* Chemistry and Biochemistry of Plant Pigments (T.W. Goodwin, ed.). Academic Press, London.

Swain, T. and Hillis, W.E. 1959. The phenolic constituents of *Prunus domestica*. I. The quantitative analysis of phenolic constituents. *J. Sci. Food Agric.* 10:63-68.

Takamura, H. and Yamagami, A. 1994. Antioxidative activity of mono-acylated anthocyanins isolated from Muscat Bailey A grape. *J. Agric. Food Chem.* 42:1612-1615.

Timberlake, C.F 1980. Anthocyanins [related to beverages]—occurrence, extraction and chemistry [coloring material]. *Food Chem.* 5:69-80.

Timberlake, C.F. 1988. The biological properties of anthocyanin compounds. *NATCOL Quarterly Bulletin* 1:4-15.

Timberlake, C.F. and Henry, B.S., 1988. Anthocyanins as natural food colorants. *Prog. Clin. Biol. Res.* 280:107-121.

Wang, H., Cao, G., and Prior, R.L., 1997. Oxygen radical absorbing capacity of anthocyanins. *J. Agric. Food Chem.* 45:304-309

Wrolstad R.E. 1976. Color and pigment analyses in fruit products. *Oregon St. Univ. Agric. Exp. Stn., Bulletin* 624:1-17.

Wrolstad, R.E., Putnam, T.P., and Varseveld, G.W. 1970. Color quality of frozen strawberries: Effect of anthocyanin, pH, total acidity and ascorbic acid variability. *J. Food Sci.* 35:448-452.

Wrolstad, R.E., Culbertson, J.D., Cornwell, C.J., and Mattick, L.R. 1982. Detection of adulteration in blackberry juice concentrates and wines. *J. Assoc. Off. Anal. Chem.* 65:1417-1423.

Wrolstad, R.E., Hong, V., Boyles, M.J., and Durst, R.W. 1995. Use of anthocyanin pigment analysis for detecting adulteration in fruit juices. *In* Methods to Detect Adulteration in Fruit Juice and Beverages, Vol. I (S. Nagy and R.L. Wade, ed.). AgScience Inc., Auburndale, Fla.

Zapsalis, C. and Francis, F.J. 1965. Cranberry anthocyanins. *J. Food Sci.* 30:396-399.

Key References

Giusti et al.,1999. See above.

Compares the molar absorptivity of many anthocyanins in different solvent systems.

Somers and Evans, 1974. See above.

Spectral methods are described for generating several color quality indices for wines.

Wrolstad et al., 1982. See above.

Description of the pH differential method for determination of total anthocyanins and indices for anthocyanin degradation as applied to fruit juices and wines.

Contributed by M. Mónica Giusti,
University of Maryland
College Park, Maryland

Ronald E. Wrolstad,
Oregon State University
Corvallis, Oregon

Separation and Characterization of Anthocyanins by HPLC

Anthocyanins are water-soluble pigments which impart the red, purple, and blue coloration of many fruits, vegetables, and cereal grains. Their analysis is useful for commodity identification since the anthocyanin fingerprint pattern is distinctive for different commodities. The authors' laboratory has demonstrated how HPLC anthocyanin analyses can be effectively applied to determine the authenticity of various anthocyanin containing fruit juices (Wrolstad et al., 1994).

Two primary analytical methods for HPLC separation of anthocyanins are described (see Basic Protocol 1 and Alternate Protocol). Basic Protocol 1 has a shorter analysis time and is appropriate for matrices containing simpler anthocyanin glycosides, while the Alternate Protocol is longer and more suitable for those with more complex, nonpolar, acylated anthocyanins. In addition, a couple of auxiliary procedures are presented that are useful for characterization of unknown anthocyanin peaks. The first of these simplifies a chromatogram through acid hydrolysis, thus removing the sugar group(s) and any attached acyl groups from the anthocyanin to form the anthocyanidin (aglycon; see Basic Protocol 2). There are only six anthocyanidins that occur in nature (Figure F1.3.1), so a chromatogram can often be greatly simplified by this treatment. It also helps to confirm the identity of the parent compound or compounds by identification of the aglycon. The final technique discussed is base saponification to remove acylating groups that may be attached to some of the anthocyanins (see Basic Protocol 3). This procedure can often be used prior to and in conjunction with acid hydrolysis to more fully characterize a particular compound. Identification of the sugars can be accomplished by GLC of their tri-methyl-silyl derivatives and acylating acids by HPLC (Gao and Mazza, 1994). For more complete identification, additional techniques (e.g., electrospray mass spectroscopy and NMR; Giusti et al., 1999) are necessary for determining the molecular weight, the nature of glycosidic linkages, and the sites of acyl and sugar substitution.

The protocols presented here allow one to analyze the anthocyanins in fruit juices, natural colorants, and extracts from various anthocyanin sources. These profiles are useful for the identification of species, varieties, and for quality assessment of commercial products. They are also used to detect misbranding or adulteration of fruit products with other anthocyanin containing fruits, juices, or colorants.

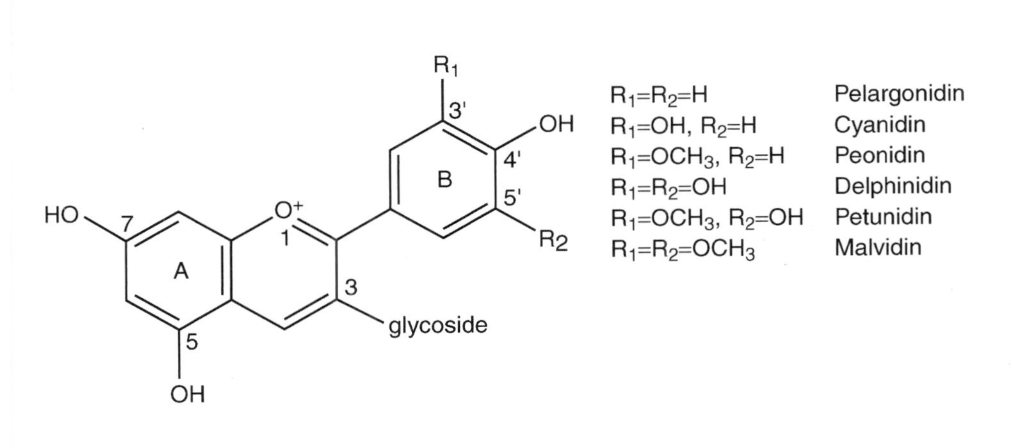

Figure F1.3.1 Generalized structure for anthocyanin pigments. A and B rings are labelled.

$R_1=R_2=H$ Pelargonidin
$R_1=OH, R_2=H$ Cyanidin
$R_1=OCH_3, R_2=H$ Peonidin
$R_1=R_2=OH$ Delphinidin
$R_1=OCH_3, R_2=OH$ Petunidin
$R_1=R_2=OCH_3$ Malvidin

Contributed by Robert W. Durst and Ronald E. Wrolstad

Anthocyanins

F1.3.1

NOTE: While a variable wavelength UV-Vis detector is sufficient for the protocols in this unit, a UV-Vis diode array detector, which collects spectra for individual peaks, greatly enhances the methods.

SAMPLE PREPARATION OF ANTHOCYANINS AND THEIR HPLC SEPARATION ON SILICA C₁₈ COLUMNS

This method is used for anthocyanin samples of lesser complexity that do not contain acylated pigments. It is the most common method used for anthocyanin analysis and is effective for most commodities; the major exceptions being red grape and cabbage containing products. This protocol describes the dilution, filtration, and reversed-phase HPLC analysis of these samples.

Materials

Acidified H₂O: add 2 to 4 drops of HCl (~0.01% final) in 500 ml H₂O in a wash bottle

4% phosphoric acid (H₃PO₄)

Acetonitrile

10% acetic acid/5% acetonitrile/1% phosphoric acid in water, filter sterilized: store up to 1 month at 25°C

Anthocyanin standards (optional): cranberry juice cocktail or purified anthocyanins (Extrasynthese or Polyphenols AS)

0.45-μm syringe filters suitable for aqueous samples

HPLC system capable of generating binary gradients, detector capable of monitoring at 520 nm, and reversed-phase C₁₈ column (e.g., Phenomenex Prodigy ODS-3 or Supelco LC-18)

1. Dilute samples with water, acidified water, or 4% phosphoric acid, as appropriate to confine the absorbance of the peaks within the limits of the detector as determined by trials. Filter through 0.45-μm filter.

   For most applications, this is all the sample preparation that is necessary. Appropriate dilution is deceptive as it will vary from one commodity to another dependent on the number of anthocyanins in the matrix. The visual appearance and absorptivity of a sample is often not a sufficient indicator for HPLC peak height/area, as samples that have many anthocyanin peaks will need to be more concentrated than a sample that has the same total anthocyanin content but only a single major peak. Most anthocyanin-containing fruit juices will give a suitable response between single strength juice and a dilution of 1:5.

2. Set the flow rate of the HPLC system to 1 ml/min across a 5-μm × 250-mm × 4.6-mm Prodigy ODS-3 column (or equivalent) at ambient temperature. Set the detector at 520 nm. Inject 50 μl sample into the HPLC system and start a gradient similar to that

Table F1.3.1 HPLC Gradient for Anthocyanin Separation on Silica C₁₈ Columns

Time (min)	Percent A[a]	Percent B[b]
0	0	100
5	0	100
20	20	80
25[c]	40	60
30	0	100

[a]Solvent A is acetonitrile (CH₃CN).

[b]Solvent B is 10% acetic acid/5% acetonitrile/1% phosphoric acid in water.

[c]Useful data ends at 25 min.

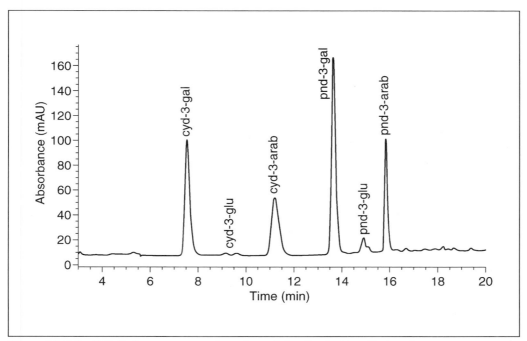

Figure F1.3.2 HPLC chromatogram of cranberry juice. Peaks identified on figure.

outlined in Table F1.3.1. Analyze data as described (see Data Analysis in Critical Parameters and Troubleshooting).

The exact gradient will have to be determined by the analyst, as variations in column brands causes significant differences in the retention times of the peaks of interest. A good material to use as a reference sample is commercial cranberry juice cocktail. It has a simple profile of four major peaks and two minor ones that cover the retention times of most peaks of interest, and will give the experimenter some assurance that the analytical system is behaving satisfactorily. See Figure F1.3.2 for a sample chromatogram with the peaks identified. Another alternative is purified anthocyanins.

Suggested changes are to increase the percentage A at the beginning if the retention time for early peaks are considerably longer than that shown in Figure F1.3.2 or decrease it at 20 min if the later peaks are not well resolved (see Critical Parameters and Troubleshooting).

The step from 20 to 25 min is included to wash highly-retained, non-polar materials from the column. If one sees significant anthocyanin peaks in this region, it may indicate that the sample contains acylated anthocyanins, and that the Alternate Protocol would be more appropriate for that particular sample.

SAMPLE PREPARATION OF ACYLATED ANTHOCYANINS AND THEIR SEPARATION ON POLYMERIC C₁₈ COLUMNS

ALTERNATE PROTOCOL

This method is used for complex anthocyanin matrices which usually contain acylated anthocyanins. It can be used for nonacylated pigments also, but because of the extra effort involved in sample preparation and the longer run times involved, it is generally not used for these compounds. Commodities appropriate for this protocol include red grapes, red cabbage, and blueberries. This protocol describes the dilution, sample preparation (including solid phase extraction), filtration, and reversed-phase HPLC analysis of samples using a polymeric based C_{18} column. Steps 1 to 4 are used to separate the anthocyanins from sugars, acids, and other water soluble compounds that are likely present in the sample.

Anthocyanins

F1.3.3

Additional Materials (*also see Basic Protocol 1*)

Acidified methanol: add 2 to 4 drops of HCl (~0.01% final) to 500 ml methanol in a wash bottle

C_{18} solid phase extraction (SPE) cartridges (e.g., C_{18} Sep-Pak; Waters Chromatography)

Syringe

Rotary evaporator and appropriate flasks

5-μm × 250-mm × 4.6-mm polymeric support reversed phase (C_{18}) column (e.g., Polymer Labs PLRP-S)

Isolate anthocyanins

1. Activate a C_{18} SPE cartridge through successive applications of 5 ml acidified methanol and 5 ml acidified water.

2. Using a syringe, apply an anthocyanin containing sample (typically 1 ml) to an activated C_{18} SPE cartridge. Wash twice with 5 ml water each time to remove sugars and acids.

   If this is a qualitative analysis, quantities are not critical, apply as much material as can be conveniently loaded on the SPE cartridge without excessive bleed. If one is performing quantitative analysis, then careful measurements of the volume applied to the SPE cartridge, and the final volume after evaporation and after solvation are necessary.

3. Elute the anthocyanins with two 5-ml aliquots of acidified methanol and collect eluate in a flask that can be put on a rotary evaporator.

4. Remove the acidified methanol from the sample on a rotary evaporator until a small drop of liquid is still present.

   Try not to evaporate the sample all the way to dryness as it will be difficult to dissolve.

   This constitutes the purified anthocyanin fraction that is subjected to further sample preparation in this protocol and elsewhere (see Basic Protocols 2 and 3).

Prepare sample

5. Dissolve the sample in 4% phosphoric acid, dilute as necessary, and pass through a 0.45-μm filter.

   Appropriate dilution is deceptive as it will vary from one commodity to another dependent on the number of anthocyanins in the matrix. The visual appearance and absorptivity of a sample is often not a sufficient indicator for HPLC peak height/area, as samples that have many anthocyanin peaks will need to be more concentrated than a sample that has the same total anthocyanin content but only a single major peak.

   Most anthocyanin-containing fruit juices will give a suitable response between single strength juice and a dilution of 1:5.

Table F1.3.2 HPLC Gradient for Anthocyanin Separation on Polymeric C_{18} Columns

Time	Percent A[a]	Percent B[b]
0	6	94
55	25	75
65[c]	25	75
70	6	94

[a]Solvent A is acetonitrile (CH_3CN).

[b]Solvent B is 4% phosphoric acid (H_3PO_4) in water.

[c]Useful data ends at 65 min.

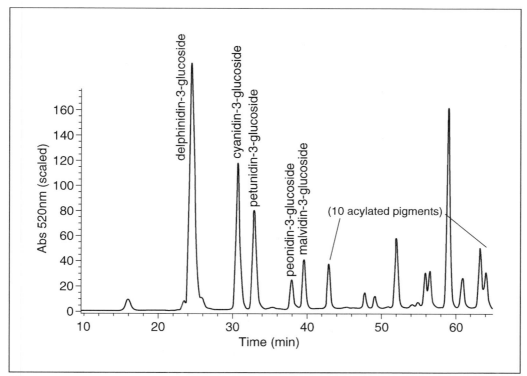

Figure F1.3.3 HPLC chromatogram of concord grape juice. Peaks identified on figure.

Analyze

6. Set the flow rate of the HPLC system to 1 ml/min across a 5-μm × 250-mm × 4.6-mm C_{18} polymeric support reversed-phase column or equivalent at ambient temperature. Set the detector at 520 nm. Inject 50 μl sample into the HPLC system and start a gradient similar to that outlined in Table F1.3.2. Analyze data as described (see Data Anaylsis in Critical Parameters and Troubleshooting).

> The exact gradient will need to be determined by the analyst, as variations in column brands cause significant differences in the retention times of the peaks of interest. A good sample to use as a working standard is concord grape juice. It has a complicated profile of 7 major peaks and up to 10 minor ones that cover the range of retention times for most peaks of interest, and will give the experimenter some assurance that the system is behaving satisfactorily. See Figure F1.3.3 for a sample chromatogram with peak identification.

> Suggested changes are to increase the percentage of A at the beginning if the retention time for early peaks are considerably longer than that shown in Figure F1.3.3, or decrease it at 55 min if the later peaks are not well resolved (see Critical Parameters and Troubleshooting).

ANTHOCYANIDINS: PREPARATION AND HPLC

This method is used to simplify a chromatogram by reducing the number of compounds in a sample to the six aglycons. This protocol describes the dilution, preparation (including solid phase extraction and acid hydrolysis), filtration, and reversed-phase HPLC analysis of the sample.

Materials

Acidified methanol and H_2O: add 2 to 4 drops of HCl (~0.01% final) to 500 ml methanol or H_2O in a wash bottle
2 N HCl
Nitrogen gas
4% phosphoric acid
Acetonitrile
10% acetic acid/5% acetonitrile/1% phosphoric acid in water, filter sterilized: store up to 1 month at 25°C
Anthocyanidin standards (optional): e.g., concord grape juice (cyanidin, delphinidin, malvidin, peonidin, and petunidin), strawberry juice (cyanidin and pelargonidin)

C_{18} solid phase extraction (SPE) cartridges (C_{18} Sep-Pak; Waters Chromatography)
Rotary evaporator and appropriate flasks
20-ml screw-top test tubes with Teflon-lined caps
Boiling water and ice baths
0.45-μm filters
HPLC system (see Basic Protocol 1)

1. Isolate anthocyanin fraction (see Alternate Protocol, steps 1 to 4).

2. Dissolve the sample in 10 ml 2 N HCl and transfer to a 20-ml screw-top test tube. Flush the tube with nitrogen gas and seal with Teflon-lined cap.

3. Place tube in a boiling water bath and allow to hydrolyze for 30 min.

4. Cool sample in an ice bath.

5. Apply sample to an activated C_{18} SPE cartridge and rinse twice with 5 ml water each time. Elute anthocyanins twice with 5 ml acidified methanol each time and collect in a flask appropriate for a rotary evaporator.

6. Remove acidified methanol from the sample on the rotary evaporator until a small drop of liquid is still present.

   Try not to evaporate the sample all the way to dryness. It makes redissolving difficult.

7. Dissolve the sample in 2 to 5 ml 4% phosphoric acid, filter through 0.45-μm filter and place in a 20-ml screw-top test tube. Flush the tube with nitrogen, seal with a Teflon-lined cap, and store on ice in the dark until injection.

   The anthocyanidins are unstable and very prone to oxidation and light induced polymerization, hence these precautions. The sample should be analyzed within 60 min or significant degradation may occur in spite of the above precautions.

8. Set the flow rate of the HPLC system to 1 ml/min across a 5-μm × 250-mm × 4.6-mm Prodigy ODS-3 column (or equivalent) at ambient temperature. Set the detector at 520 nm. Inject 50 μl sample into the HPLC system and start a gradient similar to that outlined in Table F1.3.3. Analyze data as described (see Data Analysis in Critical Parameters and Troubleshooting).

The exact gradient will have to be determined by the analyst. Variations in column brands show significant differences in the retention times of the peaks of interest. A good sample to use as a working standard is commercial concord grape juice. It contains five of the six anthocyanidins (it is missing pelargonidin). Pelargonidin can be prepared by hydrolyzing a strawberry sample, which should contain pelargonidin (~90%) and small amounts of cyanidin. See Figure F1.3.4 for a sample chromatogram with the peaks identified.

Suggested changes would be to increase the percentage of solution A at the beginning if retention times are excessively long or decrease it at 20 min if there is incomplete resolution. It is not unusual to have minor amounts of unhydrolyzed anthocyanins glycosides in the sample. If amounts are excessive, increase the hydrolysis time or decrease the amount of sample subjected to hydrolysis.

Table F1.3.3 HPLC Gradient for Anthocyanidin Separation on C_{18} Silica Columns

Time	Percent A[a]	Percent B[b]
0	5	95
20[c]	20	80
25	5	95

[a]Solvent A is acetonitrile (CH_3CN).

[b]Solvent B is 10% acetic acid/5% acetonitrile/1% phosphoric acid in water.

[c]Useful data ends at 20 min.

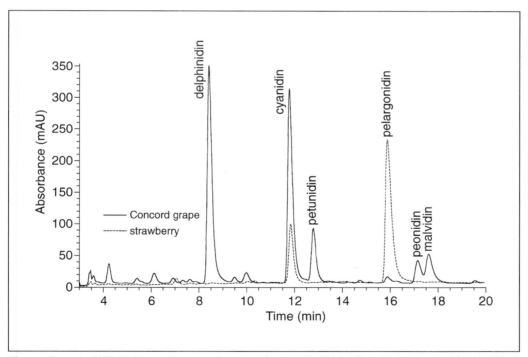

Figure F1.3.4 HPLC chromatogram of concord grape (solid line) and strawberry (dashed line) anthocyanins. Peaks identified on figure.

SAPONIFICATION OF ACYLATED ANTHOCYANINS AND THEIR HPLC SEPARATION

This method clearly shows which anthocyanin peaks in a sample matrix contain acyl substituents. This protocol describes sample dilution and preparation (including solid phase extraction and saponification). Subsequent HPLC analysis will be via one of the protocols already described for anthocyanin separation (see Basic Protocol 1 and Alternate Protocol).

Materials

Acidified methanol and H_2O: add 2 to 4 drops of HCl (~0.01% final) to 500 ml methanol or H_2O in a wash bottle
10% (w/v) KOH
2 N HCl
4% phosphoric acid
Nitrogen gas (optional)

C_{18} solid phase extraction (SPE) cartridges (e.g., C_{18} Sep-Pak; Waters Chromatography)
Rotary evaporator and appropriate flasks
20-ml screw-top test tubes with Teflon-lined caps
0.45-μm filters

1. Isolate anthocyanin fraction (see Alternate Protocol, steps 1 to 4).

2. Dissolve the sample in ~10 ml 10% (w/v) KOH in a 20-ml screw-top test tube with Teflon-lined cap and allow to react at room temperature for 8 to 10 min in the dark. Neutralize the sample with ~5 ml 2 N HCl.

3. Apply sample to an activated C_{18} SPE cartridge. Rinse twice with 5 ml water each time. Elute anthocyanins by adding 5 ml acidified methanol twice and collect in a rotary evaporator flask.

4. Remove acidified methanol from the sample on the rotary evaporator until a small drop of liquid is still present.

   Try not to evaporate the sample all the way to dryness. It makes redissolving difficult and may also induce hydrolysis of the glycosidic substituents.

5. Dissolve the sample in 2 to 5 ml 4% phosphoric acid, filter through a 0.45-μm filter and place in a 20-ml screw-top test tube with Teflon-lined cap.

   The anthocyanins are generally stable, but should be analyzed within a day or so. For samples that need to be stored longer, flush tube with nitrogen and store refrigerated or frozen.

6. Analyze the sample by an appropriate HPLC procedure (see Basic Protocol 1 or Alternate Protocol).

   The choice of which protocol to use is determined by data interpretation needs. Basic Protocol 1 may be appropriate since the sample should no longer contain any acylated anthocyanins. If one needs to compare the saponified chromatogram with that of the unsaponified sample to determine which peaks are acylated, then using the Alternate Protocol would be the better choice since direct comparisons of the two can then be made. One should see the decrease (or complete loss) of acylated peaks and the appearance of or increase in unacylated anthocyanin peaks liberated by saponification.

Background Information

Anthocyanins are the water-soluble pigments which impart the red, purple, and blue coloration of many fruits, vegetables, and cereal grains. These pigments are largely responsible for the color characteristics of raw and processed products. Their analysis is useful for commodity identification since the anthocyanin fingerprint pattern is distinctive for different foods. Figure 1.3.1 shows the generalized structure for anthocyanin pigments. There are six common anthocyanidins (aglycons liberated from anthocyanins by acid hydrolysis). Structural variations are greatly expanded through varying patterns of glycosidic substitution at the 3 and 5 positions. Additional variations can occur through acylation of the sugar substituents with organic acids. Over 300 different anthocyanin pigments have been found in nature (Strack and Wray, 1994). The anthocyanin pigment patterns exhibited by different species and even different varieties of the same species have proven useful in chemotaxonomic investigations of higher plants (Harborne and Turner, 1984). The authors have demonstrated how HPLC anthocyanin analyses can be effectively applied to determine the authenticity of various anthocyanin containing fruit juices (Wrolstad et al., 1994).

HPLC has become the method of choice for analysis of anthocyanin pigments. It offers several advantages over the traditional techniques of paper or thin-layer chromatography (TLC) such as greater resolution, shorter analysis times, and easy quantitation. Reversed-phase chromatography also has the attractive feature of predictability of elution order based on polarity; triglycosides typically elute before diglycosides, which elute before monoglycosides. An exception to this generalization are the -3-rutinosides which have longer retention times than the corresponding -3-glucosides because of the nonpolarity imparted by the C-6 methyl group of rhamnose. Glycosides of hexoses are more polar and elute earlier than glycosides of pentose sugars, e.g., -3-galactosides elute prior to -3-arabinosides. The elution order of the aglycons can be predicted on the basis of the number of hydrophilic phenolic and hydrophobic methoxyl groups. Elution order is: delphinidin (dpd), cyanidin (cyd), petunidin (ptd), pelargonidin (pgd), peonidin (pnd), and malvidin (mvd). Because of their instability and inherent difficulties for purification, the availability of anthocyanin standards has been limited and they have tended to be expensive. This has hampered the technique from progressing beyond the fingerprint or semiquantitative stages. Fortunately this does not seriously limit the utility of HPLC for qualitative analyses since the anthocyanins of most commodities are well characterized and auxiliary techniques are available for making correct peak assignments (Hong and Wrolstad, 1990a,b). Juices and extracts (e.g., prepared from cranberries, grapes, strawberries) can themselves serve as references for retention indices and spectral characteristics. Commercial sources for purified anthocyanins have recently become more available, which facilitates identification and quantitation (e.g., Extrasynthese and Polyphenols AS).

Critical Parameters and Troubleshooting

The sample preparation procedures and HPLC separations described in this unit are designed for aqueous samples. When investigating the anthocyanin composition of plant materials and solid foods, aqueous extracts can be prepared through acetone extraction as described in *UNIT F2.1*. When methanol extraction is used for anthocyanin isolation (*UNIT F1.1*), the methanol should be nearly completely removed on a rotary evaporator and replaced with water. For isolation of anthocyanins by solid-phase extraction (*UNIT F1.1*), it is critical that the sample be in aqueous solution. Injection of a dilute alcoholic solution (<15%) of anthocyanins may have relatively little effect on chromatographic behavior, since the small injection volume (typically 50 μl) will have a minor influence on the composition of the mobile phase which starts at about 5% organic solvent. Wines can be injected as is, the amount of ethanol in the injected wine sample being too small to have a significant impact on mobile phase composition.

Isolation of anthocyanins by solid-phase extraction (SPE; see Alternate Protocol) prior to injection is not necessary for most analyses. This is particularly the case for separation on Silica C_{18} columns (see Basic Protocol 1); however, in cases where the sample contains considerable degraded or polymeric pigment because of processing or storage abuse, improved chromatographic resolution will be achieved through sample clean-up. The experience of the authors has been that materials containing acylated anthocyanins (e.g., red grapes, red cabbage, radishes) and highly complex samples (e.g., blueberries) may also benefit from anthocyanin purification with C_{18} SPE.

These are relatively simple and robust procedures. Some of the factors to keep in mind when working with anthocyanins are excessive heat, light, oxidation, and sample handling, as these factors can alter or destroy these labile compounds. The preparation and HPLC separation of anthocyanins on silica C_{18} columns (see Basic Protocol 1) is the easiest and most robust of the procedures. Typical care in filtering samples and solvents prior to HPLC analysis is about all that is necessary.

Preparation of acylated anthocyanins and anthocyanidins (see Basic Protocol 2 and Alternate Protocol) requires that the SPE cartridges are not overloaded with sample, or that sample is lost because it is not adsorbed onto the SPE cartridge. Also, the sample should not be rotoevaporated to dryness or left at elevated temperature for excessive time periods.

Preparation of anthocyanidins (see Basic Protocol 3) requires that samples be handled quickly and kept cold and in the dark to prevent degradation.

Saponification of acylated anthocyanins and their HPLC separation (see Basic Protocol 3) requires that the samples be thoroughly saponified (longer time if necessary) to differentiate acylated peaks from those that are not.

Table F1.3.4 presents more information on common problems.

Data analysis

The methods described in this unit are qualitative analysis; therefore, results are typically reported as percent total peak area. Chromatographic peaks that are less than 1% of total peak area are generally ignored during data

analysis, which is justified since these small peaks do not materially affect the color or composition of a product. The analyses can be performed in a quantitative fashion if the analyst is careful to adjust and record all volumes and quantities during extraction procedures. The use of an authentic external standard such as cyd-3-glu is necessary to determine proper coefficients to use for calculations. Even though the extinction coefficients are different for each anthocyanin (varies with aglycon and sugar substituent), it would be impractical to try and quantitate each peak with it's specific standard; therefore, it is the accepted practice to use the values from a single standard (i.e., cyd-3-glu) and apply those values to all the peaks in a chromatogram. Other useful numbers that are sometimes reported include peak ratios — i.e., (% cyd-3-gal/% cyd-3-arab) or even to sum up various peaks to get the ratios by aglycon cyd/pnd — e.g., (% cyd-3-gal + %cyd-3-glu + % cyd-3-arab)/(% pnd-3-gal + % pnd-3-glu + % pnd-3-arab).

Anticipated Results

Either Basic Protocol 1 or the Alternate Protocol should first be conducted in analyzing an unknown sample. Because of its ease and simplicity, sample preparation of anthocyanins and their HPLC separation on silica C18 columns (see Basic Protocol 1) is usually the preferred choice, unless the presence of acylated anthocyanins is anticipated, in which case the protocol described for acylated anthocyanins is used (see Alternate Protocol). If the anthocyanin profile is inconsistent with previously published chromatograms, or if there are extraneous unidentified peaks, then simplification is recommended (see Basic Protocol 3). Acid hydrolysis will simplify the chroma-

Table F1.3.4   Troubleshooting Guide for HPLC Analysis of Anthocyanins and Anthocyanidins

Problem	Possible cause	Solution
All peaks come out early and/or bunched together	Differences between column elution profiles	Adjust gradient to decrease percentage acetonitrile at early run times.
Not all peaks are eluted by the end of the gradient or they are eluted during the wash out section of the gradient	Differences between column elution profiles	Adjust gradient to increase percentage acetonitrile at later run times, earlier run times, or both.
Two very early peaks, resolved, but shortly after the void volume of the column	Betalains (beet colorant) in the sample	Check the peak spectrum against the literature or standards to see that they are actually anthocyanins.

togram since there are only six anthocyanidins, and few commodities contain more than 2 or 3 of them. This protocol is very complementary to Basic Protocol 1 for investigations on fruit juice authenticity. For example, red raspberry contains cyanidin and pelargonidin but never contains delphinidin, while strawberry should have ~90% pelargonidin, 10% cyanidin, and no other anthocyanidins; cranberry contains about equal amounts of cyanidin and peonidin with traces of delphinidin but never pelargonidin nor malvidin, while grape samples always contain malvidin.

Data interpretation and peak identification are aided when one uses a diode array detector. Acquiring spectral information about the peaks as they elute from the chromatogram allows one to deduce some information about the structure of the peaks, and to determine peak purity and accurate wavelength of maximum absorption. The latter helps to determine the amount of hydroxyl and methoxyl substitution on the B ring (Fig. F1.3.1; Hong and Wrolstad, 1990a,b). The amount and type of phenolic or organic acid acylation can be inferred from the UV spectrum (Hong and Wrolstad, 1990a,b). Figure F1.3.5 shows that acylation can be determined from the $A_{acyl-max}/A_{vis-max}$ (Harborne, 1958). The $A_{acyl-max}$ is the peak in the 310 to 320 nm region. If the ratio is very low, there is no acylation. If it is 0.5 to 0.7 then there is a single acylation, while ratios of 0.8 to 1.1 indicate two acylating groups. Figure F1.3.5 also shows that -3- versus -3,5-glycosylation can be determined from the $A_{440}/A_{vis-max}$ (Hong and Wrolstad, 1990a,b). If the ratio is greater than ~0.3 the peak is -3-glycosylated. If it is less than 0.2 it is -3,5-diglycosylated.

Spectral information can be used to determine if peaks are acylated. This is determined by a peak in the 310 nm region (Hong and Wrolstad, 1990a,b). If the sample contains acylated peaks, then the longer Alternate Protocol is recommended. Simplification (see Basic Protocol 3) is advised for further characterization of the pigments. One should see elimination (reduction) of those peaks which are acylated anthocyanins, and either an increase in the glycoside peak(s), or the appearance of a new peak if the parent glycosylated pigment is not present in the original sample.

Typical HPLC chromatograms with peak assignments for several common fruits are illustrated in a book chapter by Wrolstad et al. (1994). While most of the chromatograms shown were generated using the Alternate Protocol, the pattern (elution order and relative peak areas) is essentially the same as for Basic

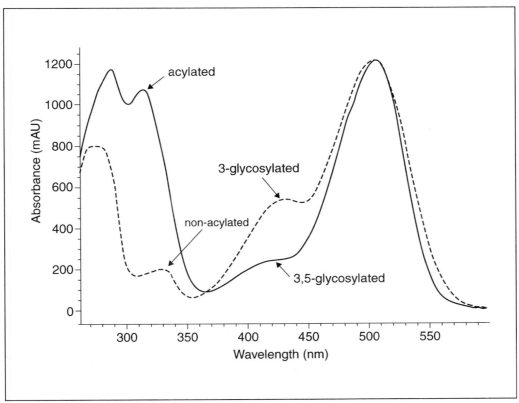

Figure F1.3.5 Spectra of -3-glycosylated (nonacylated; dashed line) and -3,5-glycosylated (acylated; solid line) pigments.

Protocol 1, except that retention times are longer. One should routinely run a sample of known composition to confirm retention times and ensure that the system is running appropriately. This is important as the analysis is quite sensitive to minor changes in solvent composition. Cranberry juice cocktail makes a good reference material that is readily available, inexpensive, and includes peaks over a broad range of retention times. Authentic standards can also be purchased. Cyanidin-3-glucoside is an appropriate choice since it is the most common occurring anthocyanin pigment in nature. Availability of authentic standards makes quantitative analysis of the anthocyanins possible using the external standard method.

Figure F1.3.4 shows HPLC chromatograms for anthocyanidins generated from acid hydrolysis of concord grape and strawberry juices. Extraneous peaks may be present because of incomplete hydrolysis, and degradation and polymerization of the labile aglycons even more of a problem. For acylated anthocyanins, higher yields of anthocyanidins will be achieved if the sample is first saponified (see Basic Protocol 3) and then subjected to acid hydrolysis (see Basic Protocol 2).

Figure F1.3.6 compares the HPLC chromatograms for radish anthocyanin extract before and after saponification. Prior to saponification there are four unique peaks present in the sample. Upon removal of the acylating group(s) from each of the four peaks by Basic Protocol 3, the same parent is formed (pelargonidin-3-sophoroside-5-glucoside), hence only a single peak in the chromatogram after saponification (Giusti and Wrolstad, 1996).

Time Considerations

In Basic Protocol 1 both the sample preparation and analysis are quick and easy. Sample preparation is a matter of a few minutes, while analysis takes 30 to 35 min between injections.

The Alternate Protocol has a longer sample preparation time (15 to 20 min) and a considerably longer analysis time of 75 min between injections.

Basic Protocol 2 takes about an hour to prepare a sample, but by staggering the start of preparation, 2 to 3 samples/hour can be prepared. The limiting factor is getting them analyzed before they degrade, since the analysis time is about 30 min between injections. Remember that it will likely be necessary to prepare a concord grape standard and/or a strawberry standard periodically to confirm retention times.

Basic Protocol 3 sample preparation takes 15 to 25 min with analysis taking either 30 to 35 min if Basic Protocol 1 is used or 75 min if Basic Protocol 2 is used for analysis.

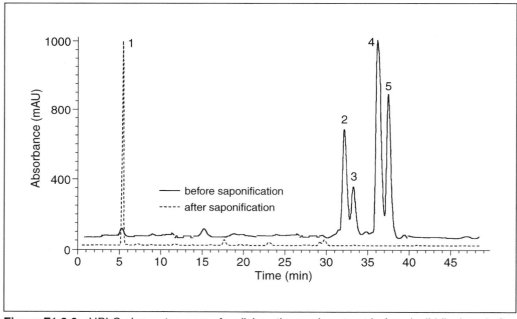

Figure F1.3.6 HPLC chromatograms of radish anthocyanin extract before (solid line) and after (dashed line) saponification. Peak 1: pelargonidin-3-sophoroside-5-glucoside; Peaks 2, 3, 4, 5: pelargonidin-3-sophoroside-5-glucoside acylated with p-coumaric (2), ferrulic (3), p-coumaric and malonic (4), or ferrulic and malonic acid (5), respectively. Note that the saponified sample has only pelargonidin-3-sophoroside-5-glucoside.

Literature Cited

Gao, L. and Mazza, G. 1994. A rapid method for complete characterization of simple and acylated anthocyanins by high performance liquid chromatography and capillary gas liquid chromatography. *J. Agric. Food Chem.* 42:118-125.

Giusti, M.M. and Wrolstad, R.E. 1996. Characterization of radish anthocyanins. *J. Food Sci.* 61:322-326.

Giusti, M.M., Rodriguez-Saona, L., Griffin, D., and Wrolstad, R.E. 1999. Electrospray and tandem mass spectroscopy as tools for anthocyanin characterization. *J. Agric. Food Chem.* 47:4657-4664.

Harborne, J.B. 1958. The chromatographic identification of anthocyanin pigments. *J. Chromatogr.* 1:473-488.

Harborne, J.B. and Turner, B.J. 1984. Plant Chemosystematics. Academic Press, New York.

Hong, V. and Wrolstad, R.E. 1990a. Use of HPLC separation/photodiode array detection for characterization of anthocyanins. *J. Agric. Food Chem.* 38:708-715.

Hong, V. and Wrolstad, R.E. 1990b. Characterization of anthocyanin containing colorants and fruit juices by HPLC/photodiode array detection. *J. Agric. Food Chem.* 38:698-708.

Strack, D. and Wray, V. 1994. The anthocyanins. *In* The Flavonoids, Advances in Research Since 1986 (J.B. Harborne, ed.) pp. 1-22. Chapman and Hall, New York.

Wrolstad, R.E., Hong, V., Boyles, M.J. and Durst, R.W. 1994. Use of anthocyanin pigment analyses for detecting adulteration in fruit juices. *In* Methods to Detect Adulteration of Fruit Juice Beverages (S. Nagy and R.L. Wade, eds.) pp. 260-286. AgScience, Auburndale, Fla.

Key References

Hong and Wrolstad, 1990a. See above.

Methods for systematic identification of anthocyanins are described.

Hong and Wrolstad, 1990b. See above.

The HPLC and spectral characteristics for 10 different anthocyanin extracts and colorants are presented.

Wrolstad et al., 1994. See above.

HPLC chromatograms with peak assignments are given for several commodities along with applications of anthocyanin analyses for determining authenticity of fruit juices.

Contributed by Robert W. Durst
   and Ronald E. Wrolstad
Oregon State University
Corvallis, Oregon

Characterization of Anthocyanins by NMR

According to the authors' unpublished files, the structures of more than 620 different anthocyanins have hitherto been elucidated. Individual anthocyanins have diverse impact on color and color stability of many fresh and processed fruits and vegetables, as well as, most probably, potential with respect to health benefits. Thus, the demand for exact structure elucidations has swelled with increased interest for these water-soluble plant pigments, which have colors ranging from salmon-pink through red, violet, and nearly black. In this context, the development of reliable superconducting magnets coupled to application of pulse technique and its associated Fourier transform have made nuclear magnetic resonance spectroscopy (NMR) the most important tool for complete structure elucidation of anthocyanins. The NMR method is especially valuable for determination of the nature of the sugar moieties and the sites of acyl and sugar substitutions. Other methods and instrumentation used for separation and characterization of anthocyanins have been described in *UNITS F1.1 & F1.2*, as well as *UNIT F1.3*.

This unit consists of a description of a sequence of NMR experiments (^{1}H, CAPT, DQF-COSY, TOCSY, HSQC, HMBC; see Background Information), which normally will be sufficient to achieve complete assignments of the proton and carbon resonances of anthocyanins (see Basic Protocol), and assumes that sufficient pure anthocyanin material (at least 1 mg) has been obtained. This sequence will also supply enough information to determine the linkage positions between the anthocyanin building blocks. Values for additional parameters typical for the HMBC, HSQC, DQF-COSY, TOCSY, NOESY, and ROESY experiments applied on anthocyanins are included in Table F1.4.1. This unit also describes preparation of anthocyanin NMR samples (see Support Protocol 1), and a method for revealing important steps during recording of a 1-D ^{1}H NMR spectrum (see Support Protocol 2). The latter may be used as a basis for recording ^{13}C NMR spectra, and for initial preparation of two-dimensional homo- and heteronuclear experiments. A discussion concerning 2-D NMR experiments (DQF-COSY, TOCSY, HSQC, HMBC, NOESY, and ROESY), chemical shifts (^{1}H and ^{13}C), and coupling constants (^{1}H-^{1}H) on selected anthocyanins (Table F1.4.3 to Table F1.4.7), as well as other information, is included in the later part of the unit (see Commentary). As an example for analysis, a detailed route for structure elucidation of the major anthocyanin cyanidin 3-O-(2″-O-β-glucopyranosyl-6′-O-α-rhamnopyranosyl-β-glucopyranoside) isolated from tart cherries, *Prunus cerasus*, is described (see Support Protocol 1). An NMR glossary including some common NMR terms is also included in the appendix at the end of this unit.

RECORDING NMR SPECTRA FOR ASSIGNMENTS OF PROTON AND CARBON SIGNALS OF ANTHOCYANINS

Structure elucidation, including NMR, depends on assignments of the proton and carbon resonances. This protocol describes a sequence of NMR experiments which, in most cases, will assign all the proton and carbon resonances of the individual building blocks of an anthocyanin, as well as giving the linkage points between these units.

Various parameters must be taken into account when an NMR experiment is performed. Different manufacturers (e.g., Bruker, Varian, Jeol) label some of these parameters differently; however, the user manuals usually contain enough details about basic acquisition and processing. For the NMR experiments included in this protocol, the authors recommend referring to Braun et al. (1998) for descriptions of the pulse programs and important acquisition and processing parameters. The nomenclature used in this protocol

---

Contributed by Øyvind M. Andersen and Torgils Fossen

Anthocyanins

F1.4.1

follows the Bruker AMX instrument using the UXNMR system; however, a glossary of abbreviations and symbols of other manufacturer dialects (Varian, Jeol) is included as an appendix in this key reference. Table F1.4.1 contains important acquisition parameters. Background information about the same NMR experiments applied on anthocyanins is provided later in the unit (see Commentary).

The proton resonance frequency, not the field strength, is usually applied to classify the NMR machines. Thus, the NMR instrument should be at least 300 MHz or stronger (e.g. the field strength must be 70 kG or stronger) to achieve constructive anthocyanin spectra. Optimal spectral resolution depends on several factors, including the quality of the NMR tubes (Table F1.4.2).

Table F1.4.1 Parameters for 2-D NMR Experiments Performed on a Typical Anthocyanin[a,b]

Experiment	Example	SW (^{13}C) ppm	SW (^1H)	Number of scans	Number of FIDs (t_1)	Data points (t_2)	Optimized coupling constants
HMBC	Fig. F1.4.7	160	8.5	2	512	2K	145 Hz ($^1J_{CH}$) 8Hz ($^nJ_{CH}$)
HSQC	Fig. F1.4.6	160	8.5	2	512	2K	145 Hz ($^1J_{CH}$)
COSY	Fig. F1.4.3		8.5	2	512	2K	7.5 Hz ($^3J_{HH}$ and $^2J_{HH}$)
TOCSY	Fig. F1.4.5		9.2	16	512	2K	
NOESY			8.5			2K	
ROESY			8.5			2K	

[a]Abbreviations: SW, sweep width.

[b]Cyanin 3-(2″-glucosyl-6″-rhamnosylglucoside) (**15**; see Fig. F1.4.1), recorded at 600 MHz for ^1H and ^{13}C, respectively, on a Bruker DMX 600 instrument.

Table F1.4.2 NMR Sample Tube Requirements for NMR Instruments with Different Field Strengths[a]

NMR instrument	Norell sample tube	Wilmad sample tube
Up to 900 MHz	No. 5020-USP	Wilmad 542
Up to 800 MHz	No. 5010-USP	Wilmad 541
Up to 600 MHz	No. 509-UP	–
Up to 500 MHz	No. 508-UP	Wilmad 535
Up to 400 MHz	No. 507-HP	Wilmad 528
Up to 300 MHz	No. 506-P	Wilmad 597, 526, 527

[a]Data provided by the NMR sample tube producers Wilmad and Norell.

Table F1.4.3 Guide to molecules used in Tables F1.4.4 and F1.4.5[a]

Identifier	Name	Reference
1	Cyanidin 3-arabinoside	Ramstad et al., 1995
2	Delphinidin 3-rhamnoside	Cabrita et al., 2000
3	Delphinidin 3-galactoside	Fossen et al., 1998
4	Pelargonidin 3-glucoside	Pedersen et al., 1993
5	Cyanidin 3-glucoside	Andersen et al., 1991a
6	Peonidin 3-glucoside	Van Calsteren et al., 1991
7	Apigeninidin 5-glucoside	Swinny et al., 2000
8	Petunidin 3-rhamnoside-5-glucoside	Catalano et al., 1998
9	Malvidin 3,5-diglucoside	Andersen et al., 1995
10	Cyanidin 3-(2″-xylosylgalactoside)	Cabrita, 1999
11	Cyanidin 3-(2″-xylosylglucoside)	Andersen et al., 1991a
12	Cyanidin 3-(2″-glucosylgalactoside)	Slimestad and Andersen, 1998
13	Cyanidin 3-(2″-glucosylglucoside)	Slimestad and Andersen, 1998
14	6-Hydroxycyanidin 3-(6″-rhamnosylglucoside)	Nygård et al., 1997
15[b]	Cyanidin 3-(2″-glucosyl-6″-rhamnosylglucoside)	Unpub. observ.
16	Delphinidin 3-(6″-acetylgalactoside)	Fossen et al., 1998
17	Cyanidin 3-(6″-malonylglucoside)	Fossen et al., 1996
18	Delphinidin 3-(2″-galloyl-6″-acetylgalactoside)	Fossen et al., 1998
19[c]	Delphinidin 3-(6″-malonylglucoside)-5-glucoside	Unpub. observ.
20	Petunidin 3-(6″-(4″″-E-p-coumaroylrhamnosyl)-glucoside)-5-glucoside	Andersen et al., 1991b
21	Malvidin 3-(6″-(4″″-E-caffeoylrhamnosyl)-glucoside)-5-glucoside.	Slimestad et al., 1999

[a]All samples dissolved in CD_3OD acidified with CF_3COOD or DCl at 25°C.

[b]See Figure F1.4.1.

[c]See Figure F1.4.4.

Anthocyanins

F1.4.3

Table F1.4.4 ^{1}H NMR Spectral Data for Selected Anthocyanins[a]

Molecule[b]	1	2	3	4	5	6	7	8	9	10	11	12	13	14	15	16	17	18	19	20	21
Aglycone	cy	dp	dp	pg	cy	pn	ap	pt	mv	cy	cy	cy	cy	OHcy	cy	dp	cy	dp	dp	pt	mv
3							8.16														
4	9.02	9.04	9.07	9.10	9.10	9.00	9.24	9.05	9.24	9.00	8.96	9.05	9.06	8.99	8.90	9.00	8.99	8.99	8.90	8.97	9.09
6	6.75	6.74	6.74	6.74	6.76	6.67	7.06	7.07	7.17	6.73	6.73	6.72	6.74		6.72	6.75	6.74	6.73	7.06	7.06	7.12
8	6.98	6.94	6.95	6.96	6.98	6.94	7.14	7.08	7.26	6.95	6.94	6.95	6.98	7.20	6.92	6.96	6.95	6.90	7.05	7.06	7.12
2'	8.16	7.65	7.88	8.61	8.14	8.13	8.38	7.67	8.12	8.11	8.05	8.13	8.11	8.07	8.00	7.87	8.07	7.64	7.77	7.94	8.12
3'				7.10			7.10														
5'	7.10			7.10	7.11	7.07	7.10	7.60		7.08	7.05	7.12	7.12	7.09	7.08		7.08			7.78	8.12
6'	8.40	7.65	7.88	8.61	8.31	8.23	8.38		8.12	8.33	8.28	8.27	8.28	8.27	8.21	7.87	8.33	7.64	7.77		
OMe						3.99		4.02	4.09											4.02	4.10
3gly/5glc	ara	rha	gal	glc	glc	glc	5glc	rha	glc	gal	glc	gal	glc	rha	glc	gal	glc	gal	glc	glc	glc
1''	5.37	5.82	5.36	5.37	5.38	5.33	5.21	5.98	5.45	5.48	5.53	5.49	5.54	5.42	5.45	5.34	5.36	5.63	5.58	5.69	5.59
2''	4.13	4.35	4.11	3.76	3.78	3.64	m	4.32	3.75	4.32	4.07	4.36	4.12	3.76	4.11	4.11	3.79	5.78	3.86	3.85	3.81
3''	3.88	4.02	3.78	3.68	3.65	3.58	m	4.02	3.65	4.01	3.90	3.99	3.85	3.66	3.83	3.78	3.58	4.07	3.78	3.75	3.69
4''	4.08	3.66	4.05	3.58	3.56	3.45	3.47	3.59	3.51	4.08	3.62	4.07	3.62	3.50	3.55	4.04	3.43	4.13	3.55	3.61	3.57
5''/5A''	4.10	3.73	3.85	3.70	3.67	3.60	m	3.60	3.61	3.95	3.71	3.90	3.65	3.83	3.77	4.15	3.88	4.27	4.01	3.97	3.92
6A''/5B''	3.87	1.37	3.86	4.05	4.02	3.91	3.97	1.37	4.03	3.92	4.03	3.90	4.00	4.14	4.11	4.43	4.65	4.52	4.56	4.16	4.12
6B''			3.86	3.86	3.82	3.71	3.76		3.79	3.92	3.85	3.90	3.81	3.68	3.67	4.37	4.37	4.44	4.45	3.85	3.82
2''xyl/6''rha										xyl	xyl			rha	rha						
1'''										4.81	4.87			4.77	4.72						
2'''										3.28	3.29			3.92	3.85						
3'''										3.43	3.43			3.78	3.68						
4'''										3.52	3.53			3.41	3.39						
5'''/5A'''										3.76	3.80			3.65	3.63						
6'''/5B'''										3.13	3.16			1.28	1.20						
5glc/2''glc								5glc	5glc			2''glc	2''glc		2''glc				5glc	5glc	5glc
1'''								5.28	5.26			4.81	4.85		4.85				5.29	5.31	5.29

continued

Table F1.4.4 ¹H NMR Spectral Data for Selected Anthocyanins^a, *continued*

Molecule^b	1	2	3	4	5	6	7	8	9	10	11	12	13	14	15	16	17	18	19	20	21
2'''								3.69	3.73			3.30	3.29		3.28				3.79	3.82	3.80
3'''								3.59	3.59			3.40	3.36		3.38				3.69	3.71	3.77
4'''								3.49	3.44			3.32	3.32		3.32				3.55	3.65	3.64
5'''								3.61	3.69			3.02	3.03		3.01				3.87	3.72	3.71
6A'''								3.99	4.01			3.50	3.54		3.54				4.06	4.06	4.02
6B'''								3.79	3.82			3.50	3.54		3.54				3.85	3.92	3.89
2''gall/4'''cou/4'''caf																		gall		cou	caf
α																				6.27	5.69
β																				7.61	6.88
2																		7.06		7.46	7.22
3																				6.88	
5																				6.88	6.86
6																		7.06		7.46	7.00
6''ace/6''mal																ace	mal	ace	mal		
2''''/2''																2.13	3.44	2.18	3.46		

^aAbbreviations: ap, apigeninidin; ara, arabinose; ace, acetyl; caf, caffeoyl; cou, coumaroyl; cy, cyanidin; dp, delphinidin; gal, galactose; gall, galloyl; glc, glucose; gly, glycoside; mal, malonyl; mv, malvidin; pg, pelargonidin; pn, peonidin; pt, petunidin; OHcy, 6-hydroxycyanidin; rha, rhamnose; xyl, xylose.
^bRefer to Table F1.4.3 for molecule identities and references.

Table F1.4.5 ^{13}C NMR Spectral Data for Selected Anthocyanins[a,b]

	1	2	3	4	5	6	7	8	9	10	11	12	13	14	15	16	17	18	20	21
Aglycone	cy	dp	dp	pg	cy	pn	ap	pt	mv	cy	cy	cy	cy	OHcy	cy	dp	cy	dp	pt	mv
2	164.59	164.49	164.49	163.85	164.36	164.19	173.8	164.14	164.5	164.15	164.29	164.20	164.34	162.28	163.03	164.38	164.30	164.51	163.79	164.56
3	145.56	144.51	145.49	145.31	145.64	145.49	112.0	145.49	146.8	145.32	145.40	145.35	145.33	145.85	144.28	145.98	145.55	145.62	146.09	146.25
4	136.46	135.51	136.61	137.08	137.03	137.35	149.9	133.90	136.2	136.23	136.34	136.58	136.80	134.39	135.01	135.68	136.69	135.34	134.08	134.99
5	159.10	159.11	159.03	157.48	159.55	159.29	158.6	156.97	157.4	159.63	159.56	159.16	159.16	141.95	158.04	159.38	159.70	159.12	156.89	156.75
6	103.42	103.38	103.29	103.49	103.50	103.42	105.5	105.43	105.9	103.46	103.63	103.34	103.34	135.62	102.67	103.30	103.43	103.30	105.54	105.61
7	170.44	170.34	170.38	170.59	170.56	170.70	172.1	169.48	169.8	170.22	170.71	nd	170.23	159.37	169.50	170.24	170.49	170.33	169.73	169.90
8	95.24	95.08	95.03	95.28	95.19	95.24	98.5	97.33	97.6	95.14	95.23	95.10	95.15	95.24	94.43	95.13	95.26	95.18	97.55	97.64
9	157.64	157.71	157.72	157.44	157.75	157.86	160.2	156.69	156.9	157.45	157.75	157.58	157.60	151.56	156.51	157.64	157.71	157.66	156.61	153.33
10	113.29	113.36	113.29	113.39	113.45	113.63	114.5	113.31	113.6	113.19	113.42	113.28	113.30	114.04	112.21	113.87	113.23	112.97	113.03	113.31
1'	121.26	120.00	120.07	120.61	121.31	121.14	121.6	119.68	119.7	121.27	121.36	121.28	121.27	121.38	120.25	120.01	121.18	119.53	119.58	119.50
2'	118.48	112.11	112.62	135.67	118.56	115.19	134.1	108.56	111.1	118.66	118.70	118.69	118.58	118.11	117.62	112.71	118.37	112.68	109.52	110.97
3'	147.51	147.73	147.56	117.91	147.41	149.51	118.9	149.84	149.9	147.37	147.79	147.36	147.40	147.19	147.62	147.57	147.44	147.38	149.76	149.89
4'	155.94	144.89	144.71	166.57	155.78	156.37	168.3	147.61	147.1	155.75	156.16	155.64	155.67	154.97	154.80	144.91	155.86	144.79	147.75	146.75
5'	117.46	147.73	147.56	117.91	117.48	117.55	118.9	145.76	149.9	117.44	117.55	117.53	117.45	117.51	116.69	147.57	117.39	147.38	146.16	149.89
6'	128.58	112.11	112.62	135.67	128.22	128.84	134.1	113.90	111.1	128.57	128.80	128.09	128.22	127.68	127.50	112.71	128.47	112.68	114.03	110.97
OCH₃						56.91		57.16	57.4										57.22	57.26
3gly/5glc	ara	rha	gal	glc	glc	glc	glc	rha	glc	gal	glc	gal	glc	glc	glc	gal	glc	gal	glc	glc
1''	104.13	102.64	104.63	103.70	103.79	103.77	103.1	102.14	102.7	102.16	101.69	102.40	101.98	103.39	100.90	104.06	103.58	102.41	102.54	102.79
2''	72.10	71.55	72.16	74.77	74.80	74.88	75.0	71.50	74.5	80.12	81.91	80.67	82.33	74.68	80.89	71.89	74.64	73.20	74.73	74.78
3''	73.71	72.31	74.87	78.13	78.13	78.11	78.2	72.36	78.7	75.08	78.47	74.75	77.84	77.98	76.79	74.60	77.90	72.74	78.23	78.25
4''	68.68	73.31	70.14	71.06	71.11	71.12	71.5	73.29	71.4	70.01	71.06	69.85	71.22	71.27	69.98	70.31	71.31	70.48	71.31	71.32
5''	66.75	72.16	77.80	78.73	78.79	78.87	79.9	72.25	77.7	77.66	78.90	77.53	77.69	77.42	76.22	75.15	75.93	75.46	77.69	77.66
6''		17.95	62.35	62.33	62.39	62.33	62.8	18.05	62.5	62.29	62.55	62.28	62.27	67.87	66.75	65.20	65.47	65.17	67.33	67.08
6'rha/2''xyl										xyl	xyl			rha	rha				rha	rha
1'''/1''''										106.04	105.89			102.10	101.18				102.15	102.13
2'''/2''''										75.78	75.98			71.85	70.95				72.09	72.09
3'''/3''''										77.85	78.17			72.32	71.50				70.39	70.34
4'''/4''''										70.94	71.25			73.84	72.90				75.40	75.28
5'''/5''''										67.13	67.37			69.80	68.95				67.88	67.85
6'''/6''''														17.90	17.09				17.85	17.84
5glc/2''glc								5glc	5glc			2''glc	2''glc			2''glc			5glc	5glc
1'''								102.50	104.4			105.35	104.98			103.82			102.69	103.07
2'''								74.59	74.9			75.97	75.87			75.00			74.84	74.78
3'''								77.82	78.5			77.68	77.98			76.75			77.88	77.90

continued

Table F1.4.5 ^{13}C NMR Spectral Data for Selected Anthocyanins[a,b], *continued*

	1	2	3	4	5	6	7	8	9	10	11	12	13	14	15	16	17	18	20	21
4''								71.04	71.1			71.14	70.82		70.20				70.95	70.97
5''								78.63	79.0			77.89	78.58		77.10				78.64	78.73
6''								62.32	62.4			62.27	62.27		61.32				62.13	62.16
2''gall/4''''cour																				
4''''caf																		gall	cou	caf
α																			114.98	114.99
β																			147.09	147.36
1																		120.82	127.16	127.73
2																		110.50	131.31	115.38
3																		146.30	116.92	146.75
4																		140.15	161.26	149.59
5																		146.30	116.92	116.50
6																		110.50	131.31	122.99
C=O																		168.14	169.09	168.98
6''acyl																ace	mal	ace		
1''''/1'''																172.76	166.65	172.83		
2''''/2'''																20.69		20.74		
3'''																	170.19			

[a]See Table F1.4.3 for definitions of anthocyanin identifiers and references, and Table F1.4.4 for abbreviations (nd, not detected).
[b]Data for **19** not included.

Table F1.4.6 Typical ^{1}H-^{1}H Coupling Constants of the Most Common Anthocyanidins[a]

	Pelargonidin	Cyanidin	Peonidin	Delphinidin	Petunidin	Malvidin
H-4	s (b)	d, 0.9 Hz	d, 0.9 Hz	d, 0.9 Hz	s (b)	s (b)
H-6	d, 1.9 Hz	d, 1.9 Hz	d, 1.9 Hz	d, 1.9 Hz	d, 1.9 Hz	d, 1.9 Hz
H-8	dd, 0.9 Hz, 1.9 Hz	dd, 0.9 Hz, 1.9 Hz	dd, 0.9 Hz, 1.9 Hz	dd, 0.9 Hz, 1.9 Hz	d (b), 1.9 Hz	d (b), 1.9 Hz
H-2'	'd', 8.7Hz	d, 2.3 Hz	d, 2.3 Hz	s	d, 2.1 Hz	s
H-3'	'd', 8.7 Hz					
H-5'	'd', 8.7 Hz	d, 8.7 Hz	d, 8.7 Hz			
H-6'	'd', 8.7 Hz	dd, 2.3 Hz, 8.7 Hz	dd, 2.3 Hz, 8.7 Hz	s	d, 2.1 Hz	s
OCH$_3$		s		s	s	s

[a]Abbreviations: *b*, broad; *d*, doublet; *'d'*, semidoublet; *s*, singlet.

Table F1.4.7 Typical ^{1}H-^{1}H Coupling Constants for the Monosaccharides Commonly Identified in Anthocyanins[a]

	Glucopyranose	Galactopyranose	Rhamnopyranose	Xylopyranose	Arabinopyranose
1″	d, 7.7 Hz	d, 7.7 Hz	d, 1.5 Hz	d, 7.7 Hz	d, 6.2 Hz
2″	dd, 7.7 Hz, 9.4 Hz	dd, 7.7 Hz, 9.6 Hz	dd, 1.5 Hz, 3.3 Hz	dd, 7.7 Hz, 9.2 Hz	dd, 6.2 Hz, 8.1 Hz
3″	t, 9.4 Hz	dd, 9.6 Hz, 3.4Hz	dd, 3.3 Hz, 9.3 Hz	t, 9.2 Hz	d (b) 8.1 Hz
4″	t, 9.4 Hz	dd, 3.4 Hz, 1.0 Hz	t, 9.4 Hz	ddd, 10.3 Hz, 9.2 Hz, 5.5 Hz	m
5A″	ddd, 2.0 Hz, 6.7 Hz, 9.4 Hz	ddd, 1.0 Hz, 3.7 Hz, 8.3 Hz	dd, 9.5 Hz, 6.2 Hz	dd, 11.4 Hz, 5.5 Hz	dd, 15.0 Hz, 4.1 Hz
5B″				dd, 11.4 Hz, 10.3 Hz	d (b), 15.0 Hz
6A″	dd, 12.0 Hz, 2.0 Hz	dd, 11.8 Hz, 8.3 Hz	d, 6.2 Hz		
6B″	dd, 12.0 Hz, 6.7 Hz	dd, 11.8 Hz, 3.7 Hz			

[a]Abbreviations: *b*, broad; *d*, doublet; *s*, singlet; *t*, triplet.

NMR instrument of 300 MHz or stronger and appropriate sample tube (Table F1.4.2)

Additional reagents and equipment for preparation of anthocyanin NMR samples (see Support Protocol 1), recording the 1-D ^{1}H NMR spectrum (see Support Protocol 2), and recording the 2-D ^{1}H-^{13}C HMBC, 2-D ^{1}H-^{13}C HSQC, 2-D ^{1}H-^{1}H DQF-COSY, 2-D ^{1}H-^{1}H TOCSY, 2-D ^{1}H-^{1}H NOESY, and 1-D ^{13}C CAPT spectrums (Braun et al., 1998)

1. Prepare the NMR sample (see Support Protocol 1).

   In this protocol, 20 mg cyanidin 3-(2″-glucosyl-6″-rhamnosylglucoside) dissolved in 0.5 ml CD$_3$OD/CF$_3$COOD (95:5 v/v) represents a typical sample. Using a relatively concentrated sample (~50 mM) facilitates shorter experiment time and better quality of the 1-D ^{13}C and 2-D NMR spectra.

2. Set the temperature (usually 298K) on the NMR instrument, and allow time for the instrument to achieve the chosen temperature.

3. Insert the NMR sample into the sample tube.

4. Read a shim file suitable for the NMR solvent and solvent volume (~40 mm) being used.

   The shimming of the NMR sample must be optimized to improve the magnetic field homogeneity (for procedure, see Braun et al., 1998). A shim file stored on the NMR computer is normally used as a starting point for the shimming procedure. This shim file should be regularly updated from the optimized shim settings of the sample. On high-field superconducting NMR spectrometers, it is essential that the probe-head be correctly tuned to the observed frequency to obtain a favorable signal-to-noise ratio. Furthermore, impedance matching of the network must also be performed. In the more recent NMR spectrometers, wobbling functions are programmed in the software, making tuning and matching a very easy process.

5. Record the 1-D ^{1}H NMR spectrum (see Support Protocol 2).

6. Record the 2-D ^{1}H-^{13}C heteronuclear multiple bond correlation (HMBC) spectrum (Braun et al., 1998, pp. 489-492).

7. Record the 2-D ^{1}H-^{13}C heteronuclear single quantum coherence (HSQC) spectrum (Braun et al., 1998, pp. 497-500).

8. Record the 2-D ^{1}H-^{1}H double quantum filtered correlation spectroscopy (DQF-COSY) spectrum (Braun et al., 1998, pp. 481-484).

9. Record the 2-D ^{1}H-^{1}H total correlation spectroscopy (TOCSY) spectrum (Braun et al., 1998, pp. 501-504).

10. Record the 2-D ^{1}H-^{1}H nuclear Overhaüser enhancement spectroscopy (NOESY) spectrum (Braun et al., 1998, pp. 405-408).

11. Record the 1-D ^{13}C compensated attached proton test (CAPT) spectrum (Braun et al., 1998, pp. 165-167).

    ^{1}H and ^{13}C NMR spectral data for various anthocyanins are given in Tables F1.4.4 and F1.4.5. Typical ^{1}H-^{1}H coupling constants for common anthocyanidins and their monosaccharides are given in Tables F1.4.6 and F1.4.7.

PREPARATION OF AN ANTHOCYANIN NMR SAMPLE

This protocol describes isolation of anthocyanins, using cherries as an example, as well as how a pure anthocyanin, cyanidin 3-(2″-glucosyl-6″-rhamnosylglucoside) (**S.15**; Fig. F1.4.1) is treated before NMR experiments are performed. In this protocol, 20 mg **S.15** is dissolved in 0.5 ml of 95:5 (v/v) CD_3OD/CF_3COOD. Refer to *UNIT F1.1* for further details on extraction, purification, and isolation of anthocyanins. Common NMR solvents for anthocyanins are given in Table F1.4.8.

Materials

Cherries (*Prunus cerasus*)
99.9:0.1 (v/v) methanol/HCl
Ethyl acetate
Amberlite XAD-7 resin (Sigma)
1% (v/v) trifluoroacetic acid (TFA) in methanol
20:79.5:0.5 and 40:59.5:0.5 (v/v) methanol/H_2O/TFA
Nitrogen gas
95:5 (v/v) CD_3OD/CF_3COOD

80 × 5–cm chromatography column
90 × 5–cm Sephadex LH-20 column
NMR sample tube (Table F1.4.2) with plug
Glass pipets
~100°C oven

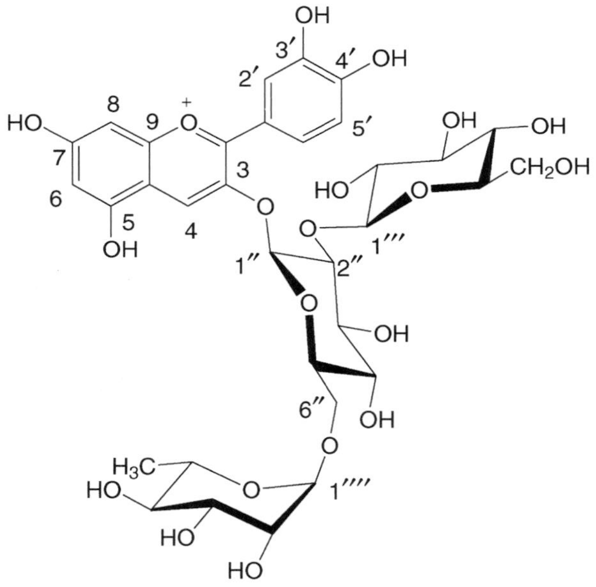

15

Figure F1.4.1   The structure of the major anthocyanin, cyanidin 3-*O*-(2″-*O*-β-glucopyranosyl-6″-*O*-α-rhamnopyranosyl-β-glucopyranoside), isolated from tart cherries, *Prunus cerasus*. Structure number **15** corresponds to ^{1}H and ^{13}C NMR data in Tables F1.4.4 and F.1.4.5.

Table F1.4.8 Common NMR Solvents for Anthocyanins

Solvent	Secondary reference for ^{1}H NMR and ^{13}C NMR
95:5 (v/v) CD$_3$OD/CF$_3$COOD	Residual solvent peak (CHD$_2$OD) at δ 3.40 and the solvent peak at δ 49.0 from TMS for ^{1}H and ^{13}C, respectively
80:20 (v/v) DMSO-d_6/CF$_3$COOD	Residual solvent peak (DMSO-Hd_5) at δ 2.49 and the solvent peak at d 39.6 from TMS for ^{1}H and ^{13}C, respectively

1. Extract anthocyanins from 700 g cherries using 900 ml of 99.9:0.1 (v/v) methanol/HCl three times.

2. Purify the concentrated extract by partitioning against an equal volume of ethyl acetate three times, followed by adsorption chromatography using an 80 × 5–cm column packed with Amberlite XAD-7 resin, which has been washed in advance with 2 liters water. Elute using 1% trifluoroacetic acid methanol.

3. Isolate pure anthocyanins on a 90 × 5–cm Sephadex LH-20 column using a solvent gradient of 20:79.5:0.5 (v/v) to 40:59.5:0.5 (v/v) methanol/water/TFA.

4. Evaporate the acidified methanol from the sample under nitrogen (2 hr).

5. Dry the glass equipment (i.e., NMR sample tube without plug and glass pipettes) for 2 hr in an ~100°C oven.

6. Dry the sample, pipet tips, and the NMR sample tube plug 2 hr in a desiccator under vacuum at room temperature.

7. Dissolve the sample in 0.5 ml of 95:5 (v/v) CD$_3$OD/CF$_3$COOD.

8. Transfer the sample to an NMR sample tube using a dry pipet.

9. Plug the NMR sample tube and seal with Parafilm.

RECORDING OF THE 1-D ^{1}H NMR SPECTRUM AND INITIAL PREPARATIONS FOR 2-D EXPERIMENTS

This protocol describes how a 1-D ^{1}H NMR spectrum is recorded. This type of spectrum may provide useful information about the identity of the aglycone, number of attached sugar units, and presence or absence of acyl moieties. This procedure may also be used as the basis for recording ^{13}C NMR spectra, and for initial preparations of 2-D homo- and heteronuclear experiments. See Basic Protocol for materials.

1. If it hasn't been done already, prepare the NMR sample, set the temperature, and read a shim file as described (see Basic Protocol, steps 1 to 4).

2. Adjust the lock-phase and check for ^{2}H lock signal saturation.

3. Tune and match the probe head's ^{1}H and ^{13}C channels, respectively.

   Tuning and matching of the ^{13}C channel is only necessary for experiments involving ^{13}C (e.g., HSQC, HMBC).

4. Define a new data set and read a suitable parameter file for the 1-D ^{1}H experiment. Record a preliminary 1-D ^{1}H spectrum with one scan and a relatively large sweep width (at least 10 ppm).

   From this preliminary spectrum, the sweep width of the 1-D ^{1}H spectrum must be defined so that all anthocyanin signals are included. The anthocyanin ^{1}H and ^{13}C signals are normally found in the spectral regions 0.5 to 9.5 ppm and 10 to 180 ppm, respectively.

SUPPORT PROTOCOL 2

Anthocyanins

F1.4.11

5. Establish the 90° pulses for ^1H and ^{13}C, respectively, by the following procedure.

    a. Perform on-resonance 90° pulse calibration by defining the middle of the spectrum (O1) at the strongest spectrum signal—e.g., the water peak or the solvent peak in the preliminary ^1H spectrum (or the solvent peak in the ^{13}C NMR spectrum).

    b. Record a new spectrum at the ~90° pulse. Fourier-transform and correct the phase.

    c. Determine the ~360° pulse by multiplying the ~90° pulse by four ($4 \times 90° = 360°$).

    Thereafter, record the spectrum with the ~360° pulse.

    d. Record a spectrum with the slightly changed ~360° pulse.

    e. Repeat substep b using the ~360° pulse. Continue repeating until the pulse is exactly 360°.

    At this point, the intensity of the selected on-resonance signal is ~0 (i.e., the positive and negative amplitudes of the signal are equal).

    f. Divide the exact 360° pulse by 4 to find the exact 90° pulse.

6. Adjust the receiver gain by using the RGA command (for a Bruker instrument).

7. Select number of scans (usually 8 to 256) for the experiment.

    The optimum number of scans for a given sample depends on several factors (e.g., sample concentration, field strength of the instrument) and must be determined individually.

8. Record the final 1-D ^1H NMR spectrum.

COMMENTARY

Background Information

The name anthocyanin (anthos meaning flower and kyanos meaning blue) was originally used to describe the pigment in blue cornflower (*Centaurea cyanus*). In a checklist from 1988 (Harborne and Grayer, 1988), 256 different anthocyanidin and anthocyanin structures were reported from a variety of plant sources with colors ranging from salmon-pink through red and violet, to nearly black. In 1999, the number of anthocyanins was increased to 453 (Harborne and Baxter, 1999), and the number today is >620 (unpub. observ.). This dramatic increase may in general reflect the fact that recent improvements in methods and instrumentation have made it easier to use smaller quantities of material to achieve results at increasing levels of precision. The progress with respect to complete structure elucidations is mainly caused by application of soft mass spectrometry techniques such as fast atomic bombardment (FAB; Self, 1987), tandem mass spectrometry (Glaessgen et al., 1992; Giusti et al., 1999), electrospray ionization (ESI; Kondo et al., 1994; Piovan et al., 1998), and matrix-assisted laser desorption/ionization (MALDI; Sporns and Wang, 1998; Sugui et al., 1999), and above all, various NMR techniques.

Today, it is possible to make complete assignments of all proton and carbon atoms in the NMR spectra of most isolated anthocyanins. These assignments are normally based on chemical shifts (δ) and coupling constants (J) observed in 1-D ^1H and ^{13}C NMR spectra (Fig. F1.4.2), combined with correlations observed as cross-peaks in various homo- and heteronuclear 2-D NMR experiments (see below for details on COSY, TOCSY, HSQC, HMBC, NOESY, and ROESY).

The advance in computing power has been an important factor for the success of the more advanced NMR techniques. Running many scans and accumulating the data may enhance weak signals, because baseline noise, which is random, tends to cancel out. One of the main advantages to be gained from signal averaging combined with the use of Fourier transform methods and high field magnets is the ability to obtain ^{13}C NMR spectra. This isotope of carbon exists in low abundance (1.108%) compared to the essentially 100% abundance of ^1H. The NMR sensitivity also depends on the magnetogyric constants, which for ^{13}C is only a quarter of the value of ^1H. Thus, the sample amount required for ^{13}C NMR spectra is about ten times that for ^1H NMR spectra, and the number of scans are normally 100 times higher.

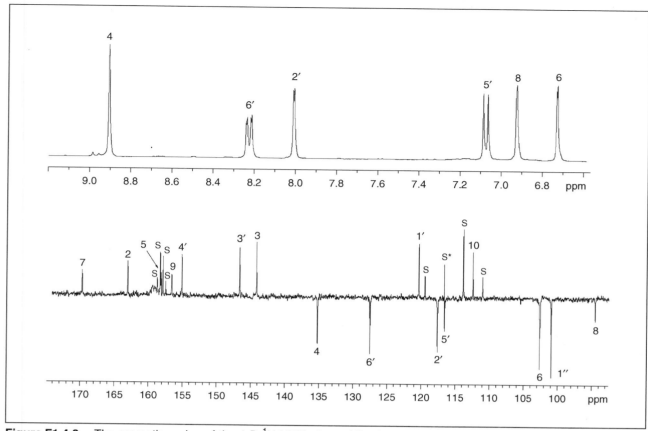

Figure F1.4.2 The aromatic region of the 1-D ^{1}H NMR of cyanidin gives rise to a characteristic splitting pattern (upper spectrum). The lower spectrum shows the aromatic region of the 1-D ^{13}C compensated attached proton test (CAPT) NMR spectrum of cyanidin. This spectrum contains all the fifteen ^{13}C resonances of the aglycone in addition to solvent signals (labeled S) and the anomeric sugar proton labeled 1″. In this spectrum, the ^{13}C nuclei which have a proton attached are represented with resonances pointing downwards, while the quaternary ^{13}C nuclei are pointing upwards. The ^{1}H NMR experiment was obtained within 25 sec, while the CAPT experiment was obtained within 1 hr 46 min.

Several experiments are suitable for recording anthocyanin 1-D ^{13}C NMR spectra. For example, in Fig. F1.4.2 the authors have used a compensated attached proton test (CAPT; Torres et al., 1993; Braun et al., 1998).

Introduction to 2-D NMR experiments

The purpose of the standard 1-D ^{1}H NMR experiment is to achieve structure-related information about sample protons (i.e., chemical shifts, spin-spin couplings, and integration data) describing the relative number of protons. Applied to anthocyanins, this information may help to identify the aglycone (anthocyanidin), number of monosaccharides present, and anomeric configuration of the monosaccharides. However, for most anthocyanins, the information gained by a standard 1-D ^{1}H NMR experiment is insufficient for complete structure elucidation. In recent years, various 2-D NMR experiments have evolved as the most powerful tools for complete structure elucidation of anthocyanins.

Two-dimensional NMR spectra are mainly produced as contour maps. These maps may be best imagined as looking down on a forest where all the trees (representing peaks in the spectrum) have been chopped off at the same fixed height. 2-D NMR spectra are produced by homonuclear (^{1}H-^{1}H) and heteronuclear (^{1}H-^{13}C) experiments.

Homonuclear ^{1}H-^{1}H-correlated NMR experiments, like the ^{1}H-^{1}H DQF-COSY and ^{1}H-^{1}H TOCSY experiments shown in Figs. F1.4.3 and F1.4.5, generate NMR spectra in which ^{1}H chemical shifts along both axes are correlated with each other. Values on the diagonal of these spectra correspond to chemical shifts, (e.g., trees; see above paragraph) that would have been shown in a 1-D ^{1}H NMR experiment. It is the off-diagonal spots called cross-peaks that present information which is new. These cross-peaks arise from coupling interactions between different ^{1}H nuclei. A cross-peak observed above the diagonal will normally also be found below the diagonal, thus producing a nearly

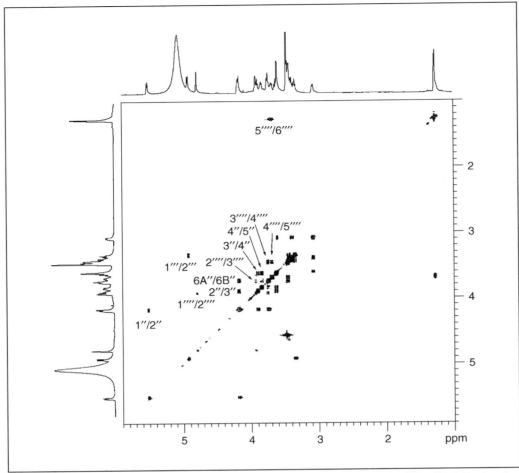

Figure F1.4.3 ¹H-¹H DQF-COSY spectrum of the sugar region of **15** (Fig. F1.4.1) showing assignments of the individual ¹H resonances of the anthocyanin sugar units by ³J_{HH} and ²J_{HH} interactions. The cross-peaks involving the rhamnose and one of the glucose units are labeled. The COSY spectrum was obtained within 39 min.

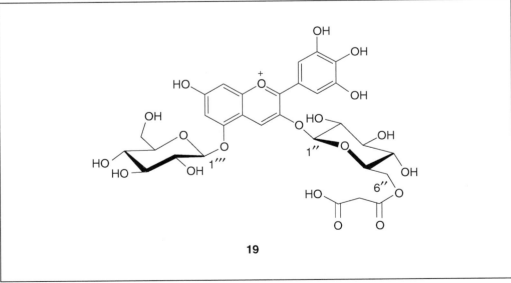

19

Figure F1.4.4 The structure of the acylated anthocyanin delphinidin 3-*O*-(6″-malonyl-β-glu-copyranoside)-5-β-glucopyranoside, isolated from *Aster novi belgii*. Structure number **19** corresponds to ¹H NMR data in Table F1.4.4.

symmetrical spectrum. The 1-D ^{1}H NMR spectra may be placed as projections along the top and left part of the 2-D NMR spectra, as shown in Fig. F1.4.3 and Fig. F1.4.5. The 1-D spectra are positioned along the axes to help assignments of the cross-peaks. The ^{1}H-^{1}H NOESY and ^{1}H-^{1}H ROESY experiments described below are other examples of homonuclear NMR experiments.

Heteronuclear NMR experiments are represented by the HSQC and HMBC spectra in Fig. F1.4.6 and Fig. F1.4.7. The ^{13}C NMR spectrum (or ^{13}C-projection) is displayed along one axis, and the ^{1}H NMR spectrum (or ^{1}H-projection) along the other. The ^{1}H-^{13}C correlations are shown as cross-peaks in the spectrum. Contrary to homonuclear NMR experiments, there exist no diagonal and only one cross-peak for each correlation.

COSY

The part of the 1-D ^{1}H NMR spectrum of cyanidin 3-O-(2″-glucopyranosyl-6′-O-α-rhamnopyranosyl-β-glucopyranoside), **15**, between 3.0 and 4.5 ppm, which represents the majority of sugar protons, is too crowded to present interpretable information. However, the double-quantum filtered correlation spectroscopy (^{1}H-^{1}H DQF-COSY) spectrum of this region (Fig. F1.4.3) makes it possible to assign all the protons of the rhamnose and the two glucose units of this compound. The DQF-COSY spectrum shows couplings between neighboring protons ($^2J_{HH}$, $^3J_{HH}$ and $^4J_{HH}$) revealed as cross-peaks in the spectrum. Interpretation may start from the anomeric H-1″ signal on the diagonal at 5.45 ppm, which is detected 1 to 2 ppm downfield to the rest of the sugar signals. A ruler can then be placed horizontally through the cross-peak labeled 1″/2″.

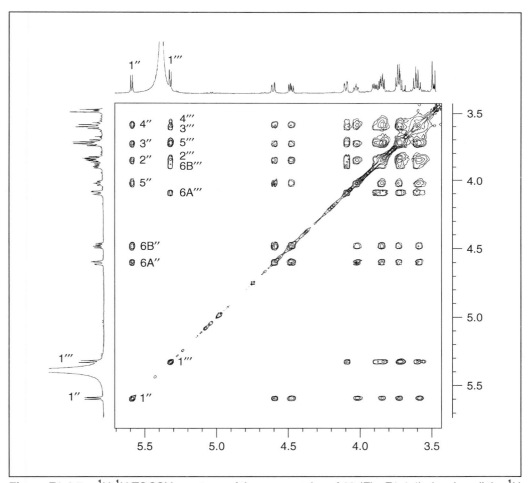

Figure F1.4.5 ^{1}H-^{1}H TOCSY spectrum of the sugar region of **19** (Fig. F1.4.4) showing all the ^{1}H resonances belonging to each spin system (e.g., to each sugar unit). In this spectrum it is possible to determine the chemical shift of all the protons belonging to each of the sugar units of **19**. The cross-peaks of each sugar unit are assigned in the spectrum. The TOCSY experiment was obtained within 5 hr 18 min.

Anthocyanins

This cross-peak represents the coupling between H-1″ and the proton located on the neighbor carbon (H-2″). The chemical shifts of H-2″ is located where the ruler intersects the diagonal (4.11 ppm). When the ruler is moved vertical through the H-2″ position on the diagonal, it intersects the cross-peak labeled 2″/3″. Thereafter the ruler is located horizontally through this cross-peak, and the intersection with the diagonal gives the chemical shift of H-3″ (3.83 ppm). In a similar manner, it is possible to assign all the protons of the monosaccharides of **15** (Table F1.4.4).

The ¹H-¹H DQF-COSY experiment is a modification of the standard ¹H-¹H COSY experiment. The chief advantage of the double-quantum filtered technique is that noncoupled proton signals are eliminated. This technique is especially useful for eliminating the water signal, which may overlap with anthocyanin sugar signals.

TOCSY

The 2-D homonuclear ¹H-¹H total correlation spectroscopy (TOCSY) experiment (Fig. F1.4.5) identifies protons belonging to the same spin system. As long as successive protons are coupled with coupling constants >5 Hz, magnetization is transferred successively over up to five or six bonds. The presence of heteroatoms, such as oxygen, usually disrupts TOCSY transfer. Since each sugar ring contains a discrete spin system separated by oxygen, this experiment is especially useful for assignments of overlapped anthocyanin sugar protons in the 1-D ¹H NMR spectrum. As seen in Fig F1.4.5, the labeled cross-peaks reveal how all the protons within each of the two sugar rings of **19** are correlated. Be aware that the cross-peak intensity is not an indicator of the distance between the protons involved, and that all expected correlations may not appear in a TOCSY spectrum due to the selected mixing time. To

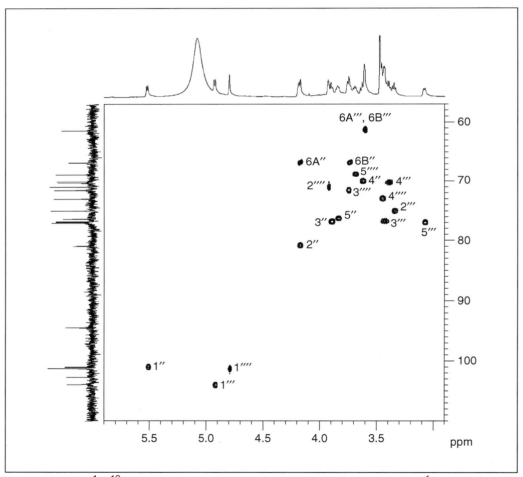

Figure F1.4.6 ¹H-¹³C HSQC spectrum of the sugar region of **15** showing all the ¹J_{CH} correlations, and thus all the ¹H and ¹³C chemical shifts of the three sugar units, but the cross-peak of the methyl group of rhamnose (H6″″/C6″″) at 1.2/17.1 ppm, which occurs beyond the presented region. The HSQC was obtained within 38 min.

avoid this latter problem, it may help to record a second spectrum with another mixing time.

In the 1-D TOCSY experiment, the resonances of one proton are selected and this signal is transferred in a stepwise process to all protons that are J-coupled to this proton. Instead of cross-peaks, magnetization transfer is seen as increased multiplet intensity. Thus, this 1-D TOCSY spectrum looks like a normal ^1H NMR spectrum including only the protons that belong to the same spin system as the chosen proton. The 1-D TOCSY experiment version is especially useful to determine coupling constants. For anthocyanins having several overlapping spin systems (e.g., several similar sugar units), additional experiments such as 2-D HSQC-TOCSY may be necessary for complete assignments.

HSQC and HMBC

The proton signals of the three monosaccharides of **15** have previously been assigned by a combination of the ^1H-^1H COSY and ^1H-^1H TOCSY experiments. The one-bond ^1H-^{13}C correlations observed in the heteronuclear single quantum coherence (HSQC) spectrum of the sugar region of **15** (Fig. F1.4.6), allow the assignment of the corresponding sugar ^{13}C signals. The one-bond ^1H-^{13}C coupling constants of anthocyanins, $^1J_{CH}$, are observed between 125 and 175 Hz (Pedersen et al., 1995).

The heteronuclear multiple bond correlation (HMBC) experiment correlates proton nuclei with carbon nuclei that are separated by more than one bond. In Fig. F1.4.7 the $^3J_{CH}$ and $^2J_{CH}$ couplings dominate. Major applications related to anthocyanins include the assignment of resonances of nonprotonated carbon nuclei of the

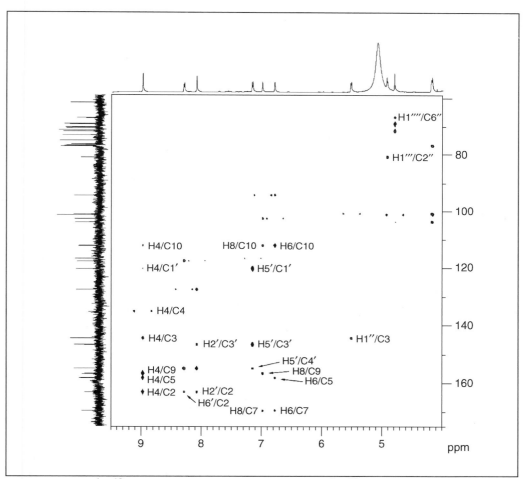

Figure F1.4.7 ^1H-^{13}C HMBC spectrum of **15** revealing all ^1H-^{13}C long-range couplings of the aglycone. Some important couplings of the sugar units are also labeled: The cross-peak H1'''/C3 (5.45/144.28 ppm), which shows the linkage between the aglycone and one of the glucose units, and the cross-peaks H1'''/C-2'' (4.85/80.89) and H1''''/C-6'' (4.72/66.75) showing the linkages between the inner glucose unit and the 2''-xylosyl and 6''-rhamnosyl units, respectively. The HMBC experiment was obtained within 40 min.

aglycone (anthocyanidin) and potential acyl group(s). Since long-range correlation of protonated carbon resonances also occur to carbon nuclei that are separated by nonprotonated carbons or other heteronuclei like oxygen, the experiment provides valuable information about the linkage points between the anthocyanin building blocks—aglycone, sugar unit(s), and acyl moieties (Fig. F1.4.7). The intensity of the cross-peaks generated by this experiment is related to the size of the multiple-bond coupling constant, $^{n}J_{CH}$, and the choice of the delay. However, there is no simple relationship between the size of $^{n}J_{CH}$ and the number of intervening bonds, which means that careful analysis of the spectrum in combination with other data is required. One common example of this problem is observed in the aromatic anthocyanin region, where some $^{2}J_{CH}$ are too small to be detected as cross-peaks, while the $^{3}J_{CH}$ are large, resulting in intense cross-peaks.

In the HSQC and HMBC experiments the ^{1}H chemical shifts constitute the natural frequency dimension (called F2), while the ^{13}C chemical shifts constitute the artificial frequency dimension (called F1). The ^{13}C dimension is made by the delays, the ^{13}C pulses and the $^{1}J_{CH}$ couplings. During ^{1}H acquisition in the HSQC experiment, ^{13}C decoupling is applied (garp decoupling). Contrary to the HSQC experiment, decoupling is not applied in the gs-HMBC experiment. In the latter experiment, $^{2}J_{CH}$ and $^{3}J_{CH}$ are favored, and $^{1}J_{CH}$ couplings are suppressed by a low pass filter in the HMBC pulse sequence (Braun et al., 1998). However, the $^{1}J_{CH}$ is not fully eliminated, and some $^{1}J_{CH}$ are observed as symmetrical doublets (Fig. F1.4.7). The one-bond filter has a delay that is matched to the inverse of $^{1}J_{CH}$. The $^{1}J_{CH}$ may also be entered as 145 Hz in this experiment. The HSQC and HMBC experiments are the modern version of the corresponding HETCOR and COLOC experiments (Braun et al., 1998). The two former experiments are more sensitive than the latter since ^{1}H FIDs are acquired, and not ^{13}C FIDs as in the two latter experiments.

NOESY

Couplings don't necessarily have to occur through bonding. Protons that are close to each other in space may be observed as cross-peaks in a nuclear Overhaüser enhancement spectroscopy (NOESY) spectrum. Thus, the more sensitive NOESY experiment proves to be an alternative technique to HMBC for determination of some linkages within an anthocyanin. When a sugar is attached to the aglycone 3-position, a cross-peak between H-4 and the anomeric proton is normally observed. Anomeric protons of monosaccharides attached to the aglycone 5- and 3′-positions show similar cross-peaks to H-6 and H-2′, respectively, while the anomeric proton of a sugar attached to the 7-position will exhibit cross-peaks to both H-6 and H-8.

Cross-peaks observed in a NOESY spectrum may reveal both intra- and intermolecular distances between anthocyanin protons (Nerdal et al., 1992). This type of information has been used for depiction of the association mechanism involving anthocyanins (Nerdal and Andersen, 1992; Houbiers et al., 1997; Gakh et al., 1998; Giusti et al., 1998). Furthermore, based on relative integration of the volume of cross-peaks in the NOESY spectrum, 3-D distance information can be estimated. Cross-peaks corresponding to two protons with a known distance are used as references for the distance calculations. Thus, intermolecular association of two anthocyanin molecules (petanin) have been evidenced by NOESY NMR experiments and distance geometry calculations (Nerdal and Andersen, 1992).

ROESY

Similar to the NOESY experiment the ^{1}H-^{1}H rotational nuclear Overhaüser effect spectroscopy (ROESY) experiment is useful for determination of the signals arising from protons which are close in space, but not necessarily connected by chemical bonds. A ROESY spectrum yields through-space correlations via the rotational nuclear Overhaüser effect (ROE). When one multiplet is irradiated, the intensities of multiplets arising from nearby nuclei are affected. Similar to the NOESY experiment, the ROESY spectrum contains a diagonal and cross-peaks. The diagonal consists of the 1-D spectrum. The cross-peaks indicate an ROE effect between two multiplets. ROESY is especially useful for cases where NOESY signals are weak, because they are near the transition between negative and positive, which often may be the case for medium-sized organic molecules like anthocyanins. ROESY cross-peaks are always negative. The ROESY (and NOESY) experiment also yields cross-peaks arising from chemical exchange. Exchange peaks and TOCSY-type artifacts are always positive so they can be told apart from ROESY correlations.

Assignments of NMR signals

To show how the structure of a relatively complex anthocyanin is completely elucidated by the above-mentioned NMR spectroscopic techniques, a detailed route applied on the major anthocyanin, cyanidin 3-*O*-(2″-*O*-β-glucopyranosyl-6″-*O*-α-rhamnopyranosyl-β-glucopyranoside), **15**, isolated from tart cherries, *Prunus cerasus,* is described below.

Assignments of the protons and carbons of the aglycone

The downfield part of the ¹H NMR spectrum of **15** (Fig. F1.4.2) shows six resonances in accordance with the aglycone cyanidin (see Table F1.4.4). The singlet at 8.90 ppm is typical for H-4 of most aglycones (Table F1.4.5). The signals at 8.00, 7.08, and 8.21 ppm are assigned to H-2′, H-5′ and H-6′, respectively, based on their coupling constants (Table F1.4.6). The signals at 8.00 and 7.08 ppm are two doublets with $^4J_{HH}$ and $^3J_{HH}$ of 1.6 and 8.6 Hz, respectively, while the signal at 8.21 ppm is a double doublet with a *meta*-coupling to H-2′ ($^4J_{HH}$ = 1.6 Hz) and an *ortho*-coupling to H-6′ ($^3J_{HH}$ = 8.6 Hz). These couplings are thereafter confirmed by their $^4J_{HH}$ and $^3J_{HH}$ cross-peaks in the DQF-COSY spectrum. The two signals at 6.72 and 6.92 ppm assigned to H-6 and H-8, respectively, have a small *meta*-coupling to each other. In many ¹H NMR spectra, this coupling is not obvious; however, it is normally recognized as $^4J_{HH}$ cross-peaks in their COSY and TOCSY spectra when their chemical shift difference is significant. It is nevertheless more common to assign H-6 and H-8 from the $^1J_{CH}$ cross-peaks in the HSQC spectrum, since C-8 normally has its chemical shift around 95 ppm, while C-6 is found around 102 ppm (Table F1.4.5).

After the chemical shifts of the protons of **15** have been assigned, the chemical shifts of the corresponding carbons are assigned from the HSQC experiment. The remaining problem of assigning the quaternary carbon atoms is addressed by the HMBC experiment, which is optimized for $^2J_{CH}$ and $^3J_{CH}$ couplings. A reasonable starting point is at H-4 (8.90 ppm), which couples to seven carbon atoms (Fig. F1.4.7). The strongest cross-peaks are the $^3J_{CH}$ couplings to C-2, C-5, and C-9, respectively. The assignment of C-2 at 163.03 ppm is deduced via its $^3J_{CH}$ couplings to H-4, H-2′ and H-6′, while the remaining signals at 158.04 ppm (C-5) and 156.51 ppm (C-9) are assigned by their additional $^2J_{CH}$ couplings to H-6 and H-8, respectively. The $^2J_{CH}$ coupling to H-4 and the $^3J_{CH}$ coupling to the anomeric proton at 5.45 ppm firmly assign C-3 to the signal at 144.28 ppm. The location of C-10 at 112.21 ppm is deduced from its $^3J_{CH}$ couplings to both H-6 and H-8. The response at 120.25 ppm is assigned to C-1′ by its $^3J_{CH}$ coupling to H-5′ and its $^4J_{CH}$ coupling to H-4. Both C-4′ and C-3′ couples with H-2′ and H-5′, however, only C-4′ (not C-3′) couples with H-6′ (Fig. F1.4.7). The remaining assignment of C-7 to 169.50 ppm is based on its $^2J_{CH}$ to both H-6 and H-8.

Assignments of the protons and carbons of the sugars

The anomeric proton and carbon signals appear considerably downfield of the other sugar resonances, and thus the three cross-peaks at δ5.45/100.90, δ4.85/103.82 and δ4.72/101.18 in the HSQC spectrum of **15** (Fig. F1.4.6) together with integration data, indicate three monosaccharides. The spectral region between δ81 and δ60 in the CAPT spectrum show fourteen resonances which, together with the three anomeric carbon resonances and the methyl signal at δ1.20/17.09, are in agreement with three hexoses. Starting from the doublet at 4.72 ppm, the observed cross-peak with the signal at 3.85 ppm in the DQF-COSY permit assignment of H-2‴′ (Fig. F1.4.3). The chain of coupled protons H-2‴′, H-3‴′, H-4‴′, H-5‴′, and H-6‴′ is thereafter assigned using the same spectrum. Subsequently, the chemical shifts of the corresponding carbon atoms (Table F1.4.5) are assigned from the HSQC experiment, which, together with ¹H-¹H coupling constants, are in agreement with a α-linked rhamnopyranosyl. Similarly, the protons of the two other sugar units are assigned by a "sequential walk" through the cross-peaks in the DQF-COSY spectrum. In cases where several protons show similar chemical shifts, the relationships are supported by the total correlation spectroscopy (TOCSY) experiment, which gives cross-peaks between the anomeric protons and all the protons in the same sugar unit. The chemical shifts and the ¹H-¹H coupling constants of the sugars of **15** (Table F1.4.4, Table F1.4.5 & Table F1.4.7) agree with one α-rhamnopyranosyl and two β-glucopyranosyl units.

Determination of the linkage points

The HMBC spectrum of **15** (Fig. F1.4.7) reveals the H-1″/C-3 cross-peak at δ5.45/144.28 ppm establishing that the glucosyl is attached to the aglycone 3-position (Fig. F1.4.1). In the same spectrum, the rhamnosyl and the other glucosyl unit are found to be attached to the 6″- and 2″-positions, respectively, by the cross-peaks at δ4.72/66.75 (H-1‴/C-6″) and δ4.85/80.89 (H-1‴/C-2″). Thus, the identity of **15** is found to be cyanidin 3-*O*-(2″-*O*-β-glucopyranosyl-6″-*O*-α-rhamnopyranosyl-β-glucopyranoside).

The binding sites of the sugars of the anthocyanidins may also be derived from the ROESY or NOESY experiments, which reveal neighborship through space. Strong cross-peaks between an anomeric proton and H-4 will, for instance, indicate that a sugar moiety is connected to the aglycone 3-position. Pronounced downfield shift effects may also confirm the linkage positions of the sugar units.

Critical Parameters and Troubleshooting

A successful structure elucidation of anthocyanins by NMR depends on several factors, including relatively high sample purity and stability. The authors recommend sample purity above 80%, as indicated by an HPLC chromatogram recorded at 280 nm, which will reveal most aromatic compounds in the sample. However, be aware that the molar absorptivities of different aromatic compounds may vary significantly. In addition to aromatic impurities, the sample may contain aliphatic impurities, which may interfere with interpretation of the sugar region of the NMR spectra in particular. The purity of many NMR samples will normally be improved by Sephadex LH-20 column chromatography (e.g., using 39.6:0.4:60 methanol/TFA/H$_2$O as the eluent).

Each anthocyanin may occur in several equilibrium forms. The use of small amounts of acids is necessary to keep the anthocyanin in the flavyllium cationic form, which is reckoned to be the most stable equilibrium form; however, it is well known that anthocyanin hydrolysis may occur under acidic conditions. The hydrolysis rate will be diminished by low storage temperatures and acid concentrations. Mineral acids (like DCl) should be replaced with organic acids (like CF$_3$COOD). In recent years it has been recognized that dicarboxylic acid moieties like malonyl may be esterified by an alcoholic solvent (e.g., methanol) under acidic conditions. Thus, the authors have experienced modification of many malonylated anthocyanins both during the isolation procedure and during storage of the NMR samples (Fossen et al., 2001). When recording multidimensional NMR spectra, a 1-D proton NMR spectrum should be recorded prior to and after each experiment to reveal potential changes within the sample. The authors recommend using the same sample for all NMR experiments despite potential modification during storage. Between the experiments, the NMR solution should be kept in a −20°C freezer. For long-term storage (i.e., months), the sample should be dried.

A common problem with respect to NMR spectroscopy on anthocyanins is the relatively slow exchange of the H-6 and H-8 aglycone protons with deuterium in acidified deuterated solvents. Especially when weak anthocyanin samples are involved, this exchange may prevent detection of correlations to C-6 and C-8 in the HSQC experiment and important long-range correlations involving the same protons in the HMBC experiment. As a rough guide, approximately half of the H-6 and H-8 protons are exchanged after 15 hr storage of an anthocyanin dissolved in 19:1 (v/v) CD$_3$OD/CF$_3$COOD, room temperature. Therefore, the authors suggest running HMBC and HSQC among the first NMR experiments.

It is very difficult to achieve ^{1}H NMR spectra of anthocyanins without an intense "water peak" around 5 ppm, which may overlap with signals commonly representing anomeric protons. After some hours of storage in the acidified deuterated solvent, this peak tends to migrate upfield ~0.4 ppm. It is therefore advised to record ^{1}H NMR spectra immediately after preparation to reveal peaks which may be hidden under the "water peak," and to repeat this procedure after several hours as well.

Anticipated Results

The proton and carbon chemical shifts of twenty-one and twenty different anthocyanins are presented in Table F1.4.4 and Table F1.4.5, respectively. These anthocyanins are chosen to illustrate the chemical shifts of the majority of anthocyanin building blocks reported. The linkage positions of the various anthocyanin building blocks may be conspicuous through shift comparison. However, be aware of shift effects caused by variation in solvent, pigment concentration and temperature. Table F1.4.6 contains typical [1]H-^{1}H coupling constants of the most common anthocyanidins.

Table F1.4.7 contains typical [1]H-^{1}H coupling constants for the monosaccharides commonly identified in anthocyanins. When the ring size of the anthocyanin monosaccharides has been reported, all but one have been reported as pyranoses (arabinofuranosyl has been identified in zebrinin; Idaka et al., 1987). When examined, the anomeric configurations of the glucosyl, galactosyl, xylosyl and glucuronyl

Table F1.4.9 Time Requirement for NMR Experiments on a 50 mM Solution of Cyanidin 3-(2″-Glucosyl-6″-Rhamnosylglucoside)

NMR experiment	Number of scans	Number of experiments	Experiment time
1-D ^{1}H	4	1	25 sec
1-D ^{13}C CAPT	2000	1	1hr 46 min
2-D ^{1}H-^{1}H DQF-COSY	2	512	39 min
2-D ^{1}H-^{1}H TOCSY	8	256	1hr 56 min
2-D ^{1}H-^{13}C HSQC	2	512	39 min
2-D ^{1}H-^{13}C HMBC	2	512	40 min

units have been reported with the anomeric β-configuration, while the arabinosyl and rhamnosyl units have the α-configuration. Be aware that although the D- and L-form of many anthocyanin monosaccharides have been published, these assignments are lacking experimental verification.

The single ^{1}H-^{13}C bond of the aglycone and monosaccharides of selected anthocyanins have been found to be between 125 and 175 Hz (Pedersen et al., 1995).

Time Considerations

The time required for each type of NMR experiment depends on many factors, including sample concentration and complexity, shimming, magnetic field strength, and the sensitivity of the individual experiments. Indication of time requirements of individual 1-D and 2-D experiments are given in the legends of Figures F.1.4.2, F.1.4.3, F.1.4.5, F.1.4.6, and F.1.4.7, as well as Table F1.4.9.

Literature Cited

Andersen, Ø.M., Aksnes, D.W., Nerdal,W., and Johansen, O.P. 1991a. Structure elucidation of cyanidin-3-sambubioside and assignments of the ^{1}H and ^{13}C NMR resonances through two-dimensional shift-correlated NMR techniques. *Phytochem. Anal.* 2:175-183.

Andersen, Ø.M., Opheim, S., Aksnes, D.W., and Frøystein, N.Å. 1991b. Structure of petanin, an acylated anthocyanin isolated from *Solanum tuberosum*, using homo- and hetero-nuclear two-dimensional nuclear magnetic resonance techniques. *Phytochem. Anal.* 2:230-236.

Andersen, Ø.M., Viksund, R.I., and Pedersen, A.T. 1995. Malvidin 3-(6-acetylglucoside)-5-glucoside and other anthocyanins from flowers of *Geranium sylvaticum*. *Phytochem.* 38:1513-1517.

Braun, S., Kalinowski, H.-O., and Berger, S. 1998. 150 and More Basic NMR Experiments. Wiley–VCH, Weinheim, Germany.

Cabrita, L. 1999. Analysis and Stability of Anthocyanins. Bergen, Norway.

Cabrita, L., Frøystein, N.Å. and Andersen, Ø.M. 2000. Anthocyanin trisaccharides in blue berries of *Vaccinium padifolium*. *Food Chem.*, 69:33-36.

Catalano, G., Fossen, T., and Andersen, Ø.M. 1998. Petunidin-3-O-α-rhamnopyranosyl-5-O-β-glucopyranoside and other anthocyanins from flowers of *Vicia villosa*. *J. Agric. Food Chem.* 46:4568-4570.

Fossen, T., Andersen, Ø.M., Øvstedal D.O., Pedersen, A.T., and Raknes, Å. 1996. Characteristic anthocyanin pattern from onions and other *Allium*. *J. Food Sci.* 61:703-706.

Fossen, T., Larsen, Å., and Andersen, Ø.M. 1998. Anthocyanins from flowers and leaves of *Nymphaéa × marliacea* cultivars. *Phytochem.* 48:823-827.

Fossen, T., Slimestad, R., and Andersen, Ø.M. 2001. Anthocyanins from maize (*Zea mays*), and reed canarygrass (*Phalaris arundinacea*). *J. Agric. Food Chem.* 49:2318-2321.

Gakh, E.G., Dougall, D.K., and Baker, D.C., 1998. Proton nuclear magnetic resonance studies of monoacylated anthocyanins from the wild carrot: Part 1. Inter- and intra-molecular interactions in solution. *Phytochem. Anal.* 9:28-34.

Giusti, M.M., Ghanadan, H., and Wrolstad, R.E. 1998. Elucidation of the structure and conformation of red radish (*Raphanus sativus*) anthocyanins using one- and two-dimensional nuclear magnetic resonance techniques. *J. Agric. Food Chem.* 46:4858-4863.

Giusti, M.M.,Rodriguez-Saona, L.E., Griffin, D., and Wrolstad, R.E. 1999. Electrospray and tandem mass spectroscopy as tools for anthocyanin characterization. *J. Agric. Food Chem.* 47:4657-4664.

Glaessgen, W.E., Seitz, H.U., and Metzger, J.W. 1992. High-performance liquid chromatography/electrospray mass spectrometry and tandem mass spectrometry of anthocyanins from plant tissues and cell cultures of *Daucus carota* L. *Biol. Mass Spectrom.* 21:271-277.

Harborne, J.B. and Baxter, H., 1999. The Handbook of Natural Flavonoids, Vol. 2. John Wiley & Sons, New York.

Harborne, J.B. and Grayer, R.J., 1988. Flavonoid checklists. *In* The Flavonoids: Advances in Research (J.B. Harborne, ed.), p. 538, Chapman & Hall, London.

Houbiers, C., Lima, J.C., Macanita, A.L., and Santos, H. 1998. Color stabilization of malvidin 3-glucoside: self-aggregation of the flavylium cation and copigmentation with the Z-chalcone form. *J. Phys. Chem. B.* 102:3578-3585.

Idaka, E., Ohashi, Y., Ogawa, T., Kondo, T., and Goto, T. 1987. Structure of zebrinin, a novel acylated anthocyanin isolated from *Zebrina pendula. Tetrahedron Lett.* 28:1901-1904.

Kondo, T., Ueda, M., Yoshida, K., Titani, K., Isobe, M., and Goto, T. 1994. Direct observation of a small-molecule associated supramolecular pigment, commelinin, by electrospray-ionization mass-spectroscopy. *J. Am. Chem. Soc.* 116:7457-7458.

Nerdal, W. and Andersen, Ø.M. 1992. Intermolecular aromatic acid association of an anthocyanin (Petanin) evidenced by two-dimensional nuclear Overhauser enhancement nuclear magnetic resonance experiments and distance geometry calculations. *Phytochem. Analys.* 3:182-189.

Nerdal, W., Pedersen, A.T., and Andersen, Ø.M. 1992. Two-dimensional nuclear Overhauser enhancement NMR experiments on pelargonidin-3-glucopyranoside, an anthocyanin of low molecular mass. *Acta Chem. Scand.* 46:872-876.

Nygård, A.-M., Aksnes, D.W., Andersen, Ø.M., and Bakken, A.K. 1997. Structure determination of 6-hydroxycyanidin-, 6-hydroxydelphinidin-3-(6″-O-α-L-rhamno-pyranosyl-β-D-glucopyra nosides) and other anthocyanins from *Alstroemeria* cultivars. *Acta Chem. Scand.* 51:108-112.

Pedersen, A.T., Andersen, Ø.M., Aksnes, D.W., and Nerdal, W. 1993. NMR on anthocyanins, assignments and effects of exchanging aromatic protons. *Magn. Reson. Chem.* 31:972-976.

Pedersen, A.T., Andersen, Ø.M., Aksnes, D.W., and Nerdal, W. 1995. Anomeric sugar configuration of anthocyanin O-pyranosides determined from heteronuclear one-bond coupling constants. *Phytochem. Analys.* 6:313-316.

Piovan, A., Filippini, R., and Favretto, D. 1998. Characterization of the anthocyanins of *Catharanthus roseus* (L.) G. Done in vivo and in vitro by electrospray ionization ion trap mass spectrometry. *Rapid Commun. Mass Spectrom.* 12:361-367.

Ramstad, B., Pedersen, A. T., and Andersen, Ø.M. 1995. Delphinidin 3-α-arabinopyranoside and other anthocyanins from flowers of *Rhododendron* cv. Lems Stormcloud. *J. Hortic. Sci.* 70:637-642.

Self, R. 1987. Fast-atom-bombardment mass spectrometry in food science. *Appl. Mass Spectrom. Food Sci.* 239-288.

Slimestad, R. and Andersen, Ø.M. 1998. Cyanidin 3-(2-glucosylgalactoside) and other anthocyanins from fruits of *Cornus suecica. Phytochem.* 49:2163-2166.

Slimestad, R., Aaberg, A., and Andersen, Ø.M. 1999. Acylated anthocyanins from petunia flowers. *Phytochem.* 50:1081-1086.

Sporns, P. and Wang, J., 1998. Exploring new frontiers in food analysis using MALDI-MS. *Food Res. Intern.*, 31:181-189.

Sugui, J.A., Wood, K.V., Yang, Z., Bonham, C.C., and Nicholson, R.L. 1999. Matrix-assisted laser desorption ionization mass spectrometry analysis of grape anthocyanins. *Am. Enol. Vitic.* 50:199-203.

Swinny, E.E., Bloor, S.J., and Wong, H. 2000. ^{1}H and ^{13}C NMR assignments for the 3-deoxyanthocyanins, luteolinidin-5-glucoside and apigeninidin-5-glucoside. *Magn. Reson. Chem.* 38:1031-1033.

Torres, A.M., Nakashima, T.T, and Mcclung R.E.D. 1993. J-compensated proton-detected heteronuclear shift-correlation experiments. *J. Magn. Reson. Ser.* A 102:219-227.

Van Calsteren, M.R., Cormier, F., Chi, B.D., and Laing, R.R. 1991. Proton and carbon-13 NMR assignments of the major anthocyanins from *Vitis vinifera* cell suspension culture. *Spectroscopy* 9:1-15.

Key References

Braun et al., 1998. See above.

A detailed description of the basic NMR experiments are presented. Refer to pages 501 to 504 for information concerning the [1]H-^{1}H TOCSY spectrum.

Agrawal, P.K. 1989. Carbon-13 NMR of Flavonoids. Elsevier, Amsterdam.

A huge compilation of ^{13}C NMR data on various flavonoid classes. Data on anthocyanins are very limited, and several assignment mistakes appear throughout the book.

Markham, K.R. and Geiger, H. 1993. [1]H nuclear magnetic resonance spectroscopy of flavonoids and their glycosides in hexadeuterodimethylsulfoxide. *In* The Flavonoids: Advances in Research Since 1986 (J. B. Harborne, ed.) pp. 441-497. Chapman & Hall, London.

An excellent compilation of ^{1}H NMR data on various flavonoids including 71 spectra. Data on seven anthocyanins are included.

Contributed by Øyvind M. Andersen
and Torgils Fossen
University of Bergen
Bergen, Norway

NMR Glossary

Chemical shift (δ): A dimensionless quantity defined as: $\delta = (\nu_{sample} - \nu_{reference})/\nu_0 \times 10^6$, where ν_{sample} is the resonance frequency of the sample, $\nu_{reference}$ is the resonance frequency of the reference, tetramethylsilane (TMS; defined as zero), and ν_0 is the observing frequency (e.g., 300 or 600 MHz). The unit for the δ scale is ppm (parts per million) and is independent of the strength of the applied magnetic field. Exact resonance frequency of a nucleus is a function of the environment (chemical/magnetic) of the observed nuclei.

Coupling constant (J in Hz): Interaction between nuclear spins mediated through chemical bonds giving rise to mutual splitting of resonance lines.

Free induction decay (FID): An oscillating voltage recorded as the magnetization vector precesses (rotates) in the laboratory frame.

Fourier transformation (FT): Mathematical operation to convert a time domain spectrum (FID) to a frequency domain spectrum (normal NMR spectrum).

Memory locations (number of data points): Number of data points stored. Proportional to the acquisition time.

Relaxation: Energy loss to the surroundings of an excited nucleus when it returns to its ground state. (1) Spin-lattice relaxation in which the relaxation process involves the entire framework or aggregate of neighbors of the high-energy nucleus. (2) Spin-spin relaxation which implies transferring the excess energy, ΔE, to a neighboring nucleus, provided that the particular value of ΔE is common to both nuclei.

Scan: Individual FID. Usually the intensity of an individual FID is so weak that the FIDs of many pulses are added together (accumulated) in the computer unit to obtain a stronger signal—i.e., better signal-to-noise ratio (S/N)—before the signal is transformed.

Signal-to-noise ratio (S/N): Increases in proportion to the square root of the number of scans (ns). The strength of the NMR signals are thus improved by increasing the number of scans.

Spectral width or sweep width (SW): The defined range of frequencies in which signals are expected to be found.

Carotenoids

F2.1   Extraction, Isolation, and Purification of Carotenoids — F2.1.1
Basic Protocol 1: Solvent Extraction and Isolation of Carotenoids — F2.1.1
Basic Protocol 2: Prepurification of Carotenoids by Crystallization — F2.1.3
Support Protocol 1: Removal of Water from Carotenoid-Containing Samples
    Using a Vacuum Oven — F2.1.4
Support Protocol 2: Removal of Water from Carotenoid-Containing Samples
    Using Alcohol — F2.1.5
Reagents and Solutions — F2.1.6
Commentary — F2.1.6

**F2.2   Detection and Measurement of Carotenoids by UV/VIS
Spectrophotometry** — F2.2.1
Basic Protocol 1: Preparation and Calibration of Individual Carotenoid
    Standards — F2.2.1
Basic Protocol 2: Measurement of Total Carotenoid Concentration in Food
    Colorants, Pharmaceuticals, and Natural Extracts — F2.2.3
Commentary — F2.2.4

F2.3   Chromatographic Techniques for Carotenoid Separation — F2.3.1
Basic Protocol 1: Isocratic Carotenoid Separation Using Wide-Pore,
    Polymeric C18 — F2.3.1
Support Protocol 1: Standards Preparation and Calibration — F2.3.3
Support Protocol 2: Sample Preparation — F2.3.5
Alternate Protocol 1: Isocratic Carotenoid Separation Capable of Simultaneous
    Separation of Retinol and Tocopherol Using Spherisorb ODS2 — F2.3.8
Alternate Protocol 2: Gradient Separation Using C 30 Carotenoid Column — F2.3.9
Basic Protocol 2: Normal-Phase Separation of Xanthophylls — F2.3.11
Commentary — F2.3.12

F2.4   Mass Spectrometry of Carotenoids — F2.4.1
Basic Protocol 1: Electron Impact and Chemical Ionization Mass
    Spectrometry of Carotenoids — F2.4.1
Basic Protocol 2: Fast Atom Bombardment, Liquid Secondary Ion Mass
    Spectrometry, and Continuous-Flow Fast Atom Bombardment of
    Carotenoids — F2.4.2
Basic Protocol 3: Matrix-Assisted Laser Desorption/Ionization Time-of-Flight
    Mass Spectrometry of Carotenoids — F2.4.3
Basic Protocol 4: Electrospray Ionization Liquid Chromatography/Mass
    Spectrometry of Carotenoids — F2.4.4
Basic Protocol 5: Atmospheric Pressure Chemical Ionization Liquid
    Chromatography/Mass Spectrometry of Carotenoids — F2.4.5
Commentary — F2.4.6

Contents

1

Extraction, Isolation, and Purification of Carotenoids

Recently, food processors and technologists have shown a great interest in the isolation, identification, and purification of natural pigments, including carotenoids, due to their nutritional value, their use as colorants, and their potential as health aids. This unit describes a practical way of extracting, isolating, and purifying carotenoids from plant materials. The method is based mainly on the natural form in which the carotenoids are found (esterified or free) and to some extent on their polarity and/or solubility in the solvents used. Common and readily available solvents and laboratory equipment are suggested.

The extraction and isolation of three groups of carotenoids of different polarity are described in Basic Protocol 1. A method for prepurifying carotenoids using crystallization is described in Basic Protocol 2. Carotenoids may be purified further by chromatographic techniques (*UNIT F2.3*) and characterized (*UNITS F2.2 & F2.4*). Support Protocols 1 and 2 describe the preparation of the sample before extraction. This process consists mainly of removing water from the sample followed by sample grinding or homogenizing.

NOTE: Carotenoid pigments should be protected from light, oxygen, and heat (see Critical Parameters and Troubleshooting).

SOLVENT EXTRACTION AND ISOLATION OF CAROTENOIDS

This protocol begins with the extraction of a dehydrated sample. It continues with a saponification scheme to initiate the isolation of the carotenoid mixture. During saponification, the esters are hydrolyzed and the free pigments released. Then, to continue the isolation, column chromatography is suggested as a simple and fast means of separating the three main groups of carotenoids based on their different polarities.

Materials

Dehydrated plant material (see Support Protocols 1 and 2)
Extractant: 1:1 (v/v) hexane/acetone or hexane alone
Saponifying solution: 40% (w/v) KOH in methanol (cool to room temperature before bringing up to volume)
Salting-out solution: 10% (w/v) Na_2SO_4
Na_2SO_4, anhydrous (powder form)
Adsorbent (see recipe)
Carotene eluant: 4% (v/v) acetone in hexane
Monohydroxy pigment (MHP) eluant: 1:9 (v/v) acetone/hexane
Dihydroxy pigment (DHP) eluant: 1:8 (v/v) acetone/hexane

500-ml extraction vessel
Explosion-proof shaft mixer (e.g. model SIU04X; Lightnin)
Whatman no. 42 filter paper
Rotary evaporator (e.g., Büchi Rotavapor, Brinkmann Instruments) attached to vacuum pump, ≤55°C
56°C water bath
125-ml separatory funnel
600 × 40–mm chromatography column
Glass wool
Vacuum filtration device
1-liter filtration flasks

Contributed by Gustavo A. Rodriguez

Carotenoids

F2.1.1

Extract carotenoids

1. Place dehydrated plant material (and filter paper if Support Protocol 2 was used) in a 500-ml extraction vessel.

2. Add 3 vol (v/w) extractant and mix 15 min using an explosion-proof shaft mixer to suspend paste in solvent.

   The type of extractant should be chosen based on the carotenoids of interest (see Commentary).

   No heating is necessary.

3. Vacuum filter mixture using Whatman no. 42 filter paper. Save filtrate.

4. Separate filtrant cleanly from filter paper, if possible, and extract a second time using 2 vol extractant. Continue extractions using smaller volumes of extractant until no appreciable color is observed in filtrate. Save and pool all filtrates.

   If filtrant cannot be separated from paper, both should be placed in the extraction vessel.

   At this point, concentrated filtrate from the sample preparation (see Support Protocol 2, step 8) may be pooled with the filtrates.

5. Concentrate extracts to ~40 ml in a rotary evaporator attached to a vacuum pump at $\leq 55°C$.

6. Add 3 ml saponifying solution and stir 45 min at 56°C.

   This is the temperature recommended for hot saponification of pigments by the AOAC (1990).

7. Transfer saponified extract to a 125-ml separatory funnel and add 1 vol salting-out solution.

8. Remove bottom layer and wash upper layer three times with 10 ml water.

9. Add 3 g anhydrous Na_2SO_4 and filter using Whatman no. 42 filter paper. Save filtrate.

Isolate carotenoids

10. Plug the bottom of a 600×40–mm chromatography column with glass wool. Mount the column on a vacuum filtration device, using a 1-liter filtration flask as a receiving vessel (Figure F2.1.1).

11. Add adsorbent to obtain a 20-cm layer while applying vacuum.

12. Level the surface of the adsorbent and place a firm 2-cm layer of anhydrous Na_2SO_4 on top.

13. Pour carotene eluant into column until eluant wets all of the adsorbent.

14. Replace receiving vessel with a clean flask and pour filtrate (step 9) into column.

15. Allow all the sample to penetrate into the adsorbent and then add carotene eluant until the first carotenoid band that separates is completely collected in the flask.

    The carotenes and esters present in the sample are contained in this fraction. Other carotenoids remain on the column.

    The chromatography process is monitored visually. The exact volume of eluant added to the column will vary depending on sample concentration and composition.

16. Elute monohydroxy pigments with MHP eluant and dihydroxy pigments and more polar pigments with DHP eluant, using a clean receiving flask for each. If necessary, store pigments ≤ 24 hr at 0° to 5°C and protect from light.

    Eluates should be processed as soon as possible.

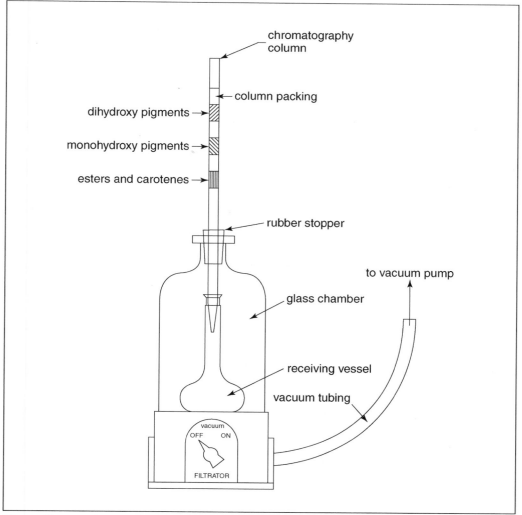

Figure F2.1.1 Rapid separation and collection of pigments using column chromatography with vacuum filtration.

PREPURIFICATION OF CAROTENOIDS BY CRYSTALLIZATION

Once the carotenoids have been isolated as described in Basic Protocol 1, they can generally be crystallized as an initial step to purification. Actually, what is most likely to happen is a co-crystallization. When working with a nonpolar fraction, α- and β-carotene may co-crystallize. In the same way, a polar fraction may yield lutein-zeaxanthin crystals. A pure carotenoid product may be obtained by crystallization of a fraction derived from a preparatory chromatographic procedure, which can be done using TLC, HPLC (*UNIT F2.3*), or in some cases column chromatography.

Materials

> Isolated carotenoid eluate (see Basic Protocol 1, step 15 or 16)
> Precipitating solvent: 1:1 (v/v) hexane/acetone or hexane alone, room temperature and cold (0° to 5°C)

> 50- or 125-ml pear-shaped flasks
> Rotary evaporator (e.g., Büchi Rotavapor, Brinkmann Instruments) attached to vacuum pump.
> Nitrogen gas tank with regulator adaptable to rotary evaporator and vacuum oven
> Whatman no. 42 filter paper
> Vacuum oven, 40°C

BASIC PROTOCOL 2

Carotenoids

F2.1.3

1. Place isolated carotenoid eluate in a 50- or 125-ml pear-shaped flask and concentrate eluate to near saturation in a rotary evaporator attached to a vacuum pump under a stream of nitrogen from a nitrogen gas tank.

   Only one carotenoid eluate (carotenes, MHPs, or DHPs) should be prepurified at a time.

   The flask size will depend on the volume of eluate to be concentrated.

2. Add precipitating solvent drop by drop until the solution starts turning cloudy, indicating that precipitation has begun.

   A solvent mixture should be selected for crystallization that will easily solubilize the carotenoid of interest and also, when added in small quantities, will force the carotenoid out of solution and start crystallization. The extractants from Basic Protocol 1 (hexane/acetone or hexane alone) are appropriate.

3. Refrigerate (0° to 5°C) solution overnight.

   Crystallization can be accelerated by placing solution at −20° to 0°C for 3 or 4 hr. Usually better crystals are produced by a slow crystallization.

4. Vacuum filter crystals using Whatman no. 42 filter paper. Wash crystals with cold (0° to 5°C) precipitating solvent.

5. *Optional:* Repeat crystallization sequence to obtain crystals of higher purity, beginning with a saturated solution of crystals (step 4) in acetone and continuing with step 2.

6. Dry crystals in a vacuum oven at 40°C, flushing drying chamber occasionally with nitrogen.

7. Pack crystals under nitrogen and store ≤6 months at −20°C if further purification work is anticipated.

REMOVAL OF WATER FROM CAROTENOID-CONTAINING SAMPLES USING A VACUUM OVEN

Plant tissues contain variable amounts of water, which is easily eliminated in order to facilitate extraction with organic solvents. Using a low-temperature vacuum oven is a good way of removing water from materials that are easily pulverized. This is true for many vegetables.

Materials

Plant tissue of interest
Vacuum oven, 60°C
Laboratory mill with 30-mesh screen

1. Weigh plant tissue of interest and record weight.

   The initial wet weight is important as, in many cases, extraction and purification of carotenoids are done to quantify pigment concentrations, which must be related to the fresh material.

   The amount of tissue used is dependent on the expected pigment concentration. Usually 100 to 500 g is appropriate.

2. Place sample inside a vacuum oven at 60°C. Maintain set temperature.

3. Turn on and control the vacuum between 12 and 20 in. Hg (305 to 508 mmHg).

4. Open oven door occasionally during drying and stir material to obtain uniform dehydration.

5. Dry to ~8% moisture content.

> *Overdrying can damage the carotenoids present in the sample.*

> *A moisture balance calibrated to indicate an 8% moisture content can be used to monitor weight loss and may be useful until experience is gained. The calibration can be performed as a gravimetric determination on a sample analyzed in parallel. This measurement can also be used to estimate an 8% moisture content relative to the initial sample weight.*

6. Grind material to 30 mesh with a laboratory mill.

7. Weigh powder for extraction.

> *The weight will also be indicative of residual moisture if the initial amount is known (see Chapter A1).*

REMOVAL OF WATER FROM CAROTENOID-CONTAINING SAMPLES USING ALCOHOL

SUPPORT PROTOCOL 2

This is a practical way of removing water from plant tissues when water is needed for homogenization or when the dry material cannot be ground to a powder due to the presence of lipids, waxes, or sugars.

Materials

Plant tissue of interest
95% (v/v) ethanol
Hexane

Explosion-proof shaft mixer (e.g., model SIU04X; Lightnin)
Whatman no. 41 filter paper
Separatory funnel
Rotary evaporator (e.g., Büchi Rotavapor, Brinkmann Instruments) attached to vacuum pump, ≤55°C

1. Weigh plant tissue of interest as described (see Support Protocol 1, step 1).

2. Mix plant tissue to a paste with an explosion-proof shaft mixer. If the material does not lend itself to mixing, add sufficient water to homogenize sample.

3. Add 2 vol of 95% ethanol to homogenate. Mix thoroughly 5 min.

4. Vacuum filter homogenized sample using Whatman no. 41 filter paper. Save filtrate.

5. If filtrate is cloudy, repeat steps 3 and 4 using 1 vol of 95% ethanol. If filtrate is clear, continue with step 6.

> *A cloudy filtrate is an indication that additional alcohol is needed.*

6. Remove and save filtrant along with the filter paper. Store ≤24 hr at −20°C.

> *The sample and the filter paper are now ready for extraction.*

7. Pool filtrates in a separatory funnel. Add 1 vol hexane.

> *Some of the more polar pigments are usually carried into the filtrate and must be recovered by extraction.*

8. Remove bottom layer and concentrate upper layer containing the lipophilic carotenoids in a rotary evaporator attached to a vacuum pump at ≤55°C. Store ≤24 hr at −20°C.

All solvents used must be good grade analytical reagents and water should be distilled. For suppliers, see SUPPLIERS APPENDIX.

Adsorbent

Mix equivalent weights of silica gel 60 GF_{254} (Merck) and diatomaceous earth (Hyflo Super-Cel; Celite, World Minerals) in a mechanical blender or in a large plastic bag for 2 hr. Store ≤3 months in a sealed container at room temperature.

COMMENTARY

Background Information

Carotenoids have been of great interest, and their importance in food coloration has been well reviewed by Kläui and Bauernfeind (1981). Extraction, isolation, and purification are described excellently by Schiedt and Liaaen-Jensen (1995).

This unit focuses on procedures for obtaining the important carotenoids that are found naturally in foods and plant material and that are of nutritional and pharmacological interest. The pigments referred to include α- and β-carotenes, β-cryptoxanthin, lutein, zeaxanthin, lycopene, capsanthin, and capsorubin, among others.

For extraction, water content is considered an important factor. It has been found that working with low-moisture samples simplifies the extraction process. Industrial extraction normally is done with dry material, which reduces complications arising from the solvents used for processing and their recovery.

The plant material as prepared in Support Protocols 1 and 2 is ready for extraction. It may contain some water (≤10%), which will not affect extraction. Freeze-drying may be another way of eliminating water with little pigment damage and may be an acceptable alternative to drying in a vacuum oven as long as the moisture content is taken to an appropriate level for efficient extraction. Freeze-drying is not adequate if sugar or other water-soluble compounds must be eliminated by physical separation (e.g., by filtration or centrifugation).

The solvent used for extraction must be chosen according to the polarity of the pigments presumably present. If this characteristic is unknown, an acetone/hexane (1:1, v/v) mixture is suitable. When it is known that the carotenoids in the sample are nonpolar or are in the ester form, hexane is a good choice for extraction. Ethanol will extract polar carotenoids, and a nonpolar solvent like hexane will promote crystallization.

In this unit, extraction is carried out using solvents. Other methods for extracting pigments are available. The use of supercritical carbon dioxide is described by Favati et al. (1988), Chao et al. (1991), and Spanos et al. (1993). Enzymes may also be used to help digest plant material and then facilitate pigment extraction by physical or chemical means (Sims et al., 1993; Thomas et al., 1998; Koch et al., 1999). Ultrasound may also be used to aid the extraction process; it is used in laboratory and industrial work for extracting pigments and other compounds. Vegetable oil may be used as an extractant and pigment carrier. However, using oils for extraction produces low pigment concentrates. By introducing organic solvents, pigments are more easily solubilized, and extracts of high purity may be obtained.

Following extraction, an efficient way of initiating the isolation of carotenoids is to saponify the extract. This removes many of the unwanted lipids present in the sample as well as chlorophyll. The saponification by-products, which to a great extent are sodium or potassium salts, are easily separated by an aqueous solution of a highly polar salt. The addition of water also helps wash off excess alkali and other water-soluble and water-complexed compounds. This procedure hydrolyzes xanthophyll esters to form the hydroxylated carotenoid.

After saponification, column chromatography is applied as recommended in the Official Methods of Analysis of the AOAC (1990), although a larger column is used to accommodate the larger sample volume. This type of chromatography is very practical for most laboratories and is very much in use in those that specialize in carotenoid studies. Efficient separation of the main groups of carotenoids is achieved using solvents of different polarity. In this standard analytical procedure, the first eluant that is put through the column is a nonpolar solvent, such as hexane or a 90:10 (v/v) hexane/acetone mixture, to elute carotenes and xanthophyll esters (all low-polarity compounds). An 80:20 (v/v) hexane/acetone solvent combination may be used to elute moderately polar monohydroxy carotenoids, and a 60:40 (v/v) hexane/acetone mixture may be used to elute the very polar fraction of dihydroxy pigments. The recommended solvents in Basic Protocols 1 and 2 are the ones found to be the most convenient for these purposes.

Crystallization as described in Basic Protocol 2 generally provides products with two or more pigments, but sometimes produces crystals of a single carotenoid.

Critical Parameters and Troubleshooting

When working with carotenoid pigments, it is difficult to overemphasize the need to protect these compounds from light, oxygen, and heat. It is therefore necessary to work in dim light and to keep intermediary products in the dark. Also, wherever possible, a vacuum or inert atmosphere must be used. High temperatures should be avoided, and samples being processed should be kept in a refrigerator or freezer when work is discontinued.

Complete dehydration before extraction is not recommended. A small amount of water is often useful when a low-polarity solvent mixture is the extractant. On the other hand, excess water may make extraction inefficient. Using the right extractant is of great importance and depends largely on which carotenoids are sought. Using the hexane/acetone mixture as the extractant is advantageous because the same pair of solvents is used later for crystallization.

Once carotenoids have been extracted and isolated, they tend to be very unstable, as natural antioxidants and protecting agents have been stripped away. An antioxidant (such as BHT, a tocopherol blend or vitamin E, rosemary extract, or other food-grade chemical) should then be added to the carotenoid. Many time-encapsulating agents or other substances are added to the carotenoid concentrates, but the antioxidants may represent ~5% of the total formulation (Vilstrup et al., 1998; Kowalski et al., 2000). The presence of metals and chlorophylls will also affect pigment stability, but this problem is largely solved in the salting-out step of Basic Protocol 1.

The salting-out may produce an interface that often has a high pigment concentration and occasionally even crystals that begin to precipitate. It is important to collect the interface and wash it with salt solution or an appropriate solvent to allow the pigments to combine with the upper phase. This interface may also contain hydrated phospholipids or some other waxy material that can be reduced or eliminated by filtration, followed by washing of the filter cake with the extractant to recover pigment. Often the use of an inert filter aid will speed up filtration and will also capture the undesirable matter.

When the sample has been passed through the chromatography column in Basic Protocol 1, the isolates are in a solvent mixture (hexane/acetone) that can be used for crystallization after saturating by evaporation and enhancing precipitation. If a different solvent pair is necessary for crystallization, the hexane/acetone may be eliminated in a rotary evaporator and the new solvent pair may be added to the concentrate.

Anticipated Results

Yields in milligrams of the pigment sought are fully dependent on the content in the raw sample. Nevertheless, these protocols typically produce isolates in the range of 50% to 70% purity. After prepurification by a single crystallization step, purity of the concentrates will range between 65% and 85%. It should be pointed out that the pure crystalline forms are normally not obtained. As mentioned above, preparative chromatography (*UNIT F2.3*) will produce fractions that can be crystallized or simply dried to a pure form.

Time Considerations

Completion of the full process will take 3 to 4 days depending on the dehydration protocol chosen. Vacuum drying (Support Protocol 1) will take a full day. Extraction, isolation, and startup of prepurification will take another day. It is advisable to crystallize overnight and to filter, wash, and dry the crystals the next day. If a second crystallization step is desirable add another day.

Literature Cited

AOAC (Association of Official Analytical Chemists). 1990. Carotenes and Xanthophylls in Dried Plant Materials and Mixed Feeds. AOAC Method 970.64. *In* AOAC Official Methods of Analysis, 15th ed. (K. Helrich, ed.) pp. 1048-1049. AOAC, Arlington, Va.

Chao, R.R., Mirlvaney, S.J., Sanson, D.R., Hsieh, F., and Tempesta, M.S. 1991. Supercritical CO_2 extraction of anatto (*Bixa orellana*) pigments and some characteristics of the color extracts. *J. Food Sci.* 56:80-83.

Favati, F., King, J.W., Friedrich, J.P., and Eskins, K. 1988. Supercritical CO_2 extraction of carotene and lutein from leaf protein concentrates. *J. Food Sci.* 53:1532-1536.

Kläui, H. and Bauernfeind, J.C. 1981. Carotenoids as food colors. *In* Carotenoids as Colorants and Vitamin A Precursors (J.C. Bauernfeind, ed.) pp. 47-317. Academic Press, New York.

Koch, L., Sandor, M., Kalman, T., Attila, P., and Victorovich, B.S. October, 1999. Natural carotenoid concentrates from plant material and a process for preparing the same. U.S. patent 5,962,756.

Kowalski, R.E., Mergens, W.J., and Scialpi, L.J. July, 2000. Process for manufacture of carotenoid compositions. U.S. patent 6,093,348.

Schiedt, K. and Liaaen-Jensen, S. 1995. Isolation and analysis. *In* Carotenoids, Vol. 1A (G. Britton, S. Liaaen-Jensen, and H. Pfander, eds.) pp. 81-108. Birkhauser Verlag, Basel, Switzerland.

Sims, C.A., Balaban, M.O., and Matthews, R.F. 1993. Optimization of carrot juice color and cloud stability. *J. Food Sci.* 58:1129-1131.

Spanos, G.A., Chen, H., and Schwartz, S.J. 1993. Supercritical CO_2 extraction of β-carotene from sweet potatoes. *J. Food Sci.* 58:817-820.

Thomas, R.L., Deibler, K.O., and Barmore, C.R. November, 1998. Extraction of pigment from plant material. U.S. patent 5,830,738.

Vilstrup, P., Jenses, N.M., and Krag-Andersen, S. September, 1998. Process for the preparation of a water dispersible carotenoid preparation in powder form. U.S. patent 5,811,609.

Key References

Bauernfeind, J.C. (ed.) 1981. Carotenoids as Colorants and Vitamin A Precursors. Academic Press, New York.

Excellent review of applications and importance in the food industry and other fields.

Britton, G., Liaaen-Jensen, S., and Pfander, H. (eds.) 1995. Carotenoids, Vol. 1A. Isolation and Analysis. Birkhauser Verlag, Basel, Switzerland.

Technical advancements in carotenoids are followed up from the original edition edited by Otto Isler.

Contributed by Gustavo A. Rodriguez
Prodemex
Los Mochis, Sinaloa, Mexico

Detection and Measurement of Carotenoids by UV/VIS Spectrophotometry

The majority of carotenoids exhibit absorption in the visible region of the spectrum, between 400 and 500 nm. Because they obey the Beer-Lambert law (i.e., absorbance is linearly proportional to the concentration), absorbance measurements can be used to quantify the concentration of a pure (standard) carotenoid (see Basic Protocol 1) or to estimate the total carotenoid concentration in a mixture or extract of carotenoids in a sample (see Basic Protocol 2). Considerations for the preparation of carotenoid-containing samples are presented in Critical Parameters (see Sampling and Sample Preparation).

NOTE: Carotenoids are easily degraded. See Critical Parameters for a discussion of suitable precautions.

NOTE: All extinction coefficients in this unit assume a 1-cm pathlength.

PREPARATION AND CALIBRATION OF INDIVIDUAL CAROTENOID STANDARDS

In this protocol, commercially purchased carotenoid standards are dissolved in a suitable solvent and the absorbance measured at its maximum wavelength (λ_{max}). Using published extinction coefficients and taking into consideration the dilution factor, the concentration of the standard carotenoid is calculated. The spectrum is also scanned in order to evaluate the fine structure (see Spectral Fine Structure in Background Information). The carotenoid solution should ideally be assayed by HPLC as described in *UNIT F2.3* to establish chromatographic purity and thus correct the calculated concentration.

Materials

~1 to 5 mg standard carotenoids (Table F2.2.1 and Table F2.2.2)
High-grade organic solvent (Table F2.2.2)
10- to 50-ml volumetric flask
Additional reagents and equipment for HPLC analysis (*UNIT F2.3*)

1. Dissolve carotenoid (e.g., ~1 to 5 mg) in a suitable solvent (see Table F2.2.2). Make to an accurate volume (e.g., 10 to 50 ml) in a volumetric flask.

   A larger volume may be used if desired.

Table F2.2.1 Commercial Sources of Carotenoids[a]

	Sigma Aldrich	Extra-synthese	Atomergic Chemical	Indifine Chemical	Fisher Scientific	Fluka Chemical	Carl Roth GmgH
β-carotene	X	X	X	X	X	X	X
α-carotene[b]							
Lycopene	X	X	X	X			X
β-crytoxanthin	X	X	X			X	
Zeaxanthin		X	X	X			X
Lutein	X	X	X	X			X

[a]The companies may have offices or distributors in other countries.

[b]At the time of writing there were no commercial suppliers of α-carotene.

Contributed by K. John Scott

It is essential that the carotenoid be completely dissolved. With crystalline samples, dissolution can be aided by initial addition of a small amount of a more effective solvent (e.g., dichloromethane) prior to making to volume for spectrophotometric measurement (for most commonly assayed carotenoids, no more than 10% of the total volume should be required; however, lycopene may require up to 100%). Where a carotenoid is supplied in a sealed ampule, dissolution can be conveniently achieved by adding successive small aliquots of the more effective solvent to the ampule, and transferring to a volumetric flask. It is not easy to assess complete dissolution visually; therefore, it is advisable to filter the solution through a suitable solvent-compatible 0.45-μm filter.

2. Dilute the solution (e.g., 1:50) in desired solvent if necessary to give ~0.3 to 0.7 AU as measured on a spectrophotometer at λ_{max}.

   It is recommended that at least two independent dilutions be made to ensure confidence in the measurement.

   If a larger final volume is used (see step 1) reduce the dilution accordingly.

3. Warm up the spectrophotometer per manufacturer's instructions.

4. Zero the spectrophotometer with solvent in a cuvette.

   NOTE: *Cuvettes must be kept scrupulously clean; avoid handling the surfaces of the cell (see Critical Parameters).*

5. Place a cuvette containing the carotenoid solution into the sample cell holder of the spectrophotometer.

6. Measure the absorbance at λ_{max}. Take reading immediately.

   See Critical Parameters concerning degradation.

7. Scan to allow measurement of the fine structure (see Table F2.2.3; also see Spectral Fine Structure in Background Information).

   The same principles can be applied to measurement of chromatographic fractions.

8. Calculate the concentration of carotenoid as shown in the example below for all *trans*-β-carotene.

$$\frac{A \times V_1}{A^{1\%}} \times C^{1\%} = \frac{0.5 \text{ AU} \times 50}{2592 \text{ AU}} \times 10 \text{ mg/ml} = 96.5 \text{ μg/ml}$$

Table F2.2.2 Data on λ_{max} and Extinction Coefficients of a Selection of Carotenoids

	MW	$A^{1\%}$	$\varepsilon^{1 \text{ mM}}$	λ(nm)	Solvent	%III/II
α-carotene	537	2710	145	445	Hexane	55
β-carotene	537	2592	139	450	Hexane	25
β-cryptoxanthin	553	2460	136	450	Hexane	25
Lutein	569	2550	145	445	Ethanol	60
Lycopene	537	3450	185	470	Hexane	65
Zeaxanthin	569	2480	141	450	Hexane	25
Natural carotenoids as food colors						
Bixin (Bixa orellana)	395	4200	166	456	Petroleum ether	
Capsanthin (paprika)	585	2072	121	483	Benzene	
Capsorubin (paprika)	601	2200	132	489	Benzene	
Synthetic food colors						
β-apo-8′-carotenal	417	2640	110	457	Petroleum ether	
Canthaxanthin	564	2200	124	466	Petroleum ether	0

Detection and Measurement of Carotenoids by UV/VIS Spectrophotometry

Where A is the absorbance reading of the diluted sample (0.5 AU), V_1 is the dilution factor (50×), $A^{1\%}$ is the absorbance of a 1% solution (i.e., the extinction coefficient; 2592 AU), and $C^{1\%}$ is the concentration of a 1% solution (10 mg/ml). Using this formula, the concentration of the original solution in this example is = 96.5 μg/ml.

9. Subsequent to measurement of the carotenoid concentration, the solution should be assayed by HPLC to establish the chromatographic purity (see UNIT F2.3).

> *For example, assuming the same values as Equation F2.2.1, if the total chromatographic area is 10000, and the area of all-trans β-carotene peak is 9500, then the chromatographic purity is 9500/10000 × 100 or 95%, and the actual concentration of β-carotene is 95.5 μg/ml × 95/100 or 91.7 μg/ml.*

MEASUREMENT OF TOTAL CAROTENOID CONCENTRATION IN FOOD COLORANTS, PHARMACEUTICALS, AND NATURAL EXTRACTS

The same principle as described above can be used for the "estimation" of the carotenoid content of extracts of food colorants, pharmaceuticals, foods, biological samples, or chromatographic fractions. This procedure employs calculations used for individual carotenoids of high purity and thus will estimate the "total carotenoids" present in a food or biological extract, where a mixture of carotenoids would be expected. Greater accuracy can be obtained as extracts are purified to contain single components (see Commentary). A spectrum scan is not employed in this procedure as the fine structure of a mix of carotenoids can only be identified after HPLC separation (see Commentary).

Materials

Sample
Appropriate solvent (Table F2.2.2)
Additional equipment and reagents for sample extraction (UNIT F2.1).

1. Prepare the sample as detailed in UNIT F2.1, taking into consideration the guidelines detailed in Critical Parameters (see Sampling and Sample Preparation). Dilute the sample appropriately using a suitable solvent.

> *Samples containing esterified carotenoids, chlorophyll, or high levels of fat may require saponification (UNIT F2.1).*

2. Warm up spectrophotometer per manufacturer's instructions.

3. Zero the spectrophotometer with solvent in a cuvette.

Table F2.2.3 Additional Spectral Characteristics of Carotenoids

Common name	Chemical name	Solvent	Absorption peaks		
β-carotene	(β,β-carotene)	Hexane	425	450	478
α-carotene	(β,ε-carotene)	Hexane	422	445	473
Lycopene	(φ,φ-carotene)	Hexane	444	470	502
β-cryptoxanthin	(3-hydroxy-β-carotene)	Hexane	428	450	478
Zeaxanthin	(β,β-carotene-3,3′-diol)	Hexane	425	450	478
Lutein	(β,ε-carotene-3,3′-diol)	Ethanol	421	445	474
Capsanthin		Petroleum ether	450	475	505
Capsorubin		Petroleum ether	445	479	510
Bixin		Petroleum ether	432	456	490
Canthaxanthin		Petroleum ether		466	
β-apo-8′-carotenal		Ethanol	405	430	460

4. Place a cuvette containing a suitably diluted (i.e., ~0.3 to 0.7 AU) extract of pharmaceutical, food, or biological material into the sample cell holder of the spectrophotometer.

5. Measure absorbance at selected λ_{max} (Table F2.2.2 and Table F2.2.3). Read immediately.

   See Critical Parameters concerning degradation.

6. Estimate the total carotenoid concentration of the sample (see Basic Protocol 1, step 9 and Equation F2.2.1).

   With the exception of individual food colorants and single carotenoid pharmaceutical products, it is unlikely that the extract will be composed of only one predominant carotenoid; therefore, a specific λ_{max} or extinction coefficient (Table F2.2.2) cannot be used. In this case it is convenient to use a λ_{max} of 450 nm and a typical $A^{1\%}$ value of 2500. Alternative values and other considerations are discussed elsewhere (see Commentary).

COMMENTARY

Background Information

The majority of carotenoids exhibit absorption in the visible region of the spectrum, mainly between 400 and 500 nm (see Table F2.2.2 for examples). A few carotenoids (e.g., phytoene) exhibit maximum absorbance in the UV region. This absorption is due to the long conjugated double bond system of carotenoids. The conjugated unsaturated part of the carotenoid molecule containing delocalized π-electrons is called the "chromophore" and is re-

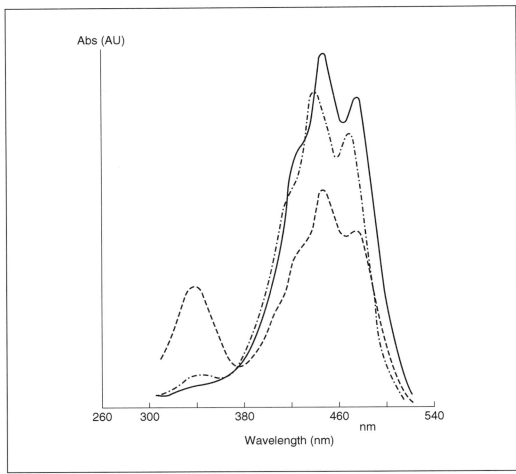

Figure F2.2.1 The spectral characteristics of *all-trans* β-carotene (solid line), 9-*cis* β-carotene (dashed and dotted line) and 15-*cis* β-carotene (dashed line).

sponsible for the absorption of light in the visible region.

The differences in the spectral characteristics of individual carotenoids are often small, but are of great importance in their identification; however, carotenoids having the same chromophore, such as β-carotene and its hydroxy-derivative zeaxanthin, have identical spectra. The spectra of *cis-* or *Z*-isomers of carotenoids, while being similar to the *all-trans* or *all-E* form, are different in as far as they exhibit a change in the λ_{max} to a shorter wavelength, a decrease in the magnitude of the absorbance, a reduction in the fine structure, and additional absorption bands in the UV region around 340 nm and 280 nm (Fig. F2.2.1). It is not the intention of the author to go into any great detail of the molecular characteristics of carotenoids in this unit, as details can be found in a excellent chapter by Britton (1995).

Extinction coefficients

Carotenoids in solution obey the Beer-Lambert law, where absorbance (*A*) equals concentration multiplied by extinction coefficient ($A^{1\%}$), where the extinction coefficient ($A^{1\%}$) is defined as the absorbance of a 1% (10 g/liter) solution of carotenoid, in a defined solvent, in a 1-cm path-length cuvette, at a specific wavelength (λ). This information can be used to quantify the concentration of a pure (standard) carotenoid (see Basic Protocol 1), or to "estimate" the total carotenoid concentration in a mixture or extract of carotenoids in a sample (see Basic Protocol 2). The extinction coefficient may also be expressed in terms of molarity.

As seen in Table F2.2.2, a 1% solution (10 g/liter) of β-carotene has an absorbance of 2592 AU and a molecular weight of 537 g/mol (i.e., 1 mM = 0.537 g/liter); therefore, the absor-

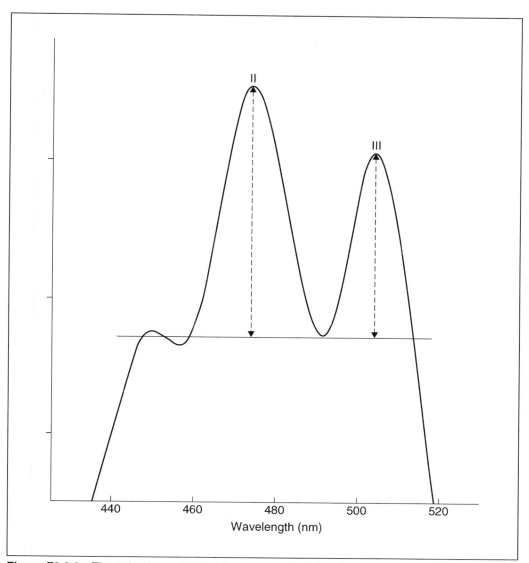

Figure F2.2.2 The calculation of %III/II from the spectral fine structure.

bance expected for a 1 mM solution is: (0.537 g/liter)/(10 g/liter) × (2592 AU) = 139 AU/1 mM β-carotene. Using this ratio, if a β-carotene solution has an absorbance of 5, then the concentration is given as: (5 AU)/(139 AU/mM) = 0.036 mM. It is important to note that due to the inherent difficulties in procedures for accurate determination of extinction coefficients, there may be a significant level of uncertainty in some published values. Small variations (e.g., 2 to 3 nm) may also occur in published data of absorption maxima. Whenever possible, the spectrum of a compound under investigation should be compared directly with an authentic pure standard. The spectra of the unknown and the standard should be identical for both the λ_{max} and the fine structure (see below).

Spectral fine structure

In addition to the absorption maxima of the carotenoids, the "shape" of the spectra provides important information for identification of purified carotenoid extracts or pure standard (while the identity of the standard is generally not in question, it is a good idea to check the purity by fine structure analysis). Fine structure is demonstrated in Figure 2.2.2, and is measured as a ratio of the absorbance maxima to one of the shoulders (i.e., %III/II). The longest wavelength band is called III and the middle absorption band II. The baseline is taken as the minimum between the two peaks, and the height of each peak is measured. Carotenoids having the same chromophore, such as β-carotene and its hydroxy-derivative zeaxanthin, have identical fine structure, while conjugated ketocarotenoids, for example canthaxanthin, have only a rounded spectrum with no fine structure (see Fig. F2.2.3), thus the %III/II is 0.

Spectral monitoring during HPLC

The spectral characteristics of a standard can be monitored during HPLC using a diode-array detector (*UNIT F2.3*). A directory of standard spectra can be stored, enabling additional identification of sample peaks. The actual absorption maxima and fine structure will be dependent on the composition of the mobile phase (see Fig. F2.2.4). Peak I may only occur as a "shoulder" with *cis*-carotenoids, while an additional peak is observed at around 340 nm (see Fig. 2.2.1).

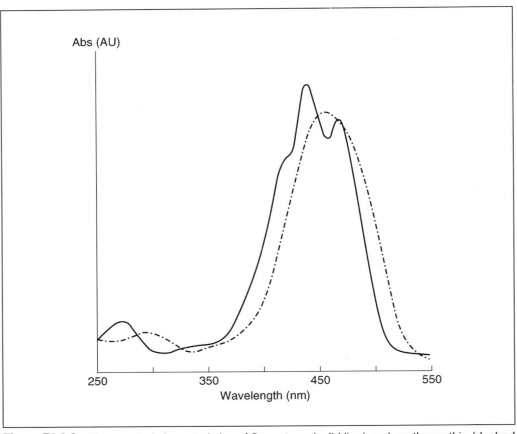

Figure F2.2.3 The spectral characteristics of β-carotene (solid line) and canthaxanthin (dashed line). Redrawn from original by Jaffé and Orchin (1962).

Critical Parameters and Troubleshooting

Good laboratory practice for UV/VIS spectrophotometry

While the procedures outlined above are fairly straight forward, in order to obtain maximum accuracy, "good laboratory practice" should be applied at all times.

1. Whenever possible the preparation and handling of carotenoid solutions should be carried out under yellow/gold fluorescent lighting to avoid light induced degradation.

2. All glassware should be scrupulously clean and reagents (e.g., solvents) should be of the highest quality.

3. The spectrophotometer should be located in a clean environment away from direct sunlight and drafts, at a reasonably even temperature, and free from electrical interference.

4. Any spillage should be cleaned up immediately.

5. When working with organic solvents it is advisable to use stoppered cells to prevent damage to cell holders and other sensitive parts of the instrument, and to avoid evaporation of the sample.

6. It is advisable to have the instrument regularly serviced (i.e., annually) by the manufacturer's engineer.

7. Once they have become familiar with the equipment, day to day maintenance can be carried out by the user; however, if in doubt, consult the manufacturer.

8. Modern equipment displays a range of error messages, often in the form of a letter and number, if the instrument fails or is not working to specification (e.g., failure of light source). Refer to the user's manual for details.

9. Cuvettes should be kept clean and the faces to be placed in the light beam should not be handled.

10. For normal measurement of carotenoids, glass cells can be used.

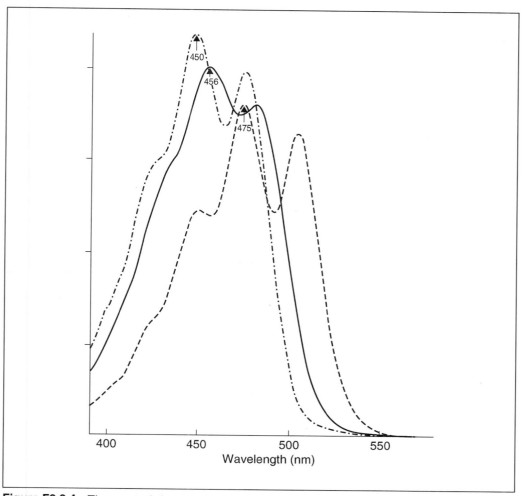

Figure F2.2.4 The spectral characteristics of β-carotene (solid line), lutein (long-dashed line), and lycopene (short-dashed line) in an acetonitrile-based HPLC solvent (75:25:5 v/v/v acetonitrile/methanol/dichloromethane).

11. It is not necessary to completely fill the cells, ~2/3 full is normally sufficient. This helps to avoid accidental spillage.

12. Before placing the cells in the cell holder, the optical faces can, if necessary, be polished carefully (avoiding any spillage) with a fine tissue.

13. For accurate results, it is advisable to use optically matched pairs or sets of cuvettes (i.e., cuvettes manufactured to a high specification to ensure they have the same optical parameters).

14. As pointed out above (see Basic Protocol 1), when preparing carotenoid standard solutions, complete dissolution of the solid material is essential; however, it is always advisable to filter the solution through a solvent compatible filter.

15. Stock solutions that have been stored (e.g., in a refrigerator), should be allowed to warm up to room temperature, refiltered, and a "new" concentration calculated prior to preparing a new working solution, which should be checked for chromatographic purity.

16. Spectrophotometric readings should be carried out immediately after the solution has been placed in the cuvette to avoid evaporation of the solvent or degradation.

17. Accurate measurement of standard solutions is a critical factor in the overall analysis of carotenoids. Inaccuracy of measurement can be a major cause of intra- and interlaboratory variation.

18. Experience will allow the analyst to recognize possible problems and anomalies, but a useful maxim is "if in doubt repeat it."

Sampling and sample preparation

Sample selection, number of samples, sample handling, and sample preparation prior to extraction are important factors effecting data quality. The type of sample (e.g., vitamin preparations, fruit drinks, supplemented foods, fruits and vegetables, biological materials), will determine to a large extent the sampling protocol and how the samples are handled. In this respect, the main considerations are the degree of homogeneity of the material and the possible variation in the vitamin content, not only between different materials, but also between different samples of the same material. Powdered, freeze-dried, and liquid materials for example, are likely to be more homogeneous with respect to vitamin distribution than fruits and vegetables. With fruits and vegetables, a good rule of thumb is that levels of vitamin in the outer part (e.g., outer leaves, skin, peel) are generally higher, considerably so in some cases, than the inner parts.

Equally with fruits and vegetables, the variety, origin, season of year, growing conditions, etc., will affect the vitamin content. In addition, consideration must be given to typical ways of preparation and cooking of the foods in the home. It is essential that the sampling protocol and the number of samples collected reflect the purpose of the exercise, be it to determine between batch variation of vitamin supplement preparations, the variation within vegetables of the same type, or to obtain a typical overall value for that vegetable. Ideally the time between sample collection and analysis should be as short as possible. The protocol should minimize any effects that may cause undesirable losses prior to analysis.

Frozen materials should be stored at −20°C, fruits and salad vegetables at around 4°C, and canned foods at room temperature. Powdered and freeze-dried materials should be stored in the dark in their original containers. Storage of fresh materials should preferably not exceed 3 days. After the initial preparation (see below), fresh or cooked materials can be conveniently stored at −20°C for a short time prior to extraction.

As indicated above, the preparation of the sample depends on the type of material and the homogeneity of that material. With dry powdered or liquid materials, the whole or parts of the samples collected can simply be thoroughly mixed prior to analysis.

Vegetables and fruits are prepared as appropriate for typical "in home" preparation methods (e.g., by removal of outside leaves, peeling, coring). Larger items such as cabbages, may be quartered, and then one quarter from each of the individual samples cut and mixed. Smaller items are cut and mixed. Further subsamples of the individual samples (e.g., 100 g depending on the number of individual samples collected initially) of the cut mixed materials are taken and these subsamples bulked and thoroughly mixed.

In order to obtain a thoroughly representative sample for extraction and subsequent analysis the following procedure has been used in the author's laboratory. Immediately after the preparation of the composite sample (raw and or after cooking as appropriate) the sample was frozen in liquid nitrogen and ground under liquid nitrogen in a Waring-type blender. Ground sample was then stored in an air-tight bottle under nitrogen at −20°C for up to 3 days prior to analysis. All manipulations were carried out under yellow/gold fluorescent lighting.

Extraction

Methods of extraction are dealt with in detail in *UNIT F2.1*. All manipulations should be carried out under gold/yellow light, avoiding exposure to daylight or artificial "white" fluorescent light. All solvents must be of a high degree of purity. Cooking procedures such as boiling may cause disruption of the cellular matrix of vegetable material making the carotenoids more readily extractable. Samples containing high levels of fat, esterified carotenoids, or chlorophylls require saponification (see *UNIT F2.1*); however, in many instances chlorophylls may be separated from the carotenoids of interest during column chromatography avoiding the need for saponification. Certain saponification conditions may cause degradation of carotenoids, particularly the xanthophylls, so the concentration of KOH, time, and temperature must be carefully assessed for the particular type of material being analyzed.

Solvents

Different solvents and the composition of the mobile phase used in HPLC may effect lmax. Generally speaking, the lmax in hexane, ethanol, and petroleum ether will show little if any change, but chloroform, for example, will show a shift to a longer wavelength. Other factors which may effect the spectral characteristics are water in water-miscible solvents, protein in carotenoproteins, and low temperatures.

Degradation of carotenoids

It must be remembered that carotenoids are sensitive to light, heat, air, and active surfaces; therefore, precautions must be taken during preparation and any subsequent storage to avoid any detrimental effects such as degradation, formation of stereoisomers, structural rearrangement, and other physicochemical reactions. The conditions of handling and any preparation prior to storage or extraction are critical to ensuring that there is no degradation of the analytes prior to analysis. Where possible, all manipulations during the preparation of standard solutions and extracts should be carried out under yellow/gold fluorescent lighting.

In certain instances, degradation can be rapid, thus it is essential that the chances of it occurring be lessened or avoided completely. This is particularly important in the case of standard carotenoid solutions. It has been reported that lycopene in particular can be rapidly degraded in chloroform from certain sources (Scott, 1992). With all standard solutions it is important not to assume that a standard has remained stable during storage.

Carotenoids in chlorophyll containing samples may be subject to chlorophyll sensitized photoisomerization, resulting in the production of significant amounts of *cis* (*Z*) isomers even during a brief exposure of an extract to light. As an example of a related problem, in acetone, *cis*-isomers can be produced as a result of the production of triplet state carotenes. The presence of O_2 in stored samples, peroxides in solvents, or oxidizing agents can rapidly lead to bleaching and carotenoid epoxides and apocarotenoids. Unsaturated lipids and metal ions can also enhance oxidative breakdown. Impurities such as plasticizers, especially phalates, can produce severe problems in spectrophotometric analysis, so all contact of samples, solvents and other reagents with plastic materials should be avoided wherever possible.

Carotenoids can be converted into mixtures of geometrical isomers under appropriate conditions, the most common being iodine catalyzed photoisomerization. This produces an equilibrium mixture of isomers, in general the *all-trans* isomers predominates. These isomers in an isomeric mixture cannot be measured separately by simple spectrophotometric determination. The usual method of subsequent measurement would be chromatographic separation, diode-array detection, and spectral analysis. In the absence of any definitive data on extinction coefficients for *cis*-isomers, they are quantified against the *all-trans* isomer. Modern procedures involve the direct synthesis of *cis*-carotenoids.

Extinction coefficients for carotenoid extracts

There are fewer problems if the extract contains essentially only one carotenoid, or if a single carotenoid has been collected from a chromatographic separation; however, most food extracts will contain at least two predominant carotenoids, and green vegetables in particular will also contain chlorophylls, which have absorption bands in the same region as the carotenoids. Plasma samples will contain a whole array of different carotenoids, although only five or six may predominate, but even these may, depending on the diet, be present at very different levels in different individuals, and some may even be absent; however, for plasmas, it may be a useful tool for screening individuals with "high" and "low" carotenoid

Carotenoids

levels. Here again it does not give any information about individual carotenoids and samples indicating a "high" level may be biased in favor of one or two carotenoids.

The problem relating to chlorophylls can be overcome to some extent by saponification of the sample, which will remove the chlorophylls; however, care must be taken in the choice of conditions, as some carotenoids, particularly the xanthophylls, may be degraded (see UNIT F2.1). On the other hand, it is possible to use an alternate wavelength for the carotenoids. For example most of the major carotenoids of interest in foods have an absorption peak around 480 nm, where any absorption of chlorophylls causes less interference; however, it is then necessary to use alternate extinction coefficients (e.g., 2180 for β-carotene).

If the extract or fraction is composed of one predominant carotenoid, it can be monitored at the appropriate wavelength and the $A^{1\%}$ for that carotenoid used. If the extract is composed of more than one carotenoid, the absorbance at 450 nm can be measured and the A1% for β-carotene used. Results can be expressed as total β-carotene equivalents. Alternatively a "typical" $A^{1\%}$ value of 2500 can be used for comparing relative quantities between various sample extracts. An $A^{1\%}$ value of 2500 is appropriate since the predominate carotenoids, with the exception of lycopene ($A^{1\%}$ = 3450), have λ between 2460 and 2710. Of course if samples of tomatoes are to be analyzed, which contain lycopene predominantly, then an $A^{1\%}$ value of 3450 would be more appropriate.

It must be stressed that the value obtained in this way for mixtures of carotenoids is only an approximation. For more definitive analysis column chromatography (UNIT F2.3) should be used.

Anticipated Results

Using the extinction coefficients given above, the example for β-carotene, a 1% solution (1 g/100 ml or 10 mg/ml) would give a theoretical absorbance reading of 2592 AU. A 2 μg/ml solution should therefore give an absorbance reading of 0.518 AU; however, as indicated earlier, the standard solution must be analyzed by HPLC to determine the chromatographic purity (see Basic Protocol 1).

Time Considerations

The time taken to carry out the procedures outlined above will depend on the knowledge and experience of the analyst and their familiarity with the equipment; however, once familiar with the procedure, the preparation and calibration of a standard solution of a carotenoid (see Basic Protocol 1) should not take >1 hr.

The time required for Basic Protocol 2 will depend on the extraction procedure.

Literature Cited

Britton, G. 1995. UV/visible spectroscopy. In Carotenoids, Volume 1B (G. Britton, S. Liaanen-Jensen, and H. Pfander, eds.) pp. 13-62. Birkhäuser, Basel, Switzerland.

Jaffé, H.H. and Orchin, M. 1962. Theory and Applications of Ultraviolet Spectroscopy. John Wiley and Sons, New York.

Scott, K.J. 1992. Observation on some of the problems associated with the analysis of carotenoids in foods by HPLC. Food Chemistry 45:357-364.

Key References

Bauerfiend, J.C. (ed.) 1981. Carotenoids as Colorants and Vitamin A Precursors: Technological and Nutritional Applications. Academic Press, New York.

A comprehensive treatise on carotenoid color technology.

Britton, G., Liaanen-Jensen, S., and Pfander, H. (eds.) 1995. Carotenoids. Volumes 1A and 1B, Spectroscopy. Birkhäuser, Basel, Switzerland.

Workbooks as well as a reference book with practical guidance and examples.

Goodwin, T.W. (ed.) 1988. Plant Pigments. Academic Press, London.

Chlorophylls and carotenoids (distribution, function and analysis).

Pfander, H. (ed.) 1987. Key to Carotenoids, 2nd ed. Birkhäuser, Basel, Switzerland.

Structural formula, common and chemical designations, and references.

Contributed by K. John Scott
Institute of Food Research
Colney, Norwich, United Kingdom

Chromatographic Techniques for Carotenoid Separation

This unit describes several liquid chromatographic techniques for separating and measuring carotenoids. The first protocol incorporates a reversed-phase separation using a wide-pore, polymerically-synthesized C18 column with visible detection at 450 nm (see Basic Protocol 1). The first alternate protocol is also isocratic C18 reversed-phase, but permits simultaneous analysis of retinol, tocopherols, and carotenoids using both a programmable UV-Vis detector and fluorescence detector, or a single diode-array detector (see Alternate Protocol 1). The second alternate protocol is oriented toward more detailed carotenoid analysis of geometric isomers (see Alternate Protocol 2); it incorporates a unique C30 "carotenoid" column with gradient separation and visible detection at 450 nm. The final basic protocol described in this unit is a normal-phase separation permitting more complete quantitation of xanthophylls and their isomers (see Basic Protocol 2). Two support protocols are described. The first details the preparation of standards for generating a calibration curve (see Support Protocol 1), while the second provides guidelines for sample preparation based on knowledge of the sample matrix (see Support Protocol 2).

ISOCRATIC CAROTENOID SEPARATION USING WIDE-PORE, POLYMERIC C18

This method requires the least sophisticated equipment and relies heavily on the unique characteristics of the column to separate the carotenoids (Craft et al., 1992; Epler et al., 1992). It incorporates the use of a polymeric C18 column, which has been shown to offer unique selectivity for structurally similar compounds such as geometric isomers. The addition of a second detector or use of a diode-array detector permits the simultaneous analysis of tocopherols, but not retinol. If the method is modified to incorporate a solvent gradient, retinol can be measured also (MacCrehan and Schonberger, 1987).

Materials

    HPLC grade methanol
    HPLC grade acetonitrile
    HPLC grade triethylamine (TEA)
    Calibration standards and (optional) internal standard (see Support Protocol 1)
    Food sample of interest (see Support Protocol 2)

    Vacuum filtration device, ultrasonicator, or inline vacuum degasser
    HPLC system:
      Column: Vydac 201 TP or 218 TP C18 column, 5 μm, 250 × 4.6 mm
        (Vydac/Separations Group; preferred), Bakerbond WP C18 (J.T. Baker), or
        HiPore RP 318 (BioRad Laboratories) and guard column containing similar
        packing material
      Data recorder: computer data system or integrator
      Detector: fixed, variable, programmable, or diode-array (DAD) UV-Vis detector
      Injector: manual or automatic
      Pump: Isocratic

Prepare mobile phase

1. Prepare mobile phase by mixing 900 ml methanol, 100 ml acetonitrile, and 1 ml of triethylamine (TEA).

   Triethylamine serves as a modifier to prevent both nonspecific adsorption and oxidation.

Contributed by Neal E. Craft

2. Degas the mobile phase via vacuum filtration, ultrasonic agitation, or inline vacuum degasser.

Set HPLC conditions

3. Set the pump flow rate at 1.0 ml/min.

4. Set UV-Vis detector at 450 nm (436 nm if using filter photometer).

5. Inject individual standards and the standard mixtures, including any optional internal standard, as described to generate a standard curve (see Support Protocol 1).

6. Inject 10 to 50 µl of sample (see Support Protocol 2) and any optional internal standard dissolved in a solvent miscible with the mobile phase (e.g., ethanol, methanol, acetonitrile).

> *The complete separation from lutein to lycopene requires ~20 min. Figure F2.3.1 illustrates the separation of carotenoid standards using this system. The elution order using this method (i.e., lutein, zeaxanthin, β-cryptoxanthin, echinenone, α-carotene, β-carotene, lycopene) differs from many other methods.*
>
> *Carotenoid retention and separation are influenced by column temperature; at temperatures above 20° to 25°C, lutein and zeaxanthin may not be well separated.*
>
> *Tocopherols can be measured simultaneously by using a diode array detector, a second UV detector set at 280 to 300 nm, or a fluorescence detector set at 296 nm excitation and 336 nm emission.*

7. Calculate the final concentrations of carotenoids in samples by multiplying the peak areas of analytes by the calibration response factors. Apply sample weight and dilution factors to arrive at the concentration of carotenoids in the original sample (i.e., initial concentration).

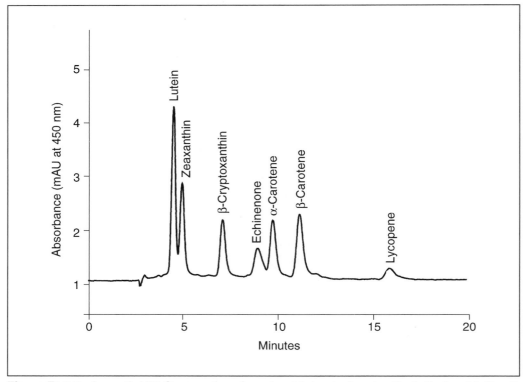

Figure F2.3.1 Isocratic HPLC separation of carotenoid standards using Basic Protocol 1. Conditions: 5-µm × 250-mm × 4.6-mm Vydac 201TP column, 90:10 methanol/acetonitrile mobile phase, 1.0 ml/min flow rate, visible detection at 450 nm, column temperature 25°C.

The response factor is the ratio of the analyte concentration to the peak area (or height) produced under the defined set of conditions during calibration (e.g., wavelength, solvent composition, column). Typically, data systems will generate response factors for each analyte from calibration data. For example, to calculate the initial concentration, where the sample weight (W1) is 0.5 g, the initial dilution (D1) is 25 ml, and the dilution factor for injection (D2) is 10×. The final concentration of the injected sample (Fc) equals the response factor multiplied by the peak area (or height). The initial concentration (Ic) equals the final concentration multiplied by the initial dilution multiplied by the dilution factor for injection divided by the sample weight (Ic = [Fc × D1 × D2]/W1 = [5.0 μg/ml × 25 ml × 10]/0.5 g = 2.5 mg/g, where Fc = Pa × Rf = 25,000 units × 0.002 = 5.0 μg/ml).

STANDARDS PREPARATION AND CALIBRATION

Analytical methods are only as good as the initial calibration; therefore, it is essential that the calibration for the accompanying HPLC methods be performed carefully. Carotenoids are labile compounds that are seldom obtained in pure form and degrade readily upon exposure to oxygen and light. Precautions should be taken to minimize standard and sample exposure to UV light and air. Given the above considerations, it is never recommended that carotenoid calibrants be prepared gravimetrically without verification using a spectrophotometer. This can lead to serious errors in quantitation (Craft et al., 1990). Many impurities contribute mass but not color to the standards. In addition, carotenoids tend to dissolve slowly in many solvents. Thus accuracy can be significantly improved by applying absorptivities ($E^{1\%}_{cm}$) and Beer's Law to filtered solutions of carotenoids to obtain concentration. The value assigned can be further refined by correcting for peak purity. This is accomplished by injecting individual standard solutions into the HPLC column while monitoring the wavelength maximum of each standard. Once the standard solution concentrations have been established, the individual carotenoid standard solutions can be mixed to form calibration solutions. The general procedure is provided below.

Materials

Crystalline carotenoid standards: lutein, zeaxanthin, β-cryptoxanthin, lycopene, α-carotene, and β-carotene
Reagent alcohol with and without 30 ppm butylated hydroxytoluene (BHT)
HPLC grade tetrahydrofuran (THF) stabilized with 250 ppm BHT

Table F2.3.1 Commonly Used Absorptivity Values for Carotenoids and Their Corresponding Wavelength Maxima

Analyte	Wavelength (nm)	Absorptivity $(E^{1\%}_{cm})^a$
α-carotene	444	2800
β-carotene	452	2592
δ-carotene	456	3290
α-cryptoxanthin	445	2636
β-cryptoxanthin	452	2386
Lutein	445	2550
Lycopene	472	3450
Neoxanthin	439	2243
Phytoene	285	1250
Phytofluene	347	1577
Violaxanthin	443	2250
Zeaxanthin	452	2350

[a]Absorptivities listed above were taken from Bauernfeind (1981). For some carotenoids, $E^{1\%}_{cm}$ is provided in petroleum ether or hexane and is not significantly different from those in ethanol.

Internal standard(s)
HPLC grade ethanol or hexane

Vacuum filtration apparatus with 0.45-μm membrane
Spectrophotometer (285 to 456 nm)
Polypropylene or PTFE membrane, 0.45 μm for vacuum filtration

Preparation of calibration standards

1. Dissolve ~1 to 2 mg lutein, zeaxanthin, β-cryptoxanthin, and other xanthophylls directly in 100 ml reagent alcohol containing 30 ppm BHT. Dissolve ~1 to 2 mg lycopene, α-carotene, and β-carotene in 10 ml THF stabilized with BHT, then dilute to 100 ml with reagent alcohol.

   For normal-phase separations, solutions should be prepared using hexane.

   Crystalline carotenoid standards are available from Sigma, Indofine Chemical, Atomergic Chemetals, Fluka Chemical, Kemin Industries, and others.

2. Vacuum filter stock solutions through a 0.45-μm membrane to remove any undissolved material.

3. Measure the absorbance of the solutions at the wavelength maximum, as described in Table F2.3.1, against an appropriate blank (i.e., reagent alcohol) on a spectrophotometer. Dilute appropriately with reagent alcohol to measure between 0.5 and 1.0 AU.

4. Inject each standard individually into the appropriate HPLC column (see Basic Protocols 1 and 2 and Alternate Protocol 1 and 2), monitoring its wavelength maximum to determine the necessary purity correction.

   For example, if the area of the analyte peak constitutes 90% of the total peak areas, then the concentration calculated using spectrophotometric absorbance is adjusted to 90%.

5. (Optional) Inject the internal standard(s) to determine if there are any degradation products or impurities that may co-elute and absorb at the wavelength of the analytes, and to determine the purity correction of the internal standard.

   An internal standard is a compound that is not present in the sample, but is chemically and physically similar to the analytes of interest. A fixed quantity is incorporated into the calibration solutions. The same concentration of internal standard is added to the samples during extraction to compensate for analyte recovery and injection variability. As seen in Figure F2.3.1, Echinenone, which is not typically found in foods, is used as the internal standard. Unfortunately, compounds which may be used as internal standards for carotenoid analysis are not readily available commercially.

6. Mix standards in the range expected for each analyte in the sample matrix (see Table F2.3.2), optionally adding a fixed amount of internal standard. Dilute to a known volume (usually 5 ml) with ethanol to provide the desired final concentration.

Table F2.3.2 Calibration Range for Carotenoids

Analyte	Range (μg/ml)
α-carotene	0.05–5.0
β-carotene	0.05–10.0
β-cryptoxanthin	0.05–5.0
Lutein	0.05–10.0
Lycopene	0.05–10.0
Zeaxanthin	0.05–5.0

A minimum of 3 concentrations should be prepared although 5 concentrations weighted toward the lower end are preferred.

Total carotenoids in any single solution should not exceed 20 μg/ml.

For normal-phase separations use hexane rather than ethanol for dilution.

7. Inject mixture (volume as recommended in the protocols for the particular column and conditions; see Basic Protocols 1 and 2 and Alternate Protocols 1 and 2) and generate calibration curves.

Most HPLC instruments include computer data systems which automatically plot peak response versus concentration to generate response factors; however, if using an older system, the standard curve can be plotted manually. The calibration curves should be linear with a correlation coefficient of >0.98 and intersect very near the origin.

After calibration of the HPLC, a food-based quality control material or reference material should be analyzed on a routine basis to validate the performance of the method. The chromatograms illustrated in the following sections of this chapter are of a food matrix created from a mixture of baby foods and infant formula; however, a similar standard food reference material (SRM 2383) may be directly purchased from the National Institute of Standards and Technology (NIST).

SAMPLE PREPARATION

Food materials are variable and complex matrices. Knowledge of the sample matrix is critical for accurate carotenoid quantification. The type and chemical form of carotenoids and the composition of the food matrix are critical to the amount of sample preparation that is necessary prior to sample analysis. Many factors regarding the food matrix must be considered for efficient carotenoid extraction. For instance, the relative content of lipid to carotenoid in the food matrix influences the method of sample preparation. If both the lipid and carotenoid content are high (e.g., margarine), it may be possible to dilute the sample in an organic solvent that is miscible with the HPLC mobile phase for direct injection; however, when the lipid content of the sample is high and the carotenoid content is low, saponification is useful to separate the lipid (primarily triglycerides) from the carotenoids. Another factor is the form of the carotenoid that is present in the sample. Carotenes (hydrocarbon carotenoids) do not form ester linkages and can be directly extracted by homogenizing in the presence of lipophilic solvents (e.g., hexane, ethyl acetate, and toluene); however, the xanthophylls frequently form esters which will readily extract into lipophilic solvents. The xanthophylls are more easily quantified in the free form, which requires hydrolysis. Saponification (alkaline hydrolysis) is necessary to remove chlorophylls that are present in many foods because they can interfere with the detection of carotenoids. In other words, there is not one given sample preparation that will apply to all foods; therefore, to provide general guidelines, sample preparation has been broken into the following three categories. 1. Oil-based food samples containing only hydrocarbon carotenoids or nonesterified xanthophylls that are visibly yellow to red in color (e.g., margarine). 2. Direct extraction for food samples that do not contain xanthophyll esters or chlorophylls, and have a low lipid and high carotenoid content (e.g., carrots). 3. Saponification for food samples containing xanthophyll esters, chlorophylls, or high lipid and low carotenoid content (e.g., spinach, eggs).

Materials

Food sample
HPLC grade tetrahydrofuran (THF) containing 250 ppm BHT
Reagent alcohol
Magnesium carbonate
50:50 methanol/THF

10% (w/v) pyrogallol in reagent alcohol
40% (w/v) potassium hydroxide (KOH) in methanol
Nitrogen gas
Saturated NaCl (~6 M)
75:25 hexane/THF
Sodium sulfate

50-ml volumetric flask
0.45-μm filter (optional)
Tabletop centrifuge and appropriate swinging bucket rotor
Vacuum filtering apparatus and Whatman filter no. 42 paper
30- to 50-ml tube with cap
Shaking water bath or ultrasonic bath, 60°C
1.5 × 12–inch glass column with sintered glass frit or plugged with glass wool
Solvent evaporation apparatus (e.g., SpeedVac, rotary evaporator, nitrogen manifold, TurboVap)

For oil-based food samples containing only hydrocarbon carotenoids or nonesterified xanthophylls that are visibly yellow to red in color (e.g., margarine)

1a. Weigh 0.5 to 5.0 g of sample.

2a. Dissolve in 25 ml of THF stabilized with 250 ppm BHT, then dilute to volume in a 50-ml volumetric flask. Dilute the sample further with reagent alcohol until the carotenoid concentration is ~5 to 10 mg/liter.

   The carotenoid concentration is approximated by diluting and measuring the absorbance at 450 nm on a spectrophotometer where 1 AU = ~4 mg/liter.

3a. (Optional) Filter through 0.45-μm filter prior to injection.

   If the sample is free of particulate, it may be directly injected into the HPLC column for analysis.

For direct extraction of food samples that do not contain xanthophyll esters, chlorophylls, or have a low lipid and high carotenoid content (e.g., carrots)

1b. Homogenize ~5 g of ground sample with 10% (w/w) magnesium carbonate and 25 ml of 50:50 methanol/THF.

   Lyophilized and dry samples should be reconstituted with water prior to extraction. They may require saponification if the samples contain chlorophylls or xanthophyll esters. Beadlet materials should be suspended in hot water, saponified, or enzymatically hydrolyzed before extraction. Follow the manufacturer's recommendations before analyzing by HPLC.

2b. Centrifuge in a swinging bucket rotor for 10 min at $6000 \times g$, 4 °C and remove the supernatant.

3b. Repeat steps 1b and 2b until the extracting solvent is colorless.

4b. Combine the extracts and vacuum filter through Whatman paper no. 42 to remove particles.

5b. Dilute to a known volume with reagent alcohol so that the THF represents <10% of the total solution (e.g., if total volume is 90 ml dilute to 500 ml).

6b. (Optional) Filter through 0.45-μm filter prior to injection.

   If the sample is free of particulate, it may be directly injected into the HPLC column for analysis.

For saponification of food samples containing xanthophyll esters, chlorophylls, or high lipid and low carotenoid content (e.g., spinach, eggs)

1c. Proceed with a direct sample extraction as instructed in steps 1b to 5b.

2c. Transfer a 5-ml aliquot to a 30- to 50-ml capped tube for saponification.

3c. Add 1 ml of 10% (w/v) pyrogallol in reagent alcohol.

4c. Add 2 ml of 40% (w/v) KOH in methanol.

> *The final concentration of reagents in the mixed solution should be ~5% to 10% KOH and ≥ 1% pyrogallol.*

5c. Flush the tubes with nitrogen gas and cap.

6c. Saponify the samples at 60°C for 1 hr in a shaking water bath or 30 min in an ultrasonic bath.

7c. After saponification, dilute the samples with 8 ml of saturated NaCl solution.

8c. Extract by vigorous mixing with 10 ml 75:25 hexane/THF.

> *Free xanthophylls, both endogenous and present in the saponified samples, are more polar and extract less efficiently into lipophilic solvents. Frequently, the addition of a polar organic solvent (tetrahydrofuran, methylene chloride, diethyl ether) is required to thoroughly extract them from the sample matrix and aqueous phase.*

9c. Remove the upper phase.

10c. Repeat the extraction step until the upper phase is colorless.

> *For some samples the phases will not separate on standing and may require centrifugation to break emulsions.*

11c. Wash the combined extract with water to remove traces of KOH and pyrogallol that may have been co-extracted.

12c. Place 3 in. solid sodium sulfate in the bottom of a 1.5×12–in. glass column plugged with sintered glass frit or plugged with glass wool. After removing the water, pass the extract through the sodium sulfate to remove any traces of water remaining.

13c. Remove the solvent from the extract using a solvent evaporation apparatus.

14c. Dissolve residue in a known volume of reagent alcohol to yield 5 to 20 mg/liter (an absorbance of 1.5 to 5.0 AU at 450 nm).

> *Samples extracted into strong organic solvents (hexane, ether, pet ether, ethyl acetate, etc.) must be transferred into a solvent miscible with the mobile phase. A small volume (e.g., 1 ml) of the organic extract should be evaporated under N_2 gas and dissolved in reagent alcohol. Further dilutions with alcohol may be necessary to obtain 5 to 10 mg/liter concentration before HPLC injection.*

15c. (Optional) filter through 0.45-μm filter prior to injection.

> *If this sample is free of particulate, it may be directly injected into the HPLC column for analysis.*

ISOCRATIC CAROTENOID SEPARATION CAPABLE OF SIMULTANEOUS SEPARATION OF RETINOL AND TOCOPHEROL USING SPHERISORB ODS2

This method is the simplest approach for simultaneous carotenoid, retinol, and tocopherol analysis. Both the column and mobile phase have been chosen to provide efficiency and selectivity for the analysis of these components without the use of a gradient. The method uses the Spherisorb ODS2 column, in which the C18 chain is monomerically bound to the silica particles (i.e., the C18 chain binds at one site to the silica particles).

Additional Materials (also see Basic Protocol 1)

Ammonium acetate
p-dioxane
Triethylamine (TEA)
Calibration standards and (optional) internal standard (see Support Protocol 1)
Food sample of interest (see Support Protocol 2)

Vacuum apparatus and 0.45-μm polytetrafluoroethylene (PTFE) or polypropylene membrane
HPLC system (see Basic Protocol 1):
  Column: 3-μm × 150-mm × 4.6-mm Spherisorb ODS2 (ES Industries, Phenomenex, or Waters) and guard column containing similar packing material

Prepare solutions

1. Dissolve 0.385 g ammonium acetate in 50 ml methanol.

2. Mix 830 ml acetonitrile, 130 ml *p*-dioxane, and 1 ml triethylamine (TEA).

   TEA serves as a modifier to prevent both nonspecific adsorption and oxidation.

3. While stirring, slowly add 40 ml methanol/ammonium acetate solution (step 1).

4. Vacuum filter through 0.45-μm PTFE or polypropylene membrane.

Set HPLC conditions

5. Set the pump flow rate at 1.5 ml/min.

6. Set UV-Vis detector at 450 nm (436 nm if using filter photometer).

7. Inject 10 to 30 μl individual standards and the standard mixtures, including any optional internal standard, as described to generate a standard curve (see Support Protocol 1).

   Column temperature should be maintained at a fixed temperature between 22° and 29°C in either a room with well regulated temperature, or a column oven for above ambient temperatures.

   If the injection solvent is of comparable solvent strength to that of the mobile phase, up to 30 ml may be used as the sample injection volume; however, if the injection solvent is more lipophilic than the mobile phase, the injection volume is limited to 10 ml.

8. Inject 10 to 30 μl standard or sample dissolved in a solvent miscible with the mobile phase (e.g., methanol, acetonitrile).

   The complete separation from retinol to β-carotene requires ~15 min. Figure F2.3.2 illustrates the separation of vitamins and carotenoids in the mixed food extract using this LC system. The elution order using this method is: lutein, zeaxanthin, β-cryptoxanthin, lycopene, α-carotene, and β-carotene.

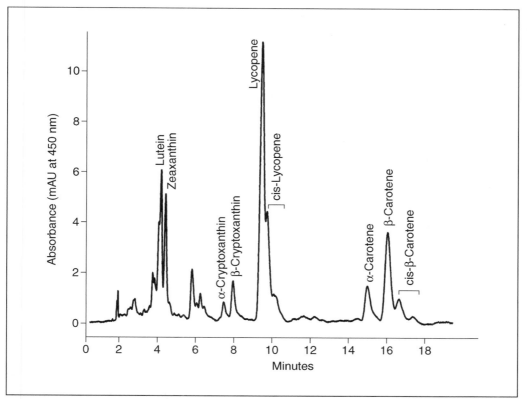

Figure F2.3.2 Isocratic HPLC separation of the food reference material carotenoids using Basic Protocol 2. Conditions: Spherisorb 3-μm × 150-mm × 4.6-mm ODS2 column, 83:13:4:0.1 acetonitrile/dioxane/150 mM ammonium acetate in methanol/TEA mobile phase, 1.5 ml/min flow rate, visible detection at 450 nm, column temperature 29°C.

Detection

9a. *For carotenoids:* Monitor 450 nm.

9b. *For retinol and carotenoids:* Monitor 325 nm for 3.2 min, then program the detector to change the wavelength to 450 nm.

10. *For tocopherols (optional):* Place fluorescence detector after UV-Vis detector and monitor excitation at 296 nm and emission at 336 nm.

11. *For diode array detectors (optional):* monitor 296 nm for tocopherols, 325 nm for retinol, and 450 nm for carotenoids.

12. Calculate the final concentrations as described above (see Basic Protocol 1, step 7).

GRADIENT SEPARATION USING C 30 CAROTENOID COLUMN

This method requires the most sophisticated equipment and yields the most detailed results by utilizing the C30 column that was created specifically for carotenoid separation. The C30 column is polymerically bonded yielding selectivity similar to the polymeric C18. The column has an intermediate pore diameter (200 Å) and a 30 carbon alkyl chain, which results in higher carbon content and therefore stronger retention of the carotenoids. Due to the strong retention of carotenoids on this column, a gradient must be employed unless the sample is previously fractionated or only contains a specific group of carotenoids. It is possible to separate a wide polarity range of carotenoids and their isomers.

ALTERNATE PROTOCOL 2

Carotenoids

F2.3.9

Additional Materials (*also see Basic Protocol 1*)

Calibration standards and (optional) internal standard (see Support Protocol 1)
Food sample of interest (see Support Protocol 2)
Mobile phase:
   Solvent A: 0.05% TEA and 50 mM ammonium acetate in methanol
   Solvent B: 0.05% TEA in isopropyl alcohol
   Solvent C: 0.05% TEA in THF stabilized with 250 ppm BHT

HPLC system (see Basic Protocol 1)
   Pump: ternary gradient
   Column: 3-μm 250-mm × 4.6-mm C30 with 3-μ guard column (Waters)
   Column oven: 35°C

Set HPLC conditions

1. Set the pump flow rate at 1.0 ml/min.

2. Set UV-Vis detector at 450 nm (436 nm if using filter photometer).

3. Inject 10 to 50 μl individual standards and the standard mixtures, including any optional internal standard as described to generate a standard curve (see Support Protocol 1).

4. Inject 10 to 50 μl sample (see Support Protocol 2) dissolved in a solvent miscible with the mobile phase (e.g., ethanol, 90% ethanol/10% isopropanol).

   The complete separation from lutein to lycopene requires ~40 min with an additional 15 min to equilibrate the column back to the initial mobile phase. Figure F2.3.3 illustrates the separation of carotenoid in the mixed food extract using this LC system. The elution order

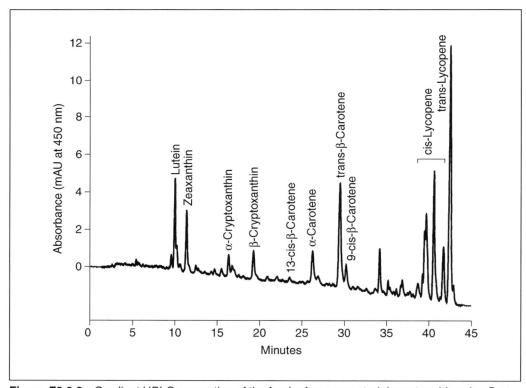

Figure F2.3.3 Gradient HPLC separation of the food reference material carotenoids using Protocol 3. Conditions: 3-μm × 250-mm × 4.6-mm Waters C30, column, 1.0 ml/min flow rate, visible detection at 450 nm, column temperature 35°C, solvent A = 50 mM ammonium acetate in methanol, B = isopropyl alcohol, C = tetrahydrofuran (all solvents contain 0.1% TEA). Flow program: 90% A/10% B linear gradient, 54% A/35% B/11% C over 24 min, linear gradient to 30% A/35% B/35% C over 11 min, hold 8 min, then return to initial conditions over 10 min.

using this method differs from many other methods: lutein, zeaxanthin, β-cryptoxanthin, 15-cis-β-carotene, 13-cis-β-carotene, α-carotene, trans β-carotene, 9-cis-β-carotene, cis-lycopene, and trans-lycopene.

Prepare gradient

5. Starting with 90% solvent A/10% solvent B, establish a linear gradient over 24 min to 54% solvent A/35% solvent B/11% solvent C, followed by a second linear gradient over 11 min to 30% solvent A/35% solvent B/35% solvent C. Hold 5 min, return to initial conditions over 10 min. Hold 5 min before next injection.

   Triethylamine in the solvent serves as a modifier to prevent both non-specific adsorption and oxidation.

Detect

6. *For carotenoids:* Monitor 450 nm.

7. *For tocopherols (optional):* Place fluorescence detector after UV-Vis detector and monitor excitation at 296 nm and emission at 336 nm.

8. *For a diode array detector (optional):* Monitor 296 nm for tocopherols, 325 nm for retinol, and 450 nm for carotenoids.

9. Calculate the final concentrations as described above (see Basic Protocol 1, step 7).

NORMAL-PHASE SEPARATION OF XANTHOPHYLLS

BASIC PROTOCOL 2

In normal-phase chromatography, polar components are more strongly retained than nonpolar components. Thus, hydrocarbon carotenes elute quickly while xanthophylls are retained and separated. This approach provides a more complete separation of polar carotenoids and their geometric isomers. This protocol is useful to the analyst that is specifically interested in the xanthophyll fraction of a sample.

Materials

Hexane
Dioxane
Indole-3-propionic acid (IPA)
TEA
Calibration standards and (optional) internal standard (see Support Protocol 1)
Food sample of interest (see Support Protocol 2)

Vacuum filtration device, ultrasonicator, or inline vacuum degasser
HPLC system (see Basic Protocol 1):
   Column: 5-μm × 250-mm × 4.6-mm Lichrosorb Si column (ES Industries, Phenomenex, or EM Science) and guard column containing similar packing material

Prepare mobile phase

1. Prepare mobile phase by mixing 800 ml hexane, 200 ml dioxane, 15 ml IPA, and 2.0 ml TEA.

   TEA serves as a modifier to prevent both nonspecific adsorption and oxidation.

2. Degas the mobile phase via vacuum filtration, ultrasonic agitation, or inline vacuum degasser.

Set HPLC conditions

3. Set the pump flow rate at 1.0 ml/min.

4. Set UV-Vis detector at 450 nm (436 nm if using filter photometer).

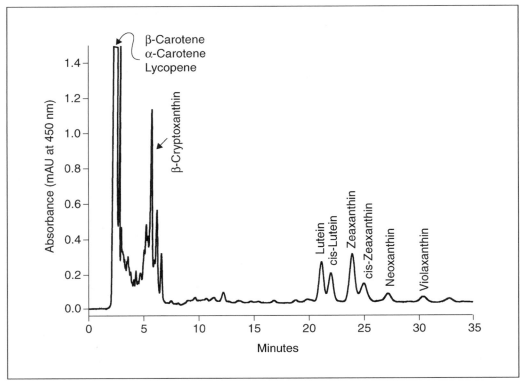

Figure F2.3.4 Isocratic HPLC separation of the food reference material carotenoids using Protocol 4. Conditions: Lichrosorb Si60, 5 μm, 250 × 4.6 mm column, hexane/dioxane/IPA/TEA (80:20:0.15:0.02) mobile phase, 1.0 ml/min flow rate, visible detection at 450 nm.

5. Inject 10 to 100 μl individual standards and the standard mixtures, including any optional internal standard as described to generate a standard curve (see Support Protocol 1).

6. Inject 10 to 100 μl of standard (see Support Protocol 1) or sample, including any optional internal standard, (see Support Protocol 2) dissolved in hexane.

> *The complete separation from β-carotene to violaxanthin requires ~35 min. Figure F2.3.4 illustrates the separation of carotenoids in a mixed food extract using this LC system. The hydrocarbon carotenes (β-carotene, α-carotene, lycopene) elute together at the solvent front. The elution order is: β-cryptoxanthin, α-cryptoxanthin, lutein, cis-lutein, zeaxanthin, cis-zeaxanthin, neoxanthin, and violaxanthin.*

7. Calculate the final concentrations as described above (see Basic Protocol 1, step 7).

COMMENTARY

Background Information

Carotenoids were among the first compounds to be separated by liquid chromatography, and actually inspired Mikael Tswett who coined the name meaning "color writing". Although the technique and instrumentation have evolved greatly since then, liquid chromatography is still the best mode of separating carotenoids. Until the 1970's all separations were performed using adsorption chromatography and low-pressure columns. The use of high-pressure pumps decreased analysis time and permitted the development of more efficient and novel column packing materials.

Carotenoid separations can be accomplished by both normal- and reversed-phase HPLC. Normal-phase HPLC (NPLC) utilizes columns with adsorptive phases (i.e., silica) and polar bonded phases (i.e., alkylamine) in combination with nonpolar mobile phases. In this situation, the polar sites of the carotenoid molecules compete with the modifiers present in the solvent for the polar sites on the stationary phase; therefore, the least polar compounds elute first while the polar analytes, like the xanthophylls, are retained longer. Reversed-phase HPLC (RPLC), which is most commonly used for carotenoid analysis, incorporates non-polar bonded phases, like the C18, or polymer phases, such as polystyrene-divinyl benzene, with polar mobile phases. The carotenoids partition between the nonpolar stationary phase and the polar mobile phase. Xanthophylls elute early due to their preference for the polar mobile phase while carotenes elute later due to their affinity for the stationary phase.

Several column factors are important for the separation of this very similar group of compounds. They include: the ligand, the ligand chain length, particle size, pore diameter, and method of ligand bonding. The ligand, which is bonded to the base silica, will determine the polarity of the column. The chain length is the number of carbon molecules in the ligand bonded to the silica and will influence analyte retention on the column. Particle size influences separation efficiency with smaller particles having a higher efficiency than larger particles. The pore diameter influences the carbon load of the column and thus, the retention of the analytes. The smaller the pore diameter, the greater the surface area and the higher the carbon load; however, if the pore diameter is too small, it will inhibit the large carotenoid molecules passage through the column. Finally, the method of ligand bonding refers to the synthesis used to bond the ligand to the base silica. The most common way of bonding the ligand is monomerically, in which the carbon chains are bonded to the silica surface using monochlorosilanes. The carbon chains extend into the pores of the silica and the analytes must pass by the individual chains on their way through the column. The second method is polymeric synthesis in which the carbon chains are bonded to the silica surface using trichlorosilanes in the presence of specific quantities of water. In this case, the carbon chains do not extend as individual finger-like projections into the path of the analytes, but the chains are cross-linked, forming a kind of irregular net that the analytes must pass through. Subtle changes from column to column in any of these factors will influence analyte separation. The reader should refer to the article by Craft (1992) for a detailed discussion of how these factors influence carotenoid separations.

The sensitivity of the method is influenced by the type of detector and lamp used in the detector. Deuterium lamps have weak energy in the wavelength range that carotenoids ab-

sorb, therefore the signal-to-noise is low. Tungsten and Xenon lamps have stronger energy in the 450-nm wavelength range and yield higher signal-to-noise ratios. Older diode-array detectors (DAD) tend to have poorer sensitivity than good programmable wavelength UV-Vis detectors due to the dispersion of light across the array of diodes. To improve sensitivity of a DAD, the bandwidth should be increased to 10 to 20 nm; however, to obtain good spectra of carotenoids, the bandwidth needs to be set at 1 to 5 nm, as the spectral difference between *cis* isomers is frequently only 2 to 4 nm. Newer DAD detectors have both deuterium and tungsten lamps providing better sensitivity for carotenoid analysis. In addition, at least one manufacturer incorporates a light pipe in the flow cell to improve sensitivity ~5-fold. Amperometric and coulometric detection of carotenoids have been used successfully (Mac-Crehan and Schonberger, 1987; Gamache et al., 1997). The advent of the coulometric array detector by ESA has permitted very low detection limits (picogram) along with a "finger print" profile. This mode of detection does require the use of a supporting electrolyte in the mobile phase (Ferruzzi et al., 1998). The addition of salts or ions to the mobile phase may alter elution profiles. Phytoene and phytofluene, which don't absorb at 450 nm, can be measured in each of the reversed-phase methods described by monitoring 285 nm and 345 nm, respectively. They elute with retention times in the area of hydrocarbon carotenoids.

Critical Parameters

Many parameters are critical to the successful and reproducible chromatographic separation of carotenoids. To assure stable baselines and consistent pump flow, solvents should be degassed before use to minimize outgassing during the gradient. Outgassing at the detector flow cell results in baseline noise. Solvents can be degassed using vacuum filtration through a 0.45-μm filter, ultrasonic agitation for 15 min, helium sparging, or use of an inline solvent degasser. Use of a backpressure restrictor after the detector to maintain ~35 to 100 psi backpressure will also help to prevent this problem.

Due to the limited solubility and slow rate of dissolution of most carotenoids in organic solvents, be vigilant to avoid the use of solutions with high concentrations of carotenoids. The relative solubility of β-carotene and lutein in different solvents is discussed in Craft and Soares (1992). Calibration standards should be filtered through 0.45-μm membranes to remove any undissolved crystals. Tiny carotenoid crystals may precipitate out of stock solutions when placed at freezer temperatures. It is essential to dissolve the standard before use or filter the sample and reassign its concentration. Dissolution can be facilitated by agitation in an ultrasonic bath or warming the container in a 60°C water bath. When protected from light and oxygen, most stock and calibration solutions are stable at −20°C for >6 months. Lycopene, however, degrades rapidly and is only stable ~1 week. Note that care should be taken to avoid directly exposing carotenoid stock solutions or samples to light, oxygen and acids since the carotenoids are labile to these conditions.

During the sample preparation, one must remember that the lipid behaves as a strong (lipophilic) solvent and in adequate concentration will cause peak broadening or doubling. In addition, dilution of these samples prior to injection must be great enough to prevent overloading the column with lipid. Thus, if the lipid content is high and the carotenoid content is low, the lipid (primarily triglycerides) must be separated from the carotenoid prior to HPLC analysis. Two general mechanisms can be used to remove the lipid: physical separation (solid phase extraction or liquid-liquid extraction) and hydrolysis (alkaline or enzymatic). Physical separation is only applicable under certain conditions since the solubility of hydrocarbon carotenoids is very similar to many lipids; however, if a sample contains high triglyceride and only polar carotenoids, the carotenoids could be partitioned into a polar organic solvent (e.g., methanol) or onto a polar solid phase extraction (SPE) cartridge (Si, NH_2 or CN) while the triglyceride remains with the nonpolar solvent (e.g., hexane). Saponification (alkaline hydrolysis), is the most common means of removing lipid from samples and has been described earlier (see Support Protocol 2). When saponification is used, it is essential to include antioxidants such as pyrogallol and ascorbic acid. Do not use BHT as it forms polymers when heated under alkaline conditions. The polymers absorb light in the visible range and co-elute with some carotenoids. Enzymatic hydrolysis requires the presence of a lipase to break down the triglyceride. Unfortunately, enzymes require an aqueous environment to function and the triglyceride is not soluble in water; therefore, the lipid forms a separate layer or oil droplets that provide limited surface area upon which the lipase can function.

Samples, even at moderate concentrations, injected into the HPLC column may precipitate in the mobile phase or at the column frit. In addition, the presence of other compounds (e.g., lipids) in the injection sample may drive the carotenoids out of solution or precipitate themselves in the mobile phase, trapping carotenoids. It is best to dissolve the sample in the mobile phase or a slightly weaker solvent to avoid these problems. Centrifugation or filtration of the samples prior to injection will prevent the introduction of particles that may block the frit, fouling the column and resulting in elevated column pressure. In addition to precipitation, other sources of "on-column" losses of carotenoids include nonspecific adsorption and oxidation. These can be minimized by incorporating modifiers into the mobile phase (Epler et al., 1993). Triethylamine or diisopropyl ethylamine at 0.1% (v/v) and ammonium acetate at 5 to 50 mM has been successful for this purpose. Since ammonium acetate is poorly soluble in acetonitrile, it should be dissolved in the alcoholic component of the mobile phase prior to mixing with other components. The ammonium acetate concentration in mobile phases composed primarily of acetonitrile must be mixed at lower concentration to avoid precipitation. In some cases, stainless steel frits have been reported to cause oxidative losses of carotenoids (Epler et al., 1992). When available, columns should be obtained with biocompatible frits such as titanium, Hastolloy C, or PEEK.

Not all C18 (ODS) columns are manufactured in the same way. As such, substitution of other columns for those listed in the above methods will probably yield inferior results. For a detailed explanation of column and solvent effects on the separation of carotenoids, see Epler et al. (1992).

Anticipated Results

During reversed-phase HPLC, carotenoids elute from the column in the order of polar to nonpolar. Reversed-phase chromatography tends to permit the elution of a wider range of components during a single isocratic separation. The three reversed-phase methods described herein should produce separations of the major carotenoids (i.e., lutein, α-cryptoxanthin, β-cryptoxanthin, lycopene, α-carotene, and β-carotene), in addition to components that are difficult to resolve, such as lutein/zeaxanthin and *cis/trans* isomers. Isocratic normal-phase chromatography is typically more appropriate for the separation of a narrow polarity range of geometric and positional isomers. The elution order is from nonpolar to polar. A wider range of analytes can also be separated using a normal-phase gradient.

Time Considerations

Calibration is time consuming when performed correctly. It may require 1 or 2 days to perform all the necessary steps (i.e., prepare stocks, filter, measure absorbance, check purity, dilute, mix, and inject calibrants). Once the stock solutions and mixed calibration solutions have been prepared, a calibration check can be performed in ~4 hr. Sample preparation, depending on the matrix, may require a few minutes or a few hours. If an autosampler is unavailable for overnight injection the extracts are typically stable overnight, refrigerated at $-20°$ to 4°C. It is prudent to maintain the autosampler tray temperature from 4° to 15°C to reduce sample degradation. HPLC analysis of the extracted sample requires 20 to 60 min. Typically one technician can extract 12 to 24 samples per day to be analyzed overnight or the next day.

Literature Cited

Bauernfeind, J.C. (ed.) 1981. Carotenoids as Colorants and Vitamin A Precursors. Academic Press, San Diego.

Craft, N.E. 1992. Carotenoid reversed-phase high-performance liquid chromatography methods: Reference compendium. *In* Methods in Enzymology, Vol. 213, pp.185-203. Academic Press, San Diego.

Craft, N.E. and Soares, J.H., Jr. 1992. Relative solubility, stability, and absorptivity of lutein and β-carotene in organic solvents. *J. Agric. Food Chem.* 40:431-434.

Craft, N.E., Sander, L.C., and Pierson, H.F. 1990. Separation and relative distribution of all *trans*-β-carotene and its *cis* isomers in β-carotene preparations. *J. Micronutrient Anal.* 8:209-221.

Craft, N.E., Wise, S.A., and Soares, J.H., Jr. 1992. Optimization of an isocratic high-performance liquid chromatographic separation of carotenoids. *J. Chromatogr.* 589:171-176.

Epler, K.S., Sander, L.C., Ziegler, R.G., Wise, S.A., and Craft, N.E 1992. Evaluation of reversed-phase liquid chromatographic columns for recovery and selectivity of selected carotenoids. *J. Chromatogr.* 595:89-101.

Epler, K.S., Ziegler, R.G., and Craft, N.E. 1993. Liquid chromatographic method for the determination of carotenoids, retinoids and tocopherols in human serum and in food. *J. Chromatogr.* 619:37-48.

Ferruzzi, M.G., Sander L.C., Rock C.L., and Schwartz S.J. 1998. Carotenoid determination in biological microsamples using liquid chromatography with a coulometric electrochemical array detector. *Anal. Biochem.* 256:74-81.

Gamache, P.H., McCabe, D.R., Parevez, H., Parvez, S., and Acworth, I.N. 1997. The measurement of markers of oxidative damage, antioxidants and related compounds using HPLC and coulometric array analysis. *In* Columetric Electrode Array Detectors for HPCL (I.N. Acworth, M. Naoi, S. Parvez, and H. Parvez, eds.) pp. 91-119. VSP Publications, Zeist, The Netherlands.

MacCrehan, W.A. and Schonberger, E. 1987. Determination of retinol, α-tocopherol, and α-carotene in serum by liquid chromatography with absorbance and electrochemical detection. *Clin. Chem.* 33:1585-1592.

Key References

Craft, 1992. See above.

A systematic overview of the principles involved in and model applications of the reverse-phase HPLC separation of carotenoids.

Epler et al., 1992. See above.

A comprehensive comparison of columns and mobile phases for use in the separation of carotenoids by reverse-phase HPLC.

Contributed by Neal E. Craft
Craft Technologies
Wilson, North Carolina

Mass Spectrometry of Carotenoids

The high sensitivity and selectivity of mass spectrometry (MS) facilitates the identification and structural analysis of small quantities of carotenoids that are typically obtained from biological samples such as plants, animals, or human serum and tissue. Structural information from the abundant fragmentation is provided by classical ionization methods, such as electron impact (EI; see Basic Protocol 1) and chemical ionization (CI; see Basic Protocol 1), but molecular ions are not always observed. Recent advances in soft ionization techniques, such as fast atom bombardment (FAB; see Basic Protocol 2), matrix-assisted laser desorption/ionization (MALDI; see Basic Protocol 3), electrospray ionization (ESI; see Basic Protocol 4), and atmospheric pressure chemical ionization (APCI; see Basic Protocol 5), have facilitated the molecular weight determination of carotenoids by minimizing fragmentation that is typical of EI and CI. Once the molecular weight of a carotenoid has been established using one of these ionization techniques, collision-induced dissociation (CID) and tandem mass spectrometry (MS/MS) can be used to augment fragmentation and obtain structurally significant fragment ions that may aid in the differentiation of structural isomers, such as differentiation of lutein from zeaxanthin, or of α-carotene from β-carotene and lycopene. CID and MS/MS parameters are independent of the ionization step so that no modifications of the sample preparation and ionization procedures are needed. Although CID and MS/MS can be used with any ionization technique, the application of this approach is illustrated using FAB ionization in Basic Protocol 2.

MS can be coupled to high-performance liquid chromatography (HPLC) to obtain separation of isomeric carotenoids or to remove interfering contaminants prior to ionization and detection. Except for MALDI, every ionization method discussed in this unit has been utilized during liquid chromatography MS (LC/MS; see Basic Protocols 1, 2, 4, and 5). However, LC-APCI-MS is now the preferred approach due to its widespread availability and ease of use. Although LC/MS and LC/MS/MS (see Basic Protocol 2) are routinely carried out, gas chromatography MS (GC/MS) is rarely used, because carotenoids typically decompose when exposed to the high temperatures of the GC process. Reviews of LC/MS and MS of carotenoids have been published by van Breemen (1996, 1997). A general introduction to mass spectrometry is given in Watson (1997).

ELECTRON IMPACT AND CHEMICAL IONIZATION MASS SPECTROMETRY OF CAROTENOIDS

Electron impact (EI) and chemical ionization (CI) MS have been used for carotenoid analysis for more than 30 years. Therefore, mass spectra of unknown carotenoids can be compared to a large number of published mass spectra to aid in identification. Unlike the newer "soft" ionization techniques in MS, EI and CI produce considerable fragmentation, and molecular ions are not always observed. Therefore, these techniques are most useful for obtaining fingerprints, or characteristic fragmentation patterns, instead of confirming or determining molecular weights of carotenoids. Alternatively, direct exposure EI (DEI) and/or direct exposure CI (DCI) MS can be used.

Materials

Carotenoid sample (1 to 100 mg/liter; 2 to 200 μM) dissolved in volatile organic solvent (e.g., hexane, tetrahydrofuran, methyl-*tert*-butyl ether, acetone), stored in an airtight glass vial

Reagent gas: methane or isobutane (for CI only)

Contributed by Richard B. van Breemen

Microsyringe

Mass spectrometer equipped with direct insertion probe and electron impact (EI) and/or chemical impact (CI) ionization

NOTE: Carotenoid solutions degrade within hours at room temperature, but they can be stored(in many cases) ≤1 month at or below −20°C and 3 months at or below −70°C.

1. Using a microsyringe, load 1 µl carotenoid sample onto a direct insertion probe of a mass spectrometer.

2. Let solvent evaporate, then insert probe into the ion source of the mass spectrometer.

3. Introduce reagent gas (for CI only) and then turn on the electron beam (typically 70 eV for EI and 200 eV for CI).

4. Heat probe to vaporize carotenoid.

5. Record mass spectrum over the range m/z (mass-to-charge ratio) 50 to 800 for EI and m/z 100 to 800 for CI.

   All known carotenoids and their major fragment ions should be included in this range. By beginning the scan at m/z 100 during CI, reagent-gas ions can be avoided.

BASIC PROTOCOL 2

FAST ATOM BOMBARDMENT, LIQUID SECONDARY ION MASS SPECTROMETRY, AND CONTINUOUS-FLOW FAST ATOM BOMBARDMENT OF CAROTENOIDS

Fast atom bombardment MS (FAB-MS) and liquid secondary ion MS (LSIMS) are matrix-mediated desorption techniques that use energetic particle bombardment to simultaneously ionize samples such as carotenoids and transfer them to the gas phase for mass spectrometric analysis. Unlike with the EI and CI techniques, molecular ions are usually abundant and fragmentation is minimal.

Materials

3-Nitrobenzyl alcohol

Carotenoid sample (1 to 100 mg/liter; 2 to 200 µM) dissolved in volatile organic solvent (e.g., hexane, tetrahydrofuran, methyl-*tert*-butyl ether, acetone), stored in an airtight glass vial

HPLC solvents for LC/MS using continuous-flow FAB (e.g., methanol, methyl-*tert*-butyl ether)

Microsyringe

Mass spectrometer or tandem mass spectrometer equipped for fast atom bombardment (FAB)–MS or liquid secondary ion (LSI)MS, with direct insertion probe and with continuous-flow ionization source (as needed)

Syringe pump or HPLC pump capable of delivering flow rates of 1 to 10 µl/min (for continuous-flow only)

Reversed-phase HPLC (typically C18 or C30) column (for continuous-flow FAB-MS or LSIMS)

NOTE: The 3-nitrobenzyl alcohol matrix is typically stable at room temperature for ~3 months. Older solutions begin to turn yellow as they decompose and should be replaced. Carotenoid solutions will degrade within a few hours at room temperature but can be stored ≤1 month at or below −20°C and in some cases ≤3 months at or below −70°C.

NOTE: For LC/MS or flow injection using continuous-flow FAB, the mass spectrometer must be equipped with a continuous-flow ionization source.

NOTE: Details about chromatography columns used for carotenoids are contained in UNIT F2.3.

109

For probe analysis:

1a. Load 1 µl of 3-nitrobenzyl alcohol onto a direct insertion probe, then use a microsyringe to load 1 µl carotenoid sample onto the surface of the liquid matrix.

2a. Let solvent evaporate and then insert probe through the vacuum interlock into the ion source of a mass spectrometer or tandem mass spectrometer.

3a. Turn on FAB-MS or LSIMS beam and record the positive ion mass spectrum over the range *m/z* (mass-to-charge ratio) 50 to 900.

> *If desired, the peaks at m/z 154 and m/z 307 for the protonated monomer and dimer of the matrix, 3-nitrobenzyl alcohol, can be eliminated by scanning from m/z 310 to 900.*

For continuous-flow analysis:

1b. Set up a syringe pump or an HPLC pump and equilibrate a reversed-phase HPLC column with appropriate eluents such as methanol and methyl-*tert*-butyl ether (e.g., 70:30, v/v).

> *A reversed-phase HPLC column (typically C18 or C30) is required for HPLC separations. Because the flow rate into the continuous-flow FAB-MS or LSIMS source must be <10 µl/min, either a capillary column must be used or else the flow must be split postcolumn. For narrow-bore HPLC columns operated at 200 µl/min, the split ratio would be 30:1. Isocratic or gradient separations may be used. A syringe pump is usually necessary for capillary columns, but standard HPLC pumps are sufficient for applications using narrow-bore columns.*

2b. Add the matrix, ~0.1% (v/v) 3-nitrobenzyl alcohol prepared in appropriate mobile phase, postcolumn at a flow rate of ~1 to 3 µl/min.

3b. Interface the continuous-flow probe to a mass spectrometer. Tune the continuous-flow FAB-MS or LSIMS ion source on the 3-nitrobenzyl alcohol dimer ion at *m/z* 307.

4b. Inject a carotenoid sample onto the HPLC column, turn on the FAB-MS or LSIMS beam, and record the positive ion mass spectrum over the range *m/z* 300 to 1000.

> *An injection volume of 1 to 5 µl is typical for narrow-bore columns.*

MATRIX-ASSISTED LASER DESORPTION/IONIZATION TIME-OF-FLIGHT MASS SPECTROMETRY OF CAROTENOIDS

BASIC PROTOCOL 3

A matrix-mediated ionization technique, matrix-assisted laser desorption/ionization time-of-flight (MALDI-TOF) MS uses an intense flash of laser light to vaporize a solid matrix containing the sample. Although usually regarded as an ionization method reserved for high-mass compounds such as proteins and polymers, MALDI has shown remarkable promise for the analysis of carotenoids. In particular, MALDI has been effective in the ionization of intact esterified carotenoids that would fragment too extensively using other ionization methods.

Materials

Carotenoid sample (1 to 100 mg/liter; 2 to 200 µM) dissolved in acetone, store in airtight glass vial
Acetone saturated with 2,5-dihydroxybenzoic acid (sample matrix)

Microsyringe
Matrix-assisted laser desorption/ionization time-of-flight (MALDI-TOF) mass spectrometer with UV laser (i.e., 337-nm nitrogen laser), MALDI probe, and optional delayed extraction and postsource decay

Carotenoids

NOTE: Carotenoid solutions degrade rapidly at room temperature (within several hours) but can be stored for at least 1 month at or below −20°C and ~3 months at or below −70°C in some cases.

1. Mix 20 µl carotenoid sample with 10 µl acetone saturated with 2,5-dihydroxybenzoic acid.

2. Using a microsyringe, load 5 to 10 µl carotenoid/matrix sample onto the target of a MALDI probe.

3. Let solvent evaporate (only a few seconds are required) and then insert probe through the vacuum interlock into the ion source of a MALDI-TOF mass spectrometer.

4. Record MALDI-TOF mass spectra in positive ion mode. Look for molecules ions and protonated molecules in the range m/z 300 to 1000.

ELECTROSPRAY IONIZATION LIQUID CHROMATOGRAPHY/MASS SPECTROMETRY OF CAROTENOIDS

Unlike other LC/MS techniques in which removal of the mobile phase and sample ionization are discrete steps, electrospray is both an ionization method and an interface between an HPLC system and a mass spectrometer. Electrospray is also one of the most universal ionization techniques for MS, as virtually every class of compound has been analyzed using this technique, including carotenoids.

For flow injection analysis of carotenoids, ~1 µl of a 0.1 to 100 µM carotenoid sample can be injected into a solvent stream and carried into the electrospray source without chromatography. This approach is particularly useful for high-throughput analysis of pure samples and for tuning and optimizing the parameters of the electrospray source. The sample should be dissolved in a volatile organic solvent (e.g., hexane, tetrahydrofuran, methyl-*tert*-butyl ether, acetone, methanol), but may contain water. The sample should be stored in an airtight glass vial. A narrow scan range (e.g., m/z 520 to 620 for compounds such as β-carotene and lutein) or selected ion monitoring of the molecular ions (e.g., m/z 536 and 568 for β-carotene and lutein, respectively) is usually used during flow-injection analysis.

When HPLC is used as part of the analysis, the mobile phase is typically a mixture of methanol and methyl-*tert*-butyl ether (i.e., 50:50, v/v), although other HPLC solvents for LC/MS using electrospray (e.g., water, tetrahydrofuran) can be used. It is important to note that entirely organic solvent systems might pose a fire hazard for some home-built ion sources and some older commercial instruments if air leaks into the ionization chamber. Therefore, water or a halogenated solvent should be added to the mobile phase postcolumn to suppress ignition. The electrospray source must always be vented outside the laboratory. The scan range is typically m/z 300 to 1000 in order to include known carotenoids and this esters.

When tuning the electrospray source of a mass spectrometer or tandem mass spectrometer, optimum sensitivity for carotenoids will be obtained using the highest possible voltage on the electrospray needle before corona discharge occurs. For example, an electrospray needle voltage of −5100 V provided excellent sensitivity in one published report (van Breemen, 1995). Because not all electrospray ion sources support voltages this high, the sensitivity might be lower in other systems. A reversed-phase HPLC C18 or C30 column is typically used for LC/MS with electrospray ionization. The flow rate into the mass spectrometer, as controlled by a syringe pump or HPLC pump, should be matched to the design of the system, but in most current commercial systems the flow rate can be varied between 1 and 1000 µl/min. Microbore, narrow-bore, or analytical-scale columns at flow rates from 1 to 1000 µl/min are compatible with most LC/MS electrospray interfaces. Details about chromatography columns used for carotenoids are contained in *UNIT F2.3*.

2,2,3,4,4,4-Heptafluoro-1-butanol can be added postcolumn to give a final concentration of 0.1% (v/v) to enhance ionization efficiency during electrospray. Typically, a 2% (v/v) solution in mobile phase is added at 50 µl/min to the HPLC column effluent at 1 ml/min. Addition of this reagent is optional.

ATMOSPHERIC PRESSURE CHEMICAL IONIZATION LIQUID CHROMATOGRAPHY/MASS SPECTROMETRY OF CAROTENOIDS

Most mass spectrometers equipped for electrospray ionization can be converted to APCI, and many commercial LC-APCI-MS instruments are equipped with both ionization techniques. During APCI, ionization takes place in an atmospheric pressure chamber when the sample molecules collide with solvent ions formed in a continuous corona discharge. Unlike electrospray, the needle used to spray the HPLC effluent is not at high voltage.

For flow-injection analysis of carotenoids, ~1 to 5 µl of a 0.1 to 100 µM carotenoid sample, which may contain water, can be injected into a solvent stream and carried into the APCI source without chromatography. This approach is particularly useful for high-throughput analysis of pure samples and for tuning and optimizing the parameters of the APCI source. The sample should be dissolved in a volatile organic solvent (e.g., hexane, tetrahydrofuran, methyl-*tert*-butyl ether, acetone, methanol), but may contain water. The sample should be stored in an airtight glass vial. A narrow scan range of *m/z* 520 to 620 is appropriate for most carotenoids such as β-carotene and lutein. Alternatively, selected in monitoring of the molecular ions or protonated or deprotonated molucules may be used.

When HPLC is used as part of the analysis, the mobile phase is typically a mixture of methanol and methyl-*tert*-butyl ether (i.e., 50:50, v/v), although other HPLC solvents for LC/MS using APCI (e.g., water, tetrahydrofuran) can be used. It is important to note that if combustible nonaqueous solvent systems are used, water or a halogenated solvent such as methylene chloride or chloroform should be added to the mobile phase postcolumn to suppress ignition in the ion source. In addition, the APCI source must be vented outside the laboratory and should not allow air into the ionization chamber. A scan range of *m/z* 300 to 1000 will include the known carotenoids and their most common esters.

A reversed-phase HPLC C18 or C30 narrow-bore column is typically used for LC/MS with APCI. Details about chromatography columns used for carotenoids are contained in *UNIT F2.3*. For most APCI systems, the optimum flow rate into a mass spectrometer or tandem mass spectrometer equipped with APCI, as controlled by a syringe pump or HPLC pump, is usually between 100 and 300 µl/min, which is ideal for narrow-bore HPLC columns. Larger diameter columns should be used with a flow splitter postcolumn to reduce the solvent flow into the mass spectrometer. For example, if a 4.6 mm i.d. column was used at a flow rate of 1.0 ml/min, then the flow must be split postcolumn ~5:1 so that only 200 µl/min enters the mass spectrometer.

COMMENTARY

Background Information

EI and CI

Most of the original structural elucidation studies of the >600 known carotenoids used EI and CI (Moss and Weedon, 1976). Although these ionization methods are usually used with GC during GC/MS, carotenoids are too thermally labile to pass through the hot oven of a GC and must be introduced into the MS using a direct insertion probe. Most EI and CI mass spectra of carotenoids have been acquired using positive ion mode, but negative ions may be formed during CI with electron capture. For example, McClure and Liebler (1995a,b) used negative ion chemical ionization to analyze oxidation products of β-carotene. EI and CI are gas-phase ionization techniques, which means that carotenoids must be volatilized prior to ionization using thermal heating. This process results in some pyrolysis prior to ionization and promotes fragmentation after ionization, which explains why molecular ions are not always observed in EI and CI mass spectra.

EI and CI have been used in combination with LC/MS in a technique called particle-beam LC/MS. During particle-beam LC/MS, the LC eluate is sprayed into a heated, near-atmospheric pressure chamber to evaporate the mobile phase, and the resulting sample aerosol is separated from the lower-molecular-weight solvent molecules in a momentum separator. Next, the sample aerosol enters the MS ion source, where the aggregates strike a heated metal surface and disintegrate into gas-phase sample molecules that are ionized by EI or CI. To minimize fragmentation and enhance sensitivity during particle-beam LC/MS, negative ion electron capture CI is usually used for carotenoid ionization. For example, Khachik et al. (1992) used particle-beam LC/MS with negative ion electron capture CI to analyze polar carotenoids extracted from human serum. They observed abundant molecular anions, $M^{-\bullet}$, and simple fragmentation patterns, such as loss of water from the molecular ion for xanthophylls like lutein.

FAB-MS and LSIMS

In widespread use since 1982 (Barber et al., 1982), FAB-MS and LSIMS are matrix-mediated techniques. Although matrices such as glycerol or thioglycerol are suitable for hydrophilic compounds such as peptides, a more hydrophobic matrix is required for the analysis of the nonpolar carotenoids. The most effective matrix for carotenoid ionization is 3-nitrobenzyl alcohol (Caccamese and Garozzo, 1990; Schmitz et al., 1992). Ionization and desorption of the carotenoid analyte occur together during the bombardment of the matrix by fast atoms (or ions) to produce primarily molecular ions, $M^{+\bullet}$. Unlike most FAB-MS and LSIMS analyses, protonated or deprotonated molecules of carotenoids are not usually observed. In addition, fragmentation is minimal compared to EI, CI, or MALDI.

Because molecular ions dominate FAB-MS and LSIMS mass spectra of carotenoids, collision-induced dissociation (CID) can be used in a tandem mass spectrometer to enhance the formation of structurally significant fragment ions. These ions provide information about carotenoid functional groups, such as the presence of hydroxyl groups, esters, or rings, or the extent of conjugation of the polyene chain. For example, the most abundant fragment ion in the tandem mass spectrum of β-carotene corresponds to loss of a neutral molecule of toluene, $[M-92]^{+\bullet}$, and indicates the presence of extensive conjugation within the molecule. This ion is also abundant in the tandem mass spectra of α-carotene, γ-carotene, lycopene, astaxanthin, neurosporene, β-cryptoxanthin, zeaxanthin, and lutein. Because the toluene molecule originates within the conjugated polyene chain and not from a terminus of these molecules, compounds such as neoxanthin, phytoene, or phytofluene, which have disruptions in their polyene chains, do not fragment to eliminate toluene. Instead, phytoene and phytofluene exhibit a fragment ion of $[M-94]^{+\bullet}$.

For example, the ion of $[M-69]^{+\bullet}$, which is observed in the tandem mass spectra of lycopene, neurosporene, and γ-carotene but not α-carotene, β-carotene, lutein, or zeaxanthin, indicates the presence of a terminal acyclic isoprene unit. Elimination of a hydroxyl group or a molecule of water, $[M-17]^{+\bullet}$ or $[MH-18]^{+}$, from carotenoids such as astaxanthin or zeaxanthin is characteristic of the presence of a hydroxyl group. Also, tandem mass spectrometry can be used to distinguish between isomeric carotenoids such as α-carotene and β-carotene, or lutein and zeaxanthin. For example, the ring of α-carotene containing the double bond that is not conjugated to the rest of the polyene chain shows unique retro-Diels-Alder fragmentation to form the ion of $[M-56]^{+\bullet}$. In a similar manner, isomeric lutein and zeaxanthin differ by the position of the same carbon-carbon double bond that distinguishes α-carotene and β-carotene. Consequently, only lutein fragments to eliminate this unusual ring and form an ion of m/z (mass-to-charge ratio) 428.

The fragmentation patterns and characteristic fragment ions for the carotenoids observed in FAB-MS and LSIMS tandem mass spectra are also observed in the tandem mass spectra obtained following ESI (see Basic Protocol 4), APCI (see Basic Protocol 5), and other methods. A detailed account of structure determination of carotenoids using FAB ionization with CID and MS/MS is presented in van Breemen et al. (1995). Finally, another advantage of MS/MS is that matrix ions formed during FAB-MS or LSIMS, and any other contaminating ions, are eliminated, which simplifies interpretation of the mass spectrum.

FAB ionization has been used in combination with LC/MS in a technique called continuous-flow FAB LC/MS (Schmitz et al., 1992; van Breemen et al., 1993). Although any standard HPLC solvent can be used, including methyl-*tert*-butyl ether and methanol, the mobile phase should not contain nonvolatile additives such as phosphate or Tris buffers. Volatile buffers such as ammonium acetate are compatible at low concentrations (i.e., ≤10 mM). Continuous-flow FAB has also been used in combination with MS/MS (van Breemen et al., 1993). The main limitations of continuous-flow FAB compared to other LC/MS techniques for carotenoids, such as ESI and APCI, are the low flow rates and the high maintenance requirements. During use, the 3-nitrobenzyl alcohol matrix polymerizes on the continuous-flow probe tip causing loss of sample signal. As a result, the continuous-flow probe must be removed and cleaned approximately every 3 hr.

MALDI-TOF

MALDI-TOF-MS facilitates the analysis of carotenoids and other natural products with detection limits that are lower than most other techniques. For example, subpicomole quantities can be detected (Wingerath et al., 1999). The enhanced sensitivity is the result of the efficiency of the pulsed ionization and detection system in which a complete mass spectrum is recorded with each laser flash. Like FAB and LSIMS, molecular ions are the most abundant sample ions, although some protonated molecules and [M-H]+ ions may be formed as well. Abundant molecular ions of carotenoid esters have been observed using MALDI-TOF-MS (Kaufmann et al., 1996; Wingerath et al., 1996),

which is significant because most other desorption ionization techniques, including FAB, ESI, and APCI, tend to produce molecular ions of esters in lower abundance with abundant deesterified fragment ions. MALDI-TOF-MS is not yet compatible with on-line chromatographic systems, so techniques such as continuous-flow FAB, ESI, or APCI should be selected when LC/MS is required.

ESI

During electrospray ionization, the HPLC eluate is sprayed through a capillary electrode at high potential (usually 2000 to 7000 V) to form a fine mist of charged droplets at atmospheric pressure. As the charged droplets are electrostatically attracted towards the opening of the MS, they encounter a cross-flow of heated nitrogen that increases solvent evaporation and prevents most of the solvent molecules from entering the MS. Although ions produced by electrospray ionization are usually preformed in solution by acid-base reactions (i.e., $[M+nH]^{n+}$ or $[M-nH]^{n-}$), carotenoid ions are probably formed by a field desorption mechanism at the surface of the droplet, which appears to be enhanced by the presence of halogenated solvents such as heptafluorobutanol (van Breemen, 1995). As a result of this unusual ionization process, electrospray ionization of carotenoids produces abundant molecular cations $(M^{+\bullet})$ with little fragmentation, and no molecular anions.

Because of the efficiency of the solvent removal in the LC/MS interface and the flexibility of the ESI process, ESI is compatible with a wide range of HPLC flow rates (from 0.1 nl/min to 1 ml/min) and with a variety of mobile phases including the methanol/methyl-*tert*-butyl ether solvent system that is ideal for separations using C30 carotenoid columns. A potential limitation of ESI for quantitation is its relatively narrow dynamic range (approximately two orders of magnitude). Aside from this narrow range of linear response during quantitative analysis, ESI shows excellent sensitivity and is compatible with HPLC using a wide range of solvents and flow rates. An example of the positive ion LC-ESI-MS analysis of carotenes in an extract of human plasma is shown in Figure F2.4.1.

APCI

APCI uses a heated nebulizer to facilitate solvent evaporation and obtain a fine spray of the mobile phase instead of a strong electromagnetic field as in electrospray. Unlike elec-

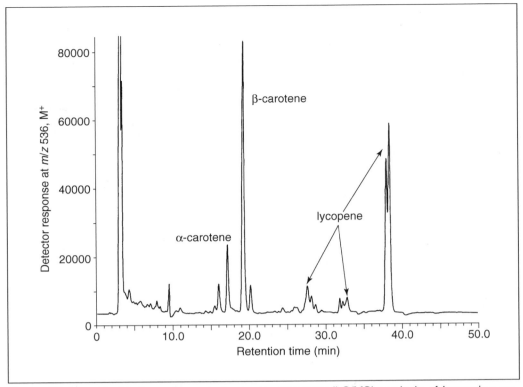

Figure F2.4.1 Liquid chromatography/mass spectrometry (LC/MS) analysis of isomeric carotenes in a hexane extract from 0.5 ml human serum. Positive ion electrospray ionization MS was used on a quadrupole mass spectrometer with selected ion monitoring to record the molecular ions of lycopene, β-carotene, and α-carotene at m/z (mass-to-charge ratio) 536. A C30 HPLC column was used for separation with a gradient from methanol to methyl-*tert*-butyl ether. The all-*trans* isomer of lycopene was detected at a retention time of 38.1 min and various *cis* isomers of lycopene eluted between 27 and 39 min. The all-*trans* isomers of α-carotene and β-carotene were detected at 17.3 and 19.3 min, respectively.

trospray, the nebulizer is not surrounded by a strong electromagnetic field. Ionization takes place by spraying the mobile phase containing analyte into a corona discharge in an atmospheric pressure chamber. The corona discharge ionizes the solvent gas (a chemical reagent gas) that ionizes the carotenoid analyte through ion-molecule reactions. Ions are then drawn into the aperture of the MS for analysis.

The main advantage of APCI compared to ESI for carotenoid analysis is the higher linearity of detector response (exceeding four orders of magnitude of carotenoid concentration), which suggests that LC-APCI-MS should be the preferred MS technique for carotenoid quantitation. Disadvantages of APCI include the multiplicity of molecular ion species, which might lead to ambiguous molecular weight determinations, and abundant fragmentation, which tends to reduce the abundance of the molecular ions. Like ESI, APCI provides high sensitivity for carotenoids with detection limits

near 1 pmol injected on-column during LC/MS. The superior linearity of the MS response during APCI of carotenoids suggests that this LC/MS technique may become the standard for carotenoid quantitation.

Critical Parameters

During EI and CI of carotenoids, the ion source of the MS should be maintained at a low temperature because carotenoids are heat labile. A typical low source temperature would be 100°C. Higher source temperatures can be used, but will result in fewer molecular ions and increased fragmentation.

The choice of matrix for FAB-MS, LSIMS, and MALDI is essential for efficient sample ionization. For example, the use of 3-nitrobenzyl alcohol instead of glycerol, thioglycerol, or most other more common matrices is essential for the formation of abundant carotenoid ions during FAB and LSIMS. Nonpolar carotenoids (e.g., the carotenes) are insoluble in polar matrices such as glycerol or thioglycerol and ionize inefficiently. During MALDI, the matrix and carotenoid sample must be dissolved in the same solvent. Usually, acetone is suitable for this purpose. Furthermore, the matrix must absorb strongly at the wavelength of the incident laser beam. Although a UV laser and UV-absorbing matrix are described, other types of lasers and matrices can be used.

The mobile phase for ESI and APCI should be volatile and can include all common HPLC solvents and compositions. Nonvolatile mobile-phase additives such as nonvolatile buffers or ion-pair agents should be avoided, because these compounds will precipitate and contaminate the ion source. During electrospray, volatile buffers such as ammonium acetate, ammonium formate, and ammonium carbonate can be used at concentrations <40 mM. At higher concentrations, even volatile buffer ions will suppress electrospray ionization. In contrast, volatile buffers at concentrations exceeding ~40 mM will not suppress ionization during APCI, although they may clog the nebulizer or block the entrance to the MS if their evaporation is too slow relative to their concentration.

Troubleshooting

Because extensive fragmentation is typical of EI and CI mass spectra, molecular ions or protonated molecules might not be observed. In order to confirm the molecular weight of a carotenoid, desorption EI or desorption CI (also known as in-beam EI and CI) can be utilized to increase the abundance of the molecular ion species. If the molecular weight of the carotenoid remains uncertain, then softer ionization techniques should be investigated, such as FAB-MS, ESI, MALDI, or APCI.

The limits of detection for carotenoids using FAB-MS and LSIMS are not as low as with most other ionization techniques (Schmitz et al., 1992). Therefore, ≥10 pmol of each carotenoid should be loaded onto the direct insertion probe per analysis. The matrix, 3-nitrobenzyl alcohol, has been effective in facilitating the ionization of all types of carotenoids. However, more polar matrices such as glycerol or thioglycerol might be useful for the FAB-MS or LSIMS analysis of polar xanthophylls such as astaxanthin. Because glycerol and thioglycerol are poor solvents for hydrophobic compounds, they are unlikely to solvate and thus facilitate the ionization of the nonpolar carotenes such as β-carotene.

Whether using FAB-MS, LSIMS, or MALDI, matrix ions might interfere with the detection of certain carotenoids. Therefore, alternate but chemically similar matrices with different molecular weights might need to be utilized on a case-by-case basis.

Anticipated Results

EI and CI

Abundant fragment ions will be observed below m/z 300 due to random cleavage of carbon-carbon bonds. Therefore, these ions are of little value in structure elucidation. Fragment ions above m/z 300 will be useful in confirming the presence of specific ring systems and functional groups. Internal cleavage to eliminate toluene, $[M-92]^{+\bullet}$, is an abundant ion for most carotenoids including β-carotene, α-carotene, and lycopene. For example, β-carotene and zeaxanthin eliminate toluene to form fragment ions of m/z 444 and m/z 476, respectively. This fragmentation pathway indicates the presence of an extensive conjugated carbon-carbon double bond system. For comparison, the incompletely conjugated carotene phytoene does not form an abundant fragment ion corresponding to $[M-92]^{+\bullet}$. As discussed above (see Troubleshooting), molecular ions or protonated molecules will usually be observed in low abundance.

FAB-MS and LSIMS

Molecular ions, $M^{+\bullet}$, will be observed with almost no fragmentation. For example, the positive ion FAB-MS mass spectrum of the carotene phytofluene is shown in Figure F2.4.2. The base peak of m/z 542 corresponds to the molecular ion. For the more polar xanthophylls, protonated molecules may be observed as well as molecular ions. During bombardment, sample ions will be continuously produced for several minutes until the sample or the matrix is consumed. During this time, exact mass measurements and high resolution measurements can be carried out to determine carotenoid elemental compositions. If structurally significant fragment ions are desired, then CID can be used to fragment the molecular ions followed by MS/MS to record the resulting product ions.

MALDI

Abundant molecular ions, $M^{+\bullet}$, will be observed as the base peaks in the MALDI-TOF mass spectra. Delayed extraction will enhance the abundance of the molecular ions relative to background noise, and the use of postsource decay will facilitate the detection of structurally

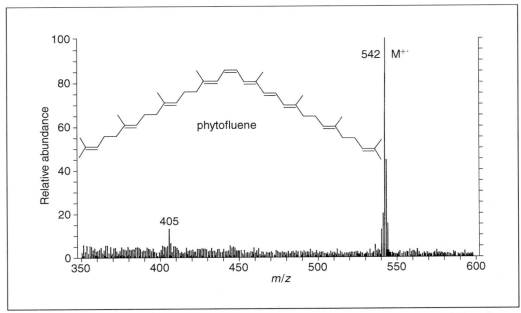

Figure F2.4.2 Positive ion fast atom bombardment (FAB-MS) mass spectrum of phytofluene isolated from blueberries. The base peak of *m/z* (mass-to-charge ratio) 542 corresponds to the molecular ion. Characteristic of FAB-MS, background signals are observed at every *m/z* value. The mass spectrum was obtained during continuous-flow FAB-MS LC/MS using a magnetic sector mass spectrometer. Although the 16-*cis* isomer of phytofluene is shown, the FAB mass spectra of the all-*trans* and other *cis* isomers are indistinguishable.

significant fragment ions that resemble the product ions observed during MS/MS with CID (Kaufmann et al., 1996; Wingerath et al., 1996). Delayed extraction and postsource decay are available on many commercial MALDI-TOF mass spectrometers. Delayed extraction is simply the introduction of a variable pause, usually milliseconds in length, between the firing of the laser during ionization and the translocation of the sample ions from the source to the time-of-flight mass analyzer. This delay allows high-energy and low-mass matrix ions and fragment ions to disperse and be quenched, leaving primarily the sample molecular ions in the ion source for subsequent analysis. Postsource decay is the process of analyzing metastable ions (or fragment ions formed outside the ion source) using a reflectron type of time-of-flight mass spectrometer. This type of analysis is similar to MS/MS, and the types of carotenoid fragment ions that are observed are analogous to those obtained using MS/MS with CID.

ESI

Electrospray ionization will produce molecular ions, $M^{+\bullet}$, with almost no fragmentation for carotenes and many xanthophylls. As the polarity of the carotenoid increases, the prob-

ability of forming protonated (positive ion mode) or deprotonated (negative ion mode) molecules will increase. Negative molecular ions of carotenoids, $M^{-\bullet}$, have not been reported using electrospray. The LC/MS analysis of isomeric carotene weighing 536 Da is shown in Figure F2.4.1. For this analysis, hydrophobic carotenoids including lycopene, α-carotene, and β-carotene were extracted from human serum using hexane and then separated on a narrow-bore C30 reversed-phase HPLC column using a mobile-phase gradient from methanol to methyl-*tert*-butyl ether. Molecular ions were formed using ESI and the signals for all ions of *m/z* 536 are shown in the computer-reconstructed mass chromatogram in Figure F2.4.1. The all-*trans* isomers of α-carotene, β-carotene, and lycopene were detected at retention times of 17.3 min, 19.3 min, and 38.1 min, respectively. The other peaks of *m/z* 536 eluting between 27 and 39 min corresponded to *cis* isomers of lycopene. In addition, *cis* isomers of α-carotene and β-carotene were observed eluting near the all-*trans* isomers between 15 and 21 min.

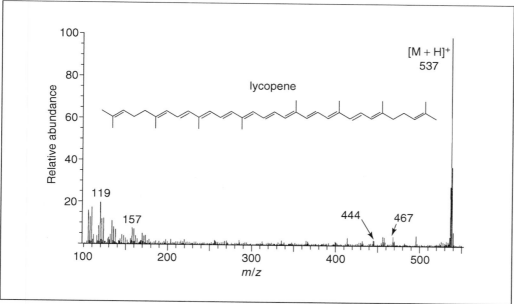

Figure F2.4.3 Flow-injection positive ion atmospheric pressure chemical ionization (APCI) mass spectrum of ~1 pmol lycopene. The carrier solvent for flow injection analysis consisted of methanol/methyl-*tert*-butyl ether (50:50; v/v) at a flow rate of 200 µl/min. The lycopene standard was isolated from tomatoes. The all-*trans* isomer of lycopene is shown, which is the most abundant isomer found in the tomato. This carotene is the familiar red pigment of the tomato.

APCI

Carotenoids form both molecular ions and protonated molecules during positive ion APCI, and molecular ions and deprotonated molecules during negative ion analysis. The relative abundances of molecular ions and protonated or deprotonated molecules vary with the mobile-phase composition (van Breemen et al., 1996). For example, polar solvents such as alcohols lead to an increased abundance of protonated carotenoids, and nonpolar solvents such as methyl-*tert*-butyl ether facilitate the formation of molecular ions. Even though APCI tends to produce more fragmentation in the ion source than either ESI or FAB-MS, these fragment ions are often not abundant. For example, the positive ion APCI mass spectrum of lycopene is shown in Figure F2.4.3. If additional fragmentation is desired for structure confirmation, then CID with MS/MS would be required.

The elemental composition and exact mass of many common carotenoids is shown in Table F2.4.1.

Time Considerations

Because MS is a rapid analytical technique, the sample throughput can exceed 120 samples per hour. Specifically, the throughput of EI, CI,

and FAB-MS is typically up to 20 samples per hour, and the rate-limiting step is usually the minute or two required to introduce the direct insertion probe or sample target containing each sample through the vacuum interlock. Although up to 60 analyses per hour may be achieved using MALDI, these rapid measurements are only possible after an array of samples has been loaded into the ion source of the mass spectrometer. Most commercial MALDI-TOF mass spectrometers allow multiple samples to be loaded onto a single sample stage, thereby increasing the efficiency of sample introduction into the mass spectrometer. Therefore, the rate-limiting step for MALDI-TOF-MS is sample preparation, which includes mixing samples with matrix solutions, depositing each solution on the MALDI target, allowing the solvent to evaporate, and then introducing the target through a vacuum interlock into the ion source of the mass spectrometer.

During chromatographic analysis, such as particle-beam LC/MS, continuous-flow FAB-MS or LSIMS, and LC/MS using ESI or APCI, the rate-limiting step is the chromatographic separation. Most analytical-scale HPLC separations require 15 to 30 min each, which limits the throughput to only two or four samples per hour. However, the high throughput demanded

Table F2.4.1 Elemental Compositions and Exact Masses[a] of Common Carotenoids[b]

Carotenoid	Elemental composition	Exact mass
β-Apo-8'-carotenal	$C_{30}H_{40}O$	416.3079
3-Hydroxy-β-apo-8'-carotenal	$C_{30}H_{40}O_2$	432.3028
α-Carotene	$C_{40}H_{56}$	536.4382
β-Carotene	$C_{40}H_{56}$	536.4382
γ-Carotene	$C_{40}H_{56}$	536.4382
Lycopene	$C_{40}H_{56}$	536.4382
Neurosporene	$C_{40}H_{58}$	538.4539
ς-Carotene	$C_{40}H_{60}$	540.4695
Phytofluene	$C_{40}H_{62}$	542.4852
Phytoene	$C_{40}H_{64}$	544.5008
2',3'-Anhydrolutein	$C_{40}H_{54}O$	550.4175
Echinenone	$C_{40}H_{54}O$	550.4175
α-Cyrptoxanthin	$C_{40}H_{56}O$	552.4331
β-Cryptoxanthin	$C_{40}H_{56}O$	552.4331
Alloxanthin	$C_{40}H_{52}O_2$	564.3967
Canthaxanthin	$C_{40}H_{52}O_2$	564.3967
Diatoxanthin	$C_{40}H_{54}O_2$	566.4124
Lutein	$C_{40}H_{56}O_2$	568.4280
Isozeaxanthin	$C_{40}H_{56}O_2$	568.4280
Zeaxanthin	$C_{40}H_{56}O_2$	568.4280
Lycopene-16,16'-diol	$C_{40}H_{56}O_2$	568.4280
4-Ketoalloxanthin	$C_{40}H_{50}O_4$	578.3760
Pectenolone	$C_{40}H_{52}O_3$	580.3916
Phoenicoxanthin	$C_{40}H_{52}O_3$	580.3916
4-Ketozeaxanthin	$C_{40}H_{54}O_3$	582.4073
Antheraxanthin	$C_{40}H_{56}O_3$	584.4229
Lutein epoxide	$C_{40}H_{56}O_3$	584.4229
7,8,7',8'-Tetradehydroastaxanthin	$C_{40}H_{48}O_4$	592.3553
7,8-Didehydroastaxanthin	$C_{40}H_{50}O_4$	594.3709
Astaxanthin	$C_{40}H_{52}O_4$	596.3866
Neoxanthin	$C_{40}H_{56}O_4$	600.4179
Isozeaxanthin bispelargonate	$C_{58}H_{88}O_4$	848.6683

[a]Exact mass is defined as the monoisotopic molecular weight of a molecule and is calculated using the mass of the most abundant isotope of each element.

[b]Moss and Weedon (1976) contains a more extensive table of carotenoids and their molecular weights.

by combinatorial chemistry has resulted in the development of HPLC methods utilizing shorter HPLC columns and higher mobile phase flow rates, such that the throughput can exceed 60 LC/MS analyses per hour using ESI or APCI. When chromatographic separations are not required, flow injection ESI- or APCI-MS analyses can be carried out with throughputs exceeding 10 samples per minute or >600 samples per hour.

Literature Cited

Barber, M., Bordoli, R.S., Elliott, G.J., Sedgwick, R.D., and Tyler, A.N. 1982. Fast atom bombardment mass spectrometry. *Anal. Chem.* 54:645A-657A.

Caccamese, S. and Garozzo, D. 1990. Odd-electron molecular ion and loss of toluene in fast atom bombardment mass spectra of some carotenoids. *Org. Mass Spectrom.* 25:137-140.

Kaufmann, R., Wingerath, T., Kirsch, D., Stahl, W., and Sies, H. 1996. Analysis of carotenoids and carotenol fatty acid esters by matrix-assisted laser desorption ionization (MALDI) and MALDI-post-source-decay mass spectrometry. *Anal. Biochem.* 238:117-128.

Khachik, F., Beecher, G.R., Goli, M.B., Lusby, W.R., and Smith, J.C. Jr. 1992. Separation and identification of carotenoids and their oxidation products in the extracts of human plasma. *Anal. Chem.* 64:2111-2122.

McClure, T.D. and Liebler, D.C. 1995a. A rapid method for profiling the products of antioxidant reactions by negative ion chemical ionization mass spectrometry. *Chem. Res. Toxicol.* 8:128-135.

McClure, T.D. and Liebler, D.C. 1995b. Electron capture negative chemical ionization mass spectrometry and tandem mass spectrometry analysis of β-carotene, α-tocopherol and their oxidation products. *J. Mass Spectrom.* 30:1480-1488.

Moss, G.P. and Weedon, B.C.L. 1976. Chemistry of the carotenoids. *In* Chemistry and Biochemistry of Plant Pigments, 2nd ed. (T.W. Goodwin, ed.) pp. 149-224. Academic Press, New York.

Schmitz, H.H., van Breemen, R.B., and Schwartz, S.J. 1992. Applications of fast atom bombardment mass spectrometry (FAB-MS) and continuous-flow FAB-MS to carotenoid analysis. *Methods Enzymol.* 213:322-336.

van Breemen, R.B. 1996. Innovations in carotenoid analysis using LC/MS. *Anal. Chem.* 68:299A-304A.

van Breemen, R.B. 1997. Liquid chromatography/mass spectrometry of carotenoids. *Pure Appl. Chem.* 69:2061-2066.

van Breemen, R.B., Schmitz, H.H., and Schwartz, S.J. 1993. Continuous-flow fast atom bombardment liquid chromatography/mass spectrometry of carotenoids. *Anal. Chem.* 65:965-969.

van Breemen, R.B., Schmitz, H.H., and Schwartz, S.J. 1995. Fast atom bombardment tandem mass spectrometry of carotenoids. *J. Agric. Food Chem.* 43:384-389.

van Breemen, R.B., Huang, C.-R., Tan, Y., Sander, L.C., and Schilling, A.B. 1996. Liquid chromatography/mass spectrometry of carotenoids using atmospheric pressure chemical ionization. *J. Mass Spectrom.* 31:975-981.

Watson, J.T. 1997. Introduction to Mass Spectrometry, 3rd ed. Lippincott-Williams and Wilkins, Philadelphia.

Wingerath, T., Stahl, W., Kirsch, D., Kaufmann, R., and Sies, H. 1996. Fruit juice carotenol fatty acid esters and carotenoids as identified by matrix-assisted laser desorption ionization (MALDI) mass spectrometry. *J. Agric. Food Chem.* 44:2006-2013.

Wingerath, T., Kirsch, D., Kaufmann, R., Stahl, W., and Sies, H. 1999. Matrix-assisted laser desorption ionization postsource decay mass spectrometry (Review). *Methods Enzymol.* 299:390-408.

Key References

Moss and Weedon, 1976. See above.

Extensive molecular ion and fragment ion abundances are listed for positive ion EI mass spectra of carotenoids.

van Breemen, R.B. 1995. Electrospray liquid chromatography-mass spectrometry of carotenoids. *Anal. Chem.* 67:2004-2009.

An important early reference describing ionization conditions for carotenoid analysis using electrospray MS.

van Breemen et al., 1993. See above.

Describes the selection and optimization of matrix and mobile-phase conditions for carotenoid analysis using LC-MS/MS with FAB-MS ionization.

van Breemen et al., 1995. See above.

Contains the most complete set of tandem mass spectra of carotenoids and their interpretation. Fragmentation patterns and characteristic fragment ions described here are common to all CID tandem mass spectra of carotenoids.

van Breemen et al., 1996. See above.

The original reference for LC-APCI-MS of carotenoids, containing essential information for carrying out C30 HPLC separations with on-line APCI mass spectrometric detection.

Wingerath et al., 1999. See above.

Describes sample preparation and matrix requirements for analyzing carotenoids using MALDI-TOF MS and discusses the use of postsource decay for obtaining structurally significant fragment ions of carotenoids during MALDI-TOF MS.

Contributed by Richard B. van Breemen
University of Illinois at Chicago
Chicago, Illinois

Miscellaneous Colorants

F3.1   Betalains   **F3.1.1**
    Basic Protocol 1: Spectrophotometric Determination of Betacyanins and
       Betaxanthins   F3.1.1
    Basic Protocol 2: Quantification of Individual Betacyanins by High-
       Performance Liquid Chromatography (HPLC)   F3.1.2
    Support Protocol: Extraction of Betalains from Beets   F3.1.4
    Commentary   F3.1.5

**F3.2   Spectrophotometric and Reflectance Measurements of Pigments of
Cooked and Cured Meats**   **F3.2.1**
    Basic Protocol 1: Measurement of Nitrosylheme Concentration   F3.2.1
    Basic Protocol 2: Measurement of Total Heme Concentration   F3.2.3
    Basic Protocol 3: Reflectance Detection of Globin Hemochromes   F3.2.4
    Reagents and Solutions   F3.2.5
    Commentary   F3.2.5

F3.3   Measurement of Discoloration in Fresh Meat   **F3.3.1**
    Basic Protocol 1: Analysis of Metmyoglobin in Ground Meat Extracts   F3.3.1
    Basic Protocol 2: Analysis of Metmyoglobin in Fresh Meat Surfaces Using
       Diffuse Reflectance Spectrophotometry   F3.3.3
    Alternate Protocol: Analysis of Fresh Meat Surface Color Using Colorimetry   F3.3.5
    Basic Protocol 3: Isolation of Total Myoglobin for In Vitro Studies   F3.3.6
    Basic Protocol 4: Preparation of Oxymyoglobin by Reduction
       of Metmyoglobin   F3.3.7
    Commentary   F3.3.8

Contents

1

Betalains

Betalains, the pigments of the red beet, have found increased use as food colorants. This unit describes a multiple-component system using visible spectrophotometry to determine the concentration of the pigment components (i.e., betacyanins and betaxanthins) calculated in terms of betanin and vulgaxanthin-I, respectively (see Basic Protocol 1). The total pigment content (betalain) is expressed as the sum of betacyanins and betaxanthins.

In the first protocol (see Basic Protocol 1), the maximum light absorption of betanin, the major betacyanin, and vulgaxanthin-I, the major betaxanthin, are measured. The betanin and vulgaxanthin contents are calculated using each pigment's 1% absorptivity value $A^{1\%}$. The method takes into account small amounts of interfering substances.

A second protocol is presented in which the betacyanin pigments are first separated into their individual components and then quantified by applying each pigment's $A^{1\%}$ (see Basic Protocol 2). This technique is particularly useful when measuring pigment content in heat-treated, partially-degraded, or stored colorant mixtures, as large amounts of degradation products will interfere with the spectrophotometric method.

In addition, a protocol describing the extraction of betalains from beets is provided (see Support Protocol).

SPECTROPHOTOMETRIC DETERMINATION OF BETACYANINS AND BETAXANTHINS

This protocol describes a spectrophotometric method for measuring the betacyanin and betaxanthin content in beet juice or beet tissue extract (see Support Protocol). Light absorption measured at 538 nm and 476 nm is used to calculate the betanin and vulgaxanthin-I concentrations, respectively. In addition, the absorption at 600 nm is measured and used to correct for small amounts of impurities. Since no prior separation of the pigments is made, the light absorption measurement at A_{538} and A_{476} includes all minor betacyanins and betaxanthins, respectively. The results are expressed as betacyanin (calculated in terms of betanin) and betaxanthin (calculated in terms of vulgaxanthin-I). The total betalain concentration is expressed as the sum of the betacyanins and betaxanthins.

Materials

0.05 M phosphate buffer, pH 6.5: 4/9.4 (v/v) 8.863 g/liter Na_2HPO_4/6.773 g/liter KH_2PO
Beet juice or tissue extract (see Support Protocol)
Spectrophotometer and appropriate recording device
1-cm path-length quartz cuvette

1. Turn on the spectrophotometer and allow the instrument to warm up for at least 30 min before taking measurements.

2. Zero the spectrophotometer at 476, 538, and 600 nm using 0.05 M phosphate buffer, pH 6.5 as the solvent blank.

3. Dilute beet juice or tissue extract with 0.05 M phosphate buffer, pH 6.5 such that the A_{538} of the sample is between 0.4 and 0.5 AU.

   See Critical Parameters for a discussion about the narrow range of absorbance units.

Contributed by Joachim H. von Elbe

4. Obtain the visible absorption spectrum of the solution between 450 and 650 nm in recorded form and read the absorption directly at 538, 476, and 600 nm.

5. Calculate the corrected light absorption of betanin and vulgaxanthin-I using the follow set of equations:

$$x = 1.095 \times (a - c)$$
$$z = a - x$$
$$y = b - z - x/3.1$$

Where

a = light absorption of the sample at 538 nm

b = light absorption of the sample at 476 nm

c = light absorption of the sample at 600 nm

x = light absorption of betanin minus the colored impurities

y = light absorption of vulgaxanthin-I corrected for the contribution of betanin and colored impurities

z = light absorption of the impurities.

The derivation of these working equations is discussed elsewhere (see Critical Parameters).

6. Calculate the betanin and vulgaxanthin-I concentrations using each pigment's $A^{1\%}$ and applying the appropriate dilution factor (step 3).

The absorptivity value ($A^{1\%}$) is the extinction coefficient representing a 1% solution (1.0 g/100 ml) and is 1120 for betanin and 750 for vulgaxanthin-I (Wyler and Dreiding, 1957; Piattelli and Minale, 1964).

As an example, if a 1-ml sample of centrifuged beet juice was diluted to 150 ml with buffer, and the absorption spectrum obtained resulted in a light absorption at 538, 476, and 600 nm of 0.422 (a), 0.378 (b), and 0.052 AU (c), respectively, then applying the above equations results in x (betanin) = 0.405, y (vulgaxanthin-I) = 0.230, and z (impurities) = 0.017. Using $A^{1\%}$ for each pigment and the dilution factor of 150×, betanin and vulgaxanthin-I concentrations in the juice of 54 mg/100 ml and 46 mg/100 ml are derived, respectively.

QUANTIFICATION OF INDIVIDUAL BETACYANINS BY HIGH-PERFORMANCE LIQUID CHROMATOGRAPHY (HPLC)

This protocol describes the separation of individual betacyanin pigments (i.e., betanin, isobetanin, betanidin, and isobetanidin) by HPLC, and the quantification of each by comparing the peak area of each pigment to the peak area of a standard curve or collecting each fraction of known volume and applying the respective $A^{1\%}$. The pigments are separated using either isocratic or gradient separation conditions.

Materials

Solvent A: 18/82(v/v) CH_3OH/0.05 M KH_2PO_4: adjust to pH 2.75 with H_3PO_4
Solvent B: CH_3OH
Sephadex G-25 (Amersham Pharmacia Biotech) water slurry, pH 2.0: adjust pH with HCl
Sample of beet juice or beet tissue extract (see Support Protocol), pH 2.0: adjust pH with HCl
1% acetic acid
Four 7.88-mm × 61-cm Bondapak C_{18}/Porasil B columns connected in series
17.8/81.2/1.0 (v/v/v) CH_3OH/0.05 M KH_2PO_4/acetic acid
0.1% HCl

Standard analytical or preparatory HPLC:
    Variable wavelength detector
    Injection loop: 10 ml (preparatory) or 20 µl (analytical)
    Column: reversed-phase Bondapak C_{18}/Porasil B (Water Associates)
1000-ml gel filtration column (Amersham Pharmacia Biotech)
Freeze drier
0.45-µm HA filter (Millipore)
Table top centrifuge 25°C

Initialize HPLC gradient

1. Set the flow rate of the HPLC at 8 ml/min at room temperature. If using a gradient, set the initial mixture to 100% solvent A, changing over 9 min to 80% solvent A, 20% solvent B.

   If an isocratic separation is desired, use 100% solvent A.

Prepare standards

2. Slurry pack a 1000-ml gel filtration column with a Sephadex G-25 water mixture, pH 2.0.

3. Apply 150 ml beet juice or tissue extract, pH 2.0.

4. Elute the column with 500 ml 1% acetic acid.

5. Collect red (betacyanin) fraction and freeze dry.

6. Dissolve freeze dried material in a minimum of double-distilled water, filter through a 0.45-µm HA-Millipore filter and chromatograph by preparatory LC—i.e., place ten

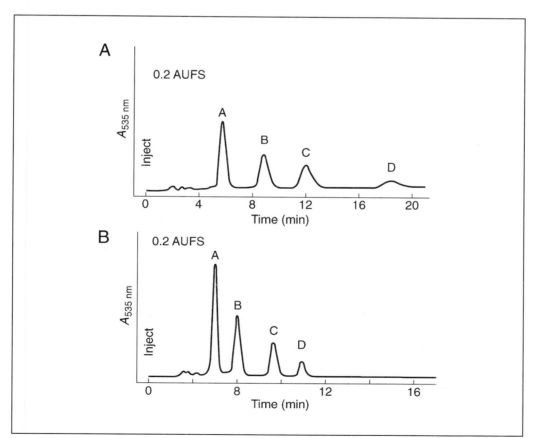

Figure F3.1.1   Isocratic gradient LC chromatogram of betacyanin pigments. Peak A: betanin; peak B: isobetanin; peak C: betanidin, and peak D: isobetanidin. Adopted from Schwartz and von Elbe (1980).

samples on four 7.88-mm × 61-cm Bondapak C_{18}/Porasil B columns connected in series. Elute columns with 17.8/81.2/1.0 (v/v/v) CH_3OH/KH_2PO_4/acetic acid using a flow rate of 8 ml/min.

7. Collect both the betanin and betanidin fraction (~50 ml each; Fig. F3.1.1). Place each fraction into a centrifuge tube and maintain at −15°C for 3 hr to initiate crystallization.

8. Thaw samples at 25°C and collect crystals by centrifugation in a tabletop centrifuge at room temperature.

9. Recrystallize both fractions by incubating in 0.1% HCl at 25°C for 4 to 5 hr to collect either betanin or betanidin·HCl.

10. Dissolve each fraction in water and adjust to known volume to give an absorbance reading between 4 and 5 AU.

11. Calculate the concentration using the light absorption at 538 nm and the respective $A^{1\%}$ for betanin and betanidin (i.e., 1120 and 1275, respectively).

See Wyler and Dreiding (1959), Wilcox et al. (1965), and Schwartz and von Elbe (1980) for more information.

Prepare standard curve

12. Separately chromatograph fraction A and C (Fig. F3.1.1), each in multiple concentrations (i.e., between 0.3 and 0.8 AU). Plot concentration versus peak area to obtain a standard curve for betanin and betanidin, respectively.

The standard curves can be used to quantify the respective isomer, since the maximum absorption peaks for the isomers are the same. It is important to use the same conditions (isocratic or gradient elution) for the sample and the standard curve. Results may be expressed in terms of betanin or betanidin, or may be added and reported in terms of total betacyanins.

Analyze sample

13. Chromatograph sample by identical analytical HPLC and determine concentration by comparison with standard curves.

Sources of samples for betacyanin analysis will vary; however, in all cases the pigments are easily extracted with water. Before chromatography, it is recommended that the extracts be filtered using a 0.45-μm HA filter. Use chromatographic conditions as described under initialize HPLC gradient (step 1).

SUPPORT PROTOCOL

EXTRACTION OF BETALAINS FROM BEETS

The analysis of betalains presently has been largely restricted to the determination of the pigment content in beet tissue or to pigment degradation studies in ideal solutions. In the case of beet tissue, the pigments are easily extracted with water, since all betalains are water soluble. This support protocol gives a method to obtain a beet extract from beet tissue. Other food samples which contain betalain can most likely be treated similarly, but interfering substances may prevent the use of the spectrophotometric method. See Critical Parameters for discussion of interfering substances.

Materials

Beet tissue
Celite (Aldrich)
Nitrogen gas (recommended)
Blender

1. Extract 50 g beet tissue with 150 ml distilled water in a blender for 1 min.

2. Mix beet puree with an equal amount of filter aid (celite), quantitatively transfer to a Buchner funnel, and filter through Whatman no. 1 filter paper using reduced pressure.

3. Wash the filter cake several times with distilled water until the extract is colorless.

   The extraction should be carried out under a nitrogen atmosphere to prevent the degradation of the pigments. Flushing the Buchner funnel with nitrogen gas is recommended.

4. Combine all extracts and make to known volume.

   Samples should be analyzed immediately, but can be stored for up to 24 hr under refrigeration if necessary. Longer storage is not recommended.

COMMENTARY

Background Information

Quantitative analysis is used to establish the total pigment content of a colorant and to determine its color strength. The relatively rapid method for the quantification of total pigment content has been very useful in the development of new beet cultivars suitable for pigment production. Present beet cultivars used in vegetable production are relatively low in total pigment. Through a selective breeding process, total pigment per gram of beet tissue has been increased several fold (Goldman et al., 1996). In the selection process it is essential to be able to rapidly determine the total betalain content as well as the betacyanin and betaxanthin content.

Critical Parameters

The multiple component system using beet juice is illustrated in Figure F3.1.2. In Figure F3.1.2 are shown four visible absorption spectra of (1) vulgaxanthin-I, (2) betanin, (3) a mixture of pure betanin and vulgaxanthin-I, and (4) beet juice. The absorption maximum for betanin is 538 nm and 476 nm for vulgaxanthin-I. Betanin also absorbs light at 476 nm and therefore will contribute to the absorption value at that wavelength. Thus, it is necessary to make a correction and subtract from the measured absorption at 476 nm the amount contributed by the presence of betanin to obtain an accurate absorption measurement for vulgaxanthin I. Since the absorption of betanin at 476 nm is not constant and varies with concentration, the ratio A_{476}/A_{538} is used in the calculations. The betanin content can be estimated directly from the absorption measurement at 538 nm, after correcting for the absorption of the impurities at 600 nm. Vulgaxanthin I does not absorb at 538 nm, and therefore no correction is needed for it at this wavelength.

A calculation of the absorption of each pigment in beet juice can be determined with the following equation set:

$$x = a - z$$
$$y = b - z - x/(A_{538}/A_{476})$$
$$z = c - x/(A_{538}/A_{600})$$

For definition of terms see the working equation set (see Basic Protocol 1, step 5).

The quotient A538/A476 for betanin is not influenced by concentration and has a mean value of 3.13 ± 0.03 in phosphate buffer at pH 6.5 over the range 0.2 to 0.8 AU; therefore, in calculations, it is rounded to 3.1. The quotient A538/A600 was found to decrease with increasing light absorption. If the sample is diluted so that the absorption at 538 nm is between 0.4 and 0.5 AU, a value for A538/A600 of 11.5 can be applied. Even if the absorption is as low as 0.2, the error when using 11.5 is within one one-thousandth of a unit. With the two quotients, the equation set above can be simplified to equation set described in the first protocol (see Basic Protocol 1, step 5). The 0.4 to 0.5 AU range is recommended to measure the light absorption, because in this range, the light absorption ratio A538/A600 is 11.5 and in this range solutions obeying Beer's law have the lowest relative error (Bauman, 1962; von Elbe and Schwartz, 1984). pH 6.5 is used because the spectra of pure pigment solutions (Fig. F3.1.1) were determined using this value. Other pH values can be used, but corrections for changes in the light absorption ratios must be made. Similarly, the relative error of readings between 0.09 and 0.7 AU can result in reasonable accuracy.

Beet juice or beet extracts are known to contain other pigments (isobetanin, prebetanin, vulgaxanthin-II) besides betanin and vulgaxanthin-I. The error that is introduced by calculating all betacyanins in terms of betanin and all be-

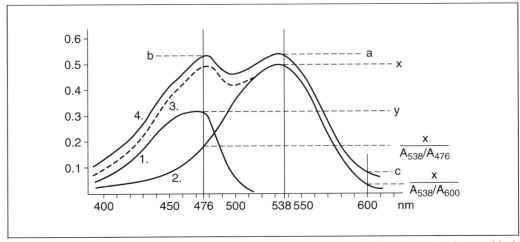

Figure F3.1.2 Calculation of betanin and vulgaxanthin-I in beet juice. Spectrum 1: vulgaxanthin-I; spectrum 2: betanin; spectrum 3: pure betanin plus vulgaxanthin-I; and spectrum 4: beet juice.

taxanthins in terms of vulgaxanthin-I is negligible, as betanin and vulgaxanthin-I comprise ~95% of the pigment concentration and the other betacyanins and betaxanthins have a maximum light absorption similar or very closely to the absorption of betanin and vulgaxanthin-I.

Anticipated Results

The spectrophotometric determination (see Basic Protocol 1) is applicable to fresh beet juice or beet tissue extracts (see Support Protocol). If samples have been exposed to conditions that could result in pigment degradation, caution must be exercised because the corrections made in the method for the presence of impurities may not be sufficient to account for the degradation products (von Elbe et al., 1983).

The results are calculated in terms of betanin, the major betacyanin, and vulgaxanthin-I, the major betaxanthin in beets. Although the minor pigment components also contribute to the light absorption at A_{538} and A_{476} expressing their concentration in terms of betanin and vulgaxanthin-I results in a small or negligible error because of the similarity of their light absorption characteristics to the major pigment components.

The high-performance liquid chromatography method (see Basic Protocol 2) allows for the separation of individual betanin and its aglycon betanidin and their isomers. The aglycon is naturally occurring in beet tissue, but in relatively small amounts. The isomers, isobetanidin and isobetanin, are easily formed when solutions of beet extract are subjected to either heat or acid (Schwartz and von Elbe, 1983). The results can be expressed in terms of the individual pigments or as total betacyanins. The HPLC separates the pigments from degradation products, making it the preferred method over the spectrophotmetric method when substantial amounts of degradation products are believed to be present.

Time Considerations

When juice samples are analyzed for total pigment content (see Basic Protocol 1), the time required per sample involves obtaining appropriate dilutions and a light absorption spectrum and should be <30 min. When tissue extraction is involved, the time required for extraction will be a minimum of 15 min.

When performing HPLC (see Basic Protocol 2) the time of analysis will depend on the conditions used—i.e., isocratic versus gradient. Isocratic separation of the individual betacyanins, as shown in Figure F3.1.2, requires 20 min, compared to 9 min using a gradient elution system.

Literature Cited

Bauman, R.P. 1962. Absorption spectrometry nomenclature. *Anal. Chem.* 51:172.

Goldman, I.L., Eagen, K.A., Breitbach, D.N., and Gabelman, W.H. 1996. Simultaneous selection is effective in increasing betalain pigment concentration but not total dissolved solids in red beets (*Beta vulgaris* L.). *J. Am. Soc. Hort. Sci.* 121:23-26.

Piattelli, M. and Minale, L. 1964. Pigments of centrospermae–I. Betacyanins from *Phyllocactus hybridus* Hort. and *Opuntia ficus-indica* mill. *Phytochemistry* 3:307-311.

Schwartz, S.J. and von Elbe, J.H. 1980. Quantitative determination of individual betacyanin pigments

F3.1.6

by high-performance liquid chromatography. *J. Agric. Food Chem.* 28:540-543.

Schwartz, S.J. and von Elbe, J.H. 1983. Identification of betanin degradation products. *Z. Lebensm. Unters. Forsch.* 176:448-53.

von Elbe, J.H. and Schwartz, S.J. 1984. Ultraviolet and visible spectrophotometry. *In* Food Analysis: Principles and Techniques (D.W. Gruenwedel and J.R. Whitaker, eds.) p. 220. Marcel Dekker, New York.

von Elbe, J.H., Schwartz, S.J., and Attoe, E.L. 1983. Using appropriate methodology to predict food quality. *Food Technol.* 37:87-91.

Wilcox, M.E., Wyler, M., Marby, T.J., and Dreiding, A.S. 1965. Die Struktur des Betanin. *Helv. Chim. Acta* 48:252-258.

Wyler, H. and Dreiding, A.S. 1957. Kristallisiertes Betanin. *Helv. Chim. Acta* 40:191-192.

Wyler, H. and Dreiding, A.S. 1959. Darstellung und Abbauprodukte des Betanidins. *Helv. Chim. Acta* 42:1966-1702.

Key References

Nilsson, T. 1970. Studies into the pigments in beetroot. *Lantbrukshoegskolans Annl.* 36:179-219.

Detailed study of the development of the spectrophotmetric method to determine betacyanin and betaxanthin content in red table beets.

Piattelli, M., Minale, L., and Prota, G. 1965. Pigments of centrospermae-III. Betaxanthins from *Beta vulgaris* L. *Phytochemistry* 4:121-125.

Reports on the conversion of betacyanins to betaxanthins.

von Elbe, J.H. 1977. The betalains. *In* Current Aspects of Food Colorants (T.E. Furia, ed.) p. 29. CRC Press, Cleveland, Ohio.

Gives quantitative data on beet pigments in red table beets and shows the effect of certain horticultural practices on the pigment content in beets.

Contributed by Joachim H. von Elbe
University of Wisconsin
Madison, Wisconsin

Spectrophotometric and Reflectance Measurements of Pigments of Cooked and Cured Meats

Color and color uniformity are important criteria for retail acceptance of both fresh and processed meats. This unit describes the methods for detection and quantitation of pigments in nitrite-cured or uncured cooked meats. Protocols for measurement of fresh meat pigments are presented in *UNIT F3.3*.

The pink cured meat pigment mononitrosylhemochrome is a complex of nitric oxide (NO), ferrous heme iron, and heat-denatured globin protein (Table F3.2.1). The pink nitrosylheme (NO-heme) moiety may be extracted from the protein in aqueous acetone and quantitated by A_{540} (see Basic Protocol 1). The percent nitrosylation may be determined from measurement of ppm NO-heme relative to ppm total acid hematin (hemin) extracted in acidified acetone (see Basic Protocol 2), since NO-heme is completely oxidized to hemin in acid solution (i.e., 1 ppm NO-heme = 1 ppm hemin).

Non-nitrosyl pink pigments, the denatured globin hemochromes, may be present in cooked meats under anaerobic (reducing) conditions, as occurs in the center of large roasts during refrigerated storage. Since the pigments consist of reduced heme iron in association with nitrogenous ligands of denatured proteins, the complex is not extractable for quantitation. Globin hemochromes may be detected, however, by their characteristic dual reflectance maxima near 528 to 530 nm and 558 nm (see Basic Protocol 3).

MEASUREMENT OF NITROSYLHEME CONCENTRATION

The measurement of cured meat pigment concentration is based on the A_{540} of the nitrosyliron(II)protoporphyrin group (also known as nitrosylheme or NO-heme; mol. wt. 646) in an extraction solution of 80% (final) acetone in water, taking into consideration the 70% water content of the meat sample. Hornsey (1956) established that only the pink NO-heme was extracted in 80% acetone. Heme groups from fresh meat pigments (Table F3.2.1) are not extractable in 80% acetone. However, upon acidification with hydrochloric acid, NO-heme in 80% acetone was completely oxidized to hemin. Thus, NO-heme concentrations could be expressed in equivalent ppm hemin.

NOTE: Ferroprotoporphyrin is the term used for reduced heme (mol. wt. 616; Windholz et al., 1976). Hematin (mol. wt. 633) is one of several terms used for the oxidized or ferric heme. Other terms include ferriporphyrin hydroxide or ferriheme hydroxide, due to the binding of a hydroxyl to the ferric heme iron. Hornsey (1956) used the term "acid hematin" (mol. wt. 652) to describe the heme oxidation product in acidified acetone solutions. The more common term "hemin" is used in this unit. It is also known as chlorohemin, due to binding of a chloride ion to the ferric heme iron.

Materials

Meat sample to be tested
Aqueous acetone (see recipe)

15×90–mm glass test tubes with screw caps
Glass stirring rod
50-mm-diameter funnels
Whatman no. 42 filter paper, 9-cm diameter
Spectrophotometer with halogen lamp and 1-cm-pathlength quartz cuvettes

---

Contributed by Daren Cornforth

Table F3.2.1 Major Pigments of Fresh Cooked, and Cured Meats[a]

Pigment	Formation	Heme status	Heme iron state	Globin status	Color
Fresh meat:					
Myoglobin	Deoxygenation of oxymyoglobin; reduction of metmyoglobin	Intact	Fe(II)	Native	Purplish red
Oxymyoglobin	Oxygenation of myoglobin	Intact	Fe(II)	Native	Bright red
Metmyoglobin	Oxidation of myoglobin	Intact	Fe(III)	Native	Brown
Cooked meat:					
Denatured globin hemochrome	Heat denaturation of myoglobin; reduction of globin hemichrome	Intact	Fe(II)	Denatured	Pink or red
Denatured globin hemichrome	Heat denaturation of metmyoglobin; oxidation of globin hemochrome	Intact	Fe(III)	Denatured	Gray or brown
Cured meat:					
Nitrosylmyoglobin (NO-Mb)	Nitric oxide complex with myoglobin	Intact	Fe(II)	Native	Red
Mononitrosyl-hemochrome	Heating of NO-Mb	Intact	Fe(II)	Denatured	Bright red or pink
Mononitrosyl-hemichrome	Oxidation of mononitrosyl hemochrome (fading in air)	Intact	Fe(III)	Denatured	Gray or brown
Verdoheme	Excess nitrite oxidizes porphyrin ring	Ring opened	Fe(III)	Absent	Green

[a]Adapted from von Elbe and Schwartz (1996) courtesy of Marcel Dekker, Inc.

NOTE: The following operations should be conducted in subdued light to reduce pigment fading during extraction.

1. Trim oxidized (faded brown or gray) surfaces of a meat sample and dice a lean portion into 2- to 3-mm cubes.

2. Weigh out duplicate 2.0 g samples and transfer to 15 × 90–mm glass test tubes containing 9.0 ml aqueous acetone.

   This 90% aqueous acetone solution should yield a final of 80% acetone and 20% water when the 70% water content of the sample is taken into account. If the sample water content deviates from 70%, the solution should be adjusted.

3. Macerate the meat mass thoroughly with a glass stirring rod (~1 min). Seal tube caps to reduce evaporation.

4. Hold at room temperature in subdued light for 10 min, then filter through Whatman no. 42 filter paper (9 cm diameter) into a clean test tube. Cap the tube to reduce evaporation.

5. Transfer filtrate to a 1-cm-pathlength quartz cuvette and read A_{540} within 1 hr against a blank cuvette containing aqueous acetone.

   Disposable plastic cuvettes should not be used, as the acetone will solubilize the cuvette and cause turbidity in the reference cell.

Cured Meat

F3.2.2

6. Calculate the concentration of NO-heme in accordance with the relationship NO-heme concentration (ppm hemin) = $A_{540} \times 289$ ppm.

This equation is derived from the equation $A_{540} = abC$, where A_{540} is sample absorbance, a is absorptivity, b is length of light path (1 cm), and C is concentration of absorbing material (in mM). Absorptivity is a constant dependent upon the wavelength of radiation and the nature and molecular weight of the absorbing material. Millimolar absorptivity, denoted by the symbol E, is the product of absorptivity and molecular weight, with units of liter $mmol^{-1}$ cm^{-1}. The E_{540}^{mM} of NO-heme in 80% aqueous acetone is 11.3 liter $mmol^{-1}$ cm^{-1} (Hornsey, 1956).

The conversion factors needed to express the concentration in ppm hemin (1 ppm = 1 μg/g) and the dilution factor must also be taken into consideration. The dilution factor is the total extraction fluid volume (ml) divided by the sample weight (g). The total extraction fluid volume includes the water content of the sample plus the amount of aqueous acetone solution. Most cooked meats have ~70% water. Thus, for a 2 g sample containing 1.4 ml water, the total extraction volume = 1.4 ml + 9 ml acetone solution, and the dilution factor = 10.4 ml/2 g sample = 5.2. Pearson and Tauber (1984) rounded the dilution factor to 5, but for best results the decimal should be retained. Also, if the sample water content is significantly different from 70%, the dilution factor should be recalculated as appropriate.

Thus, sample NO-heme concentration (ppm) = $A_{540} \times$ (1 cm mmol NO-heme/11.3 liter) $\times$ 1/cm $\times$ total extraction volume in ml/sample weight in g $\times$ 1 mol hemin/mol NO-heme $\times$ 652 g hemin/mol hemin $\times 10^6$ μg/g $\times 10^{-3}$ mol/mmol $\times 10^{-3}$ liter/ml. Substituting the dilution factor of 5 into this equation and simplifying, NO-heme concentration = $A_{540}/11.3 \times 5 \times$ 652 μg/g = $A_{540} \times 289$ ppm (as hemin). Hornsey (1956) and Pearson and Tauber (1984) rounded the concentration factor to 290 ppm.

MEASUREMENT OF TOTAL HEME CONCENTRATION

Total heme pigments in meat samples are determined after extraction with acidified acetone solution, since the heme groups of both fresh meat pigments (myoglobin, oxymyoglobin, and metmyoglobin; Table F3.2.1) and cured meat pigment (nitrosylhemochrome; Table F3.2.1) are solubilized and oxidized to hemin (Hornsey, 1956). Hemin (acid hematin; mol. wt. 652) is quantitated by its absorption peak at 640 nm. After determination of NO-heme concentration (Basic Protocol 1) and total heme (as hemin; this protocol) on the same sample, the efficiency of meat curing may be expressed as a percentage: curing efficiency (%) = (ppm NO-heme/ppm total heme) $\times$ 100. Good or acceptable pigment conversion is generally considered to be 80% to 90% of the heme pigments converted to nitrosylheme (Pearson and Tauber, 1984).

Materials

Meat sample to be tested
Acidified acetone (see recipe)

15×90–mm glass test tubes with screw caps
Glass stirring rod
50-mm-diameter funnels
Whatman no. 42 filter paper, 9-cm diameter
Spectrophotometer with halogen lamp and 1-cm-pathlength quartz cuvettes

NOTE: The following operations should be conducted in subdued light to reduce pigment fading during extraction.

1. Trim oxidized (faded brown or gray) surfaces of a meat sample and dice a lean portion into 2- to 3-mm cubes.

2. Weigh out duplicate 2.0 g samples and transfer to 15×90–mm glass test tubes containing 9.0 ml acidified acetone. To minimize acetone evaporation, use a cali-

brated pipet rather than a graduated cylinder to transfer acetone solution into glass test tubes.

This acidified 90% acetone solution should yield a final of 80% acetone and 20% water when the 70% water content of the sample is taken into account. If the sample water content deviates from 70%, the solution should be adjusted.

3. Macerate the meat mass thoroughly with a glass stirring rod (~1 min). Seal tube caps to reduce evaporation.

4. Hold at room temperature in subdued light for 1 hr, then filter through Whatman no. 42 filter paper (9 cm diameter) into a clean test tube. Cap the tube to reduce evaporation.

5. Transfer filtrate into a 1-cm-pathlength quartz cuvette and read A_{640} within 1 hr against a blank cuvette containing acidified acetone.

Disposable plastic cuvettes should not be used, since the acetone will solubilize the cuvette and cause turbidity in the reference cell.

6. Calculate total hemin concentration using the equation total hemin concentration = $A_{640} \times 680$ ppm.

This equation is derived as described for NO-heme concentration (see Basic Protocol 1, step 6). Millimolar absorptivity, E_{540}^{mM}, of hemin in 80% acetone is 4.8 liter mmol^{-1} cm^{-1} (Hornsey, 1956), giving the constant 680 in place of 289.

REFLECTANCE DETECTION OF GLOBIN HEMOCHROMES

The term globin hemochrome or denatured globin hemochrome (Table F3.2.1) is used to describe the heterogeneous mixture of pink pigments present under reducing conditions in uncured cooked meats. Globin hemochromes are pink complexes between ferrous heme iron of heat-denatured myoglobin and various nitrogenous compounds in meat, possibly including amino acids, nicotinamide, or nitrogenous side chains of other heat-denatured proteins. Since heat denaturation of proteins is necessary for their formation, they often occur in well-cooked meats (internal temperature >76°C). Pink color may intensify during anaerobic storage after cooking, as the brown hemichromes are slowly reduced. The globin hemochromes are detected in meat slices by their characteristic reflectance maxima near 528 and 558 nm.

Materials

White standard (powdered barium sulfate)
Meat sample to be tested

Recording spectrophotometer with integrating sphere attachment, with ports for sample and standard
Clear polyethylene vacuum bags (1.5-mil thickness)

1. Standardize a recording spectrophotometer to 100% reflectance from 420 to 700 nm, using the white standard (powdered barium sulfate) in both the sample and standard ports of the reflectance attachment.

2. Obtain a uniform meat slice that is 3 cm × 3 cm (sufficient to completely cover the sample port on the reflectance attachment) and >3 mm thick.

3. To exclude air and minimize fading, rapidly place the fresh slice in a clear polyethylene vacuum bag (1.5-mil thickness). Press the bag against the sample from bottom to top to remove air bubbles.

4. Place the bagged sample snugly in the sample port of the reflectance sphere, with the freshly sliced surface facing inward (toward the detector). Record reflectance (% of standard) from 420 to 700 nm.

5. Allow a control meat slice to fade in air for 15 to 30 min.

6. Bag the sample and record the reflectance spectra as described for fresh samples.

7. Subtract baseline spectrum (control slice) from sample spectrum obtained in step 4.

REAGENTS AND SOLUTIONS

*Use distilled water in all recipes and protocol steps. For common stock solutions, see **APPENDIX 2A**; for suppliers, see **SUPPLIERS APPENDIX**.*

Aqueous acetone

Add 18 ml distilled water to a 200-ml volumetric flask. Add spectrophotometric grade acetone, mix, and bring to volume with additional acetone. Store up to 6 months at 3°C.

Acidified acetone

To 4 ml concentrated hydrochloric acid, add distilled water to a total volume of 20 ml and mix. Transfer dilute hydrochloric acid to a 200-ml volumetric flask. Add spectrophotometric grade acetone, mix, and bring to volume with additional acetone. Store up to 1 month at 3°C. Discard solutions that become yellowish.

CAUTION: *For safety reasons, concentrated acids are usually added slowly to a larger volume of water. The reader may wish to modify this recipe so that the acid is added to the water.*

COMMENTARY

Background Information

Nitrosylheme, also known as NO-heme or nitrosyliron(II) protoporphyrin, is readily extracted from the cured meat pigment mononitrosylhemochrome (Killday et al., 1988) into an aqueous acetone solution for quantitation by spectroscopy. Hornsey (1956) established the following important points for accurate NO-heme quantitation.

1. Maximum NO-heme extraction occurred with an acetone/water ratio of 75% to 85% (Table F3.2.2).

2. Only the nitrosylheme derivative of the muscle pigments was extracted in 80% acetone. Heme from metmyoglobin, oxymyoglobin, or deoxymyoglobin (or hemoglobin) was not extracted.

3. The absorption spectra for extracted NO-heme exhibited a broad peak from 535 to 565 nm. Absorbance from 680 to 700 nm was <0.03.

4. The NO-heme extract from uncooked cured meat or model solutions with pure hemoglobin or myoglobin faded rapidly.

5. Extracts from cooked cured meat were stable for ≥ 1 hr, indicating that reducing substances (probably cysteine or glutathione) were also extracted, conferring stability against light-catalyzed air oxidation. (In uncooked cured meat, these compounds would also be present but at least partly oxidized, depending on the degree of oxygen incorporation during maceration and other processing steps.)

6. Fading of acetone extracts of pure NO-myoglobin solutions could be prevented by adding 1 ml of fresh 0.5% neutralized cysteine hydrochloride to 9 ml of NO-myoglobin solution.

Hornsey (1956) also found that extraction with an acidified 80% acetone solution for 1 hr gave a hemin solution, derived from oxidation of the heme moiety of both NO-heme and non-nitrosylated heme pigments. Hemin absorption spectra exhibited distinct peaks at 512 and 640 nm. Hornsey (1956) used the A_{640} of sample filtrates as a measure of the total heme pigments. Solutions of both hemin and NO-heme in 80% acetone conformed with Beer's

Table F3.2.2 Effect of Acetone Concentration on Optical Density (A_{540}) of Extracts from Lean, Cooked, Cured Pork Ham[a]

Acetone concentration (%)	A_{540}
96.5	0.324
93	0.330
86	0.360
80	0.370
75	0.360
70	0.308
65	0.270
60	0.225

[a]Adapted from Hornsey (1956) with permission from John Wiley & Sons on behalf of the Society of Chemical Industry.

Law, with straight lines passing through the origin.

In the Hornsey (1956) procedure, the sample was minced thoroughly, and a 10 g sample was mixed in a tall beaker (to prevent undue evaporation) with 10 ml of a solution of 40 ml acetone and 3 ml water. Total fluid volume was 50 ml, including the 7 ml water in the sample. This procedure was preferred over dilution of sample to 50 ml final volume, since calculations were simplified and correction for the volume of insoluble meat tissues was avoided. Later workers have modified the procedure, using smaller samples and less solvent. Pearson and Tauber (1984) used a 2 g sample and capped test tubes to prevent acetone evaporation, and Carpenter and Clark (1995) used 5 g samples.

The Hornsey (1956) procedure and its modifications have received widespread acceptance as relatively rapid measures of the adequacy of cure development in processed meats. The Hornsey procedure is also an accurate method for nutritional assessment of heme and heme iron content of meats (Carpenter and Clark, 1995), where ppm heme iron = ppm total heme/11.7. However, one caveat should be noted. The total heme pigment measurement is higher in cured meats than in similar uncured samples. Roasted turkey breast meat, for example, was reported by Ahn and Maurer (1989a) to have 23, 26, 34, and 34 ppm total pigment in samples formulated with 0, 1, 10, and 50 ppm nitrite, respectively. This effect should be considered to avoid overestimation of the heme iron content of cured meats.

Tappel (1957) was among the first to record spectral characteristics of the pink pigments of well-cooked meats, noting reflectance minima at 424, 528 to 530, and 555 to 560 nm for denatured globin hemochrome, and 423, 523 to 525, and 555 to 558 nm for nicotinamide hemochrome. The term globin hemochrome is used to describe the heterogeneous mixture of pink pigments occurring under reducing conditions in uncured cooked meats. Globin hemochromes are pink complexes between ferrous heme iron of heat-denatured myoglobin and various nitrogenous compounds in meat, possibly including amino acids, nicotinamide, or nitrogenous side chains of other heat-denatured proteins. Since heat denaturation of proteins is necessary for their formation, they often occur in well-cooked meats (internal temperature >76°C; Ghorpade and Cornforth, 1993). Pink color may intensify during anaerobic storage after cooking, as the brown hemichromes are slowly reduced. Tappel (1957) and Tarladgis (1962) used sodium dithionite to rapidly reduce the globin hemichromes of cooked meat slices to pink globin hemochromes for spectroscopic studies. Tappel (1957) also exposed dithionite-reduced slices to carbon monoxide, and noted reflectance minima at 542 and 570 nm for denatured globin carbon monoxide hemochrome. Reviewers including this author (Cornforth et al., 1991) have attributed the surface pinking of meats heated in a gas oven in part to the presence of carbon monoxide. However, more recent studies (Cornforth et al, 1998) indicate that nitrogen dioxide is the combustion gas responsible for surface pinking. Although carbon monoxide binds to myoglobin in fresh meats, it is released upon cooking (Watts et al., 1978). Carbon monoxide contributes to pinking in cooked meats only under anaerobic or reducing conditions, unusual at the surface of meat cooked in a gas oven.

Cytochrome c is more heat stable than myoglobin, and may contribute to residual pinking in cooked pork or poultry (Ahn and Maurer, 1989b; Girard et al., 1990). Cytochrome c solutions exhibit absorption maxima at 414, 520, and 550 nm. Cytochrome c is more resistant than other pink pigments to fading upon exposure of the meat surface to air (Girard et al., 1990). Pinkness of uncured cooked meat slices that fades rapidly after slicing is likely due to presence of globin hemochromes.

Critical Parameters and Troubleshooting

Incomplete pigment extraction (NO-heme and total pigments)

Incomplete pigment extraction may be due to insufficient maceration. If so, dice or grind meat into smaller particles, macerate more thoroughly with the glass rod, or homogenize the sample for 20 to 30 sec with a probe-type blender (e.g., Kinematica polytron with small head).

Incomplete pigment extraction may also be due to variable sample moisture content. Maximum pigment extraction is obtained in an extraction solution of 80% acetone and 20% water, including the water content of the sample. The procedure as outlined is for samples with 70% ± 3% water. Determine the water content of the sample, and then adjust the water content of the extraction solution as needed to obtain a final 80% acetone and 20% water in the extraction (Carpenter and Clark, 1995).

Excessive evaporation of acetone during extraction will alter the acetone/water ratio, causing incomplete pigment extraction. The tubes should be capped during extraction and handling to minimize evaporation.

Fading (NO-heme and total pigments)

Extracted samples may fade if exposed to bright light or if held for an excessive period before obtaining absorbance values. Keep sample in subdued light during extraction, and obtain absorbance values within 1 hr. Uncooked cured meat extracts or extracts of pure pigments in model systems are more prone to fading than extracts from cooked meat systems. Addition of 0.2 ml fresh cysteine solution (0.5% cysteine HCl) per 10 ml total fluid (acetone plus water plus cysteine solution) will prevent fading for several hours.

Fading (globin hemochromes)

Pink globin hemochromes tend to fade rapidly after slicing of cooked meats. Fading may be slowed by covering the exposed surface with a transparent plastic film (e.g., Saran wrap or polyethylene) to retard pigment oxidation by atmospheric oxygen, allowing more time to obtain a representative sample reflectance spectrum. Spreading a thin coating of vegetable or mineral oil over cut surfaces may accomplish the same purpose, as it has been observed that pink pigments from roasted pork are resistant to fading when the lean portion was covered by melted fat.

Some investigators have soaked cooked meat slices in freshly prepared 1% sodium dithionite solution to maintain the pigments in the reduced state. However, this is not truly representative of the fresh meat slice, since dithionite will also reduce the brown globin hemichromes, intensifying the pink color to a level greater than that actually observed after slicing.

Anticipated Results

Total heme pigments vary among species and muscles, with levels >140 ppm for cooked beef products (Pearson and Tauber, 1984). Carpenter and Clark (1995) used the acetone extraction method of Hornsey (1956) to determine heme iron content of various cooked meats. They reported heme iron levels of 21, 9, 2.2, and 1.4 ppm for cooked beef round, pork picnic, pork loin, and chicken breast, respectively. Hemin (mol. wt. 652) is 8.54% iron. Thus, these meats contained 245, 105, 25, and 16 ppm total heme, respectively (Carpenter and Clark, 1995). Ahn and Maurer (1989a) reported a value of 23 ppm total heme in cooked turkey breast.

About 80% to 90% conversion of heme pigments to nitrosylhemochrome is desirable for cured meats (Pearson and Tauber, 1984). However, beef pastrami with typical cured color may have as low as 62% conversion (94 ppm NO-heme out of 153 ppm total heme; Cornforth et al., 1998).

Globin hemochrome levels in cooked meats have not been quantitated. Their presence is indicated, however, by a reflectance spectrum with a large reflectance minimum at ~558 nm and a smaller minimum or shoulder at ~528 to 530 nm (Tappel, 1957; Ghorpade and Cornforth, 1993).

Time Considerations

Preparation of aqueous and acidified acetone solutions for NO-heme and total heme determination takes 30 min. Measurement of NO-heme concentration of a single sample takes an additional 30 min. Measurement of total heme concentration of a single sample takes 1.5 hr. Often it is desirable to make both measurements on the same sample. By starting the 1-hr extraction for total heme and then the 10-min extraction for NO-heme, both determinations may be done in 2 hr (including the 30 min for preparation of solutions).

Approximately 30 min is sufficient to obtain a reflectance spectrum for detection of globin hemochromes. Approximately 1 hr is required to install the reflectance attachment and run calibration curves.

Literature Cited

Ahn, D.U. and Maurer, A.J. 1989a. Effects of added nitrite, sodium chloride, and phosphate on color, nitrosoheme pigment, total pigment, and residual nitrite in oven-roasted turkey breast. *Poultry Sci.* 68:100-106.

Ahn, D.U. and Maurer, A.J. 1989b. Effects of sodium chloride, phosphate, and dextrose on the heat stability of purified myoglobin, hemoglobin, and cytochrome c. *Poultry Sci.* 68:1218-1225.

Carpenter, C.E. and Clark, E. 1995. Evaluation of methods used in meat iron analysis and iron content of raw and cooked meats. *J. Agric. Food Chem.* 43:1824-1827.

Cornforth, D.P., Calkins, C.R., and Faustman, C. 1991. Methods for identification and prevention of pink color in cooked meat. *Proc. Annu. Reciprocal Meat Conf., Am. Meat Sci. Assoc.* 44:53-58.

Cornforth, D.P., Rabovitser, J.K., Ahuja, S., Wagner, J.C., Hanson, R., Cummings, B., and Chudnovsky, Y. 1998. Carbon monoxide, nitric oxide, and nitrogen dioxide levels in gas ovens related to surface pinking of cooked beef and turkey. *J. Agric. Food Chem.* 46:255-261.

Ghorpade, V.M. and Cornforth, D.P. 1993. Spectra of pigments responsible for pink color in pork roasts cooked to 65 or 82°C. *J. Food Sci.* 58:51-52, 89.

Girard, B., Vanderstoep, J., and Richards, J.F. 1990. Characterization of the residual pink color in cooked turkey breast and pork loin. *J. Food Sci.* 55:1249-1254.

Hornsey, H.C. 1956. The colour of cooked cured pork. I. Estimation of the nitric oxide-haem pigments. *J. Sci. Food Agric.* 7:534-540.

Killday, K.B., Tempesta, M.S., Bailey, M.E., and Metral, C.J. 1988. Structural characterization of nitrosylhemochromogen of cooked cured meat: Implications in the meat curing reaction. *J. Agric. Food Chem.* 36:909-914.

Pearson, A.M. and Tauber, F.W. 1984. Analytical methods. *In* Processed Meats, 2nd ed. (A.M. Pearson and F.W. Tauber, eds.) pp. 360-361. AVI Publishing, Westport, Conn.

Tappel, A.L. 1957. Reflectance spectral studies of the hematin pigments of cooked beef. *Food Res.* 22:404-407.

Tarladgis, B.G. 1962. Interpretation of the spectra of meat pigments. I. Cooked meats. *J. Sci. Food Agric.* 13:481-484.

von Elbe, J.H. and Schwartz, S.J. 1996. Colorants. *In* Food Chemistry, 3rd ed. (O.R. Fennema, ed.) p. 654-655. Marcel Dekker, New York.

Watts, D.A., Wolfe, S.K., and Brown, W.D. 1978. Fate of [^{14}C] carbon monoxide in cooked or stored ground beef samples. *J. Agric. Food Chem.* 26:210-214.

Windholz, M., Budavari, S., Stroumtsos, L.Y., and Fertig, M.M. (eds.) 1976. The Merck Index, 9th ed. Merck, Rahway, N.J.

Key References

Hornsey, 1956. See above.

This paper provides the basic information for spectrophotometric determination of nitrosyl and total heme pigment levels in cured meats.

Killday et al., 1988. See above.

Mass spectroscopy, nuclear magnetic resonance (NMR) spectroscopy, and infrared spectroscopy indicated that the cured meat pigment was mononitrosylhemochrome. Contrary to previous reports, no evidence was found to indicate presence of dinitrosylheme complexes.

Pearson and Tauber, 1984. See above.

This paper is an adaptation of the Hornsey (1956) method for measurement of nitrosyl and total heme pigments in cured meats, using 2 g meat samples rather than 10 g samples. Thus, less reagent is needed and more samples may be analyzed at the same time.

Contributed by Daren Cornforth
Utah State University
Logan, Utah

The author thanks Dr. Charles E. Carpenter (Nutrition & Food Sciences, Utah State University, Logan, Utah) for time and effort spent proofreading this contribution.

Measurement of Discoloration in Fresh Meat

The discoloration of fresh meat is an important process, which is determined by the relative concentration of the three redox forms of myoglobin (deoxymyoglobin, oxymyoglobin, and metmyoglobin). Loss of the desirable cherry-red appearance with subsequent replacement by reddish browns and browns is a natural process affected by a variety of intrinsic and extrinsic parameters. Because consumers often use meat color as a basis for product selection or rejection, the loss in economic value that accompanies fresh meat discoloration can be substantial. Considerable research by the meat industry and Agricultural Experiment Stations has focused on enhancing the maintenance of desirable fresh meat color through a variety of processing and packaging techniques. In order to evaluate the efficacy of these procedures, it is critical that appropriate methods are used to objectively measure and describe changes in fresh meat color (AMSA, 1991).

In general, the approaches for measuring fresh meat discoloration are relatively straightforward and simple. Fresh meat may be merchandised in either minced (ground meat) or whole (e.g., steaks, roasts) forms. While the principles involved in objectively measuring changes in color are the same for both product types, sample preparation and instrumentation are different. Minced products are often homogenized and the pigment-containing extract is analyzed by transmission or absorbance spectrophotometry (see Basic Protocol 1). Although this is the method of choice for myoglobin and hemoglobin quantification, it does not offer an accurate representation of surface color as observed by the consumer. Whole-muscle products are analyzed with reflectance techniques or colorimetry (see Basic Protocol 2 and Alternate Protocol, respectively). With these rapid methods, repeated measurements may be recorded from the same sample over time. Fresh meat discoloration projects are generally carried out for 3 to 7 days, with color analysis being performed either every day or every other day. The time course for frozen samples is longer (e.g., 90 days), with color analysis being performed much less often (e.g., every 30 days). It is imperative to establish sampling parameters based on specific research objectives prior to the beginning of a study. Finally, many investigators find it necessary to utilize a model approach for studying the biochemistry of myoglobin in vitro. These experiments often require the isolation (see Basic Protocol 3) and redox manipulation (see Basic Protocol 4) of myoglobin.

ANALYSIS OF METMYOGLOBIN IN GROUND MEAT EXTRACTS

Ground meat generally discolors much faster than whole-muscle cuts. The mincing of meat destroys cellular integrity and liberates a variety of prooxidants, which can accelerate the discoloration process. In addition, iron from meat grinding equipment surfaces can become incorporated into the meat and serve as an oxidation catalyst (Faustman et al., 1992). The penetration of oxygen, a necessary component for prooxidant reactions, into ground meat products is also greater than in whole-muscle cuts because of the porous structure. These factors promote the formation of metmyoglobin from either oxy- or deoxymyoglobin. The method of Krzywicki (1982) can be used to quantify the total concentration of myoglobin, as well as the relative concentrations of oxidized, deoxygenated, or oxygenated forms of the pigment using absorbance values at 572, 565, 545, and 525 nm.

Materials

40 mM sodium phosphate buffer, pH 6.8 (*APPENDIX 2A*), ice cold
Ground meat sample
Control sample

Spectrophotometer able to scan visible spectrum
Blender
Whatman no. 1 filter paper

Establish baseline

1. Turn on a spectrophotometer and allow it to warm up for 30 min prior to use.

2. Pipet 1.0 ml of 40 mM sodium phosphate buffer, pH 6.8, into a disposable cuvette and place it in the sample port of the spectrophotometer.

3. Scan the buffer from 650 nm to 450 nm to establish a baseline.

Prepare samples

4. Select a 25 g sample of ground meat from an appropriate location of the meat product. Additionally, select a proper control sample to prepare in parallel.

   The exact location of the sample (surface versus deep versus surface plus deep) will depend on the experimental design and question being investigated.

   In general, replicate samples are essential when analyzing ground meat color. An efficient practice may involve the preparation of patties (~25 g) so that an entire sample is used during analysis. A small petri dish (~5 cm in diameter) serves as an excellent mold for patty formation while maintaining geometric shape, surface area, and compaction.

   Control, nontreated samples subjected to the same preparation as treated samples must be used to allow for comparisons between treatments.

5. Homogenize sample in a blender with 10 vol ice-cold 40 mM sodium phosphate buffer, pH 6.8, for 45 sec at high speed.

6. Filter homogenate using a Whatman no. 1 filter paper.

   In some situations, filtration through Whatman no. 1 filter paper may not yield a clear solution (e.g., ground meat displayed for >4 days). When this occurs, pass filtrate through a 0.40-μm filter attached to a syringe.

Determine metmyoglobin concentration

7. Pipet 1.0 ml filtrate into a fresh disposable cuvette and place it in the sample port of the spectrophotometer.

8. Scan sample from 650 nm to 450 nm and record absorbance values at 5-nm intervals. In addition, record absorbance value at 572 nm.

9. Estimate total concentration of myoglobin in the sample from the A_{525} value using an extinction coefficient of 7.6 mM^{-1} cm^{-1} and a path length of 1 cm (Bowen, 1949) as follows:

   Myoglobin (mM) = $[A_{525}/(7.6\ mM^{-1}cm^{-1} \times 1\ cm)]$.

10. Calculate the relative concentration of metmyoglobin (MetMb) using the following equation:

    % MetMb = $-2.541R_1 + 0.777R_2 + 0.800R_3 + 1.098$

    where R_1 is A_{572}/A_{525}, R_2 is A_{565}/A_{525}, and R_3 is A_{545}/A_{525} (Krzywicki, 1982).

    In general, percent metmyoglobin is reported and not percent deoxy- or percent oxymyoglobin. Formulas for the two ferrous myoglobin forms can be obtained from Krzywicki (1982).

ANALYSIS OF METMYOGLOBIN IN FRESH MEAT SURFACES USING DIFFUSE REFLECTANCE SPECTROPHOTOMETRY

The surfaces of whole muscle cuts can be analyzed using diffuse reflectance spectrophotometry, a nondestructive method that is closely related to visual assessment of meat discoloration. Unlike analysis of ground meat extracts, surface analysis allows samples to be easily repackaged and stored for future use. In addition, appropriate control samples must be utilized so that valid comparisons may be made between treatments. In the following procedure, Stewart et al. (1965) used the Kubelka-Munk equation (Judd and Wyszecki, 1963) to relate both the absorption (K) and scattering (S) of light by a meat sample at a given wavelength. The K/S ratios from different wavelengths can then be manipulated to estimate the relative presence of metmyoglobin on the meat surface.

Materials

Meat sample
Control sample

Spectrophotometer able to scan visible spectrum and equipped with integrating sphere
Barium sulfate plates (to standardize spectrophotometer)
Fresh meat PVC film

Establish baseline

1. Attach an integrating sphere to a spectrophotometer, turn on the spectrophotometer, and allow it to warm up for 30 min prior to use.

2. Place a barium sulfate plate wrapped in a single layer of fresh meat PVC film in both the reference and sample ports.

3. Establish a baseline by scanning from 650 nm to 450 nm.

   Note that the slit width for diffuse reflectance spectrophotometry must be sufficiently wide to avoid excessive noise, which will result from surface measurements. The exact slit width will depend on the instrument, but 5 nm is routinely used.

Analyze samples

4. Select an appropriately sized meat sample (at least 1 cm thick) and wrap tightly with a single layer of fresh meat PVC film. Additionally, select a proper control sample to prepare in parallel.

   Wrapping the sample will prevent contamination of the inner white surface of the integrating sphere with any meat exudate.

   Control, nontreated samples subjected to the same preparation as treated samples must be used to allow for comparisons between treatments.

5. Remove barium sulfate plate and place sample in the sample port of the spectrophotometer.

6. Scan sample from 650 nm to 450 nm and record absorbance values at 5-nm intervals. In addition, record absorbance value at 572 nm.

   If research objectives require analysis over time, repackage and store samples appropriately after recording absorbance values.

Calculate percent metmyoglobin

7. Convert absorbance to reflectance using the following equation:

$$R_a = 2 - \log R_\lambda$$

where R_a is reflectance expressed as an absorbance value (as measured in step 6) and R_λ is the calculated percent reflectance.

8. Calculate the K/S_λ ratio using the following equation: $K/S_\lambda = (1 - R_\lambda)^2/(2 \times R_\lambda)$, where K is the absorption coefficient, S_λ is the scattering coefficient at a given wavelength, and R_λ is the reflectance at a given wavelength.

9. Calculate the K/S value using the following equation: $K/S = (K/S_{572})/(K/S_{525})$, where K/S_λ is the value obtained from step 8.

Stewart et al. (1965) recognized a large difference in light absorption at 572 nm between metmyoglobin and ferrous myoglobins. They proposed a procedure by which the K/S_{572} would be calculated and divided by K/S_{525} (to account for potential differences in muscle myoglobin concentrations that could exist between different carcasses) to yield a K/S value for meat in which metmyoglobin comprised from 0% to 100% of the total pigment. In order to obtain these standards, ground beef was treated with potassium ferricyanide, an oxidizing agent, to obtain 100% metmyoglobin, or with sodium hydrosulfite, a reductant, to obtain 100% ferrous myoglobin (0% metmyoglobin). Approximately 20 samples of each of the two treatments were measured and the K/S plotted against 0% and 100% metmyoglobin. Stewart et al. (1965) assumed a linear relationship between K/S and metmyoglobin concentration. Franke and Solberg (1971) subsequently published results that supported this assumption, although they advocated a modified method for estimating percent metmyoglobin.

10. Calculate percent metmyoglobin (MetMb) using the following equation:

% MetMb = [100 − (K/S − 0.56)]/0.0084

where K/S is the value obtained from step 9.

The equation above was derived by Stewart et al. (1965) and is applicable to many situations. However, a similar equation for percent metmyoglobin may be derived by obtaining several meat samples (4 cm² × 1 cm thick) of similar composition to those used in the particular study and subjecting one half of the samples to sodium hydrosulfite and one half to potassium ferricyanide, as follows. To convert meat pigment entirely to reduced myoglobin, dissolve 0.02 g sodium hydrosulfite in 15 ml water and submerge the meat sample in the solution for ~1 min. A typical absorption spectra of reduced myoglobin will persist for at least 25 min. For conversion to metmyoglobin, dissolve 0.1 g potassium ferricyanide in 15 ml water and submerge the meat sample in the solution for ~1 min. A typical metmyoglobin spectrum is produced if this is examined immediately; however, depending on the meat sample, the ferric pigment may be rapidly reduced. Once absorption values are collected, they may be plotted against 0% to 100% metmyoglobin to obtain a straight line curve. Because a linear relationship is assumed between absorption and metmyoglobin concentration, metmyoglobin values may be extrapolated from the graph.

The relative proportion of metmyoglobin can be determined using one of various procedures. Although the authors prefer the Stewart et al. (1965) method as it provides values that correlate well with visual color assessment, Broumand et al. (1958), Van den Oord and Wesdorp (1971), Strange et al. (1974), and Bevilacqua and Zaritzky (1986) also describe procedures to evaluate meat color using reflectance spectrophotometry. Also, Krzywicki (1979) utilized similar approaches in an attempt to develop equations for calculating the relative proportions of deoxy-, oxy-, and metmyoglobin on beef surfaces. Krzywicki (1979) utilized extinction coefficients for myoglobin in solution, and attempted to account for structural impacts on light scattering by subtracting reflectance values obtained at 730 nm, a wavelength at which the author maintained that myoglobin would demonstrate no absorbance, and thus would control for structural differences between meat samples.

ANALYSIS OF FRESH MEAT SURFACE COLOR USING COLORIMETRY

Depending upon the equipment available for analysis, the surfaces of whole-muscle cuts may also be analyzed using colorimetry as opposed to diffuse reflectance spectrophotometry (see Basic Protocol 2). Because samples are not consumed during analysis, they may simply be repackaged and stored, according to study protocol, for future use. Colorimeters provide $L*$, $a*$, and $b*$ values, also referred to as Commission Internationale de l'Éclairage (CIE) Lab values (Clydesdale, 1978). Although $L*$ and $b*$ values are not extensively reported in meat discoloration studies, they give an indication of lightness and yellow/blue color, respectively. For the purposes of following fresh meat discoloration, the $a*$ value, or degree of redness, is the most useful. In general, $a*$ values will decrease with display time. Mathematical manipulation of the $a*$ and $b*$ values can be used to obtain chroma, $[(a*)^2 + (b*)^2]^{1/2}$, an indication of the saturation of a color. In addition to chroma, hue angle, $\tan^{-1}(b*/a*)$, may be plotted as a function of storage and/or be correlated with sensory assessment of appearance (Clydesdale, 1978).

Materials

Meat sample
Control sample

Colorimeter (e.g., Chromameter CR-200, Minolta)
White (or red/pink) standardization plate (should be provided with colorimeter)
Fresh meat PVC film (optional)

1. Standardize a colorimeter against a white standardization plate or, in some cases, a red/pink plate.

   The white plate is used by most investigators for standardization, but there may be instances where the colored standard is preferred. Whichever is chosen, that same plate must be used throughout the discoloration study. Additionally, the exact specifications of the standardization plate should be indicated in any scientific report.

2. Select an appropriately sized meat sample (at least 1-cm thick) to allow placement within the sample port of the benchtop colorimeter, or to permit placement of the hand-held colorimeter directly on the sample surface. Additionally, select a proper control sample to prepare in parallel.

   Control, nontreated samples, subjected to the same preparation as treated samples, must be used to allow for comparisons between treatments.

3. Wrap sample with fresh meat PVC film if needed.

   With hand-held colorimeters, the authors have found it unnecessary to wrap the meat with film; however, this may be necessary with certain benchtop instruments to prevent harming the equipment. If the meat sample is wrapped in film, the standardization plate should also be wrapped in film.

4. Make and record two to three readings of $L*$, $a*$, and $b*$ values per sample surface.

   If research objectives require analysis over time, repackage and store samples appropriately after taking measurements.

5. Average recorded values and calculate chroma or hue angle for the sample.

   An average of two to three readings per sample surface will more accurately reflect meat color.

ISOLATION OF TOTAL MYOGLOBIN FOR IN VITRO STUDIES

In order to study the autoxidation mechanism of myoglobin, one of the main factors involved in meat discoloration, the isolation and purification of myoglobin is necessary. Myoglobin (mol. wt. ~17,000 Da) can be purified readily from skeletal or cardiac muscle of meat-producing animals. It is a robust protein and its red color permits easy visualization of its progress during chromatography. A variety of purification procedures for myoglobins have been published; a straightforward approach adapted from both Wittenberg and Wittenberg (1981) and Trout and Gutzke (1996), which provides substantial yields of myoglobin using relatively inexpensive equipment, is provided.

Materials

Diced beef muscle trimmed of visible fat and connective tissue
Homogenization buffer (10 mM Tris·Cl/1 mM EDTA, pH 8.0), 4°C
Sodium hydroxide
Ammonium sulfate
Dialysis buffer (10 mM Tris·Cl/1 mM EDTA, pH 8.0), 4°C
Chromatography elution buffer (5 mM Tris·Cl/1 mM EDTA, pH 8.5), 4°C

Blender
Cheesecloth
Centrifuge capable of spinning at 20,000 × g, 4°C
Dialysis tubing (MWCO 12,000 to 14,000)
Sephacryl S-200 HR chromatography column (30 × 2.5–cm)
Peristaltic pump

Additional reagents and equipment for protein assays (*UNIT B1.1*) or for calculating concentration using extinction coefficients (see Basic Protocol 1)

NOTE: To minimize formation of metmyoglobin, homogenization and all subsequent steps should be performed at low temperature (0° to 5°C) and high pH (8.0 to 8.5).

Prepare homogenate

1. Homogenize 150 g diced beef muscle in a blender with 450 ml of homogenization buffer for 1 to 2 min at high speed.

2. Divide homogenate equally between tubes and centrifuge 10 min at 3000 × g, 4°C.

3. Pool supernatants, discard precipitate, and adjust pH of resulting supernatant to 8.0 using sodium hydroxide.

4. Filter supernatant through two layers of cheesecloth to remove lipid and connective tissue particles.

Precipitate myoglobin

5. Bring filtrate to 70% ammonium sulfate saturation (472 g ammonium sulfate/liter filtrate), adjust pH to 8.0 using sodium hydroxide, and stir for 1 hr.

6. Divide homogenate equally between tubes and centrifuge 20 min at 18,000 × g, 4°C to remove precipitated proteins.

7. Pool supernatants and discard precipitate.

8. Bring supernatant from 70% to 100% ammonium sulfate saturation (by adding an additional 228 g ammonium sulfate/liter supernatant), adjust pH to 8.0 using sodium hydroxide, and stir for 1 hr.

9. Divide homogenate equally between tubes and centrifuge solution 1 hr at 20,000 × *g*, 4°C. Discard supernatants.

Dialyze and purify myoglobin

10. Transfer precipitated myoglobin to dialysis tubing and dialyze against dialysis buffer (1 vol protein:10 vol buffer) for 24 hr at 4°C, changing buffer every 8 hr.

11. Equilibrate a Sephacryl S-200 HR chromatography column with chromatography elution buffer (3 column volumes) using a peristaltic pump.

12. Apply dialysate to column and resolve myoglobin extract with chromatography elution buffer at a flow rate of 60 ml/hr.

    Hemoglobin will elute first as a pale red/brown band. Myoglobin will follow as a readily visualized dark red band.

13. Collect myoglobin-containing fractions.

Concentrate myoglobin

14. Pool all myoglobin-containing fractions and bring solution to 100% ammonium sulfate saturation (761 g ammonium sulfate/liter solution), adjust pH to 8.0, and stir solution for 1 hr.

15. Divide solution equally between tubes and centrifuge 1 hr at 20,000 × *g*, 4°C. Discard supernatants and dialyze myoglobin as described in step 10.

    Alternatively, the myoglobin may be concentrated using ultrafiltration as described by Trout and Gutzke (1996).

    Native-PAGE (UNIT B3.1) can be used to assess the purity of the myoglobin extract, which should produce a single protein band with a molecular weight of 17.8 kDa.

16. Measure protein concentration of myoglobin solution (see Basic Protocol 1, step 9) and freeze in aliquots at −80°C.

PREPARATION OF OXYMYOGLOBIN BY REDUCTION OF METMYOGLOBIN

BASIC PROTOCOL 4

The basis for myoglobin oxidation in meat has often been studied using in vitro models. Commercially available myoglobin often exists in ≥95% of the ferric metmyoglobin form. Because many experiments are concerned with oxidation of ferrous oxymyoglobin, it is necessary to chemically reduce myoglobin as purchased. Additionally, this approach can be used to reduce myoglobin purified from muscle of meat-producing animals (see Basic Protocol 3), if necessary. The following procedure, adapted from Brown and Mebine (1969), involves chemical reduction of metmyoglobin with hydrosulfite. Unreacted hydrosulfite is subsequently removed via chromatography.

Materials

Myoglobin stock solution of isolated (see Basic Protocol 3) commercial myoglobin (e.g., Sigma)
Sodium hydrosulfite (sodium dithionite)
Sephadex G-25 desalting column

NOTE: It is extremely critical to perform all steps at 4°C to minimize oxidation of ferrous deoxy- or oxymyoglobin to metmyoglobin, especially when using a buffer at pH 5.6, the typical post-mortem skeletal muscle pH.

1. Treat a myoglobin stock solution (on ice) with sodium hydrosulfite at a rate of 0.1 mg sodium hydrosulfite to 1 mg myoglobin. Vortex 10 sec to reduce metmyoglobin to deoxy- and oxymyoglobin.

   Although the final concentration of myoglobin desired in the experiment must be considered, a concentrated myoglobin stock solution (~40 mM) is easily prepared and reduced. Additionally, 0.15 mM myoglobin is routinely used in the authors' experiments, as this represents an average concentration for myoglobin in different bovine skeletal muscles (Rickansrud and Henrickson, 1967).

   Commercial myoglobin should be made up with a buffer (e.g., phosphate or citrate buffer) that is suitable for the particular experimental conditions.

2. Bubble air through solution with a small pipet for 1 min to oxygenate myoglobin.

   A high quantity of oxymyoglobin is produced following oxygenation.

3. Using gravity flow, pass the myoglobin solution over a Sephadex G-25 desalting column to remove excess hydrosulfite.

   Alternatively, excess hydrosulfite may be removed from the myoglobin solution via dialysis against the buffer of choice (1 vol protein:10 vol buffer, 3 times, 8 hr each) or via mixed-bed ion-exchange chromatography (e.g., AG501-X8, Bio-Rad) (Brown and Mebine, 1969).

4. Use myoglobin immediately as oxidation can occur rapidly.

COMMENTARY

Background Information

Myoglobin is a heme protein found in the skeletal muscle of meat-producing animals. It provides the red color associated with meat and thus affects the appearance of meat. Two other heme-containing proteins found in meat are hemoglobin and cytochromes. These are generally present at relatively low concentrations and are not considered to substantially impact the color of meat from normal animals (Warriss and Rhodes, 1977; Ledward, 1984). Several reviews have been published on the biochemistry of myoglobin and its relevance in meat (Faustman and Cassens, 1990; Renerre, 1990; Cornforth, 1994).

The presence of myoglobin in meat impacts two important aspects of meat color. First, the total amount of coloration is directly proportional to the concentration of myoglobin present. Myoglobin concentration is affected by species (Livingston and Brown, 1981), animal age (Lawrie, 1974), diet (MacDougall et al., 1973), and muscle type (Lawrie, 1974). Muscles with a high proportion of red oxidative myofibers contain more myoglobin, and as animals age, the concentration of myoglobin in muscles increases. The marketing of white, or special fed, veal is based on close monitoring of the dietary iron intake in bovine calves, and restriction of this micronutrient yields muscles with reduced pigmentation (Bremner et al., 1976).

Myoglobin affects meat color stability as well as intensity. The color perceived by consumers on the surface of fresh meat is determined by the relative concentrations of the three redox forms of myoglobin. Deoxymyoglobin is a purplish pigment in which the heme iron is in a ferrous (Fe^{2+}) state with nothing bound at the sixth coordination site. Oxymyoglobin results when deoxymyoglobin is exposed to air. Oxygen binds to the sixth site of ferrous heme and produces cherry-red oxymyoglobin. Either of these ferrous myoglobins can autoxidize to brownish red metmyoglobin. Metmyoglobin contains heme iron in the ferric (Fe^{3+}) state with water bound at the sixth site. The spectra for each of the three forms of myoglobin are presented in Figure F3.3.1. It is important to note the isobestic point at 525 nm at which the spectra from all three myoglobin forms intersect. The estimation of myoglobin concentration in muscle extract is most easily accomplished at this wavelength, where a single extinction coefficient can be applied. Extinction coefficients have been determined for a number of wavelengths, and myoglobin forms and an extensive list of coefficients can be found in Bowen (1949). The wavelength at which metmyoglobin shows a strong absorbance relative to the ferrous myo-

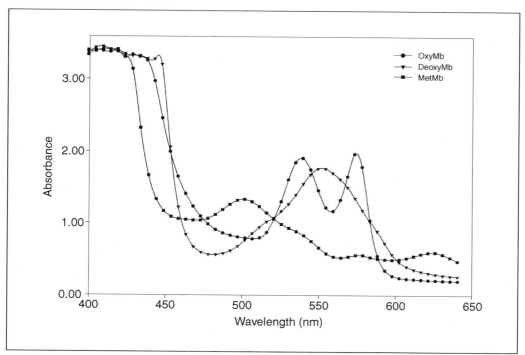

Figure F3.3.1 Absorbance spectra for deoxy-, oxy-, and metmyoglobin.

globins is 630 nm, while the wavelength at which the maximum difference in absorbance between metmyoglobin and oxymyoglobin occurs is 580 nm.

As a heme-containing protein, myoglobin shows an exceedingly intense spectral band, often called the Soret band (or *B* band), between 380 and 420 nm (Fig. F3.3.1; Gouterman, 1978). The Soret band, named after its discoverer, is attributed to the unique absorbance of the porphyrin ring, a cyclic compound formed by the linkage of four pyrrole rings through methenyl bridges (Martin, 1981).

Many factors (e.g., temperature, pH, water activity, packaging) affect the prevalence of the three myoglobin forms in meat (Faustman and Cassens, 1990) and their effects can be measured by following oxymyoglobin oxidation to metmyoglobin. For the purposes of this unit, the authors wish to emphasize the importance of both temperature and pH. Myoglobin oxidation is affected substantially by temperature with autoxidation rates higher at higher temperatures. Although myoglobin in meat is precipitated at 60°C and above (Ledward, 1971), marked denaturation of myoglobin in pure solution occurs at 65°C followed by precipitation at 70.1°C (Kristensen and Andersen, 1997). The difference in denaturation temperature can be attributed to the fact that, in meat, myoglobin coprecipitates with various other meat proteins.

The pH of a myoglobin solution is also critical, and ferrous myoglobins readily oxidize to metmyoglobin as pH is decreased from physiological to postmortem meat values. In order to minimize oxidation of deoxy- or oxymyoglobin to metmyoglobin, it is critical to keep solutions cold (0° to 5°C) at all times. In addition, a high pH (8.0 to 8.5) will discourage myoglobin autoxidation; however, meat-related research often requires an experimental pH of 5.6, the typical post-mortem skeletal muscle pH.

Critical Parameters

There are several important points that need to be stressed. First, the estimation of discoloration in meat by measurement of percent metmyoglobin is useful in a relative sense only. Proper control samples (i.e., nontreated samples subjected to identical conditions, dilutions, and analyses, as treated samples) must be in place, and color measurement must occur in the same manner with both control and treated samples. Additionally, comparisons of values for percent metmyoglobin between different laboratories are only valid when investigators use the same measurement procedure. The use of three different formulas (i.e., from three different procedures) will yield different values for percent metmyoglobin. Again, it is the relative differences between treatments, or the changes over time, that become important, and

Miscellaneous Colorants

F3.3.9

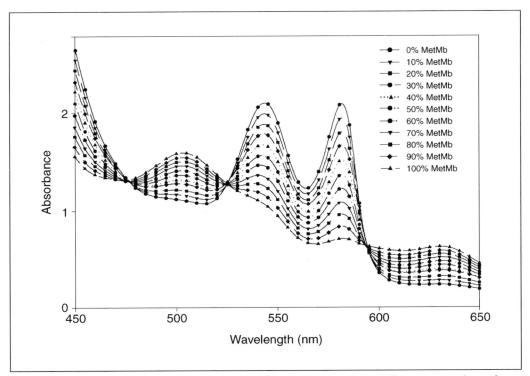

Figure F3.3.2 Absorbance spectra for myoglobin solutions containing different proportions of oxy- and metmyoglobin.

not the absolute values of metmyoglobin calculated.

As noted above, myoglobin oxidation is affected significantly by pH and temperature. At the pH of normal meat (5.6), myoglobin oxidation is more rapid than at higher pH values. It is recommended that extraction procedures utilize refrigerated conditions where possible for homogenization and filtration. The use of buffers at pH values greater than 5.6 (e.g., pH 6.8; Warriss, 1979) will also minimize any further oxidation during sample preparation. Finally, spectral analyses should be performed immediately following filtration.

Turbidity can also be problematic for some ground meat samples, as noted above. In order to avoid light scattering and erroneous results, which will occur as a result of turbidity, it is essential that proper filtration be used. In general, filtration through Whatman no. 1 filter paper is sufficient; however, when additional filtration is required, the authors have found the use of syringe-mounted filter units with 0.40-μm filters to work well.

Anticipated Results

The yield of purified myoglobin obtained with Basic Protocol 3 can vary greatly depend-

ing on the source of beef muscle used. As meat is displayed under retail display and/or storage conditions, or as myoglobin solutions are permitted to incubate, oxymyoglobin will oxidize to metmyoglobin. As such, results will be reported as metmyoglobin accumulation or oxymyoglobin loss. The units should be expressed as a percentage of the total myoglobin present rather than an absolute concentration for reasons already noted. In general, most investigators use metmyoglobin formation as a measure of meat discoloration when reporting study results. A spectrum for myoglobin containing different proportions of oxymyoglobin and metmyoglobin is presented in Figure F3.3.2. Nearly 100% oxymyoglobin is obtained if Basic Protocol 4 is followed correctly.

The use of colorimetry for evaluating discoloration in fresh meat is well established and has been reviewed elsewhere (Francis and Clydesdale, 1975; Hunt, 1980). In general, a^* values are most commonly reported, as they indicate the degree of redness in meat. Several colorimetric parameters correlate very well with sensory assessment of meat discoloration and hue angle, as shown in Figure F3.3.3.

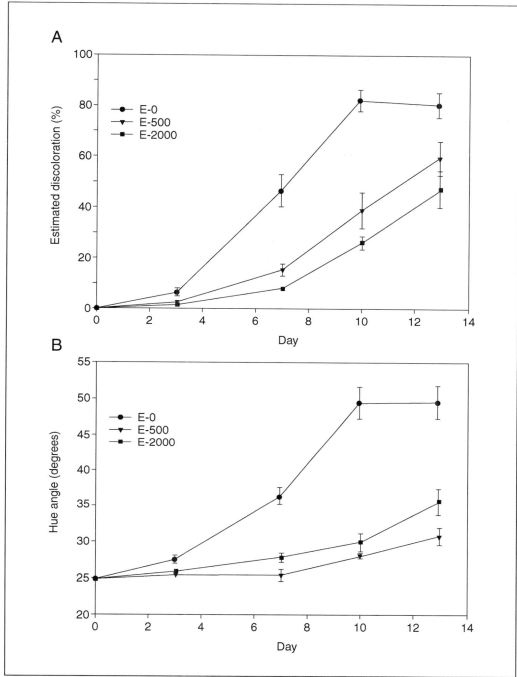

Figure F3.3.3 Comparison of subjective sensory assessment (**A**; percent discoloration) and objective colorimetric evaluation (**B**; hue angle). Beef was obtained from cattle supplemented with 0 (E-0), 500 (E-500) or 2000 (E-2000) IU α-tocopherol acetate per head per day. The α-tocopherol demonstrated a color preservation effect. Hue angle was calculated as $[\tan^{-1}(b*/a*)]*(360°/2\pi)$. Standard error bars are indicated. Adapted from Chan et al. (1995), with permission from the Institute of Food Technologists.

Time Considerations

Meat discoloration studies typically involve a maximum of 5 days, with discoloration analysis being performed every day or on alternate days. The actual experimental time involved in the objective assessment of discoloration is not extensive and depends on the number of samples being analyzed. Colorimetric measurements with hand-held colorimeters are very rapid (three measurements per meat surface in <1 min). Spectral scans of meat surfaces require 1 to 2 min. Extraction and analysis of ground meat products has the added step of homogenization and filtration prior to spectrophotometry, but relative to many laboratory procedures, this is relatively quick. Isolation and purification of preparative amounts of myoglobin requires only 2 to 3 days once appropriate preparations are made. Finally, metmyoglobin can be reduced to oxymyoglobin in 15 to 20 min.

Literature Cited

AMSA (American Meat Science Association). 1991. Guidelines for Meat Color Evaluation. *Recip. Meat Conf. Proceed. 44.* Am. Meat Sci. Assoc., Kansas City, Mo.

Bevilacqua, A.E. and Zaritzky, N.E. 1986. Rate of pigment modifications in packaged refrigerated beef using reflectance spectrophotometry. *J. Food Process. Preserv.* 10:1-18.

Bowen, W.J. 1949. The absorption spectra and extinction coefficients of myoglobin. *J. Biol. Chem.* 179:235-245.

Bremner, I., Brockway, J.M., Donnelly, H.T., and Webster, A.J.F. 1976. Anaemia and veal calf production. *Vet. Rec.* 99:203-205.

Broumand, H., Ball, C.O., and Stier, E.F. 1958. Factors affecting the quality of prepackaged meat. II. E. Determining the proportions of heme derivatives in fresh meat. *Food Technol.* 12:65-77.

Brown , W.D. and Mebine, L.B. 1969. Autoxidation of oxymyoglobins. *J. Biol. Chem.* 244:6696-6701.

Chan, W.K.M., Hakkarainen, K., Faustman, C., Schaefer, D.M., Scheller, K.K., and Liu, Q. 1995. Color stability and microbial growth relationships in beef as affected by endogenous α-tocopherol. *J. Food Sci.* 60:966-971.

Clydesdale, F.M. 1978. Colorimetry—methodology and applications. *In* Critical Reviews in Food Science and Nutrition (T.E. Furia, ed.) pp. 243-301. CRC Press, Boca Raton, Fla.

Cornforth, D.P. 1994. Color—its basis and importance. *In* Quality Attributes and their Measurement in Meat, Poultry and Fish Products (A.M. Pearson and T.R. Dutson, eds.) pp. 34-78. Blackie Academic & Professional, New York.

Faustman, C. and Cassens, R.G. 1990. The biochemical basis for discoloration in fresh meat: A review. *J. Muscle Foods* 1:217-243.

Faustman, C., Yin, M.C., and Nadeau, D.B. 1992. Color stability, lipid stability and nutrient composition of red and white veal. *J. Food Sci.* 57:302-304, 311.

Francis, F.J. and Clydesdale, F.M. 1975. Food Colorimetry: Theory and Applications. AVI Publications, Westport, Conn.

Franke, W.C. and Solberg, M. 1971. Quantitative determination of metmyoglobin and total pigment in an intact meat sample using reflectance spectrophotometry. *J. Food Sci.* 36:515-519.

Gouterman, M. 1978. Optical spectra and electronic structure of porphyrins and related rings. *In* The Porphyrins Volume III: Physical Chemistry, Part A (D. Dolphin, ed.) pp. 65-82. Academic Press, New York.

Hunt, M.C. 1980. Meat color measurements. *Recip. Meat Conf. Proceed.* 33:41-46. Am. Meat Sci. Assoc., Kansas City, Mo.

Judd, D.B. and Wyszecki, G. 1963. Color in Business, Science, and Industry, 2nd ed. John Wiley & Sons, New York.

Kristensen, L. and Andersen, H.J. 1997. Effect of heat denaturation on the pro-oxidative activity of metmyoglobin in linoleic acid emulsions. *J. Agric. Food Chem.* 45:7-13.

Krzywicki, K. 1979. Assessment of relative content of myoglobin, oxymyoglobin and metmyoglobin at the surface of beef. *Meat Sci.* 3:1-10.

Krzywicki, K. 1982. The determination of haem pigments in meat. *Meat Sci.* 7:29-36.

Lawrie, R.A. 1974. Meat Science, 2nd ed. Pergammon Press, New York.

Ledward, D.A. 1971. On the nature of cooked meat hemoprotein. *J. Food Sci. 36:883-888.*

Ledward, D.A. 1984. Haemoproteins in meat and meat products. *In* Developments in Food Proteins-3 (B.J.F. Hudson, ed.) pp. 33-75. Elsevier Applied Science Publishers, New York.

Livingston, D.J. and Brown, W.D. 1981. The chemistry of myoglobin and its reactions. *Food Technol.* 35(5):244-252.

MacDougall, B.D., Bremner, I., and Dalgarno, A.C. 1973. Effect of dietary iron on the color and pigment concentration of veal. *J. Sci. Food. Agric.* 24:1255-1263.

Martin, D.W. 1981. Porphyrins and bile pigments. *In* Harper's Review of Biochemistry (D.W. Martin, P.A. Mayes, and V.W. Rodwell, eds.) pp. 212-214. Lange Medical Publications, Los Altos, Calif.

Renerre, M. 1990. Review: Factors involved in the discoloration of beef meat. *Int. J. Food Sci. Technol.* 25:613-630.

Rickansrud, D.A. and Henrickson, R.L. 1967. Total pigments and myoglobin concentration in four bovine muscles. *J. Food Sci.* 32:57-61.

Stewart, M.B., Zipser, M.W., and Watts, B.M. 1965. The use of reflectance spectrophotometry for the assay of raw meat pigments. *J. Food Sci.* 30:464-469.

Strange, E.D., Benedict, R.C., Gugger, R.E., Metzger, V.G., and Swift, C.E. 1974. Simplified methodology for measuring meat color. *J. Food Sci.* 39:988-992.

Trout, G.R. and Gutzke, D.A. 1996. A simple, rapid preparative method for isolating and purifying oxymyoglobin. *Meat Sci.* 43:1-13.

Van den Oord, A.H.A. and Wesdorp, J.J. 1971. Analysis of pigments in intact beef samples. *J. Food. Technol.* 6:1-13.

Warriss, P.D. 1979. The extraction of haem pigments from fresh meat. *J. Food Technol.* 14:75-80.

Warriss, P.D. and Rhodes, D.N. 1977. Haemoglobin concentrations in beef. *J. Sci. Food. Agric.* 28:931-934.

Wittenberg, J.B. and Wittenberg, B.A. 1981. Preparation of myoglobins. *Methods Enzymol.* 76:29-42.

Key References

AMSA, 1991. See above.

A comprehensive guide to measuring fresh, cured, and cooked meat color published by the American Meat Science Association.

Clydesdale, 1978. See above.

An extensive review of multiple theories and analytical techniques utilized to describe food color.

Hunt, 1980. See above.

An overview of instrumental analyses used to measure meat color and their relation to human visual appraisal.

Krzywicki, 1979. See above.

Provides equations used with reflectance spectrophotometry to calculate the relative proportions of myoglobin, oxymyoglobin, and metmyoglobin on beef surfaces.

Krzywicki, 1982. See above.

Describes a method to calculate the relative concentration of myoglobin, oxymyoglobin, and metmyoglobin present in a meat extract.

Stewart et al., 1965. See above.

The procedure upon which Basic Protocol 2 is based; the researchers utilize K/S ratios to estimate the relative presence of metmyoglobin on meat surfaces.

Contributed by Cameron Faustman
    and Amy Phillips
University of Connecticut
Storrs, Connecticut

**Miscellaneous
Colorants**

F3.3.13

Chlorophylls

F4.1 Overview of Chlorophylls in Foods   **F4.1.1**
    Background and Diversity   F4.1.1
    Chlorophyll Chemistry During Food Processing   F4.1.3
    Chlorophyll Analysis   F4.1.6
    Final Considerations   F4.1.7

F4.2 Extraction of Photosynthetic Tissues: Chlorophylls and Carotenoids   **F4.2.1**
    Basic Protocol   F4.2.1
    Commentary   F4.2.4

**F4.3 Chlorophylls and Carotenoids: Measurement and Characterization
by UV-VIS Spectroscopy**   **F4.3.1**
    Absorption Maxima   F4.3.1
    Absorption Spectra   F4.3.2
    Accuracy of Spectroscopic Measurements   F4.3.3
    Quantifaction of Pigments   F4.3.5
    Determination of Total Carotenoids   F4.3.5
    Interpretation of Chlorophyll and Carotenoid Content   F4.3.6

F4.4 Chromatographic Separation of Chlorophylls   **F4.4.1**
    Strategic Planning   F4.4.1
    Basic Protocol: C18 Reversed-Phase HPLC Separation of Chlorophylls
        a and *b* and Their Nonpolar Derivatives   F4.4.2
    Alternate Protocol: C18 Reversed-Phase HPLC Separation of Polar Chlorophyll
        Derivatives and Cu^{2+} and Zn^{2+} Pheophytins   F4.4.4
    Support Protocol 1: C18 Column Cleaning   F4.4.7
    Support Protocol 2: Synthesis of Cu^{2+} and Zn^{2+} Pheophytin Standards   F4.4.8
    Commentary   F4.4.10

F4.5 Mass Spectrometry of Chlorophylls   **F4.5.1**
    Basic Protocol 1: Fast Atom Bombardment (FAB), Liquid Secondary Ion
        Mass Spectrometry (LSIMS), and Continuous-Flow FAB of Chlorophylls   F4.5.1
    Basic Protocol 2: LC/MS Using Atmospheric Pressure Chemical
        Ionization (APCI) and Electrospray Ionization (ESI)   F4.5.3
    Commentary   F4.5.4

Contents

1

Overview of Chlorophylls in Foods

BACKGROUND AND DIVERSITY

Chlorophylls are bright green natural pigments found exclusively in photosynthetic plants and select bacteria. These pigments are based on a tetrapyrrole macrocycle linked by methene bridges, a structure which is known as a porphyrin. This basic structure maintains a high degree of unsaturation, providing an extended conjugated double-bond system that has a high metal-binding capacity (Dailey, 1990). Chlorophyll's porphyrin structure is expanded by addition of a fifth isocyclic ring (ring E) joining at position 6 and γ (Figure F4.1.1). The main modifications of the basic porphyrin structure include substitutions of methyl (CH_3) groups at positions 1, 5, and 8; a vinyl ($CH_2=CH_2$) at position 2; a propionic acid group esterified to a diterpene alcohol, phytol; and a centrally bound magnesium atom (Gross, 1991). From this basic structure, five classes of chlorophylls exist naturally in plants and photosynthetic organisms—a, b, c, d, and e—with the latter being only a minor derivative. Chlorophylls a and b predominate naturally in all higher plants, while chlorophyll c, d, and e derivatives are found throughout various photosynthetic algal and diatomic species including brown, red, and yellow-green algae. Additionally, four classes of bacteriochlorophylls have been isolated in photosynthetic bacteria, with bacteriochlorophyll a and b predominating in purple bacteria while c and d are found in green and purple sulfur bacteria (Sheer, 1991; Hendry, 2000).

Chlorophylls are widely distributed among green fruits and vegetables. Generally chlorophyll a predominates over chlorophyll b by a 3-to-1 margin. While native chlorophylls function mainly as primary photosynthetic pigments, their presence is considered crucial to final food product acceptance, as the green color they impart is often associated with fresh vegetable quality. Specific distribution and content of the pigments in fruits and vegetables

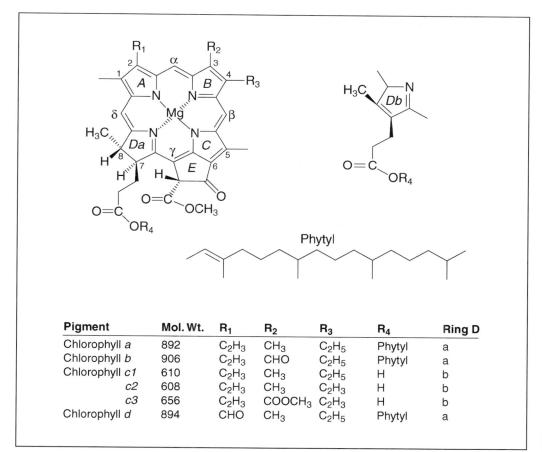

Pigment	Mol. Wt.	R₁	R₂	R₃	R₄	Ring D
Chlorophyll a	892	C_2H_3	CH_3	C_2H_5	Phytyl	a
Chlorophyll b	906	C_2H_3	CHO	C_2H_5	Phytyl	a
Chlorophyll c1	610	C_2H_3	CH_3	C_2H_5	H	b
c2	608	C_2H_3	CH_3	C_2H_3	H	b
c3	656	C_2H_3	$COOCH_3$	C_2H_3	H	b
Chlorophyll d	894	CHO	CH_3	C_2H_5	Phytyl	a

Figure F4.1.1 Structural differences between major classes of natural chlorophylls in higher plants and algae. Designation of pyrrole rings A-D and methene bridges α-δ is based on the nomenclature outlined by Fisher and Orth (1937).

Contributed by Mario G. Ferruzzi and Steven J. Schwartz

Chlorophylls

F4.1.1

Supplement 1

Table F4.1.1 Approximate Chlorophyll Content of Various Green Vegetables[a]

Tissue	Chlorophyll content (µg Chl/g vegetable tissue)		
	a	b	Total
Asparagus[b,c]			
Fresh tissue	139	74	180-300
Canned–Dry tissue	180 (Phe *a*) 110 (Pyro a)	51 (Phe a) 30 (pyro *b*)	
Beans[b,c]			
Fresh tissue	54	17	
Dry tissue		230-870	
Canned–Dry tissue	340 (Phe *a*) 260 (Pyro *a*)	180 (Phe *b*) 95 (Pyro *b*)	
Beet Leaf[d]			
Fresh tissue		1160	
Dry tissue		13100	
Broccoli[b,e]			
Floral–Fresh tissue		322	
	106	33	160
	110	43	
Stalk–Fresh tissue			
	36	2.0	
Celery[b]			
Leaves–Fresh tissue	1143	225	
Stalks–Fresh tissue	29	7	
Chinese mustard[d]			
Fresh tissue	210	28	
Collards[b]			
Fresh tissue	1009	216	
Cucumber[b]			
Fresh tissue	64	24	
Fenugreek Leaf[e]			
Fresh tissue		2010	
Dry tissue		15700	
Kale[b]			
Fresh tissue	1370	464	1870
Lettuce[b]			
Fresh tissue	334	62	
Mango Leaf[f]			
Fresh tissue:			
Amrapalli		2220	
Malaviyabhog		1880	
Langra		2660	
Dashehari		2630	
Okra[b]			
Fresh tissue	160	132	

continued

Tissue	Chlorophyll content (µg Chl/g vegetable tissue)		
	a	*b*	Total
Olive[g]			
Fresh tissue:			
Nevadillo	5480	120	5600
Hojiblanca	11610	180	11790
Marteña	6200	320	6520
Pararero	6060	120	6180
Picual 1	22320 (620 as Chl *a*)	2040 (1200 as Chl *b*)	24360
Peas[a,b,d]			
Fresh tissue	106	12	
Canned–Dry tissue	34 (Phe *a*) 33 (Pyro *a*)	13 (Phe *b*) 12 (Pyro *b*)	
Spinach[a,b]			
Fresh tissue	1380	440	1576
Dry tissue	6980	2490	
Canned–Dry tissue	830 (Phe *a*) 4000 (Pyro *a*)		

[a]Abbreviations: Chl, chlorophyll; Phe, pheophytin; Pyro, pyropheophytin.

[b]Gross (1991).

[c]von Elbe and Schwartz (1996).

[d]Negi and Roy (2000).

[e]Murcia et al. (2000).

[f]Pandey and Tyagi (1999).

[g]Gandul-Rojas and Minguez-Mosquera (1996).

is dependent on a number of factors including type of vegetable, stage of maturity, growing conditions, and commercial food processing (Gross, 1991). A sampling of chlorophyll content for a variety of fruits and vegetables is shown in Table F4.1.1. These values should be utilized only as a general guide, and not as an absolute indication of chlorophyll content in these green vegetable products, due to the wide biological variability and differences among varieties.

Chlorophyll content of food systems is normally expressed based on a reference system (e.g., wet or dry weight basis). While wet weight chlorophyll content is often reported, it should be considered in the context of the preparation's total moisture, as water content may potentially vary between varieties and preparations. Therefore, these values should be accompanied by moisture content of the tissue in order to account for this innate variability. Alternatively, chlorophyll content expressed on a dry weight basis can easily be compared across varieties and preparations. While dry weight content is preferred, it requires a drying step (preferably freeze drying) prior to analysis,

or adjustment of the wet weight values on the basis of moisture content determinations.

CHLOROPHYLL CHEMISTRY DURING FOOD PROCESSING

The orderly and programmed natural biochemical process of chlorophyll decomposition known as senescence catabolizes nearly one billion pounds of photosynthetic pigment annually (Heaton and Marangoni, 1996). This phenomenon is most evident in the fall as leaves loose their bright green colors, and has been extensively researched, but full comprehension of the complete biochemical pathways remains incomplete. Senescing tissue excluded, chlorophylls associated with the plant structure have been found to maintain excellent stability considering the harsh photodegradative environment (Hendry, 2000). The same cannot be extended to isolated chlorophylls or those subjected to food processing and preparation, including thermal treatment and acidification (Schwartz and Lorenzo, 1990).

Degradation of chlorophylls during food processing of green fruits and vegetables has been thoroughly studied and is the subject of a number of reviews (Simpson, 1985; Schwartz

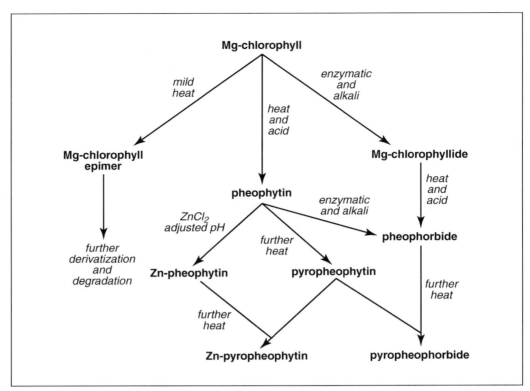

Figure F4.1.2   Major chlorophyll degradation and derivatization reactions occurring during food processing operations. Structures of major derivatives of chlorophyll *a* and *b* are depicted in Figure F4.1.3.

and Lorenzo, 1990; Heaton and Marangoni, 1996). A brief summary of the main chlorophyll reactions encountered through food processing is depicted in Figure F4.1.2, with some of the corresponding structures shown in Figure F4.1.3. Chlorophylls are extremely sensitive to physical and chemical changes encountered through food processing. These changes contribute to the perceivable discoloration of vegetable tissue from green to olive brown that is encountered during thermal processing and/or acidification. This color loss is predominantly a result of replacement of the centrally chelated magnesium atom by two atoms of hydrogen, producing metal-free pheophytin derivatives (Schwartz and Lorenzo, 1990). More severe heat treatments, as experienced in canning operations, result in the loss of C_{10} decarboxymethoxyl moeity–forming derivatives known as pyropheophytins (Schwartz et al., 1981; Schwartz and von Elbe, 1983). Pyrochlorophyll derivatives that have also been C_{10} decarboxymethoxylated but that retain the central magnesium atom have been isolated in both steamed and microwave-processed spinach leaves, but these remain less common (Teng and Chen, 1999).

Milder processing conditions also result in alteration of major chlorophyll derivatives. Formation of C_{10} epimers of the native chlorophyll structure is readily encountered during freezing, drying, and mild heating such as blanching (Katz et al., 1968; Schwartz and Lorenzo, 1990). While other conversions affect final product color, epimer formation has no detrimental effect, as the spectral properties of the epimers remain identical to those of their parent chlorophyll molecule (Sheer, 1991). Enzymatic removal of the esterified phytol by chlorophyllase results in the formation of water-soluble chlorophyllide derivatives (Schwartz and Lorenzo, 1990). These derivatives are often encountered when mild thermal treatment such as blanching is utilized, which activates chlorophyllase (Canjura et al., 1991; von Elbe and Schwartz, 1996). Further thermal processing or acidification results in formation of metal-free, water-soluble derivatives known as pheophorbides and pyropheophorbides (Schwartz and Lorenzo, 1990; Sheer, 1991).

The importance of color to final perceived quality of food products, combined with the labile nature of chlorophyll derivatives, has perpetuated numerous efforts to preserve native green vegetable appearance. Fishenbach

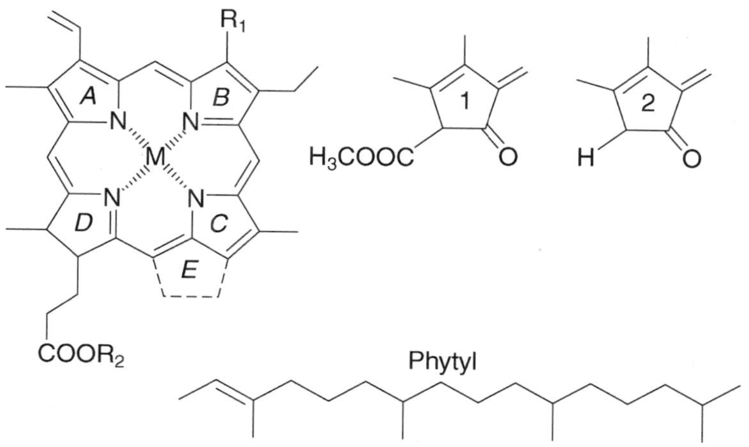

Figure F4.1.3 Structure of major chlorophyll derivatives prevalent in fresh and processed green vegetable tissue.

Derivative	Metal	R_1	R_2	Ring E
Chlorophyll *a*	Mg	CH$_3$	Phytyl	1
Chlorophyll *b*	Mg	CHO	Phytyl	1
Pheophytin *a*		CH$_3$	Phytyl	1
Pheophytin *b*		CHO	Phytyl	1
Pyropheophytin *a*		CH$_3$	Phytyl	2
Pyropheophytin *b*		CHO	Phytyl	2
Pheophorbide *a*		CH$_3$	H	2
Pheophorbide *b*		CHO	H	2
Zn-Pheophytin *a*	Zn	CH$_3$	Phytyl	1
Zn-Pheophytin *b*	Zn	CHO	Phytyl	1
Zn-Pheophorbide *a*	Zn	CH$_3$	H	2
Zn-Pheophorbide *b*	Zn	CHO	H	2
Zn-Pyropheophytin *a*	Zn	CH$_3$	Phytyl	2
Zn-Pyropheophytin *b*	Zn	CHO	Phytyl	2
Cu-Pheophytin *a*	Cu	CH$_3$	Phytyl	1
Cu-Pheophytin *b*	Cu	CHO	Phytyl	1
Cu-Pheophorbide *a*	Cu	CH$_3$	H	2
Cu-Pheophorbide *b*	Cu	CHO	H	2
Cu-Pyropheophytin *a*	Cu	CH$_3$	Phytyl	2
Cu-Pyropheophytin *b*	Cu	CHO	Phytyl	2

(1943) and Schanderl et al. (1965) first noticed formation of zinc and copper pheophytins and reported their increased thermal stability and color similarity with respect to natural chlorophylls. Since that time, great efforts have been dedicated to understanding the formation of these zinc and copper chlorophyll complexes for the purpose of color stabilization in processed foods. Zinc and copper pheophytin *a* and pyropheophytin *a* derivatives are rapidly formed by the addition of zinc and copper salts to commercially canned vegetables prior to thermal treatment, a process known as regreening (Schwartz and Lorenzo, 1990). Jones et al. (1977) studied the formation of copper and zinc derivatives, determining that complexation is favored at salt concentrations between one- and ten-fold the pigments' molar equivalence. The Crown Cork and Seal Company, Inc., has commercialized this canning technology for use with green beans under the trade name VERI-GREEN (Segner et al., 1984). von Elbe et al. (1986) investigated the composition of VERI-GREEN beans and determined that they consisted primarily of zinc-pheophytin *a* and zinc-pyropheophytin *a*. In subsequent studies it was determined that formation of zinc complexes was dependent on chlorophyll species, pH, temperature, and ion concentration (LaBorde and von Elbe, 1990; Tonucci and von Elbe, 1992). Efficient formation of zinc-pheophytins and zinc-pyropheophytins in peas subjected to a continuous flow aseptic processing system raises the possibility that color preservation by metallo-chlorophyll formation may be extended beyond simple canning to more advanced processing techniques (Canjura et al., 1999).

CHLOROPHYLL ANALYSIS

The broad range of polarity offered by the diverse array of chlorophyll pigments encountered in fresh and processed fruit and vegetable products presents a true analytical challenge. Because of the labile nature of these pigments, all procedures and sample handling should be carried out in clean glassware, under subdued light, and with fresh solvents. Extraction and analysis of chlorophyll pigments from vegetable tissue is often complicated by identification of suitable solvent systems that accommodate all chlorophyll derivatives, from the polar chlorophyllides to the more apolar pheophytins and pyropheophytins (UNIT F4.2). Organic solvents commonly used to extract lipophilic chlorophylls, including acetone and ether, will not completely extract water-soluble derivatives such as chlorophyllides and pheophorbides, as they will remain in the aqueous phase. Recovery of these water-soluble derivatives may be efficiently accomplished by homogenization of the vegetable tissue with cold methanol followed by centrifugation for solvent clarification. Methanol extracts may be collected and vacuum-dried prior to analysis. Fresh chlorophyll extracts are extremely sensitive to degradative reactions, and should therefore ideally be analyzed immediately (UNITS F4.3 & F4.4). For storage, extracts may be dried, placed at sub-zero temperatures under a nitrogen or argon atmosphere, and protected from light (Schwartz, 1998).

Spectrophotometric Assays

Spectrophotometric assessment of chlorophyll content is based on the strong electronic absorption spectra of these pigments. Arnon (1949) developed an early method measuring 80% acetone/20% water plant extracts based on the electronic absorption spectra of chlorophylls *a* and *b*. Absorbance of the extract was measured at different wavelengths, and simultaneous equations were constructed based on extinction coefficients for each derivative's unique electronic absorption maxima. Over the

last fifty years, numerous methods have appeared utilizing this approach (UNIT F4.3). White et al. (1965) developed equations for simultaneous determination of chlorophyll, pheophytin, and pheophorbide derivatives in plant tissues. Jones et al. (1977) describe methods for the spectrophotometric estimation of zinc pheophytins in complex mixtures with both chlorophylls and pheophytins. AOAC recognizes a spectrophotometric assay for the determination of chlorophyll *a* and *b* components in plant extracts (AOAC International, 1995).

Chromatographic Methods

Rapid and reliable chromatographic methods have been developed for analysis of complex pigment mixtures often encountered in fruit and vegetable tissues. Open-column methods based on gravity or vacuum-assisted flow have enjoyed wide application and success due to their low cost and excellent resolving power, allowing for both qualitative and quantitative separation. Tswett (1906a,b) was the first to accomplish column chromatography of chloroplast pigments using both calcium carbonate and sugar columns, providing basic separation of carotenoid and chlorophyll *a* and *b* pigments. Since that time, a number of different adsorbents have been used for the separation of chlorophyll derivatives, with sucrose in the form of powdered sugar (3% starch) having enjoyed the widest application (reviewed by Strain and Svec, 1969). Thin-layer chromatography (TLC) is a logical extension of open-column chromatography, often utilizing similar adsorbents, such as sucrose. TLC is based on utilization of glass plates coated with very thin layers of adsorbent material, which are developed by organic solvents.

Development of fast, accurate, and reproducible high-performance liquid chromatography (HPLC) methods has offset the use of traditional open-column and TLC methods in modern chlorophyll separation and analysis. A number of normal and reversed-phase methods have been developed for analysis of chlorophyll derivatives in food samples (UNIT F4.4), with octadecyl-bonded stationary phase (C_{18}) techniques predominating in the literature (Schwartz and Lorenzo, 1990). Inclusion of buffer salts such as ammonium acetate in the mobile phase is often useful, as this provides a proton equilibrium suitable for ionizable chlorophyllides and pheophorbides (Almela et al., 2000).

Metallo-chlorophyll derivatives such as zinc and copper derivatives present different analytical challenges as their prevalence in the food supply increases. Schwartz (1984) developed reversed-phase C_{18} methods for separation of copper and zinc pheophytins (UNIT F4.4). Reversed-phase C_{18} allows for effective resolution of zinc pheophytin in complex pigment mixtures including carotenoids and natural chlorophylls and pheophytins (Ferruzzi et al., 2001). Minguez-Mosquera et al. (1996) applied both TLC and reversed-phase HPLC for the determination of copper complexes of oxidized chlorophyll derivatives. Inoue et al. (1994) demonstrated the usefulness of an isocratic C_{18} method with methanol and acetic acid elution for separation of major components of water-soluble sodium copper chlorophyllin derivatives. These water-soluble metallo-derivatives have received interest for their potential health benefits, including anti-inflammatory, deodorizing, erythropoietic, antimutagenic, and antioxidant activities (Kephart, 1955; Sato et al., 1986; Harttig and Bailey, 1998).

Methods of Detection

Post-column detection of chlorophyll derivatives is often accomplished by ultraviolet and visible spectroscopic techniques, which take advantage of the strong electronic absorption spectra of these pigments (UNITS F4.3 & F4.4). While these methods have enjoyed wide application (Schwartz et al., 1981; Khachik et al., 1986), a major advance was made with the introduction of photodiode array (PDA) detection. Multichannel photodiode array detection allows for simultaneous monitoring of multiple wavelengths, resulting in the generation of online electronic absorption spectra of a compound as it elutes from the HPLC column. Because of the uniqueness of electronic absorption spectra of individual chlorophyll derivatives, these techniques have enjoyed extensive application for tentative identification of components from complex mixtures and extracts (UNITS F4.3 & F4.4).

In experiments where a higher degree of sensitivity and selectivity is required, fluorescence and mass-selective detectors have been applied. Picomole limits of detection offered by fluorescence makes it ideal for routine analysis requiring high sensitivity. Mass spectrometry has also proven to be both a sensitive and efficient way to identify numerous chlorophyll derivatives (UNIT F4.5). van Breemen et al. (1991) utilized both fast atom bombardment (FAB) and tandem mass spectrometry (MS/MS) for the structural characterization and mass determination of numerous derivatives including chlorophylls, chlorophyllides, pheophytins, pheophorbides, pyropheophytins, and Zn-pheophytins. Hyvärinen and Hynninen (1999) further utilized FAB-MS for identification of chlorophyll *b* allomers. HPLC with MS/MS was recently applied for identification of Cu(II)-chlorin ethyl ester in human sera (Egner et al., 2000).

FINAL CONSIDERATIONS

Analysis of chlorophylls in fruits and vegetables requires careful planning and preparation. Initial consideration must be given to sample matrix and approximate chlorophyll content. Extraction strategies will vary from tissue to tissue and with matrix status. Initial preparative steps may be required prior to extraction and analysis of samples. For example, samples of high water content such as juices may need to be dried prior to extraction. The complexity of the matrix and the nature of the analytical question will determine which techniques are most appropriate (UNIT F4.2). Spectrophotometric methods may be utilized in cases where quantitative determination of total chlorophyll content is desired (UNIT F4.3). Chromatographic techniques are appropriate when both a qualitative and quantitative chlorophyll profile is required (UNIT F4.4). Alteration of postchromatographic detection allows for flexible analyses tailored to specific application end points (UNIT F4.5). However, extreme care must be taken at every step of the analytical process to ensure that final results are free of artifacts produced from the degradation/derivatization of these labile photosynthetic pigments.

LITERATURE CITED

Almela, L., Fernández-López, J.A., and Roca, M.J. 2000. High-performance liquid chromatographic screening of chlorophyll derivatives produced during fruit storage. *J. Chromatogr. A.* 870:483-489.

AOAC (Association of Official Analytical Chemists) International 1995. Method 942.04: Chlorophyll in plants: Spectrophotometric method of total chlorophyll and the *a* and *b* components. *In* Official Method of Analysis, 16th ed. AOAC International, Arlington, Va.

Arnon, D.E. 1949. Copper enzymes in isolated chloroplast: Polyphenoloxidase in *Beta vulgaris*. *Plant Physiol.* 24:1-15.

Canjura, F.L., Schwartz, S.J., and Nunes, R.V. 1991. Degradation kinetics of chlorophylls and *chlorophyllides*. *J. Food Sci.* 56:1639-1643.

Canjura, F.L., Watkins, R.H., and Schwartz, S.J. 1999. Color improvement and metallo-chlorophyll complexes in continuous flow aseptically processed peas. *J. Food Sci.* 64:987-990.

Dailey, H.A. 1990. Biosynthesis of Heme and Chlorophylls. McGraw-Hill, New York.

Egner, P.A, Stansbury, K.H., Snyder, E.P., Rogers, M.E., Hintz, P.A., and Kensler, T.W. 2000. Identification and characterization of chlorin e4 ethyl ester in sera of individuals participating in the chlorophyllin chemoprevention trial. *Chem. Res. Tox.* 13:900-906.

Ferruzzi, M.G., Failla, M.L., and Schwartz, S.J. 2001. Assessment of degradation and intestinal cell uptake of carotenoids and chlorophyll derivatives from spinach puree using an in vitro digestion and Caco-2 human cell model. *J. Agric. Food Chem.* 49:2082-2089.

Fischbach, H. 1943. Microdeterminations for organically combined metal in pigment of okra. *J. Assoc. Off. Agric. Chem.* 26:139-143.

Fisher, H. and Orth, H. 1937. Chemie des Pyrrols. Acad. Verlag. Leipzig, Germany.

Gandul-Rojas, B. and Minguez-Mosquera, M.I. 1996. Chlorophyll and carotenoid compositon in virgin olive oils and from various Spanish olive varieties. *J. Sci. Food Agric.* 72:31-39.

Gross, J. 1991. Pigments in vegetables, chlorophylls and carotenoids. Van Nostrand–Reinhold, New York.

Harttig, U. and Bailey, G.S. 1998. Chemoprevention by natural chlorophylls in vivo: Inhibition of dibenzo[a,l]pyrene-DNA adducts in rainbow trout liver. *Carcinogenesis* 19:1323-1326.

Heaton, J.W. and Marangoni, A.G. 1996. Chlorophyll degradation in processed foods and senescent plant tissues. *Trends Food Sci. Tech.* 71:8-15.

Hendry, G.A. 2000. Chlorophylls. *In* Natural Food Colorants: Science and Technology (G.J. Lauro and F.J. Francis, eds.) Marcel Dekker, New York.

Hyvärinen, K. and Hynninen, P.H. 1999. Liquid chromatographic separation and mass spectrometric identification of chlorophyll b allomers. *J. Chromatogr., A.* 837:107-116.

Inoue, H., Yamashita, H., Furuya, K., Nonomura, Y., Yoshioka, N., and Li, S. 1994. Determination of copper(II) chlorophyllin by reversed-phase high-performance liquid chromatography. *J. Chromatogr., A.* 679: 99-104.

Jones, I.D., White, R.C., Gibbs, E., and Butler, L.S. 1977. Estimation of zinc pheophytins, chlorophylls and pheophytins in mixtures in diethyl ether or 80% acetone by spectrophotometry and fluorometry. *J. Agric. Food Chem.* 25:146-149.

Katz, J.J., Norman, G.D., Svec, W.A., and Strain, H.H. 1968. Chlorophyll diastereoisomers. The nature of chlorophylls a′ and b′ and evidence for bacteriochlorophyll epimers from proton magnetic resonance studies. *J. Am. Chem. Soc.* 90:6841-6848.

Kephart, J.C. 1955. Chlorophyll derivatives—Their chemistry, commercial preparations and uses. *Econ. Bot.* 9:3-38.

Khachik, F., Beecher, G.R., and Whittaker, N.F. 1986. Separation, identification and quantification of the major carotenoid and chlorophyll constituents in extracts of several green vegetables by liquid chromatography. *J. Agric. Food Chem.* 34:603-616.

LaBorde, L.F. and von Elbe, J.H. 1990. Zinc complex formation in heated vegetable purees. *J. Agric. Food Chem.* 28:437-439.

Minguez-Mosquera, M.I., Gandul-Rojas, B., and Garrido-Fernández, J. 1996. Preparation of Cu(II) complexes of oxidized chlorophylls and their determination by thin-layer and high performance liquid chromatography. *J. Chromatogr., A.* 731:261-271.

Murcia, M, A., Lopez-Ayerra, B., Martinez-Tome, M., and Garcia-Carmona, F. 2000. Effect of industrial processing on chlorophyll content of broccoli. *J. Sci. Food Agric.* 80:1447-1451.

Negi, P.S. and Roy, S.K. 2000. Effect of blanching and drying methods on β-carotene, ascorbic acid and chlorophyll retention on leafy vegetables. *Lebensm.-Wiss. U.-Technol.* 33:295-298.

Pandey, S. and Tyagi, D.N. 1999. Changes in chlorophyll content and photosynthetic rate of four cultivars of mango during reproductive phase. *Biologia Plantarium.* 42:457-461.

Sato, M., Fujimoto, I., Sakai, T., Aimoto, T., Kimura, R., and Murata, T. 1986. Effect of sodium copper chlorophyllin on lipid peroxidation. IX. On the antioxidative components in commercial preparations of sodium copper chlorophyllin. *Chem. Pharm. Bull.* 34:2428-2434.

Schanderl, S.H., Marsh, G.L., and Chinchester, C.O. 1965. Color reversion in processed vegetables. I. Studies on re-greened pea purees. *J. Food Sci.* 30:320-326.

Schwartz, S.J. 1984. High performance liquid chromatography of zinc and copper pheophytins. *J. Liq. Chromatogr.* 7:1673-1683.

Schwartz, S.J. and von Elbe, J.H. 1983. Kinetics of chlorophyll degradation to pyropheophytins in green vegetables. *J. Food Sci.* 48: 1303-1308.

Schwartz, S.J., Woo, S.L., and von Elbe, J.H. 1981. High performance liquid chromatography of chlorophylls and their derivatives in fresh and processed spinach. *J. Agric. Food Chem.* 29: 533-537.

Segner, W.P., Ragusa, T.J., Nank, W.K., and Hoyle, W.C. 1984. Process for the preservation of green color in canned green vegetables. U.S. Patent No. 4,473,591, September 25, 1984.

Sheer, H. 1991. The Chlorophylls. CRC Press, Boca Raton, Fla.

Simpson, K.L. 1985. Chemical changes in natural food pigments. *In* Chemical Changes in Food during Processing (T. Richerson and J.W. Finley, eds.) AVI Publishing, Westport, Conn.

Strain, H.H. and Svec, W.A. 1969. Some procedures for the chromatography of the fat-soluble chloroplast. *Adv. Chromat.* 8:118-176.

Teng, S.S. and Chen, B.H. 1999. Formation of pyrochlorophylls and their derivatives in spinach leaves during heating. *Food Chemistry* 65:367-373.

Tonucci, L.H. and von Elbe, J.H. 1992. Kinetics of the formation of zinc complexes of chlorophyll derivatives. *J. Agric. Food Chem.* 40:2341-2344.

Tswett, M. 1906a. Physikalische-chemische Studien über das Chlorophyll. Die Absorptionen. *Ber. Deut. Botan. Ges.* 24:316-323.

Tswett, M. 1906b. Absorption analyse und chromatographische Methode. Anwendung auf die Chemie des Chlorophylls. *Ber. Deut. Botan. Ges.* 24:384-385.

van Breemen, R.B., Canjura, F.L., and Schwartz, S.J. 1991. Identification of chlorophyll derivatives by mass spectrometry. *J. Agric. Food Chem.* 39:1452-1456.

von Elbe, J.H. and Schwartz, S.J. 1996. Colorants. *In* Food Chemistry (O.R. Fennema, ed.). 3rd edition. Mercel Dekker, New York.

von Elbe J.H., Huang, A.S., Attoe, E.L., and Nank, E.L. 1986. Pigment composition and color of conventional and Veri-Green canned beans. *J. Agric. Food Chem.* 34:52-54.

White, R.C., Jones, I.D., and Gibbs, E. 1963. Determination of chlorophylls, chlorophyllides, pheophytins and pheophorbides in plant material. *J. Food Sci.* 28:431-436.

Key References

Schwartz, S.J. and Lorenzo, T.V. 1990. Chlorophyll in foods. *In* Critical Reviews in Food Science and Nutrition (F.M. Clydesdale, ed.). CRC Press, Boca Raton, Fla.

This review provides an excellent overview of chlorophyll chemistry and diversity in typical food products and processing applications.

Schwartz, S.J. 1998. Pigment analysis. *In* Food Analysis (S.S. Nielson, ed.). Chapman and Hall, Gaitherburg, Md.

This chapter provides excellent insight into basic chlorophyll analysis by both spectrophotometric and chromatographic techniques.

Mario G. Ferruzzi and Steven J. Schwartz
Ohio State University
Columbus, Ohio

Extraction of Photosynthetic Tissues: Chlorophylls and Carotenoids

The extraction of chlorophylls and carotenoids from water-containing plant materials requires polar solvents, such as acetone, methanol, or ethanol, that can take up water. These extracts must then be transferred to a solvent such as diethyl ether in order to be stored stably. Samples with very high water content, such as juices and macerated plant material, are usually freeze-dried first, and can then be extracted directly with diethyl ether. After extraction, solutions are clarified and diluted to an appropriate volume to measure chlorophyll content by UV-VIS spectrophotometry. Absorption coefficients and equations needed for quantitative determination are given in UNIT F4.3.

This protocol has been developed for extraction of chlorophyll from leaf samples, but it can also be used for other plant food samples. The authors recommend, however, that the beginner initially perform the procedure using green leaf samples before extending its use to other types of samples.

NOTE: Absorption in the red and blue maxima is highest in freshly isolated chlorophyll, and then decreases with time due to formation of allomeric chlorophyll forms and possibly destruction of chlorophylls (UNIT F4.1), particularly in the presence of light. This also applies to green pigment extract solutions of leaves and other plant tissues. Therefore, chlorophyll determinations should be carried out in dim light immediately after preparing the pigment extract solution.

Materials

Green leaf samples or other greenish plant tissue samples
MgO or $MgCO_3$
100% or 80% (v/v) acetone or diethyl ether, spectrophotometric grade
Hydrophobic organic solvent: diethyl ether, light petrol, or hexane, spectrophotometric grade
Half-saturated NaCl solution
Anhydrous Na_2SO_4

Rim-sharpened cork driller (optional)
Mortar and pestle
Aluminum dishes
100°C drying oven
Freeze dryer
5-ml graduated centrifuge tubes
Explosion-proof tabletop centrifuge or a cooling tabletop centrifuge
UV-VIS spectrophotometer
1-cm-path-length cuvette
25- or 50-ml separatory funnel
Water bath set below 35°C (optional)

Prepare samples and determine water content

1. For each sample to be tested, prepare three to five replicates weighing 6 to 12 mg each and place in a mortar.

   Because values may vary between different parts of leaves and fruits, three to five replicates must be taken from each sample to obtain a significant and reliable mean value.

   For leaf samples, a rim-sharpened cork driller can be used to punch samples from the leaf. A punch area of 0.6 to 0.9 cm² will give ~6 to 12 mg fresh sample weight, equivalent to 3

Contributed by Hartmut K. Lichtenthaler and Claus Buschmann

to 6 mg dry weight. For dark green leaves, a single punch for each replicate is usually sufficient to obtain good absorbance readings (0.3 to 0.85 at 662 nm). Light green leaves may require two to three punches (12 to 36 mg fresh weight) per replicate, and yellowish green leaves may require four to five punches (24 to 60 mg fresh weight) per replicate to obtain a sufficiently high absorbance reading (>0.3 at 662 nm).

For other plant food samples, appropriate sample sizes will depend on chlorophyll content and water content. Samples with low chlorophyll content may require 100 to 200 mg per replicate. For samples with low chlorophyll content and high water content (e.g., in florescences, fruit tissues, fruit juices), it is necessary to start with 2 to 3 g fresh weight sample and freeze dry it before extraction. Frozen food samples should also be freeze dried before extraction.

2. To determine the water content of the plant sample, take another four punches (24 to 48 mg sample) and weigh them. Place them in an aluminum dish, dry ~2 hr in a 100°C drying oven, and weigh them again. Subtract dry weight from original weight to obtain water content.

   It is strongly recommended that the level of chlorophylls and carotenoids be determined on both a dry weight and a wet weight basis. The pigment values measured for a known leaf area can be converted to pigment values per gram dry weight sample with appropriate calculations. Dry weight and leaf area are reliable reference systems. Wet weight is less reliable since water content varies with storage.

Extract chlorophylls/carotenoids

3. Use the moisture content in step 2 to choose the appropriate extraction procedure, solvent, coefficients, and equations.

   For one to five leaf punches per replicate, the water content coming from the plant tissue will account for <1% to 2% of the final 5 ml extract (step 6) and can be neglected. Perform extraction as described below and use the equations for 100% acetone (UNIT F4.3).

   For 100 to 200 mg sample with very low chlorophyll, the water content in the final 5 ml acetone extract will exceed 2%. Perform extraction as described steps 4 and 5. Then, based on water content of the sample, adjust the solvent in step 6 with aqueous acetone to give a final of 5 ml of 80% aqueous acetone. Apply the equations for 80% acetone (UNIT F4.3).

   For freeze-dried samples, extract the dry sample (~0.3 to 1 g dry weight starting from 2 to 3 g fresh weight) with diethyl ether and apply the equations given for diethyl ether (UNIT F4.3). Extraction of freeze-dried plant material with diethyl ether is performed by grinding in a mortar. It is also possible to use 80% acetone or 100% methanol, but diethyl ether has proved to be an excellent solvent for quantitative extraction of chlorophylls and carotenoids from freeze-dried plant material.

4. Add 100 to 200 mg MgO or $MgCO_3$ to the sample (step 1) to neutralize plant acids and prevent pheophytin *a* formation.

   If high amounts of pheophytins are formed, the main absorption peaks near 660 and 662 nm would shift to other wavelengths and the green extract would change to a pale olive-green. A small amount of pheophytins (1% to 2%) does not significantly change the results.

5. Add 3 ml of 100% acetone (for ≤200 mg fresh sample) or diethyl ether (for freeze-dried samples with low chlorophyll content) and grind with a pestle.

   An explosion-proof motor-driven grinder or steel or glass balls can also be used to grind the sample (see Critical Parameters).

6. Transfer the turbid pigment extract comprising chlorophylls and total carotenoids to a 5-ml graduated glass centrifuge tube. Rinse the grinding device with another 1.5 ml solvent, add to the centrifuge tube, and bring to exactly 5 ml with additional solvent.

7. Centrifuge 5 min at 300 to 500 × g in an explosion-proof tabletop centrifuge at room temperature.

   Slight cooling (10° to 15 °C) can be used.

Perform spectrophotometric analysis

8. Transfer an aliquot of the clear leaf extract (supernatant) with a pipet to a 1-cm-path-length cuvette and take absorbance readings against a solvent blank in a UV-VIS spectrophotometer at five wavelengths:

   750 nm ($A_{750} = 0$ for clear extract)
   662 nm (chlorophyll a maximum using 100% acetone)
   645 nm (chlorophyll b maximum using 100% acetone)
   520 nm (for extracts from green plant tissue, A_{520} should be <10% A_{662})
   470 nm (carotenoids).

   For other solvents, use the chlorophyll maxima, absorption coefficients, and equations found in UNIT F4.3 and in Lichtenthaler (1987).

   For green to dark green leaves, extracts made from 6 to 12 mg leaf sample give A_{662} values between 0.3 and 0.85. If $A_{662} \leq 0.3$ (e.g., with light green or yellow-green leaves), the procedure should be repeated using additional sample (see step 1).

9. Apply measured absorbance values to equations given for each solvent system in UNIT F4.3 to determine pigment content (µg/ml extract solution). Multiply by 5 ml to obtain the total amounts of chlorophyll a, chlorophyll b, and carotenoids contained in the 5 ml extract.

   This represents the µg pigment in each replicate sample.

10. Determine the mean value from each set of replicates.

Store sample

11. Prepare a fresh extract from a larger amount of tissue, or pool the remainder of the replicate pigment extracts (step 7). Transfer to a 25- or 50-ml separatory funnel. Add 3 ml hydrophobic organic solvent (diethyl ether, hexane, or light petrol) and gently shake.

    In order to obtain an extract that can, under exclusion of light and water, be stored in a refrigerator for days and weeks, the pigments should be transferred to a hydrophobic organic solvent that gives a phase separation with water, such as light petrol (a mixture of hydrocarbons, boiling point 40° to 70°C), hexane, or diethyl ether.

    A 10- to 15-ml aqueous acetone extract solution (i.e., from three replicates) or a direct extract from a larger tissue sample can be extracted with 3 ml hydrophobic solvent.

    Rigorous shaking should be avoided as it causes formation of water-lipid emulsions that can only be broken down by centrifugation.

12. Add 10 to 15 ml half-saturated NaCl solution under continued gentle shaking until the hypophase (lower phase), which contains the original organic solvent (e.g., acetone), is ~50% aqueous and the chlorophylls and carotenoids are transferred to the organic epiphase.

    The half-saturated NaCl solution prevents emulsion formation.

    Under these conditions, chlorophylls a and b, carotenoids, and xanthophyll esters are transferred to the epiphase (upper phase), as described in German in Lichtenthaler and Pfister (1978) and briefly in English in Lichtenthaler (1987).

13. Transfer hypophase to a separatory funnel and extract the last traces of chlorophylls and carotenoids using 2 ml of the hydrophobic organic solvent used in step 11.

    The transfer of photosynthetic pigments from a larger extract in an aqueous organic solvent by a small amount of a hydrophobic organic solvent concentrates the pigments.

14. Combine the epiphases from the two extractions and wash once or twice with a small amount (e.g., 1 or 2 ml) of half-saturated NaCl solution.

15. Add 100 to 200 mg anhydrous Na_2SO_4 and decant the concentrated extract into a measuring flask. Close with a glass or Teflon stopper. For larger amounts of extract, concentrate the hydrophobic epiphase to a fixed volume (e.g., 5 or 10 ml) by evaporation in a water bath set below 35°C.

16. Store extract wrapped in foil at 4°C (stable for days or weeks).

> *For comparison with other plant samples, take 0.05 or 0.1 ml of this concentrated pigment extract, dilute to 5 ml with 100% acetone or diethyl ether, and determine the amount of photosynthetic pigments using the absorption coefficients and equations given in UNIT F4.3.*

COMMENTARY

Background Information

Before the extraction of photosynthetic pigments, the plant material should be well defined, i.e., one should think in advance about the reference system for the chlorophyll (Chl) and carotenoid concentration (e.g., mg Chl/g dry weight or mg Chl/m^2 leaf area). The reference systems frequently used are: fresh weight, dry weight, or leaf area. Dry weight is a reliable reference system. However, fresh weight generally is not a suitable reference system because it includes the water content of plant tissue (leaves, fruits), which is highly variable. For this reason, the dry weight of a parallel plant sample must be determined, so that the pigment content can be expressed on a dry weight basis, providing a much more reliable reference system than fresh weight. For leafy food samples such as lettuce and spinach, leaf area is also an acceptable reference system.

Critical Parameters

Plant material

Usually the plant material can be directly extracted without any pretreatment. In cases of extremely high water content (e.g., juices, macerated plant material), the sample should be freeze-dried before extraction; otherwise, the lipid-soluble pigments cannot be adequately extracted. In addition, the absorption maxima of pigments in organic solvents are shifted towards longer wavelengths when water is present, and the absorption coefficients are considerably changed with increasing water content of the extract (Lichtenthaler, 1987). This requires some precautions to apply the proper equations for pigment determination.

Extraction procedure

Extraction of photosynthetic pigments can be carried out using a mortar and pestle or a motor-driven grinder, or by shaking the plant material with glass or steel balls. When using a conventional grinder with an electric motor, one should be aware of the danger involved in using inflammable organic solvents (e.g., acetone). In this case, a mortar and pestle are safer, and they are also easier to clean. It is advisable to add small amounts of MgO or $MgCO_3$ to neutralize plant acids that cause the formation of pheophytin a from Chl a. It is advisable to start pigment extraction with a small volume of solvent. When the plant material is well homogenized, add more solvent to give a final defined volume of the extract solution. If too much solvent is used, the pigments will be too dilute and absorbance readings will no longer be possible to obtain. Thus, 6 to 12 mg leaf material can easily be extracted with 3 ml acetone, re-extracted with 1.5 ml acetone, and then brought to a 5-ml final volume. After centrifugation for 5 min, 3 ml of the clear extract solution are placed into a cuvette for quantitative determination in the spectrophotometer.

Water-containing plant materials need to be extracted with polar solvents such as acetone, methanol, or ethanol that can take up water. Freeze-dried plant tissues and freeze-dried juices can be directly extracted with diethyl ether, which contains traces of water and is more polar than light petrol or hexane. Pure light petrol or hexane are less suitable, because more polar pigments, such as Chl b or xanthophylls, are only partially extracted from freeze-dried plant samples. A few drops of acetone or ethanol added to light petrol or hexane will, however, guarantee a complete extraction. This mixture will extract Chl a, Chl b, and all carotenoids—including xanthophyll esters and secondary carotenoids that are present in many fruits and juices—from the freeze-dried plant material.

Solvent

Chlorophylls and carotenoids are generally extracted with organic solvents. One should always apply purified solvents such as spectrophotometric grades, which are commercially available. This is an essential requirement. Organic impurities in standard-grade organic solvents considerably change absorption coefficients and wavelength maxima of the pigments. Because of the relatively high water content of intact plant material, one should use a solvent that mixes with water, such as acetone, ethanol, or methanol. However, chlorophylls are unstable in these water-containing solvents. Allomeric chlorophyll types that possess different absorption characteristics can eventually form. Moreover, part of Chl a can be broken down to pheophytin a (by removal of the central Mg atom), especially in the presence of plant acids from the vacuoles of extracted plant material (UNIT F4.1). Pheophytin a exhibits quite different absorption characteristics from Chl a. In addition, light photochemically destructs chlorophylls and carotenoids. Avoid chloroform, which is sometimes proposed as an extraction solvent for chlorophylls, because it not only is poisonous, but also contains hydrochloric acid, which partially transforms Chl a into pheophytin a. The formation of pheophytin b from Chl b usually does not occur during extraction, as this requires stronger acids.

Diethyl ether (spectrophotometric grade) is the best epiphase solvent, as it has a higher capacity to solubilize photosynthetic pigments than light petrol or hexane. When concentrated, pigment extracts are stored in strongly hydrophobic solvents (such as light petrol or hexane) in the refrigerator, but small turbid flakes containing polar pigments form. This does not occur with diethyl ether, which can contain some water. If diethyl ether has been used in the epiphase for extracting pigments from the hypophase, the epiphase can be stored in a refrigerator at 4° to 6°C for 1 or 2 hr in the separatory funnel. Under these conditions, water dissolved in diethyl ether at room temperature separates from the diethyl ether and this water hypophase can be discarded.

Final preparation of the extract

For the quantitative determination of Chl a and b, it is not necessary to separate these two pigments prior to spectrophotometric measurement (UNIT F4.3), as the absorbance of the extract is measured in the red region at the wavelength positions of both Chls. From these absorbance readings, Chl a and b are calculated by a particular subtraction method. This subtraction method also applies to the determination of total carotenoids (xanthophylls and carotenes, x + c) in a total pigment extract by measuring the absorbance at 470 nm, the main absorbance region of carotenoids (UNIT F4.3).

However, for an exact determination of the pigment concentration, the extract must be fully transparent to avoid obtaining values that are too high for Chl b and total carotenoids. In most cases, the homogenized plant extract contains colorless, undissolved, very fine, solid plant material, e.g., fibers and cell wall debris. These materials make the extract turbid and scatter light, rather than absorbing light. The scattered light increases from longer to shorter wavelengths (from red region to blue). Thus, the presumed absorbance signal measured in turbid solutions by the spectrophotometer is increased differentially for individual wavelengths (see Figure F4.3.4 in UNIT F4.3). Hence, the concentration of pigments, calculated from these incorrect absorbance values, are too high. Also, the ratio of Chl a to Chl b shifts from 2.7 to 3.2 to incorrect lower values of 2.1 to 2.6.

In order to have a transparent extract without turbidity, the homogenized extract should be centrifuged (5 min at $300 \times g$) or filtered. When using a centrifuge, one should cool the extract in order to keep evaporation as low as possible and avoid problems with inflammable organic solvents, such as acetone or diethyl ether. (Several companies make explosion-proof tabletop centrifuges, which should be used.) Filtering through fine glass filters (e.g., G3) at a reduced pressure (e.g., a water-wheel pump) is another, more time-consuming procedure. If used, one should ensure that no pigments are left in the filter. In addition, glass filters are quickly plugged by the fine plant debris and/or MgO or $MgCO_3$ powder added during extraction, requiring an intricate clean-up procedure. Paper filters are not suitable, as they retain only the larger plant particles and are permeable to finely ground plant debris. Special filters that retain finer plant debris and MgO or $MgCO_3$ powder take too much time to use, and the pigments may be partially oxidized or photochemically destroyed. Thus, for routine analysis, centrifugation is the method of choice. When transferring the centrifuged, clear pigment extract solution with a pipet into the spectrophotometer cuvette, great care should be taken not to disturb the sedimented debris.

It is important to note that chlorophylls are converted to pheophytins in the presence of acids (UNIT F4.1). Formation of a significant

Chorophylls

amount of pheophytin due to unexpectedly high amounts of endogenous acids (which is very seldom the case) is observed as a change of the green pigment extract color to a pale olive-green. When this happens, add a drop of 25% HCl to the acetone extract, which will transform all chlorophyll a and b to pheophytin a and b. Then apply the absorption coefficients and equations given by Lichtenthaler (1987) for pheophytins. A small amount (1% to 3%) of pheophytins in the green extract solution does not significantly change the absorbance readings and can be tolerated. To avoid artifactual pheophytin formation, do not handle concentrated acids in the laboratory used for pigment extraction. Traces of gaseous acids in the air can convert chlorophylls to pheophytins before and during grinding and extraction.

Samples and extracts containing chlorophylls should be protected from light at all times. For storage and during analysis, extracts should be wrapped in aluminum foil.

Anticipated Results

Real pigment differences among leaf or other plant tissues of different light adaptation or developmental stages, and among fruit tissue of different maturation or senescence, occur on both a fresh and dry weight basis. Pigment differences showing up in just one reference system may not be real, as the reference system, such as dry weight (e.g., during fruit storage) or leaf area (e.g., shrinking during water stress or enlargement during leaf expansion) may have changed.

Time Considerations

One determination involving extraction with 100% acetone, centrifugation, and absorbance readings takes ~20 min. With experience, routine measurements of six samples can be completed in 40 to 60 min.

Literature Cited

Lichtenthaler, H.K. 1987. Chlorophylls and carotenoids: Pigments of photosynthetic biomembranes. *Methods Enzymol.* 148:350-382.

Lichtenthaler, H.K. and Pfister, K. 1978. Praktikum der Photosynthese. Quelle & Meyer, Heidelberg.

Contributed by Hartmut K. Lichtenthaler and Claus Buschmann
Universitaet Karlsruhe
Karlsruhe, Germany

Chlorophylls and Carotenoids: Measurement and Characterization by UV-VIS Spectroscopy

The quantitative determination of chlorophyll (Chl) *a*, Chl *b*, and carotenoids in a whole-pigment extract of green plant tissue by UV-VIS spectroscopy is complicated by the choice of sample, solvent system, and spectrophotometer used. The various plant pigments absorb light in overlapping spectral regions, depending on the system selected. This unit discusses methods used to account for such overlap by applying equations for accurate quantitative determination of Chl *a*, Chl *b*, and total carotenoids in the same pigment extract of leaves or fruits. General information on the spectroscopic characteristics of Chl *a* and Chl *b*, their specific absorption coefficients, and their quantitative determination in a whole-pigment extract of green plant tissues can be found in Šesták (1971) and Lichtenthaler (1987). For Chl structures, see *UNIT F4.1*.

ABSORPTION MAXIMA

Figure F4.3.1 shows the absorption spectrum of isolated Chl *a* and Chl *b* in diethyl ether. Chl *a* and *b* absorb with narrow bands (maxima) in the blue (near 428 and 453 nm) and red (near 661 and 642 nm) spectral ranges. The isolated yellow carotenoids have a broad absorption with three maxima or shoulders in the blue

spectral range between 400 and 500 nm (Fig. F4.3.2).

The absorption maxima of extracted pigments strongly depend on the type of solvent and, to some degree, on the type of spectrophotometer used. For example, with increasing polarity of the solvent, the red absorption maximum of Chl *a* shifts from 660 to 665 nm, and the blue absorption maximum from 428 to 432 nm. The same also applies to Chl *b*, which shifts from 642 to 652 nm and 452 to 469 nm (see, e.g., Fig. F4.3.3 and Table F4.3.1, and Lichtenthaler, 1987). These wavelength shifts of the absorption maxima are correlated with changes in the absorption coefficients used for the quantitative determination of Chls *a* and *b* and carotenoids. For these reasons, the absorbance readings of a pigment extract must be performed at the correct wavelength position, i.e., the maxima of pure Chl *a* and pure Chl *b* in a particular solvent. Moreover, the solvent-specific extinction coefficients have to be considered by applying the corresponding equations for calculation of the pigment content. Minor differences in the positions of the wavelength maxima also exist, depending on the spectrophotometer type used. Thus, the wavelength position can differ by 1.0 or 1.5 nm.

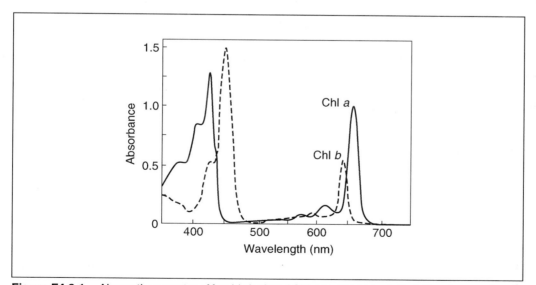

Figure F4.3.1   Absorption spectra of freshly isolated Chl *a* and Chl *b* in diethyl ether (pure solvent). The spectra were measured 40 min after extraction of pigments from leaves and 3 min after eluting the two Chls with diethyl ether from a TLC plate.

Contributed by Hartmut K. Lichtenthaler and Claus Buschmann

Chlorophylls

F4.3.1

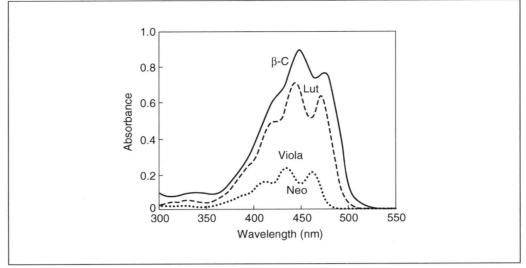

Figure F4.3.2 Absorption spectra of the major carotenoids of the photosynthetic biomembranes of green leaves of higher plants in diethyl ether (pure solvent). The carotenoids were freshly isolated from a pigment extract by TLC following Lichtenthaler and Pfister (1978) and Lichtenthaler (1987). β-C, β-carotene; Lut, lutein; Neo, neoxanthin; Viola, violaxanthin.

In order to perform spectroscopic measurements of green plant tissue extracts in the right maximum regions, one should determine the maximum red spectral position of pure Chl a and pure Chl b solutions with one's own spectrophotometer and compare them with those from the literature, given in Table F4.3.1. For a wavelength deviation of more than 1 nm, one should measure the absorbance of the pigment extract using these self-determined maxima rather than the literature values. The same equations for the particular solvent can be applied as long as wavelength positions differ by no more than 2 nm. At a deviation of >2 nm, either the spectrophotometer needs wavelength adjustment or a wrong, impure solvent has been applied. For the determination of carotenoids in the same extract solution, the wavelength position of 470 nm may be maintained, since a 1-nm shift has much less influence on the total carotenoid level than on the individual levels of Chls a and b.

ABSORPTION SPECTRA

The absorption spectrum of an extract of a green leaf containing a mixture of Chls a and b and total carotenoids (Fig. F4.3.4) is dominated by the absorption of Chl a at A_{428} (blue)

Table F4.3.1 Wavelength Maxima (A_{max}) and Specific Absorbance Coefficients (α)[a] of Chl a and b for Extracts in Different Organic Solvents

	Diethyl ether (water free)	Diethyl ether (pure)	Diethyl ether (water saturated)	Acetone (pure)	Acetone (with 20% water)	Ethanol (with 5% water)	Methanol (pure)
A_{max} Chl a [nm]	660.0	660.6	661.6	661.6	663.2	664.2	665.2
A_{max} Chl b [nm]	641.8	642.2	643.2	644.8	646.8	648.6	652.4
$\alpha_{(a)\max a}$	101.9	101.0	98.46	92.45	86.3	84.60	79.24
$\alpha_{(a)\max b}$	15.20	15.0	15.31	19.25	20.49	25.06	35.52
$\alpha_{(a)470}$	1.30	1.43	1.38	1.90	1.82	2.13	1.63
$\alpha_{(b)\max a}$	4.7	6.0	7.2	9.38	11.2	16.0	21.28
$\alpha_{(b)\max b}$	62.3	62.0	58.29	51.64	49.18	41.2	38.87
$\alpha_{(b)470}$	33.12	35.87	48.05	63.14	85.02	97.64	104.96
$\alpha_{(x+c)470}$	213	205	211	214	198	209	221

[a]Units of absorption coefficients are given in liter g^{-1} cm^{-1}. $\alpha_{(a)\max a}$ is the specific absorbance coefficient of Chl a at its red maximum; $\alpha_{(a)\max b}$ is the specific absorbance coefficient of Chl a at the red maximum of Chl b; $\alpha_{(a)470}$ is the specific absorbance coefficient of Chl a at 470 nm; $\alpha_{(x+c)470}$ is the specific absorbance coefficient of the sum of xanthophylls and carotenes at 470 nm.

and A_{661} (red). Chl b and the carotenoids absorb broadly in the blue region (400 to 500 nm).

A plant sample homogenized with an organic solvent is usually turbid and must be filtered or centrifuged to become fully transparent (see UNIT F4.2). Turbidity and light scattering lead to a higher absorption between 400 and 800 nm, with a slight but continuous increase towards shorter wavelengths (Fig. F4.3.5). Thus, measuring a turbid extract leads to an overestimation of the pigment levels, especially for Chl b and total carotenoids. Turbidity can be checked by measuring A_{750} and A_{520}. For a fully transparent leaf pigment extract, A_{750} should equal zero, since Chls a and b and carotenoids do not absorb in this region. A_{520} readings for extracts of green plant tissue should be <10% of the main Chl absorbance in the red maximum near 661 nm (diethyl ether) or 650 nm (ethanol), as shown in Figures F4.3.4 and F4.3.5.

ACCURACY OF SPECTROSCOPIC MEASUREMENTS

In order to have an exact spectroscopic measurement of absorbances, one must consider the absorbance range in which readings are made. Absorbance should be measured between 0.3 and 0.85. Leaf extracts with an absorbance <0.3 in the red region do not yield correct pigment values. There are several interfering factors, such as a base line that is not fully zeroed. Thus, values <0.3, whether read by the experimenter or given as digital values by the instrument, are not acceptable. Absorbance values >0.9 may indicate problems with the accuracy of the detector (e.g., a photomultiplier). Since the detector system examines the transmitted light of the cuvette, the absorbance is calculated from this value. When transferring the linear transmission unit to the logarithmic absorbance unit, the accuracy is exponentially reduced with rising values.

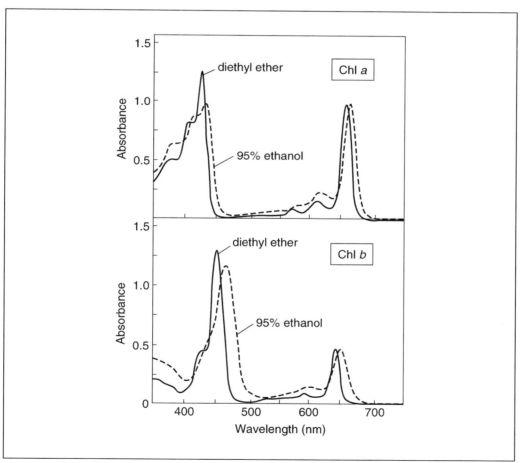

Figure F4.3.3 Differences in the absorption spectra of Chl a and Chl b in diethyl ether and 95% aqueous ethanol. For the more polar solvent (95% ethanol; broken line), the absorbance (extinction) in the blue and red absorption maxima of both Chls are decreased compared to values obtained using the less polar solvent diethyl ether (black), and the wavelength positions of the maxima are shifted to the right. For a better comparison, the absorbances in the red maxima were set at the same values.

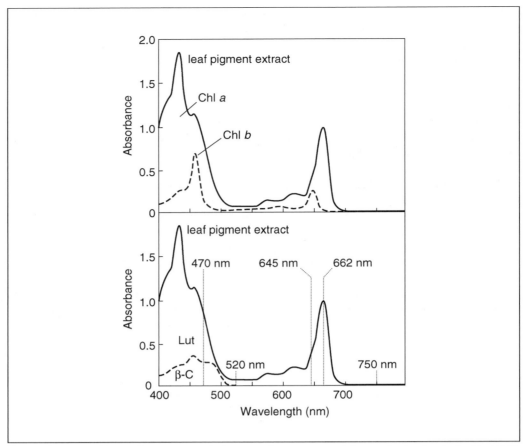

Figure F4.3.4 Absorption spectra of pigments from a green tobacco leaf extracted with 100% acetone. The leaf extract was measured directly after extracting the leaf. Chl *a*, Chl *b*, and the carotenoids β-carotene (β-C) and lutein (Lut) were measured after separation by TLC.

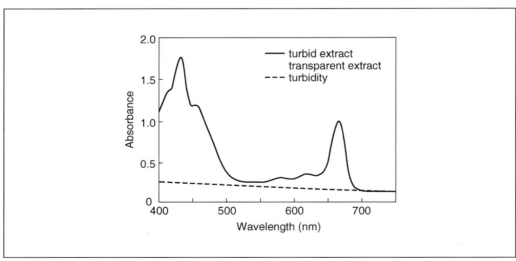

Figure F4.3.5 Absorption spectra of a leaf extract before (turbid) and after (transparent) centrifugation in 100% acetone. The difference spectrum between the two extracts represents the spectrum of turbidity.

For absorbance values <0.3, one should try to concentrate the extract (e.g., by evaporation), make a new extract using more plant material and less solvent, or extract the pigments in a separatory funnel into a small volume of a hydrophobic solvent in the epiphase. Various spectrophotometers are constructed to measure absorbance (extinction) values only up to 1.0 (i.e., a transmittance of 10%). In such cases, an absorbance >0.85 is not suitable, and the extract solution should be diluted to obtain valid Chl b and carotenoid values. In both cases, care must be taken to ensure that the final volume of the extract solution is carefully recorded and considered in the calculation of total Chls and carotenoids.

The extinction coefficients and the equations used and established by Arnon (1949) are not correct. They provide only a rough estimate of Chl a and b levels and yield inaccurate Chl b values, and, consequently, incorrect values for the Chl a/b ratio. They have been redetermined by Lichtenthaler (1987) using the extinction coefficients of Smith and Benitez (1955) for pure Chl a and Chl b in diethyl ether, which were found to be correct in the red absorption maxima at 661 and 642 nm, respectively, for purified Chls. The relative absorptions of Chl a and Chl b at other wavelengths in other organic solvents have been redetermined using modern high-resolution spectrophotometers.

To exactly determine carotenoids by measuring A_{470}, one needs to know the exact level of Chl b, which (in contrast to Chl a) also absorbs considerably at this wavelength (Fig. F4.3.1). If Chl b is overestimated, the level of total carotenoids becomes too low, and vice versa. With the redetermined extinction coefficients, the new equations permit the determination of total carotenoids in addition to Chl a and Chl b in the same green tissue extract solutions.

QUANTIFACTION OF PIGMENTS

The basis for spectroscopic quantification of pigments is the Lambert-Beer law, which defines the absorbance of a solution with respect to the specific light absorption characteristic of an individual dissolved compound:

$$A = \alpha c_w d \text{ or } A = \varepsilon c_m d$$

where A is absorbance (dimensionless), α is the specific absorbance coefficient in liter g^{-1} cm^{-1}, ε is the molar absorbance coefficient in liter mol^{-1} cm^{-1}, c_w is the weight concentration in g liter^{-1}, c_m is the molar concentration in mol

liter^{-1}, and d is the path length of the cuvette in cm, usually 1 cm.

This original Lambert-Beer law can only be applied for one isolated pigment. Absorbance coefficients taken from the literature (Table F4.3.1) are valid only for one substance (e.g., Chl a) using one solvent (e.g., 100% acetone) and one wavelength (e.g., 661.6 nm). Changes in substance, solvent, or wavelength lead to changes in the absorbance coefficient.

When the concentration of Chl a and Chl b is determined from a pigment extract containing both Chls, the equation derived from the Lambert-Beer law becomes more complex. The absorbance is then expressed as the sum of the absorbances of Chl a and Chl b. Thus, the absorbance of Chl b contributes to the absorbance of Chl a at the Chl a maximum, and vice versa:

$$A_{\max a} = A_{(a)\max a} + A_{(b)\max a} =$$
$$\left(\alpha_{(a)\max a} \times c_{ma} \times d \right) + \left(a_{(b)\max a} \times c_{mb} \times d \right)$$

$$A_{\max b} = A_{(a)\max b} + A_{(b)\max b} =$$
$$\left(\alpha_{(a)\max b} \times c_{ma} \times d \right) + \left(a_{(b)\max b} \times c_{mb} \times d \right)$$

The concentrations for Chl a (c_a) and Chl b (c_b) are then given by a different equation, where the specific contribution of Chl b to the Chl a maximum and of Chl a to the Chl b maximum are subtracted. The following equations contain the denominator z, a term formed from the four extinction coefficients of Chl a and Chl b. The light path length (usually 1 cm) is omitted here:

$$c_a = \left[\frac{\alpha_{(b)\max b} \times A_{\max a}}{z} \right] - \left[\frac{\alpha_{(b)\max a} \times A_{\max b}}{z} \right]$$

$$c_b = \left[\frac{\alpha_{(a)\max a} \times A_{\max b}}{z} \right] - \left[\frac{\alpha_{(a)\max b} \times A_{\max a}}{z} \right]$$

$$z = \left(\alpha_{(a)\max a} \times \alpha_{(b)\max b} \right) - \left(\alpha_{(a)\max b} \times \alpha_{(b)\max a} \right)$$

DETERMINATION OF TOTAL CAROTENOIDS

In an extract of plant material containing carotenoids (x + c = xanthophylls and carotenes) in addition to Chls, A_{470} (the carotenoid region) is determined as the sum of specific absorbances for Chl *a*, Chl *b*, and total carotenoids:

$$A_{470} = A_{(x+c)470} + A_{(a)470} + A_{(b)470}$$

From this follows, according to the Lambert-Beer law:

$$A_{(a)470} = \alpha_{(a)470} \times c_a \times d$$

$$A_{(b)470} = \alpha_{(b)470} \times c_b \times d$$

$$A_{(x+c)470} = \alpha_{(x+c)470} \times c_{(x+c)} \times d$$

The concentration of carotenoids $c_{(x+c)}$ is then given by the following equation, which has been reduced using $d = 1$ cm:

$$c_{(x+c)} = \frac{A_{(a)470} - \left(\alpha_{(a)470} \times c_a\right) - \left(\alpha_{(b)470} \times c_b\right)}{\alpha_{(x+c)470}}$$

The concentrations for Chl *a* (c_a), Chl *b* (c_b), and the sum of leaf carotenoids (c_{x+c}) can be calculated with the following equations given for different solvents, where the pigment concentrations are given in µg/ml extract solution.

Diethyl ether (pure solvent):

$$c_a\,(\mu g/ml) = 10.05\,A_{660.6} - 0.97\,A_{642.2}$$

$$c_b\,(\mu g/ml) = 16.36\,A_{642.2} - 2.43\,A_{660.6}$$

$$c_{(x+c)}\,(\mu g/ml) = \left(1000\,A_{470} - 1.43\,c_a - 35.87\,c_b\right)/205$$

Diethyl ether (water free):

$$c_a\,(\mu g/ml) = 9.93\,A_{660.6} - 0.75\,A_{641.8}$$

$$c_b\,(\mu g/ml) = 16.23\,A_{641.8} - 2.42\,A_{660.6}$$

$$c_{(x+c)}\,(\mu g/ml) = \left(1000\,A_{470} - 1.30\,c_a - 33.12\,c_b\right)/213$$

Diethyl ether (water saturated):

$$c_a\,(\mu g/ml) = 10.36\,A_{661.6} - 1.28\,A_{643.2}$$

$$c_b\,(\mu g/ml) = 17.149\,A_{643.2} - 2.72\,A_{661.6}$$

$$c_{(x+c)}\,(\mu g/ml) = \left(1000\,A_{470} - 1.38\,c_a - 48.05\,c_b\right)/211$$

Ethanol with 5% (v/v) water:

$$c_a\,(\mu g/ml) = 13.36\,A_{664.1} - 5.19\,A_{648.6}$$

$$c_b\,(\mu g/ml) = 27.43\,A_{648.6} - 8.12\,A_{664.1}$$

$$c_{(x+c)}\,(\mu g/ml) = \left(1000\,A_{470} - 2.13\,c_a - 97.64\,c_b\right)/209$$

Acetone (pure solvent):

$$c_a\,(\mu g/ml) = 11.24\,A_{661.6} - 2.04\,A_{644.8}$$

$$c_b\,(\mu g/ml) = 20.13\,A_{644.8} - 4.19\,A_{661.6}$$

$$c_{(x+c)}\,(\mu g/ml) = \left(1000\,A_{470} - 1.90\,c_a - 63.14\,c_b\right)/214$$

Acetone with 20% (v/v) water:

$$c_a\,(\mu g/ml) = 12.25\,A_{663.2} - 2.79\,A_{646.8}$$

$$c_b\,(\mu g/ml) = 21.50\,A_{646.8} - 5.10\,A_{663.2}$$

$$c_{(x+c)}\,(\mu g/ml) = \left(1000\,A_{470} - 1.82\,c_a - 85.02\,c_b\right)/198$$

Methanol (pure solvent):

$$c_a\,(\mu g/ml) = 16.72\,A_{665.2} - 9.16\,A_{652.4}$$

$$c_b\,(\mu g/ml) = 34.09\,A_{652.4} - 15.28\,A_{665.2}$$

$$c_{(x+c)}\,(\mu g/ml) = \left(1000\,A_{470} - 1.63\,c_a - 104.96\,c_b\right)/221$$

Methanol with 10% (v/v) water:

$$c_a\,(\mu g/ml) = 16.82\,A_{665.2} - 9.28\,A_{652.4}$$

$$c_b\,(\mu g/ml) = 36.92\,A_{652.4} - 16.54\,A_{665.2}$$

$$c_{(x+c)}\,(\mu g/ml) = \left(1000\,A_{470} - 1.91\,c_a - 95.15\,c_b\right)/225$$

INTERPRETATION OF CHLOROPHYLL AND CAROTENOID CONTENT

The concentration of Chl *a* and *b* in plant material can be quantified with different reference systems. Reference systems currently in use include mg Chl *a+b*/m² leaf area (or µg/cm² leaf area), µg Chl *a+b*/g dry weight, and mg Chl *a+b*/g fresh weight (less suitable than dry weight).

When comparing results with those of other groups or with values obtained previously, the same reference system must be applied. Changes in Chl content should be demonstrated by means of a reference that does not change, otherwise an observed variation of data may not be due to changes in Chl concentration, but instead to changes in the reference system. For instance, an increase in Chl per fresh weight (in leaves or fruits) could be solely due to a decrease in fresh weight caused by water loss. In various cases, the number of leaves, cotyledon pairs, seedlings (shoots), or fruits may be the best reference system to follow changes in pigment levels, as these numbers do not change when dry weight or leaf area vary.

The weight ratio of Chl *a* and Chl *b* (Chl *a/b* ratio) is an indicator of the functional pigment

Table F4.3.2 Leaves with High Versus Low Chlorophyll *a/b* Ratios

High *a/b* ratio	Low *a/b* ratio
Greening of etiolated leaves (4.0-10)	Fully developed green leaves (2.5-3.5)
Sun leaves (3.0-3.8)	Shade leaves (2.4-2.7)
Leaves of C_4 plants (3.0-5.0)	Leaves of C_3 plants (2.5-3.5)

equipment and light adaptation of the photosynthetic apparatus (Lichtenthaler et al., 1981). Chl *b* is found exclusively in the pigment antenna system, whereas Chl *a* is present in the reaction centers of photosystems I and II and in the pigment antenna. Whereas the light-harvesting pigment protein LHC-I of the photosynthetic pigment system PS I has an *a/b* ratio of ~3, that of LHC-II of PS II exhibits an *a/b* ratio of 1.1 to 1.3. The level of LHC-II of PS II is variable and shows a light adaptation response. Shade plants possess much higher amounts of LHC-II than sun-exposed plants and, consequently, their *a/b* ratios are lower than in sun-exposed plants (Lichtenthaler et al., 1982, 1984). Thus, a decrease in the Chl *a/b* ratio may be interpreted as an enlargement of the antenna system of PS II. Some examples for high and low Chl *a/b* ratios in leaves of different developmental stages and in fully differentiated leaves grown at low light or high light conditions are given in Table F4.3.2.

The weight ratio of Chls *a* and *b* to total carotenoids $(a+b)/(x+c)$ is an indicator of the greenness of plants. The ratio $(a+b)/(x+c)$ normally lies between 4.2 and 5 in sun leaves and sun-exposed plants, and between 5.5 and 7.0 in shade leaves and shade-exposed plants. Lower values for the ratio $(a+b)/(x+c)$ are an indicator of senescence, stress, and damage to the plant and the photosynthetic apparatus, which is expressed by a faster breakdown of Chls than carotenoids. Leaves become more yellowish-green and exhibit values for $(a+b)/(x+c)$ of 3.5, or even as low as 2.5 to 3.0 as senescence progresses. Also, during chromoplast development in ripening fruits or fruit scales, which turn from green to yellow or orange or red, the ratio $(a+b)/(x+c)$ decreases continuously and reaches values below 1.0.

Sun leaves of different trees exhibit average Chl *a+b* levels of 400 to 700 mg/m^2 leaf area (40 to 70 μg/cm^2) and shade leaves have 380 to 570 mg/m^2 leaf area (38 to 57 μg/cm^2). As sun leaves possess thicker cell walls, a lower leaf

Table F4.3.3 Examples of Chlorophyll and Carotenoid Levels and Pigment Ratios in Green Sun and Shade Leaves[a]

Leaf type		$a + b$ (mg/m^2)	x + c (mg/m^2)	$a + b$ (mg/g dw)	x + c (mg/g dw)	*a/b*	$(a + b)/(x + c)$
Fagus sylvatica (beech)	Sun leaves	510.8	126.4	6.29	1.56	3.22	4.04
	Shade leaves	450.1	85.8	12.01	2.29	2.65	5.25
Carpinus betulus (hornbeam)	Sun leaves	571.0	117.4	8.15	1.68	3.20	4.86
	Shade leaves	431.1	70.8	19.05	3.13	2.45	6.09
Populus nigra (poplar)	Dark green sun leaves	724.4	161.5	8.03	1.81	3.30	4.44
	Dark green shade leaves	568.2	109.2	12.41	2.39	2.74	5.20
	Green senescent leaves	351.5	87.4	5.00	1.24	3.08	4.02
	Yellowish-green senescent leaves	140.3	79.4	1.99	1.13	3.29	1.77

[a]Pigment levels given in mg/m^2 leaf area and in mg/g dry weight (dw). Values measured are those from fully developed leaves in June, 2000. Pigment levels within one leaf usually vary by <3%, and pigment ratios vary by <1%. Abbreviations: *a + b*; total chlorophylls *a* and *b*; x + c, xanthophylls and carotenes (total carotenoids).

F4.3.7

water content (50% to 65% fresh weight), and higher dry weight than shade leaves, they exhibit on a dry weight basis a considerably lower Chl and carotenoid content than shade leaves (Table F4.2.3). The latter, in turn, possess a higher water content (68% to 85% fresh weight) and, consequently, a lower dry weight than sun exposed leaves.

LITERATURE CITED

Arnon, D.I. 1949. Copper enzyme in isolated chloroplast polyphenoloxidase in *Beta vulgaris*. *Plant Physiol.* 24:1-15.

Lichtenthaler, H.K. 1987. Chlorophylls and carotenoids: Pigments of photosynthetic biomembranes. *Methods Enzymol.* 148:350-382.

Lichtenthaler, H.K. and Pfister, K. 1978. Praktikum der Photosynthese. Quelle & Meyer, Heidelberg (in German).

Lichtenthaler, H.K., Buschmann, C., Döll, M., Fietz, H.-J., Bach, T., Kozel, U., Meier, D., and Rahmsdorf, U. 1981. Photosynthetic activity, chloroplast ultrastructure, and leaf characteristics of high-light and low-light plants and of sun and shade leaves. *Photosynthesis Res.* 2:115-141.

Lichtenthaler, H.K., Kuhn, G., Prenzel, U., Buschmann, C., and Meier, D. 1982. Adaptation of chloroplast-ultrastructure and of chlorophyll-protein levels to high-light and low-light growth conditions. *Z. Naturforsch.* 37c:464-475.

Lichtenthaler, H.K., Meier, D., and Buschmann, C. 1984. Development of chloroplasts at high and low light quanta fluence rates. *Isr. J. Botany* 33:185-194.

Šesták, Z. 1971. Determination of chlorophylls *a* and *b*. *In* Plant Photosynthetic Production: Manual of Methods (Z. Šesták, J. Catsky, and P.G. Jarvis, eds.) pp. 672-701. Dr. W. Junk Publishers, The Hague.

Smith, J.H.C. and Benitez, A. 1955. Chlorophylls: Analysis in plant material. *In* Modern Methods of Plant Analysis (K. Paech and M.V. Tracey, eds.) pp. 142-196. Springer, Berlin.

KEY REFERENCES

Lichtenthaler, H.K. 1982. Synthesis of prenyllipids in vascular plants (including chlorophylls, carotenoids, prenylquinones). *In* CRC Handbook of Biosolar Resrouces, Vol. I, part I: Basic Principles (A. Matsui and C.C. Black, eds.) pp. 405-421. CRC Press, Boca Raton, Fla.

Presents a table (Table 7) of chloropyll, carotenoid, and vitamin E levels (in µg/g dw) of green leaf tissue, vegetables, green and red fruits (tomato, red pepper), and nongreen plant foods (carrots, cauliflower).

Lichtenthaler, 1987. See above.

Presents redetermined absorption coefficients for chlorophylls and total carotenoids, which allows the determination of all three in the same pigment extract of leaves or fruits

Šesták, 1971. See above.

Gives basic information on the measurements of chlorophylls in various spectroscopic instruments.

Contributed by Hartmut K. Lichtenthaler and Claus Buschmann
Universitaet Karlsruhe
Karlsruhe, Germany

Chromatographic Separation of Chlorophylls

Numerous methods for the analysis of chlorophyll in plant tissues have been developed. Traditionally, chromatographic techniques have always been extensively utilized for analysis of these green photosynthetic pigments. While methods such as open-column and thin-layer chromatography have been extensively utilized in the past, high-performance liquid chromatography (HPLC) has dominated separation techniques over the last 20 years. The fast, accurate, and reproducible nature of these methods makes them ideal for both research and quality assurance needs. Many normal- and reversed-phase methods have been developed for analysis of chlorophylls and their derivatives. Over the last few decades, numerous procedures employing an octadecyl-bonded stationary phase (C18) have appeared in the literature. These reversed-phase (RP) methods have predominated mainly because of their ease of use, aqueous mobile phases, and the wide commercial availability of the stationary phase, making published methodology extremely transferable and reproducible.

Major method development has concentrated on the analysis of chlorophylls and their degradation products, specifically pheophytins and pyropheophytins, since they are readily formed by thermal processing of food such as spinach and green beans. The Basic Protocol described in this unit specifically targets the separation and identification of all major chlorophyll derivatives including chlorophylls *a* and *b*, pheophytins *a* and *b*, and pyropheophytins *a* and *b*. Similar methodology has been developed for the analysis of specific polar derivatives and Cu^{2+} and Zn^{2+} derivatives, and is described in the Alternate Protocol. For a general discussion and structures of chlorophylls and their degradation products, see *UNIT F4.1*.

NOTE: Chlorophylls and their derivatives are both light and temperature sensitive. All work should be carried out under subdued lighting and at a controlled ambient temperature to avoid their degradation.

STRATEGIC PLANNING

When discussing analysis by HPLC, chlorophylls may be divided into four main groups as follows.

1. Predominant natural chlorophylls *a* and *b*.

2. Nonpolar chlorophyll derivatives such as pheophytins and pyropheophytins.

3. Polar chlorophyll derivatives such as chlorophyllides and pheophorbides.

4. Major metalloporphyrin derivatives such as Cu^{2+} and Zn^{2+} pheophytins.

The protocols described in this unit focus on the separation of these four groups. While the Basic Protocol describes a method capable of separating all constituents of groups 1 and 2, it may not be perfectly suited for analysis of all polar derivatives. Therefore, it is important to have an idea in advance of which chlorophyll compounds may be in the test sample, to allow for the application of the appropriate protocols. Prior to analysis, it is important to have an estimate of the sample extract concentration to allow for optimal use of the described protocols. Such prior knowledge allows better judgment of injection and dilution volumes, which are important factors for good resolution of chlorophyll derivatives. Finally, general laboratory conditions (e.g., subdued lighting and ambient temperature) should be carefully controlled when extracting and working with chlorophylls (*UNIT F4.2*).

Contributed by Mario Ferruzzi and Steve Schwartz

Chlorophylls

F4.4.1

C18 REVERSED-PHASE HPLC SEPARATION OF CHLOROPHYLLS *a* AND *b* AND THEIR NONPOLAR DERIVATIVES

This protocol focuses on the analysis of chlorophyll *a* and *b*, and the more nonpolar derivatives, including pheophytins and pyropheophytins. An octadecyl-bonded, reversed-phase stationary phase is used with a methanol/water mixture and ethyl acetate mobile phases in a gradient elution to provide rapid and complete separation of the major chlorophyll derivatives in 25 to 30 min. This is coupled with traditional UV/visible spectrophotometric detection at 654 nm to selectively screen these photosynthetic pigments in food and plant tissues.

Materials

HPLC-grade solvents:
  Methanol
  H$_2$O
  Ethyl acetate
  Acetone
Sample
Appropriate chlorophyll standards (Sigma-Aldrich; also see UNIT F4.2)

High-performance liquid chromatograph (HPLC) including:
  Appropriate gradient-capable solvent delivery system
  Sample injection valve
  Variable wavelength UV/VIS spectrophotometer suitable for detection at 654 nm
  Chromatographic data collection system
  Precolumn solvent filter
  Appropriate HPLC guard column and analytical column packed with octadecyl-bonded (C18) stationary phase (e.g., μBondapak C18, Waters; 201TP54, Vydac)
5-μm syringe filters
Disposable 5-ml syringes

Additional reagents and equipment for sample preparation (UNIT F4.2) and HPLC (e.g., Coligan et al., 2000)

Set up HPLC system

1. Set up an HPLC according to the manufacturer's instructions.

   Coligan et al. (2000), provides much information about general HPLC procedures.

2. Purge system fully of air and then install a guard column and an analytical column in-line.

3. Place a precolumn solvent filter between the guard column and analytical column.

   Use of a guard column is considered standard for HPLC as it greatly extends analytical column life.

4. Equilibrate analytical column ≥15 min with appropriate initial gradient conditions as described in Table F4.4.1 or Table F4.4.2.

   All solvents should be fully degassed by vacuum, helium sparging, or sonication prior to use.

   Ternary/quaternary versus binary solvent capabilities are instrument dependent.

5. Monitor baseline signal at 654 nm during equilibration to ensure that it is stable.

   For highly sensitive analyses, allow extra time (e.g., up to 1 hr) for equilibration to ensure an extremely stable baseline.

6. Check data collection system to ensure that it is operational.

Prepare sample

7. Extract chlorophylls from a sample as described in UNIT F4.2.

8. Dissolve dry extracts in 100% acetone.

   The volume of acetone is dependent on the original concentration of chlorophyll in the sample, the volume of the dried extract, and the desired concentration range for the analysis. Highly concentrated samples such as spinach may need to be diluted to ensure adherence to the linear range as prescribed by the response curves for each specific chlorophyll.

9. Pass extract through a 5-µm filter using a disposable 5-ml syringe.

Perform HPLC analysis

10. Load appropriate volume of filtered sample onto injector valve according to manufacturer's instructions.

    Injection volume will vary according to sample. For typical HPLC systems, injection volumes range between 1.0 and 100 µl. It is important that the analytical column not be overloaded. Therefore, when exploring conditions for a new sample, always start with lower injection volumes and adjust to high volumes as needed. As a rule of thumb for most analyses, injection volumes between 25 and 50 µl work best.

11. Program solvent gradient conditions as described in Table F4.4.1 or Table F4.4.2.

    As indicated in Tables F4.4.1 and F4.4.2, all gradients in this unit are simple linear gradients unless noted otherwise. This is to allow for maximum application of the described methodology to different HPLC systems.

12. Initiate chromatographic run as described by manufacturer.

13. Collect chromatographic data from specified collection system as described by manufacturer.

    Identification of chlorophylls and their derivatives can be made more efficient by use of a photodiode array detector, which allows collection of on-line spectra. However, this

Table F4.4.1 Linear Gradient Conditions for Separation of Chlorophylls and Their Nonpolar Derivatives Using a Ternary/Quaternary Solvent Delivery System

Time (min)	Flow rate (ml/min)	% Methanol	% Water	% Ethyl acetate
0.0	1.0	75.0	25.0	0.0
10.0	1.0	37.5	12.5	50.0
25.0	1.0	37.5	12.5	50.0
30.0	1.0	75.0	25.0	0.0

Table F4.4.2 Linear Gradient Conditions for Separation of Chlorophylls and Their Nonpolar Derivatives Using a Binary Solvent Delivery System[a]

Time (min)	Flow rate (ml/min)	% Solvent A	% Solvent B
0.0	1.0	100.0	0.0
10.0	1.0	50.0	50.0
25.0	1.0	50.0	50.0
30.0	1.0	100.0	0.0

[a]Solvent A, 75:25 (v/v) methanol/water; solvent B, ethyl acetate.

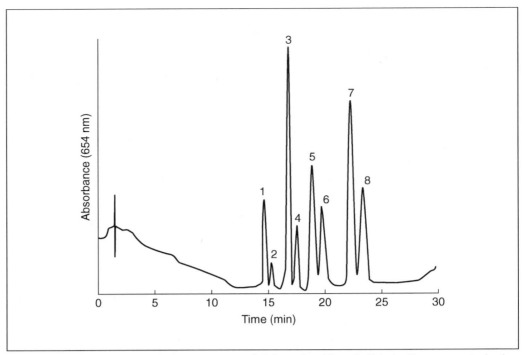

Figure F4.4.1 Typical HPLC chromatogram of eight major chlorophyll derivatives separated using the Basic Protocol. Peak identifications: 1, chlorophyll *b*; 2, chlorophyll *b'*; 3, chlorophyll *a*; 4, chlorophyll *a'*; 5, pheophytin *b*; 6, pyropheophytin *b*; 7, pheophytin *a*; 8, pyropheophytin *a*.

remains a luxury item for this method, which can still be carried out using a simple variable-wavelength detector capable of detection at 654 nm, and authentic standards for comparison and identification.

Baseline drift is a normal effect of gradient delivery (see Critical Parameters and Troubleshooting) and therefore a blank (100% HPLC-grade acetone) should be run and subtracted from the sample, if desired.

14. Run appropriate chlorophyll standards. Calculate chlorophyll concentrations from generated peak integration data according to the response curves for each specific standard.

An example of an HPLC chromatogram obtained using these conditions is given in Figure F4.4.1.

ALTERNATE PROTOCOL

C18 REVERSED-PHASE HPLC SEPARATION OF POLAR CHLOROPHYLL DERIVATIVES AND Cu^{2+} AND Zn^{2+} PHEOPHYTINS

Polar chlorophyll derivatives and metalloporphyrin derivatives such as Cu^{2+} and Zn^{2+} pheophytins can also be analyzed by C18 reversed-phase HPLC. Appropriate standards must be used; see *UNIT F4.2* for polar chlorophyll derivatives, or see Support Protocol 2 for Cu^{2+} and Zn^{2+} pheophytin standards. Gradient solvent conditions and flow rates are given in Tables F4.4.3 and F4.4.4. Otherwise, the separation is performed as described for chlorophylls and nonpolar derivatives (see Basic Protocol). Using this method, separation of polar chlorophyll derivatives can be achieved in 20 to 25 min, and separation of the metalloporphyrin derivatives in 20 to 25 min. Examples of chromatograms obtained for polar derivatives, Zn^{2+} pheophytins, and Cu^{2+} pheophytins are shown in Figures F4.4.2, F4.4.3, and F4.4.4, respectively.

Chromatographic Separation of Chlorophylls

F4.4.4

Table F4.4.3 Linear Gradient Conditions for Separation of Polar Chlorophyll Derivatives[a]

Time (min)	Flow rate (ml/min)	% Solvent A	% Solvent B
0.0	1.3	100.0	0.0
6.0	1.3	100.0	0.0
7.0	1.5	70.0	30.0
10.0	1.5	70.0	30.0
11.0	1.5	60.0	40.0
12.0	1.5	50.0	50.0
13.0	1.5	0.0	100.0
24.0	1.5	0.0	100.0
26.0	1.5	100.0	0.0

[a]Solvent A, 15:65:20 (v/v/v) ethyl acetate/methanol/water; solvent B, 60:30:10 (v/v/v) ethyl acetate/methanol/water.

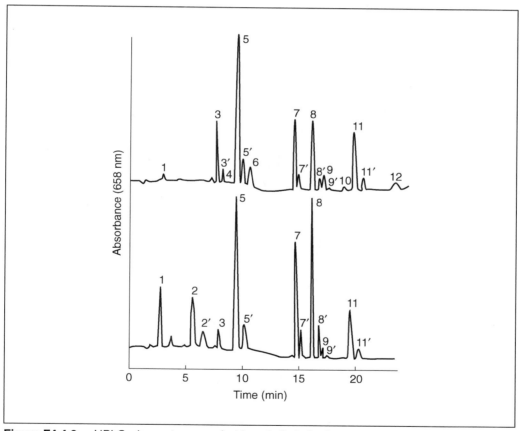

Figure F4.4.2 HPLC chromatogram of chlorophyll derivatives separated using the Alternate Protocol. Peak identifications: 1, chlorophyllide *b'*; 2, chlorophyllide *a*; 2', chlorophyllide *a'*; 3, pheophorbide *b*; 3', pheophorbide *b'*; 4, pyropheophorbide *b*; 5, pheophorbide *a*; 5', pheophorbide *a'*; 6, pyropheophorbide *a*; 7, chlorophyll *b*; 7', chlorophyll *b'*; 8, chlorophyll *a*; 8', chlorophyll *a'*; 9, pheophytin *b*; 9', pheophytin *b'*; 10, pyropheophytin *b*; 11, pheophytin *a*; 11', pheophytin *a'*; 12, pyropheophytin *a*. Reproduced from Canjura et al. (1991) with permission from the Institute of Food Technologists.

Table F4.4.4 Linear Gradient Conditions for Separation of Cu^{2+} and Zn^{2+} Pheophytins[a]

Time (min)	Flow rate (ml/min)	% Solvent A	% Solvent B
0.0	2.0	55.0	45.0
15.0	2.0	50.0	50.0
16.0	2.0	55.0	45.0

[a]Solvent A, 75:25 (v/v) methanol/water; solvent B, ethyl acetate.

Table F4.4.5 Step Gradient Conditions for C18 Column Clean-Up

Time (min)	% Methanol	% Water	% Ethyl acetate
0.0	75.0	25.0	0.0
15.0	100.0	0.0	0.0
30.0	50.0	0.0	50.0
45.0	0.0	0.0	100.0
75.0	50.0	0.0	50.0
90.0	100	0.0	0.0
105.0	75.0	25.0	0.0

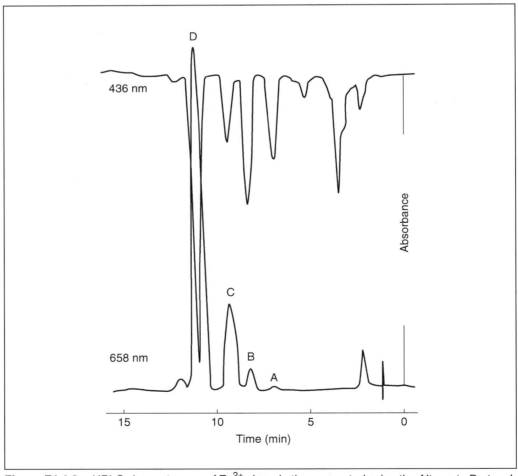

Figure F4.4.3 HPLC chromatogram of Zn^{2+} pheophytins separated using the Alternate Protocol. Peak identifications: A, allomerized Zn^{2+} pheophytin b; B, Zn^{2+} pheophytin b; C, allomerized Zn^{2+} pheophytin a; D, Zn^{2+} pheophytin a. Reproduced from Schwartz (1984) with permission from Marcel Dekker, Inc.

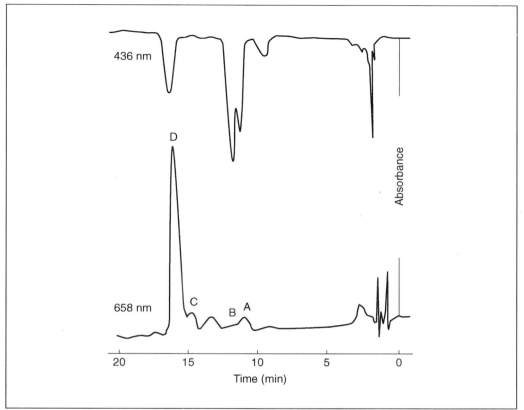

Figure F4.4.4 HPLC chromatogram of Cu^{2+} pheophytins separated using the Alternate Protocol. Peak identifications: A, allomerized Cu^{2+} pheophytin *b*; B, Cu^{2+} pheophytin *b*; C, allomerized Cu^{2+} pheophytin *a;* D, Cu^{2+} pheophytin *a.* Reproduced from Schwartz (1984) with permission from Marcel Dekker, Inc.

C18 COLUMN CLEANING

Between analyses, and at the end of each day, the HPLC analytical column should be reequilibrated with the appropriate initial gradient conditions until system pressure returns to initial operating conditions. However, on occasion, the analytical column may require a more rigorous cleaning procedure as described below in order to ensure continued optimal performance. Examples of conditions where this may be required are listed in Table F4.4.6.

1. When analysis is complete, allow system to equilibrate analytical column with appropriate initial gradient conditions as stipulated in Table F4.4.1 or Table F4.4.2 for ≥15 min.

2. Program clean-up gradient conditions as described in Table F4.4.5.

   A step gradient is used for this method whereby each of the mobile phase compositions described in Table F4.4.5 is used for 15 min, with the exception of 100% ethyl acetate, which is used for 30 min.

3. Initiate and run the clean-up gradient at a flow rate of 1.0 ml/min.

   The run will take 2 hr.

4. Upon completion of the clean-up gradient, reequilibrate analytical column with appropriate initial gradient conditions for ≥15 min.

   If the column is being prepared for storage, 100% methanol should be used for equilibration for ≥30 min prior to removal and sealing. This will help ensure longer column life.

SYNTHESIS OF Cu^{2+} AND Zn^{2+} PHEOPHYTIN STANDARDS

The use of appropriate analytical standards is important for successful chromatographic separation, identification, and quantification of chlorophyll derivatives. While chlorophyll *a* and *b* derivatives are readily available commercially (Sigma-Aldrich) both metal-free pheophytins and metalloporphyrin analogs such as Cu^{2+} and Zn^{2+} pheophytins are not. In most instances, these derivatives must be prepared from the parent Mg-chlorophyl standards prior to use. These simple synthesis techniques are based on the work of Schwartz (1984) and are to be utilized for the rapid and efficient preparation of metal-free, Cu^{2+} and Zn^{2+} pheophytin derivatives in quantities appropriate only for analytical implementation.

Additional Materials (*also see Basic Protocol*)

Concentrated hydrochloric acid (HCl; ~37% solution)
HPLC-grade diethyl ether
Zinc chloride
2.5 M $CuCl_2$ in HPLC-grade H_2O
Anhydrous sodium sulfate
Nitrogen gas
Glass vials ($\geq$11 ml; at experimentor's discretion)

1. Dissolve enough crystalline chlorophyll *a* standard in a total volume of 10 ml HPLC-grade acetone to give an absorbance of ~1.0 at 661 nm. Repeat with chlorophyll *b* (646 nm).

    Chlorophyll a and b standards can be purified from plant tissue as outlined in UNIT F4.2. Standards are also available commercially from Sigma-Aldrich.

2. Acidify each solution with 250 µl concentrated HCl.

3. Gently mix for ~10 min.

4. Extract pheophytins with 5.0 ml HPLC-grade diethyl ether. Add 1.0 ml HPLC-grade water to hasten phase separation.

5. Remove residual acid from diethyl ether layer by three successive washings with 10.0 ml water and transfer pigment-rich ether layer to a clean glass vial.

6. Dry pigment-rich diethyl ether layer by passing through approximately 50 to 100 mg of anhydrous sodium sulfate. Remove solvent under a stream of nitrogen.

7. Redissolve pheophytin *a* and *b* extracts in 4.0 ml acetone each.

8. Add 1.0 ml of 2.5 M $CuCl_2$ to a 2.0-ml aliquot of each pheophytin extract.

9. Add 0.3 g zinc chloride to a 2.0-ml aliquot of each extract.

    The reactions should go to completion within 30 min.

10. Assay the extent of complex formation and final purity by HPLC (see Alternate Protocol).

    Successful formation of Cu^{2+} or Zn^{2+} pheophytin would result in only one major chromatographic peak (>90% of total peak area). The presence of other peaks most likely results from incomplete complexation or allomerization (oxidation) of the chlorophyll molecules. In these instances further chromatographic purification (based on Alternate Protocol) may be necessary in order to achieve the desired degree of analytical purity.

11. Extract the derivatives as described in steps 4 to 6 and store dried and desiccated between −20° and −80°C until use (up to 2 to 3 months).

    Prior to use as authentic standards, the desired Cu^{2+} and Zn^{2+} pheophytins will have to be purified from other impurities stemming from the synthesis. This can be accomplished by simple preparatory or semipreparatory phase chromatography using the method described (see Alternate Protocol). If only a small amount of sample is required, collection of a few fractions from a separation using a standard analytical column may suffice.

Observed problem	Possible cause	Solution
Erratic baseline signal	Incorporation of air in mobile phase	Properly degas solvents as instructed by manufacturer
	Equilibration time too short	Allow 60 min for equilibration
Baseline drift	Normal effect of gradient delivery	Subtract blank from baseline
	Sample overload	Reduce injection volume
	Buildup of retained compounds on column	Clean column (see Support Protocol 1)
Erratic pressure reading (ΔP >100 psi)	Air trapped in pump head	Purge system as instructed by manufacturer
	Slow solvent leak in system	Check all fittings for leaks; tighten or replace if needed
Large pressure drop (ΔP >200 psi)	Solvent leak in system	Check entire system for leaks, including pump, injector, detector
Large pressure increase (ΔP >200 psi)	System fouling	Replace fouled line or filter with clean part
	Solvents prepared incorrectly	Prepare new solvents
	Precipitation of sample on column	Reequilibrate system and reduce injection volume or dilute sample
	Buildup of retained compounds on column	Clean column (see Support Protocol 1)
	Guard column failure	Replace guard column
	Analytical column failure	Replace analytical column
Precipitation of sample in injection vial	Solvent incompatibility	Reextract sample and dissolve in acetone (*UNIT F4.2*)
	Sample temperature too low	Ensure sample is at ≥20°C
	Analyte too concentrated	Dilute sample
Loss of chromatographic resolution	Buildup of retained compounds on analytical column	Clean column (see Support Protocol 1)
	Improper gradient delivery	Verify that gradient is programmed and working correctly
	Solvents prepared incorrectly	Prepare new solvents
	Injection volume too large or concentration too high	Reduce injection volume
	Column temperature too high	Reduce temperature to ≤25°C
	Degradation of analytical column	Replace analytical column
Chromatographic artifacts and apparent peak splitting	Solvent incompatibility	Reextract sample and dissolve in acetone (*UNIT F4.2*)
	Sample degradation	Carry out all work under subdued light and inject sample immediately upon being redissolved
	Detection at incorrect wavelength	Set detector to 654 nm

COMMENTARY

Background Information

The accurate measurement of chlorophylls has importance for numerous reasons ranging from simple color considerations to medical research. The most practical of these is the assessment of fruit and vegetable color quality, as chlorophylls are known to degrade rapidly when subjected to thermal processing (Schwartz et al., 1981). However, interest has also been sparked by recent literature reports that point to the possible health benefits associated with chlorophyll consumption (Hartig and Bailey, 1998).

Tswett was the first to accomplish chromatography of chloroplast pigments in 1903. Over time, advances have been made leading to the development of HPLC. With HPLC becoming the definitive method by which chlorophyll analysis is presently performed, it is not surprising that numerous methods exist in the literature. Techniques range in application from fruits and vegetables to shipboard seawater analysis. Eskins et al. (1977) developed an excellent method for chlorophyll a and b analysis in fruit. Mantoura and Llewellyn (1983) analyzed fruits and vegetables for chlorophyll and carotenoids as well as their breakdown products. However, these methods are often found to be complex, labor intensive, and time consuming, with long run times. More recent methods have focused on other pigments as well as chlorophylls. Minguez-Mosquera et al. (1992) described a method to separate chlorophylls and carotenoids from virgin olive oil that quantified seventeen pigments including seven chlorophyll derivatives.

The widespread application of HPLC methodology to chlorophyll analysis demonstrates its flexibility, effectiveness, and reliability. Methods described in this unit are based on the work of Schwartz et al. (1981). This original method allows for resolution of twelve chlorophyll derivatives in 30 min. While minor modifications were made to this method to further simplify the analysis, the final resolution and sensitivity were not compromised. Based on a commercially available reversed-phase column and an aqueous mobile phase, the method can be easily altered for specific separations. This is demonstrated in the Alternate Protocol, where the Basic Protocol was adjusted for the analysis of polar and Cu^{2+}- and Zn^{2+}-containing chlorophyll derivatives. The method for polar derivatives is based on the separation of Canjura and Schwartz (1991), while the method for

Cu^{2+} and Zn^{2+} pheophytin analysis is based on Schwartz (1984). Together, these methods allow the analyst to set up a chromatographic system based on a single stationary phase that can easily be modified for flexible and complete analysis of relevant chlorophyll derivatives. When combined with the appropriate extraction methodology, these protocols can be applied to any application relevant to chlorophyll analysis.

Critical Parameters and Troubleshooting

HPLC techniques often come with a list of potential problems. A troubleshooting guide is summarized in Table F4.4.6. One of the most important considerations is to know the limitations of the HPLC system itself. Pressure limits, solvent compatibility, gradient capability, and other parameters should all be confirmed as described in the manufacturer's instructions. However, monitoring the system pressure is a good practice that can allow for an early detection of possible problems. Generally, it is normal for system pressure to fluctuate due to changes in mobile phase composition. One should therefore establish a normal working range for this important parameter. For example, the system pressure in the Basic Protocol may be 1200 psi under initial conditions, drop to 700 psi at 20 min, and then reach 1200 psi just at 30 min. Therefore, the working range of 700 to 1200 psi should be maintained throughout the run. If the pressure increases or drops dramatically beyond this range, maintenance may be necessary (Table F4.4.6).

Anticipated Results

Typical chromatograms from all protocols described in this unit can be seen in Figures F4.4.1 to F4.4.4.

Time Considerations

The overall analysis time is dependent on a few factors, one of which is column length. With a standard analytical column length of 250 mm, run times of 25 to 30 min can be expected. Shorter, 150-mm columns may be used to shorten the run time; however, some resolution can be lost when using this approach. One must also account for preparation time for HPLC analysis. This includes steps such as sample preparation, instrument equilibration, and data analysis. This generally takes 1 hr per sample for instrumental and data analysis, plus ~1 hr for each full day's analysis for initial instrument equilibration. Finally, be sure to allocate time for sample preparation (*UNIT F4.2*).

Literature Cited

Canjura, F.L. and Schwartz, S.J. 1991. Separation of chlorophyll compounds and their polar derivatives by high performance liquid chromatography. *J. Agric. Food Chem.* 39:1102-1105.

Canjura, F.L, Schwartz, S.J., and Nunes, R.V. 1991. Degradation kinetics of chlorophylls and chlorophyllides. *J. Food Sci.* 56:1639-1643.

Coligan, J.E., Dunn, B.M., Ploegh, H.L., Speicher, D.W., and Wingfield, P.T. (eds.) 2001. Current Protocols in Protein Science. John Wiley & Sons, New York.

Eskins, K., Scholfield, C.R., and Dutton, H.J. 1977. High performance liquid chromatography of plant pigments. *J. Chromatogr.* 135:217-220.

Hartig, H. and Bailey, G.S. 1998. Chemoprevention by natural chlorophylls in vivo: Inhibition of dibenzo[a,l]pyrene-DNA adducts in rainbow trout liver. *Carcinogenesis* 19:1323-1326.

Mantoura, R.F.C. and Llewellyn, C.A. 1983. The rapid determination of algal chlorophyll and carotenoid pigments and their breakdown products in natural waters by reverse phase high performance liquid chromatography. *Anal. Chim. Acta* 151:297-314.

Minguez-Mosquera, M.I., Gandul-Rojas, B., and Gallardo-Guerrero, M.L. 1992. Rapid method of quantification of chlorophylls and carotenoids in virgin olive oil by high performance liquid chromatography. *J. Agric. Food Chem.* 40:60-63.

Schwartz, S.J. 1984. High performance liquid chromatography of zinc and copper pheophytins. *J. Liq. Chromatogr.* 7:1673-1683.

Schwartz, S.J., Woo, S.L., and von Elbe, J.H. 1981. High performance liquid chromatography of chlorophylls and their derivatives in fresh and processed spinach. *J. Agric. Food Chem.* 29:533-535.

Key References

Schwartz et al., 1981. See above.

The original method described in the Basic Protocol and modified for the Alternate Protocol. Information regarding sample preparation is also described.

Contributed by Mario Ferruzzi and
    Steve Schwartz
The Ohio State University
Columbus, Ohio

Mass Spectrometry of Chlorophylls

Mass spectra of chlorophylls and their derivatives have been obtained using a variety of desorption ionization methods including laser desorption (Tabet et al., 1985), matrix-assisted laser desorption/ionization (MALDI; Liu et al., 1999), field desorption (Dougherty et al., 1980), plasma desorption (Hunt et al., 1981), and fast atom bombardment (FAB; van Breemen et al., 1991a; Hyvarinen and Hynninen, 1999; Teng and Chen, 1999). Classical ionization techniques such as electron impact and chemical ionization have been less useful for the analysis of these thermally labile compounds. Liquid chromatography/mass spectrometry (LC/MS) has been applied to the on-line separation of chlorophylls and their degradation products, first using continuous-flow FAB mass spectrometry (van Breemen et al., 1991b) and subsequently using electrospray ionization (ESI; Zissis et al., 1999) and atmospheric pressure chemical ionization (APCI; Airs and Keely, 2000; Verzegnassi et al., 2000). FAB and APCI have been used for ionization of chlorophylls followed by tandem mass spectrometric analysis (MS/MS; van Breemen et al., 1991a; Airs and Keely, 2000). Since FAB, ESI, and APCI LC/MS have been the most widely used mass spectrometric techniques for the analysis of chlorophylls, protocols for these approaches are described in detail below.

FAST ATOM BOMBARDMENT (FAB), LIQUID SECONDARY ION MASS SPECTROMETRY (LSIMS), AND CONTINUOUS-FLOW FAB OF CHLOROPHYLLS

FAB and LSIMS are matrix-mediated desorption techniques that use energetic particle bombardment to simultaneously ionize samples like carotenoids and transfer them to the gas phase for mass spectrometric analysis. Molecular ions and/or protonated molecules are usually abundant and fragmentation is minimal. Tandem mass spectrometry with collision-induced dissociation (CID) may be used to produce abundant structurally significant fragment ions from molecular ion precursors (formed using FAB or any suitable ionization technique) for additional characterization and identification of chlorophylls and their derivatives. Continuous-flow FAB/LSIMS may be interfaced to an HPLC system for high-throughput flow-injection analysis or on-line LC/MS.

Materials

Matrix:
   3-nitrobenzyl alcohol (for static FAB or LSIMS)
   0.5% (w/v) glycerol dissolved in the mobile phase (for continuous-flow FAB or LSIMS)
Chlorophyll sample (1 to 100 mg/liter; 2 to 200 µM; see UNIT F4.2 for extraction protocols) dissolved in volatile organic solvent such as acetone, diethyl ether, or ethyl acetate (store in airtight glass vial)
HPLC mobile phase solvent (e.g., ethyl acetate, methanol, water, and/or acetone; for continuous-flow FAB/LSIMS LC/MS)

Mass spectrometer or tandem mass spectrometer (JEOL, Micromass, MAT from ThermoFinnigan) equipped with direct insertion probe and fast atom bombardment (FAB) or liquid secondary ion mass spectrometry (LSIMS); for LC/MS or flow injection using continuous-flow FAB, mass spectrometer must be equipped with continuous-flow ionization source
Microsyringe (for static FAB or LSIMS; 10-µl Hamilton or equivalent)

Contributed by Richard B. van Breemen

HPLC setup (for continuous flow FAB/LSIMS LC/MS; also see *UNIT F4.4*) including:

Syringe pump or HPLC pump capable of flow rates of 1 to 10 µl/min

Reversed-phase (typically C18) column (since flow rate into continuous-flow FAB source must be <10 µl/min, capillary column must be used or flow must be split post column)

Additional reagents and equipment for HPLC of chlorophylls (*UNIT F4.4*)

For static FAB or LSIMS analysis

1a. Load 0.5 to 1 µl of the 3-nitrobenzyl alcohol matrix onto the sample probe using a microsyringe.

> *The 3-nitrobenzyl alcohol matrix is typically stable at room temperature for ~3 months. Older solutions begin to turn yellow as they decompose and should be replaced.*

> *Although other matrices such as glycerol may be used, 3-nitrobenzyl alcohol is the most effective matrix that has been reported for static FAB or LSIMS of chlorophylls.*

2a. Load 1 µl of the chlorophyll sample onto the surface of the liquid matrix. Let solvent evaporate.

3a. Insert probe through the vacuum interlock into the ion source of the mass spectrometer.

4a. Turn on the FAB or LSIMS beam and record the positive ion mass spectrum over the range *m/z* 100 to 1000. If desired, eliminate the 3-nitrobenzyl alcohol ion at *m/z* 154 by scanning from *m/z* 200 to 1000.

> *Static FAB/LSIMS provides a continuous signal for chlorophylls and their derivatives so that exact mass measurements may be carried out to determine elemental compositions, or tandem mass spectra following CID may be recorded.*

For continuous-flow FAB/LSIMS LC/MS

1b. Set up HPLC system using reversed-phase (i.e., C18) column and interface with mass spectrometer according to the manufacturer's instructions.

> *UNIT F4.4 provides details on HPLC of chlorophylls.*

2b. Run mobile phase containing 0.5% glycerol matrix through system such that the flow rate into the mass spectrometer does not exceed 10 µl/min.

> *During LC/MS, the glycerol matrix is included in the mobile phase and does not interfere with reversed-phase chromatography or in-line absorbance detection of chlorophylls. Matrices other than glycerol might be suitable for continuous-flow FAB of chlorophylls, but none have been reported.*

3b. Load chlorophyll sample onto the HPLC injector as indicated by the manufacturer. See *UNIT F4.4* for additional details.

> *Since microbore HPLC columns are usually used for continuous flow FAB, the sample volume is usually 1-5 µl.*

4b. Acquire data by turning on the FAB or LSIMS beam and recording positive ion mass spectra over the range *m/z* 200 to 1000.

> *The scan rate should not exceed 8 sec/scan so as to preserve chromatographic resolution. Ideally, at least 8 mass spectra should be acquired per chromatographic peak. All mass spectra are stored separately in chronological order using LC/MS software, provided by the instrument manufacturer. These mass spectra may be retrieved and viewed later or used to produce computer-reconstructed ion chromatograms.*

APCI and ESI are atmospheric pressure interfaces between HPLC systems and mass spectrometers. These interfaces serve the dual functions of removing the carrier solvent from the HPLC eluate and ionizing the analyte. During APCI, the HPLC eluate is sprayed through a heated capillary into a chamber with a counter-current or cross-flow of heated nitrogen gas that facilitates the evaporation of solvent. A corona discharge in this chamber forms a steady state of solvent ions that ionize the analyte through collision processes similar to those in classical chemical ionization. During ESI, the HPLC eluate is sprayed through a capillary held at high potential relative to the surrounding chamber. As a result, electrospray droplets are smaller than those formed during APCI, and analyte ions are formed directly from the charged spray. LC/MS analyses of chlorophylls and related compounds have been reported using APCI (Airs and Keely, 2000; Verzegnassi et al., 2000) and ESI (Zissis et al., 1999).

Materials

> Chlorophyll sample (1 to 100 mg/liter; 2 to 200 μM; see *UNIT F4.2* for extraction protocols) dissolved in volatile organic solvent such as acetone, diethyl ether, or ethyl acetate (store in airtight glass vial)
> HPLC mobile phase: may include combinations of ethyl acetate, methanol, diethyl ether, methyl *tert*-butyl ether, acetonitrile, water, and/or acetone (all are compatible with APCI and ESI)
>
> HPLC system interfaced to a mass spectrometer or tandem mass spectrometer equipped with APCI or ESI (e.g., Agilent, Mircromass, ThermoFinnigan)
> C18 reversed-phase column for HPLC separations of chlorophylls: narrow-bore (2.1-mm i.d.) column at flow rate of 50 to 300 μl/min without splitting the flow, or analytical column (4.6-mm i.d.) at 1 ml/min with post-column solvent splitting of 5:1 (200 μl/min entering the mass spectrometer) for APCI; electrospray interfaces are available for use without solvent splitting over all flow rates from nl/min to 1 ml/min
> UV/VIS absorbance detector (single wavelength or diode array) placed in-line between HPLC column and mass spectrometer for additional characterization of eluting chlorophylls (optional)
>
> Additional reagents and equipment for HPLC of chlorophylls (*UNIT F4.4*)

1. Set up HPLC system and begin running mobile-phase solution at a flow rate of 200 μl/min into the APCI or ESI mass spectrometer.

   UNIT F4.4 provides details on HPLC of chlorophylls.

2. Inject a 1- to 5-μl aliquot of the chlorophyll solution onto the HPLC system using a microsyringe with a manual injector or using an autoinjector under computer control.

   HPLC separation of the chlorophyll mixture should be carried out using reversed-phase (usually C18 columns) as described in UNIT F4.4.

   Mass spectrometric or MS/MS detection of chlorophylls using APCI requires ~5 pmol of each compound per analysis (Airs and Keely, 2000).

3. *Optional:* Direct the eluate from the HPLC column through a UV/VIS absorbance detector prior to the mass spectrometer for additional sample characterization.

4. Divert the solvent front from the HPLC column, containing salts and unretained material, to waste instead of the mass spectrometer in order to minimize fouling of the LC/MS interface.

   LC/MS systems may be equipped with programmable switching valves for this purpose.

Chlorophylls

F4.5.3

5. Acquire LC/MS data by scanning the range *m/z* 200 to 1000 so that at least 8 mass spectra are acquired per chromatographic peak.

Mass spectra recorded using LC/MS software may be displayed individually, signal averaged, and background subtracted. Furthermore, these data may be used to plot computer-reconstructed selected ion or total ion chromatograms.

COMMENTARY

Background Information

FAB and LSIMS

In widespread use since 1982 (Barber et al., 1982), FAB and LSIMS are matrix-mediated techniques. The most effective matrix for static FAB/LSIMS analysis of chlorophylls and their derivatives is 3-nitrobenzyl alcohol (van Breemen et al., 1991a), whereas glycerol provides adequate sensitivity and a more robust system during continuous-flow FAB/LSIMS (van Breemen et al., 1991b). Ionization and desorption of the chlorophyll analyte occur together during the bombardment of the matrix by fast atoms (or ions) to produce molecular ions, $M^{+\cdot}$, and protonated molecules, $[M+H]^+$.

FAB ionization has been used in combination with LC/MS in a technique called continuous-flow FAB LC/MS (van Breemen et al., 1991b). Although any standard HPLC solvents may be used, including ethyl acetate, methanol, and water, the mobile phase should not contain nonvolatile additives such as phosphate or Tris buffers. Volatile buffers such as ammonium acetate are compatible. The low flow rate of continuous-flow FAB/LSIMS (<10 μl/min) is the primary limitation of this technique.

APCI and ESI

Since APCI and ESI interfaces operate at atmospheric pressure and do not depend upon vacuum pumps to remove solvent vapor, they are compatible with a wide range of HPLC flow rates. HPLC methods that have been developed using conventional detectors such as UV/VIS, IR, or fluorescence are usually transferable to LC/MS systems without adjustment. However, the solvent system should contain only volatile solvents, buffers, or ion-pair agents to reduce fouling of the mass spectrometer ion source. In the case of chlorophyll solvent systems, isocratic and gradient combinations of methanol, acetonitrile, water, acetone, and/or ethyl acetate have been used for APCI or ESI LC/MS. Unlike continuous-flow FAB/LSIMS, no sample matrix is necessary.

The APCI interface uses a heated nebulizer to form a fine spray of the HPLC eluate and to facilitate solvent evaporation. In addition, a cross-flow of heated nitrogen gas is used to

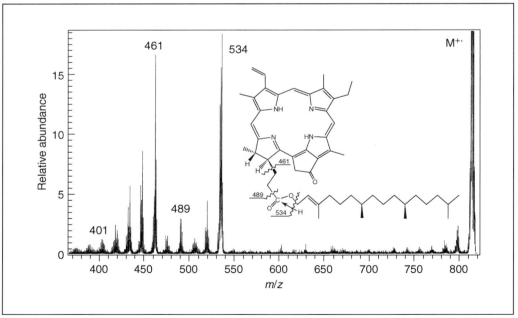

Figure F4.5.1 Positive ion fast atom bombardment (FAB) with collision-induced dissociation (CID) tandem mass spectrum of pyropheophytin *a* isolated from spinach leaves. The FAB matrix was 3-nitrobenzyl alcohol.

complete the evaporation of solvent from the droplets. The resulting gas-phase sample molecules are ionized by collisions with solvent ions, which are formed by a corona discharge in the atmospheric pressure chamber. Chlorophyll ions formed during APCI consist of cationized species such as protonated molecules, $[M+Na]^+$ and $[M+K]^+$, and positive and negative molecular ions. The relative abundance of each type of ion depends upon the solvent used, the presence of proton donors or acceptors, the levels of alkali metal ions in the sample and mobile phase, and the ion source parameters. Ions are then drawn into the mass spectrometer analyzer for measurement. A narrow opening between the mass spectrometer analyzer and the ion source helps the vacuum pumps to maintain very low pressure inside the analyzer while the APCI source remains at atmospheric pressure.

During ESI, the HPLC eluate is sprayed through a capillary electrode at high potential (usually 2000 to 7000 V) to form a fine mist of charged droplets at atmospheric pressure. As the charged droplets are electrostatically attracted towards the opening of the mass spectrometer, they encounter a cross-flow of heated nitrogen that increases solvent evaporation and prevents most of the solvent molecules from entering the mass spectrometer. ESI of chlorophylls can produce molecular ions as well as cationized species such as protonated molecules and $[M+Na]^+$ and $[M+K]^+$ (Zissis et al., 1999). The relative abundance of the cationized species depends upon the concentration of proton-donating species and trace amounts of salts in the mobile phase.

Both APCI and ESI are compatible with the same HPLC columns and solvent systems used for chlorophyll analysis. However, the main advantage of APCI compared to electrospray for chlorophyll analysis is the linearity of the detector response, which is more than two orders of magnitude larger for APCI. This suggests that APCI LC/MS might be preferred to ESI LC/MS for quantitative analysis of chlorophylls.

Critical Parameters and Troubleshooting

Chlorophylls are stabilized in vivo within chloroplasts in a complex environment of lipids, proteins, and other compounds. Once extracted from the biological matrix, chlorophylls are easily oxidized upon exposure to air and/or light to form allomers and other degradation products (*UNIT F4.1*). Therefore, care should be taken to minimize the exposure of chlorophyll samples to light and air. For example, sample

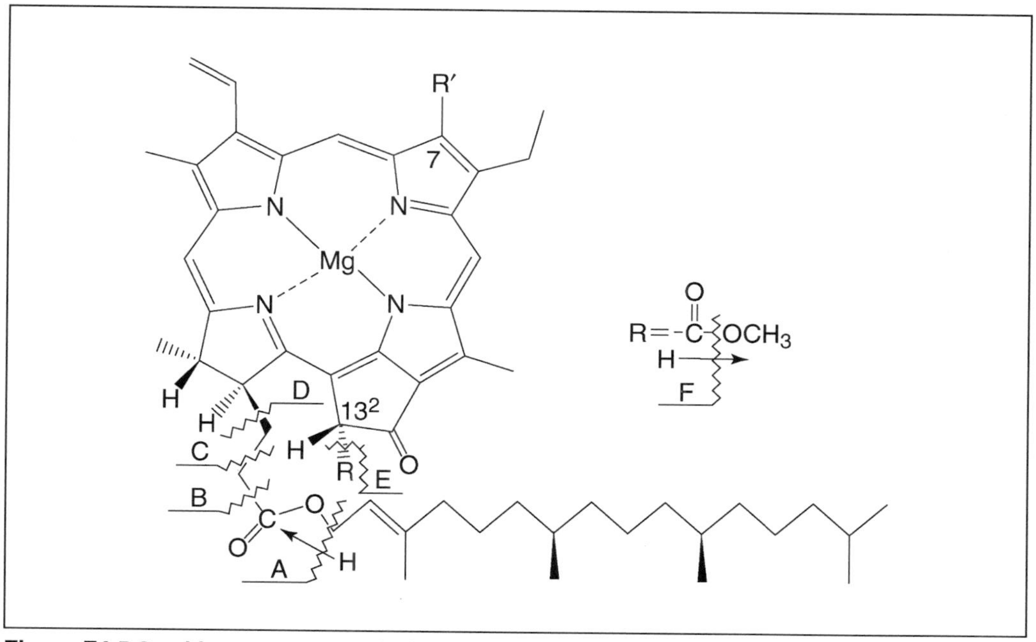

Figure F4.5.2 Mass spectrometric fragmentation scheme for chlorophylls *a* and *b* and their derivatives. Chlorophyll *a*, R = –COOCH$_3$, R′ = –CH$_3$; chlorophyll *b*, R = –COOCH$_3$, R′ = –CHO; pyropheophytin *a*, –Mg +2H, R = H, R′ = –CH$_3$; pyropheophytin *b*, –Mg +2H, R = H, R′ = –CHO; pheophytin *a*, –Mg +2H, R = –COOCH$_3$, R′ = –CH$_3$; pheophytin *b*, –Mg +2H, R = –COOCH$_3$, R′ = –CHO; chlorophyllide *a*, –phytyl chain, R = –COOCH$_3$, R′ = –CH$_3$; chlorophyllide *b*, –phytyl chain, R = –COOCH$_3$, R′ = –CHO; pheophorbide *a* (chlorophyllide *a* –Mg +2H); pheophorbide *b* (chlorophyllide *b* –Mg +2H).

vials or flasks containing chlorophylls should be amber, or else wrapped in an opaque covering to minimize the exposure of the contents to light. Furthermore, containers should be filled with inert gas such as nitrogen or argon. When opening sample vials, the room should be dark or dimly lit.

The solvents that are used for extraction of chlorophylls from leaves or other biological samples are similar to those used for HPLC (UNIT F4.4). For example, leaves are usually ground in the presence of acetone to extract chlorophylls. After filtration, the solvent is evaporated in vacuo or under a stream of nitrogen (but not air, to avoid oxidation). One additional purification step prior to HPLC (applicable to both Basic Protocols 1 and 2) is to dissolve the chlorophyll residue in hexane and then wash the hexane with water/methanol (1:1; v/v). The hexane may be removed prior to HPLC or simply diluted with mobile phase prior to analysis.

Anticipated Results

Molecular ions and protonated molecules of chlorophylls and related compounds will be formed during static FAB/LSIMS when using 3-nitrobenzyl alcohol as the matrix. Alternatively,

a glycerol matrix will facilitate the formation of protonated molecules for chlorophylls and their derivatives during continuous-flow FAB. Using static FAB/LSIMS, sample ions will be produced continuously for several minutes until the sample or the matrix is consumed. During this time, exact mass measurements, high-resolution measurements, and tandem mass spectrometric measurements may be carried out to characterize the chlorophyll sample.

As an example, the tandem mass spectrum with CID of pyropheophytin a is shown in Figure F4.5.1. Pyropheophytins can be formed from chlorophylls as a result of cooking or canning of food products. In this case, the molecular ion of m/z 812 was obtained using positive ion static FAB, fragmented using CID, and the resulting tandem mass spectrum was recorded using a magnetic sector mass spectrometer. Fragment ions were observed that confirm the presence of the phytyl chain (m/z 534, 489, and 461). The sites of cleavage within pyropheophytin a during CID are indicated in Figure F4.5.1. These and other types of fragment ions that are characteristic of chlorophylls and their derivatives are illustrated in the fragmentation scheme shown in Figure F4.5.2. The fragment ion of type A, [M-278]+, corresponds

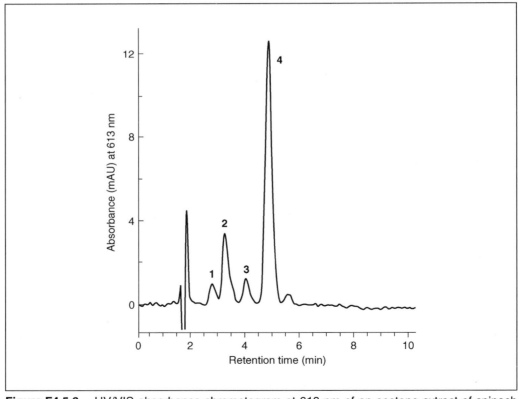

Figure F4.5.3 UV/VIS absorbance chromatogram at 613 nm of an acetone extract of spinach leaves recorded during UV/VIS LC/MS. Examples of APCI mass spectra recorded during this analysis are shown in Figure F4.5.4.

to elimination of the phytyl chain with a transfer of a hydrogen from the ion to the leaving group. Whereas ions of types A, B, and D are formed by loss of the phytyl chain and cleavage at various points along the side chain on C17, the absence of fragment ions of types E and F from the tandem mass spectrum of pyropheophytin *a* shows that this chlorophyll derivative does not contain a β-keto ester group or a methyl group at position C13^2 (R in Figure F4.5.2).

Tandem mass spectrometry following CID of molecular ions or cationized molecules may be used to enhance the formation of structurally significant fragment ions and to eliminate matrix ions and other contaminating ions from the mass spectrum. Although the example shown in Figure F4.5.1 was obtained using FAB, any other ionization method suitable for chlorophylls, such as ESI or APCI, may be used. Fragment ions provide information about chlorophyll functional groups and help to distinguish chlorophyll *a* from chlorophyll *b* and

from related pheophytins, chlorophyllides, pheophorbides, and pyropheophytins. Chlorophyll *a* may be distinguished from chlorophyll *b* by the presence of a methyl group instead of a formyl group at position C7 (R′ in Figure F4.5.2). This results in a molecular weight difference of 14 (chlorophyll *a* weighs 892 and chlorophyll *b* weighs 906 mass units). Additional details regarding structure determination of chlorophylls and their derivatives using MS/MS with CID may be found in van Breemen et al. (1991a).

The UV/VIS LC/MS analysis of an acetone extract of spinach leaves is shown in Figure F4.5.3 and Figure F4.5.4. Reversed-phase C18 chromatography was used with UV/VIS photodiode array absorbance detection followed on-line by positive ion APCI mass spectrometry. The mobile phase consisted of a 20-min linear gradient from methanol to methanol/methyl *tert*-butyl ether (91:9; v/v). The visible absorbance chromatogram at 613 nm for

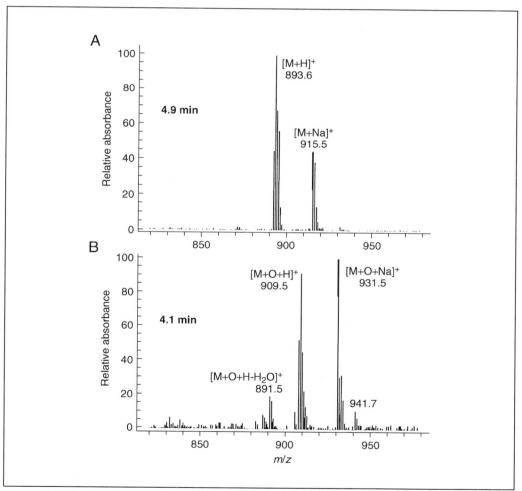

Figure F4.5.4   Positive ion APCI mass spectra of (**A**) chlorophyll *a* and (**B**) an oxidation product of chlorophyll *a*, which eluted at 4.9 and 4.1 min, respectively, during the UV/VIS LC/MS chromatogram shown in Figure F4.5.3.

this analysis is shown in Figure F4.5.3. Four peaks were detected instead of the expected two peaks for chlorophylls *a* and *b*. Diode array detection indicated that all four peaks in Figure F4.5.3 corresponded to chlorophylls. However, mass spectrometric detection was necessary to distinguish these compounds from each other. Post-column addition of 0.5% acetic acid (so that the mobile phase contained 0.05% acetic acid) was used to enhance the abundance of protonated molecules of the chlorophylls in the positive ion APCI mass spectra. Based on the mass spectra and diode array absorbance spectra, peaks 2 and 4 were identified as chlorophyll b and chlorophyll *a*, respectively, whereas peaks 1 and 3 were oxidation products of chlorophylls *b* and *a*, respectively, weighing an additional 16 mass units. The positive ion APCI mass spectra for chlorophyll a and its oxidation product (peaks 3 and 4) are shown in Figure F4.5.4.

For additional characterization of these chlorophylls, high-resolution, exact mass measurements could be carried out to confirm their elemental compositions using a magnetic sector mass spectrometer, a quadrupole time-of-flight hybrid mass spectrometer, or a Fourier transform ion cyclotron resonance mass spectrometer. In addition, tandem mass spectrometry could be used to obtain additional information regarding the structures of these compounds. A quadrupole mass spectrometer was used to obtain the data shown in Figure F4.5.4. Although an excellent instrument for high-throughput LC/MS analyses, a single quadrupole analyzer is not capable of MS/MS nor does it offer sufficient precision and resolution for exact mass measurements. Finally, HPLC could be used to purify sufficient quantities of each peak for NMR characterization. Typically, at least 1000-fold more sample is required for NMR than for mass spectrometry. Since this process would require the investment of considerably more time and sample handling, it would result in even more sample degradation.

Note that the extraction process used in the example discussed here resulted in the formation of measurable oxidation products. When this extract was reanalyzed just 24 hr later without storage under an inert atmosphere or in a freezer, no chlorophyll *a* or *b* could be detected. In summary, UV/VIS LC/MS or LC/MS/MS provides the most rapid and sensitive approach to characterizing and identifying chlorophylls and their degradation products.

Time Considerations

Since mass spectrometry is a rapid analytical technique, the sample throughput will be ~20 per hour when using FAB/LSIMS and >60 per hour when using flow injection analysis with APCI or ESI. The rate-limiting step for FAB/LSIMS is the time (1 to 2 min) required to introduce the direct insertion probe through the vacuum interlock. During LC/MS and LC/MS/MS, the slow step is the time required for chromatographic separation.

Literature Cited

Airs, R.L. and Keely, B.J. 2000. A novel approach for sensitivity enhancement in atmospheric pressure chemical ionization liquid chromatography/mass spectrometry of chlorophylls. *Rapid Commun. Mass Spectrom.* 14:125-128.

Barber, M., Bordoli, R.S., Elliott, G.J., Sedgwick, R.D., and Tyler, A.N. 1982. Fast atom bombardment mass spectrometry. *Anal. Chem.* 54:645A-657A.

Dougherty, R.C., Dreifuss, P.A., Sphon, J., and Katz, J.J. 1980. Hydration behavior of chlorophyll a: A field desorption mass spectral study. *J. Am. Chem. Soc.* 102:416-418.

Hunt, J.E., MacFarlane, R.D., Katz, J.J., and Dougherty, R.C. 1981. High-energy fragmentation of chlorophyll a and its fully deuterated analog by [252]Cf plasma desorption mass spectrometry. *J. Am. Chem. Soc.* 103:6775-6778.

Hyvarinen, K. and Hynninen, P.H. 1999. Liquid chromatographic separation and mass spectrometric identification of chlorophyll b allomers. *J. Chromatogr.* 837:107-116.

Liu, S.Q., Sun, H.R., Sun, M.Z., and Xu, J.Q. 1999. Investigation of a series of synthetic cationic porphyrins using matrix-assisted laser desorption/ionization time-of-flight mass spectrometry. *Rapid Commun. Mass Spectrom.* 13:2034-2039.

Tabet, J.-C., Jablonski, M., Cotter, R.J., and Hunt, J.E. 1985. Time-resolved desorption. III. The metastable decomposition of chlorophyll a and some derivatives. *Int. J. Mass Spectrom. Ion Processes* 65:105-117.

Teng, S.S. and Chen, B.H. 1999. Formation of pyrochlorophylls and their derivatives in spinach leaves during heating. *Food Chem.* 65:367-373.

van Breemen, R.B., Canjura, F.L., and Schwartz, S.J. 1991a. Identification of chlorophyll derivatives by mass spectrometry. *J. Agric. Food Chem.* 39:1452-1456.

van Breemen, R.B., Canjura, F.L., and Schwartz, S.J. 1991b. High-performance liquid chromatography-continuous-flow fast atom bombardment mass spectrometry of chlorophyll derivatives. *J. Chromatogr.* 542:373-383.

Verzegnassi, L., Riffé-Charlard, C., and Gülaçar, F.O. 2000. Rapid identification of Mg-chelated chlorins by on-line high performance liquid chromatography/atmospheric pressure chemical ionization mass spectrometry. *Rapid Commun. Mass Spectrom.* 14:590-594.

Zissis, K.D., Dunkerley, S., and Brereton, R.G. 1999. Chemometric techniques for exploring complex chromatograms: Application of diode array detection high performance liquid chromatography electrospray ionization mass spectrometry to chlorophyll *a* allomers. *Analyst* 124:971-979.

Key References

van Breemen et al., 1991a. See above.

The most complete reference of tandem mass spectra for chlorophylls and chlorophyll derivatives including pheophytins, chlorophyllides, pheophorbides, and pyropheophytins.

Verzegnassi et al., 2000. See above.

Class characteristic fragment ions of chlorophylls and related chlorins are summarized for both positive and negative ion APCI. Detailed instrument parameters for LC/MS using APCI are described.

Contributed by Richard B. van Breemen
University of Illinois at Chicago
Chicago, Illinois

Strategies for Measurement of Colors and Pigments

F5

F5.1	**Overview of Color Analysis**	**F5.1.1**
	Appearance and Color	F5.1.1
	Visual Color Evaluation and the Observer Situation	F5.1.2
	Measuring Tristimulus Values	F5.1.5
	CIE *XYZ* to CIELAB	F5.1.7
	Color Difference Equations	F5.1.9
	CMC and Beyond	F5.1.12
	Conclusion	F5.1.13

Contents

1

Overview of Color Analysis

APPEARANCE AND COLOR

The measurement of color is a young science, and is recognized as part of a larger field of study known as appearance science. Appearance can be defined as the phase of visual experience by which things are recognized. There are several appearance modes to identify the situation when a light source or object is recognized. The object mode encountered daily is the most familiar mode. It consists of a person observing an object illuminated by a light source. People make appearance judgments all day long without being aware of how an object is easily recognized as dull or glossy, transparent or opaque. Appearance technology classifies objects based on their distribution of incident light. Appearance attributes can be divided into two main categories—those related either to spectral (color) or to geometric (spatial) light distribution.

A strict definition of color includes (1) the object appearance that depends on light, object, and observer, and (2) the visual perception described with color names. Color is a primary attribute of appearance and it can be quantified. The measurement of color is known as colorimetry. The colorimetric principles associated with the response of the normal eye are important when reviewing color analysis. The eye-brain combination is sensitive, flexible, and able to analyze data at high speeds. Color scientists developed color (spectral) and gloss (geometric) scales founded on human color perception. They studied the response of an observer to light distributed by an object. The phenomenon of color results from this interaction of light, object, and observer. The brain processes the light distributed by the object and reaching the eye. An object has no inherent color, and so it may be stated that color exists only in the mind of the viewer. A group of trained observers will differ in their judgment and description of a perceptible color difference. The sensation of color varies even for observers with normal color vision, not to mention those without normal color vision. Approximately 8% of the male population and 0.5% of the female population are color-defective. Other factors can also influence the color experience. When identified, they lead to a method for color evaluation.

The eye, as amazing as it is, cannot measure color quantitatively. Color-order systems have been developed to specify color based on a space with coordinates. Color can be presented as an arrangement of three dimensions within a color space. One dimension relates to a lightness attribute and the other two are chromatic attributes, referred to as hue and chroma (or saturation). The human observer is not equally

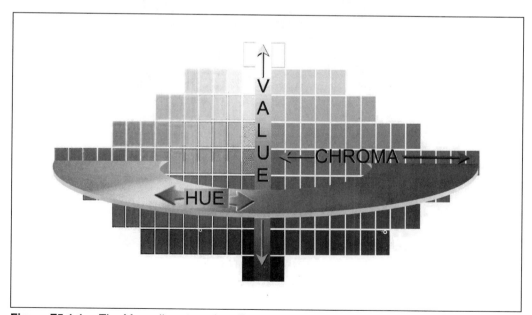

Figure F5.1.1 The Munsell system describes color in terms of hue, value, and chroma. Figure courtesy of GretagMacbeth. *This black and white facsimile of the figure is intended only as a placeholder; for full-color version of figure go to http://www.currentprotocols.com/colorfigures*

Strategies for Measurement of Colors and Pigments

Contributed by Kevin Loughrey

F5.1.1

sensitive to the three dimensions. Hue is generally the most critical, followed by chroma, and finally lightness. Munsell is a well-known color-order system that specifies surface color and has found wide acceptance in a diversity of color applications. Munsell identifies the hue, value (lightness), and chroma (saturation) of a color using color chips of equal visual spacing (Fig. F5.1.1.). The chips are arranged on pages in a book. All the chips on a single page are the same hue, but they vary in value and chroma. The observer chooses the chip that best matches the color of the object being evaluated. The chip selected will have a notation that identifies the assigned hue, value, and chroma. Visual color standards like Munsell have long been used in various food applications to specify color.

The subject of color, as it relates to the human eye and brain, is complex and fascinating. Overall, there are many factors that can affect the perceived color of a surface. Models like Munsell use an alphanumeric notation to identify the dimensions for a color and its location within the color space. A Munsell notation (hue, value, and chroma) depends on a subjective visual judgment made by a human observer. There are other color-order systems that specify color by objective three-dimensional instrument numbers. The goal of these systems is to establish a reasonable method that measures objects for color and color difference using an instrument. The value in such a method is the potential for the instrument to confirm subjective visual judgments. Successful color analysis concludes with agreement between visual assessment and instrument numbers.

VISUAL COLOR EVALUATION AND THE OBSERVER SITUATION

The experiential nature of color can be understood by examining the basic situation that results in the experience of color. The sensation of color requires three components: a light source, an object that is illuminated, and the eye-brain as a detector. Colorimetry simply refers to this three-way relationship as the observer situation. A complete understanding of this vital relationship is needed for success in the visual and instrumental evaluation of color. The regular practice for visual color appraisal must be considered carefully. Any apparatus used for examining color, such as a light booth, should incorporate illumination and viewing controls. Visual evaluations need to be precise and reproducible. A change in the viewing conditions will have a great impact on the

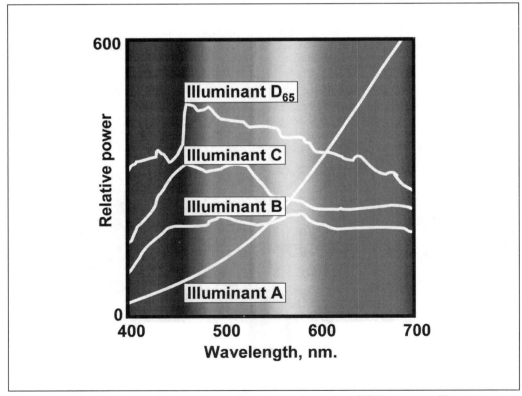

Figure F5.1.2   The spectral power distribution curves of standard CIE illuminants. Figure courtesy of GretagMacbeth. *This black and white facsimile of the figure is intended only as a placeholder; for full-color version of figure go to http://www.currentprotocols.com/colorfigures*

results of any visual appraisal. Standardized conditions for a viewing environment will greatly improve the inspection and approval of color. A number of components are critical to the success of visually judging color:

the spectral quality of the light source;
the level of illumination;
the geometric conditions;
the surround and ambient field;
the observer response functions.

A written method would define the viewing conditions. They must be the same for all observers when judging color. An excellent reference for guidance in visual color appraisal is the published standard ASTM D 1729-96 (ASTM, 2000).

The Light Source

The light source is the first important element of the observer situation. The statement has been made that without light there is no color. Sir Isaac Newton conducted experiments with an optical prism and sunlight to confirm that white light is composed of spectral colors. The human response to color is limited to the visible spectrum of light, containing wavelengths between 380 and 770 nm. The wavelengths are associated with colors, beginning with violet at the short end of the spectrum, through indigo, blue, green, yellow, orange, and finally red at the long end. The nanometer (nm) is the accepted unit of length for light waves. The light is referred to as an illuminant when

color is computed using data from a spectrophotometer. Standard illuminants for color measurement were first established in 1931 by the CIE (Commission Internationale de l'Éclairage, or The International Commission on Illumination). The CIE is the main international organization concerned with color and color measurement. Illuminants A (incandescent), B (noon sunlight), and C (overcast daylight), were recommended by the CIE for use in colorimetry. Illuminants are specified by their spectral energy distributed across the visible spectrum. Each illuminant has its own spectral power distribution curve (Fig. F5.1.2), which supports accurate identification. The curve consists of energy plotted against wavelength in nanometers. When measuring color with a spectrophotometer, the standard illuminant table is a selection in the instrument firmware or software. In 1965, the CIE proposed a D series of illuminants to better represent natural daylight. Today the illuminant most widely used in color measurement is D_{65} (average daylight). Standard illuminant tables provide a method to quantify the light as the first element of the observer situation.

The Object

The second element of the observer situation is the object. Incident light is modified when it interacts with an object. There are four main effects (Fig. F5.1.3) that may occur from this interaction:

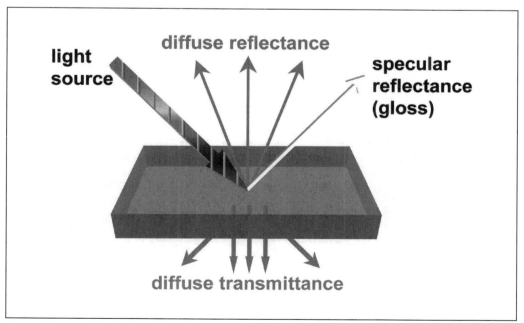

Figure F5.1.3 The interaction of light with an object modifies the light in a variety of ways. Figure courtesy of GretagMacbeth. *This black and white facsimile of the figure is intended only as a placeholder; for full-color version of figure go to http://www.currentprotocols.com/colorfigures*

Strategies for
Measurement of
Colors and
Pigments

F5.1.3

1. Specular reflection (gloss) at the surface.

2. Selective absorption within the object.

3. Diffuse reflection by scattering in the object.

4. Regular transmission through the object.

Selective absorption is the process of absorbing only certain wavelengths of light. It is the primary basis for the color of an object. A yellow color results mainly from the blue wavelengths being absorbed and the green, yellow, and red wavelengths being reflected or transmitted. In this example, the blue wavelengths of light are absorbed and never reach the eye and brain to trigger a response. The light reflected or transmitted by the object finally reaches the eye-brain of the observer and produces the perception of color. Color instruments are designed to measure this reflected or transmitted light and convert the physical data to three-dimensional numbers, corresponding to perceived color. Instrument color numbers are primarily used to assess quality in the food industry. Food color measurements determine the color quality of food ingredients. They are also used to monitor color quality during processing and storage.

Analytical instruments and techniques are employed in food technology. Certain analytical spectrophotometers are designed to measure percent absorbance at selected wavelengths. A procedure for color analysis may stipulate the measurement of the light-absorbing properties of a food. The analytical methods of color analysis differ from the colorimetric techniques. The summary of color analysis presented here is confined to the colorimetric measurement of food. A color spectrophotometer measures the light reflected or transmitted by an object employing a dispersing element such as a grating to measure at wavelength intervals of 10 or 20 nm. Data from the spectrophotometer are plotted against wavelength and displayed as a spectral curve (Fig. F5.1.4). The spectral curve is like a fingerprint of the object measured. The color of an object may be suggested if the shape of its curve relates to the curves of known colors. The spectral data for the object quantify it as the second element of the observer situation. The illuminant table first selected can now be combined with the data for the object.

The Human Observer

During the last eighty years, research has been conducted within the color community to test the response of the human eye to color stimuli. Scientists knew it was possible to match the color sensation by designing an experiment based on the mixing of three colored lights. W.D. Wright in 1928 and J. Guild in 1931 conducted the most important of these experiments. Their two experiments were inde-

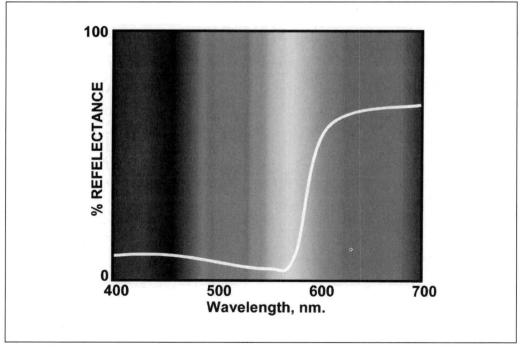

Figure F5.1.4 The spectral reflectance curve of a red object plots reflectance at each wavelength. Figure courtesy of GretagMacbeth. *This black and white facsimile of the figure is intended only as a placeholder; for full-color version of figure go to http://www.currentprotocols.com/colorfigures*

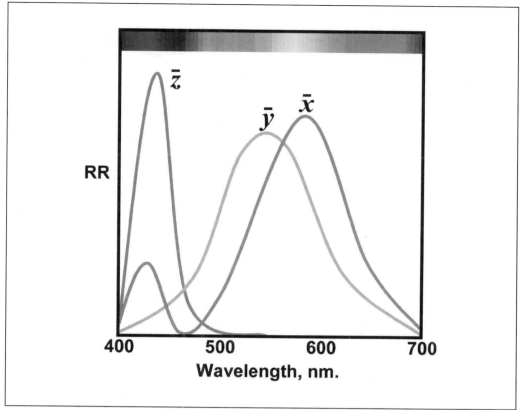

Figure F5.1.5 The 1931 CIE 2° Standard Color Observer established colorimetry. Figure courtesy of GretagMacbeth. *This black and white facsimile of the figure is intended only as a placeholder; for full-color version of figure go to http://www.currentprotocols.com/colorfigures*

pendent of each other, using different observers and different primary lights. The field of view in the experiments was 2° (similar to viewing a dime at arm's length) and was selected to involve the area of the eye known as the fovea. The fovea, located in the central region of the retina, is where the red-, green-, and blue-sensitive cones are concentrated. The cones are the receptor cells in the eye that respond to light at different wavelengths. In both experiments, normal observers visually matched monochromatic (single-wavelength) lights by mixing together three primary lights (red, green, and blue). The data curves from the experiments corresponded to the spectral response of the average human eye. The curves from the two experiments were different because Wright and Guild did not use the same set of primary wavelengths for their three lights.

The CIE transformed the original experimental curves from Wright and Guild into more useful functions that would facilitate the identification of color stimuli by numbers. The three functions resulting from the transformation represent the color-matching response functions of the average observer with normal color vision. The observer functions were standardized and adopted worldwide in 1931 as the CIE 2° Standard Color Observer (Fig. F5.1.5). Standard Observer tables are available for the original 1931 2° Observer and the more recent 1964 10° Observer. The 10° Observer was developed as an improvement of the original 2° Observer and is widely used today in many color applications. The human observer is now quantified as the third element of the observer situation. The observer joins the light and the object so the elements of the observer situation are quantified across the spectrum:

the light source as a spectral power distribution curve;

the object as a spectral reflectance or transmittance curve;

the observer as the spectral response curves.

Light, object, and observer as curves can be combined to compute numbers that agree with perceived color.

MEASURING TRISTIMULUS VALUES

The development of the observer response functions is the foundation for color measured by an instrument. The Standard Observer established a recognized method for converting

Strategies for Measurement of Colors and Pigments

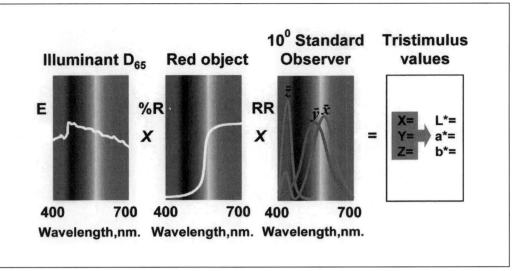

Figure F5.1.6 Light, object, and observer are combined to calculate CIE values *X*, *Y*, and *Z*. Figure courtesy of GretagMacbeth. *This black and white facsimile of the figure is intended only as a placeholder; for full-color version of figure go to http://www.currentprotocols.com/colorfigures*

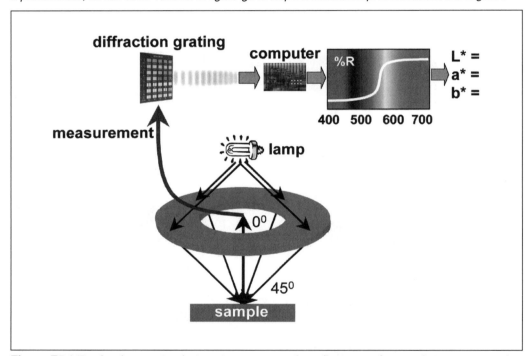

Figure F5.1.7 A color spectrophotometer measures the reflectance of a sample to compute color. Figure courtesy of GretagMacbeth. *This black and white facsimile of the figure is intended only as a placeholder; for full-color version of figure go to http://www.currentprotocols.com/colorfigures*

the spectral curve of any object into three numbers known as the CIE tristimulus values *X,Y,Z* (Fig. F5.1.6). The spectrophotometer measures the object for spectral data and the computer converts the data into the tristimulus values for a selected illuminant and observer (Fig. F5.1.7). The spectrophotometer has been referenced because the measured spectral data can be converted into tristimulus numbers that correspond to perceived color. There are other

instruments that measure color, but their design is different from a spectrophotometer.

Tristimulus colorimeters are used to measure color in certain food applications. They combine light source, filters, and photodetectors to reproduce the CIE Standard Observer response functions (Fig. F5.1.8). Colorimeters, having broad band-pass filters, do not measure the spectral data. Without spectral data they cannot offer the choice of either observer or

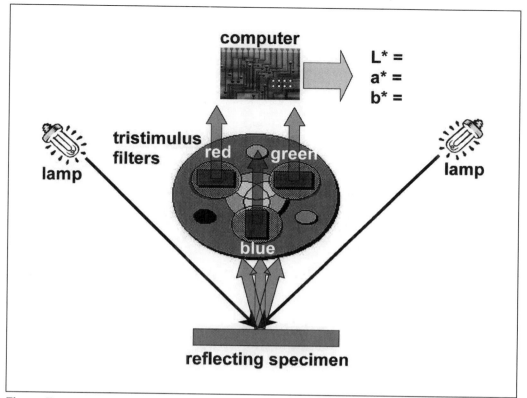

Figure F5.1.8 A colorimeter reproduces the eyes' response for one illuminant and one observer. Figure courtesy of GretagMacbeth. *This black and white facsimile of the figure is intended only as a placeholder; for full-color version of figure go to http://www.currentprotocols.com/colorfigures*

various illuminants such as A (incandescent) and F_2 (cool white fluorescent). Traditionally, their output has been limited to Illuminant C (daylight) for the 2° Standard Observer. Color spectrophotometers do not have the limitations of colorimeters because they measure the percent reflectance or transmittance (spectral data) wavelength by wavelength across the visible spectrum. The spectral data can be converted to tristimulus data for any of the illuminant-observer combinations selected. It should be noted that both spectrophotometers and tristimulus colorimeters are primarily utilized as instruments to measure the color difference between a standard and a trial, rather than as absolute devices.

CIE *XYZ* TO CIELAB

The CIE color-order system is the foundational color scale for all instrumental color scales. However, the X, Y, Z values are difficult to correlate with the perceived color of an object. In fact, the CIE scales were not intended for specifying the color of objects, nor the color difference between objects. Also, CIE scales do not have equal visual spacing for all colors. The same numerical color difference between colors does not equate to the same visual difference for all colors. Just a few years following the

adoption of the CIE system, alternate color systems became available. The new scales were created to overcome the known inadequacies of the CIE scales when measuring the colors of objects. Equations were developed that transformed the CIE data. The alternate color scales were intended to be more uniform in their visual spacing, but did use the CIE system as the base. In reality, these more uniform color spaces were at best an improvement over the CIE system.

One particular transformation of the CIE system was based on the Hering theory of color vision. The premise was that signals from the receptor cones in the eye were coded into light-dark, red-green, and yellow-blue signals as they traveled to the brain. Such color scales are known as opponent-type because they recognize three pairs of opposing signals (Fig. F5.1.9). This means that a color cannot be perceived as red and green at the same time or yellow and blue at the same time. However, a color can be green and yellow or green and blue.

The rectangular coordinates of an early opponent color system (Hunter) labeled the three dimensions of a color as L,a,b. The L coordinate represents lightness, the a coordinate represents redness ($+a$) or greenness ($-a$), and the b coordinate represents yellowness ($+b$) or blue-

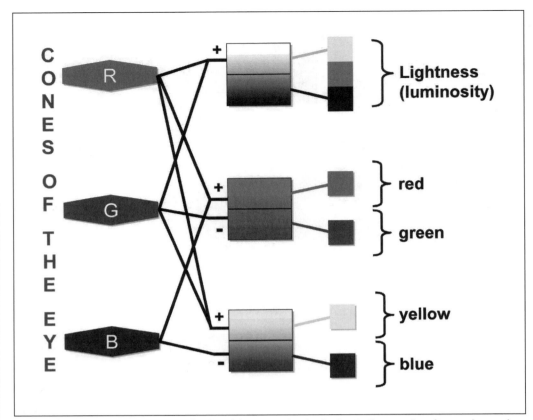

Figure F5.1.9 Hering's theory of color vision stated color is due to three pairs of opposing codes. Figure courtesy of GretagMacbeth. *This black and white facsimile of the figure is intended only as a placeholder; for full-color version of figure go to http://www.currentprotocols.com/colorfigures*

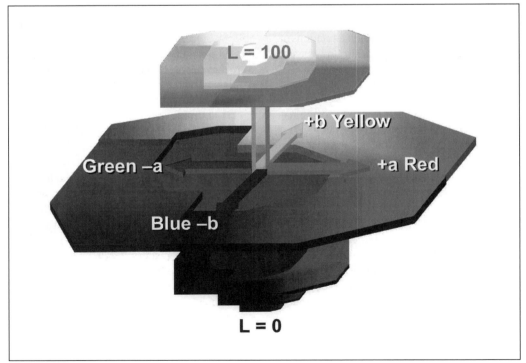

Figure F5.1.10 The Hunter *L,a,b* color space was designed for measuring color differences. Figure courtesy of GretagMacbeth. *This black and white facsimile of the figure is intended only as a placeholder; for full-color version of figure go to http://www.currentprotocols.com/colorfigures*

ness (–*b*). Using this system, an object measured for color would be assigned an *L,a,b* value to describe the color and specify its location in the three-dimensional color space. The distance between two colors, each having its own location in the color space, relates to a measurable color difference. *L,a,b* numbers correlate with perceived color because the opponent color theory was used to develop the conversion from CIE *X,Y,Z*. The Hunter *L,a,b* color space, published in 1942 (Fig. F5.1.10), became very popular because it described color and color difference in visual terms and was readily available with color instruments (tristimulus colorimeters). Today there are food applications that continue to specify color using the Hunter *L,a,b* color scale.

Work within the color community continued for the ideal, uniform color space. In 1976, the CIE recommended a more nearly uniform color space known as *L*a*b** with the official designation CIELAB. *L*a*b** is an opponent-type color space with rectangular coordinates similar to Hunter *L,a,b*. The *L** indicates lightness (0 to 100), the a* indicates redness (+) and greenness (–), and the *b** indicates yellowness (+) and blueness (–). The limits for *a** and *b** values are around –80 and +80. The two color scales do not correlate and a color located in *L,a,b* space will be in a different location in *L*a*b**.

Presently CIELAB is the most widely used color scale across the major industries, but still has limitations.

There has been a growing interest in the food industry for a color space based on a polar model. In 1976 when CIELAB was adopted, the CIE recommended an alternative color scale known as CIELCH or *L*C*H**. Of the three dimensions of color, the hue is the most critical in terms of perceptibility and acceptability for normal color observers. The *L*C*H** color space identifies the hue as one of the three dimensions. A color is located using cylindrical coordinates with *L** being the same as in CIELAB and *C** and *H** computed from *a** and *b**. The coordinates of CIELCH (also see Fig. F5.1.11) are:

L* = lightness, the same as in *L*a*b**;

C* = chroma (saturation) is the relation of a color to a neutral gray of the same lightness;

H* = hue angle is expressed in degrees, with 0° corresponding to +*a** axis (red), then continuing to 90° for the +*b** axis (yellow), 180° for –*a** (green) and finally 270° for –*b** (blue). It is important to realize that now a color is in the same location in either *L*a*b** or *L*C*H** color space. These simply represent two different methods to describe the same location for the same color. *L*C*H** has appeal for food applications because the hue is separately identified and measured.

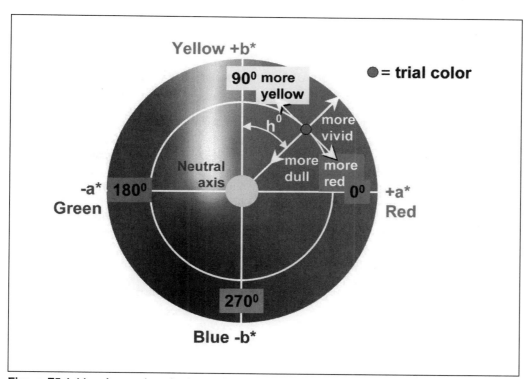

Figure F5.1.11   A sample color located using the polar coordinates of the *L*C*H** color space. Figure courtesy of GretagMacbeth. *This black and white facsimile of the figure is intended only as a placeholder; for full-color version of figure go to http://www.currentprotocols.com/colorfigures*

Strategies for Measurement of Colors and Pigments

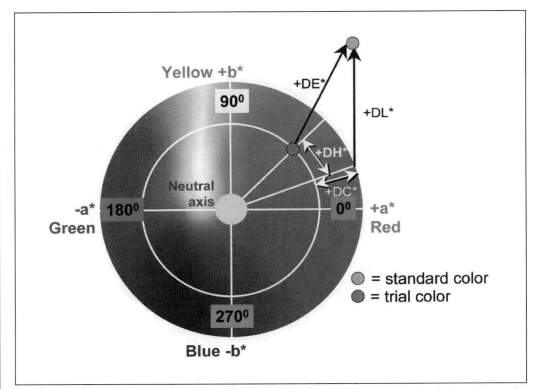

Figure F5.1.12 A difference in hue, chroma, and lightness between a standard and sample. Figure courtesy of GretagMacbeth. *This black and white facsimile of the figure is intended only as a placeholder; for full-color version of figure go to http://www.currentprotocols.com/colorfigures*

COLOR DIFFERENCE EQUATIONS

Many industrial color applications work with a color spectrophotometer to measure CIELAB color differences. CIELAB works well in specifying the color of an object, but color differences calculated with the formula have limited value for pass/fail decisions. These component color differences ($dL*$, $da*$, $db*$) are calculated by simply subtracting the $L*a*b*$ values for the standard from the $L*a*b*$ values for the trial. The $dL*$, $da*$, $db*$ values should indicate the direction of color difference from the standard (e.g., lighter or darker). They are commonly used to establish the allowable color limits from a standard. However, in applying these delta differences as limits for color acceptability, they should be adjusted for different colors. CIELAB, like Hunter $L,a,b,$ is not a visually uniform color space. The limitations of CIELAB are not well understood throughout industry and the numbers continue to be applied without being adjusted for different colors and products.

For those preferring to express differences in chroma and hue terminology, instead of $da*$ and $db*$, the following terms (also see Fig. F5.1.12) are utilized:

Delta $L*$ = the lightness difference;

Delta $C*$ = the chroma angle difference;

Delta $H*$ = the metric hue difference.

A change in hue of a color is expressed as the distance $DH*$ (capital H) along the chroma arc for the standard. A specific change in hue, $DH*$, is larger for a pair of colors (standard and trial) that are far from the neutral axis and smaller for a pair closer to the neutral axis. A negative $DH*$ indicates the trial is located clockwise in color space from the standard. A positive $DH*$ means the trial is located counterclockwise from the standard. The direction of the trial from the standard can then be interpreted as a shift in color from the standard. An example of this would be the standard having a hue angle (h°) of 90°, then a positive $DH*$ means the trial is greener, while a negative $DH*$ would mean the trial is redder than the standard. The delta differences ($DH*$ and $DC*$) in hue and chroma have been found to agree well with visual perception. However, CIELCH joins Hunter L,a,b and CIELAB as a visually nonuniform color space. Different colors require different color acceptability limits.

The Delta E (DE) is a single number widely used today for color acceptability. A traditional DE number is a measure of total color difference and is calculated from the individual component differences. When using CIELAB, the color difference equation known as *DE** can be calculated from the component or delta differences. The basic formula for *DE** is:

$$DE^* = (dL^{*2} + da^{*2} + db^{*2})^{1/2}$$

The *DE* for the Hunter *L,a,b* color scale would be calculated using the same basic equation. However, one *DE* formula cannot be converted into another formula. The Hunter *DE* and the CIELAB *DE** do not correlate, so whatever color difference equation is selected

must be specified. Different *DE* calculations have been proposed as a uniform single-number tolerance that could be applied to all colors.

Traditionally the individual delta numbers and the *DE* have served as color tolerances. The difficulty is in the application of color difference numbers when the same measured color difference does not always equate to the same visual difference. A *DE** is a good example, being well established and presently employed

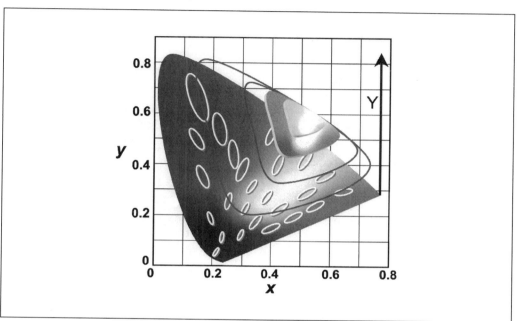

Figure F5.1.13   MacAdam ellipses (MacAdam, 1942) of differing size are plotted within the CIE chromaticity diagram (x,y). *This black and white facsimile of the figure is intended only as a placeholder; for full-color version of figure go to http://www.currentprotocols.com/colorfigures*

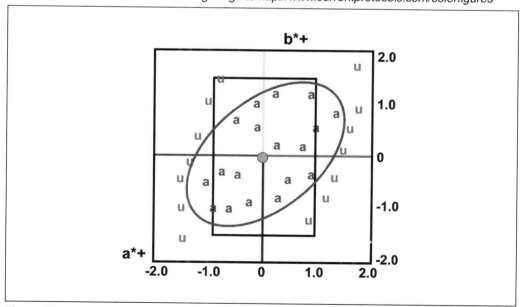

Figure F5.1.14   Elliptical tolerances versus box tolerances. Figure courtesy of GretagMacbeth. *This black and white facsimile of the figure is intended only as a placeholder; for full-color version of figure go to http://www.currentprotocols.com/colorfigures*

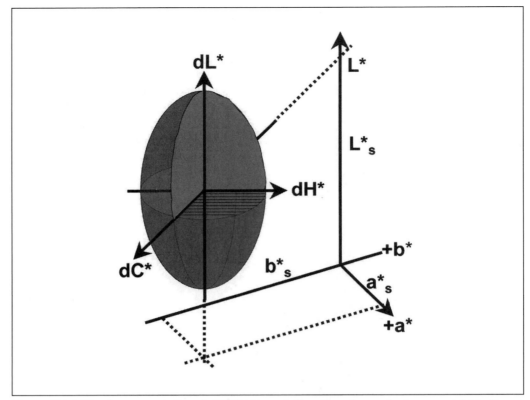

Figure F5.1.15 A representation of a CMC ellipse defines the volume of color acceptability. Figure courtesy of GretagMacbeth. *This black and white facsimile of the figure is intended only as a placeholder; for full-color version of figure go to http://www.currentprotocols.com/colorfigures*

in many color applications. Some limitations of a calculated DE^* are:

1. It represents a single number that by itself can only indicate the magnitude of color difference, but not the direction.

2. A $DE^* = 1.0$ does not translate into the same visual difference between all colors.

3. The calculated DE^* may be below the acceptable limit, but the color difference may not be acceptable visually.

There has been a long history of setting color tolerances, and even today this is not always done correctly. The criteria for color acceptability are changing and there are many factors contributing to this change. The world of specifiers and suppliers continues to involve more companies in the color-approval process. A color-control program needs to integrate a basic understanding of the difference between the perceptible limit of color and the acceptable limit. Perceptibility is the point at which a trained eye, under controlled conditions, can just begin to see a color difference.

The perceptibility of observers varies and is not the same for all colors. David MacAdam developed early color data in 1942. The Mac-Adam ellipses (Fig. F5.1.13) are plotted in the CIE 1931 chromaticity diagram. The ellipse

changes in size and shape depending upon the location in color space. These data confirm the variations in color-difference perceptibility throughout color space. Acceptability is the agreed upon limit of color difference from a standard and this depends on many factors, including the buyer and seller.

The emphasis on control in certain color applications has resulted in tighter color tolerances. Color tolerances need to be realistic so they are achievable when considering the variables introduced by people, products, and instruments. The acceptable limits should not be set at the minimum limit of what a human observer can perceive. Added to this is the visual nonuniformity that exists for both perceptible and acceptable limits. Human observers do not respond equally to all colors. Hopefully, a uniform color difference equation could be available that adjusts the perceptibility and acceptability limits for individual colors.

CMC AND BEYOND

Experience has shown that samples evaluated visually, and then measured for color difference, will not all plot within a rectangular-shaped area of acceptability. In actuality, the pattern formed by plotting the samples will be elliptical in shape. This is why CIELAB rectangular tolerances do not work well when trying to set acceptability limits that agree with the eye (Fig. F5.1.14). In practice, CIELAB accepts some samples that should not be accepted and rejects others that should be accepted.

In 1984, a new Delta E (CMC) color-difference formula was recommended for industrial pass/fail decisions. The DE_{CMC} formula defines an acceptability ellipse with the standard located in the center (Fig. F5.1.15). All trials inside the volume of the ellipse are within the CMC tolerance (usually set at $DE_{CMC} = 1.0$) and should be visually acceptable. The DE_{CMC} is calculated from the CIELCH color scale that was published in 1976 along with CIELAB. The CMC ellipse is defined by the DL^*, DC^* and DH^* coordinates. The size and shape of the ellipse will vary throughout three-dimensional color space as the standard changes in hue, chroma, and lightness. The hue is considered the most critical dimension in defining the CMC ellipse. The relationship between lightness and chroma is usually set at a 2:1 ratio, with lightness being the least critical dimension. DE_{CMC} has proven it can be applied as a single-number tolerance for all the colors of a given product.

CONCLUSION

Color scientists continue to propose newer color-difference equations. New models of color instruments are being introduced almost every year. The world of color analysis is adapting to the scientific advances that shape food technology and many other areas. Color measurement is still founded on basic principles that confirm the important relationship between the human eye and the instrument. Spectrophotometers and colorimeters assist the eye rather than replace the eye. A properly maintained color instrument is a great asset to a colorist with a trained eye.

The science and methods of colorimetry have been reviewed as a system of color analysis. The color measurement concludes successfully when the instrument data agree with the visual evaluation. The color data from the instrument provide a consistent, objective, and documented way to evaluate color. However, the final color analysis is the judgment made by the eye. Color is neither accepted nor rejected by numbers alone. What really matters in the end is how the object appears.

LITERATURE CITED

ASTM (American Society for Testing and Materials). 2000. D 1729-96, Standard Practice for Visual Appraisal of Colors and Color Differences of Diffusely-Illuminated Opaque Materials. *In* ASTM Standards on Color and Appearance Measurement, 6th ed. ASTM, West Conshohocken, Pa.

MacAdam, D.L. 1942. Visual sensitivities to color differences in daylight. *J. Opt. Soc. Am.* 32:247-274

KEY REFERENCES

ASTM. 2000. ASTM Standards on Color and Appearance Measurement, 6th ed. ASTM, West Conshohocken, Pa.

A compilation of all ASTM standards for color and appearance. An excellent resource for guidance in all aspects of color and appearance measurement.

Billmeyer, F.W. and Saltzman, M. 1981. Principles of Color Technology, 2nd ed. John Wiley & Sons, New York.

A standard work that covers the fundamental principles of color measurement. Included are practical applications of color technology.

Berger-Schunn, A. 1994. Practical Color Measurement, 1st ed. John Wiley & Sons, New York.

Provides details on computerized color measuring systems. Real-life examples are cited along with practical knowledge from the author, who has many years of industry experience.

GretagMacbeth. 1998. Fundamentals of Color and Appearance. GretagMacbeth, South Deerfield, Mass.

A short book designed to provide a basic foundation for understanding color and appearance. It is used in educational seminars intended for people involved in the evaluation and management of color and appearance where fundamental principles can be applied to solve issues.

Hunter, R.S. 1975. The Measurement of Appearance, 1st ed. John Wiley & Sons, New York.

An important textbook and reference that covers color and appearance as it relates to objects and the methods available for measurement.

Contributed by Kevin Loughrey
GretagMacbeth
South Deerfield, Massachusetts

G FLAVORS

INTRODUCTION

G1 Smell Chemicals

G1.1 Direct Sampling

G1.2 Isolation and Concentration of Aroma Compounds

G1.3 Identification and Quantitation of Aroma Compounds

G1.4 Stereodifferentiation of Chiral Odorants Using High-Resolution Gas Chromatography

G1.5 Analysis of Citrus Oils

G1.6 Solid-Phase Microextraction for Flavor Analysis

G1.7 Simulation of Mouth Conditions for Flavor Analysis

G1.8 Gas Chromatography/Olfactometry

G2 Acid Tastants

G2.1 Titratable Activity of Acid Tastants

G2.2 Liquid Chromatography of Nonvolatile Acids

SECTION G
Flavors

INTRODUCTION

It is well established that the perception of flavor is a sensory response to chemicals in foods. Taste and olfactory systems as well as the sense of touch (texture) and chemesthesis (responses detected by the trigeminal nerve receptors, i.e., the pungency of capsaicin) are the anatomical features that are involved in flavor detection and the site of the transduction process. All organisms detect extracellular chemicals by transduction processes in which a ligand binds to a receptor protein called a "seven-transmembrane G protein–activated receptor protein." The chemical analysis of flavor is the measurement of ligands that bind to these proteins. The binding of a single ligand causes the release of large numbers of intracellular "second messenger" molecules (e.g., Ca^{2+}, cAMP, IP3). These second messengers amplify the signals, which are then further modulated by the central nervous system to yield perceptions in the brain. Sensory analysis of flavor is the measurement of the perceptions induced by these ligands. The field of sensory science is concerned with the direct measurement of the flavor response using the quantification of human behavior in controlled testing environments. Academic and industrial researchers have developed a collection of sensory protocols that are used extensively to measure flavor in foods. Excellent reviews of these methods have been published over the last decade (Meilgaard et al., 1987; Burgard and Kuznicki, 1990; Stone and Sidel, 1993; Chambers and Wolf, 1996; Lawless and Heymann, 1998). This section is concerned with the use of analytical chemistry to predict sensory response, so no sensory protocols will be described beyond a brief introduction to olfactometry, as this is applied to the assessment of gas chromatographic effluents.

Flavor analysis becomes useful when the concentration values for flavorants are translated into "activity" values. The simplest of these translations is to divide the concentration obtained for a flavorant by the sensory threshold of the flavorant determined in the food. Called a "flavor unit," this ratio combines a psychophysical "constant" (although they are never constant) with a chemical measurement to produce an activity value. The activity value can then be used to predict flavor response, the ultimate goal of flavor analysis. Since the threshold values used to compute the flavor units are collected from psychometric experiments involving human subjects, they are highly variable. For the analyst, this means that the chemical measurement is likely to be much more precise and accurate than the activity values computed. A small difference in the concentration of a particular flavorant loses its significance when it is combined with a highly variable psychometric to produce an activity value. Nevertheless, a chemical measurement of flavorants, no matter how precise, is of little meaning without an appropriate activity transformation.

Activity of a flavorant is the relationship between its concentration and the perception it induces and is represented by the simple equation:

activity = concentration × coefficient (response/concentration unit).

As is the case with all biologically active chemicals (i.e., enzymes and substrates), activity (i.e., turnover number and K_m) is the relationship between a defined response and the concentration that causes it. Table G.0.1 shows some flavor activities and their definitions.

Contributed by Terry Acree

Table G.0.1 Some Flavor Activities and Their Definitions

Activity	Definition
Iso-sweetness scores	Concentration needed to yield the same sweetness as 5% sucrose solution in water
Odor activity values	The number of times an odorant is present above its threshold as measured in the food matrix
Scoville units	The number of dilutions necessary to reach the threshold of hotness for capsaicinoids and other chemesthetics

Flavor analysis is further complicated by the great range of active concentrations observed for flavorants. For example, hundreds of milligrams per gram of sucrose are required to induce sweetness while only a few picograms per gram of many odorants can induce a smell response. In addition, interpreting measured activities in terms of the sensory perceptions they invoke requires some knowledge of psychometrics. By far, the greatest impediment to the intelligent interpretation of measured flavorant concentrations is the tendency of one flavorant in a mixture to suppress the response of another in the same mixture. Although poorly documented, it is possible for some food constituents to synergistically enhance flavor of a food component. However, suppression is by far the most commonly observed interaction. Therefore, the interpretation of flavor measurements is never straightforward, but must be tempered with a knowledge of the sensory properties of the analyte in the food matrix being tested.

The challenge posed by the measurement of flavor chemicals is to quantify these chemicals in a complex food matrix at the range where they exhibit flavor activity. For glucose, a simple calorimetric test or even the measurement of refractive index is precise and sensitive enough to predict the response to sweetness, but to predict the effects of the earthy smelling compound geosmin ((4S,4αS,8αR)-octahydro-4,8α-dimethyl-4α-(2H)-naphthalenol) in drinking water requires quantification at <1 ng/ml. Therefore, the methods presented in this section were chosen to yield data that predict significant sensory response. Several reviews in the last decade summarize the theories and knowledge of the physiology of flavor and the major issues in flavor analysis (Acree and Teranishi, 1993; Murphy, 1998; Teranishi et al., 1999).

This section begins with the analysis of volatile flavors or odorants as introduced in *UNIT G1.1* with the preparation of samples for gas chromatography—the most common tool used for analysis of odorants. *UNIT G1.2* provides protocols for the isolation and concentration of samples for GC analysis. *UNIT G1.3* describes the use of isotope dilution analysis and *UNIT G1.4* describes the use of chiral chromatography for the detection of foods and flavor adulteration. *UNIT G1.8* describes the combination of gas chromatography and olfactometry (GC/O), which is the most commonly used method for odor analysis to yield odor activity values. *UNIT G1.5* provides a multitude of protocols that are used for analysis of citrus oils. *UNIT G1.6* describes the use of solid-phase microextraction as a means of sampling volatile compounds for flavor analysis. *UNIT G1.7* describes the use of mouth simulators for producing effluents with volatile ratios that are similar to those found in human exhaled breath during eating.

The following chapter (G2) covers the measurement of acid tastants, i.e., chemicals responsible for the acid and sour taste as well as the perception of astringency. *UNIT G2.1* presents methods for potentiometric and colorimetric titration of acid tastants, and *UNIT*

G2.2 presents two complementary methods for HPLC separation of nonvolatile acids. Future supplements of Current Protocols in Food Analytical Chemistry will cover the measurement of sweeteners, astringents and other chemesthetic flavorants (the chemicals that yield a flavor response in the mouth other than the simple taste system), bitter substances and chemicals that cause the "umami" response. Umami is the Japanese word for the response to monosodium glutamate, certain nucleotides, and similar compounds. In English-speaking countries umami is often translated as savory. Unfortunately, savory is also used to describe the large class of foods high in hydrolyzed protein. The use of the term "umami" makes it clear we are talking about a flavorant.

LITERATURE CITED

Acree, T.E. and Teranishi, R. 1993. Flavor Science: Sensible Principles and Techniques. American Chemical Society, Washington, D.C.

Burgard, D.R. and Kuznicki, J.T. 1990. Chemometrics: Chemical and Sensory Data. CRC Press, Boca Raton, Fla.

Chambers, E.C. and Wolf, M.B. (eds.) 1996. Sensory Testing Methods. American Society for Testing and Materials, Manual Series MNL 26. West Conshohocken, Penn.

Lawless, H.T. and Heymann, H. 1998. Sensory Evaluation of Food, Principles and Practices. Chapman & Hall, New York (now vended by Aspen Publishers).

Meilgaard, M., Civille, G.V., and Carr, B.T. 1987. Sensory Evaluation Techniques. CRC Press, Boca Raton, Fla.

Murphy, C. 1998. Olfaction and Taste XII and Int. Symp. *Ann. N.Y. Acad. Sci.* 855.

Stone, H. and Sidel, J.L. 1993. Sensory Evaluation Practices, 2nd ed. Academic Press, San Diego.

Teranishi, R., Wick, E.L., and Hornstein, I. 1999. Flavor Chemistry: Thirty Years of Progress. Kluwer Academic/Plenum Publishers, New York.

Terry Acree

Smell Chemicals

G1.1   **Direct Sampling**   **G1.1.1**
    Basic Protocol 1: Direct Headspace Sampling   G1.1.1
    Basic Protocol 2: Solvent Extraction   G1.1.3
    Support Protocol: Retention Indexing   G1.1.5
    Reagents and Solutions   G1.1.7
    Commentary   G1.1.7

G1.2   **Isolation and Concentration of Aroma Compounds**   **G1.2.1**
    Basic Protocol 1: Simultaneous Distillation Extraction   G1.2.1
    Basic Protocol 2: Volatile Traps for Concentration: Adsorption and Desorption   G1.2.3
    Commentary   G1.2.6

G1.3   **Identification and Quantitation of Aroma Compounds**   **G1.3.1**
    Basic Protocol: Identification (GC-FID; GC-MS) and Quantification of
        Volatiles (GC-FID)   G1.3.1
    Alternate Protocol: Quantification of Aroma Compounds by Isotope Dilution
        Assay (IDA)   G1.3.4
    Commentary   G1.3.6

G1.4   **Stereodifferentiation of Chiral Odorants Using High-Resolution Gas
Chromatography**   **G1.4.1**
    Basic Protocol 1: Determination of Enantiomer Composition Using Single-
        Dimensional High-Resolution Gas Chromatography   G1.4.2
    Alternate Protocol 1: Determination of Enantiomer Composition Using
        Multidimensional Gas Chromatography (MDGC)   G1.4.4
    Basic Protocol 2: Sensory Discrimination of Chiral Flavor Compounds Using
        Gas Chromatography-Olfactometry (GC-O)   G1.4.8
    Alternate Protocol 2: Sensory Discrimination of Chiral Flavor Compounds
        Using Multidimensional Gas Chromatography (MDGC)   G1.4.11
    Reagents and Solutions   G1.4.12
    Commentary   G1.4.13

G1.5   **Analysis of Citrus Oils**   **G1.5.1**
    Strategic Planning   G1.5.1
    Basic Protocol 1: Qualitative Analysis of Citrus Oils by Gas Chromatography   G1.5.2
    Basic Protocol 2: Quantitative Analysis of Citrus Compounds by Gas
        Chromatography   G1.5.3
    Basic Protocol 3: Analysis of Limonene Purity by Rapid Gas Chromatography   G1.5.5
    Basic Protocol 4: Analysis of Citrus Oils by Pycnometry   G1.5.5
    Basic Protocol 5: Analysis of Citrus Oils by Refractive Index   G1.5.5
    Basic Protocol 6: Analysis of Citrus Oils by Polarimetry   G1.5.6
    Basic Protocol 7: Quantification of Total Oil From Whole Fruit or Wet Peel   G1.5.6
    Alternate Protocol 1: Quantification of Press Cake Oil   G1.5.9
    Alternate Protocol 2: Quantification of Oil in Dry Peel and Pellet   G1.5.9
    Alternate Protocol 3: Quantification of Oil in Press Liquor and Molasses   G1.5.10
    Support Protocol 1: Determination of Moisture Content   G1.5.10
    Basic Protocol 8: Quantification of Total Aldehydes in Citrus Oils by
        Hydroxylamine Titration   G1.5.11
    Alternate Protocol 4: Quantification of Total Aldehydes in Lemon Oil by
        Acid/Base Titration   G1.5.11
    Alternate Protocol 5: Quantification of Total Aldehydes in Orange and
        Grapefruit Oil Using *N*-Hydroxybenzenesulfonamide   G1.5.12
    Basic Protocol 9: Determination of Volatile Esters   G1.5.13
    Reagents and Solutions   G1.5.14
    Commentary   G1.5.15

Contents

1

223

224

G1.6 **Solid-Phase Microextraction for Flavor Analysis** **G1.6.1**
Basic Protocol: SPME of Food Headspace for Gas Chromatography G1.6.2
Alternate Protocol: Submersion SPME in Liquid Samples G1.6.3
Support Protocol 1: Quantification of Headspace Extraction G1.6.3
Support Protocol 2: Dilution Analysis G1.6.6
Commentary G1.6.7

G1.7 **Simulation of Mouth Conditions for Flavor Analysis** **G1.7.1**
Basic Protocol: Use of the Retronasal Aroma Simulator (RAS) G1.7.2
Alternate Protocol: Use of the Model Mouth G1.7.5
Reagents and Solutions G1.7.6
Commentary G1.7.7

G1.8 **Gas Chromatography/Olfactometry** **G1.8.1**
Basic Protocol 1: Gas Chromatography/Olfactometry Using Direct Sniffing G1.8.1
Basic Protocol 2: Dilution Analysis With Gas Chromatography/
Olfactometry Using Direct Sniffing G1.8.4
Basic Protocol 3: Time Intensity Method for Gas Chromatography/
Olfactometry Using Direct Sniffing G1.8.5
Alternate Protocol 1: Detection Frequency With Gas Chromatography/
Olfactometry Using Direct Sniffing G1.8.6
Alternate Protocol 2: Posterior Intensity With Gas Chromatography/
Olfactometry Using Direct Sniffing G1.8.7
Commentary G1.8.9

Direct Sampling

Because the chemicals that stimulate the olfactory epithelium are released from the food matrix during eating, a process called retronasal smell, sampling methods for gas chromatography analysis should reflect this release; however, most analyses of aroma chemicals either reflect the composition in the food matrix or the equilibrium gas phase above the food, neither of which truly reflect the composition at the receptor site during eating. An exception is the equilibrium headspace composition, which approximates the composition at the olfactory epithelium when someone sniffs a food orthonasally (Deibler, 1999). This is important when the analyst is concerned about environmental aroma, fragrance materials, or the simple sniffing experience that can accompany eating (e.g., formal wine tasting).

The core technology used in the analysis of aroma chemicals is gas chromatography (GC); therefore, foods must be sampled so they can be introduced on to a GC column. For liquid samples it is possible to inject them into split, splitless, or on-column injectors directly. This is the preferred method for the analysis of synthetic aromas, essential oils, and aroma standards; however, solid or dilute liquid samples need to be extracted, distilled, or gas-phase generated in order to obtain useful results. This unit begins with simple direct analysis of a synthetic flavor (see Basic Protocol 1) followed by the analysis of a dilute liquid sample by solvent extraction (see Basic Protocol 2). It ends with a protocol for determining retention indices (see Support Protocol).

It is highly recommended that all retention properties be normalized to one or preferably several reference standards. Retention indexing, first described by Kováts (Kováts, 1965; see Support Protocol), is by far the most useful standardization procedure for aroma chemicals, because there are published databases (see Internet Resources) listing hundreds of aroma chemicals including their Kováts indices on several substrates.

DIRECT HEADSPACE SAMPLING

The simplest and most direct method to analyze for flavor chemicals is to separate them chromatographically and to quantify them with an appropriately sensitive detector. This is generally difficult for two reasons: (1) the chemicals are present at extremely low concentrations, and (2) the sample contains many interfering compounds. For the analyst, it is often a matter of trying a method and then altering it to minimize interferences and enhance sensitivity. Consider a simple example testing the strength of a concentrated Concord grape sample. There are two compounds that are responsible for most of the aroma of Concord grape flavor: methyl anthranilate and β-damascenone. Methyl anthranilate, generally present between 1000 and 10,000 times the concentration of β-damascenone, is the only odor-important volatile that can be detected by direct injection of a headspace sample as described in this protocol. To detect β-damascenone the sample needs to be concentrated; however, the direct extraction described in this unit (see Basic Protocol 2) is still not good enough to give convincing data for β-damascenone using FID detection. More sensitive and selective methods like those in *UNITS G1.2-G1.4* are needed to quantify very odor-active trace components.

Materials

Blank solution (e.g., ~5% ethanol)
Indexing standards solution of *n*-alkanes (C7 to C18 in 0.005% pentane; see recipe)
Essence

Contributed by Terry E. Acree

Gas chromatograph:
  Flame ionization detector (FID), 250°C
  Column oven
  Ov101 or DB5 capillary column (e.g., 20 m × 0.32–mm; f = 0.25 μm; see Table G1.1.1)
  Helium linear velocity: 36 cm/sec (~2 ml/min)
  Injector, 250°C
10-ml gas-tight syringe
100-ml flask fitted with a rubber septum

1. Program the GC oven to run isothermal for 3 min and then program to increase 2° to 6°C/min to 225°C. Hold for 10 min.

2. Inject 1 μl blank solution (e.g., ~5% ethanol) into the GC and start the program.

   This will show any artifacts or contaminants from either the injector or the column.

3. Inject 1 μl indexing solution of *n*-alkanes and start the GC program.

4. Record the retention time for each *n*-alkane to be used for indexing (see Support Protocol).

5. Place 50 ml essence in a 100-ml flask fitted with a rubber septum and allow it to equilibrate for an hour or more.

6. Withdraw 5 ml headspace gas from above the essence sample into a 10-ml gas-tight syringe. Inject slowly into the GC.

7. Record the retention time of each peak that elutes.

   Use the data from the n-alkanes to convert each time into a retention index (see Support Protocol).

8. Use a data base of retention indices (e.g., the Flavornet; see Internet Resources) to tentatively identify each peak.

9. Quantify peaks of interest (e.g., methyl anthranilate in Concord grape samples).

Table G1.1.1 GC Substrates, Polarity, and Other Phases of Similar Polarity

Phase	Polarity	Phases of similar polarity
OV-101	Nonpolar	OV-1, SP-2100, Apiezon L., DC11, DC-200, JXR Silicone, UC-W98X, SPB-1, SE-30, BP-1, CP Sil5CB, DB-1, DC-200, GB-1, OV-1CB, OV101
DB-5	Very slightly polar	SPB-5, SE-54, PTE-5, BP-5, CPSil 8CB, GB-5, 007-2, PVMS-54, Rtx-5, SE-52, SE-52CB, SE-54CB, RSL-200, Ultra-2
OV-17	Intermediate	SBP-20, SPB-35, SPB-1701, DB-1301, DC-550, Dexsil 400, 007-7, 00-17, OV-17, OV-1701, Rtx-20, SP-2250, DB17, GB-17, HP-17, PVMS-17, RSL-17, Rtx-50
C20M	Polar	Nukol, SP-1000, AT-1000, BP-20, CP-Wax51, FFAP, FFAP-CB, H-FFAP, OV-351, Stabilwax-DA, Superox, BP-20, CP Wax57CB, DB Wax, Pluronics L-121, RSL-310, SUPELCOWAX 10, Stabilwax, Superox 20M
OV-275	Very polar	SP-2330, SP-2331, SP-2380, CP SIL 84, Rtx-2330, SH-60, SH-70, Silar 9 CP, CPS1, CPS2, CP SIL 88, Rtx-2330, Rtx-2340, SH-80, SH-90, Silar 10CP

The most common method for the separation and concentration of flavor chemicals before chromatography is solvent extraction. If the aroma active components in a sample are less than a microgram/liter then solvent extraction followed by fractional distillation can be used to concentrate the analytes above 1 µg/liter. This is done for two reasons: (1) to remove the odorants from some of the interfering substances and nonvolatiles, and (2) to concentrate the sample for greater sensitivity. The choice of solvent(s) depends on a number of issues, but similar results can be obtained with many solvents. Table G1.1.2 lists a number of solvents, their polarity, and physical properties. Pentane is the least polar and ethyl acetate the most. The sample must be an aqueous or dilute sample, dissolved or slurried into water to a final concentration of 80% to 90% water. Dilute aqueous samples will present the greatest polarity difference between the solvent and the sample, driving more volatiles into the extracting solvent.

Materials

Sample
Chromatographic grade pentane
$MgSO_4$, anhydrous
Ethyl acetate (optional)

2-liter extraction flask (fermentation or Erlenmeyer flask)
2-liter separatory funnel with Teflon stopcock and ring stand
Liquid and powder funnels
250-ml rotary evaporator flask
Rotary evaporator:
  Pressure gauge
  Valve for a vacuum break
  Needle valve to control the pressure
  Waterbath to maintain temperature between 10° and 30°C
  10-ml pear-shaped rotary evaporator flask

1. Place 1 liter sample and a Teflon- or glass-coated stirring bar in a 2-liter extraction flask.

2. Gently (to prevent emulsions) add 600 ml chromatographic grade pentane.

   The amount of solvent should be more than half but less than the total sample volume. All analysis should be done at the same sample to solvent ratio (e.g., 0.6:1.0).

Table G1.1.2 Polarity, Boiling Points, and Flash Points of Solvents Used for Extraction of Volatiles from Foods

Name	Polarity	Boiling Point (°C)	Flash point (°C)
n-pentane	0.00	35–37	−40
n-hexane	0.00	68–69	−22
carbon disulfide	0.15	46.3	−30
carbon tetrachloride	0.18	76–77	NA[a]
ethyl ether	0.38	34–35	−45
methylene chloride	0.42	40–41	NA[a]
ethyl acetate	0.58	77–78	+7.2
ethanol[b]	0.88	78.5	+13
methanol[b]	0.95	64–65	+12
water[b]	large	100	NA[a]

[a]NA is not available.

[b]These solvents are used to extract volatiles from solids followed by extraction with one of the other solvents.

3. Place the extraction flask on a magnetic stirrer and adjust the speed to <1 rotation per sec (~30 rpm) for 20 min.

> NOTE: *Do not let a cyclonic vortex develop between the solvent and sample layers. This can cause an emulsion to form. A slight dent should be visible in the surface of the solvent layer as the stirring bar sets up a toroidal current in the sample layer and this is transferred to the solvent layer.*

> *Unlike many of the common extraction protocols used in organic chemistry the extraction of odorants from complex foods will result in seemingly unbreakable emulsions because of the fats and proteins they often contain; therefore, shaking the phases is seldom recommended.*

4. Set up a 2-liter separatory funnel on a ring stand as shown in Figure G1.1.1. Use a liquid funnel to help direct the liquids into the separatory funnel, and a powder funnel with shark skin–fluted filter paper containing 50 g anhydrous magnesium sulfate on a 1-liter receiver flask.

> *If the bottom phase is the aqueous phase use a receiver flask without magnesium sulfate first.*

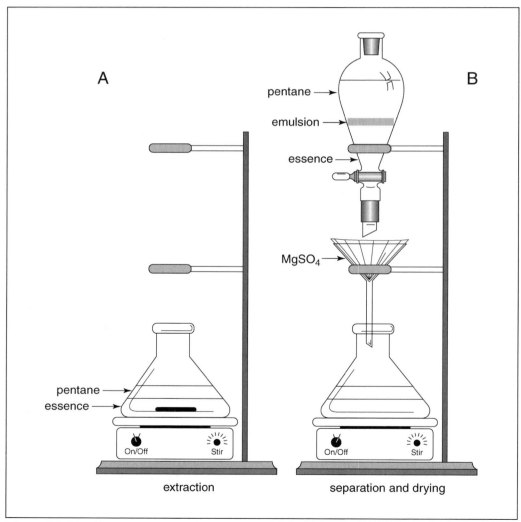

Figure G1.1.1 The apparatus used to extract liquid samples. The size of the glassware will depend on the scale of the samples. For example, a 500 ml sample will require a 2-liter fermentation flask for extraction but a 1-liter separatory funnel.

5. Remove the magnetic stirring bar and carefully pour the contents of the extraction flask into the separatory funnel. Allow the layers to separate.

   A rod can be used to direct liquids into the separatory funnel with a minimum of emulsion formation.

6. Allow the phases to separate (this may take as long as 1 or 2 hr).

   The goal is to have at least 90% of the organic phase free from any emulsion.

7. Separate the phases carefully, leaving any emulsion in the aqueous phase. Since pentane is less dense than water, allow the water phase and emulsions to drain directly into a 1-liter Erlenmeyer flask. If the odorant is highly polar, save the aqueous phase for extraction with a more polar solvent in step 9, otherwise discard.

8. Allow the organic phase to drain as quickly as possible through the magnesium sulfate.

   Don't let it overflow the powder layer and run directly through the filter paper and into the funnel. This is the nonpolar extract.

For highly polar odorants:

9. (Optional) If the sample contains highly polar odorants, set up the extraction system again and put the extracted aqueous phase back into the extraction flask. Add 600 ml ethyl acetate and repeat steps 2 through 6. This is the polar extract.

   This step is used to extract the sample with a more polar solvent.

   Uncleaned glassware from the nonpolar extraction can be used in this step.

Concentrate the sample

10. Concentrate the organic phase to ~30 ml by placing ~150 ml at a time into a 250-ml rotary evaporator flask. Place the flask in the rotary evaporator, turn on the vacuum, and close the vacuum break. Using the needle valve, allow the vacuum to increase until solvent begins to condense. Add another aliquot of the extract (150 ml total) until all the extract has been concentrated to ~30 ml.

    NOTE: *Do not distill the solvent at too-low pressure (i.e., high vacuum) or the highly volatile odorants will be lost.*

    Place concentrate in brown glass bottles with Teflon lined caps. The sample can be stored at this concentration for many months. Room temperature is fine but refrigerators often have a more constant temperature than laboratories.

11. Concentrate the sample down to 1 ml in a 10-ml pear-shaped rotary evaporator flask. Run as described (see Basic Protocol 1).

    NOTE: *When using solvents that absorb a lot of water (e.g., ethyl acetate, diethyl ether) it is necessary to dry the extract a second time with anhydrous magnesium sulfate (1 g/10 ml sample). This gives samples that are ready for gas chromatography that are 1000× more concentrated than the original sample.*

RETENTION INDEXING

The usefulness of retention data from gas chromatography can be enhanced by reporting standardized times or retention indices (RI), which involves expressing retention in terms of a ratio of the retention time (RT) of an analyte to the RT of a standard. Retention scaling based on the Kováts (1965) method requires the chromatographic separation of a homologous series of normal paraffins, esters, and others, producing an index that is the ratio of the RT of an analyte minus the RT of a less retentive standard to the RT difference between

SUPPORT PROTOCOL

Smell Chemicals

a less retentive standard and the next most retentive standard (Figure G1.1.2). Retention times are so dependent on experimental conditions that it should be a rule of analytical chemistry that all retention times be reported as standardized indices. For the last twenty years the most useful resource in laboratories using gas chromatography/olfactometry (GC/O) has been a compilation of retention indices for flavor compounds (Jennings and Shibamoto, 1980). More recently, databases published on the world wide web have become even more useful (see Internet Resources).

See Basic Protocol 1 for materials.

1. Inject 1 µl standard solution into the GC under the same conditions used to chromatograph the sample. Start the program, and determine the retention time for all paraffin standards (see Basic Protocol 1).

    For example, programming the methyl silicone OV-101 (or DB-1, SE-30, or equivalent) column to run isothermally at 35°C for 3 min, then increasing oven temperature 4°C/min to 225°C, and finally holding the temperature isothermal for 10 min, will elute all the paraffins from C7 to C18 in 30 min. These are good conditions for the analysis of fruit-juice essences.

    The exact chromatographic conditions must be determined from the properties of the samples being analyzed and the goal of the analysis. The same conditions and column must be used to run the indexing standards. Once the chromatographic conditions have been chosen, it is a good idea to test them by running the indexing standards first. The chromatogram of the standards can establish sensitivity of the detector, the performance of the column and the chromatograph before the sample is tested.

2. Calculate the retention index of all the peaks of interest in the chromatogram using the following formula.

$$RI_x = [(T_x - T_n)/(T_{n+1} - T_n)] \times (n \times 100)$$

Where:

T_x = retention time of a peak (x) to be indexed

T_n = retention time of the paraffin that immediately precedes x

T_{n+1} = retention time of the paraffin that immediately follows x

RI_x = Kováts retention index of x

n = paraffin carbon number of the paraffin that immediately precedes x

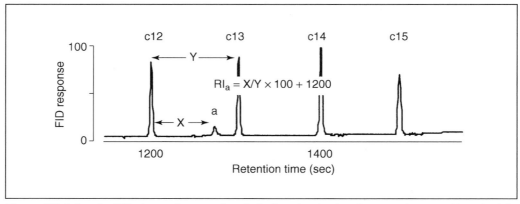

Figure G1.1.2 An FID chromatogram of an indexing standard and parameters used to calculate retention indices. For equations, see Support Protocol, step 3.

Use deionized or distilled water in all recipes and protocol steps. For common stock solutions, see APPENDIX 2A; for suppliers, see SUPPLIERS APPENDIX.

Paraffin stock solution, 1000 ppm

Using a micropipet, add one drop of each *n*-paraffin from heptane to hexadecane into a 100-ml brown glass bottle containing 50 ml of an appropriate solvent (e.g., pentane for nonpolar columns). Seal with a Teflon-lined screw cap. Store at 4°C in the refrigerator for years.

Adding 2 drops of one of the paraffins in the middle (e.g., tridecane) will yield a chromatographic pattern that makes it easy to identify all the peaks quickly. Knowledge of the exact concentration of the paraffins is not essential for indexing, but such information could be used to monitor GC performance.

Indexing standards solution, 10 ng/µl

Dilute an aliquot of the paraffin stock solution 1:100 to produce a 10 ng/µl solution. For example, add 5 ml solvent followed by 50 µl paraffin stock solution (see recipe) to a 10-ml brown bottle and seal with a Teflon-lined screw cap. Store at room temperature near the GC for months if necessary.

COMMENTARY

Background Information

Any number of extraction protocols can be used to prepare samples for gas chromatography, but in every case the goal is to simplify the chromatogram and amplify the response for the odorant. The protocol given here covers the broadest range of polarities possible using solvent extraction from an essentially aqueous sample. This and similar extraction methods are ideal for many fruit and vegetable products, homogenized meat products, and grain products ground into a slurry with water. There have been many papers comparing the virtues of different solvent extraction protocols (Leahy and Reineccius, 1984) and all have some limitation. In the protocol reported here, the use of pentane and ethyl acetate will provide an excellent result with a wide range of samples. In addition, except for their fire and explosive potentials, they are very safe solvents to use, as they have a low toxicity and no potential to accumulate in the body.

There are many samples that can be injected into a gas chromatograph in order to determine what aroma active volatile is present or to quantify a particular one (Maarse and van der Heij, 1994). Unfortunately, many of the protocols that were followed in the past did not rely on meaningful standards or retention indexing, or failed to use GC sniffing or the more formal GC/O procedures to establish that the peaks detected had odor activity; however, if the object of the analysis is to simply monitor one or two odorants known to contribute positively or negatively to the aroma of a sample, the procedure should be optimized to analyze the specific odorants accurately, precisely, and efficiently.

Critical Parameters and Troubleshooting

To do direct injections (see Basic Protocol 1) requires that the sample be low in nonvolatiles (e.g., fats, proteins, and carbohydrates). The column properties should be appropriate to achieve the objective of the analysis. For example, the ability to resolve methyl anthranilate from other analytes is critical for the assessment of flavor quality in Concord grape, but in the analysis for geosmin contamination of drinking water, sensitivity to chemicals at less than a nanogram per liter is essential. In this case a concentration step is necessary. When a sample is in a solvent that interferes with the vaporization process—e.g., ethanol, polypropylene glycol (a common carrier for synthetic flavors), and others—improved results (i.e., less peak broadening at the start of the chromatogram) can be obtained by simply adding an equal amount of a highly volatile solvent (low heat of vaporization) such as diethyl ether, preferably redistilled. This will aid in the transfer of the sample to the column and the extraction of the analytes by the substrate, and is often necessary for good splitless or on-column injections. Usually, limiting the concentration of any highly polar solvents (e.g., ethanol, polyethylene glycol) to <25% will minimize this problem.

Most if not all aroma compounds have a molecular weight of less than 350 amu and can be chromatographed within the operating temperatures of many GC substrates, especially the methyl silicones. Although the choice of substrate depends on the goals of the analysis, most flavor chemists use nonpolar substrates (e.g., SE-30, SE-54, OV-101) for samples low in polar components, and polar substrates for samples in which polar compounds (e.g., volatile acids) have a significant effect on the analysis. As pointed out in *UNIT G1.3*, the use of both substrates are required when GC retention properties are being used for identification purposes; however, it is always a good idea to use substrates that are common in the literature in order to produce data that is comparable with published data. For example, the "Flavornet" and the "Threshold" data bases on the world wide web present data for aroma compounds separated on OV-101, DB-5, OV-17, C20M, or their equivalents (see Internet Resources). These are good substrates to consider first. Table G1.1.1 list these substrates and their equivalents among the many commercially available GC stationary phases.

The types of general detectors most commonly used for the analysis of aroma compounds include flame ionization detection (FID) and electron impact mass spectroscopy (EI/MS). Although a score of special detectors can be used to detect specific aroma compounds, the three most commonly used are mass fragmentography, often called selected ion monitoring (SIM, a registered trademark of Agilent Technologies), chemical ionization mass spectroscopy (CI/MS), and GC/O. The single most commonly used among these is GC/O, since it can yield odor activity measurements directly (Acree, 1997). Furthermore,

Table G1.1.3 Troubleshooting Table

Problem	Possible cause	Solution
Compounds do not match the reported index	A damaged or contaminated column	Use a new or different column (DB-5 is the most stable column for both polar and nonpolar compounds)
Compounds do not match the index of the same standard in a different solvent	Solvent effect on the elution of the standard	Use standards dissolved in the same solvent the analytes will be dissolved in

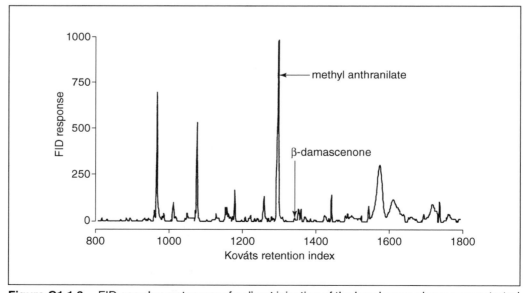

Figure G1.1.3 FID gas chromatogram of a direct injection of the headspace above concentrated extract of Concord grape essence using OV 101 substrate. Note the size of the methyl anthranilate peak and the absence of a convincing peak for β-damascenone.

when GC/O is combined with the use of authentic standards, it is the most robust method for the identification of odorants.

Solvent extraction

A number of solvents have been used to extract volatiles for aroma analysis but the optimum choice depends on a compromise. Table G1.1.2 lists the most common solvents used to extract odorants from foods. Although pentane and ethyl acetate are flammable, they have a very low toxicity, represent extremes in polarity, and a sequential extraction using these two solvents will remove most of the volatile odorants from aqueous samples (see Basic Protocol 2); however, if the desire is to do a simpler one-step extraction, then a solvent should be chosen with a polarity that will extract the volatiles of interest. For example, maltol is not extracted well with pentane, and 4-hydroxy-2,5-dimethyl-3(2H)-furanone, the smell of strawberry, is almost insoluble; therefore, the choice of the optimum solvent depends on the analyte and may require some testing to find.

Retention indexing

This method (see Support Protocol) requires a set of *n*-paraffins or *n*-ethyl esters to produce data that is comparable to that published on the internet. Table G1.1.3, the troubleshooting table below, lists some of the frequent problems encountered, their probable cause and possible solution.

Anticipated Results

Figure G1.1.3 shows a chromatogram of the headspace of Concord grape essence prepared by direct injection. At retention index 1320 is the peak caused by methyl anthranilate, one of the strongest odorants characterizing Concord grapes; however, β-damascenone, the second most potent odorant in Concord grapes, elutes at 1360 but is not visible. This is because β-damascenone is 1000× more potent (i.e., its odor threshold is 1000× lower than methyl anthranilate). This is typical result for the direct injection of headspace from natural products. Figure G1.1.4, on the other hand, shows the injection of an extract of Concord grape essence concentrated 500-fold with the β-damascenone peak large enough for quantitation.

Time Considerations

It takes ~1 hr to run a blank, and another hour to run the alkane standards by direct headspace sampling (see Basic Protocol 1). With OV-101 and similar substrates, these standards need be run only once every day or two. With more polar and less stable substrates, or with samples that contain a lot of nonvolatile material, the standards and blanks need to be run more often. In the worst case, these must be run

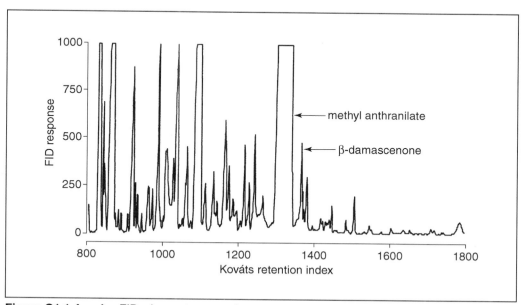

Figure G1.1.4 An FID chromatogram of concentrated extract of the same Concord grape essence shown Figure G1.1.3, drawn to display the data on a linear retention index scale. By simply comparing the index of a peak with the data listed in the flavornet, the odorants that have similar retention indices can be determined. Notice how large the methyl anthranilate peak is, but still no convincing peak for β-damascenone, even though both compounds have the same odor activity (intensity).

Smell Chemicals

G1.1.9

before and after each sample. Under these conditions, only a couple of samples can be run in a day. In the best case it should take ~1 hr to run a sample including the recycle time for the GC oven.

Literature Cited

Acree, T.E. 1997. GC/Olfactometry: GC with a sense of smell. *Anal. Chem.* 69:170A-175A.

Deibler, K.D. 1999. Gas chromatography - olfactometry (GC/O) of vapor phases. *In* Flavor Chemistry: Thirty Years of Progress (R. Teranishi, E.L. Wick, and I. Hornstein, eds.) pp. 387-395. Kluwer Academic/Plenum Publishers, New York.

Jennings, W.S. and Shibamoto, T. 1980. Qualatative Analysis of Flavor and Fragrance Volatiles by Glass Capillary Gas Chromatography, p. 472. Academic Press, New York.

Kováts, E. 1965. Gas chromatographic characterization of organic substances in the retention index system. *Adv. Chromatographia* 1:229.

Leahy, M.M. and Reineccius, G.A. 1984. Comparison of methods for the isolation of volatile compounds from aqueous model systems. *In* Anal. Volatiles: Methods Appl., Proc. - Int. Workshop (P. Schreier, ed.) pp. 19-47. Walter de Gruyter, Berlin.

Maarse, H. and van der Heij, D.G. 1994. Trends in flavour research. *Dev. Food Sci.* 35:516.

Key References

Acree, T.E. and Teranishi, R. 1993. Flavor Science: Sensible Principles and Techniques. American Chemical Society, Washington DC.

A monograph that touches upon most of the important issues in flavor clinical analysis.

Belitz, H.-D. and Grosch, W. 1999. Food Chemistry: Second Edition. Springer, Berlin.

Textbook on food chemistry with an excellent discussion of flavor chemicals.

Internet Resources

http://www.nysaes.cornell.edu/flavornet

Contains retention indicies on 4 substrates, CAS numbers, and odor qualities for over 400 chemicals identified by GC/O in food.

http://www.odor-thresholds.de

Contains thresholds and retention indices for several 100 odorants.

Contributed by Terry E. Acree
Cornell University
Geneva, New York

Isolation and Concentration of Aroma Compounds

Aroma compounds are present in minute levels in foods, often at the ppb level (μg/liter). In order to analyze compounds at these levels, isolation and concentration techniques are needed. However, isolation of aroma compounds from a food matrix, which contains proteins, fats, and carbohydrates, is not always simple. For foods without fat, solvent extraction (*UNIT G1.1*) can be used. In foods containing fat, simultaneous distillation extraction (SDE; see Basic Protocol 1) provides an excellent option. Concentration of headspace gases onto volatile traps allows sampling of the headspace in order to obtain sufficient material for identification of more volatile compounds. A separate protocol (see Basic Protocol 2) shows how volatile traps can be used and then desorbed thermally directly onto a GC column. For both protocols, the subsequent separation by GC and identification by appropriate detectors is described in *UNIT G1.3*.

SIMULTANEOUS DISTILLATION EXTRACTION

One design of the micro steam simultaneous distillation extraction apparatus is given in Figure G1.2.1. Both sample and solvent flasks are heated to their boiling points. Vapors coming from the sample and solvent flasks (Figure G1.2.1, G and F) are mixed in the

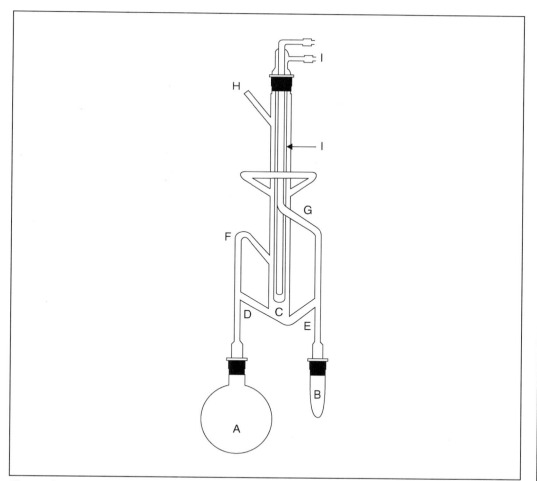

Figure G1.2.1 Simultaneous distillation extraction system. A, flask for water; B, flask for solvent; C, separation chamber; D, return tube for water; E, return tube for solvent; F, vapor tube for water; G, vapor tube for solvent; H, inlet/vent; I, cold finger. Modified from Godefroot et al. (1981).

Contributed by Deborah Roberts and Hugues Brevard

central part of the apparatus (Figure G1.2.1, C) and condensed on the cold finger (Figure G1.2.1, I) . Volatiles are transferred from the water phase to the organic solvent. The organic and water phases return to their original flasks through Figure G1.2.1, D and E tubular parts. Gradually the volatile compounds are transferred from the sample to the organic solvent. At any time during the extraction, there is no contact between the solvent and the matrix.

Materials

Deionized water
Sample
Sodium chloride (optional)
Internal standard (optional)
Buffer (optional)
Organic solvent, freshly distilled or with high purity (see Table G.1.2.1)
Drying agent (e.g., magnesium or sodium sulfate)

Simultaneous distillation extractor, which may vary in the design with respect to the solvent density (e.g., Chrompack, Alltech, or many glass-blowers)
Cooling bath
Heating bath
Stirring plate

1. Degas deionized water by vacuum, while stirring for 10 min.

   Optional: Saturate with sodium chloride, add an internal standard (e.g., 250 µl of solution containing 25 µg of internal standard), and add buffer.

   Degassing allows the elimination of residual oxygen that may modify the nature of certain chemical compounds. A better aroma recovery is obtained when the aqueous phase is saturated with salt, while an internal standard addition allows semi-quantification.

2. Cool the cold finger condenser of the simultaneous distillation extractor to 0°C, using a cooling bath.

3a. *For analytic purposes:* In the <500-ml sample flask (see Fig. G1.2.1A) containing a stir bar, place a small quantity of sample, i.e., <20 g with ~250 ml of water.

   Alternatively, an internal standard, sodium chloride, and buffer can be added in place of water.

Table G1.2.1 Commonly Used Solvents

Example of solvents	Boiling point (°C)
Diethyl ether	34.6
Pentane	35
Dichloromethane	40
Chloroform	61
Hexane	69
Trichloroethylene	87
Heptane	99
Isooctane	99
2-Pentanone	102
Trichloroethane	110
Toluene	111
Octane	125

3b. *For preparative purposes:* In 2- to 3-liter sample flask containing a stir bar, place a higher quantity of sample, i.e., 20 to 1000 g with ~250 ml of water.

4. Place adequate quantities of organic solvent (see Table G1.2.1) in solvent flask (see Fig. G1.2.1B) containing a stir bar.

   Typical quantities vary between 4 and 5 g of solvent for analytical purposes and between 40 to 50 g for preparative purposes.

   The solvent must be nonmiscible with water and should have a boiling point as low as possible. Solvents such as dichloromethane (higher density than water) and diethyl ether, pentane, or the mixture of both (lower density than water) are commonly used. Freons, chlorinated solvents, or alkanes can be used. Toxic solvent should be avoided.

5a. *For a solvent heavier than water:* Start heating the solvent first in order to fill the central part of the extractor (see Fig. G1.2.1 C) full of solvent, then start to distill the water. Heat to the boiling point of the liquids and continue the distillations.

5b. *For a a solvent with a lower density than water:* Start heating the water flask first. Heat to the boiling point of the liquids and continue the distillations.

   If sample does not contain fat or oil, the duration of the extraction should not exceed 1 hr. If it contains fats one must verify the extraction time that is necessary for an efficient aroma recovery. Depending on the matrix, one must also verify the appropriate extraction time, prior to beginning. Generally, this time does not exceed 1 hr. Solvent injection and quantification by GC/MS after the SDE is usually used to verify the minimum extraction time needed.

6. Remove the heating plate or bath from the solvent flask and let the distillation of the sample continue for 10 min.

   This will allow a better recovery of volatiles contained in tubular parts Figure G1.2.1, D and E.

7. Collect the organic phase and add a drying agent, typically magnesium or sodium sulfate.

   For 5 g of solvent, ~100 mg of drying agent is sufficient.

8. Inject the organic phase without or after concentration onto a GC column in splitless or on-column mode.

   Concentration can be performed under a gentle stream of inert gas or with a micro-concentration apparatus (e.g., Kuderna-Danish sample concentrator, Supelco, or microconcentrator). This step generates volatile losses (mainly very volatile compounds that have a boiling point lower than the solvent) and will modify the quantitative ratio.

VOLATILE TRAPS FOR CONCENTRATION: ADSORPTION AND DESORPTION

Adsorbent traps can be used to concentrate volatile compounds without solvents so that the minute quantities present in a product (often ppb levels for odor-impact compounds) can be detected. Three steps are involved: (1) concentration of headspace gas on a trap, (2) thermal desorption of the trap and transfer of volatiles to a gas chromatograph, and (3) cryofocusing of the volatiles at the beginning of the GC column. The subsequent separation by GC and identification by an appropriate detector is the subject of *UNIT G1.3*. Several of the specifics of this methodology are related to the particular thermal desorber, which may use different sizes and types of traps and have different construction designs. Although several on-line systems are available that incorporate all three steps, the methodology here will describe off-line systems.

BASIC PROTOCOL 2

Table G1.2.2 Several of the Most Popular Adsorbents for Volatile Traps

Absorbent type	Advantages	Limitations	Typical use
Tenax TA 60-80 mesh	Good thermal stability and desorption. Traps little water. Generates few artifacts.	Limited adsorption and breakthrough of very volatile compounds	For general volatile trapping. Among the most used adsorbents
Carbon molecular sieves, e.g., Carbosieves, Carboxen, and Ambersorb	Traps small volatiles well	Other volatiles not well trapped	In combination with other adsorbents
Activated charcoal	High retention of nonpolar volatiles	Water absorption. Artifact formation. Some difficulties with desorption	Less used for aroma compounds
Graphitized carbon (e.g., Carbotrap, Carbopack)	Trap hydrophobic compounds. Low affinity for water	Generally used with another adsorbent	Preferably chosen over activated charcoal for general volatile trapping

Materials

5% (v/v) dimethyldichlorosilane in toluene
Adsorbent material (see Critical Parameters and Table G1.2.2)

Silane-treated glass wool (Supelco)
Ultrasonic bath containing 1:1 (v/v) dichloromethane/methanol
100°C oven
Trap (e.g., see Figure G1.2.2) and trap caps
Vessel with connections for "purge-and-trap" (e.g., see Figure G1.2.2)
Thermal desorption system with cold trap
Gas chromatograph with detector

Prepare traps

1. Place small amount of silane-treated glass wool in trap (label this side of trap).

2. Silanize interior with 5% dimethyldichlorosilane in toluene for 5 min, rinse two times, 30 min each, in an ultrasonic bath containing 1:1 dichloromethane/methanol solution, then dry ≥2 hr in a 100°C oven.

3. Fill trap with ~150 to 250 mg adsorbent material.

   See Critical Parameters for a discussion of various adsorbent material options.

4. Place another glass wool plug at the top of the trap.

5. Precondition the traps by heating in the thermal desorber at the time recommended.

   For Tenax, 300°C for 2 hr is recommended. Thermal desorption of traps should be done with labeled side at the gas exit.

6. Keep the traps stored with their caps on in an odor-free environment such as a dessicator with positive nitrogen pressure.

   If the traps have not been used in a week, recondition them by heating for 10 min at the temperature recommended.

Isolation and Concentration of Aroma Compounds

G1.2.4

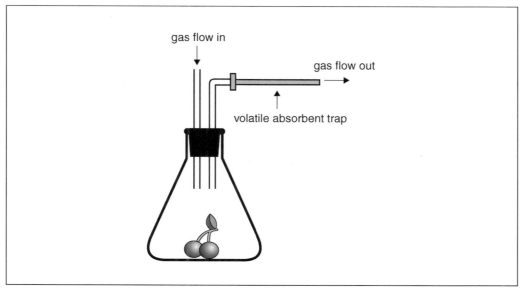

Figure G1.2.2 Typical set-up for dynamic headspace sampling using a volatile adsorbent trap.

Trap aroma compounds

A number of different set-ups can be used. Figure G1.2.2 shows one possible set-up of dynamic headspace trapping. Normally, the vessel contains the food product, an inlet for nitrogen gas, and an outlet connected to the trap. The inlet could be directed through the liquid food, often called "purge-and-trap," or above the food, often called dynamic headspace sampling.

7. Determine the time and nitrogen flow rate so that breakthrough (no more retention for a given substance) of the trap does not occur. Do this by attaching a second trap in series and analyzing this trap. For aroma compounds, smell the end of the trap to detect odor-active compound leakage. Even if the vessel is heated, the trap should be kept at room temperature.

    Trapping should be done with the labeled side at the gas entry.

8. Remove the trap and place caps on the two ends.

9. Analyze using thermal desorption system after optimization of conditions.

Desorb traps thermally and trap cryogenically

10. Place the trap in the desorber with the labeled side toward the gas exit.

11. Determine conditions of optimal desorption of traps: flow rate, temperature of traps, and transfer lines.

12. Test with a mixture of compounds of different retention indices by injecting 1 µl of compound dissolved in solvent at ~10 ppm levels onto the glass wool at the labeled side.

    For Tenax traps, a 250°C desorption temperature is recommended.

13. Adjust the cold trap to as cold a temperature as possible (temperatures between −100°C and −140°C are commonly used), and adapt as a function of the volatile compounds.

    Cryogenic trapping is used to ensure that all compounds that are desorbed from the trap begin gas chromatography at the same instant. While other types of oven cooling, for example with carbon dioxide, may retard the migration of compounds on the column, the chromatographic resolution will not be as good as with cryogenic trapping. Depending on the commercial thermal desorption system used, cryogenic trapping may be part of the instrument, or at the beginning of the column. Placing the first loop of the GC column in a Dewar of liquid nitrogen is also valid although the Dewar must be removed at the beginning of the GC run.

Cold trap heating should be done as quickly as possible for injection-like resolution.

GC analysis followed by detection as explained further in UNIT G1.3.

14. Bake the adsorbent trap at an elevated temperature with flow in the reverse direction so it is free from volatiles for the next analysis.

   Run blank trap(s) between series to ensure no carry-over contamination peaks. Also, for each sample type, check the trap to ensure that all volatiles have been desorbed during the bake-out.

15. Do a desorption of alkane standards added to the trap as in step 1 in order to establish retention indices.

   This is done in a separate experiment but in the same way as the sample analysis.

   Alkane standards can be prepared by dissolving 1 mg/liter of C5-C20 alkanes in water, and then injecting 5 µl on the trap.

COMMENTARY

Background Information

Of the two techniques described here, simultaneous distillation extraction (SDE) is a more complete volatile extraction procedure that serves to obtain quantitative information on the compounds contained in a food. Volatile trapping is a partial extraction procedure that samples the volatiles present in the headspace above a food, which are those with higher volatility. Extended trapping also induces additional volatilization of compounds initially contained in the food.

SDE

The main advantage of SDE over direct solvent extraction is the absence of contact between the matrix and the organic solvent during the extraction procedure. This allows foods of various compositions, including fat, to be analyzed. Generally, compounds that are not soluble in water are well extracted. The major drawback of such an SDE is that very soluble compounds generally stay in the water phase and are not efficiently extracted (e.g., vanillin, furaneol, phenylethylalcohol). Another drawback is the possible formation of thermal artifacts. This will not occur with thermally treated samples (i.e., coffee, cooked meat, cocoa) but will happen with fresh vegetables such as garlic and onion. Furfural, which is mentioned in many foodstuffs, is the major thermal artifact.

Different apparatus designs were proposed by several authors (Likens and Nickerson,

1964; Schultz et al., 1977; Godefroot et al., 1981). Maignial et al. (1992) developed a system for isolation of volatile compounds at room temperature by using a closed system under static vacuum. Later, Pollien and Chaintreau (1997) built a preparative device for up to 5 to 10 kg of material which is a semi-static system able to work under vacuum. Working under vacuum reduces thermal artifacts, which may occur with SDE experimentation. Samples in this case are steam distilled at ambient temperature. A solvent with a higher boiling point than that found in conventional SDE is used, which may cover up target volatile compounds during GC. This technique is less easy to handle than conventional SDE. Chaintreau (in press) is a review SDE developments.

Simple steam distillation under vacuum (Forss and Holloway, 1967; Joulain, 1986) may be used with the same goal to avoid thermal artifacts. The distillates (i.e., aromatic water phase) are further extracted by using the appropriate organic solvent. Solvents are then dried, concentrated, and injected onto a GC column.

Volatile traps

The key advantage of this headspace method, compared to static headspace analysis, is the sensitivity obtained. While static headspace shows the status at equilibrium and can measure thermodynamic constants, "purge-and-trap" or dynamic headspace can measure the kinetics of release by attaching several traps, each corresponding to a different time period (Roberts and Acree, 1995).

In flavor analysis, the most frequent use of volatile traps is in analyzing the flavor compounds in foods using "purge-and-trap" or dynamic headspace, followed by GC-MS or GCO. Additionally, the traps can be used to measure static headspace and air-matrix partition coefficients where air is pushed out of an equilibrated cell containing the sample onto a volatile trap (Chaintreau et al., 1995). Volatile traps have been also used for flavor release measurements during eating (Linforth and Taylor, 1993) or simulated eating (Roberts and Acree, 1995).

While the method described here uses thermal desorption to release the volatiles from the trap directly to the GC, another method less often used is solvent extraction of the trap. Even though this will result in a lower sensitivity as the complete trap contents are diluted in solvent, an advantage is the possibility for multiple injections. Additionally, cold traps rather than adsorbent traps are also used to trap volatiles (Badings et al., 1985) although the water and CO_2 that are also condensed can pose GC difficulties.

One of the key developments in the development of thermal desorption devices was the possibility for cryofocusing systems that have the advantage of "injection-like" samples. A short section of capillary tubing at liquid nitrogen temperatures (i.e., −160°C) traps the volatiles. When capillary columns replaced packed columns as the standard, complete flow from the desorption trap (5 ml/min minimum) to the capillary columns (~1 ml/min) was possible through the use of cryofocusing. The "split injection" interface was another development that splits the flow so that only a part of the desorbed volatiles entered the column. While this allowed the need for cryofocusing to be circumvented, sensitivity was lost due to the split.

Critical Parameters and Troubleshooting

SDE

The temperature of the baths must be precisely adjusted. It is best if these baths are dedicated to these extractions. Likewise, the cold finger condenser must be maintained at ~0°C, usually by the use of a cooling bath. The cooling temperature given by tap water (generally between 10° and 15°C) is not enough to insure a good condensation of volatile aroma, and losses may occur.

The choice of the solvent depends on the polarity of compounds contained in the product. Solvents such as dichloromethane, diethyl ether, and pentane are often chosen because they have low boiling points and are thus rapidly eluted in the GC and are easy to concentrate (Maignial et al., 1992).

In the case of foaming materials, the distillation must be carefully watched. Additional antifoaming agents can be used, but they will be coextracted and may cover some sample volatiles on the GC profile. Their use is not recommended.

Volatile traps

Sampling. The flow rate and time of trapping should be determined experimentally. Typically, flow rates in the range of 10 to 80 ml/min are used along with purge times on the order of 10 to 60 min. The flow rate can affect the type of compounds that are stripped, with highly volatile compounds being stripped with higher flow rates. An optimization study for tomato volatiles found a 20 ml/min optimum when using a 60-min trap time (Sucan and Russell, 1997). Another critical parameter to check is breakthrough. As time or flow rate is increased, compounds with higher affinity for the adsorbent material may displace other compounds. This will result in a distorted volatile profile. When performing the optimization studies, it is wise to begin with a less concentrated sample and evaluate the chromatography before testing the adsorption limits of the trap. This is to avoid system contamination. Even trace contamination of the traps and transfer lines causes large difficulties for GCO. If using pure standards, do so at dilute concentrations (i.e., ppm or lowest level needed for a reliable signal).

For the vessel purging, it is recommended that nitrogen or helium gas be used, rather than air, due to oxidation of the food product. The temperature chosen depends on the purpose of the experiment. Room temperature is often chosen to simulate volatile emanating from the product in its natural state. Slightly elevated temperatures (37°C) may better simulate flavor release from the mouth. Higher temperatures should not be used for samples that change with temperature and are normally consumed at lower temperatures, unless the purpose is to see the volatiles released upon heating. Temperatures over 60°C often result in too much water being trapped, which affects gas chromatographic separation and mass spectrometric detection when in splitless mode. If sensitivity is not an objective, split mode can be used to reduce the water on the column. Stirring is another parameter that can be included if, for example, the study's purpose is to understand the volatiles released upon chewing. After trapping, thermal desorption normally should take place immediately afterwards, although effective trap caps could add stability over time.

The sample vessel should be tested to ensure that it is leak-proof during purging with the trap attached. For example, this can be done by applying soapy water at joints. Instead of purging gas through the system, using a vacuum pump that is attached to the outlet of the trap, with a fitting allowing filtered air to enter, is another option to solve leaks (Wampler, 1997).

In some applications of purge-and-trap, foaming can be a problem. Possible solutions include using a bubble breaker or foam filtration at the top of the purging vessel (Wampler, 1997), a needle sparger, or using dynamic headspace analysis instead. Use of an antifoam emulsion can also help, although this will add some impurities and may change the flavor release, as these emulsions are oil-based.

Adsorbent choice. The choice of adsorbent material depends on the volatile compounds in the food. Of the synthetic porous polymers, the most widely used and best overall adsorbent is Tenax TA (poly-2,6-diphenyl-p-phenylene oxide) 60 to 80 mesh. While Tenax does not show an adsorption capacity for all volatiles, especially very small polar compounds such as acetaldehyde, it has good thermal stability and desorption capabilities. It also traps little water and generates very few artifacts. Table G1.2.2 shows a few limitations and advantages of various adsorbents, all of which can be purchased from chromatography suppliers. If very small volatiles are the goal, various Carbosieves could be used, or traps containing several adsorbents in series. Traps with mixed adsorbents should be desorbed immediately, before transfer between phases occurs.

Trap desorption. The choice of the thermal desorption apparatus is critical in order to avoid contamination and to be able to work with aroma compounds in a wide range of retention indices. In all systems, problems can be encountered due to reactive compounds or cold spots within the analyzer. It is recommended that all transfer lines, valves, or surfaces in contact with the volatile compounds be made of an inert material such as fused-silica or deactivated glass-lined stainless steel. Even more ideal are systems that do not have long transfer lines. Some thermal desorption systems offer a water removal system, which should not be used for recovery of flavor compounds, some of which are of low volatility and hydrophilic. Additionally, it is not appropriate to analyze thermally-labile compounds by thermal desorption of traps.

Anticipated Results

SDE

Very high sensitivity (to ppt) can be obtained with compounds that are extractable by SDE. However, highly water-soluble compounds are not extracted completely and compounds with very low boiling points, such as methanethiol, can be lost during concentration. In addition, compounds with gas chromatographic elutions simultaneously with the solvent are hidden from view. Quantification can be made by addition of an internal standard to the sample matrix.

Volatile traps

While the exact recovery depends on the volatility of the flavor compounds, most compounds can be detected with this method when present at ppb (mg/liter) concentrations. Reproducibility (CV) is between 5% and 10%. To quantify the amount trapped, an internal standard curve can be made by adding the standards in solvent directly to the trap just before thermal desorption on the side of gas entry during thermal desorption. For liquid homogeneous samples, quantification of the amount in the matrix can be done by a standard addition methodology.

Time Considerations

For SDE, a well-prepared scientist needs between 2 and 3 hr to perform the whole extraction, including the concentration. The extraction must be surveyed for ~20 min until the distillation reaches a constant rate.

For volatile trapping, preparation of the traps and conditioning requires ~1 hr. Trapping of the aroma compounds requires ~15 min. Thermal desorption of the traps followed by GC analysis requires ~1 hr.

Literature Cited

Badings, H.T., de Jong, C., and Dooper, R.P.M. 1985. Automatic system for rapid analysis of volatile compounds by purge and cold-trapping/capillary gas chromatography. *H.R.C. & C.C.* 8:755.

Chaintreau, A. Simultaneous distillation-extraction: From birth to maturity. Review. *Flavour Fragrance J.* In press.

Chaintreau, A., Grade, A., and Munoz-Box, R. 1995. Determination of partition coefficients and quantitation of headspace volatile compounds. *Anal. Chem.* 67:3300-3304.

Forss, D.A. and Holloway, G.L. 1967. Recovery of volatile compounds from butter oil. *J. Am. Oil Chem. Soc.* 44:572-575.

Godefroot, M., Sandra, P., and Verzele, M. 1981. New method for quantitative essential oil analysis. *J. Chromatogr.* 203:325-335.

Joulain, D. 1986. Study of the fragrance given off by certain springtime flowers. *In* Progress in Essential Oil Research (E. J. Brunke, ed.) pp. 57-67. Walter de Gruyter, Berlin.

Likens, S.T. and Nickerson, G.B. 1964. Detection of certain hop oil constituents in brewing products. *Am. Soc. Brew. Chem.* 5-13.

Linforth, R.S. and Taylor, A.J. 1993. Measurement of volatile release in the mouth. *Food Chem.* 48:115-120.

Maignial, L., Pibarot, P., Bonetti, G., Chaintreau, A., and Marion, J.P. 1992. Simultaneous distillation-extraction under static vacuum: Isolation of volatile compounds at room temperature. *J. Chromatogr.* 606:87-94.

Pollien, P. and Chaintreau, A. 1997. Simultaneous distillation-extraction: Theoretical model and development of a preparative unit. *Anal. Chem.* 69:3285-3292.

Roberts, D.D. and Acree, T.E. 1995. Simulation of retronasal aroma using a modified headspace technique: Investigating the effects of saliva, temperature, shearing, and oil on flavor release. *J. Agric. Food Chem.* 43:2179-2186.

Schultz, T.H., Flath, R.A., Mon, T.R., Eggling, S.B., and Teranishi, R. 1977. Isolation of volatile components from a model system. *J. Agric. Food Chem.* 25:446-449.

Sucan, M.K. and Russell, G.F. 1997. A novel system for purge-and-trap with thermal desorption: Optimization using tomato juice volatile compounds. *J. High. Resol. Chromatogr.* 20:310-314.

Wampler, T.P. 1997. Analysis of food volatiles using headspace—gas chromatographic techniques. *In* Techniques for Analyzing Food Aroma (R. Marsili, ed.) pp. 27-58. Marcel Dekker, New York.

Key References

SDE

Chaintreau, in press. See above.

Review of the evolution of SDE from the original design to the version operating under vacuum. Comparison of SDE with other extraction techniques are presented.

Forss and Holloway, 1967. See above.

Three processes were compared for volatile extraction from fat matrix: high vacuum degassing, cold-finger molecular distillation, and reduced pressure steam distillation.

Godefroot et al., 1981. See above.

The diagram of the micro steam distillation-extraction apparatus is given. The recovery yield is given for test compounds.

Maignial et al., 1992. See above.

Diagrams of SDE under vacuum and solvent consideration are given. The reader can also consult the additional literature giving other details relative to the methodology: i.e., the Chrompack user manual contains a drawing of the micro–steam distillation extraction apparatus.

McGill, A.S. and Hardy, R. 1977. Artefact production in the Likens-Nickerson apparatus when used to extract the volatile flavourous components of cod. *J. Sci. Food Agric.* 28:89-92.

Artifact formation during the extraction is demonstrated. The effect of N_2 flushing and of antioxidant addition is shown.

Picardi, S.M. and Issenberg, P. 1973. Investigation of some volatile constituent of mushroom (*Agaricus biosporus*): Changes which occur during heating. *J. Agric. Food Chem.* 21:959-962.

The authors show differences in aroma profiles and odor evaluation due to the effect of temperature.

Romer, G. and Renner, E. 1974. Simple methods for isolation and concentration of flavor compounds from foods. *Z. Lebensm. Unters. Forsch.* 156:329-335.

The design of a new apparatus is given. It is made for the extraction of larger quantities of material.

Schultz et al., 1977. See above.

The diagram of the apparatus is given. The effect of pH and the recovery yield in function of time are described.

Volatile traps

Abeel, S.M., Vickers, A.K., and Decker, D. 1994. Trends in purge and trap. *J. Chromatogr. Sci.* 32:328-338.

Extensive review article of purge-and-trap techniques geared towards environmental applications. Good discussion of choice of extraction conditions including purge rate and time.

Sucan and Russell, 1997. See above.

Example of purge-and-trap optimization of conditions.

Wampler, 1997. See above.

In depth discussion of many aspects related to headspace sampling.

Contributed by Deborah Roberts and
  Hugues Brevard
Nestle Research Center
Lausanne, Switzerland

Identification and Quantitation of Aroma Compounds

After isolation of the volatiles using different extraction and concentration procedures (*UNIT G1.2*) the samples are injected into a gas chromatograph (GC) for separation of individual compounds. Identification is based on their Kovats retention indices on two columns of different polarity, as well as comparison with mass spectroscopy (MS) spectra of the corresponding reference compounds. Quantification is then performed using one or several compounds as internal standards, which naturally do not occur in the sample, added to the sample before extraction, if possible. Depending on the type and availability of these standards, quantitation is preferably performed by gas chromatography-flame ionization detection (GC-FID; see Basic Protocol) or gas chromatography-mass spectrometry (GC-MS) using isotope dilution assays (IDAs; see Alternate Protocol). Other methods for quantification based on external standards are less accurate, and those based on multiple standard addition at different concentrations are only rarely needed.

IDENTIFICATION (GC-FID; GC-MS) AND QUANTIFICATION OF VOLATILES (GC-FID)

Volatiles are separated by GC and their elution time monitored relative to a series of *n*-alkanes that are injected under identical conditions. The most common GC detector is the flame ionization detector (FID), which is also widely used in flavor research. The detector responds to all organic compounds that burn or ionize in its flame, and is characterized by a large dynamic range (Buffington and Wilson, 1987). The temperature program chosen for the separation of compounds is dominated by mutually exclusive parameters like speed, resolution, and capacity. For separation of highly volatile odorants, sub-ambient GC-start temperatures are recommended. The steps described in this unit for the identification are limited to compounds that are listed in common data banks. Analytical methods to identify a compound that has never been reported before are not discussed, and the reader is referred to experts in natural products chemistry. Unambiguous identification requires that an authentic standard has been shown to have the same chromatographic, spectral, and olfactory properties. Anything less is tentative; however, like other problems in analytical biochemistry the biological activity (i.e., olfaction) is a powerful indication of the identity of a standard and an analyte especially when combined with chromatography.

Materials

Aroma samples
Solution of 10 to 100 µg/ml *n*-alkanes (C5 to C25) in diethylether (freshly distilled)
Reference compound
Internal standard compound

Gas chromatograph with FID and sniffing port (Hewlett Packard)
DB5 and DBWAX capillary columns (e.g., 30-m × 0.25-mm; f = 0.25 µm; J&W Scientific)
Gas chromatograph with MS detector (with EI mode and optional CI mode; Hewlett Packard)

Additional reagents and equipment for extraction methods (see *UNIT G1.2*)

Contributed by Christian Milo

Identify compounds

1. Inject a blank sample into the GC, equipped with an FID and sniffing port, under the analytical conditions chosen for the aroma sample.

   Most odorants will elute between 700 and 1600 n-paraffin retention indices from non-polar columns like the DB5.

   For aroma extracts, the blank sample is a mixture of the solvents used in the extraction, and are concentrated in the same way as the aroma isolate. Some volatiles in aroma extracts may derive from trace impurities of the solvents. For headspace techniques, a blank run is also recommended to check impurities coming from the tubings and/or adsorbents used.

2. Inject the aroma sample (0.5 μl per solvent extract) in the on-column or splitless mode and simultaneously start the GC sniffing.

   For both on-column and splitless injections of solvent extracts the initial temperature of the GC oven is held slightly above the boiling point of the solvent for 1 to 2 min. After very rapid heating to 50°C to 60°C the heating rate usually is 4°C or 6°C per minute up to the final temperature of 220°C to 240°C for DBWAX columns and up to 250° to 280°C for DB5 capillary columns.

3. Inject the mixture of 10 to 100 μg/ml *n*-alkanes.

   The injection mode (e.g., headspace or liquid injection) should be the same as for the aroma sample.

4. Determine the retention time of the alkanes and of the peaks to be identified in the flavor sample and calculate their retention indices according to the following equation (Van den Dool and Kratz, 1963):

$$RI = 100 \times n(C) + 100 \times [T_{(a)} - Tn_{(CS)}/Tn_{(CL)} - T_{(a)}]$$

   where $n(C)$ is the number of carbons; $T_{(a)}$ is the retention time of the compound of interest; $Tn_{(CS)}$ is the retention time of smaller hydrocarbons, i.e., eluting before the compound of interest; and $Tn_{(CL)}$ is the retention time of larger hydrocarbon, i.e., eluting after the compound of interest.

   This is shown in Figure G1.3.1.

5. Compare RI indices and aroma quality as assessed during GC sniffing (Acree et al., 1984; Ullrich and Grosch, 1987) with those in the literature (Kondjoyan and Berdagué, 1996), and the Flavornet (*http://www.nysaes.cornell.edu/flavornet*).

   This will give a list of possible candidates matching the RI and sniffing criteria.

6. Repeat steps 1 to 5 on a column of different polarity.

   A DBWAX and DB5 column are recommended because most literature data is available on these phases and they represent a polar column (DBWAX) and an apolar column (DB5), which separate compounds differently.

7. Inject the aroma sample on the same types of columns (i.e., DBWAX and DB5) using an MS detector.

   Focus identification on the odor-active regions as determined by GC-sniffing and RI values as published in the Flavornet (http://www.nysaes.cornell.edu/flavornet). The fragmentation pattern obtained for the compounds can be compared with those in data banks like the Wiley/NBS Registry of Mass Spectral data (McLafferty and Stauffer, 2000) or the NIST 98 library (National Institute of Science and Technology).

8. Inject reference compound.

   An aroma compound is unambiguously identified only when aroma quality, the GC retention indices, and the MS spectra are the same as for the reference compound.

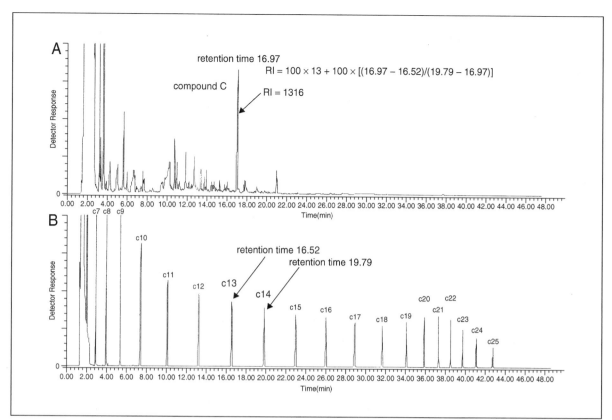

Figure G1.3.1 Determination of the retention index (RI) of an unknown compound C in an aroma extract (**A**) by comparing with a series on *n*-alkanes (**B**) analyzed under the same GC-conditions.

Quantify using GC-FID

9. Choose a compound as an internal standard.

   The standard should have similar chemical and physical properties as the compound that is to be determined (e.g., benzylthiol for quantification of furfurylthiol). It needs to be naturally absent in the sample and must not coelute with other sample components.

10. Add an appropriate amount of this standard preferably a liquid or a suspension of finely ground solid material to the sample.

    The concentration of the standard should be in the same order of magnitude as the analyte.

11. Work up sample by extraction methods (see UNIT G1.2) or headspace sampling (UNIT G1.2; Milo and Grosch, 1995; Roberts et al., 2000).

12. Inject sample into GC under the same analytical conditions as for the previously described identification steps.

13. Calculate concentration of analyte (absolute amount of *A*) according to the following equation:

$$C_{(A)} = F_{(A)}/F_{(ST)} \times C_{(ST)} \times f_{resp} \times f_{rec}$$

where $F_{(A)}$ is the peak area of analyte; $F_{(ST)}$ is the peak area of standard; $C_{(ST)}$ is the amount of standard added; f_{resp} is the FID response factor; and f_{rec} is the recovery factor.

   Sometimes the determination of FID response factors may not be necessary, especially if the standard has a similar molecular formula as the analyte of interest and therefore is close to 1.

Smell Chemicals

G1.3.3

14. Repeat steps 10 to 13 in a model system containing known amounts of analyte and standard to determine the relative recovery factors of analyte versus standard throughout the isolation procedure.

The ratio of the analyte to the standard added, divided by the ratio of the analyte to the standard measured, is the recovery factor. Multiplying the resulting measurements by the appropriate recovery factor will improve accuracy by reducing some of the matrix effects.

QUANTIFICATION OF AROMA COMPOUNDS BY ISOTOPE DILUTION ASSAY (IDA)

Aroma compounds can be quantified with high accuracy by using their corresponding isotope labeled analogs as internal standards, e.g., the concentration of dimethylsulfide in a product will be determined with d6-dimethylsulfide. Due to the almost identical chemical as well as physical properties all possible losses of the analyte during the isolation, extraction, and concentration steps are fully compensated for. Since many standards are not available solvent-free in a high enough purity, the procedure involves (1) the determination of the concentration of labeled standards, (2) homogenization of isotopically labeled standard throughout the sample, and (3) GC-MS analysis in different MS modes depending on the nature of the compound (fragmentation pattern) and sensitivity needed. Although chemical ionization in positive and negative mode may have certain advantages in terms of selectivity and sensitivity, electron impact ionization for MS is sufficient for many cases.

Materials

Isotope labeled standards: a few aroma volatiles are commercially available from CDN Isotopes and Cambridge Isotope Laboratories (CIL); custom synthesis is available from Aspen Research Laboratories and Aldrich; the standards may also be synthesized in-house
Purified aroma compounds to be quantified
Ester of high GC purity (e.g., methyloctanoate)
Organic solvents (highest purity)
Food sample to be analyzed

Extraction apparatus or headspace adsorption device
Gas chromatograph with MS detector (EI mode and optional CI mode; Hewlett-Packard)
Gas chromatograph with FID (Hewlett-Packard)

NOTE: Store standards highly diluted (ppm) in organic solvents, preferably CH_2Cl_2 or methanol (for thiols) and in a freezer. Compounds like vinylguaiacol or furfurylthiol are stable under these conditions.

NOTE: Solution of labeled standards stored in the freezer should be brought up to room temperature before opening the vials and using them. If standard solutions are in a very low-boiling solvent, e.g., pentane, use them refrigerated to facilitate pipetting.

Determine purity and concentration of labeled standard for non-commercially available standards by GC-FID

1. Inject 0.5 µl of a solution containing the labeled standard (A*) at a concentration of ~0.005% and determine the purity of the solution.

Possible impurities in the standard solutions should be identified by GC-MS. Impurities can be tolerated if they do not interfere with the quantitation.

2. Determine the GC-FID response of a known amount of the unlabeled aroma compound ($C_{(A)}$) and an ester ($C_{(E)}$) of high GC purity (e.g., methyloctanoate) that does not coelute with any impurities coming from the labeled standard solution. Calculate the response factor as follows:

$$f_{\text{resp}} = C_{(A)}/C_{(E)} \times F_{(E)}/F_{(A)}$$

where C indicates concentration and F is the peak area.

3. Take a 200-µl aliquot of labeled standard solution (A*) and 200 µl of ester (E) in approximately the same concentration and determine the exact concentration of the labeled standard by peak area comparison correcting for the previously determined response factor according to the equation:

$$C_{(A*)} = F_{(A*)}/F_{(E)} \times C_{(E)} \times f_{\text{resp}}$$

where $C_{(A*)}$ is the concentration of labeled standard; $F_{(A*)}$ is the peak area of labeled standard; $F_{(E)}$ is the peak area of the ester; $C_{(E)}$ is the concentration of the ester; and f_{resp} is the response factor [$F_{\text{resp}} = C_{(A)}/C_{(E)} \times F_{(E)}/F_{(A)}$].

The concentrations of these standard solutions should be checked regularly depending on the expected chemical stability.

Add standards

The sample should be liquid or a slurry of a finely ground solid sample.

4. Add an aliquot (10 to 500 µl) of the organic solution containing the labeled standard, keeping the ratio of analyte (A) in the food sample and its labeled analog (A*) in the range of 0.2- to 5-fold.

The volume of standard added should be kept to a minimum if headspace sampling or SPME sampling is performed. For extraction work-up procedures, the amount of solvent added via the standard may not be critical at all.

5. Stir the solution vigorously for 30 min after adding the standard and before proceeding with the work-up.

6. Transfer the sample to an extraction apparatus or headspace adsorption device for the isolation of volatiles.

Quantify by GC-MS

7. Inject 0.5 µl of a solution containing the labeled standard (A*) at a concentration of ~0.005% by GC-MS (EI mode).

8. Study the fragmentation pattern of the labeled compound and compare it with the compound to be quantified.

9. Choose selective mass traces of labeled and unlabeled compound for subsequent quantification using single ion monitoring (SIM mode).

The mass traces of the unlabeled and labeled counterparts should be specific and preferably of a high intensity. For compounds showing a strong molecular ion in MS-EI the respective molecular mass ions will be used.

In the absence of a strong molecular ion, and if the EI-spectra of labeled standard and analyte are too, other similar ionization modes need to be considered (CI positive/negative; different reactant gases).

10. Establish a calibration curve, based on selected mass traces, by injecting a series of solutions containing the same amount of labeled standard and varying amounts of the analyte over a concentration range of 0.2- to 5-fold.

To obtain the calibration curve, the amount ratio of unlabeled compound/labeled compound (x-axis) is plotted against the ratio of peak area of the mass trace of the unlabeled compound/peak area of the mass trace of the labeled compound (y-axis). The slope of the curve represents the response factor. A linear calibration curve is usually obtained within a small range of concentration (0.2- to 5-fold) and if there is no overlapping of the mass traces chosen for the unlabeled and labeled compounds. In cases of nonlinear calibration curves, linearizations can be performed and the reader is referred to the literature (Fay et al., 2000).

11. Inject the aroma extract/sample spiked with the standard, prepared in steps 4 to 6.

12. Determine the amount of analyte in the aroma sample by GC-MS via the peak areas of the selected mass traces according to the equation:

$$C_{(AS)} = C_{(A*)} \times F_{(AS)}/F_{(A*)} \times R_{(MS)}$$

where $C_{(AS)}$ is the amount of analyte in the sample; $F_{(AS)}$ is the peak area of analyte mass trace; $C_{(A*)}$ is the concentration of labeled standard; $F_{(A*)}$ is the peak area of labeled standard mass trace; and $R_{(MS)}$ is the MS response factor as determined in step 10.

COMMENTARY

Background Information

Identification (GC; GC-MS) and quantification of volatiles (GC-FID)

As perception of most aroma depends on a subtle balance of different odorants, understanding complex aroma at the molecular level means focusing on sensorially relevant odorants.

The identification of aroma-active volatiles targets the odor-active regions in the GC chromatograms. Determined by GC-sniffing and further characterized by the retention indices on columns of different polarity, not all potent odorants will give an FID signal. Their concentration in the extract may be too low to produce an FID signal but may still be sniffed due to the high sensitivity of human olfaction to certain odorants. This interferes with identification, requiring work up of higher sample quantities. Furthermore it means that the match of the retention index of an odor perception during GC-sniffing to an MS signal may be due to coelution with a seemingly pure but odorless component. Identification, as always, requires comparison with reference compounds that may need to be synthesized because they are not commercially available. Information about the occurrence of heteroatoms in a complex aroma extract can be obtained using element-specific detectors, like the flame photometric detector (FPD) for sulfur compounds and the nitrogen phosphorus detector (NPD) for nitrogen and phosphorus compounds. Besides their high selectivity for certain heteroatoms, these detectors are also more sensitive than the FID (Buffington and Wilson, 1987). A rather new

detector, the atomic emission detector (AED), allows the simultaneous recording of several elements like C, N, and S in one run, is highly sensitive and has a high linear range for S-compounds. A comparison of different detectors for the analysis of S-containing volatiles showed the strength of the AED compared to the other detectors (Mistry et al., 1994).

Reliable quantitative data are a prerequisite for evaluating the contribution of a single odorant to a positive aroma or off-flavor. They are needed to calculate odor activity values (OAV) that are defined as the ratio of the concentration to the sensory threshold of a given compound in a matrix (Rothe and Thomas, 1963; Acree et al., 1984). These values give guidance in the evaluation of the impact of an odorant to the overall aroma profile.

The main advantage of quantification using internal standards and GC-FID is the simple, inexpensive instrumental set-up as well as the availability of standards compared to IDA.

Quantitation using unlabeled compounds as internal standards and GC-FID detection lacks the high sensitivity and the high selectivity required for aroma compounds present in the ppb level. For chemically stable compounds and those in higher concentration (i.e., >1000 ppb), however, this method gives reliable data. In all other cases, using isotope labeled compounds as internal standards is the method of choice if the they are available.

Quantification of aroma compounds by isotope dilution assays

The use of stable isotope labeled compounds as internal standards requires MS detection

Table G1.3.1 Important Aroma Compounds, Their Labeled Analogs, and Literature References That Use The Technique in Flavor Research

Aroma compound	Isotope label for standard	Literature references
Acetylpyrazine	$[^2H_3]$acetyl	Schieberle and Grosch, 1987
2-Acetyl-1-pyrroline	$[^2H_{2-7}]$	Schieberle and Grosch, 1987
2-Acetyl-2-thiazoline	$[^2H_4]$	Cerny and Grosch, 1993
2-Acetyltetrahydropyridine	$[^2H_{2-5}]$	Schieberle, 1995b
2-Aminoacetophenone	$[^2H_3]$aceto	Dollmann et al., 1996
Bis(2-methyl-3-furyl)disulfide	Bis(2-$[^2H_2]$methyl)-	Sen and Grosch, 1991
2,3-Butanedione	$[^{13}C_4]$	Schieberle and Hofmann, 1997
Butanoic acid	3,4-$[^2H_{3-4}]$	Schieberle et al., 1993
Coumarin	$[^{13}C_2]$	Masanetz and Grosch, 1998a
(E)-β-damascenone	$[^2H_6]$	Sen et al., 1991
(E,E)-2,4-decadienal	3,4-$[^2H_2]$	Lin et al., 1999a
δ-Decalactone	$[^2H_5]$	Milo and Blank, 1998
Decanoic acid	4,5-$[^2H_{2-4}]$	Guth, 1997
(Z)-6-Decenal	$[^2H_{6-8}]$	Masanetz and Grosch, 1998b
2,3-Diethyl-5-methylpyrazine	$[^2H_3]$5-methyl	Cerny and Grosch, 1993
Dimethyltrisulfide	$[^2H_6]$	Milo and Grosch, 1996
(Z)-6-dodeceno-y-lactone	6,7-$[^2H_2]$	Schieberle et al., 1993
trans-4,5-Epoxy-(E)-2-decenal	4,5-$[^2H_2]$	Lin et al., 1999
trans-2,3-Epoxyoctanal	$[^2H_?]$	Guth and Grosch, 1993b
2-Ethenyl-3,5-dimethylpyrazine	$[^2H_3]$	Mayer et al., 1999
Ethyl anthranilate	$[^2H_3]$ethyl	Aubry et al., 1997
Ethyl cinnamate	$[^2H_3]$ethyl	Aubry et al., 1997
Ethyl cyclohexanoate	$[^2H_3]$ethyl	Guth and Grosch, 1993c
Ethyl dihydro cinnamate	$[^2H_3]$	Aubry et al., 1997
Ethylguaiacol	$[^2H_3]$methoxy	Semmelroch et al., 1995
5-Ethyl-3-hydroxy-4-methyl-2(5H)-furanone	$[^2H_3]$ethyl	Blank et al., 1993
5-Ethyl-4-hydroxy-2-methyl-3(2H)-furanone (homofuraneol)	$[^2H_3]$ethyl	Blank et al., 1997
Ethyl 2-methylbutanoate	2,2,2-$[^2H_3]$	Guth and Grosch, 1993c
Ethyl 3-methylbutanoate	2,2,2-$[^2H_3]$	Guth, 1997
2-Ethyl-3,5-dimethylpyrazine	$[^2H_3]$	Cerny and Grosch, 1993
3-Ethylphenol	$[1,1-^2H_2]$	Guth, 1997
2-Furfurylthiol	$[\alpha-^2H_2]$	Sen and Grosch, 1991
Geosmin	$[^2H_3]$	Palmantier et al., 1998
2-Heptanone	6,7-$[^2H_2]$	Preininger and Grosch, 1994
(Z)-4-Heptenal	$[^2H_?]$	Widder and Grosch, 1994
Hexanoic acid	$[^2H_2]$	Jagella and Grosch, 1999
Hexanal	5,6-$[^2H_2]$	Lin et al., 1999a
(E)-2-Hexenal	$[^2H_?]$	Guth and Grosch, 1993c
(Z)-3-Hexenal	$[^2H_?]$	Guth and Grosch, 1990
(Z)-3-Hexenol	3,4-$[^2H_2]$	Guth and Grosch, 1990
(Z)-3-Hexenyl acetate	$[^2H_3]$acetate	Guth and Grosch, 1993c
4-Hydroxy-2,5-dimethyl-3(2H)-furanone (furaneol)	5,6-$[^{13}C_2]$	Blank et al., 1997
3-Hydroxy-4,5-dimethyl-2(5H)-furanone (sotolone)	5,6-$[^{13}C_2]$	Blank et al., 1996
4-Hydroxy-2-nonenoic acid lactone	$[^2H_?]$	Guth and Grosch, 1993b
2-Isobutyl-3-methoxypyrazine	$[^2H_3]$methoxy	Semmelroch and Grosch, 1996

continued

G1.3.7

251

Table G1.3.1 Important Aroma Compounds, Their Labeled Analogs, and Literature References That Use The Technique in Flavor Research, *continued*

Aroma compound	Isotope label for standard	Literature references
2-sec-Butyl-3-methoxypyrazine	[$^2H_3$]methoxy	Rychlik et al., 1997
2-Isopropyl-3-methoxypyrazine	[$^2H_3$]methoxy	Masanetz and Grosch, 1998b
p-Mentha-1,3,8-triene	[$^2H_3$]	Masanetz and Grosch, 1998b
3-Mercapto-2-butanone	[$^{13}C_4$]	Schieberle and Hofmann, 1996
3-Mercapto-3-methylbutyl formate	[$^2H_6$]	Masanetz et al., 1995
4-Mercapto-4-methyl-pentan-2-one	[$^{13}C_4$]	Guth, 1997
3-Mercapto-2-pentanone	[4,5-$^2H_2$]	Sen and Grosch, 1991
Methanthiol	[$^2H_3$]	Guth and Grosch, 1994
Methional	[$^2H_3$]methyl	Sen and Grosch, 1991
4-Methoxy-2,5-dimethyl-3(2H)-furanone	[$^2H_3$]methoxy	Schieberle and Hofmann, 1997
4-Methoxy-2-methyl-2-butanethiol	[$^2H_3$]methoxy	Guth and Grosch, 1993c
2-Methoxyphenol (guaiacol)	[$^2H_3$]methoxy	Cerny and Grosch, 1993
Methylanthranilate	[$^2H_3$]	Aubry et al., 1997
3-Methylbutanal	[$^2H_2$]	Schieberle and Grosch, 1992
3-Methylbutanol	[$^2H_{2-5}$]	Schieberle, 1991
3-Methyl-2-butene-1-thiol	[$^2H_8$]	Semmelroch and Grosch, 1996
3-Methylbutyric acid	3,4-[$^2H_2$]	Guth and Grosch, 1994
5-Methyl-5H-cyclopenta(b)pyrazine	[$^2H_{1-2}$]	Schieberle and Grosch, 1987
2-Methyl-3-furanthiol	[$^2H_3$]methyl	Sen and Grosch, 1991
5-Methyl-(E)-2-hepten-4-one	6,7-[$^2H_2$]	Pfnuer et al., 1999
2-Methylisoborneol	2-[$^2H_3$]	Palmantier et al., 1998
3-Methyl-2,4-nonandione	3-[$^2H_3$]methyl	Guth and Grosch, 1990
Methylpropanal	[$^2H_7$]	Milo and Grosch, 1996
12-Methyltridecanal	12-[$^2H_{3-8}$]	Guth and Grosch, 1993a
Myristicin	[$^2H_2$]	Masanetz and Grosch, 1998b
(E,E)-2,4-Nonadienal	3,4-[$^2H_2$]	Lin et al., 1999a
(E,Z)-2,6-Nonadienal	6,7-[$^2H_2$]	Fielder and Rowan, 1999
Nonanal	3,3,4,4-[$^2H_4$]	Fielder and Rowan, 1999
(Z,Z)-3,6-Nonadienal	[$^2H_4$]	Milo and Grosch, 1993
(E)-2-Nonenal	2,3-[$^2H_2$]	Lin et al., 1999a
(Z)-2-Nonenal	[$^2H_2$]	Guth and Grosch, 1990
(Z)-1,5-Octadien-3-one	5,6-[$^2H_2$]	Lin et al., 1999b
1-Octen-3-hydroperoxide	[$^2H_2$]	Guth and Grosch, 1990
1-Octen-3-one	1,1,2-[$^2H_{2-3}$]	Lin et al., 1999b
2,3-Pentanedione	5,5,5-[$^2H_3$]	Milo and Grosch, 1993
Pentanoic acid	[$^2H_3$]	Jagella and Grosch, 1999
2-Pentylpyridine	[$^2H_4$]	Schieberle, 1996
2-Phenylethanol	1,1-[$^2H_2$]	Schieberle, 1991
2-Phenylethyl acetate	1,2-[$^{13}C_2$] acetate	Guth, 1997
2-Phenylethylthiol	[α-$^2H_2$]	Schieberle, 1996
cis-Rose oxide	[$^2H_2$]	Guth, 1997
Methylindol (skatol)	[$^2H_3$]-methyl	Preininger and Grosch, 1994
2,4,5-Trimethylthiazol	[5-methyl-$^2H_3$]-	Sen and Grosch, 1991
Vanillin	[$^2H_3$]-methoxy	Semmelroch et al., 1995
4-Vinylguaiacol	[$^2H_3$]-methoxy	Semmelroch et al., 1995
Wine lactone	[$^2H_3$]	Guth, 1997

G1.3.8

since their very similar chemical and physical properties cause them to coelute with the compound to be quantified. The isotope dilution assay is the most accurate method currently available for the quantification of labile odorants and those in low concentrations. After the first application of IDA for flavor research (Schieberle and Grosch, 1987) the method was systematically developed for >60 potent odorants (Schieberle, 1995a). The accuracy of IDAs and quantification with unlabeled internal standards was recently discussed (Preininger, 1998). Table G1.3.1 lists important aroma compounds as well as their labeled analogs and gives an extensive overview of the literature using this technique in flavor research.

One of the main advantages of IDA is that quantitative isolation of odorants from the sample is not required, provided that the internal standard is homogeneously distributed throughout the sample. Solid samples should be finely ground under liquid nitrogen in order to facilitate even penetration of the labeled standard throughout the sample. The time necessary for the standard to be evenly mixed with the analyte can be checked (Milo and Blank, 1998). Due to the high chemical and physical similarities of isotopomers, losses during the work up procedure are ideally compensated. Another strength of the method is the high selectivity and sensitivity of GC-MS, particularly in selected ion monitoring mode. Compounds that are poorly resolved by GC may not interfere in GC-MS due to different fragmentation patterns and, therefore, less sample clean-up is needed (and under CI less separation is needed).

The main hurdle to the application of IDA for quantitative measurements of aroma compounds continues to be the limited commercial availability of the labeled internal standards.

Critical Parameters and Troubleshooting

Identification and quantification of volatiles (GC-FID)

The most critical parameter for the Basic Protocol is the yield of analyte at the detector. Simulations and models are the best methods to identify the extent of the problem and to determine correction factors.

Quantification of aroma compounds by isotope dilution assays

The labeled standards used need to be isotopically stable. Isotopic stability may become an issue if the compound is labeled with deuterium in an enolizable position as is the case for the α-position of carbonyl functions. The deuterium may then be exchanged with protons from the sample during the workup and therefore falsify the result. In order to rule out such D/H exchanges, the standard may be tested under the conditions of the isolation of volatiles from the sample or, better, be replaced by a standard labeled with ^{13}C.

Furthermore the labeling of the compound should increase its molecular weight by at least 2 units, preferably 3 units, in order to minimize interferences with the natural isotope distribution of the analyte.

For non–commercially available standards, the chemical purity plays an important role. The price for a custom synthesis may be lower if only 70% chemical purity is negotiated. As long as the contaminants do not interfere with other compounds to be quantified in the sample and will not convert into the final labeled product during the workup procedure, this can be tolerated.

Although IDA is particularly useful for the quantitation of labile compounds, one needs to pay attention to possible degradation products. The possible conversion of (Z)-3-alkenals to the (E)-2 alkenals will bias results if these compounds are to be quantified simultaneously. Another example are thiols and their dimers that should be quantified in separate samples using the respective labeled standards.

The EI mode can be used in IDA, provided that characteristic ions of high intensity are available for quantification. In some cases, however, it is necessary to improve the efficiency of IDA by changing and optimizing the ionization technique as recently demonstrated for epoxy-alkenals (Blank et al., 1999).

Anticipated Results

Due to the bias that can occur in the Basic Protocol, results can be as low as 1% of the correct value for the concentration of a volatile in a sample. Most results are <80% of the correct value. This bias is usually caused by differences in the recovery between the internal standard and the analytes. Choosing an internal standard that is similar to the analyte can reduce this bias but it cannot be completely eliminated unless a separate standard is used for each analyte (see Alternate Protocol). Nevertheless, the Basic Protocol is often sufficiently accurate because the olfactory system, unlike the taste system, is compressive and insensitive to small changes in concentration (Lawless and Heymann, 1998).

Due to the high sensitivity of the IDA method quantification in the sub-ppb range, the Alternate Protocol can be applied without extensive sample cleanups. Due to the high similarity of labeled internal standard and analyte, the relative recovery of analytes from the matrix, whether oily or aqueous, approaches 100% (Guth and Grosch, 1993c; Schieberle and Hofmann, 1998); most errors come from pipetting. The coefficient of variation for compounds present in ppb quantities is between 5% and 10%.

Time Considerations

The time required for quantification of volatiles by both the Basic Protocol and the Alternate Protocol depends on the isolation/extraction procedure chosen. A complete homogenization of the labeled standards with the sample usually requires not more than 30 min and GC-MS analysis is accomplished within 1 hr. In combination with a high-throughput method like solid-phase microextraction, the GC cycle times (~1 hr) become the limiting factor in the quantification of multiple samples by IDAs.

Literature Cited

Acree, T.E., Barnard, J., and Cunningham, D.G. 1984. A procedure for the sensory analysis of gas chromatographic effluents. *Food Chemistry* 14:273-286.

Aubry, V., Etiévant, P.X., Giniès, C., and Henry, R. 1997. Quantitative determination of potent flavor compounds in Burgundy pinot noir wines using a stable isotope dilution assay. *J. Agric. Food Chem.* 45:2120-2123.

Blank, I., Schieberle, P., and Grosch, W. 1993. Quantification of the flavour compounds 3-hydroxy-4,5-dimethyl-2(5H)-furanone and 5-ethyl-3-hydroxy-4-methyl-2(5H)-furanone by stable isotope dilution assay. *In* Progress in Flavour and Precursor Studies (P. Schreier and P. Winterhalter, eds.). Allured Publishing, Wheaton, Ill.

Blank, I., Lin, J., Fumeaux, R., Welti, D.H., and Fay, L.B. 1996. Formation of 3-hydroxy-4,5-dimethyl-2(5H)-furanone (sotolone) from 4-hydroxy-L-isoleucine and 3-amino-4,5-dimethyl-3,4-dihydro-2(5)-furanone. *J. Agric. Food Chem.* 44:1851-1856.

Blank, I., Fay, L.B., Lakner, F.J., and Schlosser, M. 1997. Determination of 4-hydroxy-2,5-dimethyl-3(2H)-furanone and 2(or 5)-ethyl-4-hydroxy-5(or 2)-methyl-3(2H)-furanone in pentose sugar-based Maillard model systems by isotope dilution assays. *J. Agric. Food Chem.* 45:2642-2648.

Blank, I., Milo, C., Lin, J., and Fay, L.B. 1999. Quantification of aroma-impact components by isotope dilution assay: Recent developments. *In*

Flavor Chemistry: 30 Years of Progress (R. Teranishi, E.L. Wick, and I. Hornstein, eds.) pp. 63-74. Kluwer Academic/Plenum Publishers, New York.

Buffington, R. and Wilson, M.K. 1987. Detectors for Gas Chromatography: A Practical Primer. Hewlett Packard, Palo Alto, Calif.

Cerny, C. and Grosch, W. 1993. Quantification of character-impact odour compounds of roasted beef. *Z. Lebensm.-Unters. -Forsch. A* 196:417-422.

Dollmann, B., Wichmann, D., Schmitt, A., Koehler, H., and Schreier, P. 1996. Quantitative analysis of 2-aminoacetophenone in off-flavored wines by stable isotope dilution assay. *JAOAC (J. Assoc. Off. Anal. Chem.) Int.* 79:583-586.

Fay, L.B., Metairon, S., and Blank, I. 2000. Stable isotope dilution assay mass spectometry in flavour research: Internal standard and calibration issues. *In* Frontiers of Flavour Science (P. Schieberle and K.-H. Engel, Eds.) Deutsche Forschungsanstalt für Lebensmittelchemie, Garching, Germany.

Fielder, S. and Rowan, D.D. 1999. Synthesis of deuterated C-6 and C-9 flavour volatiles. *J. Labelled Compd. Radiopharm.* 42:83-92.

Guth, H. 1997. Quantitation and sensory studies of character impact odorants of different white wine varieties. *J. Agric. Food Chem.* 45:3027-3032.

Guth, H. and Grosch, W. 1990. Deterioration of soya-bean oil: Quantification of primary flavour compounds using a stable isotope dilution assay. *Lebensm.-Wiss. Technol.* 23:513-522.

Guth, H. and Grosch, W. 1993a. 12-Methyltridecanal, a species-specific odorant of stewed beef. *Lebensm.-Wiss. Technol.* 26:171-177.

Guth, H. and Grosch, W. 1993b. Odorants of extrusion products of oat meal: Changes during storage. *Z. Lebensm.-Unters. -Forsch. A* 196:22-28.

Guth, H. and Grosch, W. 1993c. Quantitation of potent odorants of virgin olive oil by stable-isotope dilution assays. *J. Am. Oil Chem. Soc.* 70:513-518.

Guth, H. and Grosch, W. 1994. Identification of the character impact odorants of stewed beef juice by instrumental analyses and sensory studies. *J. Agric. Food Chem.* 42:2862-2866.

Jagella, T. and Grosch, W. 1999. Flavour and off-flavour compounds of black and white pepper (*Piper nigrum* L.). *Eur. Food Res. Technol.* 209:16-31.

Kondjoyan, N. and Berdagué, J.-L. 1996. A Compilation of Relative Retention Indices for the Analysis of Aromatic Compounds. 1st ed. Laboratoire Flaveur-INRA de THEIX. Saint Genes Champanelle, France.

Lawless, H.T. and Heymann, H. 1998. Sensory Evaluation of Food, Principles and Practices. pp. 28-82. Aspen Publishers, Colorado.

Lin, J., Welti, D., Arce Vera, F., Fay, L.B., and Blank, I. 1999a. Synthesis of deuterated volatile lipid degradation products to be used as internal standards in isotope dilution assays. 1. Aldehydes. *J. Agric. Food Chem.* 47:2813-2821.

Lin, J., Welti, D., Arce Vera, F., Fay, L.B., and Blank, I. 1999b. Synthesis of deuterated volatile lipid degradation products to be used as internal standards in isotope dilution assays. 2. Vinyl ketones. *J. Agric. Food Chem.* 47:2822-2829.

Lin, J.M., Fay, L.B., Welti, D.H., and Blank, I. 1999. Synthesis of trans-4,5-epoxy-(E)-2-decenal and its deuterated analog used for the development of a sensitive and selective quantification method based on isotope dilution assay with negative chemical ionization. *Lipids* 34:1117-1126.

Masanetz, C. and Grosch, W. 1998a. Hay-like off-flavor of dry parsley. *Z. Lebensm.-Unters. -Forsch. A* 206:114-120.

Masanetz, C. and Grosch, W. 1998b. Key odorants of parsley leaves (*Petroselinum crispum* [Mill.] Nym. ssp. crispum) by odour-activity values. *Flavour Fragrance J.* 13:115-124.

Masanetz, C., Blank, I., and Grosch, W. 1995. Synthesis of [^{2}H$_6$]-3-mercapto-3-methylbutyl formate to be used as internal standard in quantification assays. *Flavour Fragrance J.* 10:9-14.

Mayer, F., Czerny, M., and Grosch, W. 1999. Influence of provenance and roast degree on the composition of potents odorants in Arabica coffees. *Eur. Food Res. Technol.* 209:242-250.

McLafferty, F. and Stauffer, D.B. 2000. Wiley/NBS Registry of Mass Spectral Data. John Wiley & Sons, New York.

Milo, C. and Blank, I. 1998. Quantification of impact odorants in food by isotope dilution assay: Strength and limitations. *In* Flavor Analysis: Developments in Isolation and Characterization (C.J. Mussinan and M.J. Morello, eds.) pp. 250-259. American Chemical Society, Washington, D.C.

Milo, C. and Grosch, W. 1993. Changes in the odorants of boiled trout (*Salmo fario*) as affected by the storage of the raw material. *J. Agric. Food Chem.* 41:2076-2081.

Milo, C. and Grosch, W. 1995. Detection of odor defects in boiled cod and trout by gas chromatography-olfactometry of headspace samples. *J. Agric. Food Chem.* 43:459-462.

Milo, C. and Grosch, W. 1996. Changes in the odorants of boiled salmon and cod as affected by the storage of the raw material. *J. Agric. Food Chem.* 44:2366-2371.

Mistry, B.S., Reineccius, G.A., and Jasper, B.L. 1994. Comparison of gas chromatographic detectors for the analysis of volatile sulfur compounds in foods. *In* Sulfur Compounds in Food (C.J. Mussinan and M.E. Keelan, eds.) pp. 8-21. American Chemical Society, Washington, D.C.

Palmantier, J.P.F.P., Taguchi, V.Y., Jenkins, S.W.D., Wang, D.T., Kim, P.-N., and Robinson, D. 1998. The determination of geosmin and 2-methylisoborneol in water using isotope dilution high resolution mass spectrometry. *Water Res.* 32:287-294.

Pfnuer, P., Matsui, T., Grosch, W., Guth, H., Hofmann, T., and Schieberle, P. 1999. Development of a stable isotope dilution assay for the quantification of 5-methyl-(E)-2-hepten-4-one: Application to hazelnut oils and hazelnuts. *J. Agric. Food Chem.* 47:2044-2047.

Preininger, M. 1998. Quantitation of potent food aroma compounds by using stable isotope labeled and unlabeled standard methods. *In* Food Flavors: Formation, Analysis and Packaging Influences (E.T. Contis, C.T. Ho, C.J. Mussinan, T.H. Parliament, R. Shahidi, and A.M. Spanier, eds.) pp. 87-97. Elsevier, Amsterdam.

Preininger, M. and Grosch, W. 1994. Evaluation of key odorants of the neutral volatiles of Emmentaler cheese by the calculation of odour activity values. *Lebensm.-Wiss. Technol.* 27:237-244.

Roberts, D.D., Pollien, P., and Milo, C. 2000. Solid-phase microextraction method development for headspace analysis of volatile flavor compounds. *J. Agric. Food Chem.* 48:2430-2437.

Rothe, M. and Thomas, B. 1963. Aromastoffe des Brotes. *Z. Lebensm. Unters. Forsch. A.* 119:302-310.

Rychlik, M., Warmke, R., and Grosch, W. 1997. Ripening of Emmental cheese wrapped in foil with and without addition of *Lactobacillus casei* subsp. casei. III. Analysis of character impact flavour compounds. *Lebensm.-Wiss. Technol.* 30:471-478.

Schieberle, P. 1991. Primary odorants of pale lager beer. *Z. Lebensm.-Unters. -Forsch. A* 193:558-565.

Schieberle, P. 1995a. New developments in method for analysis of volatile flavor compounds and their precursors. *In* Characterization of Food: Emerging Methods (A.G. Gaonkar, ed.) pp. 403-431. Elsevier Science Publishing, New York.

Schieberle, P. 1995b. Quantification of important roast-smelling odorants in popcorn by stable isotope dilution assays and model studies on flavor formation during popping. *J. Agric. Food Chem.* 43:2442-2448.

Schieberle, P. 1996. Odour-active compounds in moderately roasted sesame. *Food Chem.* 55:145-152.

Schieberle, P. and Grosch, W. 1987. Quantitative analysis of aroma compounds in wheat and rye bread crusts using stable isotope dilution assay. J. Agric. Food Chem. 35:252-257.

Schieberle, P. and Grosch, W. 1992. Changes in the concentration of potent crust odorants during storage of white bread. *Flavour Fragrance J.* 7:213-218.

Schieberle, P. and Hofmann, T. 1996. Untersuchungen zum Einfluss von Herstellungsparametern auf den Aromabeitrag intensiver Aromastoffe in Cystein/Kohlenhydrat-Reaktionsmischungen. *Lebensmittelchemie* 50:105-108.

Schieberle, P. and Hofmann, T. 1997. Evaluation of the character impact odorants in fresh strawberry juice by quantitative measurements and sensory studies on model mixtures. *J. Agric. Food Chem.* 45:227-232.

Schieberle, P. and Hofmann, T. 1998. Characterization of key odorants in dry-heated cysteine/carbohydrate mixtures-comparison with aqueous reaction systems. *In* Flavor Analysis. (C.J. Mussinan and J. Morello, eds.) American Chemical Society, Washington, D.C.

Schieberle, P., Gassenmaier, K., Guth, H., Sen, A., and Grosch, W. 1993. Character impact odour compounds of different kinds of butter. *Lebensm.-Wiss. Technol.* 26:347-356.

Semmelroch, P. and Grosch, W. 1996. Studies on character impact odorants of coffee brews. *J. Agric. Food Chem.* 44:537-543.

Semmelroch, P., Laskawy, G., Blank, I., and Grosch, W. 1995. Determination of potent odorants in roasted coffee by stable isotope dilution assay. *Flavour Fragrance J.* 10:1-7.

Sen, A. and Grosch, W. 1991. Synthesis of six deuterated sulfur containing odorants to be used as internal standards in quantification assays. *Z. Lebensm.-Unters. -Forsch. A* 192:541-547.

Sen, A., Laskawy, G., Schieberle, P., and Grosch, W. 1991. Quantitative determination of b-damascenone in foods using a stable isotope dilution assay. *J. Agric. Food Chem.* 39:757-759.

Ullrich, F. and Grosch, W. 1987. Identification of the most intense volatile flavour compounds formed during autoxidation of linoleic acid. *Z. Lebensm. Unters. Forsch. A.* 184:277-282.

Van den Dool, H. and Kratz, P.D. 1963. A generalization of the retention index system including linear temperature programmed gas-liquid partition chromatography. *J. Chromatogr.* 11:463-471.

Widder, S. and Grosch, W. 1994. Study on the cardboard off-flavour formed in butter oil. *Z. Lebensm.-Unters. -Forsch. A* 198:297-301.

Contributed by Christian Milo
Nestlé Research Center
Lausanne, Switzerland

Stereodifferentiation of Chiral Odorants Using High-Resolution Gas Chromatography

This unit describes those methods that can differentiate between enantiomers found in foods that contribute to their taste and aroma. These compounds are volatile odorants that are most easily analyzed using enantioselective high resolution–gas chromatography (HRGC). Other methods exist for the separation and analysis of chiral compounds, which include optical methods, liquid and planar chromatography, and electrophoresis, but for food volatiles, gas chromatography has evolved to the point where it is now the cornerstone for the most comprehensive analysis of volatile compounds.

Enantioselective gas chromatography can provide three quite different kinds of information: (1) the amount of each enantiomer present in a food, determined as the enantiomeric purity or the enantiomer excess, and the separation factor α for each pair of enantiomers; (2) enantiospecific sensory evaluation using gas chromatography-olfactometry (GC-O); and (3) data used as part of an authenticity determination.

The most useful results will depend upon α being >1.01 which, in turn, depends upon the selection of the correct chiral stationary phase that will provide optimum resolution of enantiomers.

The protocol selected will be determined by the application under investigation and by the instrumentation available. The following instrument configurations can be used:

1. a single oven containing a chiral column with a flame ionization detector (FID);

2. a single oven containing a chiral column with an olfactometer as the detector (GC-O);

3. a multidimensional system consisting of two columns of different selectivity. The first of these two columns, known as the precolumn, is achiral and is fitted with an FID. The second, known as the analytical column, is a chiral column, configured to receive heart-cut sections from the first column (see Alternate Protocol 1; Gordon et al., 1985; Mosandl et al., 1989; Bernreuther and Schreier, 1991; Wright, 1997) and is fitted with either an FID or an olfactometer.

    a. Two columns are enclosed in a single oven, requiring the temperature program to be the same for both columns, or for the heart-cut fraction to be trapped cryoscopically outside the oven while the temperature program is changed.

    b. Two columns are contained in separate ovens, allowing them to be operated with independent gas flows and temperature programs.

    c. Two columns are contained in separate unconnected instruments. The heart-cut fraction is collected in a cryoscopic trap and reinjected into the second column.

Basic Protocol 1 describes how to measure enantiomer ratios using single-dimensional high-resolution gas chromatography, while Alternate Protocol 1 obtains enantiomer ratios using multidimensional gas chromatography. Basic Protocol 2 gives procedures for obtaining enantiospecific sensory data using gas chromatography olfactometry, whereas Alternate Protocol 2 obtains sensory data by multidimensional gas chromatography. A specific application is chosen in each case, so that the choice of chiral column is appropriate. Once the correct operating parameters have been established, a run will take 15 to 100 min, depending upon the application. Some guidelines for the selection of operating conditions and the selection and care of chiral columns are discussed, followed by a troubleshooting section. No protocol is given for an authenticity determination,

---

Contributed by Mary G. Chisholm

because Basic Protocol 1 will usually suffice, and for this application, the determination of an enantiomer composition may be part of a more comprehensive analysis involving several instrumental methods.

DETERMINATION OF ENANTIOMER COMPOSITION USING SINGLE-DIMENSIONAL HIGH-RESOLUTION GAS CHROMATOGRAPHY

This method can be used when the enantiomers of interest are not coeluting with other compounds in the sample and when accurate quantitative information is not the highest priority of the analysis. The sample will have been prepared by an extraction method selected from those in *UNIT G1.1* and should have a concentration of 50 to 100 ppm. The identity of the components of the sample should be known from gas chromatography-mass spectrometry (GC-MS) together with their retention indices on the achiral stationary phase. Additional sample cleanup procedures may be needed to ensure the optimum results that are evaluated below:

1. online coupling with an achiral precolumn used in multidimensional gas chromatography (MDGC; see Alternate Protocol 1), part of the instrument configuration;

2. offline high performance liquid chromatography (HPLC) removes nonvolatile materials;

3. preparative gas chromatography, a concentration method, used for the examination of specific low concentration compounds of interest;

4. sampling using solid phase micro-extraction (SPME) for enantiomers with retention indices of 800 to 1400 (nonpolar column) removes all nonvolatile and semivolatile compounds, and most compounds eluting at a retention index (RI) >1400. A fast method of examining the most volatile compounds that can be used to evaluate the need for a more accurate analysis. It is very fast, but reproducibility can be poor;

5. identification of coelution by collecting a chromatogram of the sample on a column with a polar stationary phase. Examine peak areas for consistency on the two columns. Time-consuming, but the best method to check for coelution, which leads to the selection of the best column.

If the order of elution of *R* and *S* isomers is required, this must be determined by comparison of the retention times of an authentic sample with those of the analyte.

Materials

C_7-C_{20} *n*-alkane hydrocarbon standard for a nonpolar column, or C_9-C_{30} for a polar column (hydrocarbon standard; see recipe)
Sample containing an on-column concentration of 20 to 80 ng of the enantiomers of interest

High-resolution gas chromatograph fitted with the appropriate chiral column (e.g., heptakis(2,3-di-*O*-methyl-6-*O*-*t*-butyldimethylsilyl)-β-cyclodextrin in poly(14% cyanopropylphenyl/86% dimethylsiloxane)), a stationary phase widely used for the chiral analysis of many food extracts
Hydrogen and air lines
FID detector; on-column injector is preferred
Integrator or access to data collection software to record chromatograms
10-μl syringes

**Stereodifferentiation
of Chiral
Odorants
Using Gas
Chromatography**

G1.4.2

1. Use a 30-m × 0.32-mm i.d. column with 0.25-μm film thickness. Set the following operating parameters and allow to stabilize for 30 min.

   > Oven temperature: 40°C for 10 min, 2°C/min to 230°C, hold for 10 min
   > Carrier gas: H_2 at a linear velocity of 80 cm/sec
   > Injector temperature: 150°C
   > Split ratio: 100:1
   > FID detector temperature: 225°C
   > Sample size: 1 μl.

2. Enable the integrator or software for data collection.

3. Inject 1 μl of the hydrocarbon standard onto the column with a 10-μl syringe.

4. Record the chromatogram as a retention index standard for all runs performed under the same operating conditions (*UNIT G1.1*).

5. Allow the GC to reach equilibrium for the next run.

6. Inject a 1-μl sample of the analyte onto the column with a 10-μl syringe and record its chromatogram.

7. Identify the enantiomers of interest using the chromatogram from the hydrocarbon standard and record their retention times and peak areas from the integration data.

8a. Calculate the percentage of each enantiomer present from Equation G1.4.1.

$$\text{enantiomeric purity } \% = \left[\frac{R}{R+S}\right] \times 100 \text{ and } \left[\frac{S}{R+S}\right] \times 100$$

where R and S represent the areas of the R and S peaks of a pair of enantiomers in the chromatogram.

8b. Calculate the separation factor, α from Equation G1.4.2.

$$\alpha = \frac{t_{R2}}{t_{R1}}$$

where t_{Ri} and t_{R2} are the corrected retention times for peak 1 and peak 2, respectively.

> *If it suspected that the enantiomers of interest are coeluting, a correction to the areas of the affected peaks may be applied. The correction can be determined from a knowledge of the peak areas on a nonpolar and polar achiral column. If the shape of the peak and the areas are the same on both columns, then no coelution is occurring. If the areas are different, then a correction factor can be applied by comparing retention times of the separated enantiomers, and subtracting the areas of those peaks that are coeluting, which were determined from the chromatograms obtained from the achiral columns. It should be emphasized that this is not an appropriate correction to make if accurate quantitative information is required.*

DETERMINATION OF ENANTIOMER COMPOSITION USING MULTIDIMENSIONAL GAS CHROMATOGRAPHY (MDGC)

This method is used when accurate quantitative information is required and the separation produced by a single column is inadequate and cannot be rectified by reconfiguring the instrument. To justify using the more complex system, there should be a guarantee that it can deliver (1) superior resolution compared to that of a single column and (2) increased quantitative information resulting from the elimination of coelution on the chiral column.

This method requires two separate ovens that are operated independently and contain columns of different selectivities. They are connected by a switching device that can selectively transfer small portions of a chromatogram from the first column to the second. These sections of the chromatogram are known as 'heart-cut' samples. The heart-cut sample is trapped and cooled with liquid nitrogen. The first column, the precolumn, has an achiral stationary phase that should give the best possible resolution in areas of the chromatogram where enantiomers elute. The second column, the analytical column, should have a chiral stationary phase that gives the largest possible separation factor for

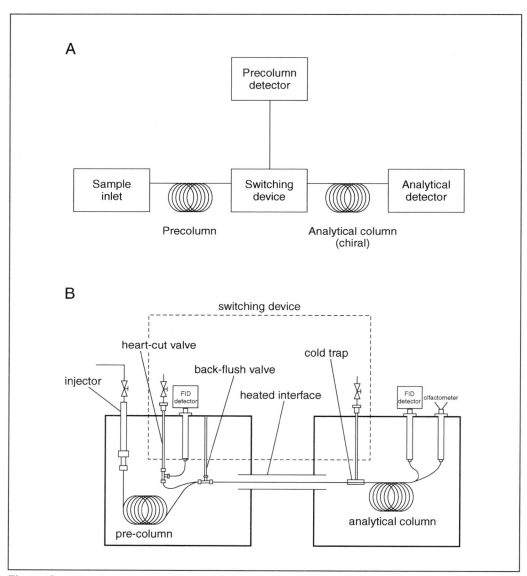

Figure G1.4.1 A two-oven configuration for MDGC. (**A**) The configuration for a basic system. (**B**) A simple system for GC-FID and GC-O showing the essential components (modified from the MDS2000 multidimensional gas chromatography system from SGE; see Table G1.4.3).

Stereodifferentiation
of Chiral
Odorants
Using Gas
Chromatography

G1.4.4

260

the chiral volatiles of interest. Details about specific instrument configurations are in the Commentary section and are shown in Figure G1.4.1. Figures G1.4.2 and G1.4.3 show chromatograms obtained from two different configurations for an MDGC system using the heart-cut technique.

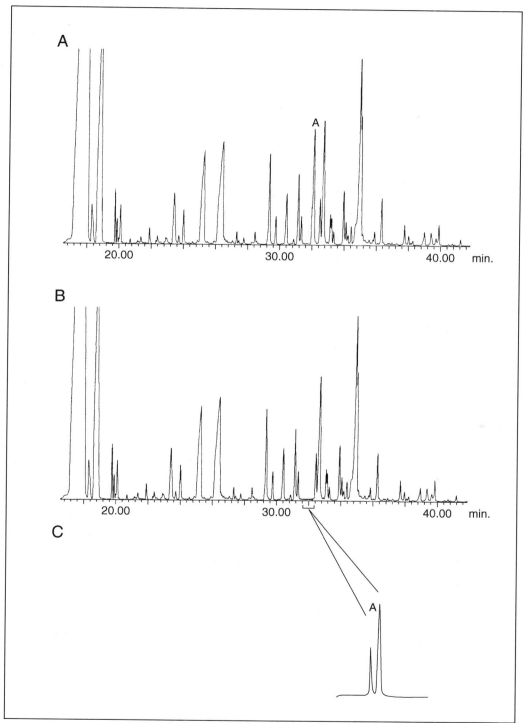

Figure G1.4.2 Chromatograms obtained using a MDGC system. (**A**) Part of a chromatogram obtained from the precolumn (nonpolar) showing the peak of interest A. (**B**) The same sample showing the result of the heart-cut on the precolumn. Peak A is missing. (**C**) The chromatogram obtained from the chiral analytical column, showing the separation of the *R* and *S* isomers.

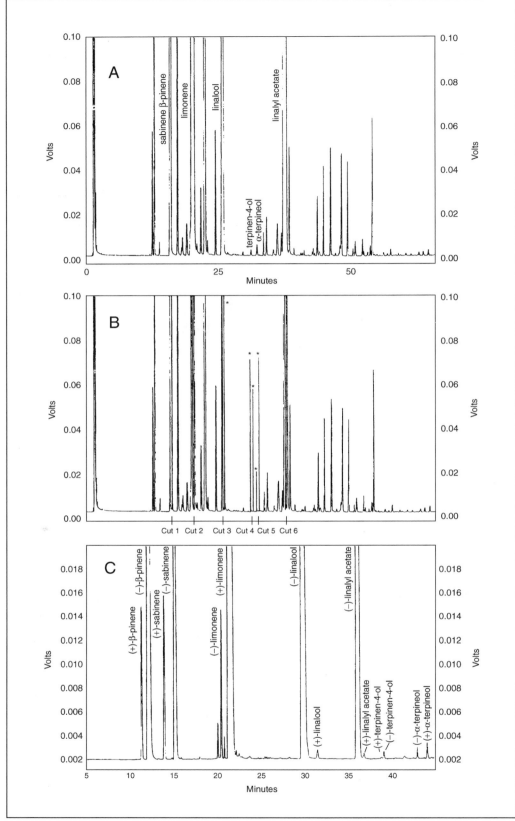

Figure G1.4.3 Chromatograms obtained using an MDGC system where 6 heart-cuts are made in one run. (**A**) Chromatogram obtained on an SE-52 precolumn, showing the peaks to be heart-cut. (**B**) The same sample showing where the cuts occurred. The asterisks are spikes caused by the valve switching. (**C**) The chromatogram obtained from the chiral analytical column showing the separation of all 6 components in the same run. Reprinted with permission from Mondello et al. Copyright 1998 American Chemical Society.

Stereodifferentiation
of Chiral
Odorants
Using Gas
Chromatography

G1.4.6

Sample preparation is similar to that of Basic Protocol 1. The precolumn can be viewed as the ideal technique for sample cleanup. No other sample cleanup is necessary. If the sample contains nonvolatile material, cleanup procedures 2 to 4 (see Basic Protocol 1) would help prolong the life of the precolumn, but it would not improve the quality of the separation on the analytical column.

Materials

Precolumn (achiral): poly(ethylene glycol) or a suitable polar stationary phase
Analytical column (chiral):
    heptakis(2,3-di-*O*-methyl-6-*O*-*t*-butyldimethylsilyl)-β-cyclodextrin in
    poly(14% cyanopropylphenyl/86% dimethylsiloxane)
C_7-C_{20} *n*-alkane hydrocarbon standard for a nonpolar precolumn, or C_9-C_{30} for a polar precolumn (hydrocarbon standard; see recipe)
Sample containing an on-column concentration of 20 to 80 ng of the enantiomers of interest
Liquid nitrogen

Multidimensional gas chromatograph with two linked independently operated ovens (first oven fitted with an on-column injector and both ovens fitted with FIDs)
Integrator for each detector or access to data collection software to record chromatograms for both detectors
10-µl syringes

Set operating conditions for an essential oil analysis or food extract

1. Set the following parameters:

    Oven 1
    Column: 30-m × 0.53-mm i.d. × 1.0-µm film thickness
    Oven temperature: 30°C for 30 sec, 40°C/min to 60°C, 3°C/min to 200°C
    Carrier gas: H_2 at a flow rate of 12 ml/min
    Injector temperature: 150°C
    FID detector temperature: 225°C
    Sample size: 2 µl

    Oven 2
    Column: 30-m × 0.25-mm i.d. × 0.25-µm film thickness
    Oven temperature: 40°C for 15 min, then 2°C/min
    Carrier gas: H_2 at a linear velocity of 80 cm/sec
    FID detector temperature: 225°C.

2. With oven 2 disabled by the switching valves or by using a separate instrument with the same column installed as the precolumn above and with the same operating conditions, carry out the following preliminary runs.

    a. Inject a 1-µl sample of the *n*-alkane hydrocarbon standard with a 10-µl syringe and record its chromatogram.

    b. Inject a 1-µl sample of the analyte with a 10-µl syringe and determine the time where each heart-cut will be taken from the chromatogram.

Use the dual oven instrument

3. Set all operating parameters and allow the instrument to stabilize for 30 min.

4. Enable the integrators, or software for data collection.

5. Set the time program for collecting the heart-cut fractions.

6. Turn on the liquid nitrogen valve to cool the trap, if it is not a part of the time program.

7. Inject a 2-μl sample of the analyte on to the precolumn with a 10-μl syringe and record its chromatogram showing the heart-cuts.

8. Record the retention time and peak areas of each heart-cut eluting from the analytical column.

9a. Calculate the percentage of each enantiomer present from Equation G1.4.3.

$$\text{enantiomeric purity } \% = \left[\frac{R}{R+S}\right] \times 100 \text{ and } \left[\frac{S}{R+S}\right] \times 100$$

where R and S represent the areas of the R and S peaks, respectively, of a pair of enantiomers in the chromatogram.

9b. Calculate the separation factor, α from Equation G1.4.4.

$$\alpha = \frac{t_{R2}}{t_{R1}}$$

where t_{R1} and t_{R2} are the corrected retention times for peak 1 and peak 2, respectively.

*BASIC
PROTOCOL 2*

SENSORY DISCRIMINATION OF CHIRAL FLAVOR COMPOUNDS USING GAS CHROMATOGRAPHY-OLFACTOMETRY (GC-O)

Determination of the odor character and intensity of enantiomers relies heavily on complete separation of the components of the sample where there is no coelution and baseline separation of enantiomers is seen (see Fig. G1.4.4). If these ideal conditions are not met, considerable errors will be incurred in making odor measurements, particularly in cases where both enantiomers have similar odors, or where one is odorless. Traces of odorants coeluting with analytes under investigation, tailing of peaks, and low resolution all seriously affect chromatographic odor data. If the retention times of two enantiomers differ by <1 min, quantitative odor data may be inaccurate.

Sample preparation is similar to that of Basic Protocol 1. The sample cleanup procedures should ensure that no extraneous odorants are eluting with chiral odorants of interest. Thus, cleanup procedure 5 is preferred, which should be modified by attaching the polar column to the olfactometer port so that unwanted odorants can be detected. If odorant coelution is found, then a new chiral stationary phase must be chosen for the sensory analysis. Cleanup procedures 2 to 4 will prolong the life of the chiral column. Cleanup procedure 1 is given as Alternate Protocol 2 of this unit.

Materials

C_7-C_{20} *n*-alkane hydrocarbon standard for a nonpolar column, or C_9-C_{30} for a polar column (hydrocarbon standard; see recipe)

Sample

For qualitative evaluation: an on-column concentration of 20 to 80 ng of the enantiomers of interest

For quantitative evaluation: a set of 6 serial dilutions of the original sample that should have an accurately known concentration of ~2.0%. The concentration can be adjusted to give the optimum odor data. Each sample is diluted by a factor of 3 from the previous sample (i.e., 1 ml sample diluted with 2 ml of solvent) and then given a random number using a "double blind" labeling method. The samples are sniffed in random order.

**Stereodifferentiation
of Chiral
Odorants
Using Gas
Chromatography**

G1.4.8

High-resolution gas chromatograph fitted with the appropriate chiral column (e.g., heptakis(2,3-di-*O*-methyl-6-*O*-*t*-butyldimethylsilyl)-β-cyclodextrin in poly(14% cyanopropylphenyl/86% dimethylsiloxane)), a stationary phase widely used for the chiral analysis of many food extracts

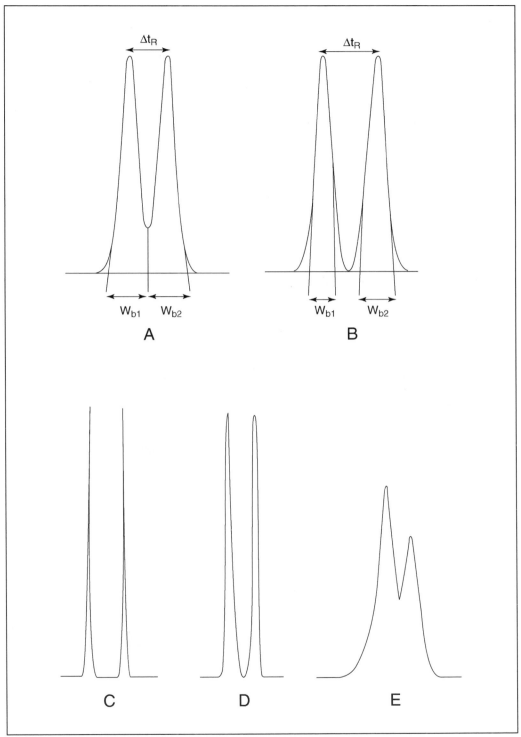

Figure G1.4.4 Resolution $R_s = 2\Delta t_R/(W_{b1} + W_{b2})$ at different separations. (**A**) sufficient separation, $R_s = 1$; (**B**) baseline separation, $R_s = 1.5$. C, D, and E have the same separation factor, but show problems cause by overloading the column. (**C**) Good resolution; (**D**) some peak broadening and tailing; (**E**) excessive tailing causing peak overlap.

FID detector installed in parallel with an olfactometer capable of delivering the effluent from the column to the human sniffer. The column is installed in either the FID port or the olfactometer port, depending upon the mode of operation. Split effluents are not recommended.

Integrator or access to data collection software to record chromatograms

Software package capable of recording timed events and odor descriptors such as Charmware (Datu)

10-μl syringe

Set operating conditions for an essential oil analysis or food extract

1. Set operating parameters:

   Column: 30-m × 0.32-mm i.d. × 0.25-μm film thickness
   Oven temperature: 40°C for 10 min, 2°C/min to 230°C, hold for 10 min
   Carrier gas: H_2 at a linear velocity of 80 cm/sec
   Injector temperature: 150°C
   Split ratio: 100:1
   FID temperature: 225°C
   Olfactometer block 225°C
   Temperature of the effluent reaching the sniffer: 35° to 40°C
   Sample size: 1 μl.

   Allow the instrument to stabilize for 30 min.

2. Enable the integrator, or software for data collection.

3. Inject 1 μl *n*-alkane hydrocarbon standard onto the column with the column in the FID port.

4. Record the chromatogram as a retention index standard for all runs performed under the same operating conditions (see Basic Protocol 1).

5. Move the column from the FID port to the olfactometer port and turn on the sniff air flows.

For qualitative analysis

6a. Inject a 1-μl sample of the sample onto the column and record the retention time and the descriptor for all odors eluting from the column by sniffing the effluent at the olfactometer port.

7a. Using the chromatogram obtained from Basic Protocol 1, operating condition 6 as a reference, identify the odors of enantiomers of interest by their retention index and determine whether contamination of odors has occurred caused by coelution or poor resolution.

8a. Assign descriptors to all resolved pairs of enantiomers where sensory discrimination has been detected.

For quantitative analysis

6b. Inject a 1-μl sample onto the column of a randomly chosen dilution from the set of serial dilutions, and record the retention time and odor descriptor for all odors eluting from the column by sniffing the effluent at the olfactometer port.

7b. Repeat step 6b until data has been collected for all 6 samples.

8b. Construct an odor chromatogram for the sample and use it to determine the odor activity value (OAV; Grosch, 1993) or the odor spectrum value (OSV; Acree, 1997) for the odorants of interest. From the odor spectrum identify the pairs of enantiomers and note their relative OSV's.

SENSORY DISCRIMINATION OF CHIRAL FLAVOR COMPOUNDS USING MULTIDIMENSIONAL GAS CHROMATOGRAPHY (MDGC)

This method is used when accurate quantitative odor data are required and the separation produced by a single column gives overlapping odors that could not be separated by adjusting the operating conditions or reconfiguring the instrument. Traces of powerful odorants coeluting with enantiomers under analysis may cause extensive distortion of odor data and be present in concentrations that are not high enough to be detected by any detector except the sniffer. Changing the precolumn may solve one set of contamination problems, only to cause new ones. The basic instrument configuration is described in Alternate Protocol 1.

Sample preparation is similar to that of Basic Protocol 1. Once it has been established that no coelution of odorants is occurring with the chiral odorants of interest (see Basic Protocol 2), then the precolumn provides the only necessary sample cleanup. Only the heart cuts reach the analytical column.

Materials

Precolumn (achiral): poly(ethylene glycol) or a suitable polar stationary phase
Analytical column (chiral):
  Heptakis(2,3-di-*O*-methyl-6-*O*-*t*-butyldimethylsilyl)-β-cyclodextrin in
  poly(14% cyanopropylphenyl/86% dimethylsiloxane)
C_7-C_{20} *n*-alkane hydrocarbon standard for a nonpolar precolumn, or C_9-C_{30} for a
  polar precolumn (hydrocarbon standard; see recipe)
Sample for either qualitative or quantitative determinations (see Basic Protocol 2)
Liquid nitrogen

Multidimensional gas chromatograph with two linked independently operated
  ovens. The first oven is fitted with an on-column injector, both ovens are fitted
  with FIDs and the second oven has an olfactometer installed in parallel with the
  FID. The analytical column is installed in either the FID port or the
  olfactometer port as described (see Basic Protocol 2).
Integrator for each detector or access to data collection software to record
  chromatograms for both detectors
10-μl syringes

Set operating conditions for an essential oil analysis or food extract

1. Set operating parameters for each oven.

   Oven 1
   Column: 30-m × 0.53-mm i.d. × 1.0-μm film thickness
   Oven temperature: 30°C for 30 sec, 40°C/min to 60°C, 3°C/min to 200°C
   Carrier gas: H_2 at a flow rate of 12 ml/min
   Injector temperature: 150°C
   Detector temperature: 225°C
   Sample size: 2 μl.

   Oven 2
   Column: 30-m × 0.25-mm i.d. × 0.25-μm film thickness
   Oven temperature: 40°C for 15 min, then 2°C/min
   Carrier gas: H_2 at a linear velocity of 80 cm/sec
   Detector temperature: 225°C

Smell Chemicals

G1.4.11

2. With oven 2 disabled by the switching valves or by using a separate instrument with the same column installed as the precolumn and with the same operating conditions, carry out the following preliminary runs.

   a. Inject a 1-μl sample of an *n*-alkane hydrocarbon standard with a 10-μl syringe and record its chromatogram.

   b. Inject a 1-μl sample of the analyte with a 10-μl syringe and determine the time where each heart-cut will be taken from the chromatogram.

Use the two-oven instrument

3. Place the analytical column in the olfactometer port and turn on the sniff air flows.

4. Set all operating parameters and allow the instrument to stabilize for 30 min.

5. Enable the integrators, or software for data collection.

6. Set the time program for collecting the previously determined heart-cut fractions.

7. Turn on the liquid nitrogen valve to cool the trap if it is not a part of the time program.

8. Proceed as described in Basic Protocol 2, qualitative analysis steps 6a to 8a; quantitative analysis steps 6b to 8b.

REAGENTS AND SOLUTIONS

Use deionized or distilled water in all recipes and protocol steps. For common stock solutions, see APPENDIX 2A; for suppliers, see SUPPLIERS APPENDIX.

Hydrocarbon standard

The hydrocarbon standard provides a universal scale (retention index, RI) for the characterization of volatile odorants. Since a single determination may require the use of more than one instrument (GC, GC-O, GC-MS), it is crucial that every time a run is made using new operating conditions, a new calibration is recorded using the hydrocarbon standard. This is the only way that RI data from one instrument can be compared with that from another provided that the stationary phase is the same. RIs do not vary with the operating conditions, while retention times do. (See more on the use of hydrocarbon standards in *UNIT G1.1*.)

The standard is composed of a series of *n*-alkanes, from C_7 to C_{30}. They should be dissolved in the same solvent that is used for dissolving the sample being analyzed. A suitable concentration is 100 μg/ml of each one, with two, e.g., C_{11} and C_{13}, at a concentration of 50 μg/ml to act as markers in the chromatogram. This stock solution should be stored at −10°C to minimize loss of low molecular weight alkanes and used as a stock solution. Dilute it by 1:10 for use as a standard. A convenient procedure for making up the stock solution follows.

Using a Pasteur pipet, determine the weight of 1 drop of *n*-decane. In a dark screw-top bottle, add 1 drop each of *n*-undecane and *n*-tridecane, 2 drops each of all other liquid alkanes, and the same weight of each of the solid alkanes. Record the weights of each one. Start by weighing the alkanes of highest molecular weight first to minimize loss by evaporation. Add the appropriate amount of pentane (or the required solvent) to make a solution of 100 μg/ml of each alkane in solution together. Store at −10°C.

continued

For nonpolar columns use a standard composed of C_7-C_{18} *n*-alkanes

For polar columns use a standard composed of C_9-C_{30} *n*-alkanes

Over a period of time, preferential evaporation of the lower molecular weight alkanes will occur. Unless known weights of each alkane are required, this is not a serious problem. Using a working solution minimizes this problem and also avoids contamination. When the quality of the standard becomes unacceptable, a new sample can be diluted from the stock solution.

COMMENTARY

Background Information

Food scientists are interested in the enantiomer distribution of chiral food odorants because enantiomers may have different odors and odor intensities. Determination of enantiomer ratios and their sensory properties can provide information about origin of food aromas and the perceived variations in the taste of foods. These data can be collected only when the enantiomers are separated using enantioselective high-resolution gas chromatography, which is the leading method for stereodifferentiation of chiral food odorants.

More recently, enantiomer ratios have been used as evidence of adulteration in natural foods and essential oils. If the enantiomer distribution of a chiral component of a natural food does not agree with that of a questionable sample, then adulteration can be suspected. Chiral GC analysis alone may not provide adequate evidence of adulteration, so it is often used in conjunction with other instrumental methods to completely authenticate the source of a natural food. These methods include isotope ratio mass spectrometry (IRMS), which determines an overall $^{13}C/^{12}C$ ratio (Mosandl, 1995), and site-specific natural isotope fractionation measured by nuclear magnetic resonance spectroscopy (SNIF-NMR), which determines a $^2H/^1H$ ratio at different sites in a molecule (Martin et al., 1993), which have largely replaced more traditional analytical methods using GC, GC-MS, and HPLC.

The requirements for analyzing food odorants are demanding. The system must have a sensitivity of a few parts per billion or less, and be capable of handling highly volatile compounds. Although many methods exist for the analysis of chiral compounds, gas chromatography is the only viable method for analyzing food odorants because many are present in amounts too low for detection by most analytical methods. The development of gas chromatography-olfactometry has helped to improve the detection of odorants with low thresholds present in trace amounts in foods.

Until chiral stationary phases were developed, the separation of enantiomers using gas chromatography was tedious, and was of little value in flavor analysis. Currently, the stationary phases most often used are a wide range of modified α-, β-, and γ-cyclodextrins dissolved in polysiloxanes. They are quite versatile and many are thermally stable up to 230°C. Some are now available where the cyclodextrin is anchored to the polysiloxane by the method of derivatization, which increases their useful lifetime. OV-1701 and SE-54 with 10% to 30% modified cyclodextrin are two of the most versatile chiral stationary phases that are commercially available. There is no systematic way to determine the most suitable stationary phase for a particular separation because the role played by molecular recognition in the mechanism of chiral separation using cyclodextrins is not yet fully understood. The online database CHIRBASE is a collection of published separations, and may be consulted to find the best stationary phase for a specific application (CHIRBASE, 1992). The literature describing applications of chiral chromatography to flavor analysis and systematic selection of chiral stationary phases is disorganized. Several companies marketing cyclodextrin columns have excellent Web sites describing the capabilities of the columns they make, including flavor applications (Restek Corporation, 1997; Supelco, 1998; also see Internet Resources for ASTEC web site). There are over a hundred chiral columns that are commercially available. See Tables G1.4.1 and G1.4.2 for cyclodextrin columns that are readily available, together with some leading suppliers.

Important terms

1. Enantiomeric purity: the measured ratio % of the detected enantiomers.

$$\text{enantiomeric purity } \% = \left[\frac{R}{R+S}\right] \times 100$$

$$\text{and } \left[\frac{S}{R+S}\right] \times 100$$

where R and S represent the areas of the R and S peaks, respectively, of a pair of enantiomers in the chromatogram.

2. Enantiomeric excess or optical purity, ee, the relative difference of the separated enantiomers.

$$\%\text{ee} = \left[\frac{R-S}{R+S}\right] \times 100$$

where $R > S$.

3. Separation factor, α:

$$\alpha = \frac{t_{R2}}{t_{R1}}$$

where t_{R1} and t_{R2} are the corrected retention times for peak 1 and peak 2, respectively.

4. Peak resolution or separation efficiency, R_s

$$R_s = \frac{2\left(t_{R2} - t_{R1}\right)}{W_{b1} + W_{b2}}$$

where W_{b1} is the peak width at base of the less retained peak, and W_{b2} is the peak width at base of the more retained peak measured at 4σ, where σ is the standard deviation for the peak.

5. Retention time, t'_R

$$t'_R = t_R - t_M$$

where t_R is the total retention time; time (or distance) from the point of injection to the point of peak maximum and t_M is the gas hold-up time; minimum time (or distance) required for the elution of a nonretained substance.

6. Retention index, RI

$$\text{RI}_i = 100n + 100\left[\frac{\log x_i - \log x_n}{\log x_{n+1} - \log x_n}\right]$$

where x_n is the corrected retention time of the n-alkane eluting before x_i.

7. Capacity ratio (partition ratio), k

$$k = \frac{t'_R}{t_M}$$

Table G1.4.1   Commercially Available Cyclodextrin Columns

2,3,6-Tri-*O*-methyl-β-cyclodextrin in OV1701[a]

2,3,6-Tri-*O*-methyl-γ-cyclodextrin in OV1701[a]

2,3,6-Tri-*O*-pentyl β-cyclodextrin[a]

2,3,6-Tri-*O*-pentyl-α-cyclodextrin

2,6-Di-*O*-pentyl-3-*O*-butyryl-γ-cyclodextrin[a]

2,6-Di-*O*-methyl-3-*O*-pentyl-β-cyclodextrin in OV1701[a]

2,6-Di-*O*-pentyl-3-trifluoroacetyl α-cyclodextrin

2,6-Di-*O*-pentyl-3-trifluoroacetyl β-cyclodextrin

2,6-Di-*O*-pentyl-3-trifluoroacetyl γ-cyclodextrin[a]

2,3-Di-*O*-methyl-6-*O*-*t*-butyldimethylsilyl-β-cyclodextrin

2,3-Di-*O*-ethyl-6-*O*-*t*-butyldimethylsilyl-β-cyclodextrin

2,3-Di-*O*-ethyl-6-*O*-*t*-butylsilyl-β-cyclodextrin

2,3-Di-*O*-acetyl-6-*O*-*t*-butyldimethylsilyl-β-cyclodextrin in OV1701[a]

2,3-Di-*O*-acetyl-6-*O*-*t*-butyldimethylsilyl-γ-cyclodextrin in OV1701[a]

O-(*S*)-2'-Hydroxypropyl-per-*O*-methyl α-cyclodextrin

O-(*S*)-2'-Hydroxypropyl-per-*O*-methyl β-cyclodextrin[a]

O-(*S*)-2'-Hydroxypropyl-per-*O*-methyl γ-cyclodextrin

2,6-Di-*O*-pentyl α-cyclodextrin

2,6-Di-*O*-pentyl β-cyclodextrin

2,6-Di-*O*-pentyl γ-cyclodextrin

2,6-Di-*O*-pentyl-3-*O*-acetyl α-cyclodextrin

3-*O*-Acetyl-2,6-Di-*O*-pentyl β-cyclodextrin

3-*O*-Butyryl-2,6-Di-*O*-pentyl γ-cyclodextrin 60% in OV 1701

[a]Widely used columns.

8. Odor activity value OAV

$$OAV = \frac{\text{concentration of odorant}}{\text{detection threshold in sample}}$$

9. Odor spectrum value OSV. Normalized odor data taken from each peak of the chromatogram that has been adjusted for odor compression using Stephen's Law (Acree, 1997; Ong et al., 1998). The odor chromatogram is redrawn as an odor spectrum. An odor spectrum value can be determined for each odorant in the sample. The values are independent of the method used for collecting odor data and of the concentration, and convey the relative importance of each odorant in the sample.

The measurement of odor intensity using OAVs is described by Grosch (1993, 1994). It requires the determination of the concentration of each odorant in the sample, and for those present in trace quantities, a stable isotope dilution assay must be used (Guth, 1997). This may make the determination of OAVs very tedious if many values are required. OSVs are normalized peak areas from an odor chromatogram and represent a more realistic representation of the importance of the odors in a sample as perceived by the nose. Their determination is described by Acree (1997).

Critical Parameters

When reliable chiral data about food odorants are required, it is often important to have good information about the nature and identity of the compounds under analysis before a chiral analysis is attempted. The information below may be essential for a good analysis, and methods for obtaining these data are described elsewhere.

1. Identity of odorants to be analyzed, determined by GC-MS.

2. A guarantee that the enantiomers are stable under the conditions of the analysis, and that no racemization occurs.

3. The order of elution of the R and S isomers, determined by comparison with a known authentic sample analyzed under the same conditions as the unknown.

4. Concentration of odorants in sample being analyzed, determined by use of internal standards, GC-MS in selected ion monitoring (SIM) mode or a stable isotope dilution assay for trace quantities (see UNIT G1.3).

5. Retention index of odorants on the stationary phase used in the chiral analysis (single oven) or on the precolumn (MDGC), and on more than one stationary phase if found to be necessary for the identification of odorants, determined by using a hydrocarbon standard, described above and in UNIT G1.1.

6. Odor descriptors of odorants to be analyzed and of those coeluting with analytes, determined by GC-O analysis. It may be important to know the odor descriptors for pairs of enantiomers in cases where contamination occurs. It may also be important in cases where one enantiomer is odorless, or where both enantiomers have similar odors but different intensities.

7. A sample that has been concentrated by preparative GC or MDGC to enable 1 and 4 to 6 described above to be determined. Concentration techniques are described in UNIT G1.2.

Table G1.4.2 Suppliers of Chiral Capillary Chromatography Columns

Advanced Separation Technologies ASTEC
Alltech Associates
Carlo Erba Reagenti
Chrompack International
Chrom Tech
CS-Chromatographie Service
J & W Scientific/Fisons
Macherey-Nagel GmbH
Mega
Restek
SGE International
Sumitomo Chemical
Supelco
Technicol
TPC Ziemer

Selection of instrumentation

The success of a chiral analysis will ultimately depend upon the instrumentation available. Single dimensional separations have become routine with a large selection of efficient chiral columns now available (see Table G4.1.1), but they have some limitations. Many powerful odorants are present in trace quantities, and are often hidden in the chromatogram by coeluting compounds, and may not show in the chromatogram, even if they can be made to elute alone. A one-oven system will provide reliable data when enantiomers are not coeluting, and the sensory data can be assumed to be for a single compound. If it is believed that coelution is occurring, a simple method to confirm that suspicion is to obtain a new chromatogram using a different stationary phase. If this is followed by a GC-O run using the new column, then the integrity of the odor data can be established. The Flavornet database (Acree and Arn, 1997), a developing source for the identification of flavor compounds by GC-O, is very useful for tracking flavor compounds by their retention indices on different stationary phases.

If more efficient chromatography is needed for improved separations and reliable quantitative data, then a multidimensional system must be used. The analysis can hardly be described as routine, since the cost of commercially available instrumentation is high, and there are very few instruments that are currently available (see Table G1.4.3). Many reported enantiospecific analyses described in the literature use an instrument that is either no longer in production (Mosandl et al., 1990; Bernreuther and Schreier, 1991) or is a sophisticated custom-built system (Mondello et al., 1999). Several designs that have been used for the switching device that transfers small portions of the first chromatogram to the second column have been described (Wright, 1997; Bertsch, 1999). The challenges confronting the flavor chemist when accurate enantiospecific data are required are described in the analysis, identification, and quantification of the sulfur odorants of passion fruit (Werkoff et al., 1998).

MDGC performed using a single oven is not a preferred configuration for the analysis of complex mixtures of food volatiles. It is important to keep in mind that no chromatographic system has yet been devised that is capable of completely resolving all the components of a complex mixture.

Selection of operating conditions

Use of the literature is strongly recommended as the starting point for determining the optimum operating conditions for a separation. The ultimate goal is for the separation factor α to be as large as possible. The chiral stationary phase, temperature program, gas flow though the column, and sample size are key parameters for a successful separation (König, 1992; Sponsler and Biederman, 1997).

Chiral columns are often operated isothermally at temperatures between 40° and 60°C, or with heating rates significantly lower than for routine analyses.

On-column injection is preferred, because it occurs at room temperature. Hot injector ports can lead to decomposition of the sample, and to racemization of chiral components. If a single oven system is used, then split injection offers some advantages because it is important not to expose the chiral column to large amounts of solvents. With an MDGC system, the heart-cutting technique removes the solvent from the chiral column, so on-column injection is preferred.

Hydrogen is the preferred carrier gas because runs are shorter and the resolution is superior. Helium will give satisfactory results unless α is close to 1.0. Linear velocities using hydrogen are usually ~80 cm/sec, which is higher than for routine separations because it results in narrower peaks.

Sample size must be kept to a minimum, 40 to 80 ng on-column concentration for chiral components, otherwise overloading, broadening of peaks, and loss of resolution will occur.

Selection and care of chiral columns

Column selection is not yet a precise science. There are no uniform theories for the separation mechanisms that operate for the resolution of enantiomers using the wide array of chiral stationary phases that are currently available. It is therefore strongly recommended that the literature is searched to determine whether the analyte under investigation has been separated and which chiral stationary phase was used. Many chiral food volatiles have now been separated and a comprehensive database has been compiled that is continually updated and contains most reported separations (Koppenhoefer et al., 1993; Roussel and Piras, 1993). There are many other sources in the literature (Anonymous, 1993a,b; König, 1993; Maas et al., 1994; Schreier et al., 1995; Juchelka et al., 1998; Miranda et al., 1998).

Modified cyclodextrins are the most versatile and widely used chiral stationary phases. The most widely used columns contain 10% to 50% cyclodextrin dissolved in either OV-101 or SE-52 polysiloxane. They are thermally stable up to 230°C but require some care in use because cyclodextrins are soluble in many solvents and can be washed off the column if they are exposed to too much solvent.

Considerations in selecting a column

To select a chiral column, first the literature should be consulted, and then the availability of the column should be assessed (see Tables G1.4.1 and G1.4.2).

Use standard lengths (25 to 30 m). To avoid overloading the column, the capacity can be increased by using a thicker film thickness and a wider diameter. This will not alter the separation efficiency. Occasionally, very short columns (2 m) give better results.

Care of CD columns

New columns should be conditioned following the manufacturer's instructions. Most are conditioned at lower temperatures than for achiral columns. The temperatures used for separation should be as low as possible, which leads to better separation and less deterioration of the sample.

The column should not be exposed to unnecessarily large volumes of solvent. Columns can be stored in a clean dry atmosphere, but they should be sealed by a flame or taped for very short periods of time. Some columns need to be stored in an atmosphere of nitrogen to avoid serious loss of selectivity. The manufacturer's instructions should be followed for column storage.

Troubleshooting

Most analytical problems arise from poor separations and coelution of analytes. Some selected problems are discussed.

If the separation factor approaches 1.0, possible causes include: chiral stationary phase is incorrect, carrier gas velocity is too high, or only one enantiomer is present in the sample.

If the separation is acceptable but the resolution is poor (see Fig. G1.4.2), possible causes: the temperature program is incorrect, carrier gas velocity is incorrect, column is overloaded, focussing of the heart-cut sample is insufficient (MDGC), the column is old and has lost efficiency, or chiral stationary phase is inappropriate.

If coelution is occurring, possible causes: an incorrect polysiloxane is in the chiral column, the temperature program is wrong, the effect on different components may vary, MDGC may be needed, or an unsuitable precolumn is used. It is easier to solve coelution problems on the precolumn than on the analytical column.

The sensory data are inconsistent and coelution is suspected. If the solutions given above fail, then the problem may be difficult to resolve because the contamination may be occurring from an odorant that is present in such small quantities that it is only detectable by GC-O. Possible causes: the separation factor α is too small, tailing of peaks is occurring, both enantiomers have the same odor, or one enantiomer is odorless. Interaction of coeluting odors may vary with small variations in the concentrations of any of the odorants; the note of an odorant can vary with concentration.

Table G1.4.3 Instrumentation: Multidimensional Gas Chromatography Systems

Currently available	
Microanalytics Instrument	Complete MDGC systems are available, suitable for chiral aroma analysis
SGE International	The switching device is sold as an accessory to attach to a single or dual oven system
Systems widely used that are no longer in production	
Siemens AG	Sichromat 2 double-oven system with a "live switching" coupling piece and independently controlled ovens
Chrompack International	Multiple Switching Intelligent Controller (MUSIC) system, based on the Deans pressure switching concept
Custom-built systems	Many are described in the literature. An example of a sophisticated system can be found in Mondello et al. (1999)

Alternative methods for overcoming problems caused by coelution and poor resolution are as follows:

1. Synthesize chiral compounds of interest and determine OSVs of enantiomeric pairs as pure authentic samples.

2. Use a trained sensory panel to collect data when separation factors approach 1.0. The odor data can then be analyzed using statistical methods.

3. Using a different achiral stationary phase will alter the order of elution. This may eliminate some coelution problems, but new ones may occur. It may alter the order of elution of the *R* and *S* isomers.

Anticipated Results

Enantiomer ratio

Values for the enantiomer ratio found in natural products can range from 0% to 100%. If the enantiomeric purity is 100%, care must be taken to determine which isomer is present by comparison of retention times with a known standard. Samples from different geographic or growing regions may show some variation in the ratio of a specific compound, while others will not deviate from a known value.

Knowledge of enantiomer ratios, particularly for compounds that occur naturally with 100% of one enantiomer, can form the basis of an authenticity determination (Kreis and Mosandl, 1992; Mosandl and Juchelka, 1997). In such analyses, deviations from known enantiomer ratios of authentic samples can indicate adulteration of the sample, where synthetic materials have been added, or the sample has been distilled although it was labeled as a natural product.

Limit of detection. A compound must give an FID response to be detected and quantified. This requires a concentration of a few ppm if the resolution is high and coelution is not occurring. Using SIM (see UNIT G1.3), odorants can be quantified at the ppb level.

Enantiospecific sensory evaluation

The integrity of the sensory data will depend upon the enantiomers being of the highest chemical and optical purity. This requires odorants to elute as single compounds without tailing of peaks with at least 30 to 60 sec between each enantiomer. These conditions may be unattainable since sensory measurements are very susceptible to the presence of trace impurities of other odorants.

If qualitative data are required, and the enantiomers differ significantly in their odor quality, then acceptable results may be obtained. If odor intensity measurements (OAVs or OSVs) or threshold values are required, then the conditions described above must be obtained if the data is to be of value. Bernreuther et al. (1997) and Koppenhoefer et al. (1994) have published the enantiospecific sensory data for a variety of chiral odorants.

Limit of detection. Any odorant present above its threshold level can be detected in a GC-O experiment. This can vary from a few ppb to a ppt level or higher. Odorants present in trace quantities are frequently below the limit of detection of an FID, so GC-O is the only method available for their identification. GC-O coupled with SIM provides the most sensitive method of identifying and quantifying trace odorants.

Time Considerations

Once the optimum operating conditions are established, individual runs can be quite short, depending upon the retention time at which enantiomers elute, from 10 to 100 min. Most enantiomers of interest can be made to elute in <30 min when a single oven system is used. The instrument has to be cooled and stabilized before another run can be carried out, so the total time from one run to the next will be ~1 hr. The run time will depend mostly upon the temperature program.

For an MDGC system, the total run time will depend upon the instrument configuration, and on how the heart-cut fraction is transferred from oven 1 to oven 2. If each enantiomer is examined separately on the analytical column, then the total run make take only 10 to 15 min longer than the time for a single oven system. However, this will require a run for each enantiomer examined. If all the heart-cuts in a sample are examined in a single run on the analytical column, then a run will take 2 to 3 times as long as the time for a single oven system and it will depend upon the temperature program for the analytical column. Heating rates are rarely >2°C/min.

On a 30-m nonpolar column, the chromatogram for a hydrocarbon standard C_7 to C_{20} with a temperature program of 2°C/min starting at 35°C will take 90 min to run with an alkane eluting every 6 to 8 min. With a rate of 6°C/min, the run will take 40 min, with an alkane eluting every 2 to 3 min.

Stereodifferentiation
of Chiral
Odorants
Using Gas
Chromatography

G1.4.18

Acree, T.E. 1997. GC/olfactometry. *Anal. Chem.* 69:170A-175A.

Anon. 1993a. Collection of enantiomer separation factors obtained by capillary gas chromatography on chiral stationary phases. *J. High Resolut. Chromatogr.* 16:338-352.

Anon. 1993b. Collection of enantiomer separation factors obtained by capillary gas chromatography on chiral stationary phases. *J. High Resolut. Chromatogr.* 16:312-323.

Bernreuther, A. and Schreier, P. 1991. Multidimensional gas chromatography/mass spectrometry: A powerful tool for the direct chiral evaluation of aroma compounds in plant tissues. II. Linalool in essential oils and fruits. *Phytochem. Anal.* 2:167-170.

Bernreuther, A., Epperlein, U., and Koppenhoefer, B. 1997. Enantiomers: Why they are important and how to resolve them. *In* Techniques for Analyzing Food Aroma (R.E. Marsili, ed.) pp. 143-207. Marcel Dekker, New York.

Bertsch, W. 1999. Two-dimensional gas chromatography. Concepts, instruments and applications—Part I: Fundamentals, conventional two-dimensional gas chromatography, selected applications. *J. High Resolut. Chromatogr.* 22:647-665.

Gordon, B.M., Rix, C.E., and Borderging, M.F. 1985. Comparison of state-of-the-art column switching techniques in high resolution gas chromatography. *J. Chromatogr. Sci.* 23:1-10.

Grosch, W. 1993. Detection of potent odorants in foods by aroma extract dilution analysis. *Trends Food Sci. Technol.* 4:68-73.

Grosch, W. 1994. Determination of potent odorants in foods by aroma extract dilution analysis (AEDA) and calculation of odor activity values (OAVs). *Flavour Fragrance J.* 9:147-158.

Guth, H. 1997. Quantitation and sensory studies of character impact odorants of different white wine varieties. *J. Agric. Food Chem.* 45:3027-3032.

Juchelka, D., Beck, T., Hener, U., Dettmar, F., and Mosandl, A. 1998. Multidimensional gas chromatography coupled online with isotope ratio mass spectrometry (MDGC-IRMS). Progress in the analytical authentication of genuine flavor components. *J. High Resolut. Chromatogr.* 21:145-151.

König, W.A. 1992. Gas Chromatographic Enantiomer Separation with Modified Cyclodextrins. Hüthig Buch, Heidelberg.

König, W.A. 1993. Collection of enantiomeric separation factors obtained by capillary gas chromatography on chiral stationary phases. *J. High Resolut. Chromatogr.* 16:569-586.

Koppenhoefer, B., Nothdurft, A., Pierrot-Sanders, J., Piras, P., Popescu, P., Roussel, C., Stiebler, M., and Trettin, U. 1993. CHIRBASE: A graphical molecular database on the separation of enantiomers by liquid, supercritical fluid, and gas chromatography. *Chirality* 5:213-219.

Koppenhoefer, B., Behnisch, R., Epperlein, U., and Holzsuch, H. 1994. Enantiomeric odor differences and gas chromatographic properties of flavors and fragrances. *Perfum. Flav.* 19:1-14.

Kreis, P. and Mosandl, A. 1992. Chiral compounds of essential oils. Part XII. Authenticity control of rose oils, using enantioselective multidimensional gas chromatography. *Flavour Fragrance J.* 7:199-203.

Maas, B., Dietrich, A., and Mosandl, A. 1994. Collection of enantiomer separation factors obtained by capillary gas chromatography on chiral stationary phases. *J. High Resolut. Chromatogr.* 17:109-115, 169-173.

Martin, G.J., Remaud, G.S., and Martin, G.G. 1993. Isotopic methods for control of natural flavours authenticity. *Flavour Fragrance J.* 8:97-107.

Miranda, E., Sánchez, F., Sanz, J., Jimenéz, M.I., and Martinéz-Castro, I. 1998. 2,3-Di-O-pentyl-6-O-*tert*-butyldimethylsilyl-β-cyclodextrin as a chiral stationary phase in capillary gas chromatography. *J. High Resolut. Chromatogr.* 21:225-233.

Mondello, L., Catalfamo, M., Proteggente, A.R., Bonaccorsi, I., and Dugo, G. 1998. Multidimensional capillary GC-GC for the analysis of real complex samples. 3. Enantiomeric distribution of monoterpene hydrocarbons and monoterpene alcohols of mandarin oils. *J. Agric. Food Chem.* 46:54-61.

Mondello, L., Catalfamo, M., Cotroneo, A., Dugo, G., Dugo, G., and McNair, H. 1999. Multidimensional capillary GC-GC for the analysis of real complex samples: Part IV. Enantiomeric distribution of monoterpene hydrocarbons and monoterpene alcohols of lemon oils. *J. High Resolut. Chromatogr.* 22:350-356.

Mosandl, A. 1995. Enantioselective capillary gas chromatography and stable isotope ratio mass spectrometry in the authenticity control of flavors and essential oils. *Food Rev. Int.* 11:597-664.

Mosandl, A. and Juchelka, D. 1997. Advances in the authenticity assessment of citrus oils. *J. Essent. Oil Res.* 9:5-12.

Mosandl, A., Hener, U., Hagenauer-Hener, U., and Kustermann, A. 1989. Stereoisomeric flavor compounds. XXXII. Direct enantiomer separation of chiral γ-lactones from food and beverages by multidimensional gas chromatography. *J. High Resolut. Chromatogr.* 12:532-536.

Mosandl, A., Hener, U., Kreis, P., and Schmarr, H.-G. 1990. Enantiomeric distribution of α-pinene, β-pinene and limonene in essential oils and extracts; Part 1. Rutaceae and Gramineae. *Flavour Fragrance J.* 5:193-199.

Ong, P.K.C., Acree, T.E., and Lavin, E.H. 1998. Characterization of volatiles in rambutan fruit (*Nephelium lappaceum* L.). *J. Agric. Food Chem.* 46:611-615.

Roussel, C. and Piras, P. 1993. CHIRBASE, a molecular database for storage and retrieval of chromatographic chiral separations. *Pure Appl. Chem.* 65:235-244.

Schreier, P., Bernreuther, A., and Huffer, M. 1995. Analysis of Chiral Organic Molecules. Methodology and Applications. pp. 132-233. Walter de Guyter, New York.

Sponsler, S. and Biederman, M. 1997. Optimization of chiral separations using capillary gas chromatography. *Am. Lab.* 24C-H.

Supelco. 1998. Chiral Cyclodextrin Capillary GC Columns. Sigma-Aldrich, Bellefonte, Penn.

Werkoff, P., Güntert, M., Krammer, K., Sommer, H., and Kaulen, J. 1998. Vacuum headspace method in aroma research: Flavor chemistry of yellow passion fruits. *J. Agric. Food Chem.* 46:1076-1093.

Wright, D.W. 1997. Application of multidimensional gas chromatography techniques to aroma analysis. *In* Techniques for Analyzing Food Aroma (R. Marsili, ed.) pp. 113-141. Marcel Dekker, New York.

Key References

Beesley, T.E. and Scott, R.P.W. 1998. Chiral Chromatography. John Wiley & Sons, Chichester, United Kingdom.

Contains excellent background and resource material.

König, 1992. See above.

Contains excellent background and resource material.

Schreier et al., 1995. See above.

Contains useful lists of applications and available chiral columns with their suppliers.

Bertsch, 1999. See above.

Gives a good review of types of instrumentation used in MDGC applications with examples from flavors and aromas as well as other areas.

Mosandl, A. 1992. Capillary gas chromatography in quality assessment of flavors and fragrances. *J. Chromatogr.* 624:267-292.

Mosandl, 1995. See above.

Werkoff, P., Brennecke, S., Bretschneider, W., Güntert, M., Hopp, R., and Surburg, H. 1993. Chirospecific analysis in essential oil, fragrance and flavor research. *Z. Lebensm. Unters. Forsch.* 196:307-328.

Internet Resources

Acree, T.E. and Arn, H. Flavornet. 1997. Gas chromatography-olfactometry of natural products. http://www.nysaes.cornell.edu/fst/faculty/acree/flavornet/index.html.

A developing database of retention indices for flavor compounds determined on OV101, DB5, OV17, and C20M. No subscription required.

http://www.astecusa.com/g_chromatography/index.htm

An overview of the capabilities of the columns made by Advanced Separation Technologies (ASTEC).

CHIRBASE, A Molecular Database for Chiral Chromatography. 1992. http://chirbase.u-3mrs.fr/.

A comprehensive database containing all published chromatographic separations using chiral stationary phases. It can be searched by the built in capabilities of Chembase and ISIS. It is continually updated. Subscription required for access.

Restek Corporation. 1997. A Guide to the Analysis of Chiral Compounds by GC. www.restekcorp.com/chiral/chiral.htm.

An overview of the capabilities of the chiral GC and the columns made by Restek. No subscription required.

Contributed by Mary G. Chisholm
Behrend College, The Pennsylvania
  State University
Erie, Pennsylvania

**Stereodifferentiation
of Chiral
Odorants
Using Gas
Chromatography**

G1.4.20

Analysis of Citrus Oils

The recovery of citrus oils goes back hundreds of years to the sale of lemon oil in Sicily. Citrus oil is recovered as cold-pressed oil, essence oil, or *d*-limonene. The term cold-pressed oil refers to citrus oil recovered during juice extraction, whereas essence oil is captured during evaporation. Citrus oils are composed of 90% monoterpenes, with orange and grapefruit oils containing 95% *d*-limonene. The quality of citrus oils determines their function and market value. Most cold-pressed oils are concentrated (folded) and then fractionated. The fractionated oil is either sold or used as a flavor precursor. Mandarin, lime, and lemon oil are typically processed for their oil and sold to flavor houses. Other oils are added back to juices to enhance flavor. Identification of volatile compounds is determined by chemical and physical parameters. The following protocols are designed to measure total available oil and to evaluate quality and composition of citrus oils. Gas chromatography characterizes the differences between cultivars, while titration techniques quantify *d*-limonene, aldehyde, and ester content.

STRATEGIC PLANNING

The following protocols are primarily conducted to ensure the quality of citrus oils during various phases of citrus processing. Citrus processors must account for quality and quantity of citrus oils sold to flavor houses. This not only includes monitoring recovered oils (i.e., cold-pressed oil, essence oil, and *d*-limonene) but also the total available oil.

Basic Protocols 1 and 2 provide flavor profiles for orange, mandarin, grapefruit, lemon, and lime oil. These profiles determine price and function depending on the composition and are gas chromatography (GC) techniques based on 30- to 45-min runs. However, verification of compounds at specific retention times is time consuming, with both the column and the flow rate effecting retention times. In addition, quantification can be a lengthy process. The generation of standard curves for quantitative analysis requires multiple runs of individual components. However, time should be minimized once verification and standard curves are established for individual compounds, provided GC conditions remain constant. Basic Protocol 3 is another GC technique researchers can use to confirm the purity of *d*-limonene. This is a 30-min assay based on an FID response.

In addition to chemical assays, Basic Protocols 4 through 6 evaluate the physical properties of citrus oils. These methods are based on the high percentage of *d*-limonene, in which citrus oils have characteristic values. Table G1.5.7 provides the physical and chemical properties defined by Food Chemical Codex (NRC, 1981).

Basic Protocol 7 applies more directly to citrus processors. It measures the total available oil. These procedures are also based on the level of *d*-limonene (>95%) in the sample. According to EPA regulation, processors must account for both recovered oil and emission vapors (*d*-limonene). The regulation permits 100 tons of volatile organic compounds per year. This is a highly accurate titration method with results slightly skewed for lemon and lime samples due to the various terpenes. However, most lemon- and lime-processing plants recover a higher percent of the total available oil and are in compliance with EPA regulations. Alternate Protocols 1 to 3 measure oils in press cakes, dry peel, and press liquor.

Basic Protocols 8 and 9 look at specific groups of compounds, aldehydes and esters, which are of more interest to flavor houses and research scientists. For flavorists, the strength of "citrus flavor" is determined from the aldehyde content, while the "fruity aroma" is attributed to the ester content. Over the years, these procedures have been modified. Basic Protocol 8 is a titration method to determine aldehyde content in citrus oils. It was originally developed for lemon oil; however, it is also applicable to other citrus oils. Alternate Protocols 4 and 5 also produce similar results. Alternate Protocol 4 is based on

Contributed by Trevor Gentry

citral levels, the predominate aldehyde in lemon oil. Alternate Protocol 5 is a spectro-photometric method, in which additional equipment and reagents are needed. Basic Protocol 9 is also a colormetric assay for determining ester content in citrus oils. It is a rapid technique with accurate results.

Basic Protocols 3 through 9 are primarily useful as quality control measures. They are rapid, usually within 30 min, given reagent preparation. The results are used to monitor the quality of a process. These results support established values for "high" quality citrus oil. Basic Protocols 1 and 2 are more involved and are better suited for research purposes. The equipment is more sensitive and also more expensive. Furthermore, the strength of the GC analysis can be enhanced by the addition of a mass spectrometer to identify either contaminates or unknown compounds present in a sample.

QUALITATIVE ANALYSIS OF CITRUS OILS BY GAS CHROMATOGRAPHY

The desired citrus flavor is the result of volatile compounds in specific proportions. Citrus oils have unique composition profiles depending on the cultivar, the processing conditions, and the storage conditions. There have been more than 200 different compounds identified in citrus oils. However, the degree of unsaturation in monoterpenes leads to rapid oxidation and unstable compounds. The reactivity and volatility of citrus oils require strict quality control protocols.

The quality of citrus oils is based on the purity of the sample as determined by a gas chromatogram (GC) profile. A small sample is injected onto a column, which separates the individual components. Separation is based on the physical interaction between the column and the sample as indicated by retention times. Figure G1.5.1 illustrates the GC setup.

Materials

Acetone
Citrus oil

Gas chromatograph including:
Flame ionization detector (FID)
Integrator
Nonpolar (DB-1, SE-30, OV-1) or weakly polar (DB-5, SE-52, Rt-5) column: 20 m × 0.10-mm i.d. with 0.1-μm film (J & W Scientific)
10-μl syringe

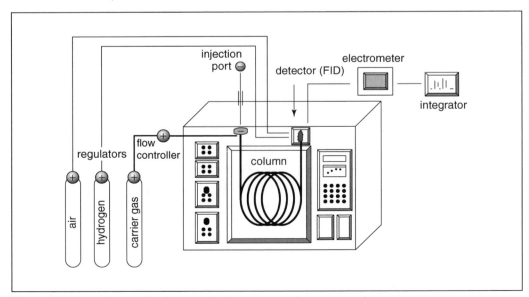

Figure G1.5.1 Schematic diagram depicting a gas chromatography setup.

1. Set up GC with the following conditions and ignite FID.

> Injector: 275°C
> Detector: FID 350°C, nitrogen make-up gas (30 ml/min)
> Carrier gas: 60 cm/sec hydrogen, split ratio 1:275.

2. Input the following temperature program for the oven:

> 70°C for 1 min
> heat to 250° at 30°C/min
> heat to 310° at 20°C/min
> hold at 310°C for 2 min
> cool to 40°C.

3. Rinse 10-µl syringe 2 to 3 times with acetone.

4. Rinse 10-µl syringe 2 to 3 times with citrus oil sample.

5. Press start and inject 1.0 µl citrus oil sample.

6. Replicate samples as needed.

QUANTITATIVE ANALYSIS OF CITRUS COMPOUNDS BY GAS CHROMATOGRAPHY

Gas chromatography provides a rapid analysis of citrus oil quality. This technique can be further enhanced to determine quantitative levels of individual compounds. Compounds can be measured based on the FID response. A standard curve with known concentrations is used to extrapolate an unknown concentration.

Materials

> Sample to be analyzed
> Standard (compound of interest)
> Solvent (e.g., ethanol, hexane)
> 100-ml volumetric flask

> Additional reagents and equipment for gas chromatography (see Basic Protocol 1)

1. Perform GC analysis of the sample as described above (see Basic Protocol 1).

2. Weigh 0.1 mg standard into a 100-ml volumetric flask, dissolve in an appropriate solvent, and bring volume to 100 ml (final 10 ppm).

   Check the solubility of the compound in solvent prior to the experiment.

3. Prepare a series of concentrations down to 0.1 ppm following the dilution scheme in Figure G1.5.2.

   This scheme generates seven standards covering a range of 0.1 to 10 ppm, which will cover the concentration of most compounds. Data must be within the linear region of the curve.

4. Press start and inject 1.0 µl of the first standard onto the column. Record response. Perform analysis of each concentration three to four times.

   Standards should be quantified in order of increasing concentration, to prevent carry-over responses.

5. Average the replicate responses at each concentration and then generate a standard curve by plotting response versus concentration.

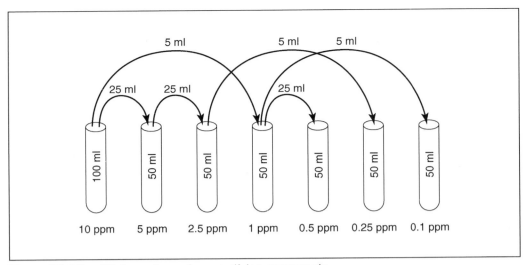

Figure G1.5.2 Dilution scheme for quantifying compounds.

6. Calculate the concentration of the unknown sample using the equation:

$$[unknown] = (response - intercept)/slope$$

This equation is derived from the basic linear regression equation, y = mx + b, where y is response, m is slope, x is concentration of the unknown, and b is the y intercept.

BASIC PROTOCOL 3

ANALYSIS OF LIMONENE PURITY BY RAPID GAS CHROMATOGRAPHY

Limonene is the major monoterpene in orange oil. This is a colorless and odorless compound at high purity. However, it rapidly oxidizes to carveol and carvone in the presence of air. Under acidic conditions, α-terpineol, β-terpineol, and γ-terpineol are also produced. Many of the impurities present in limonene have much higher odor potencies. These odor potent compounds can be perceived as "limonene odor".

Basic Protocol 3 is a rapid GC method to evaluate limonene purity. The needle is coated with a residual limonene layer in the syringe. The sample volatilizes into the column in which the sensitivity of GC detectors (FID) provides a reliable indication of purity. This is known as a "wet needle" injection.

Materials

Limonene sample
Additional reagents and equipment for gas chromatography (see Basic Protocol 1)

1. Follow the same GC/FID conditions for citrus oil analysis (see Basic Protocol 1).

2. Rinse needle with limonene sample.

3. Withdraw and then discard 1.0 µl of sample.

   This "wets" the needle, creating a limonene residue on the inside surface.

4. Press start and inject needle into septum.

5. Withdraw the plunger slightly (1 to 2 µl) and rapidly depress plunger.

   The combination of the pressure created from the plunger and temperature of the injection port will cause the sample to volatilize and enter the gas phase.

6. Determine limonene purity as the % peak area for the limonene response compared to the total area of the chromatogram.

ANALYSIS OF CITRUS OILS BY PYCNOMETRY

Specific gravity is the density (mass per volume) of the sample compared to the density of water (see Table G1.5.7 in Anticipated Results). Samples are equilibrated to the same temperature and weighed using a pycnometer. The pycnometer is a sample bottle that provides a constant volume.

Materials

Citrus oil, 25°C
25°C water bath
Pycnometer (VWR Scientific Products)
Analytical balance (Fisher)

1. Fill pycnometer with citrus oil cooled to 25°C.

    A specific gravity vial (VWR Scientific Products) can be substituted for a pycnometer.

2. Place pycnometer in a 25°C water bath and adjust citrus oil level as needed.

    Oils will expand or contract according to temperature.

3. Put cap on pycnometer and dry.

4. Weigh on an analytical balance and record.

5. Repeat procedure using the same pycnometer with water at 25°C.

6. Calculate the specific gravity (SG) using the equation:

    SG = oil weight/water weight

ANALYSIS OF CITRUS OILS BY REFRACTIVE INDEX

The refractive index is another physical property used to identify citrus oils. A refractometer is based on the law of refraction, or Snell's law, where the angle of refraction is a function of the sample. The sample is placed between two prisms. The light is trapped within the lower material (oil) and is completely reflected internally at the boundary surface. The field is then adjusted until the light and dark sections are equal. The line separating the two halves is known as the critical ray. The critical ray falls along a fixed scale. The instrument readings are calibrated to known samples (i.e., water or glass) with automatic corrections for temperature. The reliability of this procedure correlates to the high limonene content (>90%) in citrus oils.

Materials

Citrus oil
Refractometer

1. Record temperature of citrus oil.

2. Place 1 to 2 drops citrus oil on the glass prism of a refractometer.

3. Adjust the line between light and dark fields until distinct.

4. The readings are indexes of refraction. These readings are a conformation of the standard of identity for a citrus oil. The composition of these oils provides a precise reading.

    Most modern refractometers have built-in temperature corrections.

ANALYSIS OF CITRUS OILS BY POLARIMETRY

Optical rotation measures the degree that light is rotated (see Table G1.5.7 in Anticipated Results). In citrus oils, *d*-limonene is the major enantiomer in the sample. Since other optically active compounds are often present in racemic mixtures, there is no net rotation and thus they are ignored. If a compound is a racemic mixture, the polarimeter will not give a reading. Readings can be verified with known standards.

Materials

> Citrus oil
> Polarimeter with 25-mm polarimeter tube (VWR Scientific Products)

1. Adjust citrus oil to 25°C.

2. Place 25-mm polarimeter tube in trough of polarimeter between polarizer and analyzer.

3. Adjust analyzer until both halves have equal light intensities.

4. Determine direction of rotation based on this adjustment.

   The direction is positive (+) if the analyzer was adjusted clockwise, negative (−) if adjusted counterclockwise.

5. Read degrees directly. Multiply number by 4 (to normalize to a 100-mm tube standard).

QUANTIFICATION OF TOTAL OIL FROM WHOLE FRUIT OR WET PEEL

The processing of citrus oil is highly regulated. Processors must account for total available oil, recovered oil, and emitted oil. This is done primarily for environmental reasons, due to the reactive nature of terpenes. The total available oil varies depending on the cultivar.

The amount of oil in a sample is determined by Scott oil analysis (AOAC, 1990e). This is a bromination reaction previously used to determine the number of fatty acid double bonds. This titration method quantifies the recoverable oil in fruits and fruit products based on the release of Br_2 and the formation of limonene tetrabromide (Braddock, 1999). Figure G1.5.3 illustrates the chemical reaction for the bromination of limonene. Other monoterpenes (α-pinene and citral) also react; however, the method is accurate to within 10 ppm limonene (Scott and Valdhuis, 1966). For this procedure, limonene is co-distilled with isopropanol and titrated with a potassium bromide/bromate solution.

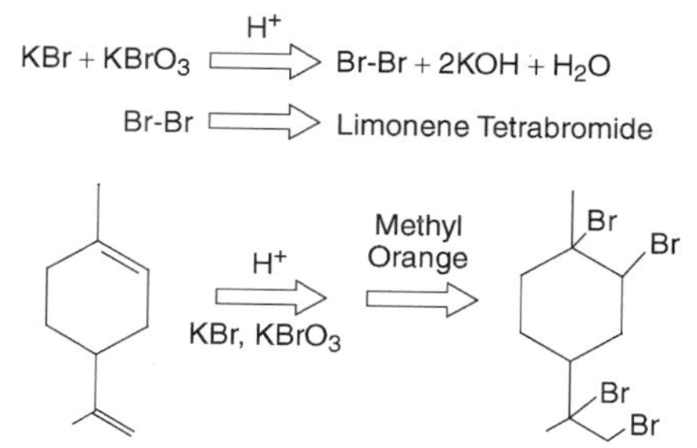

Figure G1.5.3 Chemical reaction of Scott oil analysis. Reprinted from Braddock (1999) with permission from John Wiley & Sons, Inc.

Materials

> Whole fruit or wet peel (shredded)
> Isopropanol
> Methyl orange indicator solution: 0.5% (v/v) methyl orange in 1:2 (v/v) HCl
> 0.1 N KBr/KBrO$_3$ solution (see recipe)

> 3-liter blender (e.g., Waring Blendor)
> Distillation apparatus (Fig. G1.5.4): 500 ml round-bottom flask with 24/40 neck attached to an adapter with a 28/15 ball connected to a 250-ml flask
> Glass beads
> Electric heater
> 25-ml buret

> Additional reagents and equipment for determining moisture content (see Support Protocol 1)

1. Determine moisture content of sample (see Support Protocol 1)

2. Quarter whole fruit. Weigh 500 g sample (whole fruit or wet peel) into a 3-liter blender. Add 2500 g distilled water. Mix 2 min at medium speed and 1 min at high speed.

3. Place 25 g homogenate into a distillation flask with three to four glass beads.

   Samples should be analyzed in triplicate.

4. Add 25 ml isopropanol to homogenate.

5. Begin distillation and collect 25 to 30 ml distillate.

   Figure G1.5.5 shows the distillation set-up.

6. Add 10 ml methyl orange indicator solution to distillate.

7. Titrate with 0.1 N KBr/KBrO$_3$ solution to an endpoint of pH 5.4.

8. Calculate the total oil from whole fruit or peel using the equation:

   g oil/g peel = (x ml 0.1 N titrant/25 g sample) × (3000 g homogenate/500 g peel) × (0.004 ml oil/ml titrant) × (0.84 g oil/ml oil)

   The 0.004 ml oil/ml titrant is a correction factor based on the standardization of the potassium bromide/bromate solution.

9. Convert to lb. oil/ton dry solids using the equation:

   lb. oil/ton dry solids = (lb. oil/lb. peel) × (2000 lb. peel/ton peel) × (y ton peel/ton dry solids)

   where y ton peel/ton dry solids is the inverse of % solids, measured in step 1.

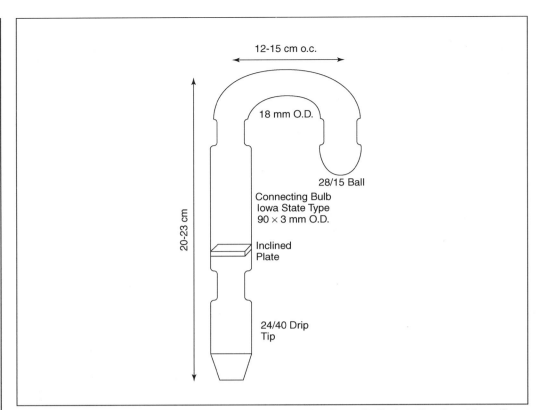

Figure G1.5.4 Depiction of connecting tube adapter for direct distillation. Reprinted from Scott and Veldhuis (1966) with permission from AOAC International.

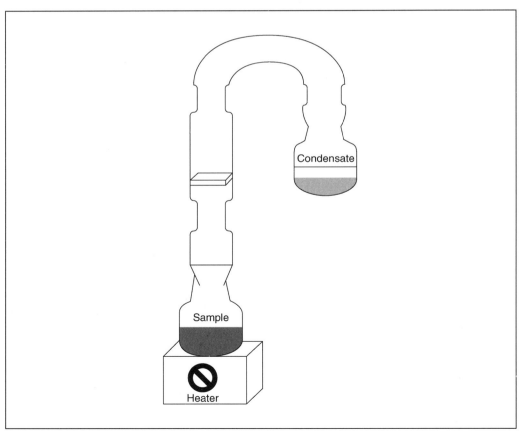

Figure G1.5.5 Schematic diagram for distillation procedure.

QUANTIFICATION OF PRESS CAKE OIL

The recovery of cold-pressed oil is not 100% effective. Most oil recovery units capture between 50% and 70%, depending on the type of unit. After the juice is extracted, the wet peel is sent to the feed mill. The peel is treated with 0.3% lime and sent through a shredder. The peel is then pressed, generating press cake and press liquor. The amount of oil in the press cake is critical to volatile organic compounds (VOC) emission levels. This quantitative method is based on the Scott oil analysis described in Basic Protocol 7.

Additional Materials (also see Basic Protocol 7)

Press cake sample

1. Perform analysis as described (see Basic Protocol 7, steps 1 to 7) but use 300 g press cake (PC) with 2700 g distilled water.

2. Calculate oil quantity from the press cake sample using the equation:

   g oil/g PC = (x ml 0.1 N titrant/25 g sample) × (3000 g homogenate/300 g PC) × (0.004 ml oil/ml titrant) × (0.84 g oil/ml oil)

3. Convert to lb. oil/ton dry solids using the equation:

   lb. oil/ton dry solids = (lb. oil/lb. PC) × (2000 lb. PC/ton PC) × (y ton PC/ton dry solids)

   where y ton PC/ton dry solids is the inverse of % solids, measured in step 1 of Basic Protocol 7.

QUANTIFICATION OF OIL IN DRY PEEL AND PELLET

The amount of oil released from the press cake during drying is at a constant rate, 0.86 kg oil/1 kg press cake (Gentry, 2001). The volatiles released from the press cake are then vented through the waste heat evaporator during concentration of the molasses. The difference in limonene concentration between the press cake and the dry peel is the maximum amount of VOC emissions. The concentration in the press cake is determined from the Scott oil analysis used in Basic Protocol 7.

Additional Materials (also see Basic Protocol 7)

Dry peel or pellets

1. Perform analysis as described (see Basic Protocol 7, steps 1 to 7) but use 100 g dry peel/pellets (DP) with 2900 g distilled water.

2. Calculate the oil from the dry peel sample using the equation:

   g oil/g DP = (x ml 0.1 N titrant/25 g sample) × (3000 g homogenate/100 g DP) × (0.004 ml oil/ml titrant) × (0.84 g oil/ml oil)

3. Convert to lb. oil/ton dry solids using the equation:

   lb. oil/ton dry solids = (lb. oil/lb. DP) × (2000 lb. DP/ton DP) × (y ton DP/ton dry solids)

   where y ton DP/ton dry solids is the inverse of % solids, measured in step 1 of Basic Protocol 7.

QUANTIFICATION OF OIL IN PRESS LIQUOR AND MOLASSES

During the concentration of the press liquor to molasses, d-limonene is distilled and recovered. The recovery of d-limonene is a function of evaporator capacity. Most citrus processors add the molasses back onto the wet peel. The difference between the press liquor and the molasses is the amount of distilled limonene. This is determined by Scott oil analysis.

Additional Materials (also see Basic Protocol 7)

0.0247 N KBr/KBrO$_3$ solution (see recipe)
Press liquor (PL) or molasses
Additional reagents and equipment for determination of °Brix (UNIT H1.4)

1. Determine the soluble solids (°Brix) with a refractometer (UNIT H1.4).

2. Perform analysis as described (see Basic Protocol 7, steps 2 to 7), but use 100 g solids and bring to 3000 g total weight, and reduce concentration of titrant to 0.0247 N.

   The normality of the titrant is lower due to the lower limonene content. It is more of an issue for molasses.

   The appropriate amounts of sample and water are determined from the desired amount of solids and the Brix measurement. For 100 g solids from an 18°Brix press liquor, the homogenate would consist of 100 g solids × (1/0.18) = 556 g press liquor and 2444 g water.

3. Calculate the oil from press liquor using the equation:

   g oil/g PL = (x ml 0.0247 N titrant/25 g sample) × (3000 g homogenate/m g PL) × (0.001 ml oil/ml titrant) × (0.84 g oil/ml oil)

4. Convert to lb. oil/ton dry solids using the equation:

   lb. oil/ton dry solids = (lb. oil/lb. PL) × (2000 lb. PL/ton PL) × (y ton PL/ton dry solids)

   where y ton PL/ton dry solids is the inverse of °Brix, measured in step 1.

DETERMINATION OF MOISTURE CONTENT

Basic Protocol 7 was developed to estimate the mass balance of d-limonene within a citrus processing plant. However, the calculations must be normalized to provide a basis of comparison. This is achieved by determining the moisture content of those samples. Samples are dried at low temperatures under a vacuum. This removes the moisture in the sample and prevents any decomposition of the remaining solid matter.

Materials

Sample
Metal weighing dishes
58°C vacuum oven

1. Record weight of metal dish (w_{dish}).

2. Place 10 g sample into metal dish. Record wet total weight (w_{total}).

3. Place overnight into a 58°C vacuum oven.

4. Weigh sample for dry total weight (w'_{total}).

5. Calculate the percent solids and percent moisture using the following equations:

   $$w_{total} - w_{dish} = w_{wet}$$

   $$w'_{total} - w_{dish} = w_{dry}$$

   $$(w_{wet} - w_{dry})/w_{wet} \times 100 = \% \text{ moisture}$$

   $$100 - \% \text{ moisture} = \% \text{ solids}$$

Table G1.5.1 Protocol Modifications Based on Different
Types of Citrus Oils[a]

Sample	Amount (g)	Time (min)	CF (g/ml KOH)
Lemon	10	15	0.0761
Lime	10	15	0.0781
Tangerine	20	30	0.0781
Orange	20	30	0.0781
Grapefruit	20	30	0.0781

[a]Data from Redd et al. (1986). CF, correction factor.

QUANTIFICATION OF TOTAL ALDEHYDES IN CITRUS OILS BY HYDROXYLAMINE TITRATION

BASIC PROTOCOL 8

Aldehyde content is considered an important flavor note included in the standard of identities for citrus oils. The flavor strength of an oil is based on the aldehyde content, where higher is better. The two major aldehydes are acetaldehyde and octanal. The quantification of aldehydes is based on the reaction of citral in the sample with a hydroxylamine solution, followed by titration with KOH in the presence of ethyl orange indicator. It is modified from AOAC Method 955.32 (AOAC, 1990c; Redd et al., 1986). This method was originally developed for lemon oils; however, it is applicable to other citrus oils.

Materials

Citrus oil
0.5 N hydroxylamine solution (see recipe)
0.5 N KOH in 60% (v/v) ethanol

1. Weigh 10 to 20 g sample into a 150-ml beaker. Record exact weight.

   For exact amounts of different citrus oils see Table G1.5.1.

2. Add 35 ml of 0.5 N hydroxylamine solution; swirl occassionally for 15 to 30 min (Table G1.5.1).

3. Titrate with 0.5 N KOH to pH 3.5.

4. Calculate the total amount of aldehydes in lemon oil using the correction factor 0.0761 g citral/1 ml 0.5 N KOH; for other oils, use the correction factor 0.0781 g decanal/1 ml 0.5N KOH.

   % aldelhyes = (x ml titrated) × CF × 100 % × (1/oil weight)

QUANTIFICATION OF TOTAL ALDEHYDES IN LEMON OIL BY ACID/BASE TITRATION

ALTERNATE PROTOCOL 4

This procedure measures citral content in lemon oil using acid/base titration. It has been modified from the hydroxylamine method, and is from AOAC Method 955.38 (AOAC, 1990d).

Materials

Lemon oil
10% (w/v) phenylhydrazine in absolute ethanol
Benzene
Methyl yellow indicator: 0.1% (w/v) *p*-dimethylaminoazobenzene in absolute ethanol
0.5 N *p*-toluenesulfonic acid in absolute ethanol

Smell Chemicals

G1.5.11

1. Weigh 15 g lemon oil sample into 125-ml flask.

2. Pipet 10 ml of 10% phenylhydrazine solution. Let stand 30 min at room temperature.

3. Add 25 ml benzene.

4. Add 0.2 ml methyl yellow indicator. Titrate with 0.5 N p-toluenesulfonic acid to an endpoint of pH 3.5.

5. Titrate a separate 10-ml sample of phenylhydrazine solution with 0.5 N p-toluenesulfonic acid to an endpoint of pH 3.5.

6. Calculate the amount of total citral using the equation:

$$\text{citral (g)} = (x \text{ ml titrant for sample} - y \text{ ml titrant for phenylhydrazine}) \times 0.076$$

and convert to % citral in the 15-g sample.

ALTERNATE PROTOCOL 5

QUANTIFICATION OF TOTAL ALDEHYDES IN ORANGE AND GRAPEFRUIT OIL USING *N*-HYDROXYBENZENESULFONAMIDE

The *N*-hydroxybenzenesulfonamide (HBS) test is a spectrophotometeric assay for determining total aldehydes in orange and grapefruit oil. The method is based on the reaction between HBS and the aldehydes in the oil. HBS contains a secondary amine that reacts with the carbonyl carbon from the aldehyde. This reaction creates an enamine that is highly conjugate and resonance stable. The addition of KOH and $FeCl_3$ prevents the removal or addition of functional groups from destabilizing the structure. The compound has a maximum absorptivity at 525 nm that is quantifiable according to the Beer-Lambert law ($A = \varepsilon cl$). A standard curve using octanal and 2-hexenal is generated. From the standard curve, the aldehyde concentration is extrapolated. This reaction is specific for aldehydes and is more accurate than the hydroxylamine protocol.

Materials

    Orange or grapefruit oil
    Isopropanol
    1.5% (w/v) hydroxybenzenesulfonamide (HBS) in 99% (v/v) isopropanol
    1.0 N KOH
    1.0% (w/v) ferric chloride ($FeCl_3$) in 2 N HCl
    Octanal
    2-Hexenal
    Spectrophotometer with 1-cm-pathlength cuvettes

Measure total aldehyde content in sample

1. Dilute 0.2 ml of orange or grapefruit oil in 100 ml isopropanol.

2. In triplicate tubes, add 1.0 ml HBS solution to 10 ml diluted sample.

3. Add 1.0 ml of 1 N KOH. Mix solution and let stand for 10 min at room temperature.

4. Add 1.0 ml of 1.0% $FeCl_3$ solution. Mix and let stand for 10 min.

5. Place sample in 1-cm-pathlength cuvette and measure absorbance at 525 nm in a spectrophotometer.

Measure standards

6. Dilute 0.25 g octanal to 250 ml with isopropanol in a volumetric flask.

7. Dilute 0.25 g of 2-hexenal to 250 ml with isopropanol in a volumetric flask.

8. Mix the two solutions in a 500-ml flask to produce a 1000 ppm aldehyde stock solution.

9. Prepare standards using the following dilution scheme:

> 1.0 ml aldehyde stock + 99 ml isopropanol = 10 ppm
> 2.0 ml aldehyde stock + 98 ml isopropanol = 20 ppm
> 3.0 ml aldehyde stock + 97 ml isopropanol = 30 ppm
> 5.0 ml aldehyde stock + 95 ml isopropanol = 50 ppm.

10. Repeat steps 1 to 5 using standards.

Calculate aldehyde content of sample

11. Average triplicate values of the standards and generate a standard curve by plotting A_{525} versus concentration.

12. Average triplicate values of the sample and calculate aldehyde concentration (in ppm) by linear regression from the standard curve:

$$[\text{aldehyde}] = (A_{525} - \text{intercept})/\text{slope}$$

13. Calculate percent aldehyde using the following equations:

$$\% \text{ aldehyde in orange oil} = \text{reading from curve} \times 529 \times 100$$

$$\% \text{ aldehyde in grapefruit oil} = \text{reading from curve} \times 522 \times 100$$

DETERMINATION OF VOLATILE ESTERS

BASIC PROTOCOL 9

Esters contribute to the "fruity" aroma of citrus oils. Esters can be converted to hydroxamic acid with alkaline hydroxyamine. This reaction complexes with ferric chloride, which can be measured colorimetrically.

Materials

> 2 M hydroxylamine hydrochloride
> Citrus oil sample
> 3 N NaOH
> 4 N HCl
> 10% (w/v) ferric chloride ($FeCl_3$) in 0.1 N HCl
> Ethyl acetate
> Spectrophotometer with 1-cm-pathlength cuvettes

Measure ester content in sample

1. In triplicate tubes, add 2.0 ml of 2 M hydroxylamine hydrochloride to 5 ml citrus oil sample.

2. Add 2.0 ml of 3 N NaOH. Mix solution and let stand for 5 min at room temperature.

3. Add 2.0 ml of 4.0 N HCl.

4. Add 2.0 ml of 10% $FeCl_3$ solution.

5. Place sample in a 1-cm-pathlength cuvette and measure absorbance at 525 nm in a spectrophotometer.

Measure standards

6. Dilute 1.000 g ethyl acetate to 1000 ml with water in a volumetric flask to make a 1000 ppm standard stock solution.

   Flasks must be sealed due to the volatility of ethyl acetate.

7. Prepare a dilution series using the following scheme:

   > 0.5 ml stock + 99.5 water = 5 ppm
   > 1.0 ml stock + 99.0 water = 10 ppm
   > 2.5 ml stock + 97.5 water = 25 ppm
   > 5.0 ml stock + 95.0 water = 50 ppm
   > 10.0 ml stock + 90.0 water = 100 ppm
   > 12.5 ml stock + 87.5 water = 125 ppm
   > 15.0 ml stock + 85.0 water = 150 ppm
   > 20.0 ml stock + 80.0 water = 200 ppm.

8. Repeat steps 1 to 5 using standards.

Calculate ester content of sample

9. Average triplicate values of the standards and generate a standard curve by plotting A_{525} versus concentration.

10. Average triplicate values of the sample and calculate the ester concentration (in ppm) by linear regression from the standard curve:

$$[ester] = (A_{525} - intercept)/slope$$

REAGENTS AND SOLUTIONS

Use deionized, distilled water in all recipes and protocol steps. For common stock solutions, see APPENDIX 2A; for suppliers, see SUPPLIERS APPENDIX.

Arsenious oxide (As_2O_3) solution

Dissolve 5 g As_2O_3 in 50 ml of 1 N NaOH. Add 50 ml of 1 N H_2SO_4 to adjust pH to ~7.0. Dilute to 1 liter with distilled water. Store up to 7 days at 8°C.

This recipe is from AOAC Method 939.12 (AOAC, 1990a).

Bromphenol blue indicator

Dissolve 0.2 g bromphenol blue in 5 ml of 0.05 N NaOH. Dilute to 100 ml with 60% (v/v) ethanol. Store up to 7 days at 8°C.

Hydroxylamine solution, 0.5 N

Dissolve 34.75 g $H_2NOH \cdot HCl$ in 40 ml of hot water and make up 1 liter with 95% ethanol. Determine pH of solution and adjust to pH 3.5. Store up to 7 days at 8°C.

Potassium bromide/bromate ($KBr/KBrO_3$) solution, 0.1 N

Dissolve 2.8 g $KBrO_3$ and 12 g KBr in boiling water and dilute to 1 liter (0.1 N solution). Dilute 125 ml 0.1 N solution with 500 ml water (final 0.0247 N).

Solutions are shelf stable at room temperature. If solutions have not been used for an extended period (6 to 8 weeks), standardize solutions prior to usage.

To standardize the solution, add 40 ml As_2O_3 solution (see recipe) to a 300-ml Erlenmeyer flask. Add 10 ml HCl and 3 drops methyl orange. Titrate with $KBr/KBrO_3$ solution.

Normality $KBr/KBrO_3$ = (ml As_2O_3 × normality As_2O_3)/(ml $KBr/KBrO_3$)

This recipe is from AOAC Method 947.13 (AOAC, 1990b).

Table G1.5.2 Market Value and Limonene Content of Various Citrus Oils[a]

Citrus oil	d-Limonene (%)	Market value (U.S. $/gallon)
d-Limonene	100	0.45
Cold-pressed Valencia	>95	0.77
Cold-pressed early-mids	>95	0.65
5× orange oil (Valencia)	90	16.00
10× orange oil	80-85	NA
25× orange oil	60-65	NA
36× orange oil	1-2	NA
Tangerine	95	14.00
Mandarin (Italian)	95	35.00
Grapefruit	93-95	14.00
Lemon (Californian)	75-80	9.50
Lemon (Italian)	75-80	22.75
Lime (Mexican)	50-55	25.50

[a]Adapted from Braddock (1999) with permission from John Wiley & Sons, Inc.

COMMENTARY

Background Information

The diversification and reactivity of oil components can decrease functionality and value. Table G1.5.2 lists the market value for most citrus oils. Prices range from $0.45 per gallon for d-limonene to $35.00 per gallon for Sicilian Mandarin. Figure G1.5.6 presents chemical structures of compounds that are important to the flavor of citrus oils.

Orange oils (early-mid/Valencia)

The desired orange flavor is the result of volatile compounds in specific proportions (Shaw, 1991). There are six major contributors to orange flavor: acetaldehyde, citral, ethyl butanoate, d-limonene, nonanal, octanal, and α-pinene with two major types of essence oils, early-mid and Valencia (Shaw, 1991). Early-mid oranges include Hamlin and Pineapple.

More than 200 different compounds have been identified in cold-pressed Valencia oil (Maarse and Visscher, 1989). Table G1.5.3 is a list of over 100 compounds identified in cold-pressed Valencia oil. d-Limonene is the predominant terpene present in citrus oil. It has a weak "citrus-like" aroma; however, at high levels (190 ppm) it is detrimental to orange flavor. It is believed to be a carrier of minor oil-soluble flavor compounds (Shaw, 1991). The second most abundant terpene is myrcene. Myrcene has long been known to impart a negative characteristic. It has been shown to give a pungency or bitterness at high levels. α-Pinene is another monoterpene that contributes to orange flavor.

However, the most notable difference between orange oils is the percentage of valencene. Valencene is a sesquiterpene hydrocarbon. In cold-pressed Valencia oil, it is 10 to 20 times higher than other orange oils (Coleman et al., 1969).

Aldehyde content in orange oil is vital. Acetaldehyde is considered the major "freshness" compound. It has a low threshold concentration (3 ppm) and is often added back to reconstituted frozen concentrate. Citral is another important aldehyde. Citral is an isomeric mixture of neral and geranial. It is typically used at 40 ppm in synthetic orange flavorings. Other aldehydes along with esters and alcohols play a significant role in orange flavor. Ethyl butanoate has a strong, pleasant, fruity aroma with a detection level of 0.4 ppm. Linalool and α-terpineol also contribute to orange flavor, but like other compounds become objectionable at high levels.

The term folded oils refers to concentrated oils. This typically involves a distillation process; however, alcohol washing can also be used. Alcohol washing is based on the insolubility of d-limonene in 60% to 70% ethanol. These processes predominately remove terpene compounds, although aldehydes (octanal) are also reduced. Oils that are more than 20-fold concentrated are called "terpeneless oils" and are more stable. Distillation is predominately used by flavor houses. Flavor houses purchase cold-pressed oil, which is concentrated and fractionated. These fractionated portions are sold for flavorings or flavor precursors.

Smell Chemicals

G1.5.15

Table G1.5.3 Volatile Compounds Associated with Cold-Pressed Valencia Oil[a]

Acids	Alcohols	Aldehydes	Esters	Ketones	Oxides	Terpenes
Acetic	Amyl alcohol	Acetaldehyde	Bornyl acetate	Acetone	*cis*-Limonene oxide	δ-Cadinene Camphene
Capric	Borneol	Citral geranial	Citronellyl acetate	Carvone	*trans*-Limonene oxide	Δ-3-Carene
Caprylic	*cis*-Carveol	Citral neral	Decyl acetate	α-Ionone		Caryophyllene
Formic	*trans*-Carveol	Citronellal	Ethyl acetate	Methyl heptenone		α-β-Copaene
	Citronellol	*n*-Decanal *n*-Dodecanal	Ethyl butanoate	6-Methyl-5-hepten-2-one		α-β-Cubebene
	n-Decanol	Dodecene-2-al-1	Ethyl 3-hydroxy hexanoate	Nootkatone		*p*-Cymene
	Dodecanol	Formaldehyde	Ethyl isovalerate	Piperitenone		β-Elemene
	Elemol	Furfural	Ethyl 2-methyl butanoate			Farnesene
	Ethyl alcohol	*n*-Heptanal	Ethyl proprionate			α-β-Humulene
	Geraniol	*n*-Hexenal	Geranyl acetate			*d*-Limonene
	Heptanol	trans-Hexen-2-al-1	Geranyl butyrate Geranyl formate			2,4-*p*-Methediene
	Hexanol-1	*n*-Nonanal	Linalyl acetate			Myrcene
	Isopulegol	*n*-Octanal (*E*)-2-Pentenal	Linalyl propionate			α-Phellandrene
	Linalool	*Perillydehyde*	1,8-*p*-Methadiene-9-yl acetate			α-β-Pinene
	cis-trans-2,8-*p*-Menthadiene-1-ol	α-Sinsensal	Methyl butanoate			Sabinene
	1,8-*p*-Methadiene-1,2-diol	β-Sinsensal	Neryl acetate			α-Terpinene
	8-*p*-Methene-1,2-diol	*n*-Undecanal	Nonyl acetate			β-Terpinene
	1-*p*-Methene-9-ol		n-Octyl acetate			α-Terpinolene
	Methyl alcohol		Perillyl acetate			α-Thujene
	Methyl heptanol		Terpinyl acetate			Valencene
	Nerol					
	n-Nonanol					
	n-Octanol					
	α-Terpineol					
	Terpinen-4-ol					
	Undecanol					

[a]Data from Kesterson et al. (1971) and Shaw (1991).

Mandarin/tangerine oil

Mandarin oranges encompass a wide variety of loose-skin citrus cultivars. In the U.S., the term tangerine is interchanged with mandarin. Tangerines are a deeper colored orange. There is a distinct aroma difference between tangerine and Sicilian mandarin oil. However, the value of mandarin oil versus tangerine oil is significant. Mandarin oil (Sicilian) is the most expensive. Similar to other citrus oils, large levels of terpenes are present. *d*-Limonene values range from 65% to 94% in mandarin oil. However, at lower levels of limonene, there is an increase in γ-terpinene.

The two major components responsible for mandarin flavor are methyl-*N*-methylanthranilate and thymol. These have been based on threshold levels for methyl-*N*-methylanthra-

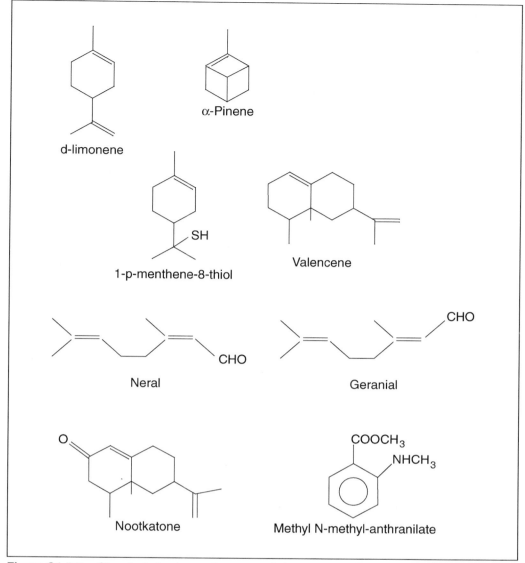

Figure G1.5.6 Chemical structures of compounds important to citrus flavor.

nilate. Methyl-*N*-methylanthranilate is five times higher in Sicilian mandarin oil than Dancy tangerine. In addition, β-pinene and γ-terpinene must be increased in tangerine oil to produce a "mandarin" flavor. Table G1.5.4 compares the major flavor components for mandarin and tangerine oil. Decanal, octanal, α-sinensal, and acetaldehyde are also important to mandarin flavor.

Grapefruit oil

Grapefruit, typically, has harsher flavor notes. Similar to other citrus oils, grapefruit oil is predominately composed of terpenes. There have been 206 volatile compounds identified in grapefruit juice. However, the flavor impact compounds are more dominant. Nootkatone and 1-*p*-methene-8-thiol are the major compo-

nents for grapefruit oil. Nootkatone is a sesquiterpene ketone that is pungent and aromatic. Sensory panels described samples with high levels of nootkatone as having a lingering bitterness with a metallic taste (Berry et al., 1967). This occurs at 9 ppm. Nootkatone has a flavor threshold in water of 1 ppm and 5 to 6 ppm in grapefruit juice. It has been suggested nootkatone be used as a quality index for grapefruit oil. However, 1-*p*-methene-8-thiol has been described as the "aroma of fresh grapefruit juice" (Demole et al., 1982). There are also significant aldehyde levels, 1.8% (Shaw, 1991).

Lemon oil

The aroma of lemon oil, like other citrus oils, is dependent upon the proportions of specific compounds. Lemon oil has a higher aldehyde

Table G1.5.4 Comparison of the Major Flavor Component for Mandarin and Tangerine Oranges[a]

Component	Tangerine oil	Mandarin oil
Methyl-*N*-methylanthranilate	0.07%	0.65%
β-Pinene	0.17%	1.80%
γ-Terpinene	1.74%	14.00%
Thymol	0.02%	0.18%

[a]Reprinted from Wilson and Shaw (1981) with permission from the American Chemical Society.

Table G1.5.5 Volatile Compounds Important to Lemon Flavor[a]

Alcohols	Aldehydes	Esters	Ethyl ethers	Terpenes
α-Bisabolol	Geranial	Geranyl acetate	Carvyl	Bergamotene
Geraniol	Neral	Methyl epijasmonate	8-*p*-Cymenyl	Caryophyllene
		Neryl acetate	Fenchyl	β-Pinene
			Myrcenyl	γ-Terpinene
			α-Terpinyl	

[a]From Shaw (1991) by courtesy of Marcel Dekker, Inc.

Table G1.5.6 Chemical Compounds Identified in Lime Oils[a]

Alcohols	Aldehydes	Esters	Ketones	Oxides	Terpenes
Borneol	Acetaldehyde	Ethyl acetate	Acetone	1,4-Cineole	α-Bergamotene
p-Cymene-8-ol	*n*-Decanal	Geranyl acetate	Piperitenone	1,8-Cineole	Camphene
Decanol	Geranial	Neryl acetate		*cis*-Linalool oxide	β-Caryophellene
Ethyl alcohol	Hexenal			*trans*-Linalool oxide	*p*-Cymene
Geraniol	2-Hexanal				Dipentene
cis-3-Hexen-1-ol	Neral				α-Elemene
Isoamyl alcohol	*n*-Nonanal				α-Humulene
Isopropyl alcohol	*n*-Octanal				β-Humulene
Lauryl alcohol	Perillaldehyde				*d*-Limonene
Linalool					Myrcene
cis-*p*-2-Menthen-1-ol					α-Pinene
Methyl alcohol					β-Pinene
2-Methyl-2-butanol					γ-Terpinene
2-Methyl-3-buten-2-ol					Terpinolene
3-Methyl-2-buten-1-ol					
Nerol					
Nonanol					
Octanol					
Terpinene-4-ol					
α-Terpineol					
β-Terpineol					

[a]From Azzouz and Reineccius (1976) with permission from the Institute of Food Technologists.

(2.0% to 13.2.%) content than orange and grapefruit oil (Shaw, 1979). High quality oils will have an aldehyde content between 4% and 5%. Lemon oil also has a lower *d*-limonene level. However, the terpene content is 1% higher than other oils with a wider variety of terpene compounds. In particular, β-pinene and γ-terpinene are generally higher in lemon oil. The predominant esters present are neryl and geranyl. In addition, there is a 5% thymol concentration (Moshonas et al., 1972). Table G1.5.5 lists volatile compounds important to lemon flavor.

Lime oil

Lime oil is similar to lemon oil with a few exceptions. Citral content is higher and octanal is the main straight-chained aldehyde. Neryl and geranyl acetate is also higher. The processing of lime oil, cold-pressed versus distilled, does contribute to noteable differences. During distillation, the oil is in contact with acidic juice for a longer time. Distilled lime has reduced levels of citral, β-pinene, and γ-terpinene and increased amounts of *p*-cymene, terpinen-4-ol, and α-terpineol (Slater and Watkins, 1964). In addition, α-thujene, neral, geranial, decanal, geranyl acetate, neryl acetate, and α and β-elemene are absent in distilled lime oil. However, there is very little varietal difference in cold-pressed lime oils, West-Indian and Mexican (Azzouz and Reineccius, 1976). Table G1.5.6 lists the 54 compounds identified in lime oils.

Total available oil

The reactivity of hydrocarbons in citrus oils has brought considerable attention to citrus processing. Processors are only allowed to emit certain levels of volatile organic compounds (VOC). For this reason, a processor must account for the total available oil. With these procedures, 90% of VOC levels can be accounted for (Gentry, 1999).

Critical Parameters

Qualitative analysis

Depending on the choice of column, flow rates and temperature program are important parameters for the qualitative analysis of citrus oils. It is also important to have the same temperature program for quantifying compounds. Replication of injections for generating a standard curve is vital. Injections should be done on the same day by the same technician. A standard curve with an R^2 value >0.9 is sufficient.

The temperature for measuring physical parameters, specific gravity, refractive index, and optical rotation is critical for obtaining accurate results.

Quantitative analysis

The solutions for measuring total available oil are stable and can be prepared in advance. The weight of the sample is important, especially for the dryer samples, press cake, and pellets. These samples absorb a lot of water. Furthermore, the concentration of the potassium bromide solution may need to be decreased if a sample has an extremely low level of limonene (i.e., press liquor or pellets).

For the colorimeteric assays, stock solutions must be fresh (1 to 3 days) and should be refrigerated. HBS solution is only good for 5 days. The standards should be run each time a series of samples is run.

Table G1.5.7 Physical and Chemical Properties of Citrus Oils[a,b]

Cultivar	SG	η	α	Aldehyde (%)	Ester (%)
Cold-pressed orange oil	0.842-0.846	1.472-1.474	+94-99	1.2-2.5	0.48
Orange essence oil	0.840-0.844	1.471-1.474	+94-99	1.0-2.5	2.12
Mandarin	0.847-0.853	1.473-1.477	+68-78	0.4-1.8	NA
Tangerine	0.844-0.854	1.473-1.476	+88-96	0.8-1.9	1.44
Grapefruit	0.848-0.856	1.475-1.478	+91-96	1.3	3.59
Cold-pressed lime (Mexican)	0.872-0.881	1.482-1.486	+35-41	4.5-8.5	6.98
Lime essence oil	0.855-0.863	1.474-1.477	+34-47	0.5-2.5	NA
Lemon (Italian)	0.849-0.855	1.473-1.476	+57-65.6	3.0-5.5	NA
Lemon (Californian)	0.849-0.855	1.473-1.476	+57-65.6	2.2-3.8	NA
Lemon essence oil	0.842-0.856	1.470-1.475	+55-75	1.0-3.5	NA

[a]Defined by the Food Chemicals Codex (NRC,1981).
[b]Optical activity (α); refractive index (η); specific gravity (SG), and aldehyde content from NRC (1981). Ester content from Kimball (1999).

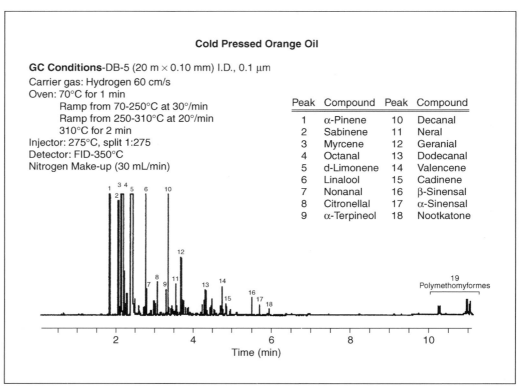

Figure G1.5.7 Typical gas chromatogram for cold-pressed orange oil. Reproduced with permission from Agilent Technologies, Inc. (see Internet Resources).

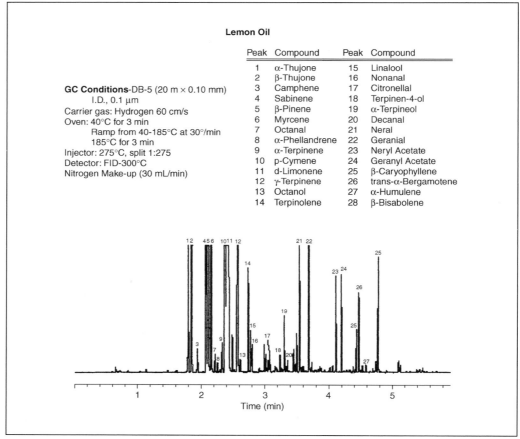

Figure G1.5.8 Chromatogram for lemon oil. Reproduced with permission from Agilent Technologies, Inc. (see Internet Resources).

The volatility of ethyl acetate requires minimal exposure when preparing standards. The same is true for limonene. Limonene should be stored in a metal canister, and the headspace should be flushed with nitrogen after usage.

Troubleshooting

Gas chromatography can be extremely machine-dependent when determining optimal settings. The key is to have reproducibility. Depending on the column, temperature programs may need to be adjusted for better peak resolution. The septa can be a problem. It should be changed regularly if used a lot (~30 injections), otherwise peaks shift retention times. In addition, artifacts can be a problem, especially if the oil has high boiling compounds. Holding the column at a higher temperature for a longer period can eliminate this. If the column continues to show artifacts, setting the column to a high temperature overnight should eliminate the problem. However, this will decrease the life-span of the column. Be sure not to exceed the manufacturer's maximum temperature.

The quantification protocols must be conducted carefully. The concentrations are extremely low. Volumetric flasks and pipets (class A) will provide the most accurate results for preparing standards. Light and oxygen exposure should also be minimized. Oils should be refrigerated in either amber or metal bottles. Plastic should never be used.

Anticipated Results

Table G1.5.7 lists the physical and chemical properties (specific gravity, SG; refractive index, η; optical activity, α) of citrus oils defined by the Food Chemicals Codex (NRC, 1981).

Figure G1.5.7 is a typical chromatogram for cold-pressed orange oil. The peaks are well

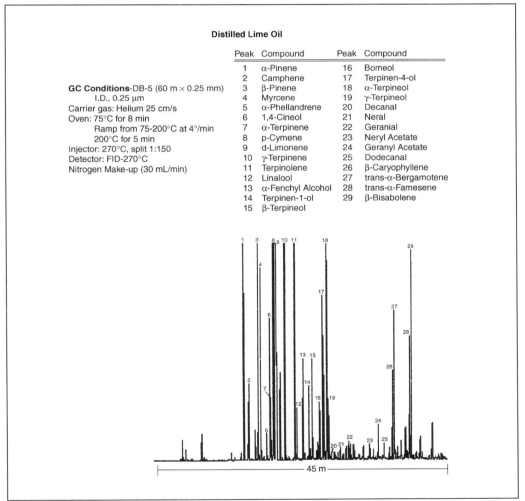

Distilled Lime Oil

GC Conditions-DB-5 (60 m × 0.25 mm)
I.D., 0.25 µm
Carrier gas: Helium 25 cm/s
Oven: 75°C for 8 min
Ramp from 75-200°C at 4°/min
200°C for 5 min
Injector: 270°C, split 1:150
Detector: FID-270°C
Nitrogen Make-up (30 mL/min)

Peak	Compound	Peak	Compound
1	α-Pinene	16	Borneol
2	Camphene	17	Terpinen-4-ol
3	β-Pinene	18	α-Terpineol
4	Myrcene	19	γ-Terpineol
5	α-Phellandrene	20	Decanal
6	1,4-Cineol	21	Neral
7	α-Terpinene	22	Geranial
8	p-Cymene	23	Neryl Acetate
9	d-Limonene	24	Geranyl Acetate
10	γ-Terpinene	25	Dodecanal
11	Terpinolene	26	β-Caryophyllene
12	Linalool	27	trans-α-Bergamotene
13	α-Fenchyl Alcohol	28	trans-α-Farnesene
14	Terpinen-1-ol	29	β-Bisabolene
15	β-Terpineol		

45 m

Figure G1.5.9 Gas chromatogram for distilled lime oil. Reproduced with permission from Agilent Technologies, Inc. (see Internet Resources).

Table G1.5.8 Identification of Peaks for Cold-Pressed Versus Distilled Lime Oil[a]

Distilled lime oil		Cold-pressed lime oil	
Peak	Compound	Peak	Compound
1	tert-Amyl alcohol	1	tert-Amyl alcohol
2	2-Methyl-3-buten-2-ol	2	2-Methyl-3-buten-2-ol
3	cis-3-Hexen-1-ol	3	n-Nonane
4	n-Nonane	4	α-Thujene
5	α-Pinene	5	α-Pinene
6	Camphene	6	Camphene
7	β-Pinene	7	β-Pinene
8	Myrcene	8	Myrcene
9	α-Phellandrene	9	D-Limonene
10	1,4-Cineole	10	1,8-Cineole
11	p-Cymene	11	γ-Terpinene
12	D-Limonene	12	Terpinolene
13	1,8-Cineole	13	Octanol
14	γ-Terpinene	14	Linaiool
15	Terpinolene	15	Terpinene-4-ol
16	Octanol	16	Decanal
17	Linalool	17	Neral
18	α-Fenchyl alcohol	18	Geranial
19	cis-β-Terpineol	19	Decanol
20	trans-β-Terpineol	20	Nerol
21	Terpinene-4-ol	21	Geraniol
22	α-Terpineol	22	α-Elemene
23	Neral	23	Thymol
24	Geranial	24	Neryl acetate
25	Decanol	25	Geranyl acetate
26	Nerol	26	β-Elemene
27	Geraniol	27	β-Caryophellene
28	α-Elemene	28	α-Bergamotene
29	Thymol	29	α,β-Humulene
30	Neryl acetate	30	Guaiene
31	Geranyl acetate	31	β-Bisabolene
32	β-Elemene		
33	β-Caryophellene		
34	α-Bergamotene		
35	α,β-Humulene		
36	Guaiene		
37	β-Bisabolene		

[a]From Azzouz and Reineccius (1976) with permission from the Institute of Food Technologists.

defined. The GC conditions are modified for this particular sample.

Hunter and Brogden (1965) generated chromatograms that provide a visual representation of the various citrus oils. The GC conditions for the orange oils are similar, giving the same retention times but different peak ratios for the various hydrocarbons. Comparisons of cold-press Valencia oil from Florida versus California shows a similar qualitative analysis; however, the valencene peak is larger in the California oil. This is not uncommon. Oil samples can show great diversity depending on the degree of oxidation, processing, and storage conditions. Fruit maturity and cultivar will also effect results. Chromatograms for early-mid orange oil also shows different response values compared to Valencia oil. Early-mid oils have

Table G1.5.9 Peel-Oil Content of Various Florida Citrus Cultivars[a]

Cultivar	Average pounds oil/ton fruit	Standard deviation (σ)
Bears lemon	15.1	2.4
Grapefruit	6.2	0.5
Hamlin	7.8	1.5
Persian lime	8.1	1.2
Pineapple	9.7	1.0
Tangerine	15.5	1.8
Valencia	13.5	1.8

[a]From Kesterson and Braddock (1975) with permission from the Institute of Food Technologists.

Table G1.5.10 Aldehyde Composition for Various Citrus Oils[a]

Oil	C_6	C_8	C_{10}	C_{12}	C_{14}	Neral	Geranial
Essence	0.5	13.8	26.7	5.5	1	6	4.2
Grapefruit	Trace	29.9	17.8	12.7	6.9	2.1	7.1
Hamlin	1.2	29.2	22.7	15.1	6.1	1.1	9.3
Pineapple	0.5	28.2	18	9.9	6.5	6.9	10.1
Tangerine	Trace	23.1	24	13.5	4.8	5	5.7
Valencia	1	27.3	30.7	9.4	5.3	4.1	4.4

[a]From Braddock and Kesterson (1976) with permission from the Institute of Food Technologists.

higher amounts of less important hydrocarbons (myrcene, α and β-copene, α and β-ylangene, β-elemene, and caryophyllene). Early-mid oils also have lower valencene peaks. However, for grapefruit oil, the two most noticeable differences are the myrcene and caryophyllene peaks. In tangerine oil, γ-terpinene, Δ, α, and β-elemene are significantly different.

Figure G1.5.8 is a chromatogram for lemon oil. The temperature program is modified compared to that of orange oil.

Figure G1.5.9 is a chromatogram for lime oil. This procedure used a longer column with lower temperature settings. The temperature also increases at a much slower rate. There were a larger number of compounds present compared to other oils. Table G1.5.8 shows peaks for lime oil samples processed under different conditions. The distilled oil has higher levels of alcohol, whereas, the cold-pressed oil has higher amounts of esters and aldehydes.

Table G1.5.9 is a summary of the total available oil for various citrus cultivars. Table G1.5.10 is the aldehyde composition for orange and grapefruit oils.

Time Considerations

Most GC temperature programs will take between 30 and 45 min. The qualitative data is simple once a library of high quality oils is collected. The quantitative data will depend on how many compounds are measured. Each standard curve will take a day to collect reliable data (minimum 12 runs).

The total available oil procedures take ~40 min. It's best to prepare samples before distillation. Samples can be titrated while the next sample is distilled. The other titration procedures should not take >20 min, provided reagents are prepared. The colorimetric assays will take ~3 hr given preparation of reagents and generation of standard curve.

Literature Cited

Association of Official Analytical Chemists (AOAC). 1990a. Standard solution of arsenious oxide. Method 939.12. *In* AOAC Official Methods of Analysis, 15th ed. AOAC, Arlington, Va.

AOAC. 1990b. Standard solution of potassium bromide-bromate. Method 947.13. *In* AOAC Official Methods of Analysis, 15th ed. AOAC, Arlington, Va.

AOAC. 1990c. Total aldehydes in lemon oil—hydroxylamine method. Method 955.32. *In* AOAC Official Methods of Analysis, 15th ed. AOAC, Arlington, Va.

AOAC. 1990d. Total aldehydes in lemon oil—Kirsten modification of the Kleber method. Method 955.38. *In* AOAC Official Methods of Analysis, 15th ed. AOAC, Arlington, Va.

Smell Chemicals

G1.5.23

AOAC. 1990e. Oil (recoverable) in fruits and fruit products. Method 986.20. *In* AOAC Official Methods of Analysis, 15th ed. AOAC, Arlington, Va.

Azzouz, M.A. and Reineccius, G.A. 1976. Comparison between cold-pressed and distilled lime oils through the application of gas chromatography and mass spectrometry. *J. Food Sci.* 41:324-328.

Berry, R.E., Wagner, C.J., and Moshonas, M.G. 1967. Flavor studies of nootkatone in grapefruit juice. *J. Food Sci.* 32:75-78.

Braddock, R.J. 1999. Handbook of Citrus By-Products and Processing Technology, pp. 149-190. John Wiley & Sons, New York.

Braddock, R.J. and Kesterson, J.W. 1976. Quantitative analysis of aldehydes, esters, alcohols, and acids from citrus oils. *J. Food Sci.* 41:1007-1010.

Coleman, R.L., Lund, E.D., and Moshonas, M.G. 1969. A research note on composition of orange essence oil. *J. Food Sci.* 34:610-611.

Demole, E., Enggist, P., and Ohloff, G. 1982. 1-*p*-Menthene-8-thiol: A powerful flavor impact constituent of grapefruit juice (*Citrus paraadisi* Macfayden). *Helv. Chim. Acta* 65:1785-1794.

Gentry, T.S. 1999. Volatile organic compounds from citrus feed mill emissions. M.S. Thesis. University of Florida, Gainesville.

Gentry, T.S., Braddock, R.J., Miller, W.M., Sims, C.A., and Gregory, J.F. 2001. Volatile organic compounds from citrus feed mills. *J. Food Proc. Engineering* 24:1-15.

Hunter, G.L.K. and Brogden, W.B. Jr. 1965. Analysis of the terpene and sesquiterpene hydrocarbons in some citrus oils. *J. Food Sci.* 30:383-387.

Kesterson, J.W. and Braddock, R.J. 1975. Total peel oil content of the major Florida citrus cultivars. *J. Food Sci.* 40:931-933.

Kesterson, J.W., Hendrickson, R., and Braddock, R.J. 1971. Florida Citrus Oils. Agricultural Experiment Stations, Institute of Food and Agricultural Sciences. University of Florida, Gainesville.

Kimball, D.A. 1999. Citrus Processing: A Complete Guide, 2nd ed., pp. 191-246. Aspen Publishers, Gaithersburg, Md.

Maarse, H. and Visscher, C.A. 1989. Citrus fruits. Products 5. *In* Volatile Compounds in Food. Quantitative and Qualitative Data, TNO-CIVO, pp. 35-99. Food Analysis Institute, Ziest, The Netherlands.

Moshonas, M.G., Shaw, P.E., and Veldhuis, M.K. 1972. Analysis of volatile constituents from meyer lemon oil. *J. Agric. Food Chem.* 20:751-752.

National Research Council (NRC). 1981. Food Chemicals Codex, 3rd ed. National Academy Press, Washington, D.C.

Redd, J.B., Hendrix, C.M., and Hendrix, D.L. 1986. Quality Control Manual for Citrus Processing Plants, Volume 1: Regulation, Citrus Methodology, Microbiology, Conversion Charts, other, pp. 63-78. Intercit, Safety Harbor, Fla.

Scott, W.C. and Veldhuis, M.K. 1966. Rapid estimation of recoverable oil in citrus juices by bromate titration. *J.A.O.A.C.* 49:628-633.

Shaw, P.E. 1979. Review of quantitative analyses of citrus essential oils. *J. Agric. Food Chem.* 27:246-257.

Shaw, P.E. 1991. Fruits II. *In* Volatile Compounds in Foods and Beverages (H. Maarse, ed.) pp. 305-327. Marcel Dekker, New York.

Slater, C.A. and Watkins, W.T. 1964. Chemical transformation of lime oil. *J. Sci. Food Agric.* 15: 657-664.

Wilson, C.W. and Shaw, P.E. 1981. Importance of thymol, methyl *N*-methylanthranilate and monoterpene hydrocarbons to the aroma and flavor of mandarin cold-pressed oils. *J. Agric. Food Chem.* 29:494-496.

Internet Resources

http://www.chem.agilent.com/scripts/
chromatograms.asp

Agilent Technologies' on-line Chromatogram Library.

Contributed by Trevor Gentry
Cornell University
Ithaca, New York

Solid-Phase Microextraction for Flavor Analysis

A compound must volatilize from a food to reach receptors in the nasal epithelium and potentially produce an aroma perception. Methods that selectively extract volatile compounds from food are advantageous for flavor analysis. Furthermore, because aroma compounds are present at very low concentrations (often below parts-per-billion levels), methods that concentrate these volatiles are necessary. Solid-phase microextraction (SPME) is an extraction method based on absorption and adsorption of compounds to polymers coated on a silica fiber, effectively concentrating analytes found in a gas or liquid phase. The absorbed compounds can subsequently be separated using gas chromatography. SPME is a nondestructive, reproducible method. A variety of fiber-coating materials are used for selective extraction of analytes. For quantification, SPME can be standardized to account for fiber selectivity. SPME can be used in conjunction with mouth simulators (*UNIT G1.7*), gas chromatography/mass spectrometry (GC/MS), gas chromatography/olfactometry (GC/O), and dilution analysis. This unit presents methods for SPME of food headspace (see Basic Protocol) and submersion SPME for liquid samples (see Alternate Protocol). In addition, quantification of headspace extraction (see Support Protocol 1) and dilution analysis (see Support Protocol 2) are presented.

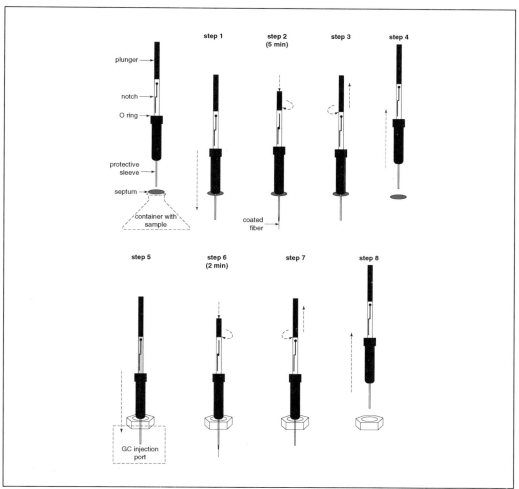

Figure G1.6.1 Illustration of procedures performed in the Basic Protocol for SPME headspace sampling.

Contributed by Kathryn D. Deibler

SPME OF FOOD HEADSPACE FOR GAS CHROMATOGRAPHY

Headspace SPME is a solventless extraction method where a silica fiber coated with adsorbant or absorbant polymer material is exposed to a gas phase to extract analytes. The food of interest is placed in a closed or open container (such as a mouth simulator). After extraction, the fiber is desorbed in a GC injection port for separation and detection of the extracted analytes.

NOTE: Figure G1.6.1 demonstrates the sequence of injection. It is important to follow this sequence carefully to avoid destruction of the fiber. Removing the fiber through a septum without retracting the fiber will strip the coating from the fiber and thus destroy it.

Materials

Food sample for analysis
Headspace vessel (any vessel that can physically support the food sample) with septum (anything that can be pierced by syringe, e.g., Parafilm, aluminum foil, cap septum)
SPME manual assembly (Supelco)
SPME fiber (Supelco), available with different coatings (Table G1.6.1)
GC instrument

1. Place food sample in vessel with septum. Insert SPME through septum.

    See Critical Parameters for discussion of sample size. Closed or open systems may be used; the container need not be sealed.

2. Expose coated SPME fiber to headspace for 5 min, monitoring precisely with a timer.

    Exposure time can be chosen arbitrarily or the time to equilibrium can be used. Time to equilibrium is determined by extracting for incremental times until a stable extracted concentration is achieved. It is imperative that the selected exposure time be used exactly and consistently for any measurements that are to be compared.

3. Retract SPME fiber.

4. Remove the SPME apparatus from the septum.

5. Inject SPME through the septum of a GC for 1.0 min with the purge off and the remaining GC parameters set to the desired analytical conditions.

    Purge flow should be set to ≥30 sec. Longer times are acceptable; 1.0 min is usually used.

6. Expose coated SPME fiber in the injection port. Leave for 2 min.

    Desorption of the analytes takes place in the first minute. The second minute is for cleaning or baking-off purposes. This time of exposure does not need to be carefully monitored and may vary widely, provided that the fiber is exposed in the injection port for at least the length of time that the purge is off.

Table G1.6.1 Selected SPME Polymer Coatings with Recommended Applications

Coating material	Recommended applications
Polydimethylsiloxane (PDMS)	Volatiles, nonpolar semivolatiles
PDMS/divinylbenzene (PDMS/DVB)	Polar volatiles
Carbowax/DVB (CW/DVB)	Polar analytes
PDMS/Carboxen (PDMS/CAR)	Trace level volatiles
Polyacrylate	Polar semivolatiles

7. Retract SPME fiber.

8. Remove SPME apparatus from the injection port.

> *Fiber degradation should be monitored after every ten sample runs (see Critical Parameters).*

SUBMERSION SPME IN LIQUID SAMPLES

As an alternative to headspace extraction, analytes can be extracted by submersion of an SPME fiber in a liquid sample such as a beverage. While the ratio of analytes in the liquid phase is different from that which would be observed in the corresponding headspace gases, the concentration of most analytes is much higher in the liquid phase. Submersion SPME is most applicable when the basic composition in the food is desired. It is often used as a replacement for solvent extraction.

Additional Materials *(also see Basic Protocol)*

   Liquid sample for analysis
   Magnetic stirrer and stir bar (or other stirring method)

1. Stir liquid sample at a high and constant rate using a magnetic stirrer and stir bar.

> *Viscous samples make mass transfer through the sample slow. Stirring reduces the time for mass transfer.*

2. Place an SPME fiber so that the tip is in the liquid.

> *For composition analysis, the depth of the fiber does not matter. For comparative analysis, it is important to use a consistent fiber depth.*

3. Expose coated SPME fiber to liquid for 30 min, monitoring precisely with a timer.

> *Exposure time can be arbitrarily chosen or the time to equilibrium can be used. Time to equilibrium can be determined by extracting for incremental time periods until a stable extracted concentration is achieved. Equilibration time is much greater for submersion extraction than for headspace extraction. It is imperative that the selected exposure time be used exactly and consistently for any measurements that are to be compared.*

4. Retract SPME fiber.

5. Remove the SPME apparatus from the liquid.

6. Inject into GC instrument for specific period of time (see Basic Protocol, steps 5 through 8).

QUANTIFICATION OF HEADSPACE EXTRACTION

Because SPME extracts compounds selectively, the response to each compound must be calibrated for quantification. A specific compound can be quantified by using three GC peak area values from solvent injection, static headspace (gas-tight syringe), and SPME. The solvent injection is used to quantify the GC peak area response of a compound. This is used to quantify the amount of the compound in the headspace. The SPME response is then compared to the quantified static headspace extraction. These three stages are necessary because a known gas-phase concentration of most aroma compounds at low levels is not readily produced. A headspace of unknown concentration is thus produced and quantified with the solvent injection. Calibration must be conducted independently for each fiber and must include each compound to be quantified.

Materials

Volatile compound to be quantified

GC-grade solvent, e.g., ethyl acetate (for solvent injection)

Low-volatizing, nonaqueous solvent: e.g., propylene glycol (for static extraction)

Liquid nitrogen or other cryogenic liquid

GC instrument

100-ml addition funnel without side arm, with standard taper ground-glass joint (24/40) at top and bottom

50-ml round-bottom flask

Exterior thread adapter (24/40) with septum and cap

5-ml gastight syringe (SGE)

500-ml Dewar flask (small enough to fit in GC oven)

Perform solvent injection

1. Prepare a 0.01% (w/v) solution of the volatile compound of interest in a GC-grade solvent.

   This solution can contain more than one compound to be tested. However, if a single compound is used, the purity of the chemical can be measured and taken into account in calculating the actual concentration. Purity of the standard chemicals is important.

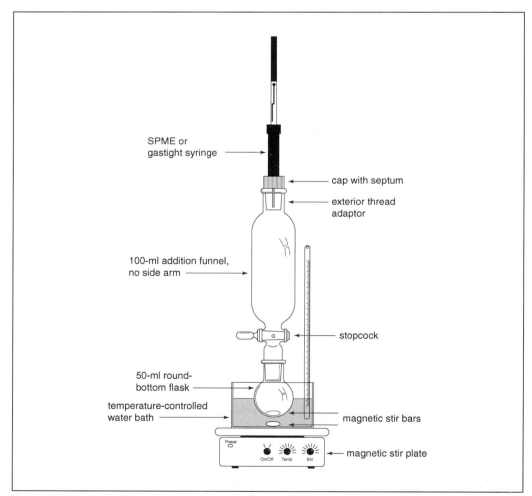

Figure G1.6.2 Apparatus for quantification of headspace extraction in Support Protocol 1.

2. Inject 1 µl of the solution into a GC and analyze according to manufacturer's instructions using the appropriate analytical parameters consistently throughout.

3. Repeat steps 1 and 2 for a total of three measurements and average the peak areas to give the solvent peak area (A_s).

   If the solution contained more than one standard compound, a separate value of A_s is obtained for each compound.

Perform static extraction

4. Set up a 50-ml round-bottom flask, 100-ml addition funnel (with no side arm), threaded adaptor, screw cap with septum, magnetic stir bar, magnetic stir plate, and temperature-controlled water bath as shown in Figure G1.6.2.

   The temperature of the liquid can be controlled with a hot water bath or any other heating or cooling method.

5. Prepare at least 300 ml of ~0.1% (w/v) solution of the volatile compound in a low-volatizing, nonaqueous solvent (e.g., propylene glycol).

   A low-volatility solvent is preferred to reduce solvent interaction in the headspace. More than one solution may be necessary if there is chromatographic overlapping of standard compounds or contaminants.

   The concentration of each volatile must be high enough for gastight syringe extraction, but low enough that the SPME fiber is not saturated and the GC detector limits are not exceeded. An acceptable concentration must be determined empirically by interative application of the two extractions.

6. Transfer 25 ml of solution to the round-bottom flask and stir with the stopcock open for at least 30 min.

7. Close the stopcock to separate the liquid phase in the flask from the gas phase in the addition funnel.

8. Extract 5 ml headspace through the septum with a 5-ml gastight syringe.

   If a cryofocusing unit is installed on the GC column, that should be used instead of the procedures in steps 9 and 11. Cryofocusing freezes the injected compounds at the immersed section, so more sample may be injected over time. Heating of that section results in the transfer or "injection" onto the column.

9. Place approximately the initial loop of the GC column in a Dewar flask containing liquid nitrogen.

10. With purge flow off, inject gas from the syringe into the column at a rate that is equal to or below the flow rate of the column.

    If the column flow rate is 2 ml/min, injection of 5 ml should take at least 2.5 min.

11. Start the GC run using the same parameters used for solvent injection (step 2) and immediately (but carefully) remove the column loop from the cryogenic liquid.

    If the peak area is too small for detection or for reasonable calculations (i.e., <10% of the peak area from sample analysis), the static extraction analysis should be repeated with a more concentrated solution.

12. Repeat steps 6 to 11 for a total of three measurements and average the peak areas to give the vapor peak area (A_v).

    If the solution contained more than one standard compound, a separate value of A_v is obtained for each compound.

Perform SPME

13. Repeat steps 6 and 7.

14. Inject the SPME fiber being calibrated through the septum and measure peak area (see Basic Protocol), but expose the fiber to the headspace for exactly 5 min.

   This is likely to be more than enough time to reach equilibrium, since the liquid phase is separated and the gas phase is relatively large. The time should be measured precisely, e.g., using a digital stopwatch that measures down to a tenth of a second.

   If another GC peak resulting from SPME of the sample or other extractions is larger than the peak observed, the fiber is not saturated. Saturation can be tested (see Critical Parameters). If the fiber is saturated, or if the GC peak is flat at the top or exceeds the detector limits, the static extraction and SPME analysis should be repeated with a less concentrated solution.

15. Repeat steps 13 and 14 for a total of three measurements and average the peak areas to give the SPME peak area (A_{SPME}).

   If the solution contained more than one standard compound, a separate value of A_{SPME} is obtained for each compound.

Analyze data

16. Calculate the calibration constant (k_{SPME}) for each compound calibrated.

   $$k_{SPME} = (C_s V_s / A_s) \times (A_v / V_v) \times (1/A_{SPME})$$

   where C_s is the concentration of compound in solvent injection (step 1), V_s is the volume of solvent injected (step 2), and V_v is the volume of gas in the gastight syringe (step 8). A_s, A_v, and A_{SPME} are defined in steps 3, 12, and 15, respectively.

   There is a unique k_{SPME} for each compound and fiber phase.

17. Calculate the concentration of standard compound extracted from a headspace by the calibrated SPME (see Basic Protocol) by multiplying the measured peak area by the calibration factor.

SUPPORT PROTOCOL 2

DILUTION ANALYSIS

Gas chromatography/olfactometry (GC/O) based on dilution analysis (e.g., CharmAnalysis or Aroma Extraction Dilution Analysis) gives an indication of what compounds are most potent in the aroma of foods. The application of SPME to GC/O dilution analysis can be achieved by varying the thickness of the fiber phase and the length of exposure, resulting in various absorbant volumes.

Additional Materials (also see Basic Protocol)

   Gas chromatography/olfactometer (GC/O; DATU, Inc.)
   SPME fibers with same coating material but varying thicknesses

1. Perform SPME (see Basic Protocol), but inject into a GC/O.

2. Repeat using a fiber with the same coating material but at approximately one-third the thickness or by exposing only one-third of the original fiber's length. For example, if SPME was originally performed by fully exposing a 1-cm long, 100-μm polydimethylsiloxane (PDMS) fiber, repeat with a 30-μm PDMS fiber or expose only 0.33 cm of fiber.

   The fiber is always fully exposed during desorption in the GC injection port. Fractional exposure can be consistently achieved by either drilling a notch in the injection holder or by restricting the plunger with an O ring. The amount exposed must be measured from the

Table G1.6.2 Comparison of Extraction Methods

Method	Equipment	Sample form	Quantitative	Concentration	Precision
SPME	Holder, fiber	Headspace or liquid	Indirectly	Yes	Very good
Solvent extraction	Solvent, glassware	Solid or liquid	Sample composition	Yes	Excellent
Porous polymer	Gas flow, trap, porous polymer, thermal desorption unit, and cryofocus or solvent	Headspace	Yes	Yes	Good
Purge and trap	Gas flow, trap, porous polymer, thermal desorption unit, and cryofocus or solvent	Liquid	No	Yes	Good
Static headspace	Gastight syringe, cryofocus	Headspace	Yes	No	Fair

fiber, not from the plunger, since the final ~0.5 cm of plunger length is not fiber. It has been observed that the holder must be tightly assembled to ensure consistent exposure amounts.

Commercial nominal thickness is often different from actual thickness. Actual thickness should be requested from the manufacturer or measured under a microscope.

3. Repeat step 1 using a fiber that is one-ninth the original thickness (e.g., 15 μm PDMS) or exposing one-third of a fiber that is one-third the original thickness (e.g., exposing 0.33 cm of 30 μm PDMS).

4. Repeat step 1 exposing one-third of a fiber that is one-ninth the original thickness (e.g., exposing 0.33 cm of 15 μm PDMS).

5. Make potency dilution analysis calculations for GC/O.

See Acree (1997), Acree and Barnard (1994), Acree et al., (1984), and Grosch (1994) for details on GC/O

COMMENTARY

Background Information

SPME

SPME is a sample-preparation technique based on absorption that is useful for extraction and concentration of analytes either by submersion in a liquid phase or exposure to a gaseous phase (Belardi and Pawliszyn, 1989; Arthur et al., 1992). Following exposure of the fiber to the sample, absorbed analytes can be thermally desorbed in a conventional GC injection port. The fiber behaves as a liquid solvent that selectively extracts analytes, with more polar fibers having a greater affinity for polar analytes. Headspace extraction from equilibrium is based on partition coefficients of individual compounds between the food and headspace and between the headspace and the fiber coat-

ing (Arthur et al., 1992; Zhang and Pawliszyn, 1993). Zhang and Pawliszyn (1993) give a thorough discussion of the theory of headspace equilibrium extraction. [It has been shown that equilibrium need not be achieved for extraction; instead, a consistent extraction time is sufficient (Ai, 1997).] Removing the solid or liquid sample from the gas phase after analytes have volatilized simplifies the sampling system to only two systems equilibrating: the SPME coating and the gas phase (Roberts et al., 2000). This can be accomplished using an apparatus such as the one shown in Figure G1.6.2.

SPME first found application in evaluating pollutants in water (Belardi and Pawliszyn, 1989). Since then, SPME has been used in an array of fields including the compositional analysis of water, air, essential oils, caffeine, apple volatiles, pharmaceutical products, insect pheromones, and botanicals (Hawthorne et al., 1992; Malosse et al., 1995; Field et al., 1996; Matich et al., 1996; Guidotti et al., 1999; Elmore et al., 2001; Lee et al., 2001). Several good reviews of SPME have been written (e.g., Eisert and Pawliszyn, 1997; Pawliszyn, 1997).

SPME has been commercially available since 1993 with various adsorbent and absorbant materials and various coating thicknesses. Table G1.6.1 lists some of these coatings and the types of compounds for which they are recommended. Fibers can be made in the laboratory using different polymer coatings (Belardi and Pawliszyn, 1989).

For a compound to contribute to the aroma of a food, the compound must have odor activity and volatilize from the food into the headspace at a concentration above its detection threshold. Since aroma compounds are usually present in a headspace at levels too low to be detected by GC, headspace extraction also requires concentration. SPME headspace extraction lends itself to aroma analysis, since it selectively extracts and concentrates compounds in the headspace. Some other methods used for sample preparation for aroma analysis include purge-and-trap or porous polymer extraction, static headspace extraction, and solvent extraction. A comparison of these methods is summarized in Table G1.6.2.

Solvent extraction

Solvent extraction gives a representation of what is present in the food; however, this is usually quite different from what volatilizes and actually comes in contact with the olfactory epithelium to induce a sensation. Because the exhaustive nature of solvent extraction does not account for volatility of compounds, compounds with low volatility may be determined as making an erroneously large contribution to the aroma if they are present at high concentration. This is often the case with vanillin, which is often present at high concentrations, but has such a low volatility that it may contribute little or not at all to the aroma. A compound with a high volatility, like ethyl butyrate, may be present in the food at a lower concentration, but contribute more to the food's aroma. Potency is also an important factor in determining a compound's contribution to aroma; this is addressed in the literature (Acree, 1997; Acree and Barnard, 1994; Acree et al., 1984; Grosch, 1994). In contrast to solvent extraction, headspace SPME is a method that is selective for compounds that have volatilized from the food.

Just as a solvent is selective in its extraction of like compounds, so is the coating in SPME. SPME has replaced solvent extraction in many industries because it is nontoxic and nonhazardous, because disposal is simple, and because it is relatively inexpensive.

Static headspace

Static headspace sampling uses a gastight syringe and a cryogenic liquid such as liquid nitrogen to cryofocus the gas sample on the GC column. The volatilized compounds are extracted in the same ratio in which they are present above the sample, and thus the headspace can be quantified. Gastight syringes must be well monitored for leakage. A precise technique must be replicated to obtain reproducible results. Limited dilution analysis can be achieved using static headspace sampling. Since aroma compounds can be present below parts per billion, the compounds cannot be detected by conventional techniques, with the exception of GC/O. SPME allows for concentration of the compounds in the headspace for analysis. Though the ratio of extraction is distorted by the selectivity of the fiber coating, this can be accounted for by the quantification method in Support Protocol 1. However, these added steps may introduce error. Static headspace is most applicable when the concentration in the headspace is high enough to be detected.

Purge and trap

By having a gas flow over a food sample and then flow through a "trap" containing a porous polymer (e.g., Tenax TA), volatiles are extracted. If the gas flows through a liquid sample, the process is called purge and trap. These headspace methods can be quantitative. Care must be taken to avoid breakthrough of compounds on a saturated polymer. Due to the small diameter of the traps, a low flow rate is usually used to eliminate or reduce back-pressure. Technical difficulties may be encountered in releasing the volatiles from the polymer, especially with thermal desorption units. Washing with solvent masks any compounds that would elute by GC at the same time, which is often the case with diacetyl. Studies have shown that the flow rate is important in effecting the ratio of compounds released in a mouth simulator (Deibler and Acree, 2000). SPME is useful in cases where a higher gas flow rate is desired, such as with a mouth simulator. SPME is quicker and simpler than porous polymer extraction, and usually produces a more reproducible result. The direct quantitative nature of porous polymer extraction is its primary advantage over SPME. Purge and trap does not quantitatively extract what is in the headspace, but does increase the amount extracted for most compounds due to bubbling through the sample.

Stir-bar sorptive extraction

Stir-bar sorptive extraction (SBSE) is carried out using a commercially available glass stir bar (Twister, from Gerstel GmbH) coated with polydimethylsiloxane (PDMS). A special thermal desorption unit is necessary to introduce the extract into a GC. It can be applied to headspace extraction, but is intended for stirring liquid samples for extraction. The same coatings used for SPME can be used for SBSE, and thus similar selectivity should be observed.

Critical Parameters

Time and temperature

Replication of time and temperature conditions is crucial. Consistent sample conditions must be carefully followed to be able to compare samples. Temperature can be controlled by placing vials in a temperature-controlled water bath or by placing the sample in a jacketed vessel. Time can be carefully monitored using a stopwatch. Consistency in sampling time is imperative; autosamplers do a good job in this respect.

Selection of fiber coating material

Since fiber coating materials have different extraction characteristics resulting in different selectivity, no single fiber is optimal for all situations. Polar coatings tend to be selective for polar compounds, while nonpolar coatings preferentially extract nonpolar compounds. Table G1.6.1 lists some of the fiber coatings that are commercially available and the types of compound for which they are recommended. For optimal fiber selection, the Basic Protocol should be used to evaluate multiple types of coated fibers that represent a range of chemical characteristics or seem appropriate for the class of compounds expected to be present. Comparison of the chromatograms from the tested fibers will allow selection of the fiber or combination of fibers that give the best extraction of the compounds of interest. It is not always important that the peak area be the highest, since this can be accounted for during quantification (see Support Protocol 1); however, greater intensity reduces absolute error.

Conditioning the fiber

Commercial fibers must be conditioned prior to initial use. Each coating requires different conditioning parameters, and the manufacturer's recommendations should always be followed. As an example, conditioning of Supelco's 100-µm polydimethylsiloxane (PDMS) fiber is performed by heating the fiber to 250°C for 1 hour. Conditioning can be carried out in the injection port of a GC or in a special conditioning apparatus.

Cleaning the fiber

Prior to each use, a SPME fiber should be cleaned by exposing it for 2 to 10 min to a temperature that is within the range recommended by the manufacturer for desorption. Since the fiber will have been cleaned during sample desorption, subsequent sampling within a few hours does not require cleaning, provided the fiber is kept retracted and away from high levels of volatiles.

Monitoring fiber degradation

Fiber degradation can result from exposure to heat and solvents, which should be minimized. Leaving a fiber in a hot injection port overnight or for several hours will shorten the life of the fiber. Fiber degradation should be monitored by periodically extracting the headspace of a standard solution containing compounds that are expected in the food sample being analyzed. It is not necessary to use all the compounds of interest. To monitor degradation, a solution containing 0.01% (w/v) of the desired compound(s) in water is placed in a glass container with a septum seal. Any amount of solution and any size container may be used, although larger containers require less time to equilibrate, thus reducing error. The solution amount and container size must be consistent throughout the monitoring process. The container with solution is held in a heated water bath (e.g., 35°C for 1 hr). Any temperature and time can be used as long as they are consistent throughout monitoring. Headspace is analyzed as in the Basic Protocol. The process is repeated after every ten sample runs and chromatograms are compared. If the peak area has been reduced by >5%, the fiber's ability to extract volatiles has been compromised. The fiber should no longer be used for comparative or quantitative purposes, or the fiber degradation effect should be calculated in subsequent sample chromatograms. Once initial monitoring has been performed, degradation can be monitored more or less frequently, depending on the level of degradation caused by the temperature and solvent conditions being used.

Enhancing qualitative extraction

If quantitative extraction is not necessary, some sample conditions may be adjusted to increase volatility of selective compounds. It is important to recognize that the selective nature of these changes in sample conditions affects the *relative* volatility of compounds, and thus distorts the amount of each compound that volatilizes. Thus, quantification of enhanced extractions is not truly representative of the sample's relative headspace.

Commonly referred to as "salting out," adding a salt (e.g., NaCl) to the sample changes the ionic environment and occupies water. Water-soluble compounds concentrate in the available water and equilibrate into the headspace at higher concentrations. Other compounds (insoluble in water) may be unaffected or may decrease in volatility. Changing the pH has a similar effect. Additionally, adjusting the ionic state of compounds will increase or decrease volatility, since ionized compounds do not volatilize.

Heating a sample will increase the total energy, increasing volatility in a manner consistent with the ideal gas law in most cases. However, it was observed that the volatility of terpenes in a beverage is decreased with increasing temperature (Deibler and Acree, 2000). Stirring a sample also adds energy to the system and thus increases volatility.

Monitoring fiber saturation

Saturation of the fiber coating is only a concern for absorptive coatings such as divinylbenzene and carboxen, and only if comparative or quantitative analysis is desired. To verify that the fiber is not saturated, a sample of higher volatile concentration is extracted and analyzed. The resulting chromatogram should show an increased peak area for the compound. If extraction of a more concentrated sample results in the same peak area, the fiber was already saturated at the lower concentration. The amount of increase is not important. Any increase in response is acceptable to show that the fiber is not saturated. When dealing with a beverage or solution, the increased concentration can easily be achieved by simply increasing the volatile ingredients. The headspace concentration can be increased using techniques described above for enhancing qualitative extraction.

Be aware that methods used to increase volatile concentration will affect the volatile ratios; however, that may be compromised when simply looking for saturation. These methods do not increase volatility for all compounds, so if no peak-area increase is observed, it could possibly be because the volatile's concentration did not increase in the headspace and not because the fiber is saturated. It is acceptable to use different sampling parameters for testing saturation (e.g., using static sampling for saturation test, while using dynamic sampling for experiment).

Sample size

In general, the larger the sample the better. When sampling from a closed system, the equilibrium state is perturbed less from a large sample (both container and food) than from a small sample. This means that the new equilibrium state will be reached more quickly.

GC desorption depth

Desorption depth in the GC injection port only needs to be determined once for the specific GC. For optimum desorption, the center of the SPME fiber should be injected to the hottest spot in the injection port. This can be determined by using expensive accessories commercially available from GC manufacturers, or by extracting a simple standard sample and injecting the SPME fiber at different depths. The depth producing the sharpest, most intense peak is the ideal injection depth. By setting the O ring at the desired point (as shown in Figure G1.6.1), the injection depth can be maintained consistently.

Accessories

New accessories are continually made available. This unit discusses the basics necessary to conduct SPME analysis for flavor analysis. An automated sampling and injection system is available from Varian. Supelco offers a manual sampling stand setup. Injection liners are available that reduce the injection port volume to presumably produce sharper peaks. Predrilled septa for the GC are available to reduce septum coring.

Troubleshooting

No response or reduced response

A blank or nearly blank chromatogram can be the result of several situations. The concentration in the headspace may be too low for detection by GC, even with the concentrating ability of SPME. If, however, extractions of similar samples or the same sample have previously given a response, something is likely to be wrong with the equipment or methodology. If there is no response, the fiber may be broken or the purge valve may be open during desorption. If there is a reduced response, the fiber may be too old (see Critical Parameters for discussion of fiber degradation), there may be a loss of fiber phase due to improper handling (e.g., leaving in injection port overnight, stripping through septum), there may be leaks in the sampling container, or the composition of the food sample may be inconsistent.

Highly variable response

SPME can be >95% reproducible. However, the following conditions must be carefully controlled to obtain reproducible results: sample temperature, exposure time to the headspace, sample equilibration time (if using a closed container), sample flow rate (if using a dynamic system), sample size (both food sample and container), stirring speed (if stirred), and composition of the sample.

Wide peaks

Peak width is usually reasonably sharp on the chromatogram. Peak width can sometimes be tightened by using cryogenic focusing, although this is usually unnecessary.

Anticipated Results

SPME can concentrate a compound in a headspace up to 300-fold, possibly more. It extracts and concentrates compounds in a selective manner, so that compound A may be concentrated 150 times while compound B is concentrated only 10 times during the same extraction. Only volatilized compounds are extracted by headspace SPME. Not all compounds found by solvent extraction will be extracted by headspace SPME, primarily because not all compounds volatilize. Immersion SPME gives similar results to those obtained by solvent extraction.

As an example, gas flow from a cola beverage was analyzed and calibration of benzaldehyde was performed as described in Support Protocol 1. A 100-μm-thick polydimethylsiloxane (PDMS) SPME fiber was used to extract the gas flow from the cola beverage. Since equilibrium with the flow concentration was determined to occur within 5 min for the slowest-equilibrating compound, the fiber was exposed to the gas flow for 5 min and then desorbed in the injection port of a GC/MS. A 0.01%

(w/v) solution of benzaldehyde in Freon 113 was analyzed by solvent injection. Purity was found to be 99.6%, and triplicate measurements gave $A_s = 6.50 \times 10^8$. A solution of 0.1% (w/v) benzaldehyde in propylene glycol was then prepared. This solution contained other compounds being quantified, but no overlapping peaks were produced. The solution (25 ml) was stirred at 45°C with the stopcock open for at least 1 hr. Triplicate measurements gave $A_v = 4.99 \times 10^7$. The same solution was analyzed by SPME, with triplicate measurements giving $A_{SPME} = 3.00 \times 10^8$. The calibration constant for benzaldehyde using this fiber was calculated as:

$$k_{SPME} = (C_s V_s / A_s) \times (A_v / V_v) \times (1/A_{SPME})$$
$$= (0.1 \text{ g/liter} \times 10^{-6} \text{ liter} \div 6.5 \times 10^8) \times$$
$$(4.99 \times 10^7 \div 5 \times 10^{-3} \text{ liter}) \times$$
$$(1/3.00 \times 10^8)$$
$$= 5.12 \times 10^{-15} \text{ g/liter/peak area}$$

The quantity of benzaldehyde in the RAS flow from cola was calculated as:

$$Q \text{ (g/liter)} = k_{SPME} \times A_{sample}$$
$$= 5.12 \times 10^{-15} \text{ g/liter/peak area} \times$$
$$3.20 \times 10^7$$
$$= 1.64 \times 10^{-7} \text{ g/liter}$$

Time Considerations

Single extraction

Once the food sample has been prepared, a single SPME analysis time depends primarily on the determined exposure time. An exposure duration of 5 min may be arbitrarily chosen. The fiber must be cleaned prior to use, taking 2 min. Longer cleaning time is necessary after extracting complex and highly concentrated samples. The GC run time is user-dependent, and usually lasts ~1 hr. The GC must cool down before the next run. During this cooling time, the SPME may be extracting another sample.

Although the number of samples that can be extracted and analyzed per day depends strongly on the sample and methods used, on the average, if one sample is processed immediately after the other, five to ten samples can be processed in an 8-hr working day.

Fiber maintenance

Fiber selection requires as many single extractions as fiber materials being evaluated (typically two to five). Fiber selection is only conducted once for extracting similar samples. Fiber conditioning must be conducted before using a commercial fiber for the first time.

Smell Chemicals

Conditioning times range from 30 min to 4 hr. Fiber degradation should be monitored after approximately every ten uses (see Critical Parameters). This involves preparing a standard solution and performing a single extraction. Testing for saturation requires the preparation of a headspace with an increased compound concentration. Two single extractions are necessary. Saturation analysis is only conducted once for extracting similar samples.

Quantification for headspace extraction

Quantification or calibration of a single fiber is composed of nine GC runs. The solvent injection takes only the time required to prepare the solution. The solution in the headspace apparatus needs ≥30 min for equilibration and 5 min for extraction by SPME or with a gastight syringe. Quantification can be conducted before or after an experiment involving the SPME or when new compounds are evaluated.

Dilution analysis

Dilution analysis time depends on the number of dilutions to be conducted. A three-dilution series with duplication is composed of six SPME single samplings and GC/O runs.

Literature Cited

Acree, T.E. 1997. GC/Olfactometry: GC with a sense of smell. *Anal. Chem.* 69:170A-175A.

Acree, T.E. and Barnard, J. 1994. Gas chromatography—olfactometry and CharmAnalysis. *In* Trends in Flavour Research (H. Maarse and D.G. van der Heij, eds.) pp. 211-220. Elsevier Science Publishing, New York.

Acree, T.E., Barnard, J., and Cunningham, D.G. 1984. A procedure for the sensory analysis of gas chromatographic effluents. *Food Chem.* 14:273-286.

Ai, J. 1997. Solid phase microextraction for quantitative analysis in nonequilibrium situations. *Anal. Chem.* 69:1230-1236.

Arthur, C.L., Potter, D., Bucholz, K., and Pawliszyn, J.B. 1992. Solid phase microextraction for the direct analysis of water: Theory and practice. *LC-GC* 10:656-661.

Belardi, R.P. and Pawliszyn, J.B. 1989. Application of chemically modified fused silica fibers in the extraction of organics from water matrix samples and their rapid transfer to capillary columns. *Water Pollut. Res. J. Can.* 24:179-191.

Deibler, K.D. and Acree, T.E. 2000. Effect of beverage base conditions on flavor release. *In* Flavor Release (D.D. Roberts and A.J. Taylor, eds.) pp. 333-341. American Chemical Society, New York.

Eisert, R. and Pawliszyn, J. 1997. New trends in solid phase microextraction. *Crit. Rev. Anal. Chem.* 27:103-135.

Elmore, J.S., Papantoniou, E., and Mottram, D.S. 2001. A comparison of headspace entrainment on Tenax with solid phase microextraction for the analysis of the aroma volatiles of cooked beef. *Adv. Exp. Med. Biol.* 488:125-132.

Field, J.A., Nickerson, G., James, D.D., and Heider, C. 1996. Determination of essential oils in hops by headspace solid phase microextraction. *J. Agric. Food Chem.* 44:1768-1772.

Grosch, W. 1994. Determination of potent odourants in foods by aroma extraction dilution analysis (AEDA) and calculation of odour activity values (OAVs). *Flavour Fragrance J.* 9:147-158.

Guidotti, M., Ravaioli, G., and Vitali, M. 1999. Total *p*-nitrophenol determination in urine samples of subjects exposed to parathion and methyl parathion by SPME and GC/MS. *J. High Resolut. Chromatogr.* 22:628-630.

Hawthorne, S.B., Miller, D.J., Pawliszyn, J., and Arthur, C.L. 1992. Solventless determination of caffeine in beverages using solid-phase microextraction with fused-silica fibers. *J. Chromatogr.* 603:185-191.

Lee, G.-H., Suriyaphan, O., and Cadwallader, K.R. 2001. Aroma components of cooked tail meat of American lobster (*Homarus americanus*). *J. Agric. Food Chem.* 49:4324-4332.

Malosse, C.P., Ramirez-Lucas, D., and Rochat, J.M. 1995. Solid phase mictroextraction, an alternative method for the study of airborne insect pheromones (*Metamasius hemipterus, Coleoptera, Curculionidae*). *J. High Resolut. Chromatogr.* 18:669-700.

Matich, A.J., Rowman, D.D., and Banks, N.H. 1996. Solid phase microextraction for quantitative headspace sampling of apple volatiles. *Anal. Chem.* 68:4114-4118.

Pawliszyn, J. 1997. Solid Phase Microextraction: Theory and Practice. Wiley-VCH, Weinheim, Germany.

Roberts, D.D., Pollien, P., and Milo, C. 2000. Solid-phase microextraction method development for headspace analysis of volatile flavor compounds. *J. Agric. Food Chem.* 48:2430-2437.

Zhang, Z. and Pawliszyn, J. 1993. Headspace solid-phase microextraction. *Anal. Chem.* 85:1843-1852.

Key References

Zhang and Pawliszyn, 1993. See above.

Gives a thorough discussion of headspace SPME theory.

Contributed by Kathryn D. Deibler
Cornell University
Geneva, New York

Simulation of Mouth Conditions for Flavor Analysis

This unit discusses the use and design of the two mouth simulators, the retronasal aroma simulator (RAS) and the model mouth, that have successfully been verified to produce an effluent with volatile ratios similar to that found in human exhaled breath during eating. Though at a glance the apparatuses seem very different, they produce relatively similar effluents. Of obvious notability is the difference in the size of the reservoir; the RAS reservoir is 1 liter and the model mouth reservoir is 70 ml. When determining which apparatus to use, carefully consider concentration needs, absorption characteristics of compounds, and shear resistance of the food.

Figure G1.7.1 is a comparison of chromatograms from a solid-phase microextraction (SPME) from a beverage in a sealed container and from the same beverage in a mouth simulator. This comparison demonstrates that a very different volatile ratio is produced from the same food under different sampling conditions. Due to these differences, it is important to use a sampling method that simulates mouth conditions when studying flavor compositions that produce a human perception. Most methods intended to increase headspace volatile concentration, such as adding salt for salting out, do not uniformly affect volatility. For some compounds,

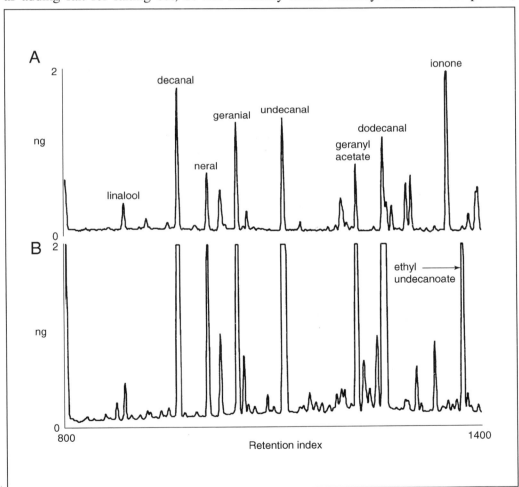

Figure G1.7.1 (**A**) A gas chromatogram of a beverage headspace sampled in the dynamic conditions of an RAS mouth simulator. (**B**) A gas chromatogram of the headspace from the same beverage under static near-equilibrium conditions. Reprinted with permission from Deibler and Acree (2000a). Copyright (2000) American Chemical Society.

Flavors

Contributed by Kathryn D. Deibler and Saskia van Ruth

G1.7.1

vaporization increases to varying degrees; for others, it decreases to varying degrees. Combining a mouth simulator with a method to increase volatility that alters the ratio of aroma compounds in the headspace defeats the purpose of using a mouth simulator.

Direct in vivo measurement of volatiles as they escape from food in the mouth is limited primarily by sensitivity and separation by contemporary analytical equipment (Roberts, 1996). Studying the dynamic volatility from a system that simulates volatility during food consumption can help predict in vivo flavor release (Dalla Rosa et al., 1992; Bakker et al., 1996).

In the study of flavors, a mouth simulator is an apparatus designed to replicate conditions in the mouth that would affect chemical partitioning into the gas phase. This partitioning is often referred to as flavor release; however, compounds not contributing to the flavor are also affected. Ideally, a mouth simulator would create an environment that from any given food would produce the same volatile ratios as the gas phase that contacts the olfactory receptors when a person eats or drinks that food.

Most mouth simulators include features that account for temperature, breath flow, mastication, and salivation in the mouth. Temperature is usually controlled using water either in a jacket or bath. Nitrogen or purified air is blown over the food to simulate breathing. Mastication forces are created by many different means, such as a stir bar, glass balls, a plunger, or blades. Salivation is often not accounted for in mouth simulators.

Mouth simulators are part of the sample preparation process. The effluent produced from a sample in a mouth simulator may be used in conjunction with several flavor analysis methods such as CharmAnalysis (Acree et al., 1984), odor identification, and determination of detection threshold. The effluent is collected and analyzed by a variety of methods.

BASIC
PROTOCOL

USE OF THE RETRONASAL AROMA SIMULATOR (RAS)

The RAS design is based on a stainless steel blender. The design is intended to be readily available, inert and easy to clean, capable of control of parameters, simple to modify for methods of headspace extraction, and, most importantly, to produce an effluent similar in ratio to the retronasal breath in humans that imparts aroma sensation (see Background Information). The RAS dimensions are such that the time to replace the gas in the reservoir per sample weight is nearly equivalent for the average human mouth. Operation, cleaning, and maintenance are rather basic. Solid or liquid food is added to the RAS, followed by artificial saliva. After the blending and air flow are started, the effluent can then be trapped on a porous polymer such as Tenax TA, sampled with SPME (UNIT G1.6), or directly analyzed, e.g., by MS-nose or sensory analysis (Roberts and Acree, 1995, 1996; Roberts et al., 1995, 1996; Roberts, 1996; Ong and Acree, 1998, 1999; Deibler et al., 2001, 2002).

The RAS has been verified to produce similar ratios of volatiles as produced by a human during consumption with four to sixteen panelists for over eight food types (Deibler et al., 2001). The comparisons were made by directly measuring the effluent or breath content with an MS-nose. The RAS has high precision (CV<5%) and sensitivity (μg/liter). Using large sample volumes, the odor-active volatiles can be collected and require less concentrating for the chemical analysis of trace components.

The RAS is not intended to simulate the size or structure of the mouth. The conditions in the mouth expected to affect volatility—i.e., temperature, breath flow, mastication, and salivation—are simulated. Temperature is controlled with a water jacket (37°C). Gas (N_2 or purified air) flow is controlled with a variable-area needle-valve flow meter (20 ml/sec). The shearing resulting from mastication is implemented with blender blades and a high-torque variable-speed motor (150 rpm). Artificial saliva can be added to the system at a 1:5 (v/w) ratio of saliva to food.

The primary components of the RAS are a 1-liter stainless-steel blender container, a stainless-steel jacket that water flows through, a variable-speed motor with controller, modified lid with inlet and outlet for gas flow, and a variable-area needle-valve flow meter. The large volume allows for the collection of sufficient volatiles to concentrate trace components for GC/MS analysis. Figure G1.7.2 shows a diagram of the RAS.

Materials

    Food being tested
    Artificial saliva (see recipe), 37°C
    Gas source (e.g., purified nitrogen, air, or humidified air)
    Retronasal aroma simulator (RAS, available from DATU) with
        temperature-controlled water jacket or water bath

    Additional reagents and equipment for headspace sampling, e.g., solid-phase
        microextraction (SPME, *UNIT G1.6*), porous polymer (e.g., Tenax TA trap),
        gas-tight syringe, MS-nose

Prepare sample

1. Bring the RAS reservoir to 37°C.

2. If the food being tested is not liquid, cut into bite-size pieces (e.g., 1.25-cm cubes of cheese).

    The RAS may not be able to break down foods with high tear resistance, such as strong gels.

3. Weigh out 150 g of food, but do not use >⅔ the volume of the RAS reservoir.

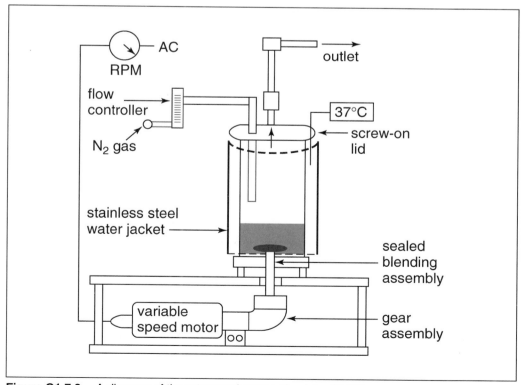

Figure G1.7.2   A diagram of the retronasal aroma simulator (RAS).

Alternatively, the amount equivalent to 30 bites or drinks may be added; then in step 5, 30 ml of artificial saliva is added (assumes food is in the mouth for 30 sec and the average simulated saliva flow rate is 2 ml/min; Bourne, 1982).

4. Add food sample to the RAS.

Analyze sample

5. Add 30 ml artificial saliva that is at ~37°C.

   More artificial saliva may need to be added to ensure that all food particles come in contact with the artificial saliva. This would be similar to an increased saliva stimulation from a dry food.

   Adding water instead of saliva may be sufficient for foods that would not be affected by the enzymes or buffering of the artificial saliva, such as soft drinks.

6. Screw lid on tightly.

7. Connect gas lines and begin flow at 20 ml/sec. Allow gas to run for ~10 sec through the system (but not through the sampling device) to purge any equilibrium or psuedo-equilibrium gas phase from the reservoir.

   The gas used may be purified nitrogen, air, or humidified air.

8. Begin blending at 150 rpm (2.5 rps).

   This produces a shear rate of ~332 sec^{-1} for a liquid, which falls within the range of shear rates in the mouth for various foods (10 to 500 sec^{-1}; Elejalde and Kokini, 1992; Roberts and Acree, 1995). Since the range of shear rates for food is large, a wide range of blending rates is acceptable (Deibler et al., 2001). If the food is not sufficiently "chewed" or broken down, a greater blending speed should be used.

   Either the blending or gas flow may be started first.

9. Begin sampling gas effluent, precisely monitoring the sampling time. Use any method of sampling headspace, including solid-phase microextraction (SPME, UNIT G1.6), absorption on porous polymers (e.g., Tenax TA trap), gas-tight syringe, MS-nose, and human sniffing. Multiple SPME fibers can be placed along the effluent path without depletion of the gas phase. A diversion of most of the gas flow may be necessary to avoid excessive back pressure when sampling with porous polymers. Alternatively, multiple outlets with porous polymer traps may be used.

   It is important to conduct steps 4 through 9 as quickly and reproducibly as is reasonable.

Shut down and clean RAS

10. Remove sampling device (e.g., SPME, Tenax TA trap).

11. Turn blending off.

12. Turn gas off and disconnect.

13. Evaluate collected sample, e.g., by gas chromatography/olfactometry (GC/O; Acree, 1997) or gas chromatography/mass spectrometry (GC/MS).

14. Remove from heated water jacket or disconnect reservoir from water supply.

15. Thoroughly clean and rinse reservoir and lid with soap and water.

    It may not be necessary to thoroughly dry the reservoir before its next use since water will be added to the system, provided the remaining dampness is minimal.

The release of aroma compounds in the mouth during eating is primarily determined kinetically, rather than thermodynamically, because of the processes occurring when food is consumed. The model-mouth system was developed to study in vitro–like aroma release and considers the bolus volume, volume of the mouth, temperature, salivation, and mastication (van Ruth et al., 1994). Volatile compounds in the effluent of the model mouth are collected on porous polymers, such as Tenax TA. Alternatively, the effluent can be measured on-line by direct mass spectrometry techniques. The model mouth can be used to study the effects of food composition and structure on aroma release, as well as the influence of oral parameters related to eating behavior.

The model mouth is composed of a sample flask and assembly, a plunger for mastication, two voltage controllers, and two variable speed motors to give precise control of vertical and circular speed of the plunger, an externally circulating temperature-controlled water bath connected to the cavity wall of the sample flask, an externally circulating temperature-controlled ethanol bath connected to a cooling coil to freeze out water, and a controlled gas supply to sweep over the food (Figure G1.7.3).

Materials

> Artificial saliva (see recipe)
> Sample
> Gas source (e.g., nitrogen or air)
> Model mouth apparatus (Fig. G1.7.3)

1. Switch ethanol bath (B in Fig. G1.7.3) and the water bath (D) on to reach their set temperatures, −10°C and 37°C, respectively.

   The ethanol bath is used to cool a coil that will freeze out water. Water may disturb the GC analysis later. The water bath is put in place to maintain the sample flask's temperature at body temperature.

2. Place sample flask in its holder and connect the warm-water pipes.

3. Screw the cooling coil on the side of the sample flask, and connect the ethanol pipes to the coil.

4. Transfer an aliquot of artificial saliva (e.g., 4 ml) to the flask.

5. Let both water and ethanol circulate through the cavity walls of the system for 10 min to allow the system to reach the set temperatures.

6. Place one spoon of sampling material (E; e.g., 6 ml) in the sample flask.

7. Place the plunger (F) in position and connect to the motors (H).

   The plunger is the mastication device and two motors regulating the vertical and circular movements control its masticatory movements.

8. Connect a trap (A) or the connection for on-line analysis to the cooling coil.

   The gas supply (nitrogen gas or air) is connected and the flow through the trap is measured. Flow rate can be between 25 and 250 ml/min, but should be consistent for all measurements. Flow rate depends on the specific sample. For short-time analysis, flow rates are usually high to account for the dead volume. For longer measurements with traps, the breakthrough through the trap depends on the total volume passing the trap, and is thus determined by flow rate and time. For direct on-line analysis of the effluent, an atmospheric-pressure chemical ionization mass spectrometer or a proton-transfer reaction mass spectrometer is connected to the cooling coil. In case of on-line analysis, no gas supply is connected; the instrument will withdraw room air through the system at a rate determined by the instrument (15 to 300 ml/min).

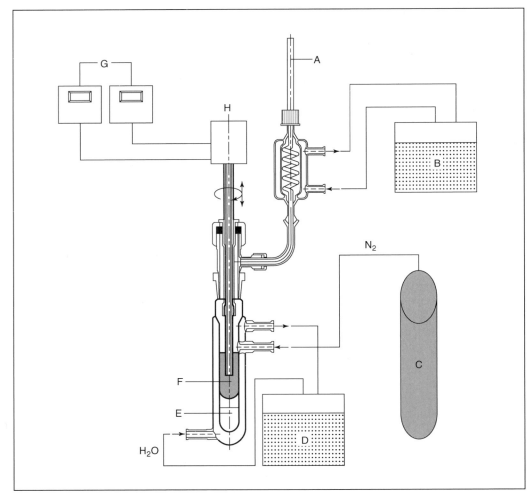

Figure G1.7.3 A diagram of the model mouth. A, trap; B, ethanol bath (−10°C); C, nitrogen gas source; D, water bath (37°C); E, sampling material (in sample flask); F, plunger; G, voltage controllers; H, motors.

9. Check the system for leaks.

10. Set the voltage controllers (G) of the variable speed motors, which regulate the vertical and circular movement of the plunger (0 to 107 cycles/min).

11. Extract the volatile compounds over a set time period. Measurement times vary between 15 sec and a few hours, and depend on the scientific requirements.

REAGENTS AND SOLUTIONS

Use deionized or distilled water in all recipes and protocol steps. For common stock solutions, see APPENDIX 2A; for suppliers, see SUPPLIERS APPENDIX.

Artificial saliva
   From Roth and Calmes (1981):
   20 mM $NaHCO_3$
   2.75 mM K_2HPO_4
   12.2 mM KH_2PO_4
   15 mM NaCl
   200 U/ml α-amylase
   Adjust pH to 7.0

From van Ruth et al. (1997):

5.208 g NaHCO$_3$
1.369 g K$_2$HPO$_4$·3H$_2$O
0.877 g NaCl
0.477 g KCl
0.441 g CaCl$_2$·2H$_2$O
0.5 g NaN$_3$
2.160 g mucin (porcine stomach mucin; Sigma-Aldrich)
200,000 U α-amylase (hog pancreas α-amylase; Sigma-Aldrich)
Bring to 1 liter with distilled water
Adjust to pH 7

These are two of many formulations for artificial saliva.

COMMENTARY

Background Information

Flavor perception results from interactions between a consumer and stimulants in a food. For the aroma part of flavor, the stimulants are volatiles that bind to receptor proteins found on the olfactory epithelium. These stimulants can reach the receptors by two routes, orthonasal or retronasal. The retronasal route is used when odorants are drawn from the mouth during eating through the nasal pharynx to produce aroma.

The composition of volatiles released from a food is different when it is sniffed (via orthonasal route) and when it is eaten (via retronasal route). This is partially due to conditions in the mouth that selectively affect volatility, thus altering the ratio of compounds that volatilize from a food system. Mouth temperature, salivation, mastication, and breath flow have all been shown to affect volatilization (de Roos and Wolswinkel, 1994; Roberts et al., 1994; Roberts and Acree, 1995; van Ruth et al., 1995c). The ideal gas law describes the effects of temperature. Saliva dilutes the sample, affects the pH, and may cause compositional changes through the action of the enzymes present (Burdach and Doty, 1987; Overbosch et al., 1991; Harrison, 1998). Mastication of solid foods affects volatility primarily by accelerating mass transfer out of the solid matrix. The gas flow sweeps over the food, creating a dynamic system. The rate of the gas flow determines the ratio of volatiles primarily based on individual volatilization rates and mass transfer.

The events of eating or drinking are dynamic processes in which equilibrium is never achieved (Castelain et al., 1994; de Roos, 1997; van Ruth and Roozen, 2000a). Volatiles travel to the nosespace simultaneous with mastication and drinking, as demonstrated by sensory tests using aroma solutions held in a dish in the mouth (Pierce and Halpern, 1996). In vivo measurements using a modified mass spectrometer detect significant concentration of volatiles from the nasal exhalation almost immediately after a food is placed in a person's mouth (Laing and Livermore, 1992; Linforth et al., 1994, 1999; Linforth and Taylor, 1998; Taylor et al., 2000). The concentration in the nosespace is greatly reduced at the instant the food is swallowed.

Equilibrium concentrations describe the maximum possible concentration of each compound volatilized in the nosespace. Despite the fact that the process of eating takes place under dynamic conditions, many studies of volatilization of flavor compounds are conducted under closed equilibrium conditions. Theoretical equilibrium volatility is described by Raoult's law and Henry's law; for a description of these laws, refer to a basic thermodynamics text such as McMurry and Fay (1998). Raoult's law does not describe the volatility of flavors in eating systems because it is based upon the volatility of a compound in a pure state. In real systems, a flavor compound is present at a low concentration and does not interact with itself. Henry's law is followed for real solutions of nonelectrolytes at low concentrations, and is more applicable than Raoult's law because aroma compounds are almost always present at very dilute levels (i.e., ppm). Unfortunately, Henry's law does not account for interactions with the solvent, which is common with flavors in real systems. The absence of a predictive model for real flavor release necessitates the use of empirical measurements.

Real foods are usually complex, having many components. To develop a theoretically accurate equation for a simple synthetic grape beverage containing water, sucrose, gum, and just two aroma compounds, one would need 17 rate equations and 10 variables to define the state. Most commercial soft drinks would require >270 rate equations and 46 variables to define the state. Since rate constants are not generally available in the literature, they would all have to be individually measured. Each constant would also have to account for the viscosity and surface tension caused by the other components. Clearly, an empirical measure of the results of all these factors might be easier to obtain and be more representative. This is the role of mouth simulators.

Development of the model mouth

The model mouth was developed as part of the doctoral program of Saskia van Ruth at the Wageningen University in the Netherlands between 1992 and 1995. The hypothesis was that only those volatile compounds released under mouth conditions are relevant for aroma analysis. The volume of the mouth, temperature of the mouth, mastication, and salivation were thought to be critical parameters. Those pa-

rameters were taken into account in the first design of the instrument, which was published in 1994 (van Ruth et al., 1994). At a later stage, the design of the instrument was optimized; design changes included assemblies that allow the removal of water vapor, the technical design and construction of the plunger motor, variable speed controllers that regulate the plunger speed, and connections for on-line measurements.

The volatiles of various foods have been analyzed using the model mouth, including French beans, bell peppers, and leeks (van Ruth et al., 1995a), cheese (Lawlor et al., 2002), bulk oils and oil-in-water emulsions (van Ruth and Roozen, 2000a), and model food systems (van Ruth et al., 2000a; van Ruth and Villeneuve, in press). Factors influencing the volatile composition were studied and included post-harvesting conditions (van Ruth et al., 1995b, 1996a) and the formation of lipid oxidation products (van Ruth et al., 1999, 2000b). Oral physiological parameters examined were mastication rate (Geary et al., 2001), saliva flow rate, and saliva

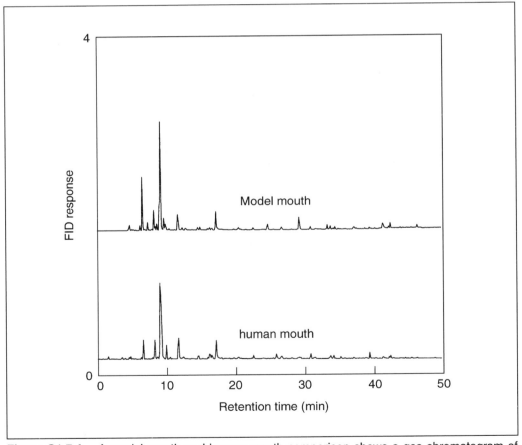

Figure G1.7.4 A model mouth and human mouth comparison shows a gas chromatogram of volatile compounds released from rehydrated French beans in the model mouth ($n = 6$; upper chromatogram) and in the mouth of assessors ($n = 12$, lower chromatogram).

composition (van Ruth et al., 1996b; van Ruth and Roozen, 2000b). Further data on the effect of oral physiological factors on volatile release can be generated with the instrument.

The model mouth was validated to be representative of volatile release in the mouth. Three dried vegetables, i.e., bell peppers, French beans, and leeks, were used for a comparison of volatile release in the model mouth and in the mouths of assessors (van Ruth et al., 1995a). An identical sample size of 1.2 g of dried vegetables and 10 ml of water were used for both techniques. In the model mouth, volatile compounds present in the effluent were extracted on the absorbent Tenax TA. In-mouth analysis was conducted during eating by withdrawing air from the mouths of twelve assessors and directing that air through a similar trap by a vacuum pump using the method described by Roozen and Legger-Huysman (1994). Gas flow rates (250 ml/min), time of isolation (12 min), chewing movements (4 movements/min), and gas chromatographic analysis of the isolated compounds were similar for model mouth and in-mouth analyses. Chromatograms of the volatile compounds extracted from the nitrogen purged through French beans in the model mouth and from oral breath are shown in Figure G1.7.4. Both qualitative and quantitative compositions of the vegetables were similar. Statistical analysis revealed no significant differences in volatile release from the three vegetables between model mouth and in-mouth analysis (Friedman two-way analysis of variance, $P < 0.05$, average CV of in-mouth 72%, of model mouth 28%). These data suggest that the model mouth mimics the volatile release in the mouth quite well.

Development of the RAS

The RAS was first designed by Roberts and Acree in 1994 using a 4-liter Waring blender (Roberts et al., 1994). The air flow rate and the blending rate were 32 ml/sec and 300 rpm, respectively. Later, a more manageable size (1-liter) Waring blender was used. This still gave the advantage of increased amount of volatiles to be concentrated. Gas flow bubbled through the liquid sample at 20 ml/sec (Roberts and Acree, 1996; Roberts et al., 1996). A chamber was added to humidify the air before it flowed through the RAS for the primary purpose of comfort for human sniffing (Deibler and Acree, 2000b). Humidification of the gas was only necessary for sensory analysis of the RAS effluent (Ong and Acree, 1998, 1999). The gas flow was redirected to no longer "purge"

through the sample, but to flow over the food as in the human system (Deibler and Acree, 2000b; Deibler, 2001). Bubbling the gas through the sample would affect the ratio and quantity of volatiles released due to the different pressure found inside the bubbles. To assure that the blending rate was not affected by the resistance of the food being studied, a high-torque variable-speed motor replaced the standard blender motor. To increase efficiency of conduction of heat to the RAS reservoir, a custom-made stainless-steal jacket was used in place of the copper coils to carry water.

Sampling methods that have been used for the RAS effluent include trapping of effluent on multiple porous polymer traps with diversion of effluent when not extracting multiple or single SPME, gastight syringe extraction, MS-nose, and human sniffing. Originally, there was just one sample port for collecting volatiles from the effluent. A branched system of stainless steel tubes with multiple septa ports permitted sampling the effluent with multiple SPME units. A flow diverter allowed for the reduction of pressure through a porous polymer trap while maintaining the high gas flow rate through the RAS.

A comparison of the aroma volatiles in the RAS effluent and the aroma volatiles in nasal expiration with several foods using the MS-nose showed a close correlation (>95%; Deibler et al., 2001). Model cheese, model chocolates, and several real foods were evaluated using four to sixteen trained panelists. The variability of the RAS was <5%, while the variability between panelists was >45%, and between analyses by a single panelist was ~25%. The variability of the panelists demonstrates the vast variety of flavor release experiences existing for humans. The precision of the RAS allows for evaluation of these different experiences. Additionally, the precision of the RAS allows for controlled chemical evaluations that would be impossible with the variation of humans. The RAS produced an increased concentration (200 times), yet with nearly the same ratio of aroma compounds as human retronasal breath. This increased concentration gives greater sensitivity to analyses. Although the actual human retronasal experience is ideal for studying flavor release, this is not always practical or desirable, such as when questionable toxicity exists, when it is not cost feasible, or when the precision and sensitivity of the RAS are desired.

Table G1.7.1 Comparison of Some Mouth Simulators

	References					
Mouth parameters simulated	Lee (1986)	Roberts et al. (1994)	van Ruth et al. (1994)	de Roos and Wolswinkel (1994)	Elmore and Langley (1996)	Springett et al. (1999)
Type of air	Helium	Air or N_2	N_2	Humid N_2	Helium	N_2
Air flow (ml/min)	50	1200	20	100	30	0-120
Mastication	Shaker with balls	Blender	Plunger/screw	Stir bar	Stir bar	Stir bar
Heat	Water bath	Water coils	Water jacket	nd[a]	Water jacket	Water jacket
Salivation	Via septum	Added	Added	Could be added	Could be added	Could be added

[a]nd, not defined.

Choice of mouth simulator

The model mouth and RAS are two examples of mouth simulators. A representative set of mouth simulators are compared in Table G1.7.1. All account for temperature, breath flow, and mastication. Only the model mouth and the RAS allow for the evaluation of solid foods and have been compared directly to human breath.

The primary difference between the output from the RAS and the model mouth is concentration. Nearly equivalent volatile ratios are produced. Loss of compounds that bind to stainless steel should be considered when using the RAS. Binding to glass (e.g., by furaneol) or silanized glass (e.g., by linalool) should be considered when using the model mouth. Food with high tear resistance (e.g., some gels) may not be broken down using the RAS. This may be compensated for by using a high blending rate with additional saliva/water or by modifying the blade shape. Both mouth simulators are acceptable for most applications where a gas flow containing concentrations similar to that found in humans' retronasal breath during eating is desired.

A mouth simulator is a valuable tool when determining what volatiles contribute to the flavor sensation during consumption of a food. This includes determining potency (e.g., CharmAnalysis), intensity (e.g., OSME; Acree and Barnard, 1994), contribution (e.g., omission tests), and effect of a compound on the flavor. Sample preparation with a mouth simulator gives a close representation of the human experience, without the expense and variability of using humans. The limitations of headspace sampling and detection sensitivity define the limits of the use of mouth simulators.

The RAS has found both commercial and academic applications (Roberts and Acree, 1995; Roberts et al., 1995, 1996; Roberts, 1996; Acree, 1997; Ong and Acree, 1998, 1999; Deibler and Acree, 2000b; Deibler et al., 2000, 2001, 2002; Deibler, 2001; Feng and Acree, 2001). The effect of the mouth parameters (i.e., temperature, flow rate, shear rate, and artificial saliva) have been evaluated (Roberts, 1996; Deibler and Acree, 2000a,b; Deibler et al., 2001). The complexity of ingredient effects on flavor release from a beverage has been demonstrated (Deibler and Acree, 2000b). Individual chemical retronasal detection thresholds have been measured for wine analyses. Retronasal aroma thresholds for calculation of odor activity units have been measured using a sensory panel and the RAS (Ong and Acree, 1998, 1999). Off flavors, aroma potency, product comparison, and product formulation have been conducted both to identify and correct problems and for the successful development of several commercial products.

Critical Parameters

RAS

Of the RAS parameters, the gas flow rate has the greatest effect on volatility and must therefore be carefully controlled (Deibler et al., 2001). The flow rate of the effluent should be periodically measured to ensure flow rate consistency. The flow rate can be measured with a simple bubble meter or an electronic flow meter. Be sure the meter used is appropriate for measuring the magnitude of the anticipated flow rate.

Upon setup, on a regular periodic basis, and anytime that the flow seems compromised, the apparatus should be scanned for leaks and any leaky connections should tightened. Leaks may be found by covering connections carrying gas with a simple soap-and-water solution or a commercial product such as Snoop Liquid Leak Detector (Nupro Company). Bubbles will be produced from any leak covered by the solution. The seals for the blade assembly and lid should be periodically tested for leaks.

If there is blockage in the gas flow path, a backpressure will develop that could potentially be dangerous. A high backpressure could cause a portion of the RAS to shoot at high speed in an unpredicted direction, potentially hurting someone. Alternatively, the backpressure could force a hole in the flow line, thus compromising the effluent being sampled. Use of a single porous polymer trap (i.e., Tenax TA) with the described flow rate will produce intolerable backpressure. This can be resolved by using a diversion flow or multiple traps.

Model mouth

Sampling time. When traps are used, the length of the time of sampling will affect the quantities of the compounds extracted. With longer extraction times, quantities will increase, but not necessarily linearly or proportionally for all compounds. As the release of aroma is a dynamic process, release rates change with time if there is significant depletion of a compound or if the release rate is zero order. The change in release rates is best observed with on-line measurements by direct mass spectrometry techniques or by interval sampling with traps. The time of extraction when traps are used is a compromise between realistic consumption times and detection limits. Breakthrough of compounds due to saturation on the porous polymer material will limit sampling time. Increasing the length and/or diameter of the sampling tube will increase the amount that the trap is able to hold, and thus increase the time that breakthrough occurs. Multiple tubes and multiple absorbent materials in series will also increase the time that breakthrough occurs. As the aim of the model mouth is to mimic mouth conditions, it is preferred to work with short isolation times (<1 min). Time of measurement should also be limited for on-line measurements.

Gas flow rates. For measurements using traps, the extraction time should be taken into account when setting the gas flow rate. With short times, gas flow rates are preferably high to account for the dead volume of the system and to reproduce in vivo breathing more accurately. If, because of detection limits (e.g., for most GC/MS), it is decided to sample for a longer time period, the breakthrough volume has to be considered. The trap will behave as an analytical column; with gas flow going through, the initially trapped compounds will move towards the end of the trap. Whether and when this happens is determined by the size of the trap, the absorbent, the volatile compound, and the total gas volume passing through the trap. The latter is determined by both the gas flow rate and time. For on-line measurements, the flow rate is usually determined by the inlet flow of the mass spectrometer. A by-pass pump may be connected to adjust the flow through the model mouth, with still the same volume going into the mass spectrometer. Again, detection limits will dictate the by-pass flow rate.

Sample size. The size of the sample will determine, to some extent, the quantity of aroma released. Generally, a larger sample size will increase the release. With relatively large samples, however, the efficiency of mastication may be affected. The obvious choice of sample size to mimic mouth conditions is the size of one bite, which varies within relatively small limits (5 to 15 g).

Sample/saliva ratio. The sample/saliva ratio is a critical parameter for volatile release in the model mouth for liquid, semi-solid, and solid foods. With liquid foods, the dilution and the change in lipid and protein concentration have an effect. With more solid types of foods, besides the effects above, saliva has an effect on the dynamics of the release. Saliva, in combination with mastication, affects the rates of mass transfer. Again, a realistic sample/saliva ratio should be chosen (e.g., in the 80:20 to 40:60 range).

Mastication rate. Mastication rates affect the extent of aroma release dramatically. A standard rate, such as 50 to 60 cycles/min, should be chosen if one is not interested in the effect of mastication rate. Usually, relatively high rates will be used (e.g., >50 cycles/min), corresponding to the chewing rates people apply when consuming solid foods. When interested in the effect of chewing behavior on flavor release, a range of mastication rates can be applied, e.g., 0, 25, 50, 75, 100 cycles/min.

Troubleshooting

RAS

Detection of no volatiles. Insufficient gas flow or insufficient blending or saliva could cause detection of no volatiles from a food with an expected high concentration of volatiles.

Reduced or absence of gas flow through the effluent outlet indicates either a leak, a blockage, or both. A blockage must be systematically discovered after any leaks have been found and repaired. If the pressure gauge (flow controller) reads the maximum reading when the flow is turned on, then the blockage is located in the system after the gauge. If the gauge registers zero while the external flow source is turned on, then the blockage or leak occurs prior to the gauge. If visual inspection does not reveal the blockage, the RAS should be taken apart starting from the outlet end, evaluating for the presence of a full-force flow after removal of each section. Once the blockage is found, it should be cleaned out and measures should be taken to prevent a blockage from re-occurring. After reassembling the RAS, check for leaks. As noted in Critical Parameters, a blockage could produce a large back-pressure that may force a part of the RAS at a high velocity in an unpredictable direction, possibly seriously injuring someone and the RAS.

If one attempts to expel the blockage using high gas/air pressure, the success depends on where and what the blockage is. The primary cause of blockage is operator error, such as a hose being clamped and forgotten, the outlet sealed by accident, or food particulates clogging the gas inlet into the RAS reservoir due to over-filling with food sample. These do not need such extreme measures as high-pressure flow to remove the blockage. Gross impurities in the gas may cause a clog. As with any equipment, the high-pressure gas should only be used on sturdy metal sections. The clogged section or valve should be removed from the system before expelling a clog with high-pressure gas. Expelling a clog with high-pressure gas is unlikely to be necessary. Teflon tubing sections not easily cleared can inexpensively and easily be replaced.

A leak can be tested for by covering gas line connections throughout the entire system with a simple soap-and-water solution or a commercial product such as Snoop Liquid Leak Detector. Bubbles will be produced from any leak covered by the solution. Tighten any leaks found and re-evaluate the flow. Some dry foods, like crackers, may require additional artificial saliva to be ground. Since dry foods often stimulate additional saliva flow, this is unlikely to compromise the flow content. Additionally, the effects of saliva have been found to have the least effect on volatility in the RAS. The shear rates resulting from mastication have a huge range (Elejalde and Kokini, 1992), thus, changing the blending rate to produce a "chewed" product may be necessary.

The detection of no volatiles from the RAS may be due to the food being tested having a concentration of volatiles below the detection limits of the sampling and analysis methods. Testing the methods with a food or model system with intense odor may indicate if the volatile concentration from the food initially being tested is just too low. A different sampling method (i.e., SPME) and/or analysis method (i.e., GC/O) may increase sensitivity enough to detect some volatiles.

Poor reproducibility. Compromised reproducibility is most likely due to something other than the RAS, such as variability of food sample, variability of the sampling method, or inconsistent application of the methodology. However, poor reproducibility can result from the RAS apparatus if there are leaks or blockages of the flow or if the unit is not properly cleaned.

Contamination in blank run. If volatiles are collected when running a blank (nothing in the RAS), there could be absorption of volatiles to the O ring that seals the lid. This can be resolved by either replacing the O ring or changing the tape that wraps the O ring. Contamination could also come from the gas source. Only pure gases should be used. Improper cleaning of the effluent sampling section is another possible source of contamination.

Saturation of sampling apparatus. Though the RAS produces an increased concentration of volatiles from foods, it is unlikely that conventional methods of sampling the effluent will cause saturation. Saturation can be indicated when a chromatogram peak is flat at the top, or when sampling of an increased concentration does not cause an increase in peak area. Changing the sampling method would be the most simple approach to eliminate saturation. Diverting some of the effluent and sampling only a fraction of the flow may eliminate saturation. Additionally, concentrations may be reduced by reducing the food sample amount; however, this change is not linear and compromises the conditions found to simulate the mouth.

Model mouth

Compounds not detected or detected in lower-than-expected concentrations. First, make sure that the problem is definitely due to a problem with the model mouth. For example, the cause of the problem may be due to the analytical equipment (e.g., gas chromatograph or mass spectrometer), inconsistencies in the food sample, and/or extraction errors. If volatile compounds are not detected or are detected in far lower-than-expected concentrations, there may be a gas leak somewhere in the system. All connections should be checked with a leak detector as described for the RAS.

If compounds are detected in lower concentrations than desirable because of detection limits and no leaks have occurred, sample size, gas flow rate, and isolation time can be adjusted to increase the total volatile quantity collected. The method can be optimized efficiently using trial runs in which one parameter is changed at a time. Be aware that volatile ratios will likely be affected by adjustments in these parameters.

Additional compounds are detected. If compounds are detected that are not expected, the purity of the gas should be checked. If the gas is not contaminated, the next step is to check the glassware and the assembly for contamination. Glassware should be cleaned with odorless detergents, rinsed with tap water three times, rinsed with distilled water three times, and dried in an odor-free drying cabinet. All water used in the model mouth should be distilled and should not have been in contact with any plastics. The artificial saliva may also be contaminated with odorous compounds. If necessary, the saliva can be prepared without the $NaHCO_3$ and can be purged with a purified nitrogen gas at a flow rate of 50 ml/min for as long as necessary. The $NaHCO_3$ can then be added and the pH adjusted afterwards. A blank run with saliva alone (i.e., no food sample) will determined if the volatile contamination has originated from the saliva. Mucin is especially known as a cause of volatile contamination.

Anticipated Results

The model mouth and the RAS will produce effluents carrying a ratio of volatile compounds that is similar to the ratio of volatiles leaving the human nose when the same food is consumed. The RAS effluent will be ~200 times more concentrated than the human breath. The RAS produces a time average representation of the retronasal breath composition.

It should be kept in mind that most analytical instruments, such as gas chromatographs and mass spectrometers, do not discriminate between volatile compounds that do or do not possess odor activity. Some form of sensory analysis must be conducted in order to select which volatile compounds contribute to the flavor of the foods. Gas chromatography-olfactometry (GC/O) is an important tool to accomplish that task.

Time Considerations

RAS

Bringing RAS to temperature requires ~20 min. To cut and measure the food takes ~5 min. Initiating RAS requires <5 min, running RAS and sampling effluent usually requires 5 to 15 min, and analyzing sample by GC/O or GC/MS takes ~1 hr. Cleaning the RAS requires ~5 min, and may be done simultaneously with analysis. The total time for GC analysis on the first sample is 1 hr and 30 min; and for subsequent samples 1 hr. The total time for direct MS analysis for the first sample is 30 min; and for subsequent samples 15 min.

Model mouth

As stated in Critical Parameters, if one is interested in the release of volatiles under mouth conditions, the time of consumption should be considered as well. In specific cases, however, detection limits may increase the time of sampling.

Bringing the model mouth to temperature requires ~20 min. Food preparation, i.e., cutting and measuring, takes ~5 min. Initiating the model mouth takes ~1 min, running the model mouth and sampling the effluent usually takes 1 min. Cleaning the model mouth requires ~10 min, and may be done concurrently with GC analysis. The time needed for GC and MS analysis is as described for the RAS.

An interesting approach is also to look into the temporal change of release of volatiles during consumption. Direct mass spectrometry techniques are able to monitor volatile concentrations in air at millisecond intervals. The time needed for GC/O, GC/MS, and direct MS analysis is as described for the RAS.

Literature Cited

Acree, T.E. 1997. GC/olfactometry: GC with a sense of smell. *Anal. Chem.* 69:170A-175A.

Acree, T.E. and Barnard, J. 1994. Gas chromatography-olfactometry and CharmAnalysis. *In* Trends in Flavour Research (H. Maarse and D.G. van der Heij, eds.) pp. 211-220. Elsevier, New York.

Acree, T.E., Barnard, J., and Cunningham, D.G. 1984. A procedure for the sensory analysis of gas chromatographic effluents. *Food Chemistry* 14:273-286.

Bakker, J., Brown, W., Hills, B., Boudaud, N., Wilson, C., and Harrison, M. 1996. Effect of the food matrix on flavour release and perception. *In* Flavour Science: Recent Developments (A.J. Taylor and D.S. Mottram, eds.) pp. 369-374. The Royal Society of Chemistry, Cambridge, U.K.

Bourne, M.C. 1982. Food Texture and Viscosity. Academic Press, New York.

Burdach, K.J. and Doty, R.L. 1987. The effects of mouth movements, swallowing, and spitting on retronasal odor perception. *Physiol. Behav.* 41:353-356.

Castelain, C., Heil, F., Caffre, I., and Dumont, J.-P. 1994. Perceived flavour of food versus distribution of food flavour compounds: Remind food texture. *In* Trends in Flavour Research (H. Maarse and D.G. van der Heij, eds.) pp. 33-38. Elsevier Science, Amsterdam.

Dalla Rosa, M., Pittia, P., and Nicoli, M.C. 1992. Influence of water activity on headspace concentration of volatiles over model and food systems. *Ital. J. Food Sci.* 4:421-432.

Deibler, K.D. 2001. Measuring the effects of food composition on flavor release using the retronasal aroma simulator and solid phase microextraction. Ph.D. dissertation. pp. 131. Cornell University, Ithaca, New York.

Deibler, K.D. and Acree, T.E. 2000a. Effect of beverage base conditions on flavor release. *In* Flavor Release (D.D. Roberts and A.J. Taylor, eds.) pp. 333-341. American Chemical Society, New York.

Deibler, K.D. and Acree, T.E. 2000b. The effect of soft drink base composition on flavor release. *In* 9th Weurman Flavour Research Symposium. Elsevier Science, Friesing, Germany.

Deibler, K.D., Acree, T.E., Lavin, E.H., Taylor, A.J., and Linforth, R.S.T. 2000. Flavor release measurements with retronasal aroma simulator. *In* 6th Wartburg Aroma Symposium, April 11, 2000, Eisenach, Germany.

Deibler, K.D., Lavin, E.H., Linforth, R.S.T., Taylor, A.J., and Acree, T.E. 2001. Verification of a mouth simulator by in vivo measurements. *J. Agric. Food Chem.* 49:1388-1393.

Deibler, K.D., Lavin, E.H., and Acree, T.E. 2002. Solid phase microextraction application in GC/olfactometry dilution analysis. *In* Analysis of Taste and Aroma (J.F. Jackson and H.F. Linskens, eds.) pp. 239-248. Springer, Berlin.

de Roos, K.B. 1997. How lipids influence food flavor. *Food Technol.* 51:60-62.

de Roos, K.B. and Wolswinkel, K. 1994. Non-equilibrium partition model for predicting flavour release in the mouth. *In* Trends in Flavour Research (H. Maarse and D.G. van der Heij, eds.) pp. 3-32. Elsevier Science, Amsterdam.

Elejalde, C.C. and Kokini, J.L. 1992. The psychophysics of pouring, spreading and in-mouth viscosity. *J. Texture Stud.* 23:315-336.

Elmore, J.S. and Langley, K.R. 1996. Novel vessel for the measurement of dynamic flavor release in real time from liquid foods. *J. Agric. Food Chem.* 44:3560-3563.

Feng, Y.-W. and Acree, T.E. 2001. Processing modulation of soymilk flavor chemistry. *In* Aroma Active Compounds in Foods (G.R. Takeoka, M. Guntert, and K.-H. Engel, eds.) pp. 251-264. American Chemical Society, Washington, D.C.

Geary, M.D., Grossmann, I., van Ruth, S.M., and Delahunty, C.M. 2001. The release of aroma compounds from oil and water model systems under varying conditions. *Irish J. Agric. Food Res.* 40:106-107.

Harrison, M. 1998. Effect of breathing and saliva flow on flavor release from liquid foods. *J. Agric. Food Chem.* 46:2727-27.

Laing, D.G. and Livermore, B.A. 1992. Perceptual analysis of complex chemical signals by humans. *In* Chemical Signals in Vertebrates VI (R.L. Doty and D. Muller-Schwartze, eds.) pp. 587-593. Plenum Press, New York.

Lawlor, J.B., Delahunty, C.M., Wilkinson, M., and Sheehan, J. 2002. Relationships between the gross, non-volatile and volatile compositions and the sensory attributes of eight hard-type cheeses. *Int. Dairy J.* 12:493-509.

Lee, W.E. III. 1986. A suggested instrumental technique for studying dynamic flavor release from food products. *J. Food Sci.* 51:249-250.

Linforth, R. and Taylor, A. 1998. Volatile release from mint-flavored sweets. *Perfumer and Flavorist* 23:47-53.

Linforth, R.S.T., Savary, I., and Taylor, A.J. 1994. Profile of tomato volatiles during eating. *In* Trends in Flavour Research (H. Maarse and D.G. van der Heij, eds.) pp. 65-68. Elsevier Science, Amsterdam.

Linforth, R.S.T., Baek, I., and Taylor, A.J. 1999. Simultaneous instrumental and sensory analysis of volatile release from gelatin and pectin/gelatin gels. *Food Chem.* 65:77-83.

McMurry, J. and Fay, R.C. 1998. Chemistry. 2nd ed. Prentice Hall, Upper Saddle River, N.J.

Ong, P.K.C. and Acree, T.E. 1998. Gas chromatography/olfactometry analysis of lychee (Litchi chinesis Sonn.). *J. Agric. Food Chem.* 46:2282-2286.

Ong, P.K.C. and Acree, T.E. 1999. Similarities in the aroma chemistry of Gewürztraminer variety wines and lychee (Litchi chinesis Sonn.) fruit. *J. Agric. Food Chem.* 47:665-670.

Overbosch, P., Achterof, W.G.M., and Haring, P.G.M. 1991. Flavor release in the mouth. *Food Rev. Int.* 7:137-184.

Pierce, J. and Halpern, B.P. 1996. Orthonasal and retronasal identification based upon vapor phase input from common substances. *Chemical Senses* 21:529-543.

Roberts, D.D. 1996. Flavor release analysis using a retronasal aroma simulator (olfactory). Cornell University, New York.

Roberts, D.D. and Acree, T.E. 1995. Simulation of retronasal aroma using a modified headspace temperature, shearing, and oil on flavor release. *J. Agric. Food Chem.* 43:2179-2186.

Roberts, D.D. and Acree, T.E. 1996a. Effects of heating and cream addition on fresh raspberry aroma using a retronasal aroma simulator and gas chromatography olfactometry. *J. Agric. Food Chem.* 44:3919-3925.

Roberts, D.D. and Acree, T.E. 1996b. Retronasal flavor release in oil and water model systems with an evaluation of volatility predictors. *ACS Symposium Ser.* 633:179-187.

Roberts, D.D., Lavin, E.H., and Acree, T.E. 1994. Simulation and analysis of retronasal aroma. In 4th Wartburg Aroma Symposium: Aroma; Perception, Formation, Evaluation (M. Rothe and H.-P. Kruse, eds.) pp. 619-626. Deutsches Institute fur Erna-hrungsforschung, Bundesrepublik, Germany.

Roberts, D.D., Elmore, J.S., Langley, K.R., and Bakker, J. 1995. The effect of viscosity on dynamic flavor release. *Colloq.-Inst. Natl. Rech. Agron.* 75:35-38.

Roberts, D.D., Elmore, J.S., Langley, K.R., and Bakker, J. 1996. Effects of sucrose, guar gum, and carboxymethylcellulose on the release of volatile flavor compounds under dynamic conditions. *J. Agric. Food Chem.* 44:1321-1326.

Roozen, J.P. and Legger-Huysman, A. 1994. Sensory analysis and oral vapour gas chromatography of chocolate flakes. *In* Aroma. Perception, Formation, Evaluation (M. Rothe and H.-P. Kruse, eds) pp.627-632. Eigenverlag Deutsches Institut fur Ernaehrungsforschung, Potsdam-Rehbruecke.

Roth, G.I. and Calmes, R. 1981. Oral Biology. C.V. Mosby. St. Louis, MO.

Springett, M.B., Rozier, V., and Bakker, J. 1999. Use of fiber interface direct mass spectrometry for the determination of volatile flavor release from model food systems. *J. Agric. Food Chem.* 47:1123-1131.

Taylor, A.J., Linforth, R.S.T., Harvey, B.A., and Blake, A. 2000. Atmospheric pressure chemical ionisation for monitoring of volatile flavour release in vivo. *Food Chem.* 71:327-338.

van Ruth, S.M. and Roozen, J.P. 2000a. Aroma compounds of oxidised sunflower oil and its oil-in-water emulsion: Volatility and release under mouth conditions. *Eur. Food Res. Technol.* 210:258-262.

van Ruth, S.M. and Roozen, J.P. 2000b. Influence of mastication and artificial saliva on aroma release in a model mouth system. *Food Chem.* 71:339-345.

van Ruth, S.M. and Villeneuve, E. 2002. Influence of α-lactoglobulin and presence of other aroma compounds on the retention of 20 aroma compounds in water. *Food Chem.* In press.

van Ruth, S.M., Roozen, J.P., and Cozijnsen, J.L. 1994. Comparison of dynamic headspace mouth model systems for flavour release from rehydrated bell pepper cuttings. *In* Trends in Flavour Research (H. Maarse and D.G. van der Heij, eds.) pp. 59-64. Elsevier Science, Amsterdam.

van Ruth, S.M., Roozen, J.P., and Cozijnsen, J.L. 1995a. Volatile compounds of rehydrated French beans, bell peppers and leeks. Part I. Flavour release in the mouth and three mouth model systems. *Food Chem.* 53:15-22.

van Ruth, S.M., Roozen J.P., and Posthumus, M.A. 1995b. Instrumental and sensory evaluation of flavour of dried French beans (*Phaseolus vulgaris*) influenced by storage conditions. *J. Sci. Food Agric.* 69:393-401.

van Ruth, S.M., Roozen, J.P., and Cozijnsen, J.L. 1995c. Changes in flavour release from rehydrated diced bell peppers (*Capsicum annum*) by artificial saliva components in three mouth systems. *J. Sci. Food Agric.* 67:189-196

van Ruth, S.M., Roozen, J.P., Hollmann, M.E., and Posthumus, M.A. 1996a. Instrumental and sensory analysis of the flavour of French beans (*Phaseolus vulgaris*) after different rehydration conditions. *Z. Lebensm.-Unters. Forsch.* 203:7-13.

van Ruth, S.M., Roozen, J.P., Nahon, D.F., Cozijnsen, J.L., and Posthumus, M.A. 1996b. Flavour release from rehydrated French beans (*Phaseolus vulgaris*) influenced by composition and volume of artificial saliva. *Z. Lebensm.-Unters. Forsch.* 203:1-6.

van Ruth, S.M., Roozen, J.P., and Legger-Huysman, A. 1997. Relationship between instrumental and sensory time-intensity measurements of limitation chocolate. *In* Flavour Perception. Aroma Evaluation (H.-P. Kruse and M. Rotne, eds.) pp. 143-151. Universitaet Potsdam, Germany.

van Ruth, S.M., Roozen, J.P., Posthumus, M.A., and Jansen, F.J.H.M. 1999. Influence of ascorbic acid and ascorbyl palmitate on the aroma composition of an oxidized vegetable oil and its emulsion. *J. Am. Oil Chem. Soc.* 76:1375-1381.

van Ruth, S.M., O'Connor, C.H., and Delahunty, C.M. 2000a. Relationships between temporal release of aroma compounds in a model mouth system and their physico-chemical characteristics. *Food Chem.* 71:393-399.

van Ruth, S.M., Roozen, J.P., and Jansen, F.J.H.M. 2000b. Aroma profiles of vegetable oils varying in fatty acid composition vs. concentrations of primary and secondary lipid oxidation products. *Nahrung* 44:318-322.

Key References

Deibler et al., 2001. See above.

Provides a description of parameter settings for the RAS that simulate mouth conditions.

van Ruth and Roozen, 2000b. See above.

Provides a description of model mouth and some effects of oral physiological parameters.

van Ruth et al., 2000a. See above.

Provides information on the effects of volatile compound character, food matrix, and temporal release profiles.

Internet Resources

http://207.150.209.95/

Web site for DATU that provides a description of the RAS and purchasing information.

Contributed by Kathryn D. Deibler
Cornell University
Ithaca, New York

Saskia van Ruth
University College Cork
Cork, Ireland

Gas Chromatography/Olfactometry

As a bioassay, gas chromatography/olfactometry (GC/O) uses human "sniffers" to assay for odor activity among volatile analytes. The core technology used in GC/O analysis is sensory testing and psychophysical measurement. In GC/O, the complex modulation of perception caused by mixture suppression is eliminated because stimulants are experienced in isolation, combined with only purified air (Acree, 1997; van Ruth, 2001a,b). Although this greatly simplifies the perceptual issues, it also means that the results cannot be used to predict the sensory properties of mixtures without supporting sensory data (Lawless and Heyman, 1999). This unit begins with the simple direct column sniffing method (Basic Protocol 1), followed by modifications for quantification, including dilution analysis (Basic Protocol 2), time intensity (Basic Protocol 3), detection frequency (Alternate Protocol 1), and posterior-intensity (Alternate Protocol 2) methods.

NOTE: Use of human subjects requires proper documentation, even for food products. For further details, see Critical Parameters.

GAS CHROMATOGRAPHY/OLFACTOMETRY USING DIRECT SNIFFING

The simplest and most direct method to detect odor-active chemicals in flavors is to separate them chromatographically and to quantify them with an appropriately selective detector: the human olfactory system. Even though the odorants are present at extremely low concentrations and the sample contains many interfering compounds, the human nose will detect the odor-active components and ignore the odorless ones. Although the method of sniffing gas chromatographic effluents is more than 40 years old, modern GC/O instruments are engineered to transfer the odorants from the GC column to a purified, humidified, thermally moderated air stream without loss of resolution or interference from oxidation or background odors. Figure G1.8.1 shows an outline for a sniff port that can be built from a laboratory Venturi vacuum pump and some simple plumbing (Acree et al., 1976). The sensory data produced with the GC/O can be recorded by a variety of methods,

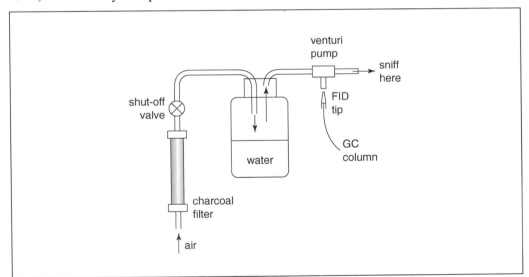

Figure G1.8.1 Diagram of the sniff port constructed from a laboratory filter (based on Acree et al., 1976; see Acree, 1997) showing the filter pump (with the check ball removed) attached to a humidifier, shut-off valve, and charcoal filter. The vacuum side of the pump is positioned over a flame ionization detector (FID) with the hydrogen gas turned off. The make-up gas helps lift the narrow (<0.2-mm-o.d.) gas chromatography (GC) effluent stream into the much larger olfactometry air stream without loss of resolution, and the 300 ml/min air combustion gas produced by the FID also prevents loss of resolution.

Contributed by Terry E. Acree and Saskia van Ruth

Flavors

G1.8.1

from simply interrupting a strip-chart recorder to creating macros for spreadsheet programs, creating scripts for database programs, or purchasing specialized software to record both detection times and sensory perceptions. All of these methods yield tables listing the time of the onset of odor detection relative to the start of the GC program, the disappearance of odor from the olfactometer air, and a descriptor for the odor.

Materials

Indexing standards solution of *n*-alkanes: 0.005% (w/v) C_7 to C_{18} in pentane (*UNIT G1.1*)

Sniffer (i.e., human subject)

Sample to be tested (select one):
  Headspace gas or solvent extract (*UNIT G1.1*) from sample of interest
  Solid-phase microextraction (SPME) fiber, containing sample of interest (*UNIT G1.6*)

OV-101 or DB5 capillary column (e.g., 20-m × 0.32-mm; f = 0.25 μm; also see *UNIT G1.1*)

Gas chromatograph (GC; e.g., Agilent 6890, Agilent Technologies) with:
  Flame ionization detector (FID)
  Sniff port (e.g., DATU, Gerstel, Microanalytics; also see Acree et al., 1976)
  Splitless injector (linear velocity 36 cm/sec or ~2 ml/min, detector temperature, 250°C)

10-ml or 10-μl gas-tight syringe, for headspace gas or solvent extract, respectively

1. Attach an OV-101 or DB5 capillary column to the FID port of a GC and program the GC oven to run as follows:

   run isothermally for 3 min
   ramp at 2° to 6°C/min up to 225°C
   hold 10 min at 225°C.

   Most people can concentrate on sniffing for ≤20 to 30 min. The program rate can be adjusted to optimize the resolution of odors and still keep the entire run to <30 min. Rate changes need to be made at the times when a hydrocarbon (i.e., a standard) is eluting so that indexing is simple. First, a new sample should be run at 6°C/min to determine the points at which odors emerge too closely to one another. Then, at the time that the nearest standard elutes prior to the time of conflicting odors, the rate should be decreased to 2°C/min. When the next standard elutes, it should be changed back to 6°C/min. Of course, some compounds will be separated only when a different substrate is used.

2. Inject 1 μl indexing standards solution of *n*-alkanes via a splitless injector and start the GC program.

3. Record the retention time for each alkane to be used for indexing (*UNIT G1.1*).

4. Move the column from the FID to the sniff port and set up the GC for another run.

   The same GC conditions should always be used for sniffing and indexing.

5. Position a sniffer so as to provide comfortable access to the sniff port for the duration of the analysis.

   Approximately 20 min will be needed to assess effluents that range between 700 and 1800 retention index.

6. Use a 10-ml or 10-μl gas-tight syringe to inject 5 ml headspace gas or 1 μl solvent extract, or insert an SPME fiber containing the sample, and start the program. Start a timer immediately after injection.

   The sniffer should not sniff until the solvent has mostly eluted.

7. Have the sniffer breathe in constant cycles at ~20 breaths/min, record the exact moment an odor is first detected at the sniff port, associate it with a descriptor, and then record the moment the odor is no longer detectable.

> *The sniffer should have been trained in advance to breathe in constant cycles and to perform the analysis.*

> *To automate the process, a simple macro can be written on a personal computer to record times and encode descriptors.*

8. Use the data from the indexing standards to convert each time into a retention index (*UNIT G1.1*).

9. Use a database of retention indices (e.g., the Flavornet; see Internet Resource) to tentatively identify each peak.

> *Verification requires sniffing an authentic standard to verify that the component and the standard have the same retention index and odor quality. Table G1.8.1 shows the result obtained when sniffing the sample used in Figure G1.8.4.*

Table G1.8.1 Single Sniff Run: Start and Stop Times of Odors Detected by GC/O[a]

Start RI	Stop RI	Odor
797	800	Sweet
800	804	Sweat
831	837	Fruity
837	839	Green
839	846	Burnt sugar
882	886	Fruity
904	910	Cat urine
948	951	Minty/fruity
952	956	Mushroom
1021	1036	Cotton candy
1037	1052	Burnt sugar
1052	1072	Caramel
1072	1079	Floral
1079	1095	Fresh
1115	1123	Meat
1130	1136	Skunky
1142	1149	Sweat
1187	1195	Sweet
1216	1233	Foxy
1239	1258	Skunky
1270	1282	Foxy
1284	1307	Vanilla
1331	1347	Apple
1348	1361	Skunky/plastic
1393	1403	Plastic
1403	1413	Plastic
1419	1428	Plastic
1428	1440	Sweet
1482	1491	Cherry

[a]Data from GC/O run shown in Figure G1.8.4 (Niagara grapes; OV-101 index). The odor descriptors were the most frequent descriptors used by the sniffer over several GC/O runs of the same sample. RI, retention index.

DILUTION ANALYSIS WITH GAS CHROMATOGRAPHY/OLFACTOMETRY USING DIRECT SNIFFING

Dilution analysis uses the sniffing of sequentially diluted samples to yield quantitative measures of potency. The samples are sniffed by a direct-sniffing GC/O protocol (see Basic Protocol 1). The analyst dilutes the samples and presents them to a sniffer who records the times at which he/she smells something. Samples are usually presented in order of decreasing concentration until no odors are detected, but this is not essential (randomization to eliminate bias is possible, if needed).

Examples of GC/O dilution analysis described in the literature include CharmAnalysis (Acree, 1997) and AEDA (Aroma Extract Dilution Analysis; Acree, 1997). These methods differ in the way the data are analyzed and presented. At the end of the analysis, the lowest concentration (the highest dilution) at which an odor was detected is the flavor dilution (FD) value. Although the experimenter can write or buy software to simplify the recording and analysis, a simple graphical procedure is to set up a plot of retention index versus dilution number and draw a horizontal line that extends from the start of odor detection to its stop for each dilution tested. Simply dropping a perpendicular at the beginning or end of an area of the graph where no odor was detected will define the odor-active regions in the chromatogram. For AEDA quantification, the highest dilution detected in an odor-active region is used as the FD value. To calculate a charm value, the FD value is multiplied by the width of the odor-active region. The advantage of charm values is that they are based on peak areas and not just peak heights; this is important for a proper assessment of polar compounds, which tend to tail during chromatography.

All materials needed for dilution analysis are listed in Basic Protocol 1.

1. Prepare sequential dilutions of a liquid extract sample in the same solvent used to prepare the extract (e.g., pentane).

   It is easier to apply dilution analysis to liquid extracts than to headspace samples.

   With a 100-fold extract of grape juice diluted at 1/3, about seven or eight dilutions are required before the last odor disappears (i.e., over a 1000-fold dilution).

2. Run indexing standards and the first sample (highest concentration) as described (see Basic Protocol 1, steps 1 to 8).

3. Repeat steps 6 to 8 of Basic Protocol 1 with sequential dilutions of the sample until no odor is detected.

4. Set up a graph of RI on the abscissa versus dilution number on the ordinate and draw a horizontal line that extends from the start of an odor to its stop for each dilution tested. Drop a perpendicular at the beginning or end of an area of the graph where no odor was detected to define the odor-active regions in the chromatogram.

5. For AEDA quantification, use the highest dilution detected in an odor-active region as the FD value.

   For example:

   $FD = 2^7 = 128$ *for the seventh dilution of a 1:1 (1/2) dilution series*

   $FD = 3^7 = 2187$ *for the seventh dilution of 1:2 (1/3) dilution series.*

6. To calculate a charm value, multiply the FD value by the width (in index units) of the odor-active region.

   For example:

   charm value = 128×4.23 index units = 541.44 charm units.

7. Use a database of retention indices (e.g., the Flavornet; see Internet Resource) to tentatively identify each peak.

Verification requires sniffing an authentic standard to ensure that the component and the standard have the same retention index and odor quality.

TIME INTENSITY METHOD FOR GAS CHROMATOGRAPHY/OLFACTOMETRY USING DIRECT SNIFFING

The time intensity method, also described as OSME (Sanchez et al., 1992), quantifies odor in terms of the psychophysical perception of intensity. It uses a cross-modal matching device (a potentiometer or a mouse on a personal computer) to instantly relate force or hand position to a perception of the changing odor intensity as stimulants rise and fall in a GC/O air stream. OSME does not measure potency (i.e., the number of times an odorant concentration is above threshold). Data are generally averaged over several sniffers to produce chromatograms with peaks that relate to perceived intensity at the sniff port. It uses the number of assessors simultaneously detecting an odor in the GC effluent (detection frequency) as a measure for the intensity of a compound. A group of assessors (six to twelve subjects) records the beginning and the end of an odor. The perceived intensities of the individual detections are combined for a specific sample to cumulate the intensity and yield a chromatogram. Data taken from a sniffing chromatogram made from the eight compounds of a reference mix for eight assessors are shown in Figure G1.8.2, where a single sniffer (HW) is shown on the left and the panel average is shown on the right. Usually, the effluent is split for two sniff ports and a flame ionization detector (FID). Thus, two assessors sniff the effluent simultaneously (without seeing each other, to avoid bias). One analysis using a group of eight assessors requires four identical gas chromatographic runs. The detection-frequency (van Ruth and Rozen, 1994; see Alternate Protocol 1) and posterior-intensity (van Ruth et al., 1996; see Alternate Protocol 2) methods of GC/O differ only in the way the data are collected and analyzed.

Materials

Reagents and equipment for GC/O with direct sniffing (see Basic Protocol 1), with additional apparatus as needed for desired number of sniffers
Cross-modal matching device: potentiometer or personal computer with mouse and in-house software for indicating perceived odor intensity

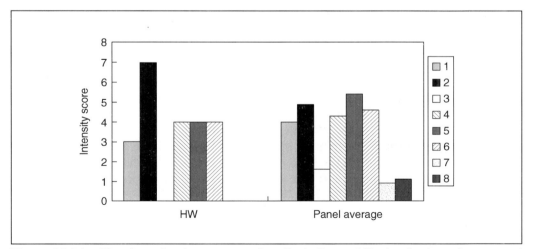

Figure G1.8.2 Time intensity chromatogram (see Basic Protocol 3), showing intensity scores of eight volatile compounds in a reference mix for a single sniffer (HW) and for the panel average (*n* = 8). Notice that sniffer HW showed no response to compounds 3, 7, and 8. This kind of specific anosmia is not uncommon, requiring the use of multiple sniffers or testing to eliminate anosmics.

1. Set up the necessary apparatus for the desired number of sniffers (see Basic Protocol 1, step 1). For a typical setup, split the effluent to two sniff ports and an FID so that two assessors can sniff the effluent simultaneously (without seeing each other).

   One analysis using a group of eight assessors requires four identical gas chromatographic runs.

2. Run indexing standards and inject sample as described (see Basic Protocol 1, steps 2 to 6).

3. Have the two sniffers breathe in constant cycles and record data (see Basic Protocol 1, step 7). At the same time, have the sniffers record the perceived change in intensity by using a cross-modal matching device to indicate a slide bar position or a cursor position along the length of a line.

   The software for the computer setup is not commercially available. Because published details are lacking, most users of this method write software for a personal computer to create a line (generally a thermometer widget available in higher-level languages such as Realbasic) to create a visual scale that the sniffer uses during his/her sniffing experience. The scale has been shown to be most precisely used if it is horizontal, and the ends are labeled "background" on the left and "very strong" on the right. The data generated are usually averaged and integrated to yield an average perceived intensity score (not a potency score) to quantify the group perception.

4. Repeat steps 2 and 3 for the remaining sniffers, using additional chromatographic runs.

5. Determine the intensity score of each odor for each sniffer. Determine the panel averages for each odor. Plot intensity scores of each odor.

6. Use retention indices to tentatively identify each peak (see Basic Protocol 1, steps 8 and 9).

ALTERNATE PROTOCOL 1

DETECTION FREQUENCY WITH GAS CHROMATOGRAPHY/OLFACTOMETRY USING DIRECT SNIFFING

The detection frequency method uses a number of sniffers to quantify an odor in the GC effluent from a single concentration. The underlying assumption is that any random sample of sniffers will functionally express a range of sensitivities, so that some sniffers will detect an odor and others will not. The conclusion is that the fraction of a group that detects an odor is related to the group potency of the odor, a notion that can be supported by the large diversity in odor thresholds observed in humans.

The sniffers (six to twelve subjects) record the beginning and end of detection for an odor. Data are collected in exactly the same way that they are collected during a single sniff run of Basic Protocol 1. The duration of the individual detections are combined for a specific sample to cumulate the number of detections and yield a chromatogram. An example of a sniffing chromatogram for the eight compounds of a reference mix for eight assessors is shown in Figure G1.8.3. Sometimes the effluent is split for two sniff ports and a flame ionization detector (FID). In this case, two assessors sniff the effluent simultaneously (without seeing each other, to avoid bias; see Basic Protocol 3).

The materials are the same as those listed for Basic Protocol 3.

1. Set up the necessary apparatus for the desired number of sniffers (see Basic Protocol 1, step 1). If needed, split the effluent to two sniff ports and an FID (see Basic Protocol 3, step 1).

2. Collect odor detection data as described (see Basic Protocol 1, steps 2 to 7).

3. Repeat using several (about eight) trained sniffers, with each sniffer repeating the assay about four times.

   Multiple sniffers are used to assess intensity, and repeated measures are used to reduce noise in the data.

4. Set up a graph of retention index on the abscissa versus dilution number on the ordinate and draw a horizontal line that extends from the start of an odor to its stop for each replicate tested. Drop a perpendicular at the beginning or end of an area of the graph where no odor was detected to define the odor-active regions in the chromatogram. Use the number of subjects that responded in each odor-active region as the frequency response (e.g., seven of eight subjects, for a frequency response of seven).

5. Use a database of retention indices (e.g., the Flavornet; see Internet Resource) to tentatively identify each peak.

   Verification requires sniffing an authentic standard to verify that the component and the standard have the same retention index and odor quality. Figure G1.8.3 shows a typical detection-frequency chromatogram.

POSTERIOR INTENSITY WITH GAS CHROMATOGRAPHY/OLFACTOMETRY USING DIRECT SNIFFING

ALTERNATE PROTOCOL 2

The posterior-intensity method involves the recording of the odor intensity on a scale after a peak has eluted from the column. A linear relationship has been shown to exist between the logarithm of the stimulus at the sniff port and the average posterior intensity of a panel of eight sniffers (van Ruth et al., 1996). As is frequently the case in GC/O, large variability was observed between the sniffers. The task for the sniffer is moderately complicated in the posterior-intensity method, resulting in differences in scale usage among sniffers. To overcome variation, a group of sniffers should be used and the end of the scales should be anchored with references (in theory). References can be provided only during training GC/O runs; it is practically impossible to provide a reference during an actual GC/O run. Despite good correlations between numbers of assessors and intensities at the sniff port and intensities of sensory attributes for a number of compounds, it is a drawback that the method is not based on real intensities.

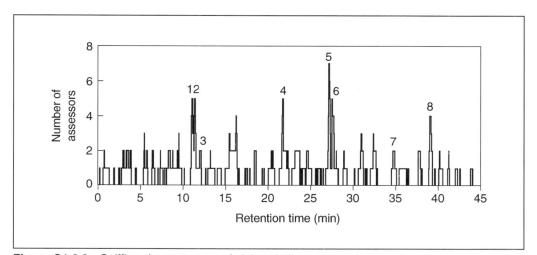

Figure G1.8.3 Sniffing chromatogram of eight volatile compounds in a reference mix obtained by the detection frequency method using eight assessors (see Alternate Protocol 1). Compounds: 1, 100 ng 2-butanone; 2, 20 ng diacetyl; 3, 500 ng ethyl acetate; 4, 100 ng 3-methyl-1-butanol; 5, 20 ng ethyl butyrate; 6, 100 ng hexanal; 7, 100 ng 2-heptanone; 8, 500 ng α-pinene.

Flavors

The materials are the same as those listed for Basic Protocol 3.

1. Set up the necessary apparatus for the desired number of sniffers and collect odor detection data as described (see Basic Protocol 1, steps 1 to 7).

2. Repeat using several (about eight) trained sniffers, with each sniffer repeating the assay about four times.

   Multiple sniffers are used to assess intensity, and repeated measures are used to reduce noise in the data.

3. Set up a graph of retention index on the abscissa versus perceived intensity on the ordinate and draw a horizontal line that extends from the start of an odor to its stop for each odor detected. Drop a perpendicular at the beginning or end of an area of the graph where no odor was detected to define the odor-active regions in the chromatogram. Use the area (or just the height) of each odor-active region as the posterior intensity (PI) response (e.g., for a peak with a width of 4.35 and a height of 3, PI = $4.35 \times 3 = 13.05$ *or* PI = 3). For a group response, average the values from individual sniffers.

4. Use a database of retention indices (e.g., the Flavornet; see Internet Resource) to tentatively identify each peak.

   Verification requires sniffing an authentic standard to verify that the component and the standard have the same retention index and odor quality. Figure G1.8.4 shows a typical posterior-intensity chromatogram.

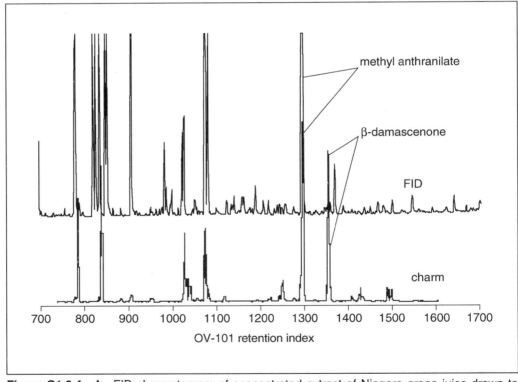

Figure G1.8.4 An FID chromatogram of concentrated extract of Niagara grape juice drawn to display the data on a linear retention index scale where the *y* axis is flame ionization response (upper trace). Below it is the charm chromatogram, where the *y* axis is dilution value. By simply comparing the index of a peak with the data listed in the Flavornet (see Internet Resource), it is possible to determine which odorants have similar retention indices. Notice how large the methyl anthranilate peak is, whereas there is no convincing peak for β-damascenone in the FID chromatogram, even though both compounds have the same potency in the charm chromatogram.

Gas Chromatography/ Olfactometry

Background Information

In the early history of gas chromatography/olfactometry (GC/O; *UNIT G1.1*), the goal of GC/O analysis was to determine when an odor elutes from a GC in order to identify it. The analysis yielded a list of times and, with appropriate standards, retention indices. When combined with other chemical analysis methods, such as mass spectrometry (MS), a name for a particular odorant could be proposed. Comparing both the chemical and sensory properties of the odorant with those of authentic standards allowed researchers to identify the odorant with considerable certainty. The number of odorants that are detected, however, is determined by a number of factors, including the design of the olfactometer, the fraction of the extract injected, and, as we now suspect, the genetics of the sniffer.

Furthermore, as an extract of a natural product is concentrated, the number of odorants detected increases indefinitely. Clearly, most of the odorants in a natural product are below their odor threshold, and it is only the most potent compounds that are involved in generating the flavor response. An odorant can be very potent at extremely low concentrations if it has an extremely low odor threshold. (*UNIT G1.1*). In practice, early GC/O analysts attempted to concentrate the sample as far as possible to identify as many potential odorants as possible. Compositional studies combined with threshold studies were then used to sort out the "important" odorants from the ones that did not contribute to the flavor experience. Rothe's odor units (OU = concentration in sample/threshold in sample) were an early attempt to rank odorants by potency. The process of determining OU values for a food required a lot of chemical and psychophysical analysis. Dilution analysis was developed to produce an OU-like value directly from GC/O without the need to know the identity of the odorant. In fact, the real value of dilution analysis is that it can tell the analyst which compounds to identify.

At a meeting in Germany in 1983, the idea of using repeated sniffs of sequentially diluted samples, now generally called dilution analysis, was proposed (Acree and Barnard, 1984). This led to the publication of CharmAnalysis in 1984 and Aroma Extract Dilution Analysis (AEDA) in 1987, both of which were based on the idea of quantifying potency by dilution to threshold. Potency here is similar to the concept of titer or the amount of dilution necessary to eliminate a detectable biological response, in this case odor.

It is important to recognize that GC/O methods have no direct meaning in terms of the potency of a compound as it is experienced during eating. Extracting an odorant from a food, injecting an aliquot of the extract into a GC/O, and sniffing it as it elutes in a Gaussian distribution of concentration produced by the chromatographic process and isolated from all the other odorants, is not the same experience as eating the food. What is obtained from GC/O dilution analysis is the odor potency of each component under conditions of maximum sensitivity. By expressing the data in relative terms, the analyst hopes to prepare a priority list of the odorants in terms of their potential to contribute to the odor experience, and to avoid time wasted on the study of components that have little likelihood of contributing to the flavor experience.

Two other quantitative GC/O methods have emerged during the last 20 years: OSME and frequency response methods (e.g., GC-SNIFF). Neither of these use dilution analysis directly, but both approximate it in similar ways. Although OSME replaced potency with perceived intensity as the measured quantity, it also averaged the results of several sniffers. It is well established that humans are highly variant in their odor thresholds, and averaging the GC/O response of a number of sniffers at a single concentration will produce higher values for the most potent odorants. In a sense, the result of an OSME analysis is the average perceived intensity of an odorant at a single concentration modulated by human diversity.

The dynamic range of OSME and GC-SNIFF data is generally less than a factor of ten, whereas dilution analysis frequently yields data that cover three or four powers of ten. It has been determined, however, that compressive transforms (log, root 0.5, and so on) of dilution analysis data are needed to produce statistics with normally distributed error (Acree and Barnard, 1994). Odor Spectrum Values (OSVs) were designed to transform dilution analysis data, odor units, or any potency data into normalized values that are comparable from study to study and are appropriate for normal statistics. The OSV is determined from the equation:

$$OSV = 100 \times \sqrt{P_i / P_{max}}$$

where P_i and P_{max} are the potency values of the ith component and the most potent component, respectively.

Critical Parameters

Exposing humans to chemicals in experimental protocols requires special procedures and documentation, even if the chemicals are extracted from foods. Human Subjects Committees must be informed of the intent to use humans in laboratory experiments, and special consent forms must be completed by each subject (referred to as a sniffer in GC/O). An example of a sniffer consent form is found in Figure G1.8.5.

Anticipated Results

Figure G1.8.4 shows a graphical display of dilution analysis data and Table G1.8.1 shows the data in tabular form. Notice that, as also shown for concord grapes in Figure G1.1.3, the methyl anthranilate peak in the FID is ten times

Jane Doe, P.I.
(111) 666-9999
Human Subjects Concerns.

1. Subjects. This research involves the testing of human response to odors they detect sniffing whole foods (cola beverages and cheeses), synthetic models of foods or extracts of foods. 1. Subjects will be trained to judge the perceived smell of flavors by sniffing them as they are separated from an instrument (a gas chromatograph-olfactometer, GCO). 2. Subjects will consist of healthy young adults recruited from Gotham City. The only criteria for exclusion will be current upper respiratory illness (colds), any known food allergies or other unusual reactions to foods, and any dietary restrictions beyond self-imposed weight control.

2. Materials. Research materials will consist of food grade flavors and natural products obtained from commercial food and flavor vendors.

3. Recruitment. Subjects will be recruited by advertisement (posters displayed in town) as well as from existing files of participants in other taste tests conducted at Gotham City. A consent form, approved by the Gotham City institutional review board, will be signed before participation, indicating the nature of the study, materials to be tasted, time involved, right to withdraw at any time without prejudice, risks, benefits and reimbursement.

4. Risks. There are few, if any, risks to the subjects. All materials will be food grade chemicals or natural products. They will be presented at dose levels equal to or less than those found in commercial foods.

5. Safeguards. All materials will be tested at levels no higher than those found in natural products, e.g., beta-damascenone occurs in apples at 100 times its threshold and will not be used at a level higher than 100 times its average reported threshold. Total intake will be limited to 1 ng per day (less than in an apple) in situations where swallowing is necessary. Rinse water and crackers will be provided to help dissipate any unpleasant residual sensations. Exposure will be limited to six sessions per day.

6. Risks/benefits. Risks to the subjects are minimal and are significantly outweighed by the scientific value of the information to be gained by the proposed studies. The only benefits to the subjects are monetary remuneration and satisfaction from participating in scientific studies.

7. Copy of the INFORMED CONSENT form to be signed by each subject that participates in the experiment is attached.

Figure G1.8.5 *(Above and at right)* Example of first and second pages of an informed consent form.

the damascenone peak, but the damascenone dilution value is the same size as for methyl anthranilate. The biggest issue for the analyst is interpretation. The simplest approach is to sort the data by charm value, FD value, OSV, and so on to see which compounds are the most potent, and then to develop compositional analyses for these compounds (*UNIT G1.6*).

Time Considerations

The biggest issue with GC/O analysis is time. It takes ~1 hr to run a blank and another hour to run a standard. With OV-101 and similar substrates, these standards need to be run only once every day or two. In this case, a complete dilution analysis of one sample could be com-pleted in 2 days. If it takes 2 days to obtain data on one sample, the number of samples that can be analyzed is very limited, especially if replicates are analyzed for statistical comparisons. With more polar and less stable substrates, or with samples containing nonvolatile materials, the standards and blanks need to be run more often. In the worst case, the standards must be run before and after each sample. Under these conditions, a complete dilution analysis could take a week for one sample. Clearly, this is not good for any routine analysis. The results are used, however, to develop a routine GC/MS or GC/FID method to measure the most potent components.

INFORMED CONSENT

This research involves the testing of human response to odors they detect sniffing whole foods (cola beverages and cheeses), synthetic models of foods or extracts of foods. Standardized sets of odorants designed to stimulate all odor receptors in the subjects will be used in a device called a GC/O to test for sensory acuity. You will be asked to sit in front of a gas chromatograph combined with an olfactometer and sniff purified humidified air in an isolated environment. The experiment will consist of four sniffing sessions conducted on different days. The maximum number of samples you will be asked to sniff in any one day is six. Each session will take 30 to 45 minutes. You will receive ____ per hour (or any fraction of an hour) or ____ for the entire experiment for participating. If you are a student, no class credit is involved.

Your participation is strictly voluntary. You have the right to leave the experiment at any time you wish, without any penalty or hard feelings. Such a decision will not influence any other relationship that you may have to the experimenters (Jane Doe) in any way. There are no right or wrong answers in these tests. It is your unique ability to detect odors that we are interested in. After the experiment, your data will be kept in a locked file cabinet. In any electronic records, you will be identified only by a code number. Your personal data will never be displayed in any presentation or publication with your identity revealed by name or initials.

Please ask any questions you have about the study at this time.

By signing below, I indicate that I am participating in this study voluntarily. I understand that I have the right to withdraw from the experiment at any time, without penalty. I also indicate that, to the best of my knowledge, I have a normal sense of taste and smell, and that I have none of the following conditions: respiratory disease such as a cold or asthma, respiratory allergies such as hay fever, food allergies, and that I am not to the best of my knowledge pregnant or breast feeding. I have no dietary restrictions or my only dietary restrictions involve self-imposed caloric restriction for weight control. All my questions about the experiment have been answered to my satisfaction. I am between the ages of 18 and 55, inclusive.

Name (Print) _____ Date _____

Signature _____

Literature Cited

Acree, T.E. 1997. GC/olfactometry: GC with a sense of smell. *Anal. Chem.* 69:170A-175A.

Acree, T.E. and Barnard, J. 1984. The analysis of odor active volatiles in gas chromatographic effluents. *In* Analysis of Volatiles (P. Schreier, ed.) pp. 251-267. Walter de Gruyter, Berlin.

Acree, T.E. and Barnard, J. 1994. Gas chromatography-olfactometry and CharmAnalysis. *In* Trends in Flavour Research (H. Maarse and D.G. Van Der Heij, eds.) pp. 211-220. Elsevier, Amsterdam.

Acree, T.E., Butts, R.M., Nelson, R.R., and Lee, C.Y. 1976. Sniffer to determine the odor of gas chromatographic effluents. *Anal. Chem.* 48:1821-1822.

Lawless, H.T. and Heymann, H. 1999. Sensory Evaluation of Food: Principles and Practices. Food Science Text Series (D.R. Heldman, ed.) Chapman & Hall, New York.

Sanchez, N.B., Lederer, C.L., Nickerson, G.B., Libbey, L.M., and McDaniel, M.R. 1992. Sensory and analytical evaluation of beers brewed with three varieties of hops and an unhopped beer. *Dev. Food Sci.* 29:403-426.

van Ruth, S.M. 2001a. Aroma measurement: Recent developments in isolation and characterisation. *In* Focus on Biotechnology, Vol. 7, Physics and Chemistry: Basis of Biotechnology (M. De Cuyper and J.W.M. Bulte, eds.) pp. 305-328. Kluwer Academic, New York.

van Ruth, S.M. 2001b. Methods for gas chromatography-olfactometry: A review. *Biomed. Eng.* 17:121-128.

van Ruth, S.M. and Roozen, J.P. 1994. Gas chromatography/sniffing port analysis and sensory evaluation of commercially dried bell peppers (*Capsicum annuum*) after rehydration. *Food Chem.* 51:165-170.

van Ruth, S.M., Roozen, J.P., Hollmann, M.E., and Posthumus, M.A. 1996. Instrumental and sensory analysis of the flavour of French beans (*Phaseolus vulgaris*) after different rehydration conditions. *Z. Lebensm. Unters. Forsch.* 203:7-13.

Key References

Acree, T.E. and Teranishi, R. 1993. Flavor Science: Sensible Principles and Techniques. American Chemical Society, Washington, D.C.

A general discussion of flavor analysis, including both historical material and specific applications, in which GC/O is described in the context of the most common methods of flavor analysis.

Belitz, H.-D. and Grosch, W. 1999. Food Chemistry, 2nd ed. Springer, Berlin.

An excellent summary of food analysis in which flavor chemistry is given detailed coverage with many examples.

Leland, J.V., Schieberle, P., Buettner, A., and Acree, T.E. 2001. Advances in gas chromatography-olfactometry. American Chemical Society (ACS) Symposium Series 782. ACS, Washington, D.C.

A monograph with chapters prepared by most of the leading researchers and users of GC/O in the world.

Internet Resource

http://www.nysaes.cornell.edu/flavornet

Contains retention indices on four substrates, CAS numbers, and odor qualities for >800 chemicals identified by GC/O analysis in food.

Contributed by Terry E. Acree
Cornell University
Geneva, New York

Saskia van Ruth
University College Cork
Cork, Ireland

Acid Tastants

G2.1   Titratable Activity of Acid Tastants   **G2.1.1**
       Basic Protocol: Potentiometric and Colorimetric Acidity Titrations   G2.1.1
       Support Protocol 1: pH Measurement   G2.1.4
       Reagents and Solutions   G2.1.6
       Commentary   G2.1.6

G2.2   Liquid Chromatography of Nonvolatile Acids   **G2.2.1**
       Basic Protocol 1: HPLC of Nonvolatile Acids Using a C18 Column   G2.2.2
       Basic Protocol 2: HPLC of Nonvolatile Acids Using an HPX-87H Column   G2.2.4
       Support Protocol: Preparation of Samples for HPLC of Organic Acids   G2.2.6
       Reagents and Solutions   G2.2.7
       Commentary   G2.2.9

Contents

1

Titratable Activity of Acid Tastants

When measuring acidity in foods and beverages there are two common units of measurement: titratable acidity and pH. There is no direct relationship between pH and titratable acidity, therefore, both must be measured experimentally. Titratable acidity (TA), also referred to as total acidity, measures the total acid content in a food or beverage system and is determined by titration of the acids in the food system with a standard base.

The term pH is used to express the concentration of free H_3O^+ in a sample and results from dissociation of the acids present. pH is defined as the negative logarithm of the hydrogen ion concentration, shown below, and can span a range of 14 orders of magnitude.

$$pH = -\log [H^+]$$

A lower pH value indicates a more acidic sample due to more free H_3O^+, and a higher pH value indicates a more basic sample. For example, lemon juice has an acidic pH (pH ~2), whereas egg whites have a basic pH (pH ~9). An accurate pH measurement is usually determined instrumentally with a pH meter.

This unit describes the methods for determining the titratable acidity in foods and beverages. To successfully determine the TA, one must have a basic understanding of pH measurements. In the Basic Protocol, titratable acidity is determined by titrating the sample with sodium hydroxide, both the potentiometric and colorimetric methods of titration are described. In order to determine the TA potentiometrically, one must know how to measure the pH of a sample. The Support Protocol describes calibrating the pH meter and pH measurement of a sample.

POTENTIOMETRIC AND COLORIMETRIC ACIDITY TITRATIONS

The choice of titration protocol to determine the TA of a sample is dependent primarily on the color of the sample. The colorimetric titration uses phenolphthalein indicator solution to determine the endpoint of the titration. Phenolphthalein indicator solution turns from colorless to pink upon reaching the endpoint; therefore, if the color of the sample interferes with this color change, the potentiometric titration is the best method. However, when using colorimetric titration, it is also common to titrate to an endpoint of pH 8.2, the endpoint of the phenolphthalein solution.

Materials

Sample
Deionized, distilled water (ddH₂O)
0.1 N sodium hydroxide solution, normalized (see recipe)
1% phenolphthalein indicator solution

pH meter (Corning, Orion)
pH electrode
250-ml beakers (potentiometric titration)
250-ml Erlenmeyer flasks (colorimetric titration)
Small magnetic stir bars
Blender (optional; if the sample needs to be macerated)
Magnetic stir plate
50-ml buret
Buret clamp
Clamp support
Stopcock
Small glass funnel

Contributed by Jane E. Friedrich

Acid Tastants

G2.1.1

Titrate by potentiometric acidity titration

1a. Assemble apparatus as shown in Figure G2.1.1A.

2a. *Optional*: If sample is a solid, add equal parts ddH$_2$O and macerate in blender at 100 rpm for 2 min. If the sample is a semi-solid (i.e., fruit), macerate in blender at 100 rpm for 2 min.

 If the sample is carbonated, degas prior to analysis.

3a. *Optional*: If sample contains suspended particles, carefully centrifuge for 5 min at 2500 rpm at room temperature or filter through neutral paper.

4a. Adjust the sample to room temperature or the appropriate temperature as indicated on the 10-ml volumetric pipet.

5a. Pipet 10 ml sample into a clean 250-ml beaker. Add 100 g degassed ddH$_2$O. Add magnetic stir bar. Cover with watch glass. Stir on magnetic stirrer until dispersed.

6a. Fill 50-ml buret with 0.1 N NaOH and slowly drain to the 0.0 starting point.

 Be sure that there is no air remaining in the outlet capillary.

7a. Carefully titrate with 0.1 N NaOH solution to the endpoint of pH 8.2.

 Add NaOH at a slow, uniform rate until the endpoint is approached. Then add NaOH dropwise, swirling between drops, until the endpoint is stable (e.g., 5 to 10 sec).

8a. Record buret reading for the milliliters of 0.1 N sodium hydroxide used.

9a. Calculate titratable acidity in terms of a standard acid using the equation below (see Table G2.1.1 to choose the standard acid).

$$\text{TA (g/100 ml)} = \frac{(V)(N)(\text{meq. wt.})(100)}{(1000)(v)}$$

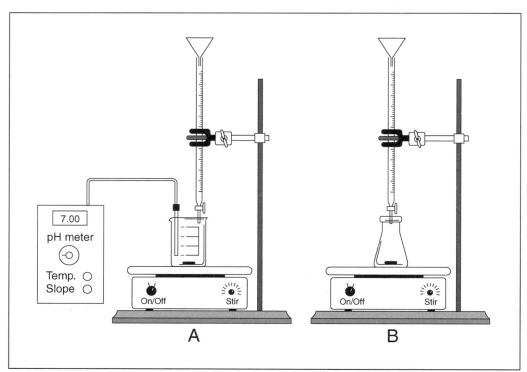

Figure G2.1.1 (**A**) Potentiometric and (**B**) colorimetric titratable acidity apparatus.

Where V is volume of sodium hydroxide solution used for titration (ml); N is normality of sodium hydroxide solution; meq. wt. is milliequivalent weight of the standard; and v is sample volume (ml).

Perform analysis in triplicate.

Titrate by colorimetric acidity titration

1b. Assemble apparatus as shown in Figure G2.1.1B.

2b. *Optional*: If sample is a solid, add equal parts ddH$_2$O and macerate in blender at 100 rpm for 2 min. If the sample is a semi-solid (i.e., fruit), macerate in blender at 100 rpm for 2 min.

  If the sample is carbonated, degas prior to analysis.

3b. *Optional*: If sample contains suspended particles, carefully centrifuge for 5 min at 2500 rpm at room temperature or filter through neutral paper.

4b. Adjust the sample to room temperature or the appropriate temperature as indicated on the 10-ml volumetric pipet.

5b. Pipet 10 ml of sample into a clean 250-ml Erlenmeyer flask. Add 100 g degassed ddH$_2$O.

6b. Add 5 drops of 1% phenolphthalein indicator solution into the sample. Add magnetic stir bar and stir on magnetic stir plate.

7b. Fill 50-ml buret with 0.1 N NaOH and slowly drain to the 0.0 starting point.

  Be sure that there is no air remaining in the outlet capillary.

Table G2.1.1 Standard Acids of Some Foods

Food	Standard acid[a]
Apple	Malic
Apricot	Malic
Banana	Malic
Blueberry	Citric
Cherry	Malic
Cranberry	Citric
Grapefruit	Citric
Grape	Tartaric
Lemon	Citric
Lime	Citric
Orange	Citric
Peach	Malic
Pear	Malic
Pineapple	Citric
Plum	Malic
Raspberry	Citric
Strawberry	Citric
Tomato	Citric
Wine	Tartaric

[a]meq. wt. values of the acids are: acetic acid, 60; citric acid, 64; lactic acid, 90; malic acid, 67; tartaric acid, 75; sulfuric acid, 49.

8b. Carefully titrate to the endpoint, a faint but definite pink color.

Add 0.1 N NaOH at a slow, uniform rate until the endpoint is approached. Then add NaOH dropwise until the color does not fade, i.e., the endpoint is stable (5 to 10 sec).

9b. Record buret reading for the milliliters of 0.1 N sodium hydroxide used.

10b. Calculate titratable acidity in terms of a standard acid in the sample using the equation below (see Table G2.1.1 to chose the standard acid).

$$\text{TA (g/100 ml)} = \frac{(V)(N)(\text{meq. wt.})(100)}{(1000)(v)}$$

Where V is volume (ml) of sodium hydroxide solution used for titration; N is normality of sodium hydroxide solution; meq. wt. is milliequivalent weight of the standard; v is sample volume (ml).

Perform analysis in triplicate.

SUPPORT PROTOCOL 1

pH MEASUREMENT

The pH can be measured instrumentally using a pH meter, a potentiometer, and an ion-selective electrode. The pH meter should be calibrated using buffers obtained from commercial sources. All pH meters come with model-specific instructions for calibration. The following protocol gives a basic overview of the calibration procedure and subsequent measurement for all pH meters. Review the instructions of the specific model in your laboratory to insure proper calibration.

Materials

Standard buffer solutions pH 4.00 and 7.00 (Fisher)
Deionized, distilled water (ddH$_2$O)
Sample

pH meter (Corning, Orion)
pH electrode
Electrode storage solution (Fisher)
25-ml beakers
50-ml beaker
Small magnetic stir bars
Magnetic stir plate
Blender (optional)
Pasteur pipet

Calibrate pH meter (2-point calibration)

1. Pour a small amount of each standard buffer solution into each of two 25-ml beaker (sufficient to cover the electrode bulb and porous plug) and place a small stir bar into each beaker.

2. Label each beaker with its pH value.

3. Allow buffers to equilibrate to room temperature.

4. Measure the temperature of the standard buffer solutions.

This temperature should be close to the calibration temperature (indicated on the reagent bottle).

5. Adjust the temperature dial on the pH meter to the temperature of the equilibrated standard buffer solutions.

6. Set sensitivity or slope control to 100% position (ignore this step for pH meters that do not have the 100% indicated).

7. Remove the electrode from the storage solution and rinse with ddH$_2$O.

8. Place the pH 7.00 standard buffer solution on the magnetic stir plate and stir slowly.

9. Immerse the electrode in the pH 7.00 standard buffer solution.

10. Adjust the buffer control dial so that the display reads pH 7.00.

11. Remove the electrode from standard buffer solution and rinse with ddH$_2$O into a 50-ml waste beaker.

12. Place the pH 4.00 standard buffer solution on the magnetic stir plate and stir slowly.

13. Immerse the electrode in the pH 4.00 standard buffer solution.

14. Adjust the sensitivity or slope control dial so that the display reads pH 4.00.

15. Remove the electrode from standard buffer solution and rinse with ddH$_2$O into the 50-ml waste beaker.

16. Immerse the electrode in the pH 7.00 standard buffer solution.

17. Adjust the buffer control dial so that the display reads pH 7.00.

18. Rinse the electrode with ddH$_2$O into waste beaker.

Analyze sample
19. If the sample is a solid, macerate the sample in a blender at 100 rpm for 2 min.

20. Pour a small amount of the sample into three 25-ml beakers (sufficient to cover the electrode bulb and porous plug) and place a small stir bar in each.

21. Equilibrate the sample to the temperature of the standard buffer solutions.

22. Rinse the electrode with a small portion of the sample to be analyzed.

   Use a Pasteur pipet to rinse the electrode with the sample.

23. Place the beaker containing the sample on the magnetic stir plate and stir slowly.

24. Allow the dial to stabilize (i.e., 15 to 20 sec).

25. Record the pH value of the sample.

26. Repeat steps 4 through 7 to perform measurement in triplicate for each sample analyzed.

27. Once completed, rinse the electrode with ddH$_2$O.

28. Store the electrode in electrode storage solution.

REAGENTS AND SOLUTIONS

Use deionized, distilled water (DDH₂O) in all recipes and protocol steps.

NaOH solution, 0.1 N

Boil 1 liter ddH₂O while stirring on magnetic stir plate. Cool ddH₂O to room temperature. Place ddH₂O in beaker with magnetic stir bar on magnetic stir plate. Weigh out 4.2 g NaOH salt. Add NaOH to water while stirring on magnetic stir plate until dissolved.

Standardized NaOH solution, 0.1 N

Oven dry ~6 g of potassium acid phthalate (KHP) at 120°C for 2 hr. Cool to room temperature in dessicator. Weigh out 5 g of potassium acid phthalate in 250-ml beaker. Record exact weight. Add 100 g (100 ml) ddH₂O. Add magnetic stir bar. Cover with watch glass. Stir on magnetic stirrer until dissolved. Assemble apparatus as shown in Figure G2.1.1A. Titrate with the NaOH solution to endpoint of pH 8.2. Record ml of NaOH solution used. Calculate the normality (N) of the NaOH using the following equation:

$$N = \frac{(\text{g of KHP}) \times 1000}{(\text{ml of NaOH}) \times (\text{meq. wt. of KHP})}$$

Perform analysis in duplicate (repeat titration). Store at room temperature in a glass reagent bottle sealed with a rubber stopper. The standardized NaOH may be stored up to two months. A white swirling precipitate often appears in the standardized NaOH at the end of its shelf life, at this point the normality is still stable but this can be used as a visual sign of the end of the shelf life.

COMMENTARY

Background Information

The taste of foods and beverages often depend on the concentration and type of acid they contain. Two common concepts deal with acidity in foods: titratable acidity and pH. Each of these must be determined experimentally as each has its own impact on food quality. Over the years it has been shown that titratable acidity and pH contribute to the acid taste. However, the acid taste of food high in organic acids is dependent primarily on titratable acidity and secondarily on pH (Plane et al., 1980). The following equation has been proposed by Plane et al. to determine an acid taste index, i.e., acidity index (I_a), of a food sample:

I_a = TA (g/liter) – pH

Two methods are commonly used to determine the endpoint of an acidity titration. The potentiometric method titrates to a predetermined pH and the colorimetric method uses an indicator that changes color at a particular pH to determine the endpoint. Other methods define the endpoint as the inflection of a titration curve, i.e., plots of pH value versus milliliter of NaOH used (Sadler and Murphy, 1998; Iland et al., 1993). However, the increased precision of these methods adds little to the prediction of acid taste.

Using an endpoint of pH 8.2 as the equivalence point for potentiometric determination of TA produces results comparable to colorimetric titrations using phenolphthalein, the classic indicator for TA determinations of organic acids. Of the methods discussed here, titrating to pH 8.2 using the potentiometric titration method is the most commonly used approach and therefore the suggested method of analysis. To calculate the TA of a sample, the milliequivalent weight of a standard acid is used to expresses all of the acids present in the sample. The standard acid is often chosen to be the predominant acid in the sample (Ough and Amerine, 1988). Individual acids must be measured by more specific chemical methods, such as liquid chromatography or capillary electrophoresis.

Critical Parameters

One of the most critical parameters in the measurement of titratable acidity is the sample. The composition of sample will affect the first steps in the procedure for determining TA. For example, if the sample to be analyzed is carbonated, e.g., soft drink, beer, champagne, it must be degassed before measuring the titratable acidity to remove the carbonic acid present in the sample. It is important to remove carbonic acid from the sample so that it will not interfere with the analysis. If the carbonic acid is not removed from the sample, the TA will be over-emphasized. High fat samples, fat content >4%, will also need to be modified prior to analysis. These samples can simply be diluted so that the total fat content is <4% prior to analysis. For further information on sample preparation, refer to Sadler and Murphy (1998). When determining TA, it is important that once the endpoint of the titration is approached, the sample must be swirled after each drop of NaOH and be allowed to reach equilibrium before further titration.

When using a pH meter, it is imperative that the meter is calibrated, and that all steps were followed to calibrate the meter, i.e., temperature was set and the slope was used to calibrate to pH 4.00. Also, the temperature of the standard buffer solutions and sample must be the same. It is recommended that the sample should be diluted prior to analysis if the sample is >15% alcohol.

Anticipated Results

The typical percent acid for the Basic Protocol will depend on the standard acid chosen. For example, if sulfuric acid is chosen as the standard acid, the percent acid will range from 0.1% to 2.4%. However, if lactic acid is chosen as the standard acid, the percent acid will range from 0.15% to 4.3% acid. Therefore, the limits of this analysis are dependent on the standard acid chosen. In general, it is hard to measure titratable acidity below 0.1% and above 4% using this method. If the titratable acidity is

above 4% the sample should be diluted before analysis.

Time Considerations

The 0.1 N NaOH must be made and normalized prior to the determination of titratable acidity, ~3 hr. The first titration will take the longest due to the fact that the amount of titrant, NaOH, to be used is unknown. After the initial titration of a sample, each additional titration will take ~20 min (including calculating the titratable acidity).

Literature Cited

Iland, P., Ewart, A., and Sitters, J. 1993. Techniques for Chemical Analysis and Stability Tests of Grape Juice and Wine, pp. 22-25. Patrick Iland Wine Promotions, Campbelltown, Australia.

Ough, C.S. and Amerine, M.A. 1988. Acidity and individual acids. *In* Methods for Analysis of Musts and Wines, 2nd ed. pp. 50-79. J. Wiley, New York.

Plane, R.A., Mattick, L.R., and Weirs, L.D. 1980. An acidity index for the taste of wine. *Am. J. Enol. Vitic.* 3:265-268.

Sadler, G.D. and Murphy, P.A. 1998. pH and titratable acidity. *In* Food Analysis, 2nd ed. (S.S. Nielsen, ed.) pp. 99-118. Aspen Publishers, Inc., Gaithersburg, Md.

Key References

Sadler and Murphy, 1998. See above.

Detailed discussion of titratable acidity and pH, gives a good overview of the theory and some applications of each analysis. Also addresses sample preparation issues.

Contributed by Jane E. Friedrich
Cargill Incorporated
Minneapolis, Minnesota

Liquid Chromatography of Nonvolatile Acids

Organic acids play an important role in maintaining the quality and nutritional value of a variety of foods. For example, the organic acid content of a fruit diminishes following harvest due to senescence. Organic acids also play a significant role in the overall flavor perception of foods. The sourness of cider is attributed to the content of lactic acid present in the cider. Analysis of nonvolatile acids in juices is widely used in industry for quality control and to test for adulteration. Adulteration of fruit juices is determined by looking at not only the individual organic acid concentrations but also the ratio of concentrations of the acids. The major organic acids present in cranberry juice are quinic, malic, and citric acids. Quinic acid is the most important of these acids in cranberry juice and is also the most uniform. Thus, it is often used as a marker to determine the content of cranberry juice in cranberry beverages.

Because of their importance, organic acids are one of the most commonly analyzed components of a food system. Many methods have been used to determine organic acids in foods, including volumetric, electrochemical, enzymatic, and chromatographic (paper, thin-layer, gas-liquid, or high-performance liquid chromatography) methods. Of the methods listed, high-performance liquid chromatography (HPLC) has long been used as the industry standard for the analysis of organic acids in a food sample and requires the least sample pretreatment.

This unit describes the standard method for determining the organic acids in a sample. The organic acid composition of a prepared sample can be analyzed using HPLC with a C18 column (see Basic Protocol 1) or an HPX-87H column (see Basic Protocol 2). These two methods have been shown to give comparable results. Basic Protocol 1 separates components using reversed-phase chromatography. This method uses a nonpolar stationary phase and a relatively polar mobile phase. Components will elute in order of decreasing polarity (i.e., the most polar component will elute from the column first). Basic Protocol 2 separates components using ion-exchange chromatography. Components will elute in order of increasing pKa. This method of chromatography is based on exchange equilibria between ions in solution and ions of like sign on the surface of an essentially insoluble, high-molecular-weight solid. As the two protocols use completely different mechanisms of separation, they provide complementary methods of analysis; however, it is not necessary to use both for each sample. The method of choice is dependent on the type of sample analyzed. Based on the mode of separation, Basic Protocol 1 is best when the acids expected are rather nonpolar, whereas Basic Protocol 2 would be the method of choice when there are several polar acids present. However, both methods are capable of separating a large range of organic acids. Only when questionable data (e.g., co-elution, poor peak resolution) are obtained is it necessary to apply both methods to the same sample. As both methods work fairly well for most sample types, other considerations when choosing a method would be time and cost. Basic Protocol 1 requires more time for analysis than Basic Protocol 2. The cost for columns in Basic Protocol 1 is slightly greater than in Basic Protocol 2, as only one analytical column is needed for the latter. All methods of analysis of organic acids in foods require sample pretreatment to dissolve the analytes for analysis. A Support Protocol describes methods of sample pretreatment that must be performed prior to instrumental analysis.

Contributed by Jane E. Friedrich

Acid Tastants

G2.2.1

HPLC OF NONVOLATILE ACIDS USING A C18 COLUMN

The method of high-performance liquid chromatography (HPLC) used may change depending on the organic acids present in the sample. For example, if one is analyzing the organic acids involved in the Krebs cycle in dairy products, the method of Doyon et al. (1991) has been optimized for exactly this type of analysis. However, the method described below is the most commonly used method by individuals analyzing organic acids in various foods and beverages in both academia and industry. This method is also the AOAC-tested and -approved method for determination of nonvolatile acids in apple juice and cranberry juice cocktail (Coppola and Star, 1986). This method utilizes a C18 column for HPLC analysis and has been modified to use two columns (the analytical column and a guard column) in series for the separation of organic acids (Coppola et al., 1995). The main purpose of the guard column is to increase the life of the analytical column by removing particulates and contaminants from the solvents. In this case, the guard column also helps in the analysis of products containing both tartaric and quinic acid. The use of two columns increases the resolution of the quinic acid peak.

In this procedure, the instrument is set up and standards are prepared, and then the sample is prepared as described elsewhere (see Support Protocol). Addition of an internal standard to the sample is also important for the analysis of organic acids. This provides a means not only for determining whether the analysis is working, but also for quantitating the percent recovery of the method. The sample is then run and concentrations are calculated.

Materials

0.05 M KH$_2$PO$_4$, pH 2.40 (see recipe)
HPLC-grade water
Organic acid standard solutions (e.g., see recipe) suitable for sample type
Internal standard
Sample
10% to 20% (v/v) acetonitrile (for storage of column)

High-performance liquid chromatograph (HPLC; e.g., Waters Chromatography) equipped with column heater, solvent pump, UV detector (set at 210 nm), integrator, autosampler, and (for manual injection) a 10-μl sample loop
15 × 0.46–cm YMC-ODS-AQ analytical column (AQ12S031546WT, Waters Chromatography)
YMC-ODS-AQ guard column (AQ12S05G304WTA, Waters Chromatography)
10-μl syringe (Hamilton Company, for manual injection only)

Additional reagents and equipment for sample preparation (see Support Protocol)

Set up HPLC instrument

1. Set an HPLC UV detector at 210 nm and allow to warm up for ~1 hr prior to analysis.

2. Set eluent flow rate to 1 ml/min.

3. Equilibrate a 15 × 0.46–cm YMC-ODS-AQ analytical column and a YMC-ODS-AQ guard column with mobile phase (0.05 M KH$_2$PO$_4$) for ≥30 min prior to the first injection.

   This two-column method uses a second analytical column as the guard column.

4. Obtain baseline of HPLC instrument.

5. Inject and run an HPLC-grade water sample as a blank. For manual injection, use a 10-μl syringe here and throughout procedure.

6. Prepare organic acid standard solutions using acids that are found in the sample.

   For each sample type, a literature review must be done prior to analysis to determine the appropriate organic acids and their concentrations. Typical acids include acetic, butyric, citric, formic, hippuric, isobutyric, isovaleric, lactic, malic, oxalic, phenylacetic, propionic, pyruvic, tartaric, uric, and valeric acids. The recipe for organic acid standard solutions (see Reagents and Solutions) describes standards that can be used for a number of fruit juices.

7. Inject standards in duplicate, record peak shapes for each acid component, and integrate results.

8. Add an internal standard to a sample. Use an acid that is not found in the sample, at a concentration that is within the range of acids that are found in the sample. Set aside an equivalent amount of internal standard in duplicate to run by itself.

   Once again, a literature review must be done to determine which acid to use as an internal standard. The internal standard will be used to calculate the percent recovery, which quantitates organic acid losses during sample preparation.

9. Prepare sample for injection (see Support Protocol).

   The sample may be prepared in advance. The internal standard should be added to the sample (step 8) before the sample is prepared (step 9) to control for losses during sample preparation.

10. Inject and run sample in duplicate. Also run internal standard alone in duplicate. Record peaks and integrate results.

11. Thoroughly flush the HPLC system and column with HPLC-grade water. For storage, flush with 10% to 20% acetonitrile to prevent microbial growth.

Calculate percent recovery

12. Determine the percent recovery of the internal standard by using the equation percent recovery $= A_{IS}/A_I \times 100$, where A_{IS} is the area of the internal standard in the sample measured in abundance units (AU), and A_I is the area of the internal standard run by itself.

Calculate acid concentrations

13. Calculate the response factor (RF) for each acid in the standard using the equation $RF = C/A$, where C is the concentration (ppm) of the organic acid in the standard solution and A is the peak area generated (AU).

14. Calculate the concentration of each acid in the sample using the equation $C_s = RF \times A_s$, where C_s is the concentration (ppm) of the organic acid in the sample and A_s is the area of the peak (AU) generated in the sample.

15. When analyzing drinking juices, correct for °Brix to relate the results to single-strength juice using the equation $C_{°Brix} = C_s \times (11.5/°Brix$ of sample$)$, where $C_{°Brix}$ is the concentration of an organic acid in the sample at 11.5 °Brix.

    The value 11.5 °Brix is used in the above equation as an example. The °Brix used should correspond to the juice of the product (Table G2.2.1). One needs to correct for °Brix only when reporting results as standard °Brix.

Table G2.2.1 Standard °Brix for Several Common Juices[a]

Juice	Standard °Brix
Apple	11.5
Apricot	11.7
Banana	22
Blackberry	10
Black currant	12.4
Boysenberry	10
Cranberry	7.5
Grape	16
Grapefruit	10
Lemon	7
Lime	7
Orange	11.8
Passionfruit	14
Peach	10.5
Pear	12
Pineapple	12.8
Prune	18.5
Raspberry	9.2
Sour cherry	14.7
Strawberry	8
Sweet cherry	20
Tangerine	11.8
Tomato	5

[a]From Nagy et al. (1993).

HPLC OF NONVOLATILE ACIDS USING AN HPX-87H COLUMN

As stated in Basic Protocol 1, the method of HPLC used may change depending on the organic acids present in the sample. This method uses a Bio-Rad Aminex HPX-87H column for HPLC analysis. An internal standard is important for the analysis of organic acids. This provides a means of not only determining if the analysis is working but also quantitating the percent recovery of the method.

Materials

0.005 M H_2SO_4 (see recipe)
HPLC-grade water
100 ppm organic acid standards (see recipe)
Internal standard
Sample
10% to 20% (v/v) acetonitrile (for storage of column)

High-performance liquid chromatograph (HPLC; e.g., Waters Chromatography) equipped with column heater, solvent pump, UV detector (set at 214 nm), integrator, autosampler, and (for manual injection) 10-μl sample loop
Aminex HPX-87H column (Bio-Rad)
Aminex HPX-87H guard column (Bio-Rad)
10-μl syringe (Hamilton Company; for manual injection only)

Additional reagents and equipment for sample preparation (see Support Protocol)

Set up HPLC instrument

1. Set an HPLC UV detector at 214 nm and allow to warm up for ~1 hr prior to analysis.

2. Set eluent flow rate to 0.6 ml/min.

3. Set column temperature to 65°C.

4. Set run time to 20 min.

5. Equilibrate an Aminex HPX-87H column and guard column with mobile phase (0.005 M H_2SO_4) for ≥30 min prior to the first injection.

6. Obtain baseline of HPLC instrument.

7. Inject and run an HPLC-grade water sample as a blank. For manual injection, use a 10-µl syringe here and throughout procedure.

Generate standard curves

8. Dilute 100 ppm organic acid standards in HPLC-grade water to prepare a series of standards ranging from 100 ppm to 10 ppm in 10 ppm increments. Use acids that are found in the sample.

   For each sample type, a literature review must be done prior to analysis to determine the appropriate organic acids and their range of concentrations for the standards. Typical acids include acetic, butyric, citric, formic, hippuric, isobutyric, isovaleric, lactic, malic, oxalic, phenylacetic, propionic, pyruvic, tartaric, uric, and valeric acids.

9. Inject each standard dilution in duplicate. Record peak shapes and integrate the results.

10. Generate a standard curve for each acid by plotting area versus concentration of organic acid for each solution. To determine the concentration of organic acid present in the sample, calculate y for the line $y = mx + b$, where y is the peak area (A) measured in abundance units (AU) produced by the organic acid in the stock solution, m is the slope of the standard curve, x is the concentration of organic acid in the solution, and b is the y intercept.

Prepare and run sample

11. Add an internal standard to a sample. Use an acid that is not found in the sample, at a concentration that is within the range of acids that are found in the sample. Set aside an equivalent amount of internal standard in duplicate to run by itself.

    Once again, a literature review must be done to determine which acid to use as an internal standard. The internal standard will be used to calculate the percent recovery, which quantitates organic acid losses during sample preparation.

12. Prepare sample for injection (see Support Protocol).

    The sample may be prepared in advance. The internal standard should be added to the sample (step 11) before the sample is prepared (step 12) to control for losses during sample preparation.

13. Inject and run sample in duplicate. Also run internal standard alone in duplicate. Record peaks and integrate results.

14. Thoroughly flush the HPLC system and column with HPLC-grade water. For storage, flush with 10% to 20% acetonitrile to prevent microbial growth.

Calculate percent recovery

15. Determine the percent recovery of the internal standard by using the equation percent recovery = $A_{IS}/A_I \times 100$, where A_{IS} is the area of the internal standard in the sample and A_I is the area of the internal standard run by itself.

Calculate acid concentrations

16. Determine the concentration of each organic acid in the sample (x in the equation from step 10) by using the sample peak area (A_s) for y.

17. When analyzing drinking juices, correct for °Brix to relate the results to single-strength juice using the equation $C_{\circ Brix} = C_s \times (11.5/°Brix$ of sample), where $C_{\circ Brix}$ is the concentration of an organic acid in the sample at 11.5 °Brix.

> *The value 11.5 °Brix is used in the above equation as an example. The °Brix used should correspond to the juice of the product (Table G2.2.1). One needs to correct for °Brix only when reporting results as standard °Brix.*

PREPARATION OF SAMPLES FOR HPLC OF ORGANIC ACIDS

All of the methods currently used to analyze organic acids in foods require some sample preparation. This is because the measurement of organic acids in a sample is actually a measurement of the analytes present. Therefore, all samples must be pretreated to dissolve the analytes for analysis. Of the various methods used to assay organic acids, HPLC calls for the least sample pretreatment on account of its high sensitivity and selectivity; this is another reason why this method is commonly used in industry. Some researchers purify samples prior to analysis to improve resolution; this is easily done with solid-liquid chromatographic techniques such as C18 Sep-Pak cartridges (*UNIT F1.1*), which will separate nonvolatile acids from anthocyanins and polyphenolics. Anion-exchange resins can be used to separate sugars from acids (*UNIT E1.2*). If the resolution of the organic acid peaks in the sample is poor, removal of the sugars, polyphenolics, and anthocyanins will improve resolution.

Materials

    Liquid, semi-solid, or solid food sample
    HPLC-grade water
    70% to 80% (v/v) ethanol or 70% to 80% (v/v) acetonitrile (for semi-solid and
        solid foods only)

    Refractometer
    Rotary evaporator (for liquid samples <11 °Brix only)
    20-ml screw-top vials
    0.45-µm syringe filters (for liquid samples)
    4-ml autosampler vials
    Blender (for semi-solid and solid foods only)
    Whatman no. 4 filter paper (for semi-solid and solid foods only)
    100-ml flasks

For liquid food samples:

1a. Place a few drops of a sample on the lower prism of a refractometer and measure and record °Brix. If the sample is in the range of 11 to 12.5 °Brix, continue with step 5a. If the sample is >12.5 °Brix (i.e., it is too concentrated), continue with step 2a. If the sample is <11 °Brix, concentrate the sample using a rotary evaporator and remeasure °Brix.

> *If the sample is a carbonated beverage it must first be degassed for 5 min in an ultrasonic bath prior to step 1a.*

2a. Weigh ~2 g sample into a 20-ml screw-top vial.

3a. Add 10 ml HPLC-grade water, cap vial, and mix well.

4a. Measure and record °Brix. If diluted sample is still too concentrated (>12.5 °Brix), add additional water, mix well, and remeasure °Brix. If diluted sample is too dilute (<11 °Brix), add one to two drops original sample, mix well, and remeasure °Brix.

Properly diluted samples should be in the range of 11 to 12.5 °Brix.

5a. Filter sample through a 0.45-μm syringe filter into duplicate 4-ml autosampler vials and cap vials. Fill vial and discard any remaining sample.

Samples may be stored up to 24 hr at 4°C.

For semi-solid and solid food samples:

1b. Macerate ~25 g sample for 3 to 5 min with equal parts HPLC-grade water and 70% to 80% ethanol or 70% to 80% acetonitrile in a blender at 100 rpm.

The choice of ethanol or acetonitrile depends upon the type of sample; however, most samples can be analyzed using either solvent. The main difference would be that ethanol will extract the more highly polar acids better than acetonitrile.

Alternate methods of preparing aqueous extracts, especially the use of a C18 Sep-Pak cartridge, are described in UNIT F1.1.

2b. Stir slurry for 2 hr.

3b. Measure and adjust °Brix as necessary (steps 1a to 4a).

4b. Filter sample through Whatman no. 4 filter paper, washing filtrate with ~5 ml HPLC-grade water, and collect in a 100-ml flask. Fill duplicate 4-ml autosampler vials with sample, cap vials, and discard any remaining sample.

REAGENTS AND SOLUTIONS

Use HPLC-grade water in all recipes and protocol steps unless otherwise noted. For common stock solutions, see APPENDIX 2A; for suppliers, see SUPPLIERS APPENDIX.

H_2SO_4, 0.005 M

Weigh out 0.4904 g H_2SO_4 and dissolve in 1 liter water, swirling to mix. Filter through a 0.45-μm filter. Degas solution by stirring under vacuum at 635 mmHg for ≥3 hr. Store up to 1 month at room temperature (~23°C).

KH_2PO_4, 0.05 M, pH 2.40

Weigh out 6.81 g monobasic potassium phosphate and quantitatively transfer to a 1-liter beaker with water. Add additional water to ~900 ml and stir. Adjust pH to 2.40 with concentrated phosphoric acid. Quantitatively transfer solution to a 1-liter volumetric flask and bring to volume with water. Filter through a 0.45-μm filter. Store up to 1 month at room temperature (~23°C).

Organic acid standard solutions

Stock solutions: Prepare stock solutions A to E using the amounts indicated below. Measure each component to the nearest 0.1 mg and add to a 100-ml volumetric flask. Bring to volume with HPLC-grade water and mix well. Store up to 1 month at 4°C.

Stock solution A: 200 mg tartaric acid, 150 mg each acetic, isocitric, and lactic acid.

Stock solution B: 1.000 g L(−)-malic acid.

continued

Acid Tastants

Table G2.2.2 Acid Concentrations (in g/liter) in Organic Acid Standards for Selected Juices

Acid	Apple	Black currant	Grape	Grapefruit	Lemon	Orange	Pineapple
Acetic	0.15	0.15	0.15	0.15	0.15	0.15	0.15
Citric	0.25	8.0	0.50	8.0	8.0	5.0	3.0
Galacturonic	0.20[a]						
Isocitric	0.15	0.15	0.15	0.15	0.15	0.15	0.15
Lactic	0.15	0.15	0.15	0.15	0.15	0.15	0.15
Malic	2.0	0.50	2.0	0.20	0.50	0.80	0.80
Quinic	0.20[a]						
Succinic	0.20[a]						
Tartaric	0.20	0.20	1.20	0.20	0.20	0.20	0.20

[a]If organic acid stock solution E is included.

Stock solution C: 1.000 g citric acid.

Stock solution D: 200 mg tartaric acid.

Stock solution E: 200 mg each galacturonic, quinic, and succinic acid.

Working standard solutions: To prepare standards, combine stock solutions A to E as appropriate for the sample. Add stock solutions to a 10-ml volumetric flask, bring to volume with HPLC-grade water, and mix well. Store up to 2 weeks at 4°C. For selected juice samples, use the amounts indicated below. For final concentrations of acids in these solutions, see Table G2.2.2.

Apple juice: 1 ml stock solution A, 2 ml stock solution B, 0.25 ml stock solution C. Addition of 1 ml stock solution E is optional.

Black currant and lemon juices: 1 ml stock solution A, 0.5 ml stock solution B, 8 ml stock solution C.

Grape juice: 1 ml stock solution A, 2 ml stock solution B, 0.5 ml stock solution C, 5 ml stock solution D.

Grapefruit juice: 1 ml stock solution A, 0.2 ml stock solution B, 8 ml stock solution C.

Orange juice: 1 ml stock solution A, 0.8 ml stock solution B, 5 ml stock solution C.

Pineapple juice: 1 ml stock solution A, 0.8 ml stock solution B, 3 ml stock solution C.

NOTE: *The standard solutions listed above are for the most common acids found in several common juices. These are not all the acids found in each juice. For example, isocitric and citric acid are commonly found in raspberry juice, and the ratio of the two is an important factor for determining adulteration of the juice.*

Organic acid standards, 100 ppm

Weigh 0.100 g each desired organic acid standard to the nearest 0.1 mg in a 10-ml volumetric flask. Bring to volume with water and mix well. Store in amber vials up to 1 month at 4°C.

Background Information

As stated earlier, organic acids play an important role in maintaining the quality and nutritional value of foods. This is because organic acids are in a constant state of flux during postharvest and tend to diminish during senescence via the Krebs cycle and the shikimic acid pathway (Haard, 1985). Organic acids are present in foods as the result of biochemical processes or from the activity of some microorganisms (mainly yeast and bacteria). They are also present due to their addition as acidulants, stabilizers, or preservatives. Organic acids possess a wide range of sensory properties. They can either play a direct role in the taste of the food, for example the sour taste of cider due to lactic acid, or have an indirect effect, as displayed by malic and acetic acid with respect to the negative correlation to sweet taste and scented flavors (Gomis and Alonso, 1996). Quinic acid is the most important of the organic acids in cranberry juice and is often used as a marker to determine cranberry juice content in cranberry beverages. Tartaric acid is regarded as a qualitative marker for the presence of grape juice, but unlike quinic acid in cranberry juice, its levels fluctuate in grape juice. Other useful qualitative markers include isocitric acid in raspberries and other berries, and the citric acid/isocitric acid ratio in berries.

Many methods have been used to determine organic acids in foods, including volumetric, electrochemical, enzymatic, and chromatographic (paper, thin-layer, gas-liquid, or HPLC) methods. However, most of these are not able to assay organic acids comprehensively; for example, the enzymatic methods are specific kits for individual organic acids (i.e., they only detect one of the acids present). Therefore, to analyze the sample comprehensively using enzymatic methods would be extremely time consuming and costly, as the analysis would have to be run separately for each acid and would require several kits.

Of the methods listed above, HPLC is the best for analyzing all organic acids in a sample, as it is able to quantitatively identify several different acids in one run. HPLC provides a method that is fast, sensitive, and reliable, which is an advantage over spectrophotometric and volumetric methods. HPLC methods are also very reproducible; the retention times of the compounds are fairly constant. However, there will be differences in retention times between columns and there will invariably be some retention time shifting on the same column over time. What is important to note is that the two different methods described in Basic Protocols 1 and 2 will give different retention times for the standards. This is because they are using different modes of separation, and as a result the retention order of the standards tends to be reversed. Of all the methods mentioned, HPLC is the most commonly used method for the analysis of organic acids.

Critical Parameters

It is imperative that the HPLC instrument, including the detector, is working correctly. The easiest way to check this is by first running a blank. If there is no response, one can move onto injecting the standards. If there is a response to the blank, the column may have been overloaded prior to this run. Refer to a troubleshooting guide for the specific HPLC system. The internet is also an invaluable source for troubleshooting (e.g., see Internet Resources). Keep in mind that the source of the problem may not be the system but may in fact be the column.

It is important to perform a comprehensive literature review prior to analysis. This is not only for the purpose of preparing standards and the addition of internal standards but also because there can be interactions between sugars and acids. For example, it has been documented that lactose can interfere with the measurement of lactic acid, which will overestimate the concentration of lactic acid in the sample. Thus, it is good to have prior knowledge of the concentrations of the acids typically found in the sample. If the resolution of the organic acid peaks is poor, it will be necessary to separate the nonvolatile acids from sugars (*UNIT E1.2*), anthocyanins, and polyphenolics.

Anticipated Results

The typical concentrations of organic acids found in samples vary greatly depending on the sample being analyzed. For example, tartaric acid is usually present in wine at 2000 to 7000 ppm, but only at 1000 to 2000 ppm in grapes, and at 75 to 200 ppm in pineapple juice. It is important to do a comprehensive literature review prior to analysis to verify that measured results are within the range of those found in the literature. For further information on the concentrations of organic acids commonly found in fruit juices, refer to Nagy et al. (1993) and Nagy and Wade (1995). Table G2.2.3 lists

Acid Tastants

G2.2.9

Table G2.2.3   Concentrations of Organic Acids Commonly Found in Various Fruit Juices[a]

Juice	Citric acid	Isocitric acid	Malic acid	Quinic acid	Tartaric acid
Apple	0.005-0.04 g/100 g	—	0.2-1.3 g/100 g	—	—
Apricot	0.4 g/100 g	70-200 mg/kg	3.5-19 g/kg	62-76 mg/100 g	—
Cranberry	1.0 g/100 g	—	0.7 g/100 g	1 g/100 g	—
Grape	0.02-0.05 g/100 g	—	0.3-0.7 g/100 g	—	0.5-1.1 g/100 g
Grapefruit	0.9-2.0 g/100 g	—	0.02-0.1 g/100 g	—	—
Lemon	4.2 g/100 g	200 mg/liter	0.26 g/100 g	—	—
Orange	0.4-1.5 g/100 g	>44 ppm	0.1-0.3 g/100 g	—	—
Passionfruit (purple)	13.1 meq/100 g	—	13.1 meq/100 g	—	—
Passionfruit (yellow)	10.55 meq/100 g	—	10.5 meq/100 g	—	—
Pineapple	0.7-0.9 g/100 g	162 mg/liter	0.1-0.2 g/100 g	1 g/100 g	—
Raspberry	1.6 g/100 g	144 mg/liter	0.4 g/100 g	15 mg/100 g	—
Sour cherry	30 mg/100 ml	1 mg/100 ml	0.8 g/100 g	—	—
Strawberry	0.67-0.94 g/100 g	49.3 mg/kg	0.1 g/100 g	10-80 mg/100 g	—

[a]From Nagy et al. (1993) and Nagy and Wade (1995). A dash (—) indicates data not included in these reports.

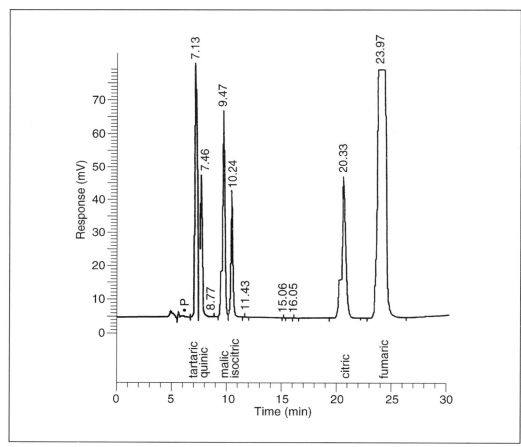

Figure G2.2.1   A typical HPLC chromatogram of organic acid standards using Basic Protocol 1. Reprinted from Coppola et al. (1995) with permission from AgScience, Inc.

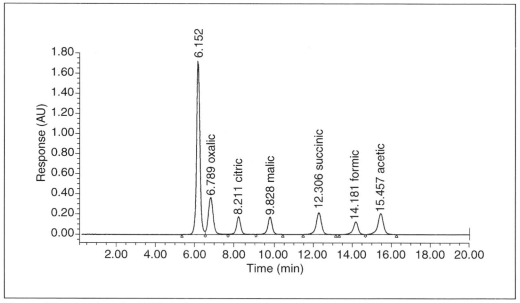

Figure G2.2.2   A typical HPLC chromatogram of organic acid standards using Basic Protocol 2.

the concentrations of organic acids commonly found in several different types of fruit juices. Figure G2.2.1 shows the separation of a standard test mix of organic acids using Basic Protocol 1. Figure G2.2.2 shows the separation of a standard test mix of organic acids using Basic Protocol 2.

Time Considerations

It will take ~2 hr to make and degas the mobile phase and prepare the standards. This must be done prior to analysis. If the sample is a semisolid or solid food, the sample pretreatment will take ~3 hr. If the sample is a liquid or beverage, the sample pretreatment will take ~1 hr. During this time it is advised to warm up the UV detector and run a blank to make sure that the HPLC unit is ready for injection of the standards.

The HPLC analysis for Basic Protocol 1 requires ~50 min per run whereas Basic Protocol 2 requires ~15 to 20 min per run. Once the standard solutions have been run for Basic Protocol 1 (~1 hr) or the standard curve has been made for Basic Protocol 2 (~2 hr), the sample can be analyzed. Sample analysis should be made in duplicate. After the HPLC analysis, the integration of the results and concentration calculations require ~15 to 30 min per organic acid.

Literature Cited

Coppola, E. and Star, M.S. 1986. Liquid chromatographic determination of major organic acids in apple juice and cranberry juice cocktail: Collaborative study. *J. Assoc. Off. Anal. Chem.* 69:594-597.

Coppola, E., English, N., Provost, J., Smith, A., and Speroni, J. 1995. Authenticity of cranberry products including non-domestic varieties. *In* Methods to Detect Adulteration of Fruit Juice Beverages, Vol. I (S. Nagy and R.L. Wade, eds.) pp. 287-309. AgScience, Auburndale, Fla.

Doyon, G., Gaudreau, G., St.-Gelais, D., Beauliea, Y., and Randall, C.J. 1991. Simultaneous HPLC determination of organic acids, sugars and alcohols. *Can. Inst. Food Sci. Technol. J.* 24:87-94.

Gomis, D.B. and Alonso, J.J.M. 1996. Analysis for organic acids. *In* Handbook of Food Analysis, Vol. 1, Physical Characterization and Nutrient Analysis (L.M.L. Nollet, ed.) pp. 715-743. Marcel Dekker, New York.

Haard, N.F. 1985. Characteristics of edible plant tissues. *In* Food Chemistry (O.R. Fennema, ed.) pp. 857-911. Marcel Dekker, New York.

Nagy, S. and Wade, R.L. (eds.) 1995. Methods to Detect Adulteration of Fruit Juice Beverages, Vol. I. AgScience, Auburndale, Fla.

Nagy, S., Chen, C.S., and Shaw, P.E. 1993. Fruit Juice Processing Technology. AgScience, Auburndale, Fla.

Key References

Bio-Rad Life Sciences Group. Bulletin 1928. US/EG REVA. Guide to Aminex HPLC Columns for Food and Beverage, Biotechnology, and Bio-Organic Analysis. Bio-Rad Life Sciences, Hercules, Calif.

Provides a good description of the Aminex HPX-87H column (see Basic Protocol 2) and its applications.

Coppola et al., 1995. See above.

Discusses HPLC methodology and gives sample calculations.

Internet Resources

http://www.waters.com

Includes an online troubleshooting guide.

Contributed by Jane E. Friedrich
Cargill, Incorporated
Minneapolis, Minnesota

H TEXTURE/RHEOLOGY

INTRODUCTION

H1 Viscosity of Liquids, Solutions, and Fine Suspensions

H1.1 Overview of Viscosity and Its Characterization

H1.2 Measuring the Viscosity of Non-Newtonian Fluids

H1.3 Viscosity Determination of Pure Liquids, Solutions and Serums Using Capillary Viscometry

H1.4 Measuring Consistency of Juices and Pastes

H2 Compressive Measurements of Solids and Semi-Solids

H2.1 General Compressive Measurements

H2.2 Textural Measurements with Special Fixtures

H2.3 Texture Profile Analysis

H3 Viscoelasticity of Suspensions and Gels

H3.1 Dynamic or Oscillatory Testing of Complex Fluids

H3.2 Measurement of Gel Rheogy: Dynamic Tests

H3.3 Creep and Stress Relaxation: Step-Change Experiments

SECTION H
Texture/Rheology

INTRODUCTION

Rheology is the science of deformation and flow of matter. In food rheology, the matter of interest is food, and the importance of its deformation and flow relate to several important properties. Of these, texture is the most important. Texture is one of four quality factors of foods; the others are flavor, appearance, and nutrition. In the food industry, there are other properties and processes in which rheology has an important role. They include formulation, manufacturing, transportation, and shelf stability. The measurement of the rheological properties of foods provides the food scientist and engineer with critical information necessary for the successful development and delivery of formulated foods to the consumer.

In the definition of rheology there are two processes, deformation and flow. Deformation suggests the presence of solid-like behavior and flow suggests the presence of fluid-like behavior. Many foods have both solid and fluid properties. The objective of this section is to provide methods for the evaluation of the rheological properties of foods. In Chapter H1, flow properties of foods are the focus. In Chapter H2, deformation properties are the focus.

Chapter H1 relates to measurement of flow properties of foods that are primarily fluid in nature. *UNIT H1.1* surveys the nature of viscosity and its relationship to foods. An overview of the various flow behaviors found in different fluid foods is presented. The concept of non-Newtonian foods is developed, along with methods for measurement of the "complete flow curve." The quantitative or fundamental measurement of "apparent shear viscosity" of fluid foods with rotational viscometers or rheometers is described. *UNIT H1.2* describes two protocols for the measurement of non-Newtonian fluids. The first is for time-independent fluids, and the second is for time-dependent fluids. Both protocols use rotational rheometers. *UNIT H1.3* describes a protocol for simple Newtonian fluids, which include aqueous solutions or oils. As rotational rheometers are new and expensive, many evaluations of fluid foods have been made with empirical methods. Such methods yield data that are not fundamental but are useful in comparing variations in consistency or texture of a food product. *UNIT H1.4* describes a popular empirical method, the Bostwick Consistometer, which has been used to measure the consistency of tomato paste. It is a well-known method in the food industry and has also been used to evaluate other fruit pastes and juices as well.

Chapter H2 describes the measurement of textural properties of solid-like foods. The first unit in that chapter, *UNIT H2.1*, describes a general procedure commonly used to evaluate the texture of solid foods. This method involves the compression of the food material between two parallel plates. There are a number of empirical textural parameters which can be evaluated with this technique. Simple compressive measurements do not provide a complete textural picture of some foods; *UNIT H2.2* presents variations to the parallel plate compression method with the use of special fixtures. For example the use of a puncture probe or a wire cutting device provide data that may relate more directly to the consumer's evaluation of texture for products like apples and cheese. *UNIT H2.3* describes a general protocol for the evaluation of a number of sensory texture parameters. This protocol is called texture profile analysis, and was originally developed in the 1960s. Its application to the evaluation of several types of foods is presented.

Chapter H3 describes the measurement of viscoelasticity of foods. *UNIT H3.1* and *UNIT H3.2* describe the use of dynamic or oscillatory application of stress and strain. *UNIT H3.1* is a general protocol designed for a broad class of foods, and *UNIT H3.2* is specifically designed for food gels. *UNIT H3.3* describes the use of the creep technique for viscoelastic measurements of a broad class of foods. Oscillatory testing has become the more widely used method, but creep measurements are important for certain properties, as will be discussed by the author.

Charles F. Shoemaker

This Section H is dedicated to Professor Marvin Tung who passed away during the development of this manual. Marvin was a main contributor to Chapter H2. I have always considered Marvin as one of the principal researchers and teachers of modern food rheology. His contributions to food rheology research are evident in the literature. And to those of us who had the pleasure to know him, his special style of teaching and mentoring will be greatly missed.

Viscosity of Liquids, Solutions, and Fine Suspensions

H1.1 Overview of Viscosity and Its Characterization   **H1.1.1**

What Is Viscosity?   H1.1.1

Apparent Viscosity and Shear Thinning Versus Shear Thickening   H1.1.1

Yield Stress in Materials   H1.1.2

Time-Dependent Behavior   H1.1.2

Simple and Complex Shear   H1.1.2

The Range of Viscosity and Shear Rate   H1.1.3

Rheometers: Controlled Rate and Controlled Stress   H1.1.4

Viscometer Data: The Rheogram   H1.1.4

Rheometer Data: The Complete Flow Curve   H1.1.6

H1.2 Measuring the Viscosity of Non-Newtonian Fluids   **H1.2.1**

Strategic Planning   H1.2.1

Basic Protocol 1: Measuring Viscosity with Nonequilibrium Ramped or
Stepped Flow Tests   H1.2.4

Basic Protocol 2: Measuring Viscosity with Equilibrium Flow Tests   H1.2.6

Commentary   H1.2.8

**H1.3 Viscosity Determination of Pure Liquids, Solutions and Serums Using
Capillary Viscometry**   **H1.3.1**

Basic Protocol: Using Capillary Viscometry to Determine the Viscosity of Pure
Liquids and Solutions   H1.3.2

Alternate Protocol: Using Capillary Viscometry to Determine the Viscosity of
Serums, Fruit Juices, or Pastes   H1.3.3

Commentary   H1.3.4

H1.4 Measuring Consistency of Juices and Pastes   **H1.4.1**

Basic Protocol   H1.4.1

Commentary   H1.4.2

Contents

1

Overview of Viscosity and Its Characterization

WHAT IS VISCOSITY?

At this point it is appropriate to define viscosity and how it is measured. In the text that follows, the shear viscosity is referred to as viscosity. It is important to realize that fluids also have an extensional or elongational viscosity when stretched; however, this property is not dealt with here. The shear mode of deformation is assumed henceforth, such that the motion in the fluid is similar to the action of the blades of a pair of shears.

The word viscosity comes from the Latin word for mistletoe, viscum. Anyone familiar with this plant is aware that it exudes a viscous sticky sap when harvested. Viscosity is defined after Isaac Newton in his *Principia* as the ratio of stress to shear rate and is given the symbol η. Stress (σ) in a fluid is simply force/area, like pressure, and has the units of pascals (Pa; S.I. units) or dynes/cm^2 (c.g.s.). Shear rate or strain rate ($\dot{\gamma}$ or $d\gamma/dt$) is the differential of strain (γ) with respect to time. Strain is simply the change in shape of a volume of fluid as a result of an applied stress and has no units. The shear rate is in fact a velocity gradient, not a flow rate. It has the bizarre units of 1/time (sec^{-1}) and is the velocity at a given point in the fluid divided by the distance of that point from the stationary plane.

If one considers fluid flowing in a pipe, the situation is highly illustrative of the distinction between shear rate and flow rate. The flow rate is the volume of liquid discharged from the pipe over a period of time. The velocity of a Newtonian fluid in a pipe is a parabolic function of position. At the centerline the velocity is a maximum, while at the wall it is a minimum. The shear rate is effectively the slope of the parabolic function line, so it is a minimum at the centerline and a maximum at the wall. Because the shear rate in a pipe or capillary is a function of position, viscometers based around capillary flow are less useful for non-Newtonian materials. For this reason, rotational devices are often used in preference to capillary or tube viscometers.

APPARENT VISCOSITY AND SHEAR THINNING VERSUS SHEAR THICKENING

For a Newtonian fluid, viscosity is a constant at a given temperature and pressure. Viscosity is resistance to flow and is analogous to electrical resistance in a wire. The equation for viscosity correlates with Ohm's Law such that stress equals voltage, shear rate equals current, and viscosity equals resistance. Viscosity results from energy dissipation within a fluid as it is forced to flow. A fluid under laminar flow can be thought of as having many infinitesimally thin layers, called lamellae, all moving relative to one another. The friction between these layers leads to the generation of heat, and it is this phenomenon that generates fluid viscosity. Most fluids are in fact non-Newtonian, so their *apparent* viscosity (a distinction for non-Newtonian fluids) varies as a function of flow rate. This means that the energy dissipation within a fluid is either increasing (i.e., leading to an increase in viscosity) or decreasing (i.e., leading to a decrease in viscosity). The majority of non-Newtonian fluids are in fact shear thinning, in that their viscosities decrease with increasing flow. The structure of most fluids lends itself to this behavior because their components do one of the following:

1. Anisotropic particles align with the flow streamlines to reduce their hydrodynamic cross-section.

2. Aggregates of particles tend to break apart under shear forces, again minimizing hydrodynamic disturbance.

3. Polymer molecules existing as random coils elongate in the streamlines.

4. Emulsion droplets deform to become more streamlined.

5. Particles arrange themselves in formations to reduce energy, much like racing cars line up behind each other to take advantage of the leading car.

Some materials will increase in viscosity as the flow rate increases, and, although this is relatively rare, it is crucial to be aware of this if processing is to be effective. Shear thickening occurs as a result of one or more of the following situations:

1. Adjacent fluid lamellae lose their boundaries and interfere with each other, causing laminar flow to change to transitional or turbulent flow. Energy dissipation increases dramatically under these circumstances.

2. At moderate volume fractions of dispersed phase (ϕ), particles or polymers can collide or entangle, again increasing energy dissipation.

**Viscosity of
Liquids,
Solutions, and
Fine Suspensions**

Contributed by Peter Whittingstall

H1.1.1

3. Very high values of ϕ force the particles into various close-packing arrangements that can themselves be disrupted at high shear rates.

YIELD STRESS IN MATERIALS

Certain materials maintain their shape against gravity or other small stresses without flow and are therefore said to have an *apparent* yield stress. Examples of this are ketchup (the bottle must be shocked to start flow) or toothpaste (the tube must be squeezed and the extruded toothpaste maintains its cylindrical shape on the brush). It is important to remember that the lack of flow is deceptive; the viscosity is not suddenly infinite below a specific yield stress. In fact, the viscosity of such materials under a range of small stresses is simply very high (e.g., >100,000 Pa), and this means that any flow that occurs is so slow as to be undetectable over a short period of time (i.e., minutes). Over sufficiently long time scales *all* matter will flow (e.g., asphalt roads will rut, medieval lead piping will sag), so to a purist the concept of a yield stress is artifactual. To a pragmatist (e.g., a production engineer), however, if a material takes too long to flow it might just as well take forever! For this reason, the measurement of yield stress has preoccupied many industries over the last 20 years or so. It is a task that is fraught with problems because the method used to measure yield stress strongly affects the value obtained. Controlled rate devices such as laboratory viscometers will tend to give a yield stress value in proportion to the lowest shear rate they can set, while controlled stress rheometers will give much lower values because they do not force the sample to move. To this end, protocols for the measurement of yield stress are included in this chapter.

TIME-DEPENDENT BEHAVIOR

If the response of a sample to a change in shear is reversible and essentially instantaneous within the time frame of the measurement, it is said to be time independent. Most shear-thinning or shear-thickening materials are time independent. Alternatively, the sample can take seconds, minutes, hours, or longer to reach steady state; such materials are said to be time dependent. This delayed response to a change in applied shear can have a significant impact on processing considerations.

The response of time-dependent samples to reversed or looped flow measurement reveals a hysteresis. If the two steps of the flow test overlap, then the sample is able to accommodate changes of shear stress or shear rate within the time frame of the experiment. This will be the case for time-independent materials unless the ramp is so quick that inertial artifacts of the drive system are seen. In time-dependent materials, the kinetics of viscosity recovery as the shear field decreases are different from the kinetics of viscosity breakdown as the shear field is increased. Thus, the two lines no longer overlap and a thixotropic loop is seen.

The most common form of time-dependent behavior is thixotropy ("changed by touch" is the literal translation). This term used to be applied to shear-thinning properties in the paint and coatings industry, and so confusion can exist in the literature. Thixotropic materials recover structure after shearing and thus "set up" when at rest over time. This can cause problems in a manufacturing facility that has down time, because the material resting in the tanks and pipes can be difficult to pump upon startup. A very rare type of time-dependent behavior is antithixotropy or negative thixotropy. In these materials, the viscosity increases during periods of shear, and slowly declines at rest.

Thixotropy is most commonly seen in materials that are applied at high shear rates, but must then recover to a high viscosity to stay in place (e.g., nondrip paints). While such behavior is undoubtedly of value in industry, it presents significant challenges to the person responsible for characterizing the material as it is manufactured. The reason for this difficulty is the inherent sensitivity of thixotropic materials to their shear history. The effects of recent handling and differences in operator technique on the measured viscosity can be substantial. Certain flow techniques are designed specifically to allow the sample to reach steady state at each point in the test and thus have an advantage over quicker but less precise tests.

SIMPLE AND COMPLEX SHEAR

It is clear then, that the measurement of non-Newtonian materials presents special challenges for a viscometer. Many industrial viscometers designed to give a single point determination have a deceptively simple operating principle. Examples include the speed at which a liquid flows out of a container through a known orifice, a bubble rises in a column of fluid, or a ball falls in a column of fluid. These simple devices are actually very complex in terms of the shear field that is generated. The shear field is the variation of shear stress or shear rate as a function of position within the measuring system. In measuring non-Newtonian materials, such devices will derive an average apparent viscosity giving relative data only.

To successfully measure non-Newtonian fluids, a known shear field (preferably constant) must be generated in the instrument. Generally, this situation is known as steady simple shear. This precludes the use of most single-point viscometers and leaves only rotational and capillary devices. Of these, rotational devices are most commonly used. To meet the criterion of steady simple shear, cone and plate, parallel plates, or concentric cylinders are used (Figure H1.1.1).

THE RANGE OF VISCOSITY AND SHEAR RATE

Rotational viscometers generally have a fixed range of measurement because of the limitations of their drive systems (motors). Viscometers typically generate flow viscosity data over a linear range of one to three orders of magnitude of shear rate. One specialized form of viscometer is called a rheometer. This device can measure viscoelasticity in a fluid but can also measure viscosity alone. Such devices are more expensive than viscometers, but can measure viscosity over wider ranges of shear rate or shear stress. This means that the processing engineer or formulator has access to more information.

The viscosity of non-Newtonian materials can vary by many orders of magnitude, and it is important to know as much of this range as possible. Differences in food stability can be seen at ultra-low shear rates (<0.01 sec^{-1}), while differences in consumption are seen at moderate shear rates (~ 50 sec^{-1}), and differences in application of the product (e.g., spreading peanut butter) are seen at high shear rates (>100 sec^{-1}).

A rheometer generates more information about a material over much wider ranges of

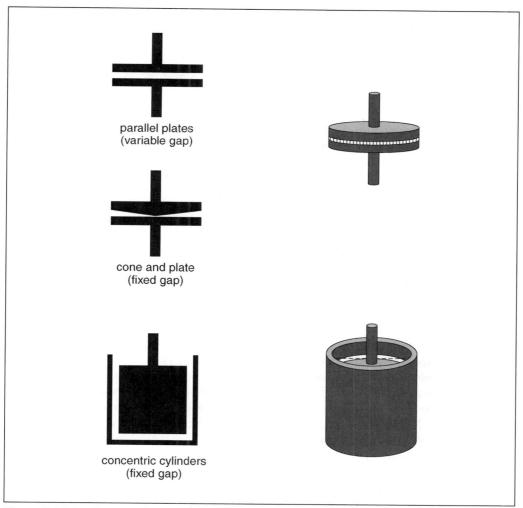

parallel plates
(variable gap)

cone and plate
(fixed gap)

concentric cylinders
(fixed gap)

Figure H1.1.1 Common fixtures used for viscosity measurements with rotational rheometers. Cross-sectional views are shown on the left and external views are shown on the right.

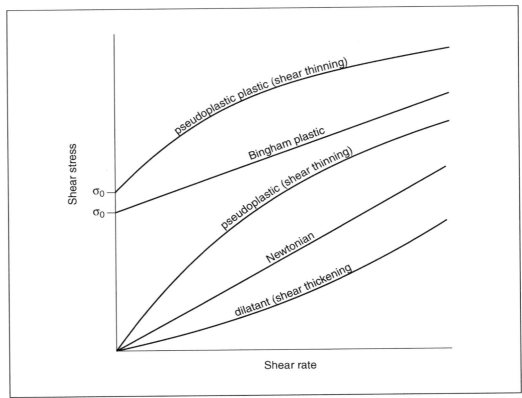

Figure H1.1.2 Rheograms or flow curves for five time-independent fluids. The Newtonian fluid yields a straight line that emanates from the origin. The other four examples are non-Newtonian fluids. σ_0 represents a yield stress point, which is common for plastic fluids.

shear rate than a viscometer. In that sense, a viscometer is to a rheometer as a hand lens is to a microscope. Certain types of rheometers can cover ranges of shear rate of six to nine orders of magnitude.

RHEOMETERS: CONTROLLED RATE AND CONTROLLED STRESS

The drive system of a rotational rheometer is likely to be optimized in one of two ways depending upon its preferred mode of operation. The most common form of rheometer is a controlled-rate (controlled-speed) device. This configuration is also used in most viscometers and has been around for decades. A shear rate is applied to a rotor by the motor controlling the viscometer's speed. The rotor is normally a flat plate or cylindrical cup. The stator is thus a cone or plate for the first two geometries or a cylindrical bob for the third (Figure H1.1.1). The stator is linked to the rotor via the sample, which acts to couple the input signal like an automobile transmission. Thus, the torque on the stator when measured by a transducer is used to derive the shear stress in the sample.

In the second type of rheometer, the controlled-stress device, the motor is optimized to

apply a torque to the rotor by an induction motor and measure its resultant displacement or speed by an optical encoder. The geometry dimensions allow for the stress to be derived from the applied torque, while the gap allows for the shear rate to be derived from the displacement or speed. These novel rheometers are growing very rapidly in popularity from their introduction in the mid-1980s, owing to their ability to probe the behavior of a sample in the region where it yields or begins to flow. By applying a stress the sample can respond naturally without being forced to flow (an inevitable consequence of setting a shear rate).

VISCOMETER DATA: THE RHEOGRAM

Data from viscometers are often presented as a linear plot of shear stress versus shear rate, sometimes called a rheogram (Figure H1.1.2). This type of plot allows the viewer to see directly if there is Newtonian behavior because the plot will take the form of a straight line through the origin. A non-Newtonian response is, by definition, nonlinear and may or may not pass through the origin. If the sample has an apparent yield stress, then the line or curve will

Newtonian	$\sigma = \eta\,\dot{\gamma}$	
Pseudoplastic	$\sigma = k\,\dot{\gamma}^{n}$	$n < 1$
Dilatant	$\sigma = k\,\dot{\gamma}^{n}$	$n > 1$
Bingham	$\sigma = \sigma_0 + \eta_{PL}\dot{\gamma}$	
Casson	$\sigma^{1/2} = \sigma_0^{1/2} + \eta_c^{1/2}\dot{\gamma}^{1/2}$	
Herschel-Bulkley	$\sigma = \sigma_0 + k\,\dot{\gamma}^{n}$	

Figure H1.1.3 Empirical models commonly used for fitting rheogram data. They all rely on the fact that the data were acquired in a linear fashion. See text for definition of variables.

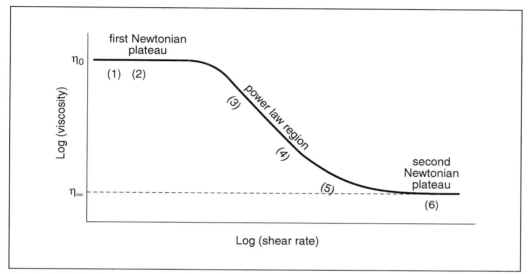

Figure H1.1.4 A complete flow curve for a time-independent non-Newtonian fluid. η_0 and η_∞ are the viscosities associated with the first and second Newtonian plateaus, respectively. Regions (1) and (2) correspond to viscosities relative to low shear rates induced by sedimentation and leveling, respectively. Regions (3) and (4) correspond to viscosities relative to the medium shear rates induced by pouring and pumping, respectively. Regions (5) and (6) correspond to viscosities relative to high shear rates by rubbing and spraying, respectively.

have some positive y-axis intercept. Several empirical models are available for fitting the data, but they all rely on the fact that the data were acquired in a linear fashion. If the range of the data is very wide (more than three orders of magnitude), then they will not fit the data well. The linear models are as follows (Figure H.1.1.3):

1. Newtonian model:
   stress = viscosity × shear rate
2. Power Law model:
   stress = consistency coefficient × (shear rate)^n

where the consistency coefficient is equivalent to apparent viscosity and n is the flow behavior index. This model allows for a change in slope (viscosity); if $n < 1$, the material is shear thin-

ning (i.e., pseudoplastic), and if $n > 1$, the material is shear thickening (i.e., dilatant).

3. Bingham model:
   stress = yield stress + (plastic viscosity × shear rate)

This is a straight-line fit shifted up the y axis to accommodate a yield stress.

4. Casson model:

   (stress)$^{1/2}$ = (yield stress)$^{1/2}$ +

   [(Casson viscosity)$^{1/2}$ × (shear rate)$^{1/2}$]

where Casson viscosity is the viscosity derived from fitting the Casson model. This slight variation of the Bingham model is often used for dispersions and emulsions.

Viscosity of Liquids, Solutions, and Fine Suspensions

H1.1.5

| Carrean-Yasnda Model (full curve) | $\dfrac{\eta - \eta_\infty}{\eta_0 - \eta_\infty} = \dfrac{1}{[1 + (k_1\dot{\gamma})^2]^{m_{1/2}}}$ |
| Cross Model (full curve) | $\dfrac{\eta_0 - \eta}{\eta - \eta_\infty} = (K\dot{\gamma})^m$ |

For portions of the Cross equation that predict portions of the complete flow curve:

Power Law (predicts power law region)	$\eta = k\dot{\gamma}^{n-1}$
Williamson (predicts first Newtonian plateau and power law region)	$\eta = \eta_0 - k\dot{\gamma}^{n-1}$
Sisko (predicts power law region and second Newtonian plateau)	$\eta = \eta_\infty + k\dot{\gamma}^{n-1}$

Figure H1.1.5 Empirical models that are used to predict the complete flow curve of non-Newtonian fluids or portions of the complete curve. In the full-curve models, K is a constant with time as its dimension and m is a dimensionless constant. See text for definition of other variables in equations.

5. Herschel-Bulkley model:

stress = yield stress +

[consistency coefficient × (shear rate)^n]

This is the most versatile of all the above models, allowing for curvature and a *y*-axis shift.

RHEOMETER DATA: THE COMPLETE FLOW CURVE

When a rheometer is used to collect viscosity data, the range covered is typically logarithmic and is therefore much wider than data from a viscometer. The data are generally plotted in a different fashion, as a graph of viscosity versus shear rate on a log-log plot. These curves have a typical shape represented by an initial plateau at low shear rate called the zero-shear viscosity (η_0), a final plateau at high shear rate called the infinite-shear viscosity (η_∞), and a linear portion linking them called the power law region. If all these characteristics are present, then the data can be considered to be a complete flow curve. Often, however, some of the high-shear or low-shear data will be absent. Typically, the controlled-rate rheometer will have problems probing the zero-shear viscosity, while the controlled-stress instrument may not reach the infinite shear viscosity (see section on rheometers; Figure H.1.1.4).

If some or all of this curve is present, the models used to fit the data are more complex and are of two types. The first of these is the Carreau-Yasuda model, in which the viscosity at a given point (η) as well as the zero-shear and infinite-shear viscosities are represented. A Power Law index (m_1) is also present, but is not the same value as *n* in the linear Power Law model. A second type of model is the Cross model, which has essentially the same parameters, but can be broken down into submodels to fit partial data. If the zero-shear region and the power law region are present, then the Williamson model can be used. If the infinite shear plateau and the power law region are present, then the Sisko model can be used. Sometimes the central power law region is all that is available, and so the Power Law model is applied (Figure H.1.1.5).

KEY REFERENCE

Barnes, H.A., Hutton, J.F., and Walters, K. 1989. An introduction to Rheology. Elsevier, New York.

Covers all the basics of experimental rheology and includes a brief section on constitution equations and other theoretical concepts.

Contributed by Peter Whittingstall
ConAgra Grocery Products
Fullerton, California

Measuring the Viscosity of Non-Newtonian Fluids

For simple fluids, also known as Newtonian fluids, it is easy to predict the ease with which they will be poured, pumped, or mixed in either an industrial or end-use situation. This is because the shear viscosity or resistance to flow is a constant at any given temperature and pressure. The fluids that fall into this category are few and far between, because they are of necessity simple in structure. Examples are water, oils, and sugar solutions (e.g., honey; UNIT H1.3), which have no dispersed phases and no molecular interactions. All other fluids are by definition non-Newtonian, so the viscosity is a variable, not a constant. Non-Newtonian fluids are of great interest as they encompass almost all fluids of industrial value. In the food industry, even natural products such as milk or polysaccharide solutions are non-Newtonian.

Two protocols are presented for non-Newtonian fluids. Basic Protocol 1 is for time-independent non-Newtonian fluids and is a ramped type of test that is suitable for time-independent materials. The test is a nonequilibrium linear procedure, referred to as a ramped or stepped flow test. A nonquantitative value for apparent yield stress is generated with this type of protocol, and any model fitting should be done with linear models (e.g., Newtonian, Herschel-Bulkley; UNIT H1.1).

Basic Protocol 2 is for time-dependent non-Newtonian fluids. This type of test is typically only compatible with rheometers that have steady-state conditions built into the control software. This test is known as an equilibrium flow test and may be performed as a function of shear rate or shear stress. If controlled shear stress is used, the zero-shear viscosity may be seen as a clear plateau in the data. If controlled shear rate is used, this zone may not be clearly delineated. Logarithmic plots of viscosity versus shear rate are typically presented, and the Cross or Carreau-Yasuda models are used to fit the data. If a partial flow curve is generated, then subset models such as the Williamson, Sisko, or Power Law models are used (UNIT H1.1).

Basic Protocols 1 and 2 both require a significant amount of common methodology that is described in Strategic Planning.

STRATEGIC PLANNING

This section describes common steps designed to measure the viscosity of non-Newtonian materials using rotational rheometers. The rheometer fixture that holds the sample is referred to as a geometry. The geometries of shear are the cone and plate, parallel plate, or concentric cylinders (Figure H1.1.1). The viscosity may be measured as a function of shear stress or shear rate depending upon the type of rheometer used.

If multiple procedures are performed, a qualitative measure of time dependence can be gained. This is achieved by linking two ramps in an up-and-down sequence and examining the area of hysteresis. A second option is to connect three segments as an up ramp, a peak hold step, and a down ramp sequence; this option has the effect of altering the area of hysteresis because the sample is broken down more effectively before the down step.

Before attempting to use either type of flow protocol, the appropriate geometry must be selected and the sample must be loaded into the device. There are certain guidelines that should be followed when selecting a geometry and when setting conditions for sample loading.

Viscosity of Liquids, Solutions, and Fine Suspensions

Contributed by Peter Whittingstall

Choosing a Geometry Type

If the sample is a simple (unfilled) liquid (i.e., a transparent or translucent liquid without particles or fibers), a simple cone-and-plate geometry should be selected for testing. Examples of such materials are honey, gum solutions (e.g., xanthan, gellan, carboxy methyl cellulose or CMC), filtered fruit juices, oils, corn syrups, weak gels, and other simple protein or polysaccharide solutions (e.g., albumen).

A filled material, such as ketchup, mustard, unfiltered juices, pastes, emulsions, or dispersions, has particles or droplets dispersed throughout its bulk. The size of such entities needs to be known if a cone-and-plate geometry is to be used, because the gap set in such a geometry is fixed, ranging from ~10 μm for a low-angle cone to ~120 μm for a high-angle cone. The rule in such circumstances is that the gap used must be approximately ten times the size of the particles or droplets.

Aggregates or flocculates must be accounted for when considering the size of such entities. If the above conditions cannot be satisfied, then a parallel-plate or concentric-cylinder system must be selected. Parallel-plate geometries may be set to a user-defined gap ranging from 1000 μm to values determined by the ability of the sample to support its own mass without slumping (normally 2000 to 3000 μm). For concentric-cylinder systems, the gap is defined by the radius ratio of the rotor and the stator.

Solid samples such as chocolate, cheese, or dough naturally lend themselves to testing on a parallel plate because they can be molded or pressed into a sheet. The sheet in turn may be conveniently punched into discs that are the same size as the plates.

Stiff gels that are unfilled are best categorized as solids and should be cut into discs from a cylindrical sample taken by means of a cork borer.

Coarse slurries, such as concentrates of fruit or vegetable pulp and meat or fish pastes with a high water content, are often best tested in a concentric-cylinder system. This is especially true if there is a question about settling or stability. Slip films could be formed at the rotor face if parallel plates are used under these circumstances. (Samples that exhibit synersis such as yogurt or high-fat-content materials are also cause for concern.)

Choosing a Geometry Size

The choice of geometry size depends upon a coarse viscosity grading of the sample. High-viscosity materials need a small-surface-area geometry, whereas low-viscosity materials require a large-surface-area geometry. Thus, if multiple sizes of geometry are available, the size should be selected based on the ease with which a sample moves in its container. Honey is normally a medium- to high-viscosity material and so is suitable for a medium (~4-cm diameter) or small (~2-cm diameter) cone. A large geometry (~6-cm diameter) should be used to measure apple juice, with its low viscosity. It is important to bear in mind that many materials appear to be immobile but will flow quite readily when disturbed. Materials with this property include tomato ketchup or paste. This type of behavior is associated with an apparent yield stress.

If the sample viscosity is very low (i.e., close to the appearance of water) and a large cone-and-plate or parallel-plate geometry is not available, an alternative is to use a concentric-cylinder arrangement. The size of this system in terms of its surface area should also be at a maximum. Therefore, the use of large-diameter cup-and-bob systems or the so-called double concentric-cylinder system is advisable. A hybrid system called the Mooney-Ewart geometry is also a possible candidate as long as the sample is unfilled. This geometry is a concentric cylinder with a cone and plate as the base. The design has similar advantages to the double concentric cylinder in terms of increased surface area.

Several materials are typically used to manufacture measuring geometries for rheometers. Steel resists solvents and functions under a wide temperature range, but has high inertia. Acrylic or transparent plastic can be sensitive to solvents and has very low inertia, but functions within a limited temperature range. Aluminum, either plain or anodized, resists some solvents, functions within a wide temperature range, and has a low inertia. Composites have similar attributes to acrylic, but are not transparent and function within a wider temperature range. Exotic metals, such as titanium, have similar attributes to steel, but have a lower inertia. In general, the low-inertia materials are the most flexible, because they do equally well for low-viscosity materials and weak gels as for rigid samples. Large steel geometries cause problems in measurement due to their density (inertia); aluminum, composites, and titanium are better choices. Acrylic is an ideal geometry material in many respects, but will not tolerate temperatures >40°C. Its transparency also makes acrylic a good choice for transparent samples.

Other Factors in Selecting a Geometry

If a sample contains volatiles such as solvents, it can change in viscosity as they are lost to the atmosphere. To prevent this, geometries often are supplied with accessories to cover the free sample edge and minimize evaporation. Generally referred to as solvent-trap covers, they should be used if there is a likelihood of drying. Another strategy to eliminate loss of volatiles is to paint a thin film of an inert liquid, such as a silicone oil, on the exposed face or meniscus of the sample. As long as the oil has a low viscosity and the sample is not oil soluble, this should not affect the results. This technique is successfully used with dough samples.

To eliminate sample slippage at the interface, certain styles of geometry will have machined grooves or cross-hatching on their shearing surface. These types of geometries are useful but add uncertainty to the absolute value of the gap. This means that they should be used only if slip is anticipated as a problem.

Setting Up a Geometry

Rheometers will have individual means of attaching a geometry to the drive system or torsion transducer. The manufacturer's instructions should be followed for setup, such that the geometry dimensions are loaded or manually typed into the software for calculating stress and strain (rate) from the raw parameters of torque and displacement (speed). Setting a reference point or bias or mapping the air bearing may also be required. Finally, a value for the geometry inertia must be calibrated or entered. These procedures should be followed so that all the required geometry information is in the software.

Next, the zero-gap reference position (the contact point where the stator and the rotor just touch) should be set using the software routines for bringing rotor and stator into contact. After this, the gap used for the test itself can be set. Generally, the zero-gap referencing procedure should be done at the temperature of interest for testing. Some instruments control temperature using an environmental chamber or oven, whereas others will heat or cool the stator. If the latter is the case, then for temperatures that are well away from ambient, the gap can be referenced at ambient and the temperature of interest set with the sample in place to aid thermal conductivity. This is reasonable if the instrument has a gap temperature compensation factor. Most "autogap" instruments will have this facility built into the firmware that controls the rotor position with respect to the stator.

There are several methods for referencing the zero gap. The deceleration of the rotor as it spins is used by some instruments; others use an electrical circuit between the rotor and the stator, which indicates the point of contact, or a normal force transducer, which senses the force of contact. The deceleration technique is the most widely available and should be used as a default method unless otherwise indicated in the instrument manual.

Loading the Sample

After setting the gap relative to the zero-gap reference, the sample is loaded into the gap. The separation of the rotor and stator are set to a large value to allow for easy access. For a cone-and-plate or parallel-plate geometry, an aliquot of sample is transferred by pipet or spatula, depending upon its fluidity, to a position centered around the axis of the rotor on the stator. For concentric-cylinder geometries, the sample should be dispensed into the bottom of the cup.

A pipet should be used for dispensing simple liquids. A syringe should not be used because of the high-shear environment in the contracted flow at its tip. Disposable plastic pipets are well suited for this task, as they may be modified to generate a uniform cross-section by removing the tip with scissors. As a general rule, it is better to overestimate the amount of sample needed than to underestimate. Whereas excess sample is relatively easy to remove, an artifact can be generated by repeating the loading process to add more sample. A significant amount of shear is generated in the sample when the gap is closed to its measurement, and thus repeating this process should be avoided.

Many instruments equipped with automatic gap setting will allow the operator to impose a gradual gap closure profile, using a linear speed decrement or even an exponential speed decrement. The latter is especially useful for weak gels (e.g., yogurt) or structured materials suspected of being thixotropic. Instruments with normal force sensors often allow the gap to be closed until a low normal force value is exceeded. This results in a slow stepwise closure for as long as the sample is capable of relaxing to accommodate the narrowing gap. For samples that are gel-like, this technique is less suitable because the sample is unlikely to relax. While using any of these techniques, the sample should be left undisturbed until the gap reaches its expected value.

Sometimes a two-step gap closure is an effective way to achieve reproducible sample loading. In this technique, a trim gap is set whose value is a small increment of the set gap (e.g., gap plus 10%). When this gap is attained, the sample is trimmed around the edge of the geometry using a flat-bladed spatula. It is important to restrict the movement of the rotor while trimming. After trimming the sample, set the final gap, which is specified by the geometry, such that the sample has a slight bulge around the perimeter. This technique is rarely necessary for a concentric-cylinder geometry, because the annulus of the cup is more readily accessible.

The last step, after the sample is loaded in place, is to cover the sample with a solvent-trap cover or silicone oil film if necessary. Once this is done, the first test can be performed.

BASIC PROTOCOL 1

MEASURING VISCOSITY WITH NONEQUILIBRIUM RAMPED OR STEPPED FLOW TESTS

This test is best for time-independent non-Newtonian fluids and can be used to gain a reconnaissance of the sample's performance, for example, where flow begins (apparent yield stress) and whether the sample appears to be time dependent or independent. This protocol is also suitable for determining whether a sample is Newtonian or non-Newtonian. The measured apparent yield-stress value is test dependent, so changing the ranges of shear covered also changes the apparent onset of flow. The hysteresis seen with time-dependent samples is also qualitative and will change with a different test range.

Measuring the Viscosity of Non-Newtonian Fluids

Use these tests carefully! Often it is best to perform these tests as a first approximation of a sample before performing steady-state tests (see Basic Protocol 2).

Materials

Sample, stored in an appropriate clean container (sealed if solvent loss is an issue)
Controlled-rate or controlled-stress rotational rheometer with
    Computer control and appropriate software for instrument control, data
        acquisition, and model fitting
    A selection of fixtures with various geometries of different sizes (e.g.,
        diameters, cone angles, and gaps)

Set up and run test

1. Load a sample on a controlled-rate or controlled-stress rotational rheometer and equilibrate to the test temperature (see Strategic Planning).

2. Program the software for a start point of 0 sec^{-1} and an end point of 100 to 500 sec^{-1}.

   If the range of the geometry is exceeded, a software warning will normally occur when these values are entered.

   Controlled-rate operation is also available for most controlled-stress rheometers. The stress range measured with this protocol can subsequently be used to set up exact stresses of interest. If controlled-rate operation is not available, a start point of 0 Pa and an end point of 100 Pa should be programmed. These values will not be suitable for all samples (e.g., an end point of <1 Pa should be used for samples like water, and 100 Pa should be used for samples like ketchup).

3. Set the linear- or log-ramp toggle, if present, to linear. Program the software for the desired test temperature.

   This test is usually carried out at or close to the ambient temperature. If a different temperature is desired, the software should be programmed with an equilibration step prior to the shear step.

4. Set up a time for covering this continuous ramp range, generally 2 to 5 min for simple liquids and 5 to 15 min for complex fluids.

   If a stepped ramp is used, the number of points and a point-time interval should be entered. This is calculated to take the same time as above (e.g., 120 to 300 points at one point per second or 240 to 600 points at two points per second). The exact timing of the test is irrelevant unless the experiment is intended to reproduce work done previously, in which case the previous protocol should be adhered to as closely as possible.

5. *Optional:* If a looped test is desired, add a second ramp that is the reverse of the first. Do not change the conditions if a symmetric hysteresis loop is required.

6. Save the software settings and set up the on-screen plotting capability to give appropriate information.

   Typically, a graph of shear stress versus shear rate is used (linear axes). If multiple y axes are an option, viscosity should be added as a second y axis.

7. Run the test, save the data, and transfer the file to whatever analysis software is to be used.

   Generally the rheometer manufacturer's software is adequate, although files may be exported to spreadsheet programs or calculation packages (e.g., Mathematica, Wolfram Research).

8. Clean the rheometer and geometry according to the manufacturer's instructions.

Viscosity of Liquids, Solutions, and Fine Suspensions

H1.2.5

9. Analyze the data, treating each ramp separately. Apply linear models (e.g., Newtonian, Power Law) based on the appearance of the curve(s).

   A straight line suggests the Newtonian model; a straight line with a positive y intercept suggests the Bingham model. A curve suggests the Power Law model and the presence of a positive y intercept suggests the Herschel-Bulkley model (UNIT H1.1).

10. Note down or print out the values obtained for viscosity or *k* (consistency coefficient), yield stress, and the flow behavior index, and make note of any calculated loop area.

    The calculated loop area is in proportion to the amount of time-dependent behavior and is not to be used for direct quantification. Samples with controlled loading conditions and the same test protocols can be compared.

MEASURING VISCOSITY WITH EQUILIBRIUM FLOW TESTS

Equilibrium flow tests are designed to generate as much of the complete flow curve (*UNIT H1.1*) as possible. They are logarithmic stepped ramps with steady-state criteria built into each step. Thus, a step does not conclude until these criteria are met or a certain maximum time is exceeded. These tests are therefore much longer in duration than the ramps in Basic Protocol 1. Equilibrium tests are preferred for time-dependent samples because they run at a rate determined by the sample's own natural response time. At low stresses and shear rates, each point will take minutes to perform. As the sample shear thins, the step times get faster. The response of even a time-dependent sample is converted into a time-independent response.

Materials

Sample, stored in an appropriate clean container (sealed if solvent loss is an issue)
Controlled-rate or controlled-stress rotational rheometer with
   Computer control and appropriate software for instrument control, data
      acquisition, and model fitting
   A selection of fixtures with various geometries of different sizes (e.g.,
      diameters, cone angles, and gaps)

Set up and run test

1. Load a sample on a controlled-rate or controlled-stress viscometer or rotational rheometer and equilibrate to the test temperature (see Strategic Planning).

2. Program the software for a procedure with a start point of the nearest order of magnitude to the lowest shear rate or stress that can be set for that geometry.

   Typically, this value is ~10^{-3} sec^{-1} for a controlled-rate instrument or ~0.01 Pa for a controlled-stress instrument.

3. Set the linear- or log-ramp toggle, if present, to logarithmic for the ramp and program the software for the desired test temperature.

   This test is usually carried out at or close to the ambient temperature. If a different temperature is desired, the software should be programmed with an equilibration step prior to the shear step.

4. Enter an endpoint of 100 to 1000 sec^{-1} for a controlled-rate instrument or 1000 times the start stress for a controlled-stress instrument.

   The exact start and end points of the ramp can be decided from the ramp performed in Basic Protocol 1. A stress of 1/10th to 1/100th of the apparent yield stress to the nearest order of magnitude should be set.

For controlled-rate devices, the lower limit is less critical. The high-shear end of the ramp can be used to determine the upper limit of stress or shear rate.

Some samples will tend to leave the gap at high rotational speeds.

5. Determine the number of steps to be used.

   Five to ten steps should be used for each order of magnitude of stress or shear rate.

6. Set the steady-state criteria for each step and a step maximum time.

   Typically a running average of the data is taken repeatedly during each step. The averaging time should be on the order of 5 to 30 sec, and at least 50 such segments should be included in each step. Thus a single step could last 250 to 1500 sec if steady-state is not achieved. The most recent average is compared with the previous average and the percent difference is calculated. An acceptable tolerance for this percent difference, somewhere between 5% and 10%, should be set. The final part of the acceptance criterion is the number of times the percent difference is within tolerance before steady state is declared for that point. Three consecutive iterations is a good figure.

   For most samples, a step maximum time of 5 to 25 min should be set, so as to interrupt the step if steady state is unlikely to be achieved. All of the acceptance parameters can be considered as a sliding scale. A fast equilibrium flow test can be not much better than a continuous ramp. If the sample is time dependent with slow rebuild kinetics, then the times should be pushed to their longest limits. Sample stability is an issue, so if the sample is likely to dry or gel at the temperature of interest, the analysis should be carried out quickly (i.e., with shorter step maximum times).

7. Save the software settings and set up the on-screen plotting capability to give appropriate information.

8. Run the test, save the data, and transfer the file to whatever analysis software is to be used.

   Generally the rheometer manufacturer's software is adequate, although file may be exported to spreadsheet programs or calculation packages (e.g., Mathematica, Wolfram Research).

9. Clean the rheometer and geometry according to the manufacturer's instructions.

Analyze data

10. Plot the data as viscosity versus shear rate on logarithmic axes.

11. Choose a model to analyze the data, using the one that gives the best (lowest) standard error.

    The models used are typically either the Cross model or the Carreau-Yasuda model (UNIT H1.1), if a complete curve is generated. A complete curve has both plateaus present (zero and infinite shear; see Figure H1.1.4).

12. To assess the yield behavior of a sample, plot the data as viscosity versus stress.

    If a sample shows a very sharp drop in viscosity over a narrow stress range, then it can be considered to have an apparent yield stress.

    Comparison of materials is often more successful on a plot of viscosity versus stress.

13. If the data at low shear are noisy, examine the data table.

    If the first data points do not reach equilibrium but exceed the maximum time allowed, then the test may be too demanding for the instrument as it is set up. The run should be repeated with a more sensitive geometry (larger surface area) or a more sensitive transducer (controlled-strain instruments only), which may improve the data. Prolonging each step may or may not improve data.

Viscosity of Liquids, Solutions, and Fine Suspensions

H1.2.7

Background Information

For non-Newtonian fluids that are time independent, the use of linear ramps and the models mentioned in Basic Protocol 1 can be useful in obtaining a reasonably broad picture of a material's performance, and this method certainly is useful in predicting how well a liquid might pour or pump. These tests become inadequate if the sample is strongly time dependent. The faster the ramp is performed, the greater the error will be. The problem with time-dependent behavior is that the rate of structure breakdown is a function of the change in shear (test controlled), whereas the rate of structure buildup is a function of the kinetics of structure assembly (sample controlled) and the change in shear (test controlled). Thus, the faster the test, the greater the likely discrepancy between the breakdown of structure and its rebuilding.

A logarithmic ramp addresses some of these problems by increasing the data density in the critical zone where flow begins. A typical logarithmic ramp spans from 0.01 to 100 Pa or 0.01 to 100 sec^{-1}, or so as to give round numbers that are orders of magnitude apart. The same time frames for continuous and stepped ramps apply here, but the effect is to magnify the low-shear end of the curve. This can be programmed in the software by switching the linear- or log-ramp toggle (see Basic Protocol 1, step 3). A yield stress measured here will still be in error, but it will probably be seen as a smaller value. Thixotropic loop test hysteresis will also be somewhat reduced but will still be imperfect and qualitative only.

If a logarithmic ramp is performed, then the data should not be fit with linear models (*UNIT H1.1*). These data should be plotted as viscosity versus shear rate on logarithmic axes and the Carreau-Yasuda or Cross models (or subsets) should be used instead. It is unlikely that the zero-shear plateau will be seen in these types of tests. For a complete flow curve, the equilibrium tests described in Basic Protocol 2 should be used.

The linear ramps (continuous or stepped) have the advantage of speed and work best for simple shear-thinning or shear-thickening materials. They should compare with single-point determinations from other equipment if the sample is Newtonian. If the sample is non-Newtonian then there may be discrepancies. This will occur if the single-point device uses a complex shear field (e.g., Zahn or Ford cup).

Data from a rotational rheometer using cone-and-plate or concentric cylinders should agree within limits of ~2% to 5% for Newtonian samples and ~10% for non-Newtonian time-independent samples. Time-dependent samples can be orders of magnitude apart.

The equilibrium flow test in Basic Protocol 2 is very useful for measurement of time-dependent samples. The nonequilibrium tests in Basic Protocol 1 are much quicker, but can give results that are hard to reproduce if the sample is time dependent. The equilibrium flow test is far longer but may only need to be performed once.

The complete flow curve allows for the response of the test sample to be predicted at rest (zero-shear behavior), when being poured out of a container (early Power Law region), when being pumped (middle Power Law region), or when being processed or consumed. Each part of the curve is important in assessing a sample's suitability.

There are additional benefits to the measurement of the complete flow curve with a controlled-stress rheometer. The low stresses give data at extremely low shear rates, allowing for good characterization of the zero-shear viscosity (η_0). This parameter is proportional to the molecular weight of polymers in solution or as melts. As the M_n or M_w (number and weight average molecular weight, respectively) increases, so does the value of η_0. The radius of curvature of the plot of viscosity versus shear stress or shear rate as it leaves the zero-shear plateau and begins to shear thin is related to the molecular weight distribution of the polymer in the sample. Monodisperse samples will give a small radius of curvature, whereas polydisperse samples will give a large radius of curvature. The same principles are true for particle size and size distribution or droplet size and size distribution in dispersions and emulsions.

Critical Parameters and Troubleshooting

The rheometer and the geometry should be in good condition (i.e., clean and free from visible damage). Results can be affected by pitting and dings. When attached to the instrument, the geometry should be concentric to the axis of rotation and should hang vertically so as to rotate without eccentric oscillation. Temperature control devices should be calibrated with a thermocouple and corrected for any offset. Prior to running a sample that is not a standard, the instrument should be calibrated with a traceable standard oil. The measured viscosity of the standard at a given temperature should be within 2% to 5% of the certified value.

The best way to detect problems occurring during the test is to visually inspect the sample rather than to focus on the data appearing on the monitor. One typical problem resulting from time-dependent behavior is irregular changes in rotor speed as the sample breaks down. Switching from a nonequilibrium to an equilibrium test can improve this problem, especially if a controlled-stress device is used.

The same type of behavior can be seen if slip is occurring, but usually it is apparent that only part of the sample is being sheared, rather than the entire bulk. Some workers mark the sample with a notch or a pen to observe the change during shear. To test for slip, a series of experiments using a parallel plate geometry and varying gaps should be run. If the data are not consistent, then slip is occurring. At that point it is necessary to use a serrated geometry or to modify the smooth surface. Slippage can be reduced by gluing fine-grade sandpaper to the surface of the parallel plates.

If the sample is highly filled or elastic it may edge fracture and be expelled at high shear rates. This is usually very easy to spot, and there is little to be done about it in a rotational instrument. Highly elastic samples are measured in a capillary device at high shear rates where edge fracture is not a problem.

Anticipated Results

The common method for graphic display of data generated using Basic Protocol 1 is a graph of shear stress on the y axis versus shear rate on the x axis. Most fluid foods will be shear thinning and their resulting graph will appear similar to a pseudoplastic or pseudoplastic plastic curve as shown in Figure H1.1.2. In some cases two curves are recorded for each sample. The first is with an increasing shear rate or stress (up curve). After the completion of this step the shear rate or stress is then decreased by the same rate to the initial value (down curve). A comparison of the up and down curves will usually match one of the following cases. In the first case, both curves will be superimposed. This suggests that the fluid is time independent. In the second case, the up curve will be above the down curve. This appears as a loop on the shear stress versus shear rate graph. This loop is called a thixotropic loop and the area inside the loop is dependent on the degree of time dependency of the material. This area is also dependent on the rate of increase of the shear rate (up curve) and the rate of decrease of the shear rate (down curve). Because the loop area is not solely dependent on the material, it is not often measured. Its appearance is simply taken as an indication that the fluid has time-dependent flow properties. In this case the fluid is often measured with the method described in Basic Protocol 2.

The common method for graphic display of data generated using Basic Protocol 2 is a log-log plot of viscosity versus shear rate (Figure H1.1.4). However, the covered shear-rate range is often not wide enough to characterize the complete curve as shown in Figure H1.1.4. When using a controlled-stress rheometer, the first Newtonian plateau and power region are measured. When controlled-rate instruments are used, only the power region is usually measured. The second Newtonian plateau region is usually not obtainable with rotational instruments. Instruments such as a high-pressure capillary rheometer are needed for these high–shear rate measurements.

Time Considerations

Analysis of a sample using Basic Protocol 1 will take 15 to 30 min. Using the equilibrium test in Basic Protocol 2 will take 25 to 60 min per sample, with longer times required as the number of data points increases.

Key References

Pettitt, D.J., Wayne, J.E.B., Renner Nantz, J.J., and Shoemaker, C.F. 1995. Rheological properties of solutions and emulsions stabilized with xanthan gum and propylene glycol alginate. *J. Food Sci.* 60:528-531, 550.

This work reports rheological measurements of food polymer suspensions and emulsions. These are typical of a large number of foods and their rheological behaviors.

Contributed by Peter Whittingstall
ConAgra Grocery Products
Fullerton, California

Viscosity Determination of Pure Liquids, Solutions and Serums Using Capillary Viscometry

This unit describes a method for measuring the viscosity (η) of Newtonian fluids. For a Newtonian fluid, viscosity is a constant at a given temperature and pressure, as defined in *UNIT H1.1*; common liquids under ordinary circumstances behave in this way. Examples include pure fluids and solutions. Liquids which have suspended matter of sufficient size and concentration may deviate from Newtonian behavior. Examples of liquids exhibiting non-Newtonian behavior (*UNIT H1.1*) include polymer suspensions, emulsions, and fruit juices. Glass capillary viscometers are useful for the measurement of fluids, with the appropriate choice of capillary dimensions, for Newtonian fluids of viscosity up to 10 Pascals (Newtons m/sec^{-2}) or 100 Poise (dynes cm/sec^{-2}). Traditionally, these viscometers have been used in the oil industry. However, they have been adapted for use in the food industry and are commonly used for molecular weight prediction of food polymers in very dilute solutions (Daubert and Foegeding, 1998). There are three common types of capillary viscometers including Ubelohde, Ostwald, and Cannon-Fenske. These viscometers are often referred to as U-tube viscometers because they resemble the letter U (see Fig. H1.3.1).

Capillary viscometers are ideal for measuring the viscosity of Newtonian fluids. However, they are unsuitable for non-Newtonian fluids since variations in hydrostatic pressure during sample efflux results in variations in shear rate and thus viscosity. This unit contains protocols for measuring the viscosity of pure liquids and solutions (see Basic Protocol) and serums from fruit juices and pastes (see Alternate Protocol).

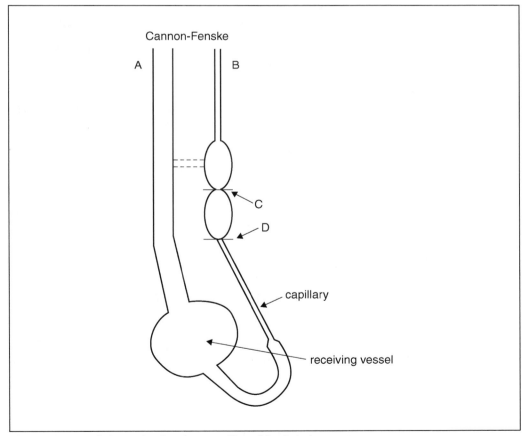

Figure H1.3.1 Schematic of a glass capillary (U-tube) viscometer.

Contributed by Jody Renner-Nantz

USING CAPILLARY VISCOMETRY TO DETERMINE THE VISCOSITY OF PURE LIQUIDS AND SOLUTIONS

This protocol describes a method for measuring the viscosity of pure liquids and solutions by capillary viscometry. The sample is loaded into a Cannon-Fenske viscometer. The time required for the sample to flow between two time points on the viscometer is used to calculate the kinematic viscosity or viscosity.

Materials

Cleaning solvent appropriate for sample
Pure liquid or solution sample

Temperature-controlled 30°C water bath consisting of:
    4.5-gallon glass tank
    Temperature-controlled immersion water circulator
    Apparatus to hold viscometer vertically in water bath
Thermometer
Capillary viscometer tube (Cannon-Fenske)
Solvent wash bottle
10-ml syringe fitted with a syringe filter holder (25-mm diameter)
Nylon membrane filters (25-mm diameter and 0.8-μm pore size)
Tubing with ID that fits over tube B of viscometer (see Figure H1.3.1)
Rubber suction bulb that connects to tubing
Faucet-adapter type aspirator
Stopwatch (±0.2 sec)

Load and equilibrate viscometer

1. Turn on the temperature-controlled water bath and set the temperature to 30°C. Allow the water bath to heat and equilibrate for 30 min before measuring the viscosity of any fluids. Verify the water bath temperature with a thermometer before proceeding.

2. Clean the Cannon-Fenske viscometer by placing the appropriate solvent into a solvent wash bottle. Continuously flush and aspirate the solvent through the viscometer until it is clean.

3. Pass dry filtered air through the viscometer to remove final traces of solvent. Mount viscometer in water bath.

4. To remove lint, dust, or other solid material from the sample, filter sample using a 10-ml syringe fitted with a syringe filter holder and a nylon filter.

5. To charge the sample into the viscometer discharge the sample from the syringe with filter directly into tube A of the viscometer. Discharge enough sample to fill the receiving vessel (see Figure H1.3.1) to about ¾ full.

    The same sample volume should be used for all subsequent measurements.

6. Make sure that the viscometer is in a vertical position in the bath.

7. Equilibrate the sample in the water bath for ~10 min.

Make the measurement

8. With the suction bulb connected to one end of the tubing, connect the other end of tubing to tube B and apply suction to draw the sample through the capillary and to a level above the etched line noted as C.

9. Remove the tubing and measure the efflux time using a stopwatch by allowing the sample to flow freely down past mark C, then measure the time it takes for the meniscus to pass from mark C to mark D in seconds.

Viscosity
Determination of
Pure Liquids,
Solutions and
Serums Using
Capillary
Viscometry

H1.3.2

Table H1.3.1 ASTM[a] Sizes, Viscosity Ranges, and Constants for Cannon-Fenske Type Viscometer Tubes

ASTM size	Viscosity range (cSt)[b]	Approx. constant
25	0.5 to 2	0.002
50	0.8 to 4	0.004
75	1.6 to 8	0.008
100	3 to 15	0.015
150	7 to 35	0.035

[a]American Society for Testing and Materials.

[b]A centistoke is a unit of kinematic viscosity equal to the kinematic viscosity of a fluid having a dynamic viscosity of 1 centipoise and a density of 1 g/cm^3.

Calculate the viscosity

10. Calculate the kinematic viscosity (η') of the sample in centistokes (cSt with units of cm^2/sec) by multiplying the efflux time in seconds by the viscometer constant (Table H1.3.1). Repeat measurements as necessary.

$$\eta' = At$$

where η' is the kinematic viscosity, A is the viscometer constant (cSt/sec), and t is the efflux time in seconds.

If the viscometer constant A is unkown, it may be determined by measuring the efflux time of a fluid with a known kinematic viscosity and calculating the viscometer constant A with the equation above.

11. The viscosity (mPa or cP) can be determined with the kinematic viscosity and density of the fluid, ρ (g/cm^3), with the following equation.

$$\eta = A\eta' = A\rho t$$

USING CAPILLARY VISCOMETRY TO DETERMINE THE VISCOSITY OF SERUMS, FRUIT JUICES, OR PASTES

This method is an adaptation of the Basic Protocol for measuring the viscosity of pure liquids and solutions. The °brix (*UNIT H1.4*) of the sample is adjusted to a desired value by dilution. In many protocols, a nominal value of 5 °brix is the accepted target value for dilution. The sample is then filtered to remove particles that would plug the capillary tube of the viscometer, and the serum viscosity is measured in a Cannon-Fenske viscometer.

Additional Materials (also see Basic Protocol)

Refractometer with automatic temperature compensation feature
Stomacher lab blender with sample bags (Fisher)
Laboratory centrifuge
Side-arm Erlenmeyer flask
Buchner funnel
Whatman no. 1 filter paper

1. Turn on refractometer.

2. Determine the initial °brix of the fruit juice, serum, or paste.

3. Calculate the amount of water needed to adjust the °brix to 5.0 using the following equation where C_I is the °brix of the original sample, C_F is the desired °brix (5.0

ALTERNATE PROTOCOL

Viscosity of Liquids, Solutions and Fine Suspensions

H1.3.3

°brix), W_I is the weight of the original sample, and W_F is the final weight of the diluted sample:

$$W_I = C_F W_F / C_I$$

4. Add water to dilute the sample to 5.0 °brix.

5. Blend the sample in the stomacher for 2 min or until the °brix measurement of the sample is consistently 5.0 ± 0.1 °brix.

6. Centrifuge the sample for 15 min at $10,000 \times g$, 25°C. Decant the supernatant into a beaker, measure the °brix, and record.

7. If there are remaining any large particles, vacuum filter the sample using a side-arm Erlenmeyer flask fitted with a Buchner funnel and Whatman no. 1 filter paper.

8. Filter the sample three times using the syringe filtering apparatus described in the Basic Protocol.

9. Repeat steps 4 through 9 of the Basic Protocol.

COMMENTARY

Background Information

The principle of operation of capillary viscometers is described by the Poiseuille equation where the rate of liquid flow (V/t) through the viscometer can be determined by the following (Steffe, 1996):

$$V/t = (\rho'gl + \Delta\rho)\Pi R^4/8\eta l$$

where ρ' is the density of the fluid, g is the force due to gravity, l is the length of the capillary tube, and $\Delta\rho$ is the pressure difference between the entrance and exit of the capillary tube. The equation can be rearranged and further simplified to account for all the constants that characterize the viscometer. This also assumes that the difference in height of the two liquid columns is relatively constant during the time required for flow. Thus, the only pressure difference across the liquid is due to the weight of the liquid. With these conditions:

$$\eta = A\rho t$$

where A is a constant that incorporates all the parameters that characterize a viscometer.

A simple way to evaluate the viscosity of an unknown (η_2) is to compare its viscosity to that of a known (η_1) such that:

$$\eta_2 = \{\rho_2 t_2/\rho_1 t_2\}\eta_1$$

where ρ_1 and ρ_2 are the densities of the known and unknown, respectively.

The viscosity of a fluid corrected for its density is also known as the absolute or kinematic viscosity.

Critical Parameters and Troubleshooting

There is a range of capillary tubes available for use depending on the viscosity range of the samples to be measured. Typical Cannon-Fenske capillary tube sizes and viscosity ranges are listed in Table H1.3.1. The viscometers are available calibrated or uncalibrated. Uncalibrated viscometers require the user to calibrate them using fluids with know viscosities.

Since viscosity is temperature dependent, the temperature-controlled water bath should be checked daily for accuracy.

Table H1.3.2 Viscosities of Sucrose Solutions Measured by a Capillary Viscometer[a]

Sucrose (%)	Viscosity × 10² (dynes s/cm²)	Standard deviation
5	1.086	0.012
10	1.217	0.011
20	1.542	0.011
30	2.107	0.008

[a]Values from Young and Shoemaker (1991).

The volume of sample used for each measurement should be kept constant.

Anticipated Results

Viscosity will vary depending on the sample. Typical results for viscosity of sucrose solutions are shown in Table H1.3.2.

Time Considerations

Once the constant-temperature water bath has reached 30°C, measuring the viscosities of pure solutions like standard oils requires ~15 min per sample. The most time-consuming part of the measurement is cleaning and drying the viscometer between samples. Preparing fruit pastes and juices for viscosity determination will take ~1 hr for two to four different samples and then an additional ~15 min/sample for the viscosity measurements.

Literature Cited

Daubert, C.R. and Foegeding, A.E. 1998. Rheological principles for food analysis. *In* Food Analysis, 2nd ed. (S.S. Nielson, ed.) pp. 558. Aspen Publishers, Gaithersburg, Md.

Steffe, J. 1996. Rheological Methods in Food Process Engineering, 2nd ed. pp. 125-127. Freeman Press, East Lansing, Mich.

Young, S.L. and Shoemaker, C.F. 1991. Measurement of shear-dependent intrinsic viscosities of carboxymethyl cellulose and xanthan gum suspensions. *J. Appl. Polymer Sci.* 42:2405-2408.

Key Reference

Barnes, H.A., Hutton, J.F., and Walters, K. 1989. Viscometers for measuring shear viscosity. *In* An Introduction to Rheology, Rheology Series, 3, 1st ed. pp. 32-34. Elsevier Science Publishing, New York.

Detailed discussion about using capillary viscometry.

Contributed by Jody Renner-Nantz
University of California
Davis, Cailfornia

Measuring Consistency of Juices and Pastes

UNIT H1.4

BASIC
PROTOCOL

An empirical method for measuring the consistency of various coarse suspensions of juices, preserves, jams, pastes, and other highly viscous products is described in this protocol. In previous units, protocols were given for the measurement of viscosity of fluid foods (*UNITS H1.2 & H1.3*). These are fundamental measurements in that the actual viscosity is measured. Viscosity is a property of the fluid and its value should not depend on the method of measurement. In this unit a popular empirical test is described. The Bostwick consistometer is used to determine sample consistency by measuring the distance in centimeters that a material flows under its own weight during a given time interval, which in this case is 30 sec at 25°C. Other temperatures can be used as long as all measurements are made at the same temperature. Viscosity and consistency are very temperature dependent (Eley, 1995; Weaver, 1996).

Consistency will also be influenced by the natural variation of soluble solids in fruit juices. Because the soluble solids level changes during processing, consistencies are evaluated among juices at a fixed level of soluble solids. The soluble solids content is determined by refractometry, which measures the percent of sugar (sucrose) by weight at the temperature indicated on the instrument. The scale used is the Brix scale, which is equivalent to percent sucrose concentration. A common Brix level used by the tomato industry to evaluate tomato paste consistency is 12 °Brix. Soluble solids (measured in degrees Brix) are often used as an indication of sugar content and maturity level in fruits, vegetables, and their industrial products (Marsh et al., 1990).

Materials

Fruit juice or paste
Brix refractometer
Stomacher laboratory blender with sample bags (e.g., Thomas Scientific)
Bostwick consistometer (Fisher Scientific)

1. Apply a drop of a fruit juice or paste sample to a Brix refractometer, hold the refractometer perpendicular to a light source, and determine the initial Brix (percent soluble solids) value for the sample.

    Adding bubbles to the sample, which will give incorrect readings, should be avoided.

2. Calculate the amount of water needed to adjust the °Brix value to 12 using the following equation:

    $$W_i = (C_f \times W_f)/C_i$$

    where W_i is the initial weight of the sample, W_f is the final weight of the diluted sample, C_f is the final desired Brix value, and C_i is the initial Brix value of the sample.

3. Blend the sample and added water in a stomacher laboratory blender fitted with a sample bag until the °Brix value of the sample is consistently 12 ± 0.1.

    A minimum of 250 g sample should be prepared for each reading.

    Samples with lower Brix values can also be measured as long as their concentrations are consistent from experiment to experiment.

    The stomacher is used to evenly blend the water into the paste without incorporating air into the sample. Other blenders of this type could also be used.

4. Adjust sample temperature to 25°C.

Viscosity of
Liquids,
Solutions, and
Fine Suspensions

Contributed by Montana Camara Hurtado

H1.4.1

5. Close the gate of a Bostwick consistometer and adjust leveling screws until the leveling bubble indicates that the consistometer is level.

6. Fill the consistometer reservoir to the point of overflow (e.g., ~300 g of 12 °Brix paste) and remove excess from top of reservoir using a spatula.

7. Release the gate and let the sample flow for 30 sec. At the end of this period, measure the distance the paste flowed from the gate to the leading edge of the sample in centimeters and record this as the Bostwick value.

8. Rinse consistometer with water and dry.

9. Make at least three replicated readings with fresh samples.

COMMENTARY

Background Information

In the food industry it has often been difficult to obtain true viscosity measurements (*UNIT H1.1*) of complex fluid foods such as coarse fruit suspensions. These are usually non-Newtonian suspensions. Fruit concentrates are dispersions of solid particles (pulp) in aqueous media (serum). Their rheological properties are of interest in practical applications related to processing, storage stability, and sensory properties. Expensive rheometers are often not available in quality control and product development laboratories. However, viscosity is nonetheless an important quality factor of these products.

In order to obtain an estimate of viscosity, many empirical tests have been developed in the food industry. Each test characterizes the viscosity of a product by some type of empirical measurement, which is largely influenced by product viscosity. The value produced by such measurements may be influenced by other rheological properties of the food as well as the instrument used for the measurement. As long as the same instrument is used in all measurements of the food products the measured values may be ranked in order of their viscosity. Generally such empirical measurements of fluid foods are said to yield consistency values. In the food industry, E. P. Bostwick of the U.S. Department of Agriculture, Canned Fruits and Vegetable Service developed a widely used empirical measurement of consistency using the Bostwick consistometer. In theory, as the Brix value of a sample increases, the Bostwick consistency value should decrease. However, if during food processing additives such as corn starches or syrups are included in the food formulation, the correlation between the Brix concentration and Bostwick consistency values could be lost. Here the term consistency may be considered as related to viscosity, although the exact numerical conversion may not be known.

Critical Parameters and Troubleshooting

Because consistency is strongly affected by temperature and concentration of soluble solids, both parameters should be carefully controlled during sample preparation and measurement. Filling the Bostwick reservoir, leveling the product, and releasing the gate must be done in a timely manner to prevent separation of serum and changes in temperature.

Anticipated Results

The consistency of a sample will vary depending on a number of factors including the ripening stage of the initial product and processing conditions. Some typical °Brix values and their standard derivations are: tomato juice, 5.5 ± 0.2; tomato puree, 7.3 ± 0.1; ketchup, 12.0 ± 0.01; and tomato paste, 12.0 ± 0.1. Bostwick values for the same products are: 8.5 ± 0.2, 9.4 ± 0.09, 22.0 ± 0.08, and 6.5 ± 0.02, respectively. Both Brix and Bostwick values can vary among similar products depending on their individual processing conditions and formulations. For both product consideration and consumer appeal, products with a high consistency or viscosity (i.e., with low Bostwick values for the same Brix concentration) are very valuable.

Time Considerations

The largest time factor in this procedure is associated with adjusting the Brix level of the samples and equilibrating their temperatures. The time required for sample preparation and measurement is estimated to be 15 min per sample.

Literature Cited

Eley, R.R. 1995. Rheology and viscometry. *In* ASTM Manual 17: Paint and Coating Testing Manual (J.V. Koleske, ed.) pp. 333-368. American Society for Testing and Materials, Philadelphia.

Marsh, G.L., Buhlert, J.E., and Leonard, S.J., 1980. Effect of composition upon Bostwick consistency of tomato concentrates. *J. Food Sci.* 45:703-706.

Weaver, C. 1996. Equipment guide. *In* The Food Chemistry Laboratory: A Manual for Experimental Foods, Dietetics, and Food Scientists. p. 97. CRC Press, Boca Raton, Fla.

Key References

McCarthy, K.L. and Seymour, J.D. 1994. Gravity current analysis of the Bostwick consistometer for power law foods. *J. Texture Stud.* 25:207-220.

Compared experimental measurements for Newtonian and power law fluids to theoretical predictions and showed that the apparent viscosity predicted by the Bostwick measurement must be correlated with flow behavior during processing and thus could be very useful to incorporate into food process design and control.

Contributed by Montana Camara Hurtado
Universidad Complutense de Madrid
Madrid, Spain

Compressive Measurements of Solids and Semi-Solids

H2.1	**General Compressive Measurements**	**H2.1.1**
	Basic Protocol	H2.1.1
	Commentary	H2.1.5
H2.2	**Textural Measurements with Special Fixtures**	**H2.2.1**
	Basic Protocol 1: Textural Measurements Using a Puncture Probe or a Cone Penetrometer	H2.2.1
	Basic Protocol 2: Textural Measurements Using Warner-Bratzler, Kramer, or Wire Cutting Fixtures	H2.2.5
	Basic Protocol 3: Textural Measurements Using Back Extrusion	H2.2.8
	Commentary	H2.2.9
H2.3	**Texture Profile Analysis**	**H2.3.1**
	Basic Protocol: Measuring the Texture of Potatoes and Carrots	H2.3.1
	Alternate Protocol 1: Measuring the Texture of Apples	H2.3.2
	Alternate Protocol 2: Measuring the Texture of Cooked Ground Beef	H2.3.2
	Alternate Protocol 3: Measuring the Texture of Gelatin Gel	H2.3.3
	Commentary	H2.3.3

Contents

1

General Compressive Measurements

The compressive measurement of a food material is important to understanding issues related to materials handling (e.g., bruising) as well as some textural properties. This unit describes mechanical testing of food specimens and the necessary sample preparation required to obtain specimen geometries generally subjected to compressive tests. In this protocol, a mechanical testing machine equipped with parallel plates is used to achieve uniaxial compression at a desired loading rate. This measurement provides an avenue to determine force and deformation up to the point of rupture (yield), and from these data determine the stiffness of the material under investigation. From a stress/strain curve, further details can be gathered on the state of the specimen at failure, as well as the apparent elastic modulus or Young's modulus of elasticity. Young's modulus is the ratio of the extensional (tensile) stress divided by the corresponding extensional strain of an elastic material, measured in uniaxial extension. It is represented by the symbol E in units of pascals (Pa). All of this information may be derived by physical analysis of the curves or by using applicable computer software.

The mechanical testing machine applies a load to a sample by compressing the food material between two parallel plates. The rigidity of the material dictates which type of load cell should be used in making measurements; however, samples can range from rigid solids to soft biological materials such as gels or apple parenchyma tissue. This protocol will focus on strongly structured food tissue systems and weakly structured gel materials. For a discussion of sample geometry selection, see Commentary.

Aspect ratios of certain types of prepared specimens vary from author to author. In addition, no standardized deformation rates have been established for many of the food specimens subjected to compressive measurements. Thus, comparison of results between different laboratories is difficult if test conditions vary. As a result, valid comparisons of data may only be possible internally.

NOTE: Safety protocols from the manufacturer of the specific mechanical testing machine should be followed closely. For example, limit switches should be adjusted to prevent any overtravel, up or down, by the cross-head.

Materials

Material specimen (prepared or intact)
Razor blades, cork borer, wire cutting device
Mechanical testing machine linked to either a chart recorder or computer (with data recording and analysis software)
Parallel plates, stainless steel or Teflon
Lubricating oil (recommended)

Prepare sample

NOTE: Preparation of specimens requires that dimensions be measured precisely. It is assumed that the dimensions of the compressed specimen are small with respect to the dimensions of the contacting surfaces (plates) to ensure that the sample remains in contact with the plates. Occasionally, it is desirable to compress a sample in its natural state (i.e., no specimen cutting).

1a. *For cubic or rectangular solid samples:* Cut flat surfaces using a razor blade or a parallel-wire cutter. For agricultural specimens (e.g., fruits and vegetables), prepare samples with an aspect ratio (height to diameter) of ~1 and a sample height of one to several centimeters. Be sure that samples have smooth parallel sides.

Contributed by Marvin A. Tung and Michael D.H. Rogers

1b. *For cylindrical solid samples:* Use a cork borer or similar cutting tool of appropriate diameter. If necessary, apply a lubricant (e.g., mineral oil) to the cork borer to facilitate removal of the cut specimen. Cut samples to the desired length using a razor blade or wire cutting device. For agricultural specimens (e.g., fruits and vegetables), prepare a sample with an aspect ratio (height to diameter) of ~1 and a sample height of one to several centimeters. Be sure that the ends of the cylindrical specimen are smooth and parallel.

Readily available cork borers range in diameter from 7.1 to 22.2 mm.

1c. *For gel samples:* Pour gelling solution into a mold with known dimensions and an aspect ratio (height to diameter) of ~1. Allow gel to form by cooling or heating, as appropriate.

2. Allow the specimen to equilibrate to the temperature at which it is most likely stored or used (e.g., refrigerated, ambient, or elevated temperature).

This assists in mimicking the conditions most commonly encountered by the specimen.

Set up testing device

3. Choose a load cell that has the capability to manage the expected maximum load exerted on the specimen.

4. Turn on the mechanical testing machine and attach parallel plates.

5. Turn on the chart recorder or computer.

If using computer software, procedures specific to the software will need to be followed.

6. For test equipment driven by computer software, enter required information—e.g., specimen dimensions, cross-sectional area, desired calculations, and the rate at which the sample will be deformed (cross-head speed or deformation rate).

Table H2.1.1 Nomenclature and Corresponding Units

Symbol	Parameter	S.I. unit
A_0	Initial cross-sectional area	m^2
D	Mean product diameter	m
$d\,(\Delta L)$	Deformation (change in length)	m
E	Young's modulus of elasticity	Pa
F	Force, load	N
$L\,(h)$	Length (height) of compressed specimen	m
$L_0\,(h_0)$	Original length (height) of specimen	m
t	time	s
u_z	Velocity in z direction (deformation rate)	$m\ s^{-1}$
ΔX	Change in width of specimen	m
X_0	Original width of specimen	m
$\dot{\varepsilon}_B$	Biaxial extensional strain rate	s^{-1}
ε_{eng}	Engineering strain	dimensionless
ε_h	Hencky strain	dimensionless
ε_f	Maximum strain at failure	dimensionless
η_B	Biaxial extensional viscosity	Pa·s
σ_B	Biaxial stress	Pa
σ_{eng}	Engineering stress	Pa
σ_h	True (corrected) stress	Pa
σ_f	Maximum stress at failure	Pa
μ	Poisson's ratio	dimensionless

7. Apply lubricating oil to specimen surfaces that contact the plates.

    This will minimize friction between the contact surfaces.

8. Place the specimen on the lower flat plate and move the upper plate to a position near the top surface of the specimen. Measure the distance between the plates to obtain the sample height (*h*).

    This may be required, depending on the computer software.

9. Zero the position of the upper plate and zero the load on the load cell.

10. Select a desired deformation rate and select a point at which to stop the test.

    Generally, testing will cease at any point beyond rupture (maximum stress) or at a chosen % strain.

11. Start the test.

    CAUTION: *The cross-head will begin to move down. Keep hands clear!*

12. Record the force (*F*) versus deformation (*d*) data for the test.

    See Table H2.1.1 for variables and units used in these calculations.

13. At any point during the test, calculate engineering stress (σ_{eng}) and engineering strain (ε_{eng}) from the force/deformation data set or plot:

$$\sigma_{eng} = \frac{F}{A_0} \qquad \varepsilon_{eng} = \frac{d}{L_0}$$

    where *F* is the applied force, A_0 is the initial cross-sectional area of the specimen, and L_0 is the initial length of the specimen (i.e., the sample height, *h*, in step 8).

Analyze data

14. For large deformations (e.g., *d* > 25%; Peleg, 1985), convert engineering stress and strain to true stress (σ_h) and Henky strain (ε_h) as follows:

$$\sigma_h = \sigma_{eng}(1 - \varepsilon_{eng}) \qquad \varepsilon_h = \ln(1 + \varepsilon_{eng})$$

15. Determine sample strength by measuring the stress at failure (σ_f). To do this, determine the force at failure via a force/deformation plot (see Figure H2.1.1) and convert force to either engineering stress or true stress. Also use deformation at failure (*d*) to calculate the strain at failure (ε_f).

$$\sigma_f = \frac{F}{A_0} \qquad \varepsilon_f = \frac{d}{L_0}$$

16. Determine Young's modulus of elasticity (*E*), which is the slope of the linear portion of the stress versus strain curve. Calculate the stiffness of the specimen from the force/deformation curve.

$$E = \frac{\sigma_h}{\varepsilon_h} \qquad \text{stiffness} = \frac{F}{d}$$

To correct deformation for an initial curved region of the force/deformation curve (toe region) due to specimen irregularities at the contact surfaces of a specimen with a Hookean

Compressive Measurement of Solids and Semi-solids

H2.1.3

region (where stress is proportional to strain and a linear relationship between stress and strain can be shown), the curve may be continued through the zero stress (force) axis to provide a corrected zero strain (deformation). For specimens with no Hookean region, the zero strain point is determined by a tangent line through the maximum slope at the inflection point of the toe region.

It should be noted that the Young's modulus for convex bodies is calculated differently. It is sometimes desirable to determine the mechanical properties of a specimen (e.g., an apple, potato, grape) in its natural state. This can provide valuable information for those interested in materials handling. These materials have convex surfaces in contact with flat loading plates; thus, the assumptions made with cube and cylindrical geometries no longer hold. Using the Hertz model (Mohsenin, 1970), Young's modulus of elasticity can be calculated for these convex bodies provided Poisson's ratio is known or estimated. In this case, Young's modulus is calculated as follows:

$$E = \sqrt{\frac{16(1-\mu^2)^2 F^2}{14.285 \times Dd^2}}$$

where μ is Poisson's ratio, F is the applied force, d is the deformation, and D is the mean product diameter.

17. Where no linear region is discernable from a force/deformation or a stress/strain curve, use a secant modulus. Construct a secant line from a desired stress (force) point and extend it to the zero strain (deformation) point. Use the slope of this secant line to give the secant modulus.

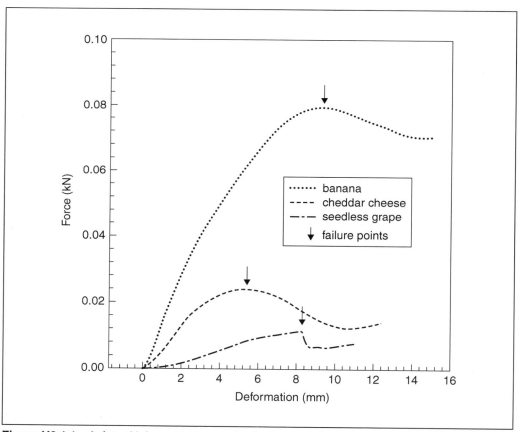

Figure H2.1.1 A force/deformation curve illustrating specimen fracture of a banana (2.5 cm length, 3.0 cm diameter, 5 mm/min deformation rate), cheddar cheese (2.0 × 2.0 × 2.0 cm, 10 mm/min deformation rate), and a seedless grape (2.2 cm length, 1.7 cm diameter, 2 mm/min deformation rate) under uniaxial compression at room temperature.

18. Analyze lubricated squeezing flow to determine biaxial extensional viscosity (η_B), which is calculated from biaxial stress (σ_B) and biaxial extensional strain rate ($\dot{\varepsilon}_B$).

$$\sigma_B = \frac{F}{A} = \frac{Fh}{A_0 h_0} \qquad \dot{\varepsilon}_B = \frac{u_z}{2(h_0 - u_z t)} \qquad \eta_B = \frac{\sigma_B}{\dot{\varepsilon}_B}$$

σ_B and $\dot{\varepsilon}_B$ are determined from force-deformation curves for materials which exhibit squeezing flow behavior (e.g., peanut butter, processed cheese).

COMMENTARY

Background Information

The study of the mechanical properties of a food is important for determining its strength, texture, and deformation characteristics. The geometry, size, and shape of a sample should conform to standards such as those set by the American Society for Testing and Materials (ASTM), or should meet assumptions for use in mechanical tests for formula development (Mohsenin, 1970).

Studies of food rheology require an understanding of stresses, strains, and rates of strain. A normal stress is one that is perpendicular to the surface it is acting on and is either compressive or tensile (Peleg, 1987). A normal stress can be expressed as an engineering stress (magnitude of applied force/initial cross-sectional area of specimen) or a true stress (magnitude of applied force/cross-sectional area of deformed specimen) (Peleg, 1987). Additionally, there are shear stresses, which are parallel to the surface it acts on. Shear stress has the same dimensions as normal stress, but the area is parallel to its direction (Peleg, 1987).

Engineering (apparent) strain is defined as the ratio between the deformation of the specimen and the initial length of the specimen. Deformation is described as the absolute elongation or length decrease in the direction of the applied force (Peleg, 1987). Engineering strain can be considered to be true strain when the deformation is small; however, when the deformation is large, this is no longer true. In this case, the most appropriate definition of true strain would be that defined by Hencky (Peleg, 1987).

The failure characteristics of a food or food material can be measured using compression, tension, or torsion. Of all the available deformation tests, possibly the most common is uniaxial compression (Lelievre et al., 1992). Bulk compression is another type of compression test, but it is seldom used due to the difficulty in applying force by means of hydraulic pressure (Bourne, 1982). The experimental data obtained from the compression of a food material (Figure H2.1.1) is often acquired using instruments such as a mechanical testing machine (Bagley et al., 1988).

A specimen can fail in compression, tension, or shear. A specimen that fractures at 45% to the longitudinal axis has failed in shear (Lelievre et al., 1992). Tensile failure results if the specimen fractures at 90% to the longitudinal axis, and compressive failure is represented by fracture along the longitudinal axis (Lelievre et al., 1992).

Uniaxial compression of a cylindrical shape is a method of measuring the behavior of a Hookean solid. Hooke's law is represented by the following relationship:

$$\sigma_{eng} = E\varepsilon_{eng}$$

where E is Young's modulus of elasticity, σ is stress, and ε is strain. Hooke's law is usually valid for solids under small strains (e.g., <0.25; Steffe, 1996). Under large deformations, Hencky strain (ε_h) should be used. This assumes that the sample maintains a cylindrical shape during compression (Steffe, 1996).

Critical Parameters

There are three principal geometries used for samples subjected to compressive measurements. The two most common are cubic and cylindrical shapes. Cubic samples are commonly used when examining cheese specimens (Ak and Gunasekaran, 1992), while cylindrical samples are common in the compression of polymer gel samples (Tang et al., 1996) and apples (Rebouillat and Peleg, 1988). The third geometry is a convex body such as a sphere. For potatoes and apples, which approximate a spherical shape, it may be possible to compress the intact specimen rather than a cut specimen (Morrow and Mohsenin, 1966). This may be preferred because of the difficulty of correlating compression data of cut specimens with that of whole specimens.

Compressive Measurement of Solids and Semi-solids

H2.1.5

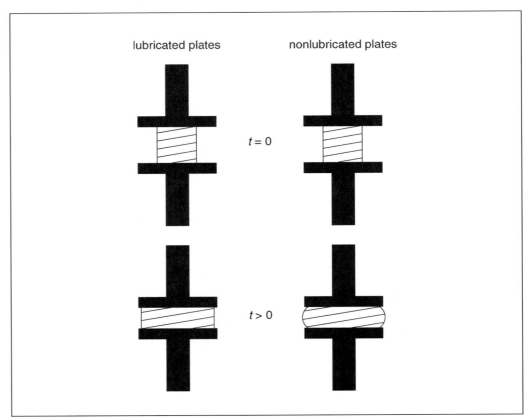

lubricated plates   nonlubricated plates

$t = 0$

$t > 0$

Figure H2.1.2 Specimen compression between parallel plates that are lubricated and nonlubricated.

Certain assumptions are often presented when considering the compression of a specimen. These include homogeneity and isotropy of the sample, the use of contacting plates that are infinitely large, and the use of a specimen whose radius of curvature is large compared to the length of the contacting area (Morrow and Mohsenin, 1966). For normal sizes and shapes of convex agricultural products, these assumptions are generally valid (Morrow and Mohsensin, 1966).

When performing a compression test, lubricated contact surfaces between the specimen and loading plates are generally preferred to nonlubricated surfaces, as lubrication prevents barreling (Figure H2.1.2). Barreling results from friction between the specimen and the contacting surfaces.

Compression of a weakly structured food between parallel plates may achieve squeezing flow (Steffe, 1996). When lubricated parallel plates are used, the result is a form of biaxial extension. Biaxial extension may be used to measure biaxial viscosity, which is a reflection of resistance to radial stretching flow in a plane. Lubricated squeezing flow of a semi-solid

specimen (Figure H2.1.3) supports the assumption that specimen deformation is homogeneous and thus extensional viscosity can be evaluated (Rao, 1992).

Ideally, the stress/strain relationship should be independent of the dimensions of a specimen in uniaxial compression (Peleg, 1987). Often, samples that have a height-to-diameter ratio (aspect ratio) ≤ 1 do not follow this rule. This occurs either from frictional forces or because the food material resists having the fluid squeezed from the matrix. Friction at the contact surfaces can increase the apparent strength of the sample and change its deformation behavior (Peleg, 1987). In fluid-filled samples, resistance to pressure dissipation depends on the length and nature of the path and on the density and microstructure of the sample (Peleg, 1987). Therefore, the greater the diameter of the sample (greater fluid path), the greater the apparent firmness and strength of the specimen.

In uniaxial compression, one dimension of a cylindrical sample is compressed while the area in contact with parallel plates increases (Steffe, 1996). The ratio between the increase

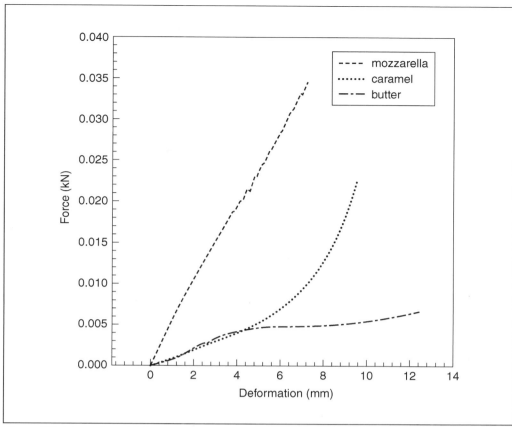

Figure H2.1.3 A force/deformation curve illustrating the lubricated squeezing flow of mozzarella cheese (2.3 cm length, 1.8 cm width, 1.8 cm height, 10 mm/min deformation rate), butter (2.1 cm diameter, 2.4 cm height, 5 mm/min deformation rate), and caramel (2.2 cm diameter, 1.9 cm height, 2 mm/min deformation rate) under uniaxial compression at room temperature.

in diameter (transverse strain) and the decrease in height (axial strain) is called Poisson's ratio (μ):

$$\mu = \frac{\Delta X / X_0}{d / L_0}$$

Compressive measurements provide a means to determine specimen stiffness, Young's modulus of elasticity, strength at failure, stress at yield, and strain at yield. These measurements can be performed on samples such as soy milk gels (Kampf and Nussinovitch, 1997) and apples (Lurie and Nussinovitch, 1996). In the case of convex bodies, where Poisson's ratio is known, the Hertz model should be applied to the data in order to determine Young's modulus of elasticity (Mohsenin, 1970). It should also be noted that for biological materials, Young's modulus or the apparent elastic modulus is dependent on the rate at which a specimen is deformed.

When evaluating force/deformation (Figure H2.1.1) or stress/strain curves, an artifact caused by the seating of the specimen, or a takeup of slack is sometimes observed. This "toe" compensation must be corrected to obtain a zero strain or zero deformation point.

Anticipated Results

Rigid specimens (e.g., apple, cheddar cheese) often exhibit a sudden decrease in force (stress) after a certain amount of deformation (maximum strain). At this point the specimen has fractured. Maximum stress and strain values may vary depending on the chosen specimen. Specimens that are weakly structured and tend to flow under lubricated compression (e.g., mozzarella cheese, marshmallow) demonstrate squeezing flow. As a result, the force (stress) continually increases as the specimen deformation (strain) increases. These materials do not fracture, but continue to stretch radially while under compression. Both rigid and soft specimens of the same material may exhibit varying characteristics depending on the deformation rate and the aspect ratio of each specimen.

Compressive
Measurement of
Solids and
Semi-solids

H2.1.7

Time Considerations

Sample preparation may require as little as 5 min (e.g., cutting a sample cube of cheese) or an hour or more (e.g., preparation of a food polymer solution and pouring in a mold). The time to run a test will depend on the deformation rate and the degree of deformation chosen for the specimen of interest. Typically, uniaxial compression of a specimen can be performed in 5 to 20 min.

Literature Cited

Ak, M.M. and Gunasekaran, S. 1992. Stress-strain curve analysis of cheddar cheese under uniaxial compression. *J. Food Sci.* 57:1078-1081.

Bagley, E.B., Christianson, D.D., and Martindale, J.A. 1988. Uniaxial compression of a hard wheat flour dough: Data analysis using the upper convected Maxwell model. *J. Texture Studies* 19:289-305.

Bourne, M.C. 1982. Food Texture and Viscosity: Concept and Measurement. Academic Press, New York.

Kampf, N. and Nussinovitch, A. 1997. Rheological characterization of carrageenan soy milk gels. *Food Hydrocoll.* 11:261-269.

Lelievre, J., Mirza, I.A., and Tung, M.A. 1992. Failure testing of gellan gels. *J. Food Eng.* 16:25-37.

Lurie, S. and Nussinovitch, A. 1996. Compression characteristics, firmness, and texture perception of heat treated and unheated apples. *Int. J. Food Sci. Tech.* 31:1-5.

Morrow, C.T. and Mohsensin, N.N. 1966. Consideration of selected agricultural products as viscoelastic materials. *J. Food Sci.* 31:686-698.

Mohsenin, N.N. 1970. Physical Properties of Plant and Animal Materials, Vol. 1. Gordon and Breach Science Publishers, New York.

Peleg, M. 1985. A note on the various strain measures at large compressive deformations. *J. Texture Studies.* 15:317-326.

Peleg, M. 1987. The basics of solid food rheology. *In* Food Texture: Instrumental and Sensory Measurement (H.R. Moskowitz, ed.) pp. 3-33. Marcel Dekker, New York.

Rao, M.A. 1992. Viscoelastic properties of cheeses. *In* Viscoelastic Properties of Foods (M.A. Rao and J.F. Steffe, eds.) pp. 173-184. Elsevier Science Publishers, New York.

Rebouillat, S. and Peleg, M. 1988. Selected physical and mechanical properties of commercial apple cultivars. *J. Texture Studies.* 19:217-230.

Steffe, J.F. 1996. Rheological Methods in Food Process Engineering, 2nd ed. Freeman Press, East Lansing, Mich.

Tang, J., Tung, M.A., and Zeng, Y. 1996. Compression strength and deformation of gellan gels formed with mono- and divalent cations. *Carbohydrate Polymers* 29:11-16.

Key References

Bourne, 1982. See above.

A good discussion of the information that can be obtained from uniaxial compression data.

Steffe, 1996. See above.

A helpful analysis of lubricated squeezing flow with problem solving examples and results for selected materials.

Contributed by Marvin A. Tung and
    Michael D.H. Rogers
University of Guelph
Guelph, Ontario, Canada

The editorial board is grateful to Dr. Ian Britt (University of Guelph) for his assistance in preparing this contribution.

Textural Measurements with Special Fixtures

This unit describes the measurement of textural properties of food using special testing fixtures. For the purposes of the unit, the focus will be on puncture probes, cone penetrometers, shear cells, wire cutting devices, and back extrusion fixtures, all of which lend themselves to textural testing in compression. Using a punch such as a Magness-Taylor puncture probe or a cone penetrometer (Basic Protocol 1) allows one to measure the firmness of a food material. The cone penetrometer, in particular, provides a means for determining a yield value. Shear cells such as those from Warner-Bratzler and Kramer (Basic Protocol 2) allow for different characteristics to be analyzed, as these fixtures combine resistance to cutting with shear, tension, and compression forces on the sample. For some structured foods, such as cheese, a wire cutting device is used to measure hardness. In addition, back extrusion cells (Basic Protocol 3) are useful for determining the force of extrusion in a form of compression-extrusion test. An understanding of human sensory interpretation of texture is essential for selecting appropriate fixtures for textural measurements.

TEXTURAL MEASUREMENTS USING A PUNCTURE PROBE OR A CONE PENETROMETER

Puncture probes and cone penetrometers provide a simple means for determining the textural properties of many structured foods (see sample results in Figure H2.2.1).

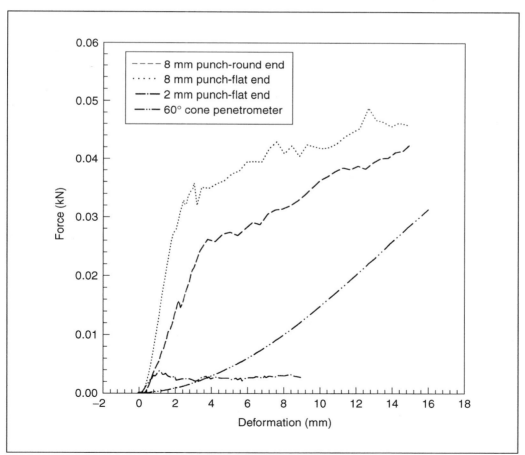

Figure H2.2.1 Force/deformation curves illustrating three puncture probe tests (50 mm/min deformation rate) of an apple specimen and a cone penetrometer test (10 mm/min deformation rate) of cheddar cheese, all at room temperature.

Contributed by Marvin A. Tung and Michael D.H. Rogers

Puncture probes are commonly used for fruits and vegetables, and allow for the determination of force at rupture of the cellular structure. The procedure outlined below is adapted from the method of Bourne (1979). Cone penetrometers are commonly employed for determining firmness and yield value for foods such as margarine and butter, which may be a reflection of the product's spreadability. Quite often it is desirable to use a testing system that provides a constant deformation rate. Additionally, a mechanical testing machine allows for production of a force/deformation curve to further analyze the data.

Materials

Material specimen (prepared or intact)
Mechanical testing machine linked to either a chart recorder or a computer with data recording and analysis software (optional; puncture probes are often hand held)
Testing fixture (select as appropriate):
Puncture probe
Cylindrical probe (either a hand-held instrument or a fixture for a mechanical testing machine), flat or rounded end (Magness-Taylor)
Cone Penetrometer (cone angle and geometry may vary)

Prepare sample

1. Prepare specimen so that the diameter and gauge length of the material are sufficient to meet the requirements for depth of penetration.

   The sample size should be large when compared to the size of the particular fixture in use so to ensure that the entire surface of the specimen remains in contact with the fixture.

   Puncture probe diameter should be held constant during a given series of tests.

2. Allow the temperature of the specimen to equilibrate to the temperature at which it is most likely stored or used (e.g., refrigerated, ambient, or elevated temperature).

   This assists in mimicking the application conditions most commonly encountered by the specimen.

Perform measurements

For hand-held puncture probes:

3a. With one hand, hold the structured food against a rigid surface (e.g., a laboratory bench).

4a. Hold puncture probe perpendicular to the surface to be tested.

5a. With the puncture probe in the other hand and the side of this hand resting on your hip, lean into the puncture probe to provide a uniform rate of force application.

   Try to avoid uneven movements during testing.

6a. For food products such as freshly harvested apples and ripe fruit, continue penetration until probe depth reaches the inscribed line on the probe.

   Often raw vegetables will require that penetration continue until contact with the instrument's splash guard.

   In the case of specimens such as apples, it may be appropriate to remove the epicarp (skin) in the region being tested.

7a. Record the maximum force from the instrument.

3b. Choose a load cell that has the capability to manage the expected maximum load to be exerted on the specimen.

4b. Turn on the mechanical testing machine and attach the appropriate fixture (puncture probe or cone penetrometer).

5b. Turn on the chart recorder or computer.

If using computer software, procedures specific to the software will need to be followed.

6b. For test equipment driven by computer software, enter required information—e.g., specimen dimensions, cross-sectional area, desired calculations, and the rate at which the sample will be deformed (cross-head speed or deformation rate).

7b. Place the specimen on the lower flat base plate and move the fixture, attached to the moving cross-head, to a position near the top surface of the specimen. Measure the distance between the lower plate and the fixture to obtain the sample height, which may be required by the computer software and in further calculations.

8b. Zero the position of the fixture in relation to the sample surface and the zero the load on the load cell.

9b. Select a desired deformation rate (dh/dt) and select a distance at which to stop the test.

Testing should cease at the appropriate depth of penetration or at rupture.

10b. Start the test.

CAUTION: *The cross-head will begin to move down. Keep hands clear!*

11b. Record the force/deformation data from the test.

12b. For puncture probes, determine sample strength by measuring the force or stress at rupture. For flat-ended probes, calculate the force engineering stress from the area of the contact surface of the probe (A) and the force at rupture (F) from a force/deformation plot.

$$\sigma_f = \frac{F}{A}$$

It is important to note that the ultimate strength of a specimen may not be the point of rupture. The rupture point may occur prior to the point illustrating the ultimate strength of a material (see Figure H2.2.2).

See Table H2.2.1 for variables and units used in these calculations.

**Compressive
Measurement of
Solids and
Semi-solids**

H2.2.3

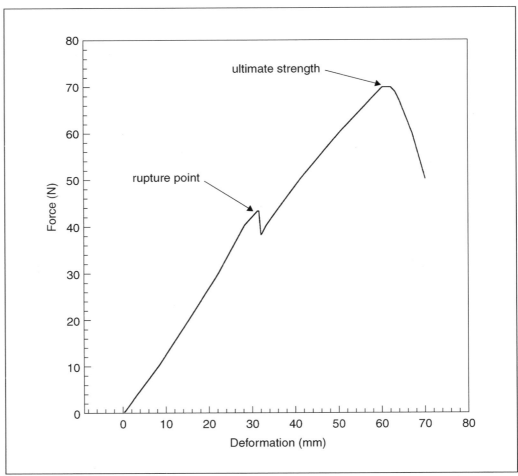

Figure H2.2.2 A force/deformation curve illustrating the potential difference between the rupture point and the ultimate strength of a food specimen (adapted from Mohsenin, 1970).

Table H2.2.1 Nomenclature and Corresponding Units

Symbol	Parameter	S.I. unit
A	Area	m^2
A'	Actual area where F' is applied	m^2
$c\,(K)$	Constant	Dimensionless
F	Force	load
F'	Actual force applied for sample	N
f	Firmness of butter	Pa
h	Height of cone	cm
K_c	Compression coefficient	Dimensionless
K_s	Shear coefficient	Dimensionless
l	Length of cone	cm
p	Perimeter	m
r	Radius of cone	cm
dh/dt	Speed of cone penetrometer	cm/s
η_{app}	Apparent viscosity	Pa·s
σ_f	Maximum stress at failure	Pa
θ	Cone angle	Degrees

13b. Calculate the yield value (*f*) for materials such as margarine and butter, as determined by cone penetrometry, using the following equations from Tanaka et al. (1971).

$$F' = \frac{F}{\cos(\theta/2)}$$

$$A' = \frac{\pi r l}{\cos(\theta/2)}$$

$$r = h\tan(\theta/2)$$

$$l = \frac{h}{\cos(\theta/2)}$$

$$F'/A' = \eta_{app}(dh/dt) + f = \frac{F\cot(\theta/2)\cos(\theta/2)}{\pi h^2}$$

If F'/A' (penetration stress) is plotted versus dh/dt (penetration speed), the slope is the apparent viscosity (η_{app}) and the intercept (penetration stress at dh/dt = 0) is the yield value (f).

TEXTURAL MEASUREMENTS USING WARNER-BRATZLER, KRAMER, OR WIRE CUTTING FIXTURES

Warner-Bratzler, Kramer, and wire cutting fixtures provide another method of measuring textural components of some food materials. The Warner-Bratzler fixture is useful for measuring the shear stress associated with cutting of meat products. The Kramer shear fixture (Figure H2.2.3) has important applications for textural measurements of products such as peas and peaches. In this case, the force at rupture is a result of both cutting and compression forces. Finally, wire cutting devices (sectilometers; Figure H2.2.4) are important for cheese and butter products. The firmness of the food material is measured as the maximum force encounter on the wire during cutting.

BASIC PROTOCOL 2

Materials

Material specimen (prepared or intact)
Mechanical testing machine linked to either a chart recorder or computer (with data recording and analysis software)
Testing fixture (select as appropriate; see Background Information):
    Warner-Bratzler test cell
    Kramer shear/compression cell (single-blade or multiblade cells)
    Wire cutting device (sectilometer; single- or multiwire)

Prepare specimen

1. Prepare speciman as appropriate for the testing device to be used:

   a. *For Warner-Bratzler measurements:* Cut specimen so that it is cylindrical in shape.

   b. *For Kramer shear cell measurements:* Cut into small cubes or chunks (i.e., 1 cm lengths) if excessively large.

   c. *For sectilometer measurements:* Cut so that specimen has a width less than the length of the wire(s) on the instrument.

2. Allow the temperature of the specimen to equilibrate to the temperature at which it is most likely stored or used (e.g., refrigerated, ambient, or elevated temperature).

   This assists in mimicking the application conditions most commonly encountered by the specimen.

Compressive Measurement of Solids and Semi-solids

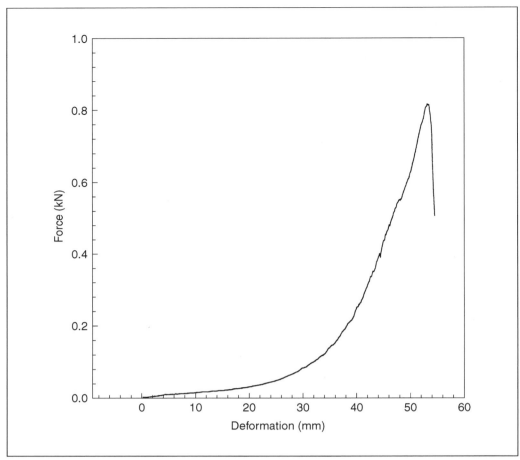

Figure H2.2.3 A force/deformation curve illustrating a compression-extrusion test (10 mm/min deformation rate) for canned green peas using a Kramer shear cell (multiblade) at room temperature.

Perform measurements

3. Choose a load cell that has the capability to manage the expected maximum load to be exerted on the specimen.

4. Turn on the mechanical testing machine and attach the appropriate fixture (Warner-Bratzler, Kramer, wire cutting fixture).

5. Turn on the chart recorder or computer.

   If using computer software, procedures specific to the software will need to be followed.

6. For test equipment driven by computer software, enter required information such as specimen dimensions, cross-sectional area, desired calculations, and the rate at which the sample will be deformed (cross-head speed or deformation rate).

7. Place the specimen on the two metal parts of the lower portion of the Warner-Bratzler fixture or the Kramer shear cell and fill the lower fixture to about half capacity. In the case of a wire cutting device, place the specimen on the lower plate.

8. Move the top fixture to a position near the top surface of the specimen. Measure the distance between the lower plate and the fixture to obtain the sample height, which may be required by computer software or in future calculations.

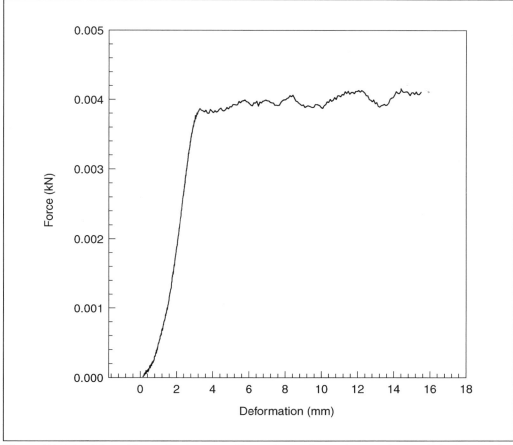

Figure H2.2.4 A force/determination curve illustrating a wire cutting test (10 mm/min deformation rate) for cheddar cheese using a single-wire sectilometer at room temperature.

9. Zero the position of the fixture in relation to the sample surface and zero the load on the load cell.

10. Select a deformation rate of 22.86 cm/min (9 inches/min) for the Warner-Bratzler test. For the Kramer shear cell test or the wire cutting test, select a desired deformation rate.

    Testing should cease at the appropriate depth of penetration or at rupture.

11. Start the test.

    CAUTION: *The cross-head will begin to move down. Keep hands clear!*

12. Record the force/deformation data for the test.

    Measurements made using a Warner-Bratzler fixture, a Kramer shear/compression cell, or a wire cutting device may demonstrate rupture prior to the reaching the ultimate strength of the specimen (Figure H2.2.2). The force at rupture is the most important factor required for textural analysis using these fixtures. A food material being tested using a Kramer cell may fail in either compression, tension, shear, or some combination of the three. Measurements obtained through sectilometer tests allow for the determination of the firmness of the sample. Firmness is the maximum force needed to push the wire through the food.

**Compressive
Measurement of
Solids and
Semi-solids**

H2.2.7

TEXTURAL MEASUREMENTS USING BACK EXTRUSION

Back extrusion is an important method for determining the yield force required for food materials that are homogeneous and that flow. The yield force is the point where flow is initiated. Using a plunger and a cylinder open at one end (Figure H2.2.5), it is possible to extrude food material back through the annular gap. Extrusion occurs beyond the rupture point of the food. This method is commonly performed with cooked materials such as peas and rice.

Materials

Material specimen (prepared or intact)
Mechanical testing machine linked to either a chart recorder or computer (with data recording and analysis software)
Back extrusion cell (plunger and a container with an open top having a certain annular gap)

Prepare sample

1. Prepare a food specimen that can flow (e.g., fruits, vegetables, liquids, and gels) and is homogeneous in size.

2. Allow the temperature of the specimen to equilibrate to the temperature at which it is most likely stored or used (e.g., refrigerated, ambient, or elevated temperature).

   This assists in mimicking the conditions most commonly encountered by the specimen.

Perform measurements

3. Choose a load cell that has the capability to manage the expected maximum load exerted on the specimen.

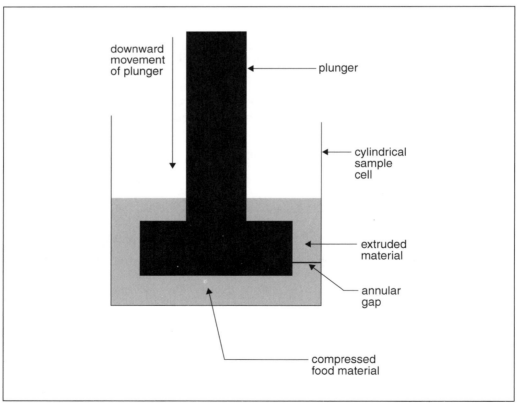

Figure H2.2.5 A diagram illustrating the operation of a back extrusion fixture being used to measure the textural properties of a flowable food material.

4. Turn on the mechanical testing machine and attach the appropriate fixture (back extrusion cell).

5. Turn on the chart recorder or computer.

   If using computer software, procedures specific to the software will need to be followed.

6. For test equipment driven by computer software, enter required information such as specimen dimensions, cross-sectional area, desired calculations, and the rate at which the sample will be deformed (cross-head speed or deformation rate).

7. Place the specimen in the container (fill to about ¾ capacity) and mount on the lower plate of the mechanical testing machine.

8. Move the top fixture (i.e., the plunger) to a position near the top surface of the specimen. Measure the distance between the lower plate and the fixture to obtain the sample height, which may be required by computer software or in future calculations.

9. Zero the position of the fixture in relation to the sample surface and the load on the load cell.

10. For the back extrusion test, select a desired deformation rate.

    Testing should cease at the appropriate depth of compression of the packed sample.

11. Start the test.

    CAUTION: *The cross-head will begin to move down. Keep hands clear!*

12. Record the force/deformation data for the test.

    The rupture force may or may not be the represented as the maximum point on the force/deformation curve (see Figure H2.2.2). At rupture the food material begins to flow back up through the annular gap.

COMMENTARY

Background Information

deMan (1976) defined texture as "the way in which the structural components of a food are arranged in a micro and macro structure and the external manifestations of this structure." The external manifestations of structure refer to the behavior of a food in a mechanical testing machine. Texture can be subdivided into two categories: rheology or physical qualities (e.g., Young's modulus, Poisson's ratio) and haptaesthesis or perceived qualities (e.g., mouthfeel, gumminess; Bourne, 1982). These qualities can be measured using a variety of different instruments.

There are four main procedures for measuring the texture of food: empirical, imitative, fundamental, and ideal (Bourne, 1982). Of these procedures, empirical tests are the most frequently used in textural measurement; however, they are poorly defined (Bourne, 1982). The protocols provided in this unit are all empirical tests.

Perceived qualities of texture require an understanding of the way in which humans interact with food (Bourne, 1982). Mastication is the process of chewing and grinding food so that it may be swallowed. Humans masticate for flavor release, contentment, and to break food into smaller pieces (Bourne, 1982). Texture can also be sensed by using the hand and fingers. Squeezing a food material between the thumb and forefinger is an excellent method for evaluating texture (Bourne, 1982).

Cone penetrometers

The cone penetrometer was first developed for measuring the firmness or yield point of solid fats such as butter or margarine (Bourne, 1982). Assuming that the cone angle, mass of the cone assembly, temperature, and time for penetration are held constant, results from cone penetrometry can be interpreted based on the depth of penetration (Dixon and Parekh, 1979). Dixon and Parekh (1979) applied a cone penetrometer for measuring the firmness of butter.

The properties of the test specimen, the weight and height of the cone, and the cone angle are determinants for the depth of pene-

Compressive Measurement of Solids and Semi-solids

H2.2.9

tration of a cone penetrometer (van Vliet, 1999). Should the specimen yield, elongational flow and shear flow results (van Vliet, 1999).

Puncture probes

Puncture probe testing involves measuring the maximum force required to drive a metal probe into a structured food (deMan, 1976). On a force/deformation curve, the force at rupture is seen as the maximum point on the curve. These instruments can be hand operated or attached to a force/deformation measuring instrument (e.g., mechanical testing machine). Puncture probes may have either a flat or a rounded end (Bourne, 1979). Puncture probes of the Magness-Taylor type are often used for firmness measurements of fruits and vegetables. Lurie and Nussinovitch (1996) found that a good correlation existed between the Magness-Taylor measurement of firmness and the compression testing of both strength and stiffness of apple cultivars. In firm apples, compression testing appeared to provide results most closely related to the properties perceived by humans (Lurie and Nussinovitch, 1996). Whether a puncture-type test or a compression-type test is more appropriate for testing texture may depend on the nature of the food.

The yield force reading obtained from a puncture test was found to depend on the perimeter and area of the punch (Bourne, 1979), as illustrated in the equation (Bourne, 1982):

$$F = K_c A + K_s p + c$$

where K_c and K_s are compression and shear coefficients, respectively, p is the perimeter of the sample, and c is a constant. K_c can be determined as the slope of a force/area curve, and K_s can be determined from the slope of a force/perimeter curve. It has been demonstrated that if the perimeter of the puncture probe is held constant, the force on a material is directly proportional to the area of the punch (Bourne, 1979). When the area of the punch is held constant, the force is directly proportional to the perimeter of the punch (Bourne, 1979). In cases where the puncture probe has a circular cross-section, the area and perimeter terms are replaced by diameter.

Of great importance, when using a handheld puncture probe, is the type of product being examined. Bourne (1982) described three categories for food materials. Products such as freshly harvested apples require a continual increase in force to continue penetration after the yield point. However, food materials such as ripe fruit require no such increase, and food materials such as raw vegetables require pushing the puncture probe into the material until it contacts the splash collar on the instrument.

Cutting instruments

For conducting shear testing, as in cutting a food sample, the Warner-Bratzler fixture is commonly used. The Warner-Bratzler blade has a V-shaped notch cut from it (Bourne, 1982). Warner-Bratzler fixtures are most often employed for measuring the tenderness or toughness of meat products. To measure this property, the blade forces the meat into the V notch until it is cut though (Bourne, 1982). As the specimen fills the area of the V notch, it is cut or sheared; however, it should be noted that the stresses induced in this action are complex, involving tensile and compressive stresses as well as shear.

Normally, meat samples are oriented perpendicular to the blade (Voisey, 1976). This allows for measurement of the force required to cut across the muscle fibres. Cylindrical specimens ranging in size from 0.5 to 2.5 cm in diameter are most often used (Voisey, 1976). A true Warner-Bratzler test requires a deformation rate of 22.86 cm/min (Voisey, 1976).

Stanley (1976) found that, for a Warner-Bratzler shear cell, the firmness of a meat product may be reflected by measurements in compression, whereas tenderness may be reflected by the tensile strength of the meat. He further indicated that firmness and tenderness may or may not be dependent on each other. For this reason, results from such tests may be difficult to interpret.

The Kramer Shear Press can be equipped with test fixtures designed to measure shearing force. It was developed originally for testing the tenderness or toughness of peas (Stanley, 1976). The multiblade Kramer shear cell consists of ten blades that are driven down through a sample (Stanley, 1976). The sample rests in a box with slits on the bottom. As the blades enter the box, the sample is compressed, sheared, and extruded through the bottom of the box (Rao, 1992). Food materials are cut into cubes or chunks and the test cell is filled to half capacity (Voisey, 1977). According to Voisey, failure occurs in two manners; by compression and by cutting. The deformation required for failure depends on the resistance of the specimen to compression and cutting.

Structured foods, such as cheddar cheese and some fruits and vegetables, are often examined using wire cutting fixtures known as sectilometers. In this case, a steel wire is moved through a sample at a controlled rate and the force on the wire is measured (deMan, 1976). Multiwire sectilometers are sometimes used to improve the precision of the testing procedure (Voisey and deMan, 1976).

Compression-extrusion fixtures

Finally, compression-extrusion testing involves an extrusion cell commonly used for weakly structured, homogeneous food products. This apparatus consists of a piston that is forced into a cylinder open at one end and containing the product (Figure H2.2.5). Beyond the point of rupture of the food, the compressed material is forced to flow back through the annular space between the piston and the cylinder (Bourne, 1976; Edwards, 1999). The gap between the piston and the cylinder is called the annulus (Bourne, 1982). Variation in the annulus width results in variation in the force required for extrusion (Bourne, 1982).

Anticipated Results

Textural measurements using a puncture probe should show a point on the force/deformation diagram where the force is a maximum. The maximum force encountered by the puncture probe, for a given sample, will vary depending on the geometry and size of the probe and the deformation rate. Cone penetrometer measurements will vary depending on the mass and angle of the cone, as well as the nature of the sample and deformation rate. Cutting instruments (i.e., Warner-Bratzler, Kramer shear, sectilometer) all make measurements based on the maximum force encountered by the fixture. The maximum force depends on the nature of the sample (e.g., cheese versus butter for a sectilometer) as well as deformation rate. Note that variable deformation rate is not a factor with Warner-Bratzler tests. Finally, the back extrusion fixture results should show a point of rupture for the food material, beyond which the food should be extruded. This point will depend on the nature of the sample, the deformation rate, the width of the annular gap, and the geometry of the plunger.

Time Considerations

Sample preparation may require as little as 5 min (e.g., removing the skin from an apple) or up to an hour or more (e.g., preparation of a food polymer solution and pouring into a mold). The time to run a test will depend on the deformation rate and the desired amount of penetration into the food material. Typically, a test using any of these special fixtures can be accomplished in 5 to 20 min.

Literature Cited

Bourne, M.C. 1976. Interpretation of force curves from instrumental texture measurements. *In* Rheology and Texture In Food Quality (J.M. deMan, P.W. Voisey, V.F. Rasper, and D.W. Stanley, eds.) pp. 355-381. AVI Publishing Company, Westport, Conn.

Bourne, M.C. 1979. Theory and application of the puncture test in food texture measurement. *In* Food Texture and Rheology (P. Sherman, ed.) pp. 95-142. Academic Press, New York.

Bourne, M.C. 1982. Food Texture and Viscosity: Concept and Measurement. Academic Press, New York.

deMan, J.M. 1976. Texture of fats and fat products. *In* Rheology and Texture In Food Quality (J.M. deMan, P.W. Voisey, V.F. Rasper, and D.W. Stanley, eds.) pp. 355-381. AVI Publishing Company, Westport, Conn.

Dixon, B.D. and Parekh, J.V. 1979. Use of the cone penetrometer for testing the firmness of butter. *J. Texture Studies* 10:421-434.

Edwards, M. 1999. Vegetables and fruit. *In* Food Texture: Perception and Measurement (A.J. Rosenthal, ed.) pp. 259-281. Aspen Publishers, Gaithersburg, Md.

Lurie, S. and Nussinovitch, A. 1996. Compression characteristics, firmness, and texture perception of heat treated and unheated apples. *Int. J. Food Sci. Tech.* 31:1-5.

Mohsenin, N.N. 1970. Physical Properties of Plant and Animal Materials, Vol. 1. Gordon and Breach Science Publishers, New York.

Rao, V.N.M. 1992. Classification, description and measurement of viscoelastic properties of solid foods. *In* Viscoelastic Properties of Foods (M.A. Rao and J.F. Steffe, eds.) pp. 3-47. Elsevier Science Publishers, New York.

Stanley, D.W. 1976. The texture of meat and its measurement. *In* Rheology and Texture In Food Quality (J.M. deMan, P.W. Voisey, V.F. Rasper, and D.W. Stanley, eds.) pp. 405-426. AVI Publishing Company, Westport, Conn.

Tanaka, M., deMan, J., and Voisey, P.W. 1971. Measurement of textural properties of foods with a constant speed cone penetrometer. *J. Texture Studies* 2:306-315.

van Vliet, T. 1999. Rheological classification of foods and instrumental techniques for their study. *In* Food Texture: Measurement and Perception (A.J. Rosenthal) pp. 65-97. Aspen Publishers, Gaithersburg, Md.

Voisey, P.W. 1976. Engineering assessment and critique of instruments used for meat tenderness evaluation. *J. Texture Studies* 7:11-48.

Voisey, P.W. 1977. Interpretation of force-deformation curves from the shear-compression cell. *J. Texture Studies* 8:19-37.

Compressive
Measurement of
Solids and
Semi-solids

Voisey, P.W. and deMan, J.M. 1976. Applications of instruments for measuring food texture. *In* Rheology and Texture In Food Quality (J.M. deMan, P.W. Voisey, V.F. Rasper, and D.W. Stanley, eds.) pp. 142-243. AVI Publishing Company, Westport, Conn.

Key References

Bourne, 1976. See above.

A useful look at the analysis of force/deformation curves.

Bourne, 1982. See above.

A good overview of the methods available for textural measurements.

Morrow, C.T. and Mohsenin, N.N. 1966. Consideration of selected agricultural products as viscoelastic materials. *J. Food Sci.* 31:686-698.

A helpful paper on the use of puncture probes for textural measurements of agricultural products.

Contributed by Marvin A. Tung
  and Michael D.H. Rogers
University of Guelph
Guelph, Ontario, Canada

The editorial board is grateful to Dr. Ian Britt (University of Guelph) for his assistance in preparing this contribution.

Texture Profile Analysis

This unit describes an instrumental texture profile analysis (TPA) using mechanical compression of a foodstuff. TPA was developed in the early 1960s to study the mechanical properties of foods and their relationship to the texture of foods. The original TPA was carried out using the General Foods (GF) Texturometer by Friedman et al. (1963). A newer TPA was developed using the Instron Universal Testing Machine by Bourne (1968) with the following testing conditions: (1) a circular plate 150 mm in diameter was used to compress the specimens, so that samples were not subjected to any shearing; (2) the plunger was set to compress the 10-mm sample to 2.5 mm (25% of its original height), then rise back to its starting position and move down again to compress the sample to 2.5 mm (25% of its original height); and (3) the Instron cross-head was set to cycle with a vertical reciprocating movement at a constant speed of 50 mm/min.

Currently, there are many compressive instruments that are suitable for TPA. In the following protocols, the Instron Universal Testing Machine is used. In the Basic Protocol, the textures of vegetables, both raw and cooked to different temperatures, are measured. For Alternate Protocol 1, the textures of different fresh apple cultivars are tested. The texture of cooked ground beef samples with varying fat content is measured in Alternate Protocol 2. Lastly, in Alternate Protocol 3, gel textures are tested among samples with varying gelatin concentrations.

MEASURING THE TEXTURE OF POTATOES AND CARROTS

Texture is an important characteristic of vegetables, and it changes during thermal processing due to the breakdown of cellular material. Although both potato and carrot are root plants, the potato has a relatively uniform granular tissue structure and high starch composition, while the carrot has a fibrous tissue structure and low starch composition. These structural and compositional differences can affect the textural changes of potatoes and carrots with thermal processing.

Materials

Eight potatoes of uniform size (~50-mm square-mesh grading)
Eight carrots of uniform size (~280 mm long and 125 mm thick at the larger end)

Cork borer, size 4
Sharp knife
Instron Universal Testing Machine (Instron)

1. Set aside a potato and a carrot to use as a raw standard.

2. Cook each of the remaining seven samples of potatoes and carrots to an internal temperature of 30°, 40°, 50°, 60°, 70°, 80°, and 90°C.

   A calibrated digital cooking thermometer is preferable for measuring sample temperature.

3. Collect from each sample (including the raw standards) four cylindrical specimens measuring 10 mm high and 12 mm in diameter using a cork borer and a sharp knife.

4. Set an Instron Universal Testing Machine to compress the specimen twice, compressing each time to 50% of its original height.

5. Test specimens at 10 mm/min cross-head speed at room temperature (~22°C).

6. Analyze the four specimens taken from each sample at each cooked temperature and raw state.

7. Obtain TPA parameters for each sample (see Commentary).

Contributed by Yoshi Mochizuki

MEASURING THE TEXTURE OF APPLES

Apple cultivars have different textures due to their internal variability of structure and composition. Some apples resist boiling and do not readily sauce. Others may undergo ready cell separation. This wide range of textural behavior illustrates the complexity due to pectins and other cell-wall materials. Select apple cultivars according to the desired processing qualities.

Additional Materials (*also see Basic Protocol*)

Mature apple for each cultivar to be tested (all of a similar size, age, and ripeness)
Cork borer, size 6

1. Slice an apple at ~15 mm above and below its equator.

2. Collect five cylindrical specimens measuring 10 mm high and 15 mm in diameter from the central slice using a cork borer and a sharp knife. Avoid the core region.

3. Set an Instron Universal Testing Machine to compress the specimen twice, compressing each time to 80% of its initial height.

4. Test specimens at 50 mm/min cross-head speed at room temperature (~22°C).

5. Analyze the five specimens taken from each cultivar.

6. Obtain TPA parameters for each cultivar (see Commentary).

MEASURING THE TEXTURE OF COOKED GROUND BEEF

Ground beef has a typical fat level of 20% to 30%. Consumers can select retail ground beef with decreased levels of fat; however, they usually perceive leaner grinds as being less palatable. To some extent this is true. In order to assure qualities such as texture, mouthfeel, tenderness, juiciness, flavor, appearance, and overall acceptability, a certain fat content is necessary in ground beef. The fat level can affect the texture of cooked ground-beef patties.

Additional Materials (*also see Basic Protocol*)

Ground beef of various fat content (~250 g of each)
Electric skillet, preheated to 150°C

1. For each type of ground beef, form two uniform patties (~113 g/patty) measuring 15 mm high and 60 mm in diameter.

2. Cook patties to an internal temperature of 77°C using a preheated (150°C) electric skillet. Cool to room temperature.

   A calibrated digital cooking thermometer is preferable for measuring sample temperature.

3. Cut four specimens as slices (10 cm × 30 cm × 10 mm) from each cooked patty.

4. Set an Instron Universal Testing Machine to compress the specimen twice, compressing each time to 30% of its initial height.

5. Test specimens at 200 mm/min cross-head speed at room temperature (~22°C).

6. Make eight measurements for each type of ground beef (four slices from each of the two cooked patties).

7. Obtain TPA parameters for each type of ground beef (see Commentary).

In the food industry, particularly the confectionery industry, gelatin is commonly used for processing gelled products. Gelatin is a soluble polypeptide derived from insoluble collagen, and it shows a reversible sol-gel change with temperature. The rigidity of the gel is one of the most important properties of gelatin and correlates with the gelatin concentration. In this protocol, the gelling quality of a gelatin is measured by measuring the gel strength as a function of gelatin concentration.

Additional Materials (also see Basic Protocol)

Gelatin
19 × 31–cm enamel pans, >2 cm in depth
Wire-cutting grid to cut 2-cm cubes

1. Mix 22, 24, 28, 35, and 45 g gelatin in five separate beakers each containing 1 liter boiling water and stir 10 min so that gelatin is dissolved.

2. Cool samples to 40°C and pour them into 19 × 31–cm enamel pans to a height of 2 cm.

3. Cover enamel pans with aluminum foil and refrigerate 24 hr at 10°C.

4. Cut three 2-cm cubes from each concentration of gelatin gel using a wire-cutting grid.

5. Set an Instron Universal Testing Machine to compress the specimen twice, compressing each time to 30% of its initial height.

6. Test specimens at 200 mm/min cross-head speed at room temperature (~22°C).

7. Analyze the three specimens taken from each gelatin concentration.

8. Obtain TPA parameters for each gelatin concentration (see Commentary).

COMMENTARY

Background Information

The instrumental TPA was developed by a group at the General Foods Corporation Technical Center (Szczesniak and Kleyn, 1963). The parameters obtained from the resulting force-time curve correlate well with sensory evaluations of the same parameters (Friedman et al., 1963). Later, the Instron Universal Testing Machine was adapted to perform a modified TPA (Bourne, 1968, 1974). A typical Instron TPA curve is shown in Figure H2.3.1.

The difference between the GF Texturometer and the Instron arises from the different compression motions. The plunger of the GF Texturometer is driven at sinusoidal speed and follows the arc of a circle. The resulting TPA curve shows rounded peaks. The plunger of the Instron, however, is driven at constant speed and shows rectilinear-shaped peaks on its TPA curve. Therefore, one disadvantage of the GF Texturometer is that the contact area between the sample and the plunger is not constant during the test; because the plunger moves through the arc of a circle, one edge of the plunger contacts the food first and, as the downstroke continues, the area of the plunger pressing on the food increases until the entire plunger area is in contact with the food at the end of the downstroke (Bourne, 1968). Another disadvantage of the GF Texturometer is that the sample platform is flexible and bends a little as the load is applied. This can result in confounding data. On the other hand, the Instron is so rigid that bending of the instrument does not occur, resulting in data that are precise. All modern TPA measurements are now carried out with two-cycle uniaxial compression instruments such as the Instron Universal Testing Machine.

The most common parameters derived from the TPA curve are shown in Table H2.3.1 (Friedman et al., 1963; Bourne, 1968). The peak force during the first compression cycle is defined as hardness. Fracturability (originally called brittleness) is defined as the force at the first significant break in the curve during the first compression cycle. The ratio of the positive force area during the second compression cycle to that during the first compression

Compressive Measurement of Solids and Semi-Solids

H2.3.3

cycle (area 2/area 1; Figure H2.3.1) was originally defined as cohesiveness when using the GF Texturometer. When the Instron is used, cohesiveness is obtained from the areas under the compression portion (downstroke) only and excludes the areas under the decompression portion (upstroke) instead of using the total area under positive force. The negative force area of the first compression cycle (area 3), adhesiveness, is defined as the work necessary to pull the plunger away from the sample. The length to which the sample recovers in height during the time that elapses between the end of the first compression cycle and the start of the second compression cycle is defined as springiness (originally called elasticity). Gumminess is defined as the product of hardness times cohesiveness, and chewiness is defined as the product of hardness times cohesiveness times springiness.

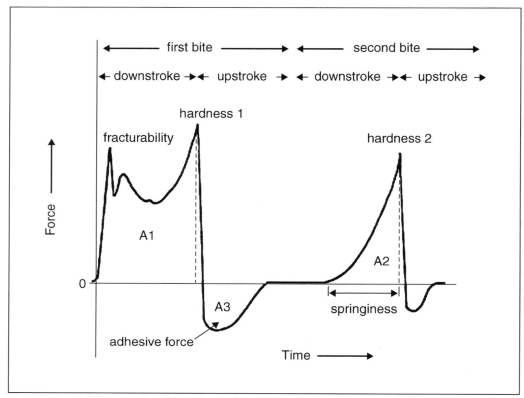

Figure H2.3.1 A generalized texture profile analysis curve from the Instron Universal Testing Machine. A1, area 1; A2, area 2; A3, area 3. The dotted lines indicate the times at which hardness is measured. See text for further discussion. Reprinted from Pons and Fiszman (1996) with permission from Food and Nutrition Press.

Table H2.3.1 Parameters and Units of Instrumental Texture Profile Analysis[a]

Mechanical parameter	Measured variable	S.I. units
Adhesiveness	Work	$N \times mm$ (mJ)
Chewiness	Work	$N \times mm$ (mJ)
Cohesiveness	Ratio of forces	Dimensionless
Fracturability	Force	N
Gumminess	Force	N
Hardness	Force	N
Springiness	Distance	mm

[a]From Bourne, 1982.

Critical Parameters

Consistent preparation of samples is probably the most critical step in obtaining good data. Another important step is the cutting and slicing of test samples from the prepared samples. Having the same size for each test sample is very critical. Generally, it is also important to avoid edge effects in a test sample. For example, for cooked samples, a test sample taken from an exterior surface will generally give different readings than one taken from the interior. Unless otherwise specified, avoid taking test samples close to the exterior surfaces of prepared samples. It is also important to have consistent aging effects on the texture of prepared samples between the time they are prepared and tested.

For a TPA measurement, make sure that the degree of compression, plunger size, and crosshead speed are the same among tests that are to be compared. Adjust the operating characteristics of the machine when testing other types of samples. For example, if the samples are harder than those used in these protocols, reduce the degree of compression and/or the cross-head speed.

Anticipated Results

Tables H2.3.2 and H2.3.3 show TPA parameters of raw and cooked potatoes and carrots prepared at some of the temperatures described in the Basic Protocol. In general, hardness, fracturability, cohesiveness, springiness, gumminess, and chewiness decreased with increasing cooking temperatures. Thus, the major changes of texture during thermal processing were due to the breakdown of cellular material. These results suggest that TPA parameters are suitable to monitor the cooking of potatoes and carrots (Mittal, 1994).

Table H2.3.4 shows an example of cohesiveness data from apple cultivars as assayed in Alternate Protocol 1. The cohesiveness discriminates between Golden Delicious and Stark Delicious. The apples of extra quality showed the lowest values of cohesiveness. These results suggest that TPA parameters can differentiate between apple cultivars (Paoletti et al., 1993).

In Table H2.3.5, hardness, springiness, and cohesiveness of cooked ground beef patties as measured in Alternate Protocol 2 are given. The hardness for 5% and 10% fat levels was higher than for 15% or higher fat levels. Beef patties containing 25% and 30% fat had lower springiness values than all other patties. The cohesiveness for beef patties containing 5% fat was higher than others. However, cooking temperature had no effect on springiness and cohesiveness (Troutt et al., 1992).

Figure H2.3.2 shows the maximum and yield force (i.e., hardness and fracturability) of

Table H2.3.2 Changes in the Texture of Potatoes Cooked to Various Center Temperatures[a]

Temperature (°C)	Chewiness (mJ)	Cohesiveness	Fracturability (N)	Gumminess (N)	Hardness (N)	Springiness (mm)
20	38.5	0.087	107	12.6	145	3.05
40	29.4	0.071	104	10.5	148	2.81
60	16.2	0.049	121	6.6	135	2.45
80	2.8	0.047	26	1.4	37	2.10

[a]From Mittal, 1994.

Table H2.3.3 Changes in the Texture of Carrots Cooked to Various Center Temperatures[a]

Temperature (°C)	Chewiness (mJ)	Cohesiveness	Gumminess (N)	Hardness (N)	Springiness (mm)
20	73.7	0.104	22.3	215	3.30
40	39.6	0.067	15.3	228	2.59
60	22.0	0.048	8.9	186	2.46
80	16.0	0.065	5.9	91	2.69

[a]From Mittal, 1994.

Compressive Measurement of Solids and Semi-Solids

H2.3.5

Table H2.3.4 Cohesiveness Values from TPA Measurements of Different Apple Cultivars[a]

Apple cultivar	Cohesiveness
Annurea	0.0192
Emperor	0.0171
Golden Delicious Extra	0.0135
Golden Delicious I	0.0299
Granny Smith	0.0520
Renetta	0.0090
Stark Delicious Extra	0.0156
Stark Delicious I	0.0205

[a]From Paoletti et al., 1993.

Table H2.3.5 TPA Values of Cooked Ground-Beef Patties with Different Fat Levels[a]

TPA parameter	Fat level (%)					
	5	10	15	20	25	30
Cohesiveness (%)	47.5	41.0	37.8	38.0	34.5	36.0
Hardness (N)	.96	.84	.69	.59	.59	.52
Springiness[b]	86.2	85.8	82.9	82.2	77.9	78.1

[a]From Troutt et al., 1992.

[b]Springiness was calculated as a ratio.

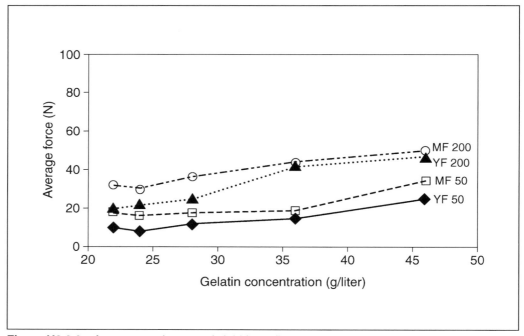

Figure H2.3.2 Average maximum and yield force (i.e., hardness and fracturability) of gelatin gels at different concentrations. Six samples for each concentration were analyzed under compression to 85% with a flat-plate probe. MF, maximum force at 50 and 200 mm/min; YF, yield force at 50 and 200 mm/min. Reprinted from Muñoz et al. (1986) with permission from Food and Nutrition Press.

gelatin gels at different concentrations. The results show that hardness and fracturability increased as the gelatin concentration increased. However, the values of hardness and fracturability were dependent on the crosshead speed. Also, the cohesiveness for gelatin gels was not significantly correlated with the concentration of gelatin (Muñoz et al., 1986).

Limitations of the instrumental TPA when compared to sensory TPA include sensitivity and complexity. However, the instrumental TPA parameters are repeatable. Also, TPA parameters correlate well within sensory TPA parameters. The best correlation and data greatly depend on the measuring conditions, the plunger, the precise sample shape, and, most importantly, the careful and skillful operation of the experimenter.

Time Considerations

The time required for one TPA test of a sample is ~5 to 10 min, which includes sample positioning and instrument cleaning. It is usually the case that sample preparation (e.g., cooking) requires much more time than TPA measurements. Most uniaxial compression instruments offered to the food industry are computer controlled and have software to immediately calculate the TPA parameters.

Literature Cited

Bourne, M.C. 1968. Texture profile of ripening pears. *J. Food Sci.* 33:223-226.

Bourne, M.C. 1974. Textural changes in ripening peaches. *J. Can. Inst. Food Sci. Technol.* 7:11-15.

Bourne, M.C. 1982. Principles of objective texture measurement. *In* Food Texture and Viscosity: Concept and Measurement, pp. 114-117. Academic Press, San Diego.

Friedman, H.H., Whitney, J.E., and Szczesniak, A.S. 1963. The Texturometer–A new instrument for objective texture measurement. *J. Food Sci.* 28:390-396.

Mittal, G.S. 1994. Thermal softening of potatoes and carrots. *Lebensm.-Wiss. Technol.* 27:253-258.

Muñoz, A.M., Pangborn, R.M., and Noble, A.C. 1986. Sensory and mechanical attributes of gel texture. I. Effect of gelatin concentration. *J. Texture Stud.* 17:1-16.

Paoletti, F., Moneta, E., Bertone, A., and Sinesio, F. 1993. Mechanical properties and sensory evaluation of selected apple cultivars. *Lebensm.-Wiss. Technol.* 26:264-270.

Pons, M. and Fiszman, S.M. 1996. Instrumental texture profile analysis with particular reference to gelled systems. *J. Texture Stud.* 27:597-624.

Szczesniak, A.S. and Kleyn, D.H. 1963. Consumer awareness of texture and other foods attributes. *Food Technol.* 17:74-77.

Troutt, E.S., Hunt, M.C., Johnson, D.E., Claus, J.R., Kastner, C.L., Kroph, D.H., and Stroda, S. 1992. Chemical, physical, and sensory characterization of ground beef containing 5 to 30 percent fat. *J. Food Sci.* 57:25-29.

Key References

Breene, W.M. 1975. Application of texture profile analysis to instrumental food texture evaluation. *J. Texture Stud.* 6:53-82.

Reviews developments and applications of the instrumental TPA using the GF Texturometer and the Instron.

Bourne, 1982. See above.

Describes food texture and viscosity including the definition of TPA characteristics using the GF Texturometer and the Instron.

Pons and Fiszman, 1996. See above.

Reviews testing conditions and TPA terminology of the instrumental TPA.

Contributed by Yoshi Mochizuki
University of California
Davis, California

Viscoelasticity of Suspensions and Gels

H3.1   **Dynamic or Oscillatory Testing of Complex Fluids**   **H3.1.1**
     Procedure 1: Determining if a Sample Is at Steady State   H3.1.3
     Procedure 2: Determining the Linear Viscoelastic Region   H3.1.4
     Procedure 3: Obtaining a Mechanical Spectrum or Fingerprint of a Sample   H3.1.6
     Special Procedure: Time-Temperature Superposition   H3.1.7
     Procedure 4: Probing Changes in Structure and Chemorheology   H3.1.7
     Commentary   H3.1.9

H3.2   **Measurement of Gel Rheoogy: Dynamic Tests**   **H3.2.1**
     Basic Protocol   H3.2.1
     Commentary   H3.2.2

H3.3   **Creep and Stress Relaxation: Step-Change Experiments**   **H3.3.1**
     Basic Protocol   H3.3.1
     Commentary   H3.3.4

Contents

1

Dynamic or Oscillatory Testing of Complex Fluids

Real world materials are not simple liquids or solids but are complex systems that can exhibit both liquid-like and solid-like behavior. This mixed response is known as viscoelasticity. Often the apparent dominance of elasticity or viscosity in a sample will be affected by the temperature or the time period of testing. Flow tests can derive viscosity values for complex fluids, but they shed light upon an elastic response only if a measure is made of normal stresses generated during shear. Creep tests can derive the contribution of elasticity in a sample response, and such tests are used in conjunction with dynamic testing to quantify viscoelastic behavior.

This unit describes four procedures for evaluating the viscoelastic properties of complex foods. These procedures should be used for samples with unknown viscoelastic properties. The instrument used for this analysis is a rotational rheometer, which is described in detail in *UNITS H1.1 & H1.2*. References to rheometers that are commercially available can be found in Schoff and Kamarchik (1996). To measure viscosity, a steady shear stress or shear rate is applied, and the sample response is measured in terms of shear rate or shear stress, respectively. A dynamic or oscillatory test is the most sensitive measure of viscoelastic structure; the sample is perturbed with a sinusoidal input (stress or strain), and the sinusoidal response is measured. The relative amplitudes of the two signals and their phase difference (δ; Barnes et al., 1989) are used to calculate modulus and viscosity (Figure H3.1.1). Thus the ratio of stress/strain defines the modulus (G) at a given moment. These parameters are complex, however, because of the sinusoidal nature of the test, and each one may be resolved (i.e., split mathe-

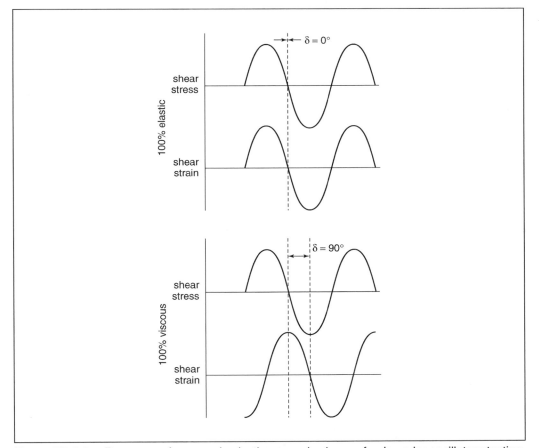

Figure H3.1.1 Responses from purely elastic or purely viscous foods under oscillatory testing. The controlled parameters of the oscillatory applied stress or strain are amplitude, frequency, temperature, and duration of measurement. The measured parameters are amplitude and phase shift (δ) of the strain or stress response.

Contributed by Peter Whittingstall

matically) into an in-phase and an out-of-phase component. The mathematics of this process are similar to vector analysis, where a single vector is resolved into two components at right angles to each other (Figure H3.1.2). From the pythagorean theorem:

$$G* = \sqrt{(G')^2 + (G'')^2} = \frac{\text{stress}}{\text{strain}}$$

The complex modulus, $G*$, is resolved into a real component, G', and an imaginary component, G'':

$$G* = G' + iG''$$

where i is the square root of −1. The real component (G') is completely in phase. It rep-

resents pure elasticity or energy storage, and is thus called the storage modulus. The remainder of the signal (G'') is considered out of phase and imaginary in terms of the elasticity. The energy that is not stored must be dissipated or lost, and thus G'' is called the loss modulus, representing energy lost through viscous dissipation. It is not, however, a measure of viscosity. For that, the complex viscosity, $\eta*$, is used, which also comprises two components, η' (dynamic viscosity) and η'' (which is not normally named, but represents undissipated energy and thus is connected to elasticity):

$$\eta* = \eta' - i\eta''$$

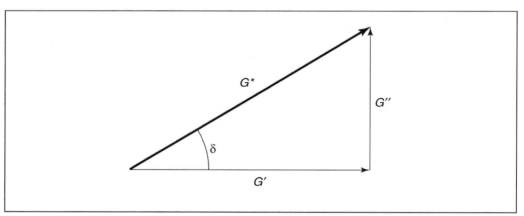

Figure H3.1.2 The oscillatory response of a food material that possesses both elastic and viscous properties can be represented by a complex variable $G*$. This variable has two components that can be expressed as either the Cartesian coordinates G' and G'' or the polar coordinates $|G*|$ and δ. $|G*|$ is the magnitude of the imaginary $G*$ and is measured as the ratio of the amplitudes of stress and strain.

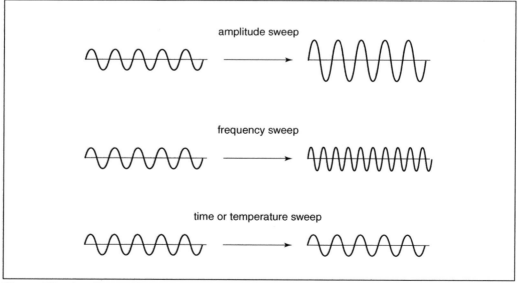

Figure H3.1.3 For oscillatory (sweep) testing, four control parameters can be varied: amplitude, frequency, time, and temperature.

For a complete evaluation of an unknown sample, all four of the following procedures should be performed in the order presented. First, it must be determined whether the sample is at steady state before rheological measurements are undertaken. Second, the linear viscoelastic (LVE) region must be determined. Third, a mechanical spectrum (or fingerprint) of the sample must be obtained using the frequency sweep test. Finally, changes in structure and chemorheology should be evaluated using a temperature sweep test. These procedures require a significant amount of common methodology that has been described in the Strategic Planning section of *UNIT H1.2*. For some samples that have strong solid-like behavior, special preparation is required, the nature of which will depend on the specific sample being analyzed (e.g., gels in *UNIT H3.2*). For the purpose of this discussion, it is assumed that all necessary choices have been made in terms of geometry selection, and that the rheometer has been set up to accept the sample for testing. Schematics of the various test types are shown in Figure H1.1.1. Schematics of the various test variables are shown in Figure H3.1.3. For general rheometry artifacts, see *UNIT H1.2*. Common artifacts such as sample drying, aging, or structure rebuilding are best detected using Basic Protocol 1 of that unit.

PROCEDURE 1: DETERMINING IF A SAMPLE IS AT STEADY STATE

The use of the dynamic (oscillation) time sweep (Figure H3.1.3) is used to monitor changes in viscoelastic structure. In Figure H3.1.4, a grease sample is used as an example of a slow recovery process.

The sample is loaded onto the instrument and the time reference is noted by starting a timer or resetting a timer in the software. A dynamic test for viscoelastic structure is then used to monitor changes in the sample that could result from mechanical relaxation, drying, or thixotropy. A time sweep test is usually performed at a constant temperature. The test is also run at a constant frequency that is comparable to real-time observation (typically 1 Hz) or at a constant angular frequency (10 rad/sec or 1.6 Hz).

The amplitude of stress or strain is held constant at a value that is within the LVE region of the sample. For an unknown material, it can be difficult to determine a reasonable value to use. If the sample is changing with time, however, it is impractical to measure the extent of the linear region until a steady state is reached. To overcome this, a controlled strain amplitude of 0.1% to 1.0% should be set before starting the test. The rheometer will operate directly if it is a controlled-strain instrument or through a feedback loop if it is a controlled-stress instrument. (*UNIT H1.1* discusses the differences between these two rheometer types.) The choice of the actual value is dictated by the type of sample. Polymer solutions and simple liquids can tolerate high strains, so ~1% strain can be used. Gels and pastes are more sensitive, so 0.1% to 0.5% strain should be used. If the stress signal generated by the transducer in a controlled-strain instrument is too low (i.e., in the range of noise), the sample must be re-evalu-

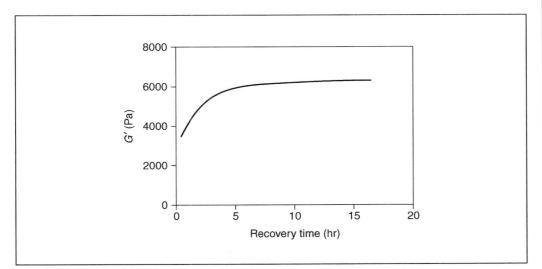

Figure H3.1.4 Structure recovery after loading a sample of a grease or fat on a cone-and-plate fixture on a rotational rheometer. G' was measured over a period of 17 hr at constant oscillatory stress, frequency, and temperature.

Viscoelasticity of Suspensions and Gels

H3.1.3

ated with a more sensitive transducer or a larger geometry.

There may be other settings in the test method, such as timings for data collection, and these should normally be left at the default values. The time for the experiment should typically be set to ~30 min. This is normally adequate, but if steady state occurs more quickly, it is not a problem, because most tests can be interrupted. The instrument should be set to display the following information in real time: storage modulus (G'), loss modulus (G''), corrected phase angle (δ), and the actual percent strain. If the instrument has the capability to display the waveforms visually, this should be switched on.

As the test progresses, the values of G' and, if possible, the stress required to achieve the commanded strain should be recorded. (These data should be used to design subsequent experiments so as to avoid exceeding the stress at which LVE behavior becomes non-linear.) When G' has reached a steady state, its value should be approximately constant (within ±5%). Beware of slow but consistent increases in G' (~1% per data point), as this may be indicative of thixotropic behavior, which will require hours for the system to reach steady state, or of slow consistent drying. The solvent trap cover should be used to eliminate the possibility of slow consistent drying in the experiment. During the test, the stress measured by the transducer, or used by the motor to achieve the commanded strain, should be recorded. This value is important in constructing the next test to evaluate the width of the LVE region.

Once steady state is achieved, the test can be halted and the data saved.

PROCEDURE 2: DETERMINING THE LINEAR VISCOELASTIC REGION

In this experiment, the same frequency and temperature used in the determination of steady state (procedure 1) are used, and the method is otherwise modified for analysis of G' as a function of stress (for controlled-stress instruments) or strain (for controlled-strain instruments). It is generally better to ramp through the stress or strain in logarithmic intervals, although a linear ramp is acceptable. Using the value for the stress needed to achieve a given strain, construct the range as follows. If, for example, a stress of 5 Pa was used to attain 1% strain, the initial stress should be set at 1 Pa or 0.1 Pa (expected strain of ~0.2% or 0.02%, respectively). This is based on the assumption that the preceding test was performed in the linear region so that the stresses and strains are in proportion. If a logarithmic test is to be performed, it is convenient to use decades (0.1%, 1.0%, 10%). A controlled-strain instrument allows for this range to be set directly. The upper limit of stress or strain is ultimately determined by the sample itself, but two to three decades will typically exceed the LVE limit of most materials. Thus, 10% or 100% strain is normally the upper limit for even flexible polymer systems.

Figure H3.1.5 shows the LVE region as a function of strain for a fat product (shortening) at 10 rad/sec. The LVE region is extremely narrow; strains of 0.02% are outside the linear

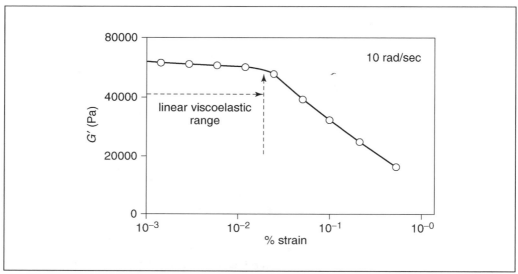

Figure H3.1.5 Determination of the linear viscoelastic range with respect to strain for a food shortening. The linear range appears to exist for strains of ≤0.02%.

zone. These data were acquired on a controlled-stress instrument and would probably be difficult to acquire on typical controlled-strain devices.

The same parameters should be displayed in real time to allow for visual determination of the point at which the linear region ends. Many materials will show a stable G' plateau, and the drop in G' that occurs beyond the LVE region is obvious. Some materials, especially highly filled ones, may always show a decline in G', and this decline is more pronounced in nonlinear behavior. Under such circumstances, it is usual to attribute the limit as occurring at $G' = 90\%(G'_{max})$. Other key indications of nonlinear behavior include a concomitant increase in G'' and δ as G' declines, as well as a change in the shape of the output signal waveform. This asymmetry is usually caused by an increase in the third harmonic in proportion to the fundamental.

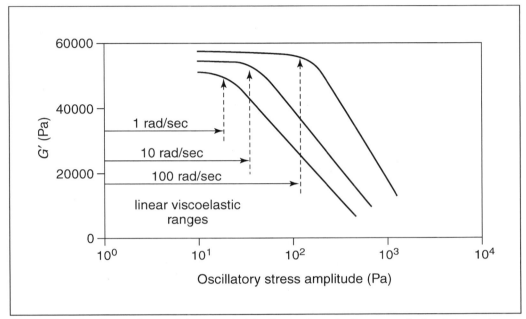

Figure H3.1.6 Dependence of the extent of the linear viscoelastic range of a food shortening on the frequency of the applied sinusoidal shear stress. As the stress frequency decreases, the range of linear behavior also decreases.

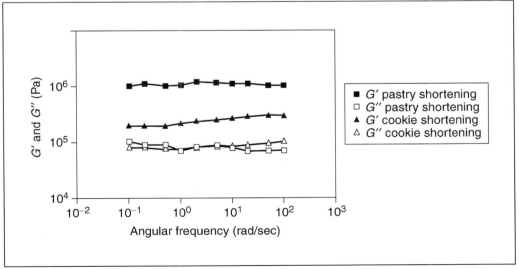

Figure H3.1.7 The use of a mechanical frequency spectrum to distinguish between cookie and pastry shortenings. Although both shortenings have similar G'' responses across different frequencies, their G' responses differ significantly.

Once a range of stresses or strains has been defined for linear behavior at one frequency, the method should be repeated at any other frequencies of interest. For example, if a frequency sweep test is to be run covering 0.1 to 100 rad/sec, then it makes good sense to determine the linearity at the lowest and highest frequencies to avoid obtaining nonlinear data in the test. Typically, a material will show a narrowing of linear behavior in terms of stress as the frequency decreases, but a broadening in terms of strain. Figure H3.1.6 shows the stress effect as a function of angular frequency for a shortening sample. At 1 rad/sec the linear region appears to end at ~15 Pa, at 10 rad/sec it ends at ~30 Pa, and at 100 rad/sec the end occurs at ~80 Pa.

PROCEDURE 3: OBTAINING A MECHANICAL SPECTRUM OR FINGERPRINT OF A SAMPLE

In the frequency sweep test, the idea is to obtain LVE data from the test material over the widest possible (or realistic) range of frequencies. The lower limit of testing is never difficult for a rheometer to achieve physically, but it may be impractical to explore. Typically, the time required to obtain data at frequencies of <0.01 rad/sec or 0.006 Hz is impractical for a laboratory schedule. (At 0.006 Hz, each data point would take 167 sec for a single iteration; most rheometers perform at least two or three iterations.) Furthermore, samples may change or degrade in nonsterile conditions over extremely long tests (i.e., hours). If it is desirable to obtain

low-frequency data, the optimal approach may be to use alternative tests such as the creep test (for controlled-stress instruments) or the stress-relaxation test (for controlled-rate instruments), which are described elsewhere (Barnes et al., 1989). Figure H3.1.7 shows a comparison of two types of shortening. Both samples have the same type of response, but they can be distinguished from one another based on the significant difference in the magnitude of G'. In the figure, both samples appear to be predominantly elastic, as $G' > G''$, and insensitive to frequency. This is to be expected for a soft solid or gelled system.

The rationale behind a frequency sweep is to compare the behavior of a sample at short times (high frequencies) that mimic high shear rates with its behavior at long times (low frequencies) that mimic low shear rates. High shear is typically used to mimic food application (e.g., spreading mayonnaise) or processing, while low shear is typically used when assessing food stability. Another point to remember is that elastic behavior is favored at high frequency, whereas viscous behavior is favored at low frequency. Figure H3.1.8 shows a frequency sweep for an elastic liquid used in a soft drink, in which candy gel beads are suspended without sedimentation. At low frequencies, the values for G' reach a plateau. This is consistent with a sample that has an apparent yield stress. The steep rise in η^* at low frequencies also indicates that the system is stable and will be able to support the beads for prolonged periods.

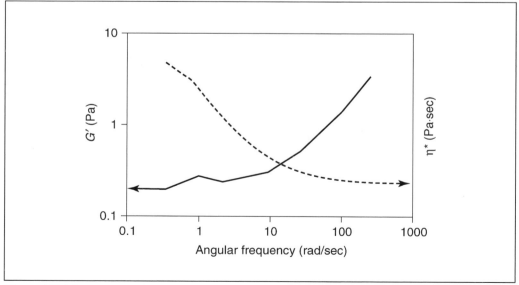

Figure H3.1.8 The characterization of a soft drink that shows a high complex viscosity, η^*, at low frequencies. This corresponds to a high viscosity at low shear rates, which gives the drink the ability to prevent the sedimentation of suspended gel beads. Solid line, G'; dashed line, η^*.

The frequency sweep can be performed at a constant stress or a constant strain amplitude, and care should be taken in selecting a value for either parameter. Viscoelastic materials, especially polymeric materials, may change their linear behavior quite drastically as a function of frequency. If the dynamic range of the rheometer has a limitation (e.g., stress transducer range in a controlled-strain instrument), then it may be feasible to test only two to three decades of frequency. At low frequencies the stress signal will drop dramatically for a given strain, presenting problems. In some sense the same issue is seen in a controlled-stress instrument at high frequencies, because the position (strain) data get very small for a given stress. To obtain wider ranges of data under these circumstances, the testing should be repeated at the new frequencies using different strains and stresses.

To cover the widest range of frequencies in a single test, a controlled-stress device operating in a controlled-strain mode, or the equivalent test for a controlled-strain device, will provide the optimal configuration because the stresses or strains, respectively, are boosted to stay within the dynamic range of the instrument. If the results for LVE behavior have not been generated, then a fixed strain of 0.1% to 1.0% is likely to be adequate. As before, lower stresses should be used if the sample is a dispersion or emulsion, and higher stresses should be used if it is polymeric in form.

Once a stress or strain has been selected, the range of frequencies is specified, usually in a logarithmic sense. Thus, if 10 rad/sec was used in the initial test to find the LVE region, then a frequency sweep from 1 to 100 rad/sec would be useful. A broader sweep of 0.1 to 1000 rad/sec (or the upper limit of the instrument) would give proportionately more information.

SPECIAL PROCEDURE: TIME-TEMPERATURE SUPERPOSITION

To obtain as much information as possible on a material, an empirical technique known as time-temperature superposition (TTS) is sometimes performed. This technique is applicable to polymeric (primarily amorphous) materials and is achieved by performing frequency sweeps at temperatures that differ by a few degrees. Each frequency sweep can then be shifted using software routines to form a single curve called a master curve. The usual method involves horizontal shifting, but a vertical shift may be employed as well. This method will not work for all materials. The method requires that the sample be thermorheologically simple, which means that its relaxation times change in a regular fashion as a function of temperature. Frequency sweep data are often analyzed using viscoelastic models. Generally, the parameters analyzed are G' and η', the in-phase components of elasticity and viscosity, respectively.

PROCEDURE 4: PROBING CHANGES IN STRUCTURE AND CHEMORHEOLOGY

The dynamic temperature sweep has many applications. First, it is used to follow changes in state, such as a melting phenomenon in chocolate, a glass transition in confectionary, or the baking process in a dough or batter. Second, it is used to follow irreversible gelation or curing in a sample. Many samples can be made to form a gel under the right conditions of temperature, and it is of great interest to follow the kinetics of structure buildup in real time, without disrupting the process itself. The latter test is normally done in a parallel-plate geometry, whereas the former could use cone-and-plate, parallel-plate, or concentric-cylinder geometries (Figure H1.1.1). Other examples include phase changes in lipid systems or emulsions.

A single frequency is traditionally used (1 Hz or 10 rad/sec) and a fixed strain or stress is applied. To obtain data points more rapidly it is advisable to use higher frequencies. The changes in structure can be very large, as many gels start out as weak liquids. For such samples, it is best to use a fixed strain (e.g., 0.1% to 1.0%). The duration of the test depends on the sample and the rate of the temperature ramp. It is important to ramp the temperature at a meaningful rate without introducing temperature gradients (normally, 1° to 5°C/min), with the lower ramp rates being preferred if run time is not a limitation. Figure H3.1.9 depicts the heating of cookie dough and its softening and then stiffening as it cooks.

Certain processes in a material are accelerated by increasing the temperature, such as phase separation in emulsions. A convenient way of predicting formulation stability is to run a temperature sweep and observe the response of G' or $\tan(\delta)$. Samples that are stable will tend to show almost no major change in these parameters, whereas unstable materials will change sharply at a critical temperature. Figure H3.1.10 shows comparative data for four emulsions that are divisible into two pairs: one pair that is likely to be stable and another that is

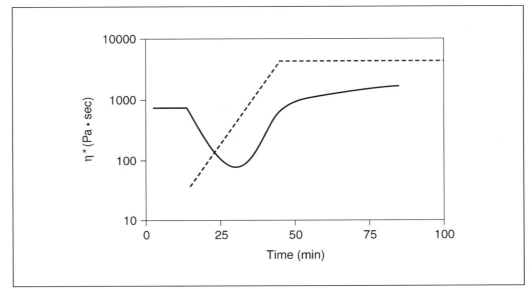

Figure H3.1.9   The characterization of the baking of a cookie dough by the measurement of the complex viscosity (η^*, solid line) as a function of temperature (°C, dashed line).

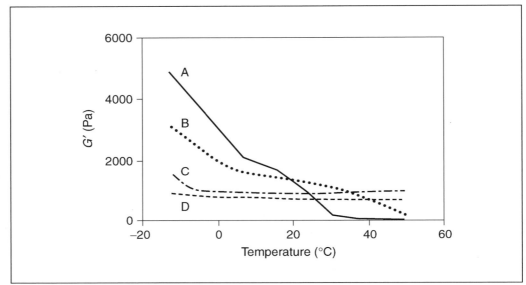

Figure H3.1.10   The temperature stability of four emulsions. Emulsions A and B show a high dependence of rheological stability on temperature. Emulsions C and D have rheological properties that are relatively independent of temperature.

likely to be unstable. Samples A and B clearly undergo transitions and lose elastic structure, whereas C and D are relatively unaffected. Consequently, it is easy to distinguish between the formulations to predict the more successful ones (i.e., C and D).

Another option is to use a multiwave technique, whereby the fundamental frequency and several other frequencies (harmonics) are added together into a single complex wave. Each one of these multiwave iterations can be deconvoluted into its components after the test is complete. Thus, each complex data point can be multiplied into several data points. Setting up such procedures involves assigning stresses or strains to each of the components, and there is a danger that the linear viscoelastic region of the sample will be exceeded because these stresses and strains may not add in a linear fashion. Often a series of preliminary experiments at each of the desired frequencies must be done in order to configure the multiwave test correctly. Because instruments vary in how they perform this type of procedure, it is beyond the scope of this work to describe this test in detail.

Because the sample may undergo significant changes in volume during the test, it is an advantage to use an instrument with an automatic feature to maintain a gap in response to the normal forces applied to the plates. Such a feature needs to be able to maintain a constant normal force within a window and to modify the gap in real time to prevent the sample from either overflowing the plates (i.e., by increasing the gap as normal force increases) or shrinking (i.e., by decreasing the gap as normal force decreases). This is normally a feature of more expensive systems because it requires a normal-force sensor. Parallel-plate systems are necessary for this work; a cone-and-plate system cannot generally be used. In general terms, the instrument should be set to keep the normal force within 10% to 20% of its start value and also to allow changes in the gap of ~10% to 20%. Note that this feature works in addition to the gap temperature compensation present in most automatic gap-setting machines.

Data obtained over a wide temperature range for a polymeric material that is not thermosetting will reveal features around the glass transition (T_g; also known as the glass-to-rubber transition) such as peaks (G'' and tan[δ]) and onsets (places where a plateau gives way to a new linear region with a negative slope). These features and the temperatures at which they occur are often of great interest to the investigator. These are analyzed by the rheometer software in a systematic way to identify the temperature at which they occur.

COMMENTARY

The mathematics for deriving these parameters are fairly simple as long as the rules of linear viscoelasticity are obeyed. For this, the stress and strain used to do the calculation must be sufficiently small that they are in linear proportion. For example, if the applied stress amplitude is doubled, then the measured strain amplitude should also double, resulting in a relatively constant value for G^*. For polymers, this region of behavior is very long, because the nature of polymers is to withstand relatively large-scale deformations. Dispersions and emulsions are less rugged, as each component is relatively small and naturally sits in its own energy minimum. Any displacement from this position is therefore likely to be unfavorable.

Although the magnitude of the stresses defining the LVE region of behavior is important, it is not intuitive to the average person. What does a stress of 20 Pa feel like? Strains are perhaps more accessible. If we were to consider

extension as a mode of deformation, we could consider the rack of Middle Ages renown as a form of tension tester used to test the resolve of its victims. A strain of 0.1% to 1.0% from such a device might be uncomfortable, but is essentially survivable as our bodies compress by that much 1-in. (2.5 cm) in a 72-in. (183-cm) adult just from walking around all day. In the same sense, most materials will accept such strains without irreversible damage (flow). Similarly, just as we would not care to contemplate being stretched by 6 in. (15 cm) in height, most samples will deform irreversibly at strains of ~10%. Polymers, which can withstand up to 100% strain on occasion, are the exception. The presence of other materials (filler) will reduce this propensity.

All samples have to be loaded into the rheometer for testing, and this process will deform the sample in some way. For simple liquids (e.g., oils, sugar solutions), this is not a problem because they will recover instantly. In contrast, structured materials such as gels or pastes will be damaged. Changes in structure may result from molecular rearrangements or other actions such as drying through solvent loss. Some samples may recover (e.g., if they are thixotropic), whereas others may not (e.g., if they are gels and cross-links are broken). Whatever the sample, the loading process is traumatic, and thus the first step is to see if the sample is at equilibrium after being loaded (procedure 1). If it is still changing with time, then further experiments must wait until an equilibrium level of structure is reached. Structure is most conveniently defined as the magnitude of G', because this property is the most susceptible to damage during shear. Thus, measuring the changes in G' as a function of time (time sweep) will allow for accurate assessment of sample stability.

The torque sweep (or strain amplitude sweep in controlled-rate instruments) is the logical and necessary next step in assessing any material. Using increasing amplitudes, the sample is deformed in a reversible fashion (i.e., no net flow occurs), and the dynamic test is thus a sensitive mechanical real-time probe. This is especially useful if a sample is changing because of temperature effects i.e., it is melting or going through a T_g. Alternatively, an isothermal change such as gelation or curing can be accurately tracked without the test itself interfering with the change of state. Thus, a time-sweep test or a temperature-sweep test is likely to be the method of choice for materials undergoing such physicochemical changes. Such

Viscoelasticity of Suspensions and Gels

H3.1.9

tests are a logical conclusion to the characterization tests, time sweep, torque or amplitude sweep, and frequency sweep.

A high priority for measurement is to define the LVE region, where the sample obeys the mathematical rules for the calculation of the various parameters. Data obtained outside of this region are qualitative only. The type of test used here is an amplitude sweep. The stress amplitude or the strain amplitude are varied at a constant frequency, and the region of linearity is clearly seen in a plot of G' (the most sensitive index of structure) versus the stress or strain. Increasing the stress or strain allows for the limit of viscoelastic behavior to be seen at a given frequency. Other methods rely on examining the sine waves themselves, as nonlinear data often result in visually deformed or asymmetric waveforms. An analysis of the third harmonic of the output is also insightful.

It is important to define the LVE region at all frequencies and temperatures of interest in subsequent tests. The length of the LVE region is frequency dependent; as the frequency decreases, the LVE region will shorten (narrow) as a function of stress and widen as a function of strain. The final aim of these preparative tests is often to examine the sample at different frequencies, performing a kind of mechanical spectroscopy upon it, and the data must remain within the linear response region at all times. It is important to remember that the response of a sample is a strong function of time. Elasticity can manifest itself at very short time intervals, but viscosity by definition is a time-consuming response. Thus, a test that scans frequencies will favor an elastic response at short times (high frequency) and a viscous response at long times (low frequency).

It is important to remember that all materials have a characteristic response time, varying from picoseconds for simple liquids like water to years for more traditional solids. If a sample appears to be a mobile liquid as it is disturbed in its container, then it will have a characteristic time of well under a second. If a sample appears to be an immobile solid, then it will have a characteristic time of several minutes or hours. Increasing the temperature of the sample will speed up molecular motion and thus decrease (shorten) the characteristic response time. Cooling the sample will have the opposite effect. If heating the sample is not an option, the only recourse is a long experiment!

Another aspect of measurement is the perception of the sample response. The simple polymer polydimethylsiloxane (PDMS),

which is used to make Silly Putty, is a material that is an excellent model of viscoelastic behavior. Its characteristic response time does not vary at a constant temperature, although its apparent properties certainly do. When the container is opened, Silly Putty is seen to have found its own level just like a liquid, albeit a very viscous one. The moment it is subjected to a short impact, however, it bounces like an elastic solid. The difference is the amount of time allowed for the test; a long test is likely to reveal some liquid-like character, whereas a short test is likely to reveal only solid-like properties.

When taken to extremes, one can imagine even traditional liquids seeming to perform like a solid if the experiment is short enough. Consider the following thought experiment. From a boat, it is a pleasant experience to jump into a body of water. Imagine the same test repeated from an aircraft at high altitude without the benefit of a parachute. The water will not change but your perception of it most definitely will! Similarly, a sharp tug on an adhesive bandage is well known as the method of choice for its removal if a few body hairs are to be spared. The short time prevents the adhesive from responding like a liquid and wetting the skin or hairs as effectively. Otherwise, an extremely slow pull would allow the adhesive to flow away from the skin and hairs. Materials are tacky to the touch if they can flow into the crevices of the skin in a reasonably short time frame, but cooling the material inevitably makes it a little easier to handle. That is why pastry chefs recommend handling dough at low temperatures and with a protective coating of flour.

Samples are mechanically brittle at sufficiently low temperatures, normally below the T_g. They also are seen to be brittle at very short times. Thus, a short sharp impact can shatter a material at room temperature if it is naturally below its T_g or if the material has been frozen. Materials that are rubbery at room temperature can be sufficiently brittle at low temperatures to fail catastrophically. In the space shuttle Challenger, an explosion was caused by uncombusted fuel escaping when a rubber O-ring was rendered brittle by low-temperature weather. The key in all of these examples is the speed of molecular motion. For energy to be dissipated, a stress can cause a local increase in molecular motion. If that route is denied by time constraints (e.g., a fast shock) or temperature control (molecular immobility), then a crack is the only way energy can leak out. Cold toffee is easily broken into bite sized pieces but will flex and flow into messy strings when left out to soften.

A frequency sweep can reveal these behavior patterns if it is collected over a sufficiently wide range. Low-frequency data are not usually difficult to obtain (provided that time is not a problem), but there are mechanical limitations to obtaining high-frequency data. At frequencies >10 Hz, it becomes impossible to correct for the inertia of the moving parts in the rheometer.

Other dynamic tests are more pragmatic in application, as they form a means of quantitatively monitoring the viscoelasticity of a material as it changes in real time or as a function of temperature. This means that melting, crystallization, gelation, and curing can all be followed without the test itself affecting the results.

Finally, oscillation data as a function of angular frequency can be used to extend the operating range of a flow test in unfilled polymer melts or solutions. These materials are often so elastic as to tear themselves out of the geometry gap during steady shear, even at relatively low shear rates (~1 Hz). An oscillation test can generate a value for complex viscosity (η^*) at a particular angular frequency that corresponds exactly to the value for η at the same shear rate. This empirical rule is known as the Cox-Merz rule and generally will not work for dispersions, emulsions, or filled polymer systems. Oscillation data can easily be generated at ~100 rad/sec, thereby extending the apparent range of the combined data by two orders of magnitude.

LITERATURE CITED

Barnes, H.A., Hutton, J.F., and Walters, K. 1989. An Introduction to Rheology. Rheology Series, Vol. 3. Elsevier, Amsterdam.

Schoff, C.K. and Kamarchik, P. 1996. Rheological Measurements. *In* Kirk-Othmer Encyclopedia of Chemical Technology, 4th ed., Vol. 21, Recycling, Oil, to Silicon (J.I. Kroschwitz and M. Howe-Grant, eds.) John Wiley & Sons, New York.

KEY REFERENCES

Barnes, H.A. 2000. A Handbook of Elementary Rheology. The University of Wals Institute of Non-Newtonian Fluid Mechanics, Aberysthyth, U.K.

Larson, R.G. 1999. The Structure and Rheology of Complex Fluids. Oxford University Press, Oxford.

Rao, M.A 1999. Rheology of Fluids and Semisolid Foods: Principles and Applications. Aspen Publishers, Gaithersburg, Md.

Steffe, J.F. 1996. Rheological Methods in Food Process Engineering. Freeman Press, East Lansing, Mich.

The above references provide an excellent general reading list for more information on complex fluids.

Contributed by Peter Whittingstall
ConAgra Foods
Irvine, California

Measurement of Gel Rheology: Dynamic Tests

This unit describes dynamic methods for monitoring the gelation process of protein dispersions and for determining rheological properties of the final gels. In a dynamic test, a test material is subjected to a controlled oscillation, the frequency of which determines the rate of deformation of the sample. The magnitude of deformation is kept very small to prevent destruction of gel structures during measurement. Two basic parameters reflecting elastic (G'; storage modulus) and viscous (G''; loss modulus) components of rheological properties are obtained at various frequencies. This frequency range will depend on the sample type and the purpose of the study (see Critical Parameters and Troubleshooting). The measuring instruments are operated by computers and the experimental procedure is simple. *UNIT H3.1* discusses much of the theory behind dynamic rheometer tests.

Materials

Sample protein solution or gel
Chemical substance to induce gelation, if needed
Immiscible reagent for preventing solvent evaporation (e.g., mineral or silicone oil), if needed

Dynamic rotational rheometer with appropriate test fixture
Computer with software package to control rheometer

1. Attach an appropriate test fixture to a dynamic rotational rheometer and connect rheometer inline with a computer.

    The appropriate text fixture type (or geometry) and size depend on the substance that is being analyzed. UNIT H1.2 includes a discussion on choosing a geometry.

    The amount of sample needed will depend on the rheometer and test fixture that are used. Ideally, there should not be excess sample (e.g., below or above the inner cylinder or outside the upper plate). Usually, sample below or above the inner cylinder does not contribute significantly to the results because of its smaller contact area compared with that of the wall of the cylinder. For steady rotational viscometry, it is often critical to cut the sample outside the upper plate. Thus it is also recommended to do so in dynamic tests. Changes in sample volume that often occur during gelation must also be considered.

2a. *For a gel induced by a chemical substance:* Mix a sample protein solution with a chemical substance to induce gelation, place the mixture in the test fixture, and cover it with an immiscible reagent.

    The chemical substance required to induce gelation, as well as the concentrations of chemical and protein, will need to be determined empirically.

2b. *For a gel induced by heating:* Place a sample protein solution in the fixture, cover it with an immiscible reagent, and use the computer to apply the appropriate temperature to the fixture for an appropriate length of time.

    The time and temperature of the incubation, as well as the protein concentration, will need to be determined empirically. Depednding on the purpose of the study, a temperature ramp or constant temperature can be used here and in steps 3 to 5.

2c. *For a preformed gel:* Place a sample gel in the fixture and cover it with an immiscible reagent, if necessary. Continue with step 4.

3. Set the computer to display G' and G''. Monitor G' and G'' during gel formation at, for example, 1 Hz and a maximum strain of 0.01.

    For a given gel, preliminary tests are required to determine an appropriate frequency and strain.

Contributed by Shinya Ikeda and E. Allen Foegeding

4. Measure the frequency dependence of G' and G'' after completion of the gelation.

For a further discussion of a frequency sweep test, see UNIT H3.1. The range of frequencies used here and in step 5 will depend on both the sample and the rheometer type.

5. Measure the strain dependence of G' and G''.

COMMENTARY

Background Information

Recent advances in dynamic rotational rheometers are of growing importance in food analyses for several reasons. (1) The measurement minimizes the destruction of the material. (2) The time required for a measurement is reasonably short in comparison with chemical or physical changes in the material. (3) The viscoelasticity of gels is characterized by determining G' and G'' in the linear viscoelastic (LVE) region; no other method gives dynamic moduli values.

Protein gelation is induced by many factors such as temperature, pH, or additives. Protein molecules in a medium become reactive under appropriate conditions and, at a certain probability, will stick to one another upon collision. The resultant aggregate diffuses and forms larger aggregates by linking to other molecules or ag-

gregates. If the protein concentration is sufficiently high, a network structure percolating in the entire system is eventually formed. This overall process is called gelation. Because protein gels generally show combinations of ideal rheological behavior (i.e., the behavior of both an ideal elastic solid and an ideal viscous fluid), protein gels are considered viscoelastic materials.

The relative magnitudes of elasticity and viscosity for a viscoelastic material depend on the scale of the observation time. If the rate of a deformation is very slow, a material may behave more like a viscous fluid. The faster the deformation occurs, the more elastic the material appears. Therefore, in the measurement of viscoelasticity, the rate of deformation should be chosen based on a practical situation of interest (e.g., the rate of processing or mastica-

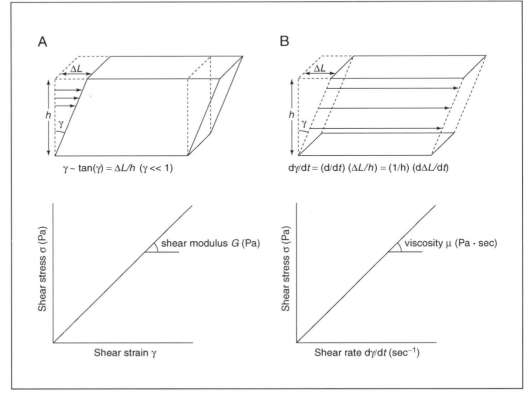

Figure H3.2.1 Deformation pattern of a substance in response to shear. (**A**) An ideal elastic solid subjected to shear. (**B**) An ideal viscous fluid subjected to shear. h, height; ΔL, displacement in length.

tion) to obtain a meaningful result. From another point of view, a material can be characterized by determining its mechanical response over a wide range of deformation rates, which is also within the scope of a dynamic test.

An elastic solid has a definite shape. When an external force is applied, the elastic solid instantaneously changes its shape, but it will return instantaneously to its original shape after removal of the force. For ideal elastic solids, Hooke's Law implies that the shear stress (σ; force per area) is directly proportional to the shear strain (γ; Figure H3.2.1A):

$$\sigma = G\gamma$$

Equation H3.2.1

where the proportional constant G (Pa) is the shear modulus, a measure of the elastically stored energy (see also *UNIT H3.1*). In contrast, a viscous fluid has no definite shape and flows upon application of an external force. Applying a shear stress to an ideal viscous fluid (Newtonian fluid) results in a homogenous layer flow (Figure H3.2.1B). A material can be characterized by a parameter reflecting an internal resistance to flow. For a Newtonian fluid, the flow rate, expressed as shear rate ($d\gamma/dt$), is directly related to shear stress:

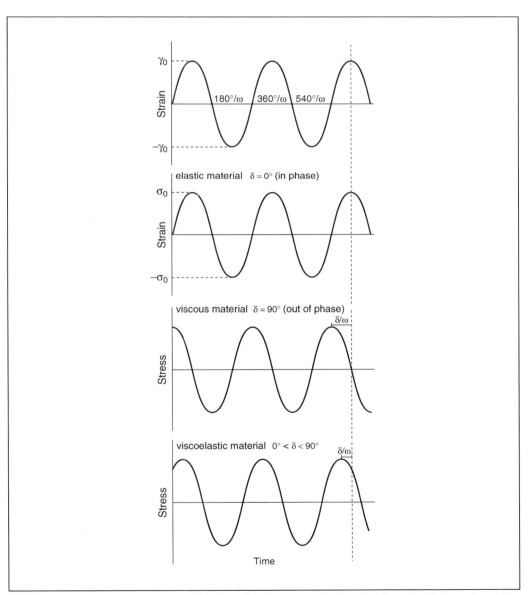

Figure H3.2.2   Responses of an ideal elastic, viscous, and viscoelastic material to a sinusoidal deformation. δ, phase angle; γ, shear strain; ω, angular frequency; σ, shear stress.

Viscoelasticity of Suspensions and Gels

H3.2.3

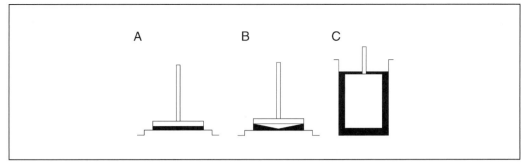

Figure H3.2.3 Test fixture geometries. (**A**) Parallel plates. (**B**) Cone and plate. (**C**) Concentric cylinders. The black regions correspond to sample material.

$$\sigma = \mu \frac{d\gamma}{dt}$$

Equation H3.2.2

where the proportional constant μ (Pa·sec) is Newtonian viscosity.

A viscoelastic material is represented by a combination of elastic and viscous bodies. If a controlled sinusoidal small strain (γ) is applied to a viscoelastic material as follows:

$$\gamma = \gamma_0 \sin(\omega t)$$

Equation H3.2.3

then the stress response (σ) is:

$$\sigma = \sigma_0 \sin(\omega t + \delta)$$

Equation H3.2.4

where γ_0 is the strain amplitude (maximum strain), σ_0 (Pa) is the stress amplitude (maximum stress), ω is the angular frequency (which

is equal to $2\pi f$, where f is the frequency), and δ is the phase angle, reflecting the phase shift between γ and σ due to viscoelasticity. Equation H3.2.4 can also be expressed as the sum of in-phase and 90°-out-of-phase components with the strain:

$$\sigma = G'\gamma_0 \sin(\omega t) + G''\gamma_0 \cos(\omega t)$$

Equation H3.2.5

where G' (Pa) is the storage modulus and G'' (Pa) is the loss modulus. When comparing Equation H3.2.5 with Equations H3.2.1 and H3.2.2, one notices that the first and second terms of Equation H3.2.5 are the same form as Equations H3.2.1 and H3.2.2, respectively. That is, the term $G'\gamma_0\sin(\omega t)$ is in the form of a constant (G') multiplied by strain ($\gamma_0\sin(\omega t)$), which is the same as Equation H3.2.1. The term $G''\gamma_0\cos(\omega t)$ is in the form of a constant (G''/ω) multiplied by shear rate ($\omega\gamma_0\cos(\omega t)$), which is the same as Equation H3.2.2. Note that the

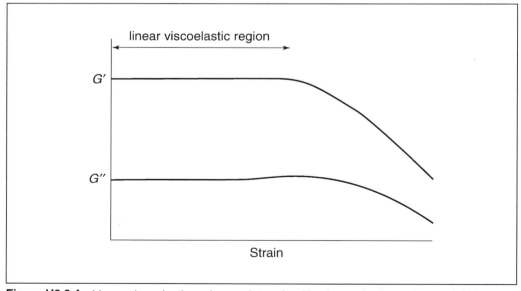

Figure H3.2.4 Linear viscoelastic region as determined by the strain dependence of G' (storage modulus) and G'' (loss modulus).

shear rate is given by differntiating Equation H3.2.3 with respect to time, $d\gamma/dt = \omega\gamma_0\cos(\omega t)$. Therefore, the storage modulus G' reflects the elastic component of viscoelasticity, namely, a measure of the elastically stored and recovered energy per cycle of deformation, and the loss modulus G'' reflects the viscous component, namely, a measure of the energy dissipated as heat. For an ideal elastic solid, the entire applied energy is stored (i.e., $G'' = 0$), and thus the strain and the stress are in phase. For an ideal viscous fluid, all the applied energy dissipates as heat (i.e., $G' = 0$), and thus the strain and the stress are 90° out of phase. These relationships are schematically illustrated in Figure H3.2.2.

By comparing Equations H3.2.4 and H3.2.5, the following equations are obtained.

$$G' = \frac{\sigma_0}{\gamma_0}\cos(\delta)$$

Equation H3.2.6

$$G'' = \frac{\sigma_0}{\gamma_0}\sin(\delta)$$

Equation H3.2.7

$$\frac{G''}{G'} = \tan(\delta)$$

Equation H3.2.8

The term $\tan(\delta)$ is referred to as the loss tangent, a measure of the relative magnitude of the viscous to the elastic component, or the relative magnitude of lost to stored energy per cycle deformation. Predominately elastic (solid-like) material has a $\tan(\delta) < 1$ ($G' > G''$), whereas a fluid-like material has a $\tan(\delta) > 1$ ($G' < G''$). Additional parameters can be defined, such as the complex modulus G^* ($G' + iG''$; where i is the imaginary unit).

$$\sigma = G^* \gamma$$

Equation H3.2.9

$$|G^*| = \sqrt{G'^2 + G''^2}$$

Equation H3.2.10

Hooke's Law, which states that a proportional relationship exists between stress and strain, usually holds for a viscoelastic material at a small strain. This phenomenon is called linear viscoelasticity (LVE). Within the LVE region, the viscoelastic parameters G' and G'' remain constant when the amplitude of the applied deformation is changed. Consequently, parameters measured within the LVE region are considered material characteristics at the observation time (frequency).

Various rheological instruments have been developed. Some instruments are designed for a specific purpose and give only instrument-specific parameters. A gel rheology test using this type of instrument is called an empirical test, which is of value if a correlation with a property of interest is found. A fundamental test has the advantage over an empirical test in determining true (i.e., material-characteristic and instrument-independent) physical properties. Some fundamental rheological tests apply a large deformation up to the point at which the material fractures, and they correlate well with

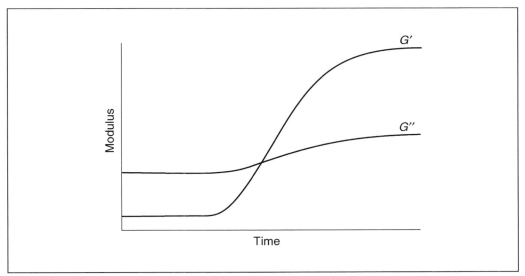

Figure H3.2.5   Development of moduli during gelation. G', storage modulus; G'', loss modulus.

results of sensory texture analyses (Montejano et al., 1985). A large deformation, however, can push the material beyond its LVE region, at which point viscoelasticity becomes a function of the magnitude of the deformation as well as of time. If this is the case, a mathematical or quantitative interpretation of the results is very complex, if it is even possible. Therefore, a fundamental test applying a small deformation is more suitable for analyzing the general rheological properties of food gels.

Critical Parameters and Troubleshooting

Most rheometers that are commercially available for a dynamic test are designed so that a controlled strain or stress is applied to the test material (Bohlin et al., 1984). A sample is placed between the two parts of a test fixture (i.e., two plates or a cone and plate). For most strain-controlled instruments, the rotationally controlled piece transmits an input to the sample and the response is detected by the stationary piece. The common types of test fixtures are the cone and plate, parallel plates, and concentric cylinder (Shoemaker et al., 1987; Figure H3.2.3). Cone-and-plate and parallel-plate fixtures are easy to set up and facilitate the calculation of dynamic moduli, whereas the concentric cylinder fixture is preferred for preventing water loss because the sample surface can easily be covered with immiscible fluids.

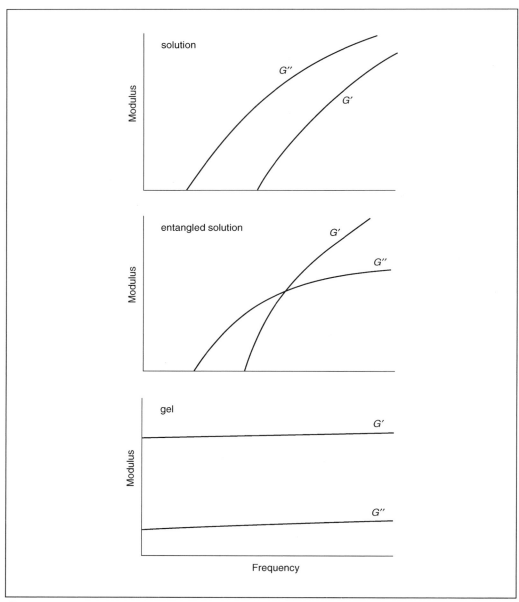

Figure H3.2.6   Three types of materials that are rheologically distinguishable based on a frequency sweep test. G', storage modulus; G'', loss modulus.

The size of the fixture is chosen based on the testing material and the sensitivity of the instrument. (Usually, various sizes and types of fixtures are available for an instrument.) Generally, responding torque becomes larger as the diameter of the fixture is increased or as the gap between the fixture surfaces decreases.

Before starting a gelation process, the operator needs to select the magnitude and frequency of the applied strain or stress. Theoretically, smaller applied deformations and lower frequencies are less likely to damage the gel network structure. A lower limit of deformation exists, however, to gain a good signal-to-noise ratio. At the same time, the magnitude of input should be chosen to ensure LVE as follows. The LVE region of the testing sample is established by running stress or strain sweeps on the material. In each case, the magnitude of the input is increased while moduli are determined (Figure H3.2.4). The LVE region is observed at a low deformation region where the moduli remain constant. As the magnitude of the input is increased, the moduli will suddenly start to decrease or increase at a certain strain or stress value, which is the limit of LVE. If various frequencies are used in the test, strain or stress sweep measurements should be done at least at the upper and lower extremes of the frequency range to ensure linearity at all frequencies. The strain-controlled measurement is recommended for testing materials that show a drastic change in rheological properties during the measurement (e.g., from sol to gel), because a stress necessary for measuring a strong gel would cause a substantial strain when applied to a weaker gel, which may exceed the LVE region.

In the case of heat-induced gelation, a step increase in temperature may not be available for some instruments. Alternatively, the temperature is raised at a fixed rate while the rheological parameters are monitored. Various kinetic phenomena such as denaturation and aggregation of proteins take place as the temperature increases, and thus the results obtained during heating cannot be considered absolute, as they depend at least on the heating rate. Attention should be paid to the heating rate or the size of the sample when comparing the results with other analytical measurements such as differential scanning calorimetry (DSC).

Anticipated Results

Figure H3.2.5 schematically shows changes in G' and G'' when a protein solution is placed under a constant high temperature. Prior to gelation, the material shows a typical fluid-like behavior ($G' < G''$). If the size of protein aggregates becomes large enough, G' increases rapidly, and after some time, a cross-over point ($G' = G''$) is observed. This point and the corresponding time are often referred to as the gel point (gelation point) and the gel time (gelation time), respectively (Clark and Ross-Murphy, 1987; Djabourov, 1988; Clark, 1992). As gelation progresses, G' becomes dominant, showing characteristics of solids ($G' > G''$). In some test systems, a cross-over of G' and G'' may not be observed. As G' is initially almost zero

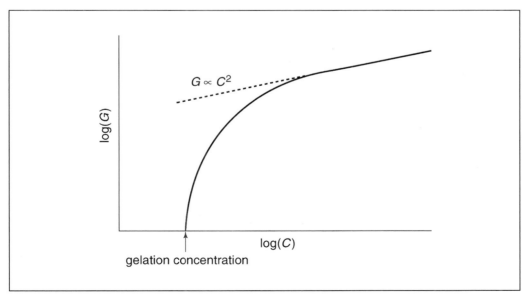

Figure H3.2.7 Concentration dependence of gel modulus (G). G can be either the storage modulus, G', or the complex modulus, G^*, as both usually coincide within experimental error. C, concentration.

because of the fluid-like nature of the material, the gelation point can be defined as the point where G' begins to increase rapidly. This point may be determined by an extrapolation depending on the sensitivity of the measuring instrument. It should be noted that the gel point determined by this procedure usually depends on the measuring frequency. Theoretical aspects of determination of the gel point have been considered in the literature (Winter, 1987; Scanlan and Winter, 1991). Comprehensive reviews regarding theoretical treatments of the gelation process of biopolymers are also available (Clark and Ross-Murphy, 1987; Ziegler and Foegeding, 1990). Although the fundamental theories of gelation were developed for describing the gelation process of synthetic polymers via linking by covalent bonds (Stockmeyer, 1943; Flory, 1953), modern progress in physics, such as advances in percolation theories, has allowed their application to the analysis of a sol-gel transition in protein gels (de Gennes, 1979).

A strain or stress sweep is used to establish the LVE region (Figure H3.2.4). The LVE region is a characteristic of a material. While the strain value at the limit of LVE rarely exceeds 0.1 for colloidal gels, a larger LVE region with a strain of up to 1 or more is usually observed for biopolymer gels (Clark and Ross-Murphy, 1987).

Although it is almost impossible for a protein gel to reach an equilibrium state (G' continues to increase gradually for hours or days), G' becomes approximately constant a few hours after gelation time (Figure H3.2.5). A gel that has experienced an aging process can be called a cured gel, and a characterization of a cured gel is also in the scope of a dynamic test. Although available frequencies are usually limited to a range of 0.001 to 10 Hz, frequency sweep measurements within the LVE region reveal the type of material (Djabourov, 1988; Scanlan and Winter, 1991; Ross-Murphy, 1994).

Typical patterns of G' and G'' produced from frequency sweep experiments are classified into three categories (Figure H3.2.6). A dilute solution of macromolecules shows $G' < G''$ at most frequencies. With increasing concentration, interactions among molecules become pronounced in what is called an entangled or semi-dilute solution. An entangled solution has a distinguishing feature of a cross-over point where $G' = G''$ or $\tan(\delta) = 1$. The material is fluid like at a frequency below the cross-over point, but solid like at a frequency above the cross-over point. The frequency at the cross-over point is thus also referred to as the relaxation frequency. If a gel is formed, G' is predominant ($G' > G''$) and both G' and G'' are relatively independent of frequency. If ideal cross-links are formed by permanent covalent bonding, the moduli are completely independent of frequency. A protein gel, however, usually shows a slight frequency dependence and is called a physical gel. When $\log(G')$ is plotted against $\log(f)$, the slope is slightly greater than zero and is typically less than 0.1. More elastic gels have lower slope values, whereas more viscous gels have higher slope values.

There have been attempts to widen the frequency scale based on a procedure called time-temperature superposition (TTS; Tokita et al., 1983; Dea et al., 1984; *UNIT H3.1*). Because this procedure assumes that only the number and the strength of cross-links are dependent on temperature, it must be determined whether the material undergoes physical or chemical changes following temperature changes. It may be possible to widen the frequency range using a few different methods or instruments based on different principles (Nishinari, 1976), although a substantial change in the shape of the spectrum may not be expected (te Nijenhuis, 1981).

The concentration (C) dependence of G' has generated much interest not only from a practical point of view but also from a scientific standpoint. When $\log(G')$ is plotted against $\log(C)$, a power law relationship is often observed. The slope of the power law is dependent on the concentration range: the slope is large (5 to 7) when the concentration is close to the critical gelation concentration, becomes smaller with increasing concentrations, and then converges to a value close to 2 (Figure H3.2.7). Regarding the relationship between G' and C, various models (Hermans, 1965; Oakenful, 1984; Clark and Ross-Murphy, 1987), including the recent introduction of fractal geometry (Bremer et al., 1990; Shih et al., 1990), have been developed.

Time Considerations

Because the duration for one measurement is very short (e.g., with a 1-Hz input, a cycle is completed in 1 sec), a dynamic test is suitable for gaining information in a short time frame or for monitoring time-dependent changes in gel network properties. When monitoring the gelation process at a fixed frequency, it usually takes a few hours for G' to become approximately constant. The constancy can be judged by a constant value of G' at a fixed frequency during a subsequent frequency or strain sweep test, which usually takes several minutes.

Literature Cited

Bohlin, L., Hegg, P.O., and Ljusberg-Wahren, H. 1984. Viscoelastic properties of coagulating milk. *J. Dairy Sci.* 67:729-734.

Bremer, L.G.B., Bijsterbosch, B.H., Schrijvers, R., van Vliet, T., and Walstra, P. 1990. On the fractal nature of the structure of acid casein gels. *Colloids Surf.* 51:159-170.

Clark, A.H. 1992. Gels and gelling. *In* Physical Chemistry of Foods (H.G. Schwartzberg and R.W. Hartel, eds.) pp. 263-305. Marcel Dekker, New York.

Clark, A.H. and Ross-Murphy, S.B. 1987. Structural and mechanical properties of biopolymer gels. *Adv. Polym. Sci.* 83:57-192.

Dea, I.C.M., Richardson, R.K., and Ross-Murphy, S.B. 1984. Characterisation of rheological changes during the processing of food materials. *In* Gums and Stabilisers for the Food Industry, 2 (G.O. Phillips, D.J. Wedlock, and P.A. Williams, eds.) pp. 357-366. Pergamon Press, Oxford.

de Gennes, P.G. 1979. Scaling Concepts in Polymer Physics. Cornell University Press, Ithaca, N.Y.

Djabourov, M., Leblond, J., and Papon, P. 1988. Gelation of aqueous gelatin solutions. II. Rheology of the sol-gel transition. *J. Phys. (France)* 49:333-343.

Flory, P.J. 1953. Principles of Polymer Chemistry. Cornell University Press, Ithaca, N.Y.

Hermans, J. 1965. Investigation of the elastic properties of the particle network in gelled solutions of hydrocolloids. I. Carboxymethyl cellulose. *J. Polym. Sci. A* 3:1859-1868.

Montejano, J.G., Hamann, D.D., and Lanier, T.C. 1985. Comparison of two instrumental methods with sensory texture of protein gels. *J. Texture Stud.* 16:403-424.

Nishinari, K. 1976. Longitudinal vibrations of high elastic gels as a method for determining viscoelastic constants. *Jpn. J. Appl. Phys.* 15:1263-1270.

Oakenful, D. 1984. A method for using measurements of shear modulus to estimate the size and thermodynamic stability of junction zones in noncovalently cross-linked gels. *J. Food Sci.* 49:1103-1104, 1110.

Ross-Murphy, S.B. 1994. Rheological methods. *In* Physical Techniques for the Study of Food Biopolymers (S.B. Ross-Murphy, ed.) pp. 343-392. Blackie Academic & Professional, Glasgow, U.K.

Scanlan, J.C. and Winter, H.H. 1991. Composition dependence of the viscoelasticity of end-linked poly(dimethylsiloxane) at the gel point. *Macromolecules* 24:47-54.

Shih, W.-H., Shih, W.Y., Kim, S.-I., Liu, J., and Aksay, I.A. 1990. Scaling behavior of the elastic properties of colloidal gels. *Phys. Rev. A* 42:4772-4779.

Shoemaker, C.F., Lewis, J.I., and Tamura, M.S. 1987. Instrumentation for rheological measurements of food. *Food Technol.* 41:80-84.

Stockmeyer, W.H. 1943. Theory of molecular size distribution and gel formation in branched-chain polymers. *J. Chem. Phys.* 11:45-55.

te Nijenhuis, K. 1981. Investigation into the ageing process in gels of gelatin-water systems by the measurement of their dynamic moduli. I. Phenomenology. *Colloid Polym. Sci.* 259:522-535.

Tokita, M., Futakuchi, H., Niki, R., Arima, S., and Hikichi, K. 1983. Dynamic mechanical properties of milk and milk gel. *Biorheology* 20:1-10.

Winter, H.H. 1987. Can the gel point of a cross-linking polymer be detected by the G'-G'' crossover? *Polym. Eng. Sci.* 27:1698-1702.

Ziegler, G.R. and Foegeding, E.A. 1990. The gelation of proteins. *Adv. Food Nutr. Res.* 34:203-288.

Key References

Clark and Ross-Murphy, 1987. See above.

Comprehensively reviews the theoretical aspects of gelation of biopolymers.

Rao, M.A. and Steffe, J.F. (eds.) 1992. Viscoelastic Properties of Foods. Elsevier Applied Science, London and New York.

Contains principles and experimental results of viscoelastic behaviors of various food materials.

Ziegler and Foegeding, 1990. See above.

Extensively reviews the rheological properties of various protein gels. Also includes an introduction to gelation theories.

Contributed by Shinya Ikeda
Osaka City University
Osaka, Japan

E. Allen Foegeding
North Carolina State University
Raleigh, North Carolina

Creep and Stress Relaxation: Step-Change Experiments

Rheometers can impose a rapid step change using their drive systems and measure the response of the sample as a function of time. These output signals can be analyzed to extract viscoelastic information, including relaxation times. Initially, when elapsed time is small, the response is analogous to high-frequency oscillation data; however, as the elapsed time increases and the material reaches steady state, the response correlates with very low-frequency data. If steady state is reached, the data can sometimes be mathematically converted to ultra-low-frequency information (*UNIT H3.1*). Thus, these techniques are often a shortcut used to obtain so-called terminal zone data (data that reflect the longest relaxation times that can be measured). These types of test are easy to assess visually because the shape of the curve is directly related to the behavior of the sample. The more linear the output, the more it is dominated by liquid properties. The more curved the output, the more it is dominated by solid-like behavior.

A controlled-stress instrument imposes a step stress, and such experiments are referred to as creep tests (Fig. H3.3.1). Typically there are one or two steps in a creep test: the imposition of a stress (retardation step) and its removal (relaxation or recovery or recoil step). The output signal is displacement, often displayed as strain, % strain, or compliance, J (strain/stress). The latter signal is displayed "as is" in the creep step, but in the recoil phase is calculated by subtraction from the previous step (zero stress would cause a nonsensical quotient), and is thus more properly called the recoverable compliance. When a creep step is analyzed, the latter linear part of the curve is associated with a viscous response, so a viscosity at a given stress (applied) and shear rate (measured) is generated. Creep can be considered as the precursor to flow, since experiments can be performed in either the linear viscoelastic region of a sample or the nonlinear region where irreversible deformation (i.e., flow) begins to dominate. Thus, creep tests can be used to probe the zero shear plateau or maximum apparent viscosity of a material as well as the onset of shear thinning. If this onset is sharply defined, the material is said to have a "yield stress," and the creep test is perhaps the most reliable way of evaluating this behavior.

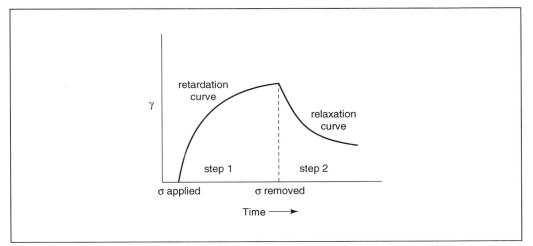

Figure H3.3.1 A creep experiment where a small constant stress (σ) is applied to a food sample (step 1) for a period of time. Afterwards the applied stress is removed (step 2). The degree of deformation (strain, γ) is measured during the experiment, and a typical response is shown.

Contributed by Peter Whittingstall

Controlled-strain instruments do not perform creep tests well because their natural mode of operation is the rapid imposition of position changes or speed changes. The equivalent controlled-strain step-change test is called a stress relaxation test. In this test, the drive system rapidly changes the position of the rotor through a certain angular displacement (strain). The output signal is torque, often displayed as stress or modulus, G (stress/strain). The detector picks up the sample's response to this change in position as a sharp rise in stress (torque). For a liquid sample this stress will decay rapidly as energy is dissipated within the fluid. The response shape will thus resemble a spike. A solid, on the other hand, can store energy, and so the rise in stress will be maintained as a plateau. Real materials are viscoelastic and thus give an exponential decay in stress. The exponential time constants are related to the viscoelastic parameters seen in the creep curves and have the same significance (Fig. H3.3.2).

The viscoelastic samples to be tested by this method may be in different forms. The simplest to work with is a soft or liquid-like viscoelastic material such as mayonnaise or other food emulsions. These are easy samples to work with terms of sample loading. More solid-like samples such as cheese or food gels are more difficult to load onto the instrument in a consistent matter. The degree of compression of soft samples should ideally be controlled using a normal force measure or force rebalance system. Slippage is also a concern and roughened plates or even adhesives may be needed if slip is an issue. As this protocol is a general one, it is assumed that the sample is already loaded on the rheometer and has achieved equilibrium in terms of temperature and viscoelastic structure (time-dependent behavior).

Materials

> Sample (prepared or intact)
>
> Rotational rheometer (*UNIT H1.1*; e.g., Bohlin Instruments, Chandler Engineering): controlled stress (for applied step shear stress) or controlled strain (for applied step shear strain) with appropriate software for rheometer control, data acquisition, and data analysis
>
> Appropriate testing fixtures (Fig. H1.1.1): parallel plates are generally preferred for these tests

1. Load sample (*UNITS H3.1 & H3.2*), then select a stress (creep test) or strain (stress relaxation) for use in the test.

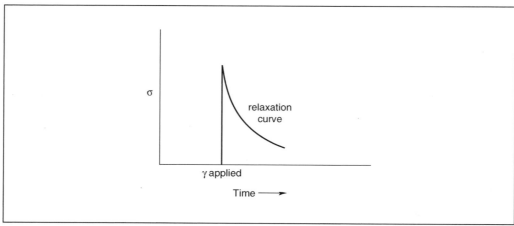

Figure H3.3.2 A stress relaxation experiment where a small constant strain (γ) is applied to a sample. After the strain is applied, the stress (σ) of the material is measured as a function of time.

The stress or strain selected should be in the linear viscoelastic region of the sample measured at a frequency of ~1 Hz or 10 radians/sec (1.6 Hz; UNIT H3.1; also see Background Information). Choose a value that is not close to the upper limit of the linear region, as a sample may leave the linear region of behavior during a prolonged step change test. Occasionally, a specific stress or strain value that the sample will face during use may be applied, where it is of interest to see if the sample recovers or deforms/flows irreversibly in response. If the linear viscolelastic region is not known, consider running an oscillation amplitude sweep using stress or strain as appropriate for the instrument. Sometimes this option is not available, so trial and error is also an alternative.

2. Apply several stresses or strains, starting low and working up. If possible, monitor the output signals of the transducer or position sensor while applying test input signals. When the signal is relatively noise free, repeat that step as a complete test.

   It is often advisable to load a new sample at this point.

   Most instruments have a mode that allows output monitoring.

   The critical phase of either a creep test or a stress relaxation test is in the first few seconds. It is here that the transducer can be overloaded or the optical encoder can give a noisy response. Most controlled-strain instruments will give an audible error signal if the transducer is overloaded. For controlled-stress instruments, the noise level is determined by the resolution of the optical encoder.

3. Once the step value is determined, apply it for at least 5 min, preferably 15 min or more.

   The sample must have reached steady state before cessation of the test or the application of a second step. Steady state in a creep test is seen as a constant slope in the strain curve. A constant slope in the stress curve may also be seen in a stress relaxation test, but often the signal is lost in the noise. A material that is liquid-like in real time will need a test period of ~5 to 10 min. A stress relaxation test is likely to be somewhat shorter than a creep test since the signal inevitably decays into the noise at some point. A creep test will last indefinitely but will probably reach steady state within an hour. For a material that is a solid in real time, all experiments should be longer as molecular motion is, by definition, slower. Viscoelastic materials will lie in between these extremes. Polymer melts can take 1 hr or more to respond in a creep test, but somewhat less time in a stress relaxation test.

   If the software offers an equilibrium tolerance option to stop the test prior to the allotted time, disable it. It is better to compare tests that have been run for the same amount of time.

4. *For creep tests:* Perform a recoil or recovery step using a segment of zero applied stress. Use a step duration that is at least as long as the retardation step.

   The recoil step is sometimes deliberately set to be longer than the retardation step in order to ensure complete recovery to steady state.

   Creep tests typically have a recoil step, unless they are of a multistep variety. For instance, some workers will use multiple retardation steps of increasing stress to visually determine the apparent yield stress or onset of flow. This is seen as a sharp biphasic response in the strain. The same yield behavior can be seen by plotting several creep curves on top of each other. When plotting compliance curves for the retardation step, the data will superimpose if the sample is in the linear region (stress is proportional to strain by definition), due to the normalizing effect of dividing the measured strain by the applied stress. As soon as the linear region is exceeded, the curves will no longer lie on top of each other.

   Stress relaxation tests need not have a second step, although some workers recommend a second step in the opposite direction. The Boltzmann superposition principle for polymers allows for multiple step-change tests of both types (stress or strain) as long as the linear limit of the polymer is not exceeded (Ferry, 1980).

Viscoelasticity of Suspensions and Gels

H3.3.3

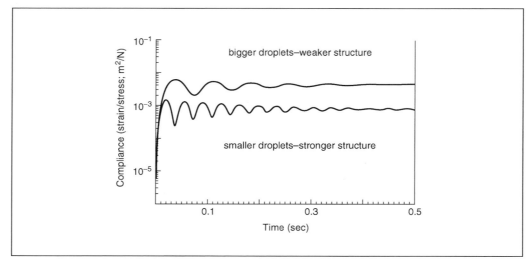

Figure H3.3.3 The immediate response to an imposed stress upon two emulsions produces two different responses with respect to size of the emulsion droplets.

5. Analyze data.

The data from step-change tests is modeled mathematically to fit one or more exponential time constants depending upon the curvature of the line. Each exponential time constant is related to a relaxation time in the sample response. Most rheometers have software that automatically fits the data to these models.

It is often necessary to ignore the first second of data from a step change test when using such models. The first few data points of a creep test often reflect a "ringing" in the sample if there is a gel-like structure present. An example is shown in Figure H3.3.3, where two emulsions are mixed at different speeds with the same formulation (in terms of, for example, stabilizers). The sample mixed at lower speed has larger droplets and therefore a lower surface area for the emulsifiers to work with. The difference is clearly seen in the pattern of damped oscillations. This damped oscillation is very useful because it can provide insight into the elastic properties of the sample, but it cannot be modeled successfully by the Burgers model. It must therefore be excluded.

The first second of a stress relaxation step can also show this type of ringing, but it is generally caused by the transducer itself. Thus, the first part of the data may be electronically filtered to remove the transducer ringing by setting a filter cutoff frequency of ~40% of the value for the resonant frequency of the transducer and geometry. Some rheometers allow for the measurement of transducer resonant frequency when measuring the geometry inertia.

Other mathematical transformations of the data allow for low-frequency oscillation data to be calculated from step change data, which is desirable since step change tests are generally quicker than low-frequency studies. This technique is outside the scope of these protocols, but interested readers are referred to Ferry (1980).

COMMENTARY

Background Information

Creep tests

The stresses used in a creep test are chosen in two ways. First, a value is chosen from oscillatory tests (specifically stress or strain amplitude sweeps at 1 Hz or 10 rad/sec; UNIT H3.1) to define the linear region. Using two to five different values, the sample is taken from a linear viscoelastic response to the onset of flow. This may require stresses to be chosen over an order of magnitude or more. Flow tests are then used to further characterize the sample.

The second method is to calculate a stress that is appropriate to a particular situation of interest. An example of this would be the stress acting on a drop of material due to its own weight as it rests on a support medium. The force of gravity tends to make the drop spread out into a film, while its surface tension tends to resist this process. The resulting force can be converted to a stress by knowing the footprint (surface area) of the drop. In the confectionary industry, the ability to dispense shaped candy is important for product consistency. The creep test at the calculated stress will predict how well different batches of material will perform.

All creep curves can be used to derive a viscosity value. This value is obtained from the slope of the output once steady state has been reached, and the exercise should not be attempted if the output is still changing in terms of its slope. On occasion, the only way to be sure that further changes will not occur is to analyze a particular curve and then repeat the experiment for a slightly longer time and re-analyze it. If the same results are obtained, the sample has reached steady state. It typically takes between 5 min and 1 to 2 hr for a material to achieve steady state. More viscous samples will take more time than less viscous samples. At low stresses the values for the viscosity can be very large since the measured shear rate is often very small. If the data are analyzed in the linear viscoelastic region, it is a good method for deriving zero shear viscosity.

The normal method for analyzing creep curves is to use the Burgers model (Barnes, 2000; Figure H3.3.4), where a spring and a dashpot represent the very short-time and steady-state long-time behavior, respectively. In between these extremes are as many as four Kelvin-Voigt units comprising a spring and a

dashpot linked in parallel with each other so that they experience the same strain. These viscoelastic elements each have a relaxation time (or more properly a retardation time) associated with them. A typical data fit will therefore list an initial compliance (or strain) associated with the single spring (J_0) and a steady-state or Newtonian viscosity (η) and shear rate associated with the single dashpot (Fig. H3.3.5). After that will be listed the compliance, viscosity, and relaxation time for as many Kelvin-Voigt units as are needed to fit the line. Finally, some kind of goodness of fit will be described (e.g., standard error). Remember that the viscosities from creep tests in the linear viscoelastic region will equate to the zero shear viscosity. Differences in zero shear viscosity and/or J_e^0 (recoverable compliance) are sometimes related to differences in molecular weight distribution (MWD) in polymers or particle/droplet size distribution in dispersions.

The steady-state viscosity is a pragmatic way of predicting certain key properties of a sample. If the stresses used in the creep experiment are well chosen they will reflect the stresses applied to the sample by the action of gravity. Thus, the viscosity under these circumstances will help predict the ability of the material to resist sagging on a vertical surface (coatings). Other uses include prediction of sedimentation velocity or creaming velocity in two-phase dispersions. The ability of a paint to level out and therefore remove brush marks by

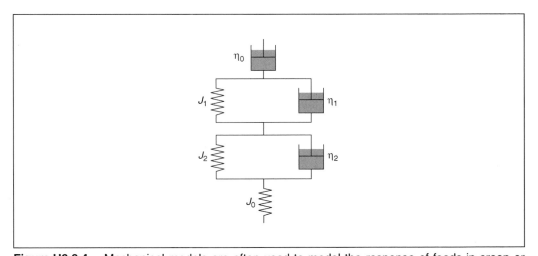

Figure H3.3.4   Mechanical models are often used to model the response of foods in creep or stress relaxation experiments. The models are combinations of elastic (spring) and viscous (dashpot) elements. The stiffness of each spring is represent by its compliance (J = strain/stress), and the viscosity of each dashpot is represent by a Newtonian viscosity (η). The form of the arrangement is often named after the person who originally proposed the model. The model shown is called a Burgers model. Each element in the middle—i.e., a spring and dashpot arranged in parallel—is called a Kelvin-Voigt unit.

the action of surface tension is also easily determined by a creep test.

Stress relaxation tests

A typical stress relaxation curve will be plotted as a modulus function (G = stress/strain). The curve of $Gf(t)$ (modulus as a function of time) is an exponential decay, and there is no equivalent to the recoil part of the creep test unless another step is programmed. The model fitting routines involve Maxwell elements (Barnes, 2000; Ferry, 1980) that comprise a spring and a dashpot in series, such that they both experience the same stress. Each Maxwell element has a relaxation time associated with it, as well as a modulus (or its inverse, compliance, represented as a spring) and a viscosity (dashpot). Rather than there being a single model equivalent to the Burgers model for fitting stress relaxation data, a series of Maxwell elements is generally used giving a relaxation spectrum.

A disadvantage of the stress relaxation technique is the fact that the signal is always decaying until it is indistinguishable from the noise. If this occurs before the sample reaches steady state, then the test is difficult to analyze. In both step-change techniques, one is interested in generating information over a very wide range of time frames. Thus, the first second of data from either technique is representative of the high-frequency data from a frequency sweep (mechanical spectrum). The steady-state information is, of course, connected to the very-low-frequency data from a sample, and it is often more convenient to access this information via a step-change test than to physically set a very low frequency and wait for the data. As a final point, the mathematical manipulation required to obtain a zero shear viscosity and/or J_e^0 is significant if stress relaxation is the source data (Ferry, 1980).

The choice of strain step to use is guided by the same principles as the choice of stress to use in a creep test. Ideally the strain chosen should come from known data gathered using an oscillation test to define the linear region. Sometimes a particular strain is chosen from known application data, and therefore may be outside the linear response of the sample. When dealing with a torque transducer, there is a fixed dynamic range to consider. The choice of geometry size and transducer range is therefore important in order to avoid overloading the transducer. A stiff sample and too large a geometry will result in a high torque. If a sample is not rigid, the output signal may be too close to the noise.

Transducers that use torsion bars or springs of known compliance may also oscillate at the beginning of the test as the step is imposed. This ringing is symptomatic of the detector and not the sample response, and thus needs to be removed by signal filtering. In a creep test this kind of behavior only occurs if the sample is

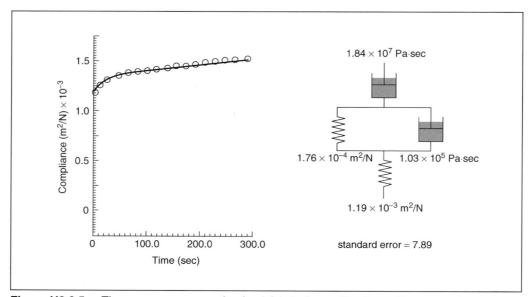

Figure H3.3.5 The creep response of a food (circles) was fitted to a Burger model with one Kelvin-Voight unit. The goodness of fit is shown as the continuous curve and the standard error. The values of compliance and viscosity of the respective springs and dashpots were outcomes of the fitting process.

"ringing." Analysis of this behavior can give insight into the elasticity of the sample.

Critical Parameters and Troubleshooting

A general point to remember when dealing with step change tests when compared to oscillation tests is that the linear response of a sample is quite often different for the two types of test. This is because the physical displacement of the structure of the material is much smaller in an oscillation test, with its forward and backward motion. A step change test is unidirectional and thus may cause more dislocation of a molecule, droplet, or particle, which in turn results in greater "damage."

Noisy data are generally undesirable and should prompt the investigator to try higher values of stress or strain. Remember that the data may be noisy to begin with in a creep test because the sample is barely responding, but as the test progresses the data often leave the noise far behind. Thus, a final curve may be generated that is still easy to analyze. Noise in a stress relaxation test should also prompt the investigator to increase the strain step, although the dynamic range of the transducer may limit the ability to recover steady-state data.

A good diagnostic for creep and stress relaxation tests is to plot them on the same scales as a function of either compliance (J) or modulus (G), respectively. If the curves superimpose, then all the data collected is in the linear region. As the sample is overtaxed, the curves will no longer superimpose and some flow is said to have occurred. These data can still be useful as a part of equilibrium flow. The viscosity data from the steady-state part of the response are calculated and used to build the complete flow curve (see equilibrium flow test in *UNIT H1.2*).

As tests are performed at larger stresses and strains, the results may show decreasing compliance or increasing modulus, respectively. If this is the case, there is a problem with the sample. Generally the cause of this behavior is time dependency. The sample is rebuilding structure and becoming more rigid as subsequent tests are performed. It is very important that the sample be allowed to reach equilibrium before performing step-change tests. Typically, the shape of the curve is distorted at medium to long times because the sample has built up structure during the test. Occasionally a negative stress or strain will appear as internal rearrangements of structure manifest themselves during the test. Again, this is indicative of the test being initiated before the sample is at equilibrium.

In general, it is important to choose a measurement geometry that optimizes sensitivity while minimizing artifacts due to slip. Parallel plate geometry is often the preferred choice as it is relatively easy to cut samples into discs. For liquids or melts, cone and plate or even concentric cylinders may be used, but they are potentially more invasive during sample loading. Care should be taken to allow all stresses from loading to dissipate before starting the test. Normal force sensors are often useful in this regard, or an oscillation time sweep can provide similar guidance.

Anticipated Results

The shape of the creep and recoil curves provide insight about the sample, as does the stress relaxation response. A very strongly curved retardation response resembling a square wave is indicative of dominantly elastic behavior. A stress relaxation response for an elastic solid gives an almost horizontal line with very little decay, because almost all the energy is stored. A creep recoil curve that almost returns to the initial state of zero compliance is indicative of a very elastic system, because stored energy is almost completely recovered.

A linear output is associated with a liquid-like or viscous response. The retardation response of a Newtonian liquid is a straight line through the origin, where the slope of the line is inversely proportional to the viscosity. A stress relaxation response for a Newtonian liquid is a narrow spike, as energy is dissipated very quickly. Typically the recoil part of a creep curve will be almost horizontal if there is little elasticity in the sample.

Real (viscoelastic) materials give an intermediate response that is an exponential curve. The exponential time constants associated with the curve are used to approximate the relaxation times of the material itself. Thus, the shape of the output curve is analyzed to give viscoelastic information, although this model fitting is only strictly legitimate in the linear viscoelastic region. Workers have shown that the mechanical parts of the models (springs and dashpots) can be associated with specific parts of a food's makeup.

Time Considerations

It is clear that these tests take a significant amount of time to perform. Generally, 15 min to 1 hr should be allowed for each test, once the appropriate stresses or strains have been found.

It is advisable to perform replicate tests (three or more) to be sure of reproducibility. For some solid-like materials with long relaxation times, the loading process may take as long as the test.

Literature Cited

Barnes, H.A. 2000. A Handbook of Elementary Rheology. Institute of Non-Newtonian Fluid Mechanics, University of Wales, Aberystwyth, U.K.

Ferry, J.D. 1980. Viscoelastic Properties of Polymers, 3rd ed. John Wiley & Sons, New York.

Key References

Barnes, 2000. See above.

Barnes, H.A., Hutton, J.F, and Walters, K. 1989. An Introduction to Rheology. Elsevier, New York.

Ferry, 1980. See above.

Steffe, J.F. 1996. Rheological Methods in Food Process Engineering, 2nd ed. Freeman Press, East Lansing, Mich.

These texts deal with creep and stress relaxation in some detail and provide a good background to the interested reader. However, some are difficult to obtain (Ferry, 1980, is out of print).

Contributed by Peter Whittingstall
ConAgra Foods
Irvine, California

I   BIOACTIVE FOOD COMPONENTS

INTRODUCTION

I1   Polyphenolics

I1.1   Determination of Total Phenolics

I1.2   Extraction and Isolation of Polyphenolics

I1.3   HPLC Separation of Polyphenolics

I1.4   Proanthocyanidins: Extraction, Purification, and Determination of Subunit Composition by HPLC

I1.5   Identification of Flavonol Glycosides Using MALDI-MS

I1.6   Analysis of Isoflavones in Soy Foods

SECTION I
Bioactive Food Components

INTRODUCTION

The stimulus for adding this new section to *Current Protocols in Food Analytical Chemistry* (CPFAC) was the identification and measurement of polyphenolics in foods, which constitutes a very active area of research because of the possible health benefits that these dietary antioxidants may provide. We wanted to give priority for a number of units concerning the analysis of these compounds, but we did not have a section in CPFAC that was an obvious fit. Our organizational format was based on food composition (water, carbohydrates, lipids, proteins, enzymes) and functionality (color, flavor, texture). Anthocyanin pigments (Section F) are in fact phenolic compounds, and methods for their analysis have much in common with methods for analysis of polyphenolics. However, since most polyphenolics are colorless, it seemed inappropriate to include them in a section on pigments, even if many of them are converted to brown pigments with enzymic oxidation. Because of their antioxidant properties, Section D on lipids was another possible placement, but we agreed that that was a bit of a stretch. A new section on nutraceuticals was ruled out because we also wanted to accommodate constituents such as alkaloids that are harmful to human health. A section on phytochemicals was considered but rejected because there are compounds of animal origin that have biological activity and are potential candidates for future units. Thus, we finally agreed that bioactive food components was an appropriate and useful title for a new section. We realize that this section is not all-inclusive, since there are already several units in CPFAC for analysis of compounds that have biological activity, such as carotenoids, chlorophyll, and cholesterol. Cross-references between this section and other appropriate units will be made wherever it is deemed useful.

The first chapter in this section deals with the analysis of polyphenolics. The term polylphenolics has come into widespread and popular usage in recent years. Presumably the word means "many phenolics," and one interpretation would be a compound having many phenolic groups. The more correct meaning is a multiplicity of different phenolic compounds. The term most likely arose when analysts learned that a very large number of the peaks they found in reversed-phase HPLC chromatograms of alcoholic plant extracts were phenolic compounds. Thus, the expression polyphenolics encompasses several classes of phenolic compounds. More than 8000 phenolic compounds have been identified in plants. Flavonoids are the largest group of phenolics, with more than 4000 structures identified to date. There are several different classes of flavonoids that differ in the oxidation state of the C ring. Structures of the flavonoid classes are shown in *UNIT II.3*. Anthocyanin pigments are an important class of flavonoids, and units describing their analysis are found in Section F. Phenolic acids along with their esters and glycosides constitute another major group of polyphenolics. Structures for these compounds as well as condensed tannins and procyanidins are found in *UNITS II.2, II.3, & II.4*.

At one time polyphenolics were regarded as "antinutrients" since they can complex with proteins and reduce protein quality. Today's heightened interest in these compounds is due to possible health benefits that are attributed to their antioxidant properties. The possible health benefits include reduced risk of coronary heart disease, cancer, stroke, and diabetes. In addition to these health-related findings, it has been long recognized that

Bioactive Food Components

polyphenolics have several important functional roles in food quality. They can impart bitterness and astringency to foods and beverages, and they are important to mouth-feel and textural quality. As substrates for polyphenoloxidase, they can generate brown color. Polymerization and condensation with proteins can result in formation of sediment and haze in beverages. Polyphenolics serve several beneficial functions in plants. They provide a protective effect against harmful UV radiation. In this regard, it is advantageous to have compounds with different absorbance maxima so there is protection throughout the UV region. When plant tissue is injured, polymerization of phenolics produces scar tissue that assists in wound healing. The antimicrobial and antiviral properties of phenolics help to protect against invading pathogens.

UNIT II.1 presents two methods for determining total phenolics by UV-visible spectroscopy. The Folin-Ciocalteau assay was specifically developed for measuring phenolics in grapes and wines. The method is widely accepted and is currently being applied to many different plant materials and foods. A second method based on spectral analysis is useful primarily for process monitoring. Methods for extraction of polyphenolics and for subsequent fractionation by solid-phase extraction are described in *UNIT II.2*. Reversed-phase HPLC of polyphenolics is described in *UNIT II.3*, with an additional method for neutral and acid fractions that have been isolated by solid-phase extraction. *UNIT II.4* deals with procyanidins, which are flavonoid polymers with flavan-3-ol subunits and have particular importance to the astringency, bitterness, and mouth-feel of wines. Methodology is described for extraction and purification, as well as for HPLC analysis of subunit composition. While HPLC and UV-visible spectroscopy are invaluable for characterizing and measuring concentrations of polyphenolics, mass spectroscopy gives more definitive identification. *UNIT II.5* describes the analysis of flavonoid glycosides using MALDI-MS. *UNIT II.6* covers the extraction and HPLC analysis of isoflavones, another class of flavanoids. There is considerable evidence that consumption of soybean products has a positive impact on human health, and that the biological activity is associated with isoflavones. Soy products are the major dietary source of isoflavones.

Ronald Wrolstad

Polyphenolics

I1.1 Determination of Total Phenolics — E1.2.1

Basic Protocol 1: Determination of Total Phenolics by Folin-Ciocalteau Colorimetry — E1.1.1

Alternate Protocol: Microscale Protocol for Folin-Ciocalteau Colorimetry — E1.1.2

Basic Protocol 2: Determination of Total Phenolics by Spectral Analysis — E1.1.3

Reagents and Solutions — E1.1.5

Commentary — E1.1.4

I1.2 Extraction and Isolation of Polyphenolics — I1.2.1

Basic Protocol 1: Ultrasound-Assisted Aqueous Methanol Extraction of Polyphenolics — I1.2.1

Alternate Protocol 1: Homogenizer-Assisted Methanol Extraction of Polyphenolics — I1.2.2

Basic Protocol 2: Separation of Anthocyanin and Non-Anthocyanin Fractions — I1.2.3

Alternate Protocol 2: Neutral and Acidic Fractionation of Polyphenolics — I1.2.5

Commentary — I1.2.6

I1.3 HPLC Separation of Polyphenolics — I1.3.1

Basic Protocol: Reversed-Phase HPLC Analysis of Polyphenolics Separated into Nonanthocyanin and Anthocyanin Fractions — I1.3.1

Alternate Protocol: Reversed-Phase HPLC Analysis of Polyphenolics Separated into Neutral and Acidic Fractions — I1.3.4

Commentary — I1.3.6

I1.4 Proanthocyanidins: Extraction, Purification, and Determination of Subunit Composition by HPLC — I1.4.1

Basic Protocol 1: Extraction of Proanthocyanidins — I1.4.1

Basic Protocol 2: Purification of Proanthocyanidins — I1.4.2

Basic Protocol 3: Determination of Subunit Composition by HPLC — I1.4.4

Reagents and Solutions — I1.4.6

Commentary — I1.4.7

I1.5 Identification of Flavonol Glycosides Using MALDI-MS — I1.5.1

Basic Protocol 1: Extraction of Flavonol Glycosides from Plant Sources — I1.5.1

Basic Protocol 2: Purification of Flavonol Glycosides for MALDI-MS Analysis — I1.5.2

Basic Protocol 3: MALDI-MS Identification of Flavonol Glycosides — I1.5.3

Commentary — I1.5.4

I1.6 Analysis of Isoflavones in Soy Foods — I1.6.1

Basic Protocol 1: Solvent Extraction of Isoflavones in Their Natural Forms from Soy Foods — I1.6.2

Support Protocol 1: Sample Preparation for Extraction — I1.6.5

Basic Protocol 2: Preparation of Individual Isoflavone Standards — I1.6.5

Basic Protocol 3: Gradient Separation and Identification of Isoflavones Using a C18 Reversed-Phase Column — I1.6.7

Support Protocol 2: Sample Preparation for HPLC Analysis — I1.6.10

Support Protocol 3: Converting Glycosidic Isoflavones to Their Aglycones Using Enzymes — I1.6.10

Commentary — I1.6.12

Contents

1

Determination of Total Phenolics

The phenols or phenolics in wine are important to both red and white wines. In red wines, this class of substances contributes to the astringency, bitterness, and other tactile sensations defined as structure or body, as well as to the wine's red color. In white wines, higher levels of phenolics are generally undesirable, as they contribute to excessive bitterness and to the tendency of the wine to brown when it is exposed to air. Phenolics in grapes and wines include many different substances: phenolic acids (e.g., hydroxybenzoic acids such as gallic acid, hydroxycinnamic acids found in grape juice), three classes of flavonoids found in the skins and seeds (the red anthocyanins, the flavonols, and the abundant flavan-3-ols, which comprise the monomeric catechins), oligomeric proanthocyanidins, and polymeric condensed tannins. (For details on phenolics classes and compound structures, refer to *UNITS II.2 & II.3*.) White wine is made by immediately pressing off the skins and seeds after harvesting, and thus contains only small quantities of flavonoids. In contrast, red wine is a whole-fruit extract made by fermenting with the skins and seeds, and the alcohol thus produced is an excellent solvent for these substances.

Measuring these different substances and reporting meaningful values in a single number is an analytical challenge. There are many different procedures for analyzing different classes of phenolic substances, but few are used in wine analysis except for anthocyanin or color measures. HPLC methods (*UNIT II.3*) that give specific information on individual substances are not widely used in wineries, but are becoming more common as the significance of particular phenolic substances becomes better understood.

There are two widely used methods for the analysis of total phenolics in wine. The Folin-Ciocalteau method (Basic Protocol 1 and the Alternate Protocol) has the advantage of a fairly equivalent response to different phenols, with the disadvantage of responding to sulfur dioxide and sugar. The direct spectral absorbance analysis (Basic Protocol 2) is quick and simple, making it suitable for process monitoring. This method, however, responds differently to the various phenolic classes, making comparisons between different wine types problematic, and also gives significant interference for sorbate.

Wine, of course, is not the only food that contains phenolics. Phenolics are found in all foods, though at low levels in most. Notable foods that are high in phenolics include coffee and tea, chocolate, fruits and derived products, some oils, spices, and some whole grains. Although the following methods were developed for—and first applied to—analysis of wines and grapes, they can be adapted for other foodstuffs (also see Commentary).

DETERMINATION OF TOTAL PHENOLICS BY FOLIN-CIOCALTEAU COLORIMETRY

Folin-Ciocalteau (FC) colorimetry is based on a chemical reduction of the reagent, a mixture of tungsten and molybdenum oxides. Singleton adapted this method to wine analysis (Singleton and Rossi, 1965) and has written two major reviews on its use (Singleton, 1974; Singleton et al., 1999). The products of the metal oxide reduction have a blue color that exhibits a broad light absorption with a maximum at 765 nm. The intensity of light absorption at that wavelength is proportional to the concentration of phenols. The FC method has been adopted as the official procedure for total phenolic levels in wine; the Office International de la Vigne et du Vin (OIV), the one international body that certifies specific procedures for wine analysis, accepts the FC method as the standard procedure for total phenolic analysis (OIV, 1990). An earlier variation was the Folin-Denis procedure, but the FC method has displaced it except in a few historical cases of official procedures that have not been updated (AOAC International, 1995).

Contributed by Andrew L. Waterhouse

Color development is slow but can be accelerated by warming the sample. With excessive heating, however, subsequent color loss is quite rapid, and timing the colorimetric measurement becomes difficult to reproduce. The reagent is commercially available, but can be prepared (Singleton and Rossi, 1965). The resulting solutions are treated as hazardous waste, and the scale of the original procedure creates a lot of waste. Fortunately, modern liquid-measuring equipment now allows for microscaling the reaction to the volume of a UV-Vis cuvette, reducing the cost of the reagent and waste disposal (see Alternate Protocol).

Materials

Sample, e.g., white wine or 10% (v/v) red wine in water
Gallic acid calibration standards (see recipe)
Folin-Ciocalteau (FC) reagent (Sigma; also Singleton and Rossi, 1965), stored in the dark and discarded if reagent becomes visibly green
Sodium carbonate solution (see recipe)

100-ml volumetric flask
Spectrophotometer set to 765 nm, with 1-cm, 2-ml plastic or glass cuvettes

1. Place 1 ml sample, a gallic acid calibration standard, or blank (deionized or distilled water) in a 100-ml volumetric flask.

    Samples and standards should be analyzed in triplicate.

    If any sample has an absorbance reading above that of the 500 mg/liter standard, it must be diluted adequately and remeasured. White wine can typically be analyzed without dilution. Red wine must be diluted with water (usually ten-fold) to fall into the range of the standards.

2. Add ~70 ml water, followed by 5 ml FC reagent. Swirl to mix and incubate 1 to 8 min at room temperature.

    The incubation must not be >8 min (see Critical Parameters, discussion of reaction time and temperature).

3. Add 15 ml sodium carbonate solution.

4. Add water to the 100-ml line, mix, and incubate 2 hr at room temperature.

5. Transfer 2 ml to a 1-cm, 2-ml plastic or glass cuvette and measure its absorbance at 765 nm in a spectrophotometer.

6. Subtract the absorbance of the blank from all readings and create a calibration curve from the standards.

7. Use this curve to determine the corresponding gallic acid concentration of the samples. Be sure to multiply by any dilution factor for the correct concentration (i.e., by ten for red wines). Report values in gallic acid equivalents (GAE) using units of mg/liter (see Critical Parameters, discussion of standardization).

ALTERNATE PROTOCOL

MICROSCALE PROTOCOL FOR FOLIN-CIOCALTEAU COLORIMETRY

This protocol is adapted for small sample volumes. The reaction is performed directly in a 2-ml cuvette. For a list of materials needed, see Basic Protocol 1.

1. Put 20 µl sample, a gallic acid calibration standard, or blank (deionized or distilled water) into a 1-cm, 2-ml plastic or glass cuvette.

Determination of Total Phenolics

2. Add 1.58 ml water, followed by 100 μl FC reagent. Mix thoroughly by pipetting or inverting and incubate 1 to 8 min.

> *The incubation must not be >8 min (see Critical Parameters, discussion of reaction time and temperature).*

3. Add 300 μl sodium carbonate solution, mix, and incubate 2 hr at room temperature.

> *A final volume of 2 ml must fill the cell adequately for a reading.*

4. Measure sample absorbance at 765 nm and analyze as described (see Basic Protocol 1, steps 6 to 7).

DETERMINATION OF TOTAL PHENOLICS BY SPECTRAL ANALYSIS

BASIC PROTOCOL 2

Phenolic substances all absorb UV light, and all of them have some absorbance at 280 nm. This property can be used to determine phenolics by spectral analysis. One problem with this method is that each class of phenolic substances has a different absorptivity (extinction coefficient, e) at 280 nm. Thus, the results cannot be related to any specific standard and are reported directly in absorbance units (AU). This also means that disparate wines (or other disparate samples) are difficult to compare with this method, as they are likely to have very different compositions.

The value of this method is that it is extremely simple and rapid, requiring only filtration and, in some cases, dilution. It is very suitable for monitoring wines during various stages of processing (e.g., fermentation) and for comparing similar wines (e.g., a single grape variety from different vineyards, or wines from a particular vineyard over different vintages).

Materials

Sample, e.g., red or white wine
Filter membrane, e.g., polytetrafluoroethylene (PTFE)
Cuvettes, transparent at 280 nm (e.g., quartz or methacrylate)
Spectrophotometer, set to 280 nm

1. Filter a sample or blank (deionized or distilled water) with a PTFE filter membrane or other material to achieve clarity.

> *Nylon or other membranes that absorb phenolics should not be used. Membranes can be tested for phenolic absorption by comparing absorbance after single and double filtration.*

2. Transfer an appropriate volume of sample to a quartz or methacrylate cuvette and measure absorbance at 280 nm in a spectrophotometer. If absorbance is not within the acceptable precision of the spectrophotometer (usually $A < 2$ AU), dilute sample as necessary and repeat.

3. Subtract absorbance of blank, and correct absorbance to original concentration and a 1-cm cuvette path length. Subtract 4 AU to report final value.

> *For instance, if a sample is diluted ten-fold with water and a reading of 0.85 AU is observed with a 2-mm cell, the correction would be as follows:*
>
> *total phenol = [A280 × DF × (1 cm/b)] - 4*
>
> *= [0.85 × 10 × (1 cm/0.2 cm)] - 4 = 38.5 AU*
>
> *where DF is the dilution factor, b is the cell path length, and 4 is an arbitrary correction for nonphenolic absorbance (see Critical Parameters, discussion of spectral analysis).*

Use deionized or distilled water in all recipes and protocol steps. For common stock solutions, see APPENDIX 2A; for suppliers, see SUPPLIERS APPENDIX.

Gallic acid calibration standards

Dissolve 0.5 g gallic acid in 10 ml ethanol and then dilute to 100 ml with water (5 g/liter final). Dilute 1, 2, 5, and 10 ml to 100 ml with water to create standards with 50, 100, 250, and 500 mg/liter concentrations, respectively. Store up to 2 weeks at 4°C.

Standards will retain 98% of their potency for 2 weeks if kept closed under refrigeration (4°C), but this potency is retained for only 5 days at room temperature.

Commercial gallic acid is usually adequately pure, but can be recrystallized from water if desired.

Sodium carbonate solution

Dissolve 200 g anhydrous sodium carbonate in 800 ml water and bring to a boil. After cooling, add a few crystals of sodium carbonate and let sit 24 hr at room temperature. Filter through Whatman no. 1 filter paper and add water to 1 liter. Store indefinitely at room temperature.

COMMENTARY

Background Information

There are many phenolic substances in plants and thus in foods. Rich dietary sources of phenolics include fruits, tea, coffee, cocoa, and processed foods derived from these, such as wine. At high levels, and in particular when sugar levels are low, phenols impart an astringency, bitterness, and color to foods. In red wine, unsweetened tea, and chocolate products, the taste is heavily influenced by the presence of phenolics. Therefore, an assessment of phenolic content in food is of great importance.

Folin-Ciocalteau method

The Folin-Ciocalteau (FC) procedure is one of the standard procedures in wine analysis, as well as in tea analysis (Wiseman et al., 2001). One drawback in interpretation is that different classes of phenolics have varying taste attributes, and tests for chemical astringency based on precipitation of proteins have been recently developed (Adams et al., 1999). In addition, if the food product contains sugar, it can mask the bitterness and astringency, as observed in ripe fresh fruit, sweetened chocolates, and tea.

The differential sensory effect of phenolics aside, a major advantage of the FC procedure is that it has a fairly equivalent response to different phenolic substances in wine, making it suitable for measuring accurate mass levels of total phenolic substances. Among the abundant phenolics in wine, the mass response factor relative to gallic acid ranges from 0.87 for caffeic acid to 1.10 for epicatechin based on values from Singleton (1974). The glucosides give lower values, but their mass is increased by the nonphenolic glycosidic substituent. Their response appears to be similar on a phenolic fraction basis. Monohydroxyphenolics such as coumaric acid also give low values, but these constitute a small fraction of the phenolics in wine. In general, the response of a phenolic is due to the number of phenolic groups, and Singleton (1974) describes the controlling factors in detail.

The FC method has also been applied to other foods. One example of particular use is for analysis of tea (Wiseman et al., 2001). The method has also been applied to vegetables (Kaur et al., 2002) and fruit (Pearson et al., 1999; Vinson et al., 2001), although in one instance the only corrected interference was ascorbate. For analysis of foodstuffs other than wines and grapes, the analyst must be aware of potential interferences. In other fields, the method has been used for analysis of medicines (Sadler and Jacobs, 1995), trees in wood chemistry (Yu and Dahlgren, 2000), and fresh waters (Thoss et al., 2002).

The FC procedure was automated some time ago (Slinkard and Singleton, 1977). Although there have been no more published reports on contemporary automation, there are many laboratories that have adapted the procedure to clinical analyzers (G. Burns and T. Collins, pers. comm.). It seems likely that the procedure could be adapted to other analyzers as well.

There are few direct comparisons of total phenol values and antioxidant measurements, although some do exist (e.g., Baderschneider et al., 1999). In general, the response of total phenol tests is comparable to antioxidant tests, with better correlations for antioxidant tests based on aqueous systems as opposed to those based on lipid media.

Spectral analysis

Phenolic substances can also be quantified by measuring absorbance at 280 nm. The applicability of this method is far more limited, however, because absorbance properties of different phenolics vary and cannot be related to a specific standard. Because of this, and because of the extreme ease of the method, spectral analysis is well suited for using total phenolic content for process monitoring.

Sample preparation

The preparation of extracts from solid foods is not trivial. In general, 70% (v/v) acetone is used to extract proanthocyanins and condensed tannins, and aqueous methanol is typically used for other classes of phenolics (*UNIT F1.1*). However, the acid levels used in some reported extraction procedures can have a devastating effect on the recovery of particular flavonoids, and this has been carefully studied for HPLC analysis of flavonoids (Merken et al., 2001). Although some transformations simply hydrolyze glycosides, which would not significantly change the total phenolic content, other degradations may significantly decrease the amount of phenolics present. In the absence of a well-developed procedure for the extraction of a particular compound from a similar matrix, re-extractions and recovery must be validated for any extraction protocol. For solid foodstuffs, the final results would be expressed in mg/100 g or mg/kg.

Critical Parameters

Folin-Ciocalteau method

Interferences. Because the color formation of the Folin-Ciocalteau reaction is based on chemical reduction of the reagent, this reaction is general enough to allow for interference from a number of sources. In wine, the principal interfering compounds are sulfur dioxide and ascorbate, though high levels of sugar indirectly enhance the readings of other analytes. Because there can be many other interferences in non-wine samples, however, it is necessary to thoroughly investigate the use of this method for different samples. The most problematic interference may well be sugar, as it is not mentioned in the original report (Singleton and

Rossi, 1965), but can be found at very high levels in, for instance, fruits. Another non-wine issue comes from the fact that the Lowry method for protein analysis is based on the reaction of the FC reagent with tyrosine phenolic groups, so samples with high protein levels will not be suitable for phenolic analysis by FC. In addition, it should be noted that phenolics will interfere with protein analysis by the Lowry method. Some investigators have applied the use of the FC procedure to analyze phenolic levels in human blood fractions, such as plasma (Nigdikar et al., 1998). It is likely, however, that this method will respond to changes in the large amount of constitutive redox-active substances, such as ascorbate, urate, and tocopherol, as well as any proteins, making it difficult to attribute changes in response to the appearance of phenolics in the blood.

The FC method does not respond to sulfur dioxide alone, but does respond to sulfur dioxide in the presence of phenolic compounds. Presumably, the phenols are oxidized by the FC reagent and then reduced by the sulfur dioxide, creating an interfering response by a type of catalytic cycle. Unfortunately, the magnitude of the interference is not constant (Saucier et al., 1999), so it is not possible to suggest an accurate correction factor, though approximate mass correction factors of 0.1 to 0.2 have been suggested (thus, 10 mg/liter sulfur dioxide would yield a response of 1 to 2 mg/liter in the FC assay; Singleton, 1988). Generally, the case where the sulfur dioxide interference is significant is in white wines, which have a lower range of total phenolic levels and a high level of sulfur dioxide. It has been suggested that this interference renders the method unusable (Somers and Ziemelis, 1980), but that conclusion has not been accepted by others.

Ascorbate is present at very low levels in wine unless it has been added, which is legal but rarely done in the United States, although it is common in some other countries. In other food samples, especially fresh fruits, ascorbate levels can be very high. For kiwifruit, a correction of 1 mg/liter per 1 mg/liter ascorbate must be applied.

The interference of sugar is easily corrected, and is only necessary with sweet or semisweet wine (>2% [w/v] sugar). With fruit samples, the sugar levels can be very high, and it is not clear if adequate corrections can be applied. Generally, it is not advisable to analyze must for phenolic levels by the FC method because of the complexity of the sugar correction. Sin-

Table I1.1.1 Approximate Correction to Folin-Ciocalteau Results for Wines Containing Invert Sugar[a]

Apparent phenol content (mg GAE/liter)	Invert sugar content		
	2.5%	5.0%	10%
100	5	10	20
200	10	15	40
500	20	30	50
1000	30	60	100
2000	60	120	200

[a]The correction value should be subtracted from the apparent phenol content for an accurate value. This correction applies to analyses conducted at room temperature. Abbreviation: GAE, gallic acid equivalents. Reproduced from Slinkard and Singleton (1977), with permission from the American Society for Enology and Viticulture.

gleton suggests conducting the analysis of the standards with the same level of sugar as the sample, but this is a complex issue because different sugars yield different intereferences (i.e., fructose has a higher response than glucose). Singleton suggests the corrections in Table I1.1.1 when sugar is not added to standards (Slinkard and Singleton, 1977) and describes additional correction factors for warmer temperatures.

These correction factors have not been directly applied to other foods, but it would be prudent to test for the presence of interfering substances, in particular sugars and ascorbate in fresh fruits, and to assess their level of interference in the assay. Correction factors for other interferences may need to be specifically developed.

Limits of detection and quantitation. Because wines that have total phenol levels lower than 50 mg/liter are quite rare, this is not a significant issue for wine. For other sample types, however, a limit of quantitation of ~0.027 AU or 20 mg/liter would be expected, based on a sample-to-sample variance of 0.003 AU (Singleton and Rossi, 1965).

Standardization. Because the analysis reveals the presence of many different substances in one result (in the case of wine, this could mean thousands of substances if one takes into account the diversity of proanthocyanidins), the only practical standard is a single substance. In the cases of wine and tea, the accepted standard is gallic acid. It is a particularly good standard because it is relatively inexpensive in pure form and is stable in its dry form. Other substances have been used, and in principle any phenol could be used, but gallic acid is strongly recommended for the above reasons and the fact that the use of a single standard makes it easier to compare data.

Gallic acid is quite stable in dry form, but will oxidize once it is in solution. This reaction is enhanced at higher temperatures. The author has found that the standard solutions should be stored in full or nearly full bottles that are kept tightly sealed between uses and are stored under refrigeration. The author has observed that >5% potency is lost after ~1 week at room temperature, but this same loss takes ~2 weeks with refrigeration. The authors would expect that greater air exposure will also accelerate decomposition.

Because a single substance is used, the result must be reported as a response equivalent to the amount or concentration of that substance. As wine is analyzed on a concentration basis, the result is reported in gallic acid equivalents (GAE) using units of mg/liter. For any standard, the results must always be reported on an equivalent basis to avoid the perception that one is measuring the amount of the standard substance.

Reaction time and temperature. The oxidative reaction caused by the FC reagent is slow and is not complete when the reading is taken. In addition, the colored product is unstable. The measurement is based on the kinetics of the process. Thus, the time of the colorimetric reading is most critical, and the temperature will affect the extent of the reaction and degradation. This is one of the reasons that standards are run each time, to accommodate for changes in room temperature, timing, and reagent condition. At higher temperatures, the colored product has a very limited lifetime. For a full discussion, see Singleton et al. (1999).

The FC reading time can be accelerated by heating the solution and halving, as a rule of thumb, the reaction time for each temperature increase of 10°C. Thus, at 40°C, readings can be taken after 30 min. Decreasing the time further by higher temperatures becomes problematic for manual measurements because the timing of each reading becomes more critical with each increase in temperature. The degradation of the component that absorbs at 765 nm, in particular, becomes very rapid at higher temperatures, so analyses that run just slightly too long will yield very poor results (Singleton, 1999). Thus, 40°C is usually the highest recommended temperature for manual use, but higher temperatures are described for automated analyses (Slinkard and Singleton, 1977).

Precision. Because the FC method relies on reaction kinetics and not stoichiometric conversion, it is not very precise, and variations of ~5% are typical for replicates, depending on the temperature control and timing precision of the reagent additions and spectral measurements.

Spectral analysis

Interferences. In wine the principal interference is sorbic acid. This is a preservative typically added to sweetened wines that have residual sugar. Because yeast would continue to ferment this sugar, spoiling the product in the bottle, such wines must be rendered sterile at bottling. This is usually done by sterile filtration, sometimes combined with the addition of sorbic acid. Sorbic acid has a very strong absorbance at 280 nm and will result in exceptionally high readings (up to 36 AU). There appears to be some dispute about the ease of dealing with the interference, as the author of this method claims that it is easy to remove the sorbic acid by isooctane extraction (Somers and Ziemelis, 1985). It has been reported, however, that an impractically large volume of isooctane would be needed to remove the sorbate (Tryon et al., 1988). Although it is likely that serial extractions would be very effective, routine multiple extractions of all samples would make the method significantly more complex when analyzing unknown wines that may contain sorbate.

Background. The method also attributes 4 AU to nonphenolic substances in all samples, and so this value is subtracted to reflect the true absorbance due to phenols. The origin of this background absorbance is not clear, but is thought to be due to protein and nucleotides. The magnitude of the background absorbance does vary with a standard deviation of 1 AU (Somers and Ziemelis, 1985). Thus, for typical white wines with an uncorrected absorbance of 8 AU, the subtraction of 4 AU leaves 4 AU with an expected standard deviation of 1 AU or 25% due to the variance in the correction factor.

Standards. There is no standard or calibration with standards, so these issues are moot.

Anticipated Results

The level of total phenolics in white wines varies from ~100 to 300 mg/liter by the FC method. The levels will be on the low end of the scale if the must was subjected to oxidative treatment and the pressing was very light. Higher levels will be observed when harder pressing of the solids is utilized or if the wine was aged in new oak barrels. By spectral analysis, white wines have an average corrected absorbance of 4 AU, with a range of 1 to 11 AU.

The FC results will be compromised by high levels of sulfites, but because sulfite levels are almost always measured in wines, the possibility of sulfite interference in a wine should be anticipated. The spectral method will be compromised by the presence of sorbate. In known wines this can easily be anticipated and, because its use is limited to sweet wines, it may be possible to check for it selectively. Another clue to its presence would be a very high absorbance. Sorbate can also be measured in wine (Caputi and Stafford, 1977).

Red wines have total phenolic levels of 1 to 3 g/liter, with typical average of ~1.8 g/liter. Differences are due to differing amounts of phenolics in grapes based on variety and growing conditions, with moderate to cooler climates yielding higher levels. Production techniques can have a secondary effect, and longer contact times or higher temperatures will increase the amount of phenolics extracted from the grape solids into the wine. Red wines have a reported range of 23 to 100 AU, with an average of 54 AU. Because levels are so high, interferences other than sorbate do not introduce significant error in the spectral method.

Time Considerations

When the FC analysis is carried out manually, there are limits to the number of samples that can be handled at once because of the need to time the reagent mixing and spectral readings. In the author's laboratory, single analysts can run 20 samples per day, divided into 2 sets. Since each sample is run in duplicate and 4 standards are included in each run, 50 individual results are generated in a day.

Literature Cited

Adams, D.O. and Harbertson, J.F. 1999. Use of alkaline phosphatase for the analysis of tannins in grapes and wine. *Am. J. Enol. Vitic.* 50:247-252.

AOAC (Association of Official Analytical Chemists) International. 1995. Tannin in Distilled Liquors. AOAC Official Method 952.03. *In* Official Methods of Analysis of AOAC International, 16th ed., (P. Cuniff, ed.) ch. 26, p. 16. AOAC Int., Arlington, Va.

Baderschneider, B., Luthria, D., Waterhouse, A., and Winterhalter, P. 1999. Antioxidants in white wine (cv. Riesling): I. Comparison of different testing methods for antioxidant activity. *Vitis* 38:127-131.

Caputi, A. Jr. and Stafford, P.A. 1977. Ruggedness of official colorimetric method for sorbic acid in wine. *J.A.O.A.C.* 60:1044-1047.

Kaur, C. and Kapoor, H.C. 2002. Anti-oxidant activity and total phenolic content of some Asian vegetables. *Int. J. Food Sci. Technol.* 37:153-161.

Merken, H.M., Merken, C.D., and Beecher, G.R. 2001. Kinetics methods for the quantitation of anthocyanidins, flavonols, and flavones in foods. *J. Agric. Food Chem.* 49:2727-2732.

Nigdikar, S.V., Williams, N.R., Griffin, B.A., and Howard, A.N. 1998. Consumption of red wine polyphenol reduces the susceptibility of low-density lipoproteins to oxidation in vivo. *Am. J. Clin. Nutr.* 68:258-265.

OIV (Office International de la Vigne et du Vin). 1990. Indice de Folin-Ciocalteau. *In* Recueil des Méthodes Internationales d'Analyses des Vins et des Moûts, pp. 269-270. OIV, Paris.

Pearson, D.A., Tan, C.H., German, J.B., Davis, P.A., and Gershwin, M.E. 1999. Apple juice inhibits human low density lipoprotein oxidation. *Life Sci.* 64:1913-1920.

Sadler, N.P. and Jacobs, H. 1995. Application of the folin-ciocalteu reagent to the determination of salbutamol in pharmaceutical preparations. *Talanta* 42:1385-1388.

Saucier, C.T. and Waterhouse, A.L. 1999. Synergetic activity of catechin and other antioxidants. *J. Agric. Food Chem.* 47:4491-4494.

Singleton, V.L. 1974. Analytical fractionation of the phenolic substances of grapes and wine and some practical uses of such analyses. *In* Chemistry of Winemaking (A.D. Webb, ed.) pp. 184-211. American Chemical Society, Washington, D.C.

Singleton, V.L. 1988. Wine phenols. *In* Wine Analysis (H.F. Linskens, and J.F. Jackson, eds.) pp. 173-218. Springer-Verlag, Berlin.

Singleton, V.L. and Rossi, J.A. 1965. Colorimetry of total phenolics with phosphomolybdic-phosphotungstic acid reagents. *Am. J. Enol. Vitic.* 16:144-158.

Singleton, V.L., Orthofer, R., and Lamuela-Raventos, R.M. 1999. Analysis of total phenols and other oxidation substrates and antioxidants by means of Folin-Ciocalteu reagent. *Methods Enzymol.* 299:152-178.

Slinkard, K. and Singleton, V.L. 1977. Total phenol analysis: Automation and comparison with manual methods. *Am. J. Enol. Vitic.* 28:49-55.

Somers, T.C. and Ziemelis, G. 1980. Gross interference by sulphur dioxide in standard determinations of wine phenolics. *J. Sci. Food Agric.* 31:600-610.

Somers, T.C. and Ziemelis, G. 1985. Spectral evaluation of total phenolic components in *Vitis vinifera*: Grapes and wine. *J. Sci. Food Agric.* 36:1275-1284.

Thoss, V., Baird, M.S., Lock, M.A., and Courty, P.V. 2002. Quantifying the phenolic content of freshwaters using simple assays with different underlying reaction mechanisms. *J. Environ. Monit.* 4:270-275.

Tryon, C.R., Edwards, P.A., and Chisholm, M.G. 1988. Determination of the phenolic content of some French-American hybrid white wines using ultraviolet spectroscopy. *Am. J. Enol. Vitic.* 39:5-10.

Vinson, J.A., Su, X.H., Zubik, L., and Bose, P. 2001. Phenol antioxidant quantity and quality in foods: Fruits. *J. Agric. Food Chem.* 49:5315-5321.

Wiseman, S., Waterhouse, A., and Korver, O. 2001. The health effects of tea and tea components: Opportunities for standardizing research methods. *Crit. Rev. Food Sci. Nutr.* 41:387-412 Suppl.

Yu, Z. and Dahlgren, R.A. 2000. Evaluation of methods for measuring polyphenolics in conifer foliage. *J. Chem. Ecol.* 26:2119-2140.

Key References

Singleton and Rossi, 1965. See above.

The original description for the use of the FC method for wine analysis.

Singleton et al., 1999. See above.

A thorough review of the FC method's use in wine analysis by the original author.

Somers and Ziemelis, 1985. See above.

The key compilation of spectral methods.

Contributed by Andrew L. Waterhouse
University of California, Davis
Davis, California

Extraction and Isolation of Polyphenolics

Polyphenolics constitute a wide range of chemical compounds composed of aromatic ring(s) with one or more hydroxyl substituents, including their functional derivatives. Methods for extraction and isolation of polyphenolics from plant material are described in this unit. Extraction and isolation are the first important steps for separation, characterization, and quantification of polyphenolics from plant material. Polyphenolics are often most soluble in organic solvents less polar than water. The solubility is dependent on the polar properties of the polyphenolics. The correct selection of the extracting solvent is not as simple as it may seem. Aqueous methanol is a popular choice of solvent (see Background Information).

Basic Protocol 1 describes the extraction of polyphenolics from freeze-dried powdered plant material using aqueous methanol and an ultrasound bath. The cavitation produced by ultrasound when used in aqueous methanol extraction assists to increase the mass transfer rate and allows higher product yield with reduced extraction time and solvent usage. Alternate Protocol 1 describes a simple sample preparation for the extraction of polyphenolics from fresh raw plant material using methanol and homogenization. Basic Protocol 2 describes an effective solid-phase extraction for isolation/purification of polyphenolics into non-anthocyanin and anthocyanin fractions. Interfering compounds such as sugars and organic acids are eliminated in this procedure. Alternate Protocol 2 describes the fractionation of polyphenolics into neutral and acidic fractions based on the fact that polyphenolic acids are completely ionized at pH 7.0 and un-ionized at pH 2.0.

ULTRASOUND-ASSISTED AQUEOUS METHANOL EXTRACTION OF POLYPHENOLICS

In this method, 80% aqueous methanol and ultrasound are used to extract polyphenolics from freeze-dried powdered plant material. Ultrasound-assisted extraction is a rapid and efficient method for the extraction of polyphenolics. Low-frequency, high-energy, high-power ultrasound in the kHz range has the advantage of significantly reducing extraction time and enhancing extraction yield. The use of fine powdered plant material maximizes polyphenolics extraction by increasing the surface area of the sample and facilitating the disruption of biological cell walls.

Materials

>   Freeze-dried powdered plant material, ground with a blender (e.g., Waring) or laboratory mill (e.g., Thomas-Wiley) and stored at −18°C under nitrogen gas, protected from light
>   80% (v/v) aqueous methanol
>   Nitrogen gas
>   Absolute methanol
>
>   500-ml Erlenmeyer flask
>   Ultrasonic cleaning bath
>   Whatman no. 2 filter paper
>   Buchner funnel
>   1000-ml round-bottom evaporating flask
>   Rotary evaporator with water aspirator or vacuum pump, 40°C water bath

CAUTION: Sonication, filtration, and transfer should be performed in a fume hood. Personal protection equipment—laboratory coat, gloves, and safety glasses—must be used. Ear protection should be utilized during sonication.

Contributed by Dae-Ok Kim and Chang Y. Lee

1. Mix 10 g of ground freeze-dried samples (accurately weighed) with 100 ml of 80% aqueous methanol in a 500-ml Erlenmeyer flask.

   The powdered plant material maximizes polyphenolic extraction because of its high contact area with solvent and easy destruction of biological cell walls. Ethanol may be employed instead of methanol in routine extractions to avoid the toxic properties of methanol. The main advantage of methanol over ethanol is its lower boiling point.

2. Under subdued light, immerse the flask into the ultrasonic bath and sonicate for 20 min at room temperature with continual nitrogen gas purging and periodic shaking. Do not allow the temperature to rise.

   To prevent oxidation of polyphenolics during extraction, a constant stream of nitrogen gas is introduced into the Erlenmeyer flask to provide an oxygen-free environment. The temperature in the ultrasonic bath during solvent extraction should be maintained at room temperature or lower because cavitational effects are reduced and possible degradation of polyphenolics takes place at increased temperatures. The temperature in the ultrasonic bath can be controlled by cooling systems or ice. The intense ultrasound achieves higher production yields, despite reduced contact time and solvent consumption, due to enhanced mass transfer rate.

3. Filter the mixture through Whatman no. 2 filter paper by vacuum suction using a chilled Buchner funnel. Rinse the filter cake with 50 ml absolute methanol.

4. Re-extract the residue with 100 ml of 80% aqueous methanol by repeating steps 2 and 3.

   Two extractions are often satisfactory for most plant material.

5. Transfer filtrates to a 1000-ml round-bottom evaporating flask with 50 ml of 80% aqueous methanol.

6. Evaporate methanol in a rotary evaporator under vacuum at 40°C until the volume of extract is reduced to 10 to 30 ml.

   Solvent evaporation should be done under reduced pressures at low temperatures to minimize the degradation of extracted polyphenolics. Due to the likely occurrence of hydrolysis, isomerization, and polymerization at higher temperatures, temperatures for the evaporation process should be maintained at 40°C or below.

7. Resolubilize the concentrate to a 100-ml volume with deionized distilled water. Flush with nitrogen gas to prevent oxidation. Store at −4°C until analysis (up to 1 day).

ALTERNATE PROTOCOL 1

HOMOGENIZER-ASSISTED METHANOL EXTRACTION OF POLYPHENOLICS

In this alternative procedure, absolute methanol rather than aqueous methanol is used to extract the polyphenolics from fresh plant material instead of freeze-dried samples. Raw plant material is washed, dried, and then immediately extracted. This protocol employs a blender to first macerate the plant material to increase the sample surface area for better contact of the solvent and the sample.

Additional Materials (also see Basic Protocol 1)

Fresh plant material
Ascorbic acid

500-ml beaker
Blender (e.g., Waring Blendor), chilled to 4°C
Polytron homogenizer (or equivalent)
500-ml filter flask

Extraction and Isolation of Polyphenolics

I1.2.2

1. Mix 50 g fresh plant material (net weight) with 1 g ascorbic acid and 100 ml absolute methanol in a 500-ml beaker.

   Ascorbic acid is incorporated into the extraction system to prevent oxidation of polyphenolics during extraction. Since fresh plant material contains considerable quantities of water, the use of absolute methanol rather than aqueous methanol provides a more effective extraction.

2. Transfer contents of beaker into chilled (4°C) blender and immediately macerate by blending at high speed for 3 min.

   The speed of homogenization should start slowly and then be gradually increased. This work should be performed in the fume hood using proper personal protection equipment (PPE).

3. Transfer ground/crushed material back into its original beaker with 50 ml of 80% aqueous methanol and place in an ice bath.

4. Homogenize sample in ice bath using a Polytron homogenizer set at 7 for 2 min.

   This work also should be performed in the fume hood using proper PPE. The sample should be kept in an ice bath to prevent an increase of temperature that may cause possible degradation of polyphenolics during homogenization.

5. Filter the homogenous sample through a Whatman no. 2 filter paper into a chilled 500-ml filter flask by water-aspirated vacuum suction using a chilled Buchner funnel. Pour the homogenate slowly into the center of the filter paper and allow to advance to the edges before applying the vacuum.

   In the case of extremely viscous samples or samples with high pectin contents, centrifugation is often employed. The sample is centrifuged for 20 min at 12,000 × g, 4°C, and then the supernatant is filtered.

6. Transfer the residue from the filter paper back to the Polytron homogenizer with 100 ml methanol and repeat steps 4 and 5. Combine filtrates and discard the extracted plant material.

7. Transfer filtrates to a 1000-ml round-bottom evaporating flask with 50 ml of 80% aqueous methanol and evaporate off methanol in a rotary evaporator under vacuum at 40°C to a volume of 10 to 30 ml.

8. Dissolve the extract to a volume of 100 ml with deionized distilled water. Flush with nitrogen gas to prevent oxidation. Store at −4°C until analysis (up to 1 day).

SEPARATION OF ANTHOCYANIN AND NON-ANTHOCYANIN FRACTIONS

BASIC PROTOCOL 2

Due to the large number of structurally similar polyphenolics in plants, analysis of individual polyphenolics is relatively difficult and complicated. Considerable amounts of interfering material can be extracted with the polyphenolic extraction procedure. An isolation/purification step is often required to eliminate components that may interfere with the analysis. In this protocol, a simple fractionation of the polyphenolic extract is performed using preconditioned C18 Sep-Pak cartridges to separate anthocyanins from non-anthocyanin polyphenolics (Figure I1.2.1; also see *UNIT F1.1*). This fractionation technique, using solid-phase extraction, makes the analysis of individual polyphenolics by high-performance liquid chromatography possible.

Materials

Ethyl acetate
Methanol with and without 0.1% (v/v) HCl
0.01 N aqueous HCl
Aqueous extract (sample) of polyphenolics (see Basic Protocol 1 or Alternate Protocol 1)
Nitrogen gas

Polyphenolics

I1.2.3

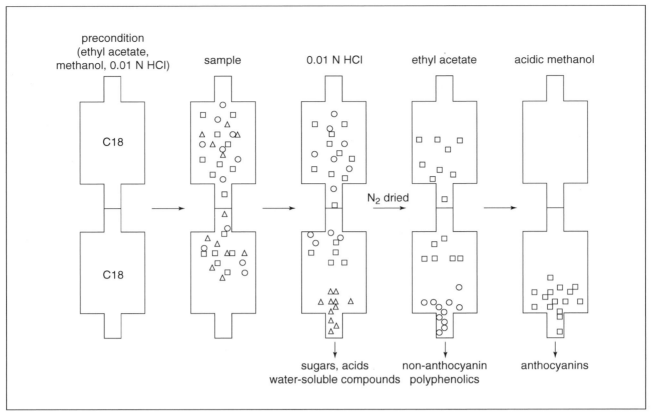

Figure I1.2.1 Fractionation of polyphenolics into non-anthocyanin and anthocyanin fractions using C18 cartridges (circles: non-anthocyanin polyphenolics; squares: anthocyanins; triangles: sugars, acids, and water-soluble compounds).

C18 Sep-Pak cartridges (Waters Chromatography)
0.45-µm polyvinylidene fluoride (PVDF) syringe-tip filter
50- and 100-ml round-bottom flasks
Rotary evaporator with water aspirator or vacuum pump, at 40°C

1. Connect two C18 Sep-Pak cartridges and precondition by sequentially passing 10 ml ethyl acetate, 10 ml absolute methanol, and 10 ml of 0.01 N aqueous HCl through the cartridges.

    A single cartridge can be used depending on solute concentration.

2. Filter aqueous polyphenolic extract (sample) through a 0.45-µm PVDF filter. Load a known volume of filtered extract onto cartridges.

3. Wash cartridges with 6 ml of 0.01 N aqueous HCl to remove sugars, acids, and other water-soluble compounds.

4. Dry cartridges by allowing a current of nitrogen gas to pass through the connected Sep-Pak cartridges for 10 min.

5. Rinse cartridges with 40 ml ethyl acetate to elute polyphenolic compounds other than anthocyanins and collect in a 100-ml round-bottom flask.

6. Elute the adsorbed anthocyanins from the cartridges with 6 ml acidic methanol and collect in a separate 50-ml round-bottom flask.

7. Remove the solvents of the non-anthocyanin fraction and anthocyanin fraction using a rotary evaporator under reduced pressure. Evaporate ethyl acetate at 20°C and methanol at 40°C or under nitrogen.

8. Dissolve each fraction in 5 ml deionized distilled water. Flush with nitrogen gas to prevent oxidation. Store at −4°C until analysis (up to 1 day).

To obtain a sufficient volume for subsequent analyses, several individual separations may be performed and then combined.

NEUTRAL AND ACIDIC FRACTIONATION OF POLYPHENOLICS

In this protocol, polyphenolics are fractionated into neutral and acidic fractions to prevent interference among polyphenolics in HPLC analysis. Phenolic acids are completely ionized at pH 7.0 and un-ionized at pH 2.0. This property allows for fractionation of neutral polyphenolics at pH 7.0 and acidic polyphenolics at pH 2.0. Two individually preconditioned C18 cartridges, one for neutral polyphenolics and the other for acidic polyphenolics, are used for this separation (Figure I1.2.2).

Additional Materials (also see Basic Protocol 2)

> Methanol
> 5 N NaOH
> 0.01 N and 1 N HCl (aqueous)

1. Precondition a C18 Sep-Pak cartridge for neutral polyphenolics with 2 ml methanol followed by 2 ml deionized distilled water.

2. Filter the aqueous polyphenolic extract with a 0.45-μm PVDF filter and adjust the pH to 7.0 with 5 N NaOH.

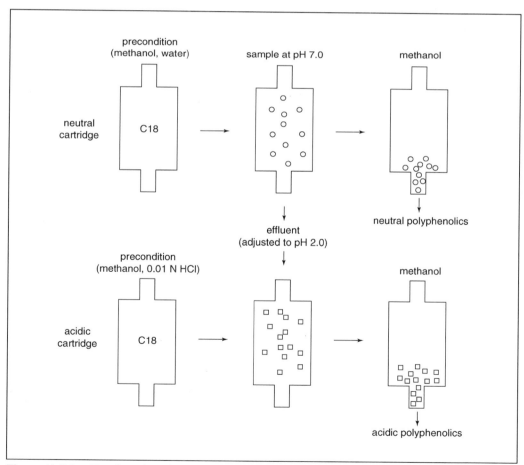

Figure I1.2.2 Fractionation of polyphenolics into acidic polyphenolics and neutral polyphenolics using C18 cartridges (circles: neutral polyphenolics; squares: acidic polyphenolics).

3. Pass extract onto the preconditioned C18 Sep-Pak cartridge to absorb the neutral polyphenolics and collect the effluent in a 50-ml round-bottom flask.

4. Precondition a second C18 Sep-Pak cartridge for acidic polyphenolics with 2 ml methanol followed by 2 ml of 0.01 N HCl.

5. Adjust the pH of the effluent portion from the first cartridge to pH 2.0 using 1 N HCl and pass through the second (acidic) cartridge to absorb the acidic polyphenolics.

6. Elute both neutral and acidic polyphenolics from their respective C18 Sep-Pak cartridges with 5 ml absolute methanol.

7. Remove the solvent of both neutral and acidic fractions using a rotary evaporator under vacuum at 40°C, or under nitrogen.

8. Resolubilize the fractions in 5 ml deionized distilled water. Flush with nitrogen gas to prevent oxidation. Store at –4°C until analysis (up to 1 day).

COMMENTARY

Background Information

Types of polyphenolics

Widely distributed in nature, polyphenolics are important aromatic secondary metabolites of plants. They contribute to the sensory quality (i.e., color, flavor, and taste) as well as the antioxidant activity of fruits, vegetables, beverages, and grains. Generally, fruits have higher polyphenolics than vegetables. Strawberries, raspberries, plums, oranges, and others are known fruits having higher concentrations of polyphenolics. Proteggente et al. (2002) reported that red plums, strawberries, and raspberries are categorized as anthocyanin-rich fruits; oranges and grapefruits are classified as flavanone-rich fruits; green grapes, onions, leeks, lettuce, broccoli, spinach, and green cabbage are flavonol-rich fruits and vegetables; and apples, pears, tomatoes, and peaches are hydroxycinnamate-rich fruits. More than 8000 natural polyphenolics are currently known to occur in plant sources. Flavonoids and their derivatives are the most common and the largest group of polyphenolics, with >4000 known structures (Middleton and Kandaswami, 1994). The term polyphenolics can be defined as compounds composed of aromatic benzene ring(s) substituted with hydroxyl groups, including functional derivatives. Polyphenolics range from structures that are very lipophilic (e.g., tangeretin) to those that are very hydrophilic (e.g., quercetin-3-sulfate). Most phenolic acids and flavonoids are compounds of relatively low molecular weight.

Polyphenolics are classified into three important groups: phenolic acids, flavonoids, and tannins. Phenolic acids include hydroxybenzoic (C_6-C_1), hydroxyphenylacetic (C_6-C_2), and hydroxycinnamic (C_6-C_3) acids (Figure I1.2.3; also see Figure I1.3.3). Hydroxycinnamic acids are most widely distributed in plant tissues. The important hydroxycinnamic acids are *p*-coumaric, caffeic, ferulic, and sinapic

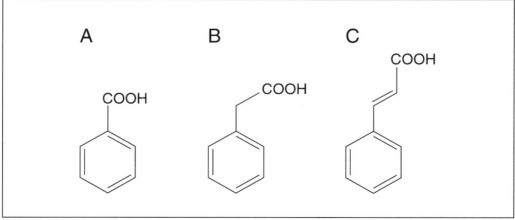

Figure I1.2.3 Structures of the hydroxybenzoic acids (**A**), hydroxyphenylacetic acid (**B**), and hydroxycinnamic acids (**C**). Also see *UNIT I1.3*.

acids. Most hydroxycinnamic acids are rarely found naturally in the free state, but occur partly as glucose esters, and most frequently as quinic acid esters (Herrmann, 1989; Möller and Herrmann, 1983). Chlorogenic acid, commonly distributed in fruits and vegetables, consists of a combination of quinic and caffeic acids.

Flavonoids have the common structure of diphenylpropane (C_6-C_3-C_6), consisting of two aromatic rings joined together by three carbons that are usually formed into an oxygenated heterocyclic ring (Figure I1.2.4). The structures and numbering systems of the various classes of flavonoids are shown in *UNIT I1.3* (see Figures I1.3.4 to I1.3.9). The oxidative state of this three-carbon chain determines the various flavonoids' classes. Flavonoids include anthocyanins, flavanols (catechins), flavonols, flavones, isoflavones, flavanones, and their derivatives. The most common flavonoids are flavones, flavonols, and their derivatives. The anthocyanins are water-soluble pigments responsible for the red, purple, and blue colors in fruits and flowers. Anthoxanthins refer to flavones, flavanones, flavonols, and isoflavones, which are colorless or white to yellow molecules (King and Young, 1999). Isoflavones, structural isomers to flavonoids, are found mainly in soy and soy-based products (Choi et al., 2000). Most flavonoids, except flavan-3-ols, are usually glycosylated, with some of the phenolic hydroxyl groups linked to sugar residues such as glucose, rhamnose, galactose, and arabinose (Lee, 2000).

Tannins are polyphenolic compounds of high molecular weight that occur naturally and react with proteins. They can form insoluble complexes with proteins, which are responsible for the taste known as astringency that is caused by precipitation of mouth proteins. Tannins are typically classified into three groups: (1) the condensed tannins, (2) the hydrolyzable tannins, and (3) the phlorotannins (Figure I1.3.10).

Condensed tannins (also referred to as procyanidins or proanthocyanidins) are oligomers or polymers of two or more catechins or epicatechins. Hydrolyzable tannins are polymers of gallic or ellagic acid, which are easily hydrolyzed by acid, alkali, or enzymes. Phlorotannins are made up of phloroglucinol subunits. Tannins having redox potentials similar to those of related simple phenolic compounds are 15 to 30 times more effective than Trolox at scavenging peroxy radicals (Hagerman et al., 1998).

Extraction of polyphenolics

The extraction is the main step in the recovery and isolation of polyphenolics and the evaluation of individual and total polyphenolics from various plant-based materials. The objective of solvent extraction is to remove all polyphenolics from the solid plant material by transferring them into a liquid phase. Polyphenolics are often most soluble in solvents less polar than water. An effective extraction of plant material depends on a proper solvent selection, elevated temperatures, and mechanical agitation to maximize polyphenolics recovery. Some polyphenolics are subject to degradation during sample preparation and extraction due to light and oxidation. Using ascorbic acid (Lee and Jaworski, 1987), sodium metabisulfite (Donovan et al., 1998), or nitrogen gas purging (Kim et al., 2002) during sample preparation will prevent oxidation of polyphenolics. It is recommended that samples be analyzed within 1 day of extraction.

Choice of solvents

The most common solvents used for the extraction of polyphenolics from plant material are methanol, ethanol, acetone, ethyl acetate, and their aqueous solvents. Aqueous methanol is a popular choice of solvent because it is efficient, has a high boiling point, and is eco-

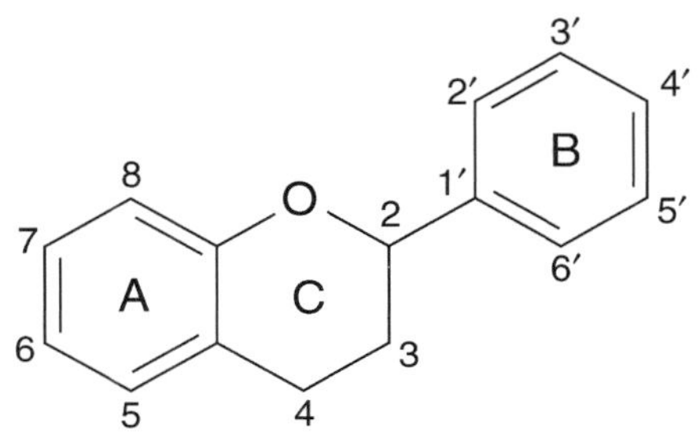

Figure I1.2.4 Structure and numbering system of flavonoids. Also see *UNIT I1.3*.

nomical. Methanol has been successfully used in extracting polyphenolics from various vegetables and fruits (Hertog et al., 1992a), such as apples (Oleszek et al., 1988), grapes (Lee and Jaworski, 1990), onions (Park and Lee, 1996), prunes and prune juice (Donovan et al., 1998), and banana bracts (Pazmiño-Durán et al., 2001). Acetone has been used to extract polyphenolics from apples (Eberhardt et al., 2000), prunes (Nakatani et al., 2000), and blackberries (Stintzing et al., 2002). Acetone extraction of anthocyanins is presented in UNIT F1.1. Ethanol has been used to extract polyphenolics from apples (Coseteng and Lee, 1987), grapes (Jaworski and Lee, 1987), and buckwheats (Kreft et al., 1999). Polyphenolics in wine and olive oil residues have been efficiently extracted with ethyl acetate (Lesage-Meessen et al., 2001; Salagoïty-Auguste and Bertrand, 1984). Polyphenolics such as catechin, epicatechin, and procyanidins are traditionally extracted from teas by brewing with hot water (Wang et al., 2000a,b).

Rutin extraction from buckwheat was optimally performed using between 50% and 60% ethanol with a 3-hr maceration (Kreft et al., 1999). Repeating the extraction two times was sufficient to extract most of the rutin. Lie et al. (2000) demonstrated the optimization of influencing factors such as extraction temperatures, extraction times, and solvent concentrations in a conventional extraction of dry powdered material of *Hypericum perforatum*. Applied extraction conditions caused variations to favor different components in the mixture. Higher extraction efficiency was achieved with moderately polar solvents to extract flavonoids such as rutin, isoquercitrin, and quercetin. The optimum extraction conditions for dry *Hypericum perforatum* leaf powder were determined as 44% to 69% ethanol in acetone with a 5.3- to 5.9-hr maceration at 55°C.

In order to extract anthocyanins in plant material, Basic Protocol 1 and Alternate Protocol 1 can be used if the extraction system is applied at a low pH (<3.0). The acidic environment is generally provided by the addition of hydrochloric acid or other organic acids to the solvent (Donner et al., 1997; Chandra et al., 2001; Pazmiño-Durán et al., 2001). Extraction of anthocyanins in an acidic environment is desirable because the red flavylium cation is then more stable. During the evaporation process, however, anthocyanins in extracts containing weak organic or mineral acids may be subject to acid hydrolysis of labile acyl groups and substituted sugars. To avoid such acid hydrolysis, anthocyanins can be extracted only by solvents without the addition of acids. For more detailed information, refer to UNIT F1.1, which extensively discusses and reviews anthocyanin extraction.

Sonication and homogenization

The traditional extraction method usually requires a long extraction time (Nakatani et al., 2000). Recently, ultrasound was applied effectively and rapidly to extract polyphenolics from plant material. The low frequency of ultrasound (in the kHz range) has the advantage of aiding extraction with significantly reduced extraction times and enhanced extraction yield. The efficient extraction of polyphenolics by ultrasound is ascribed to the cavitational phenomenon, which is the rapid formation and collapse of countless microscopic bubbles in a solvent produced by the alternating low- and high-pressure waves generated by ultrasonic sound. Bubbles filled with solvent vapor are produced at low pressure and then are compressed and finally implode at high pressure, leading to a powerful shock wave and enhanced mixing in the solvent extraction system (Paniwnyk et al., 2001). High-power ultrasound improves solvent extraction from plant material mainly due to its mechanical effects (Mason et al., 1996). These effects via cavitational collapse of bubbles enhance mass transfer rate and solvent penetration into cellular materials. In addition, the disruption or damage of biological cell walls by ultrasound results in facilitated release of the intracellular contents.

The extraction with methanol using a sonicator under strict exclusion of light at a low temperature is rapid and efficient, with >99% of the major constituents extracted from St. John's Wort in two 30-min steps of sonication (Li and Fitzloff, 2001). Methanolic extraction with ultrasound gave a significant increase of maximum yield and reduction of extraction time compared to conventional reflux extraction (Paniwnyk et al., 2001). In any extraction procedure, the solubility of the extracted components at a given extraction temperature should be considered when selecting the solvent. Ultrasonic extraction (three times for 1.5 hr each time) with unregulated temperature was comparable to water-bath extraction (three times for 5.6 hr each time at 55°C). The former showed an advantage of being less time consuming even with the drawback of the fluctuating temperature of an ultrasonic bath (Liu et al., 2000). Ultrasonic extraction with aqueous methanol to extract phenolic phytochemicals from apples under a nitrogen environment was effectively employed (Kim et al., 2002). Ultrasound-assisted extraction used at uniformly low temperatures can avoid thermal degradation and reduction of cavitational effects. It is recommended that ultrasound-assisted extraction also be used with additional agitation or shaking to avoid standing waves or the formation of solid-free regions for the preferential flow of the ultrasonic waves (Vinatoru et al., 1997).

Using both blender and homogenizer (see Alternate Protocol 1) has the advantage of immediate maceration in absolute methanol after briefly washing fresh plant material. This enhances the prevention of oxidative degradation during sample preparation for fresh plant-based materials. In addition, reducing agents such as ascorbic acid and sodium metabisulfite should be incorporated into the extraction procedure to prevent oxidation and browning reactions. Absolute methanol is used instead of aqueous methanol due to the water content of fresh, raw plant material.

Isolation/purification

Since considerable amounts of potential interfering materials can be extracted along with the polyphenolics, an isolation/purification step is often required to eliminate components that may interfere with analysis. The fractionation techniques presented in Basic Protocol 2 and Alternate Protocol 2, using solid-phase extraction to minimize the effects of sample preparation/cleanup on the integrity of the extract, will make possible the identification and quantification of individual polyphenolics by HPLC (*UNIT I1.3*), MS, and NMR.

Solid-phase extraction using a small, disposable Sep-Pak C18 cartridge has been popularized as the preferred method for cleaning and fractionating phenolic acids and flavonoids in order to remove interfering substances from crude polyphenolics extracts. The C18 solid-phase extraction technique has been used to isolate polyphenolics from vegetables, grapes, cherries, and apples. One of the major problems involved in the separation of polyphenolics is their similarity in chemical characteristics (Jaworski and Lee, 1987). A simple fractionation method was developed by Oszmianski and Lee (1990b) using a preconditioned Sep-Pak C18 cartridge to separate non-anthocyanin polyphenolics from anthocyanins for the successful analysis of non-anthocyanin polyphenolics. In this method, polyphenolics were absorbed onto the solid phase. Sugars, organic acids, and other polar components in the loaded cartridge were eluted with acidified water and the cartridge was dried by means of a continual current of nitrogen gas for several minutes. Polyphenolics other than anthocyanins were eluted from the cartridge using ethyl acetate. Anthocyanins still bound to the solid phase were eluted with acidified methanol (0.1% [v/v] HCl/methanol). A crucial step in this fractionation process is the removal of water from the cartridge using nitrogen gas. The recoveries from red grapes is as follows: ~100% for caffeoyl tartrate, 86% to 91% for procyanidin B3, 77% for rutin, and 98% to 99% for anthocyanins (Oszmianski and Lee, 1990b).

Phenolic acids are completely ionized at pH 7 and completely un-ionized at pH 2 (Salagoïty-Auguste and Bertrand, 1984). This characteristic allows for solid-phase extraction of neutral polyphenolics (catechins, procyanidins, flavonols, isoflavonoids) at pH 7 and acidic phenolics (caffeoyl tartrate, p-coumaroyl tartrate) at pH 2. Jaworski and Lee (1987) recovered >90% of acidic phenolics, catechin, and epicatechin, 104% of procyanidin B2, and 118% of procyanidin B3 by passing deproteinated grape juice through a Sep-Pak C18 cartridge. [Recoveries >100% might commonly take place due to experimental error (5% to 10%) from each experiment.] This simple, easy-to-use fractionation method also improved the resolution of many polyphenolic peaks by HPLC analysis. It was thus suggested that it facilitates the collection of unknown peaks individually, making possible their analysis and identification. However, the degradation of sample integrity can result from changes in pH, which may result in hydrolysis of acyl and glycosidic linkages, oxidative reaction of polyphenolics, and isomerization of cinnamic acids or esters (Roggero et al., 1990).

In a polyphenolic extract, anthocyanins can interfere with other polyphenolics such as procyanidins during HPLC analysis and hence should be removed prior to analysis. Anthocyanins from crude polyphenolic extracts can be removed as described in Basic Protocol 2. The ethyl acetate used for elution of phenolic compounds other than anthocyanins is removed using a rotary evaporator at 20°C. The non-anthocyanin polyphenolics are dissolved in deionized distilled water and the pH is adjusted to 7.0 with NaOH as described in Alternate Protocol 2 or the method developed by Oszmianski and Lee (1990a). In the latter method, polyphenolics were fractionated into three groups: neutral fraction A (flavanols and other polar phenolics), neutral fraction B (flavonols), and acidic phenolics. Polyphenolic extracts were adjusted to pH 7.0 with NaOH

Polyphenolics

I1.2.9

and passed through two C18 Sep-Pak cartridges connected together and preconditioned with 5 ml of methanol and 5 ml of deionized distilled water. The acidic fraction was washed with 5 ml of deionized distilled water, and 5 ml of 0.01 N HCl. Neutral fraction A absorbed on the solid phase was eluted with 5 ml of 10% acetonitrile solution acidified to pH 2.0. To elute neutral fraction B, 5 ml of 40% acetonitrile was used. The acidic fraction was adjusted to pH 2.0 with HCl and passed through a C18 cartridge preconditioned with 5 ml methanol and 5 ml of 0.01 N HCl. The absorbed phenolics were eluted with 5 ml of 40% acetonitrile. Prior to solid-phase extraction, the polyphenolic extract was subjected to phase separation with hexane to remove carotenoids and other nonpolar compounds.

Besides solid-phase extraction, column chromatography is also often used for cleanup and purification of polyphenolics from plant material. Ionic adsorbants (polyvinylpyrrolidone or PVP, polyamides, and Sephadex LH-20) and Amberlite XAD-2 resin have been used to isolate and purify polyphenolics from crude extracts. For the separation of polyphenolics from plant material, column chromatography using Sephadex LH-20, a gel-filtration matrix, is often used with various eluting solvents (Park and Lee, 1996). The most widely used solvents for column chromatography are aqueous methanol and aqueous ethanol.

Critical Parameters and Troubleshooting

The sonication time, repetitions of extraction, solvent polarity, and solvent selection may depend on the particular plant material. Before using Basic Protocol 1, the optimization of these parameters must be determined in order to efficiently extract polyphenolics from plant materials. The use of finely ground samples enhances polyphenolics recovery due to the increased surface area and easy disruption of biological cellular walls. During extraction, a continual supply of nitrogen gas should be applied to the extraction vessel to provide a nitrogen environment, thus avoiding any possible oxidative degradation.

For an accurate quantification of phenolic compounds extracted from plant material, the starting plant material should be washed, dried, and weighed prior to extraction, and the amount of the final extract should be recorded. The extracted polyphenolics are generally diluted and then concentrated by evaporation of solvents. Evaporation should be performed at 30° to 40°C under reduced pressure. When extract-

ing anthocyanins using acidic conditions for their stability (see Background Information), it should be kept in mind that acid hydrolysis of polyphenolics may take place during the concentration process. The final extract can be quantified based on gallic acid or catechin equivalents using spectrophotometric measurement before HPLC (Singleton and Rossi, 1965).

Ultrasound tends to increase the temperature of the ultrasonic water bath, which may cause pigment degradation in the extract. Without the use of a temperature control device, a temperature increase from 23° to 65°C was reported during 90 min of ultrasonication (Liu et al., 2000). The surrounding temperature and air flow inside a fume hood may also affect the temperature in the ultrasonic bath. The temperature of an ultrasonic water bath should be controlled to prevent degrading reactions during extraction at room temperature or at lower temperatures. This can be achieved by using a continuous cooling system or by periodically adding ice to the water bath. In addition to the ultrasonic bath, agitation or shaking is desirable to avoid standing waves or the formation of solid-free regions for the preferential flow of the ultrasonic waves.

After filtering the mixture of sample and organic solvent, the hose connected from the water aspirator to the flask under the Buchner funnel should first be carefully disconnected before shutting off the water source to prevent possible backflow of tap water into the flask, which would contaminate the sample. Prior to evaporation, the solvent waste reservoir should be empty, clean, and dry. Evaporation should be started slowly at reduced pressure to avoid the boiling over of solvent and samples. If by chance the polyphenolic extract boils over into the waste reservoir, immediately release the applied vacuum and transfer the extract back to the original boiling flask. If this is not satisfactory and contamination is a possibility, extraction should be performed again with a new sample.

The final extract after evaporation may be partitioned with hexane in a separatory funnel to remove chlorophyll, lipids, carotenoids, and other fat-soluble materials (e.g., in avocado, olives, seeds), if considerable amounts of these compounds are present. However, possible loss of lipophilic polyphenolics should be checked. Performing the liquid-liquid extraction after the evaporation of solvent may reduce the amount of hexane required. The extract should be flushed with nitrogen gas and stored in the dark at 4°C to avoid possible loss of sample due to oxidation and other degrading reactions. If the analysis is prolonged, the extract should be stored at −18°C or below in a freezer-resistant container. Flavones and flavonols in methanol proved to be stable for >3 months at 4°C (Hertog et al., 1992b). Extracts at −18°C might be stable at least 3 months.

C18 solid-phase extraction is used to fractionate polyphenolics for their identification and characterization. This technique can eliminate interfering chemicals from crude extracts and produce desirable results for HPLC or other analytical procedures. To obtain a sufficient volume for all analyses, several separations by solid-phase extraction may be performed. The individual fractions need to be combined and dissolved in solvents appropriate for HPLC analysis. In Basic Protocol 2, the application of a current of nitrogen gas for the removal of water from the C18 cartridge is an important step in the selective fractionation of polyphenolics into non-anthocyanin and anthocyanin fractions. After the collection of non-anthocyanin polyphenolics, no additional work is necessary to elute anthocyanins bound to the C18 solid phase if anthocyanins are not to be determined.

Anticipated Results

Ultrasound-assisted extraction provides efficient extraction in a shorter processing time than is needed for conventional extraction. The aid of ultrasound will result in a higher extraction yield and reduce solvent consumption. The extract may exhibit a wide range of colors (pale yellow, brown, red). For anthocyanin extraction in an acidic environment, the extract will be deep red, pink, or purple. The extract may contain considerable amounts of lipophilic compounds (e.g., chlorophyll, carotenoids, lipids). Prior to solid-phase extraction, those compounds can be eliminated from the extracts using liquid-liquid extraction.

Time Considerations

The ultrasound-assisted extraction of freeze-dried plant-based materials normally takes ~2 hr. If the extract is evaporated to dryness, a total of ~3 hr is necessary. As an additional sample preparation step, 2 to 4 days should be allotted for freeze-drying fresh plant materials, depending on the quantity of the material. Homogenizer-assisted extraction of fresh fruit takes <4 hr.

Isolation/purification using solid-phase extraction is a fast step, generally needing ≤3 hr. Prior to this solid-phase extraction, if it is necessary to remove chlorophyll, carotenoids, and other lipophilic compounds from the final extracts, liquid-liquid partitioning with hexane takes an additional 90 min, depending on the particular plant material.

Literature Cited

Chandra, A., Rana, J., and Li, Y. 2001. Separation, identification, quantification, and method validation of anthocyanins in botanical supplement raw materials by HPLC and HPLC-MS. *J. Agric. Food Chem.* 49:3515-3521.

Choi, Y.-S., Lee, B.-H., Kim, J.-H., and Kim, N.-S. 2000. Concentration of phytoestrogens in soybeans and soybean products in Korea. *J. Sci. Food Agric.* 80:1709-1712.

Coseteng, M.Y. and Lee, C.Y. 1987. Changes in apple polyphenoloxidase and polyphenol concentrations in relation to degree of browning. *J. Food Sci.* 52:985-989.

Donner, H., Gao, L., and Mazza, G. 1997. Separation and characterization of simple and malonylated anthocyanins in red onions, *Allium cepa* L. *Food Res. Int.* 30:637-643.

Donovan, J.L., Meyer, A.S., and Waterhouse, A.L. 1998. Phenolic composition and antioxidant activity of prunes and prune juice (*Prunes domestica*). *J. Agric. Food Chem.* 46:1247-1252.

Eberhardt, M.V., Lee, C.Y., and Liu, R.H. 2000. Antioxidant activity of fresh apples. *Nature* 405:903-904.

Hagerman, A.E., Riedl, K.M., Jones, G.A., Sovik, K.N., Ritchard, N.T., Hartzfeld, P.W., and Riechel, T.L. 1998. High molecular weight plant polyphenolics (tannins) as biological antioxidants. *J. Agric. Food Chem.* 46:1887-1892.

Herrmann, K. 1989. Occurrence and content of hydroxycinnamic and hydroxybenzoic acid compounds in foods. *Crit. Rev. Food Sci. Nutr.* 28:315-347.

Hertog, M.G.L., Hollman, P.C.H., and Venema, D.P. 1992a. Optimization of a quantitative HPLC determination of potentially anticarcinogenic flavonoids in vegetables and fruits. *J. Agric. Food Chem.* 40:1591-1598.

Hertog, M.G.L., Hollman, P.C.H., and Katan, M.B. 1992b. Content of potentially anticarcinogenic flavonoids of 28 vegetables and 9 fruits commonly consumed in the Netherlands. *J. Agric. Food Chem.* 40:2379-2383.

Jaworski, A.W. and Lee, C.Y. 1987. Fractionation and HPLC determination of grape phenolics. *J. Agric. Food Chem.* 35:257-259.

Kim, D.-O., Lee, K.W., Lee, H.J., and Lee, C.Y. 2002. Vitamin C equivalent antioxidant capacity (VCEAC) of phenolic phytochemicals. *J. Agric. Food Chem.* 50:3713-3717

King, A. and Young, G. 1999. Characteristics and occurrence of phenolic phytochemicals. *J. Am. Diet. Assoc.* 99:213-218.

Kreft, S., Knapp, M., and Kreft, I. 1999. Extraction of rutin from buckwheat (*Fagopyrum esculentum* Moench) seeds and determination by capillary electrophoresis. *J. Sci. Food Agric.* 47:4649-4652.

Lee, C.Y. 2000. Phenolic compounds. *In* Encyclopedia of Food Science and Technology (F.J. Francis, ed.) pp. 1872-1881. John Wiley & Sons, New York.

Lee, C.Y. and Jaworski, A. 1987. Phenolic compounds in white grapes grown in New York. *Am. J. Enol. Vitic.* 38:277-281.

Lee, C.Y. and Jaworski, A.W. 1990. Identification of some phenolic in white grapes. *Am. J. Enol. Vitic.* 41:87-89.

Lesage-Meessen, L., Navarro, D., Maunier, S., Sigoillot, J.-C., Lorquin, J., Delattre, M., Simon, J.-L., Asther, M., and Labat, M. 2001. Simple phenolic content in olive oil residues as a function of extraction systems. *Food Chem.* 75:501-507.

Li, W. and Fitzloff, J.F. 2001. High performance liquid chromatographic analysis of St. John's Wort with photodiode array detection. *J. Chromatogr. B* 765:99-105.

Lie, F.F., Ang, C.Y.W., and Springer, D. 2000. Optimization of extraction conditions for active components in *Hypericum perforatum* using response surface methodology. *J. Agric. Food Chem.* 48:3364-3371.

Liu, F.F., Ang, C.Y.W., Heinze, T.M., Rankin, J.D., Beger, R.D., Freeman, J.P., and Lay, J.O. Jr. 2000. Evaluation of major active components in St. John's Wort dietary supplements by high-performance liquid chromatography with photodiode array detection and electrospray mass spectrometric confirmation. *J. Chromatogr. A* 888:85-92.

Mason, T.J., Paniwnyk, L., and Lorimer, J.P. 1996. The uses of ultrasound in food technology. *Ultrason. Sonochem.* 3:S253-S260.

Middleton, E. and Kandaswami, C. 1994. The impact of plant flavonoids on mammalian biology: Implications for immunity, inflammation and cancer. *In* The Flavonoids: Advances in Research Since 1986 (J.B. Harborne, ed.) pp. 619-652. Champman & Hall, London.

Möller, B. and Herrmann, K. 1983. Quinic acid esters of hydroxycinnamic acids in stone and pome fruit. *Phytochemistry* 22:477-481.

Nakatani, N., Kayano, S.-I., Kikuzaki, H., Sumino, K., Katagiri, K., and Mitani, T. 2000. Identification, quantitative determination, and antioxidative activities of chlorogenic acid isomers in prune (*Prunus domestica* L.). *J. Agric. Food Chem.* 48:5512-5516.

Oleszek, W., Lee, C.Y., Jaworski, A.W., and Price, K.R. 1988. Identification of some phenolic compounds in apples. *J. Agric. Food Chem.* 36:430-432.

Oszmianski, J. and Lee, C.Y. 1990a. Inhibitory effect of phenolics on carotene bleaching in vegetables. *J. Agric. Food Chem.* 38:688-690.

Oszmianski, J. and Lee, C.Y. 1990b. Isolation and HPLC determination of phenolic compounds in red grapes. *Am. J. Enol. Vitic.* 41:204-206.

Paniwnyk, L., Beaufoy, E., Lorimer, J.P., and Mason, T.J. 2001. The extraction of rutin from flower buds of *Sophora japonica*. *Ultrason. Sonochem.* 8:299-301.

Park, Y.-K. and Lee, C.Y. 1996. Identification of isorhamnetin 4'-glucoside in onions. *J. Agric. Food Chem.* 44:34-36.

Pazmiño-Durán, E.A., Giusti, M.M., Wrolstad, R.E., and Glória, M.B.A. 2001. Anthocyanins from banana bracts (*Musa X paradisiaca*) as potential food colorants. *Food Chem.* 73:327-332.

Proteggente, A.R., Pannala, A.S., Paganga, G., van Buren, L., Wagner, E., Wiseman, S., van de Put, F., Dacombe, C., and Rice-Evans, C.A. 2002. The antioxidant activity of regularly consumed fruits and vegetables reflects their phenolic and vitamin C composition. *Free Radic. Res.* 36:217-233.

Roggero, J.-P., Coen, S., and Archier, P. 1990. Wine phenolics: Optimization of HPLC analysis. *J. Liq. Chromatogr.* 13:2593-2603.

Salagoïty-Auguste, M.-H. and Bertrand, A. 1984. Wine phenolics—Analysis of low molecular weight components by high performance liquid chromatography. *J. Sci. Food Agric.* 35:1241-1247.

Singleton, V.L. and Rossi, J.A. Jr. 1965. Colorimetry of total phenolics with phosphomolybdic-phosphotungstic acid reagents. *Am. J. Enol. Vitic.* 16:144-158.

Stintzing, F.C., Stintzing, A.S., Carle, R., and Wrolstad, R.E. 2002. A novel zwitterionic anthocyanin from evergreen blackberry (*Rubus laciniatus* Wild). *J. Agric. Food Chem.* 50:396-399.

Vinatoru, M., Toma, M., Radu, O., Filip, P.I., Lazurca, D., and Mason, T.J. 1997. The use of ultrasound for the extraction of bioactive principals from plant materials. *Ultrason. Sonochem.* 4:135-139.

Wang, H., Helliwell, K., and You, X. 2000a. Isocratic elution system for the determination of catechins, caffeine and gallic acid in green tea using HPLC. *Food Chem.* 68:115-121.

Wang, L.-F., Kim, D.-M., and Lee, C.Y. 2000b. Effects of heat processing and storage on flavanols and sensory qualities of green tea beverage. *J. Agric. Food Chem.* 48:4227-4232.

Key References

Jaworski and Lee, 1987. See above.

An improved analytical method for the fractionation of polyphenolics into neutral and acidic groups by passing deproteinated grape juice through a preconditioned C18 Sep-Pak cartridge.

Oszmianski and Lee, 1990b. See above.

A method for fractionation of polyphenolics into non-anthocyanin and anthocyanin fractions from red grapes by solid-phase extraction using disposable C18 cartridges.

Contributed by Dae-Ok Kim and
  Chang Y. Lee
Cornell University
Geneva, New York

HPLC Separation of Polyphenolics

The polyphenolics, ubiquitous phytochemicals in the plant kingdom, are important aromatic secondary metabolites of plants. Diverse combinations of polyphenolics are found in plant-based materials. Polyphenolics are important because they are responsible for the color and flavor of fresh and processed products. Some have strong antioxidant and anticancer activities. They are routinely consumed in the human diet in significant quantities.

In this unit, methods for reversed-phase high-performance liquid chromatography (HPLC) are described for the analysis of polyphenolics. HPLC analysis can be employed in an easy and fast manner to obtain an accurate elucidation and quantification of individual polyphenolic compounds found in plant-based materials. The separation of each polyphenolic is based on the polarity differences among polyphenolics with structural similarities and uses various combinations of mobile and stationary phases.

The Basic Protocol describes the reversed-phase HPLC analysis of polyphenolic compounds isolated into nonanthocyanin and anthocyanin fractions by solid-phase extraction. The Alternate Protocol describes the HPLC separation of acidic and neutral polyphenolic fractions. Fractionated samples are used because significant amounts of interfering compounds are extracted along with polyphenolics from plant materials. Solid-phase extraction with C18 Sep-Pak cartridges (*UNIT I1.2*) is used to selectively eliminate undesired components from crude extracts, and may minimize the effects of sample cleanup or preparation on the integrity of polyphenolics. The isolation and purification step using solid-phase extraction of polyphenolics will make possible the efficient analysis of individual polyphenolics by reversed-phase HPLC.

NOTE: Deionized and distilled or HPLC-grade water should be used in these protocols. Reagents (solvetns) should also be HPLC grade.

REVERSED-PHASE HPLC ANALYSIS OF POLYPHENOLICS SEPARATED INTO NONANTHOCYANIN AND ANTHOCYANIN FRACTIONS

In this protocol, polyphenolics isolated as nonanthocyanin and anthocyanin fractions after a sample cleanup are analyzed by reversed-phase HPLC in order to obtain an accurate measurement of individual polyphenolic constituents.

Materials

Crude polyphenolic sample (*UNIT I1.2*)
Methanol
Acidified aqueous methanol or acidified water (0.1% [v/v] HCl)
Mobile phases:
   50 mM $(NH_4)H_2PO_4$, pH 2.6 (pH adjusted with orthophosphoric acid)
   80:20 (v/v) acetonitrile/50 mM $(NH_4)H_2PO_4$, pH 2.6
   200 mM H_3PO_4, pH 1.5 (pH adjusted with ammonium hydroxide)
Polyphenolic standards (see Table I1.3.1 for suppliers)

0.45-μm poly(tetrafluoroethylene) (PTFE) syringe-tip filters
High-performance liquid chromatography (HPLC) system equipped with:
   Quaternary pump
   Diode array detector
   Vacuum degasser

Contributed by Dae-Ok Kim and Chang Y. Lee

Table I1.3.1 Elution Order of Polyphenolic Compounds Separated by HPLC of Nonanthocyanin and Anthocyanin Fractions[a]

Compound[b]	Retention time (min)
280 nm:	
Gallic acid	8.739 ± 0.155
Protocatechuic acid	12.948 ± 0.161
Epigallocatechin	15.637 ± 0.215
Catechin	17.998 ± 0.209
Vanillic acid	20.247 ± 0.229
Syringic acid	21.612 ± 0.283
Epicatechin	24.697 ± 0.391
Epigallocatechin gallate	25.710 ± 0.285
Epicatechin gallate	33.209 ± 0.110
Naringin	34.626 ± 0.039
320 nm:	
Chlorogenic acid	17.594 ± 0.175
Caffeic acid	21.553 ± 0.331
p-Coumaric acid	30.315 ± 0.281
Sinapic acid	32.266 ± 0.108
Ferulic acid	32.301 ± 0.167
Resveratrol	38.022 ± 0.121
Cinnamic acid	40.953 ± 0.088
370 nm:	
Rutin	32.226 ± 0.149
Rhoifolin	34.395 ± 0.040
Quercitrin	34.668 ± 0.071
Myricetin	36.478 ± 0.079
Luteolin	39.335 ± 0.097
Quercetin	39.752 ± 0.115
Apigenin	41.913 ± 0.048
Kaempferol	42.548 ± 0.194
520 nm:	
Kuromanin	19.525 ± 0.307
Malvin	19.855 ± 0.519
Keracyanin	20.879 ± 0.139
Delphinidin	26.590 ± 0.284
Oenin	28.130 ± 0.059
Cyanidin	30.523 ± 0.085
Pelargonidin	32.626 ± 0.041
Peonidin	33.160 ± 0.165
Malvidin	33.213 ± 0.056

[a]Retention time will depend on the specific column used. All results are expressed as mean ± SD for at least twenty replications, and were obtained on a Symmetry (Waters Chromatography) analytical column.

[b]Sigma is an appropriate supplier of all 280-, 320-, and 370-nm compounds; 520-nm compounds can be purchased from Extrasynthese. Other suppliers include Indo Fine Chemical and Polyphenols AS.

5-μm × 250-mm × 4.6-mm C18 reversed-phase column (e.g., Symmetry; Waters Chromatography)
Guard column (e.g., Symmetry Sentry; Waters Chromatography)
1- to 100-μl sample injection loop (20-μl loop is recommended)

Additional reagents and equipment for fractionating crude polyphenolics by solid-phase extraction into anthocyanin and nonanthocyanin fractions (*UNIT I1.2*)

1. Fractionate a crude polyphenolic sample into anthocyanin and nonanthocyanin fractions by solid-phase extraction as described in *UNIT I1.2*.

    For quantitative analysis by HPLC, accurate weights of the raw materials used and final volumes of extracts should be recorded. The sample volume applied to the disposable C18 cartridge, the final sample volume after evaporation, and the dissolving solvent volume must also be known.

    This fractionation step may be optional. Some samples can be directly analyzed by HPLC after filtration (step 2) without solid-phase extraction. Anthocyanins that can be detected at 280 nm can interfere with the separation of some polyphenolics. If the analyst is interested in nonanthocyanin polyphenolics, and especially if plant materials containing high levels of anthocyanins are being analyzed, this fractionation technique should be utilized.

2. Dilute anthocyanin and nonanthocyanin fractions 1:1 (v/v) with methanol and filter each through a 0.45-μm PTFE syringe-tip filter. For the anthocyanin fraction, use acidified aqueous methanol or acidified water as the solvent.

    See UNIT F1.3 for more detailed information about anthocyanin separation by HPLC.

3. Set up an HPLC system with a 5-μm × 250-mm × 4.6-mm C18 reversed-phase column and guard column. Set flow rate to 1.0 ml/min at constant room temperature (e.g., 23°C).

    Installation of a guard column having the same packing material as the main column lengthens the life of the often expensive analytical column.

    The instrument room should be maintained at constant temperature for greater reproducibility during HPLC analysis. The operating temperature and flow rate may be altered depending on the particular HPLC system.

4. Set detector at 280 nm for catechins (flavan-3-ols), naringin, and benzoic acid derivatives; 320 nm for chlorogenic acid, resveratrol, and hydroxycinnamic acids; 370 nm for flavones and flavonols; and 520 nm for anthocyanins and anthocyanidins (see Table I1.3.1).

    Detection of the nonanthocyanin fraction at 280, 320, and 370 nm can be done simultaneously. For direct injection after filtration without fractionation, or for analysis of combined nonanthocyanin and anthocyanin fractions, all four wavelengths can be used simultaneously.

5. Load a 20-μl fractionated polyphenolic sample into the HPLC system.

    Although the use of a 20-μl injection loop is recommended, the injection volume applied to the HPLC system may depend on the samples used. Injection volumes between 1 and 100 μl are generally used (Merken and Beecher, 2000).

6. Run mobile phases as described in Table I1.3.2.

    To enhance peak resolution, the user may modify the solvent gradient. The linear-gradient mobile phases used here are modified from Lamuela-Raventós and Waterhouse (1994).

7. Use the polyphenolic standards to generate characteristic UV-Vis spectra and retention times and identify individual polyphenolics in the sample.

Table I1.3.2 Solvent Gradient for Reversed-Phase
HPLC Analysis of Polyphenolics[a]

Time (min)	Solvent A (%)	Solvent B (%)	Solvent C (%)
0	100	0	0
4	92	8	0
10	0	14	86
22.5	0	16.5	83.5
27.5	0	25	75
50	0	80	20
55	100	0	0
60	100	0	0

[a]Solvent A, 50 mM $(NH_4)H_2PO_4$, pH 2.6; solvent B, 80:20 (v/v)
acetonitrile/solvent A; solvent C, 200 mM H_3PO_4, pH 1.5.

The elution order of commercially available standards is presented in Table I1.3.1. Retention times of the peaks will depend on the specific column used.

8. Analyze polyphenolic standards at a minimum of three concentrations to generate calibration curves. Analyze data and calculate the quantity of each polyphenolic compound.

    A range of 0 to 10 ppm should be used for the concentrations of the standards, which should produce a linear relationship.

ALTERNATE
PROTOCOL

REVERSED-PHASE HPLC ANALYSIS OF POLYPHENOLICS SEPARATED INTO NEUTRAL AND ACIDIC FRACTIONS

Phenolic acids are ionized at pH 7.0 and are un-ionized at pH 2.0. This property allows for solid-phase extraction of neutral polyphenolics at pH 7.0 and acidic polyphenolics at pH 2.0 to prevent interference. In this protocol, polyphenolics isolated as neutral and acidic fractions using pH adjustments are analyzed by reversed-phase HPLC in a thermostatically controlled environment.

Additional Materials *(also see Basic Protocol)*

    Mobile phases:
        5:95 (v/v) acetic acid/water (for acidic fractions)
        5:95 (v/v) acetic acid/water and 40:60 (v/v) acetonitrile/water (for neutral
        fractions)
    Binary pump
    8×100-mm C18 Radial-Pak column (Waters Chromatography) and guard column

    Additional reagents and equipment for fractionating crude polyphenolics into
        acidic and neutral fractions (UNIT I1.2)

1. Fractionate a crude polyphenolic sample into acidic and neutral fractions as described in *UNIT I1.2*.

    For quantitative analysis by HPLC, accurate weights of raw materials used and final volumes of extracts should be recorded. The sample volume applied to the disposable C18 cartridge, the final sample volume after evaporation, and the dissolving solvent volume must also be known.

    This particular fractionation step may be optional. Some samples can be directly injected after filtration (step 2) without solid-phase extraction. This technique, however, will improve the resolution of many of the HPLC polyphenolic peaks and will allow their analysis and identification.

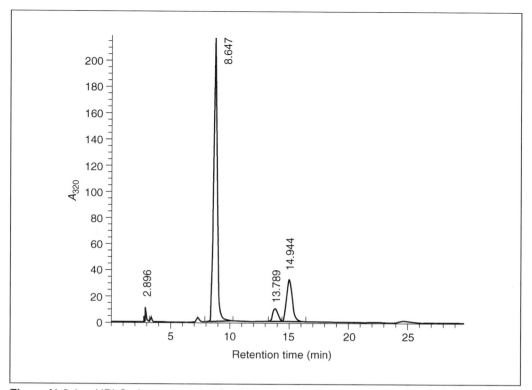

Figure I1.3.1 HPLC chromatogram of acidic polyphenolics isolated from Niagara grapes and detected at 320 nm. Retention time: 8.647 min, *trans*-caftaric acid; 13.789 min, *cis*-coutaric acid; 14.944, *trans*-coutaric acid. Reproduced from Lee and Jaworski (1987) with permission from the American Society for Enology and Viticulture.

2. Dilute acidic and neutral fractions 1:1 (v/v) with methanol and filter through a 0.45-µm PTFE syringe-tip filter.

3. Set up an HPLC system (see Basic Protocol, steps 3 and 4), but use a binary pump and an 8 × 100–mm C18 Radial-Pak column, and set the detector at 320 nm for acidic polyphenolics or 280 nm for neutral polyphenolics. Load column as in step 5 of the Basic Protocol.

4a. *For acidic polyphenolics:* Elute isocratically with 5:95 acetic acid/water for 30 min.

> *Retention times of the peaks are subject to the particular type of column. The acidic fraction from solid-phase extraction consists of phenolic acids such as cis-coutaric, trans-coutaric, and trans-caftaric acids. Isocratic elution is suitable because of the limited number of compounds found in the acidic fraction. Analysis of the acidic fraction is completed within 30 min. See Figure I1.3.1 for an HPLC chromatogram of the acidic polyphenolics isolated from Niagara grapes.*

Table I1.3.3 Solvent Gradient for Reversed-Phase HPLC Analysis of Neutral Polyphenolics[a]

Time (min)	Solvent A (%)	Solvent B (%)
0	100	0
1	100	0
50	0	100
55	100	0
60	100	0

[a]Solvent A, 5:95 (v/v) acetic acid/water; solvent B, 40:60 (v/v) acetonitrile/water.

Polyphenolics

I1.3.5

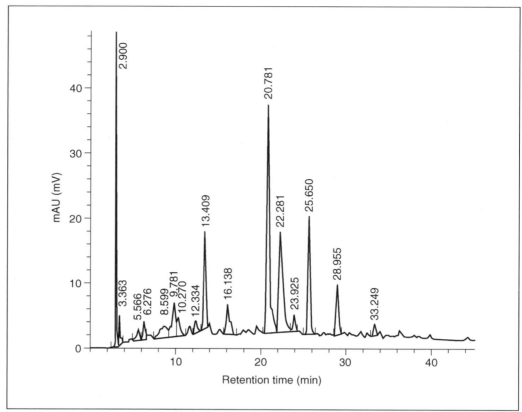

Figure I1.3.2 HPLC chromatogram of neutral polyphenolics found in Niagara grapes detected at 280 nm. Retention time: 8.599 min, procyanidin B3; 9.781 min, procyanidin B1; 13.409 min, catechin; 16.138 min, procyanidin B2; 20.781 min, epicatechin; 22.281 min, catechin-catechin-gallate; 23.925 min, catechin-catechin-gallate isomer; 28.955 min, catechin-gallate. AU, absorbance units. Reproduced from Lee and Jaworski (1987) with permission from the American Society for Enology and Viticulture.

4b. *For neutral polyphenolics:* Elute using 5:95 acetic acid/water and 40:60 acetonitrile/water over 60 min, as presented in Table I1.3.3.

> *Retention times of the peaks are subject to the particular type of column. For enhanced resolution of the peaks, the solvent gradient may be altered. See Figure I1.3.2 for a chromatogram of neutral polyphenolics found in Niagara grapes.*

5. Analyze polyphenolic standards and quantify polyphenolics in the sample as described (see Basic Protocol, steps 7 and 8).

COMMENTARY

Background Information

Polyphenolic phytochemicals are ubiquitous in the plant kingdom. These important aromatic secondary metabolites of plants are consumed in significant amounts in daily life. Their occurrence among plant-based materials is frequently varied. They contribute to sensory qualities such as color, flavor, and taste of plant-based materials. The composition of polyphenolic phytochemicals is influenced by maturity, cultivar (Lee and Jaworski, 1987), cultural practices, geographic origin, climatic conditions, storage conditions, and processing procedures (Spanos and Wrolstad, 1990). Some phytochemicals are known as nutraceuticals, which provide health benefits because of their biological activities (Dillard and German, 2000). Research on phytochemicals has been driven in recent years by their beneficial health effects, including antioxidant, anticarcinogenic, and antimutagenic activities (Huang and Ferraro, 1992) and their ability to reduce the risk of coronary heart disease (Hertog et al., 1993).

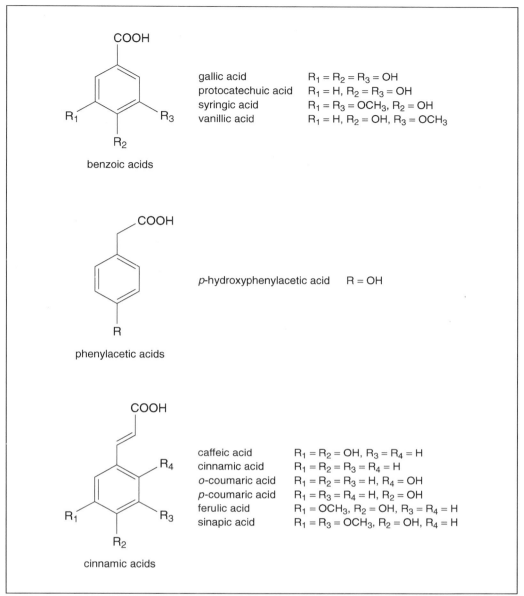

Figure I1.3.3   Structure of phenolic acids.

Polyphenolic phytochemicals are classified into three major groups: phenolic acids, flavonoids, and tannins. Phenolic acids include hydroxybenzoic, hydroxyphenylacetic, and hydroxycinnamic acids (Figure I1.3.3). Hydroxycinnamic acids are the most widely distributed of the phenolic acids in plant tissues. The important hydroxycinnamic acids are *p*-coumaric, caffeic, ferulic, and sinapic acids. Most hydroxycinnamic acids are rarely encountered in the free state in nature. They occur as glucose esters and, more frequently, as quinic acid esters (Herrmann, 1989). Phenolic acids are usually detected at wavelengths between 210 and 320 nm. In general, the polarity of phenolic acids is increased mainly by the hy-droxyl group at the 4 position, followed by those at the 3 and 2 positions. Methoxyl and acrylic groups substituted into the aromatic ring reduce polarity and increase the retention times (Torres et al., 1987). When using the Basic Protocol, the elution order for benzoic acids is gallic acid, protocatechuic acid, vanillic acid, and syringic acid (Rodríguez-Delgado et al., 2001; Table I1.3.1). The elution order for hydroxycinnamic acids is caffeic acid, *p*-coumaric acid, sinapic acid, ferulic acid, and cinnamic acid (Table I1.3.1).

The flavonoids are the largest and most important group of plant phenolics. They have the common structure of diphenylpropane ($C_6C_3C_6$; Figure I1.3.4A). The flavonoids are

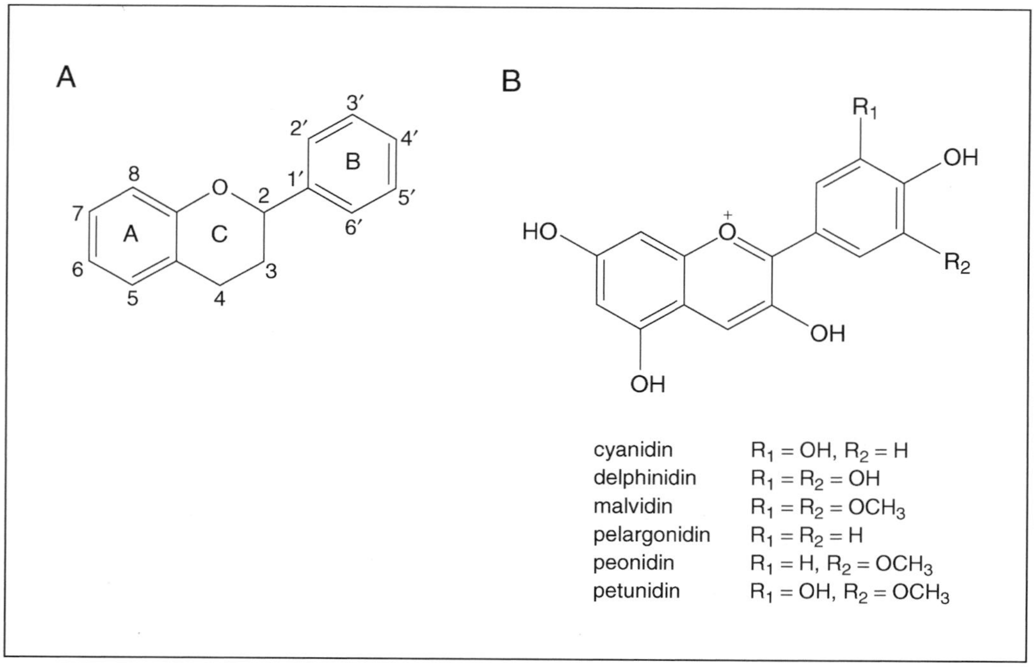

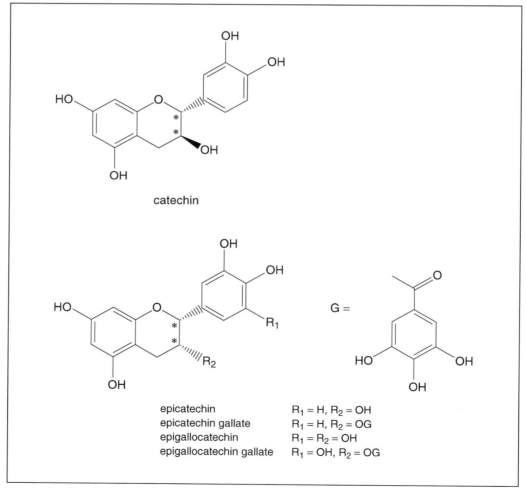

Figure I1.3.4 **(A)** Generic structure and numbering convention of flavonoids, which have a common diphenylpropane ($C_6C_3C_6$) structure. **(B)** Structure of anthocyanidins.

cyanidin	R_1 = OH, R_2 = H
delphinidin	R_1 = R_2 = OH
malvidin	R_1 = R_2 = OCH_3
pelargonidin	R_1 = R_2 = H
peonidin	R_1 = H, R_2 = OCH_3
petunidin	R_1 = OH, R_2 = OCH_3

catechin

epicatechin	R_1 = H, R_2 = OH
epicatechin gallate	R_1 = H, R_2 = OG
epigallocatechin	R_1 = R_2 = OH
epigallocatechin gallate	R_1 = OH, R_2 = OG

Figure I1.3.5 Structure of flavan-3-ols. Chiral centers indicated by asterisks. G, galloyl.

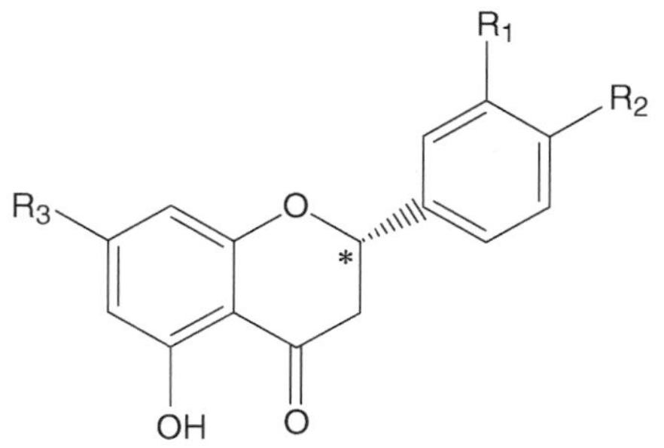

Figure I1.3.6 Structure of flavanones. Chiral center indicated by an asterisk.

eriocitrin	$R_1 = R_2 = OH$, $R_3 = O$-rutinoside
eriodictyol	$R_1 = R_2 = R_3 = OH$
hesperetin	$R_1 = R_3 = OH$, $R_2 = OCH_3$
hesperidin	$R_1 = OH$, $R_2 = OCH_3$, $R_3 = O$-rutinoside
naringenin	$R_1 = H$, $R_2 = R_3 = OH$
naringin	$R_1 = H$, $R_2 = OH$, $R_3 = O$-neohesperidoside
narirutin	$R_1 = H$, $R_2 = OH$, $R_3 = O$-rutinoside
neohesperidin	$R_1 = OH$, $R_2 = OCH_3$, $R_3 = O$-neohesperidoside

composed of six subgroups: anthocyanidins, catechins (flavan-3-ols), flavonols, flavones, flavanones, and isoflavones. Anthocyanidins and anthocyanins (glycosides or acylglycosides of anthocyanidins) are water-soluble pigments that are mainly responsible for the red, pink, purple, or blue colors in plant materials. They are visible to the human eye and are universal plant colorants. They are usually detected in the visible region of the spectrum between 500 and 530 nm, and are separated in the acidic environment of mobile phases as their flavylium cations. There are six anthocyanidins commonly present in plant-based materials: pelargonidin, cyanidin, delphinidin, peonidin, petunidin, and malvidin (Figure I1.3.4B). When using the Basic Protocol, the elution order for anthocyanins and anthocyanidins is kuromanin, malvin, keracyanin, delphinidin, oenin, cyanidin, pelargonidin, peonidin, and malvidin (Table I1.3.1).

Flavanols such as catechin, epicatechin, and epigallocatechin (Figure I1.3.5) are chiefly found in teas (green tea, oolong tea, black tea). Flavanols occur only as aglycone forms, contrary to the fact that most flavonoids exist in plants as glycosides, in which some hydroxyl groups are linked to sugar residues such as glucose, rhamnose, galactose, and arabinose (Lee, 2000). Teas are usually extracted by boiling. Filtration is often the only sample cleanup necessary for HPLC analysis. A typical wavelength for the detection of flavanols is often 210

or 280 nm. When using the Basic Protocol, the elution order for flavanols is epigallocatechin, catechin, epicatechin, epigallocatechin gallate, and epicatechin gallate (Table I1.3.1).

Flavanones (Figure I1.3.6) with a saturated heterocyclic C ring are mainly found in citrus fruits. Natural flavanones have the 2S configuration and usually occur as glycosides, frequently neohesperidosides (2-O-α-L-rhamnosyl-D-glucosides) and rutinosides (6-O-α-L-rhamnosyl-D-glucosides; Tomás-Barberán and Clifford, 2000). Their glycosylation occurs at the 7 position. Flavanones are generally detected at 280 nm, and possibly at 252, 285, 290, and 365 nm. In reversed-phase liquid chromatography, the elution order of flavanones (by decreasing mobility) is hesperidin, naringin, hesperetin, and naringenin (Swatsitang et al., 2000).

Flavones (Figure I1.3.7) are less commonly found in plants, but often occur in citrus. Polymethoxylated flavones, which include nobiletin, sinensetin, and tangeretin, are the characteristic features in citrus plants. Flavones are generally detected at 360 or 370 nm. The elution order, when analyzed as described in the Basic Protocol, is rhoifolin, luteolin, and apigenin (Table I1.3.1).

Among the flavonoids, flavonols (Figure I1.3.8) are the most prevalent in the plant kingdom. Flavonols have a hydroxyl at the C3 position. They are usually found as glycosides, of which glycosylation occurs at the 3 position

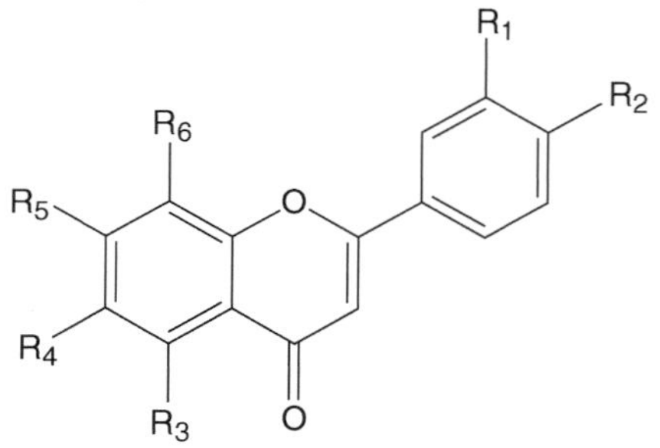

Figure I1.3.7 Structure of flavones.

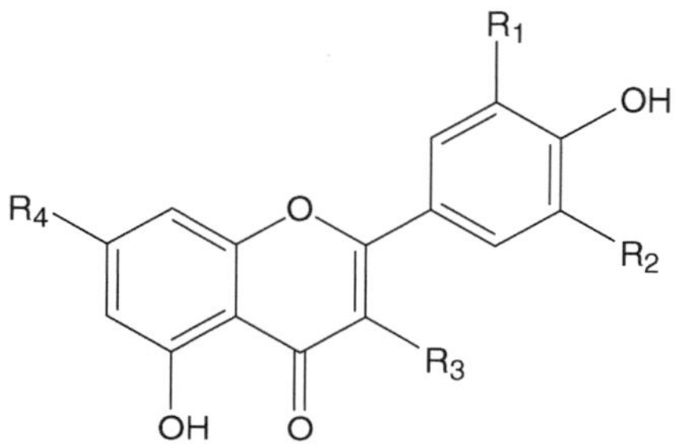

apigenin $R_1 = R_4 = R_6 = H$, $R_2 = R_3 = R_5 = OH$
luteolin $R_4 = R_6 = H$, $R_1 = R_2 = R_3 = R_5 = OH$
nobiletin $R_1 = R_2 = R_3 = R_4 = R_5 = R_6 = OCH_3$
rhoifolin $R_1 = R_4 = R_6 = H$, $R_2 = R_3 = OH$, $R_5 = O$-neohesperidoside
sinensetin $R_1 = R_2 = R_3 = R_4 = R_5 = OCH_3$, $R_6 = H$
tangeretin $R_2 = R_3 = R_4 = R_5 = R_6 = OCH_3$, $R_1 = H$

Figure I1.3.8 Structure of flavonols.

kaempferol $R_1 = R_2 = H$, $R_3 = R_4 = OH$
myricetin $R_1 = R_2 = R_3 = R_4 = OH$
quercetin $R_1 = R_3 = R_4 = OH$, $R_2 = H$
quercitrin $R_1 = R_4 = OH$, $R_2 = H$, $R_3 = O$-rhamnoside
rutin $R_1 = R_4 = OH$, $R_2 = H$, $R_3 = O$-rutinoside

in plant materials. The three most common aglycones in flavonols are kaempferol, quercetin, and myricetin. Glycosides of quercetin are usually predominant in plants. Flavonols and their glycosides are generally detected at 270, 365, and 370 nm. When analyzed using the Basic Protocol, the elution order for flavonols is rutin, quercitrin, myricetin, quercetin, and kaempferol (Table I1.3.1).

Isoflavones (Figure I1.3.9), which are structural isomers of the previously described flavonoids, are found almost exclusively in soy and soy-based products. The aromatic ring B is linked to the 3 position of the heterocyclic six-membered ring. Isoflavones can be divided into four groups: aglycones, glycosides, 6″-*O*-acetyl-glycosides, and 6″-*O*-malonyl-glycosides. Daidzein, genistein, glycitein, and their glycosides (daidzin, genistin, glycitin) are the major isoflavones. Isoflavones are detected at 260 nm, and possibly between 230 and 280 nm. The elution order for isoflavones by reversed-phase HPLC analysis is daidzin, glycitin, genistin, daidzein, glycitein, and genistein (Griffith and Collison, 2001; Gu and Gu, 2001).

Tannins (Figure I1.3.10) are polyphenolic polymers of high molecular weight that occur naturally and react with proteins. Tannins are

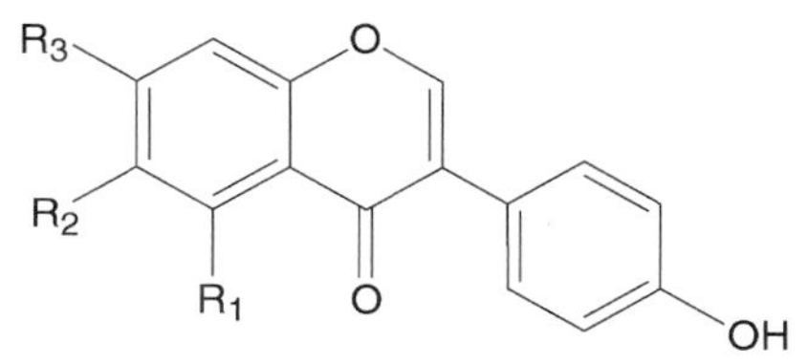

Figure I1.3.9 Structure of isoflavones.

daidzein	$R_1 = R_2 = H$, $R_3 = OH$
daidzin	$R_1 = R_2 = H$, $R_3 = O$-glucoside
genistein	$R_2 = H$, $R_1 = R_3 = OH$
genistin	$R_1 = OH$, $R_2 = H$, $R_3 = O$-glucoside
glycitein	$R_1 = H$, $R_2 = OCH_3$, $R_3 = OH$
glycitin	$R_1 = H$, $R_2 = OCH_3$, $R_3 = O$-glucoside

typically classified into three groups: condensed tannins, hydrolyzable tannins, and phlorotannins. Condensed tannins (often referred to as proanthocyanidins or procyanidins) are oligomers or polymers of flavan-3-ols (e.g., catechins and epicatechins). Hydrolyzable tannins are polymers of gallic or ellagic acid, producing gallotannins or ellagitannins, respectively. Phlorotannins consist of phloroglucinol subunits. HPLC analysis is a useful technique for quantitative or qualitative evaluation of these groups of tannins, although their analysis using HPLC is generally more difficult and complicated than HPLC analyses of phenolic acids and flavonoids. HPLC analysis of these tannins is briefly reviewed elsewhere (Mueller-Harvey, 2001; Schofield et al., 2001).

Because polyphenolics show chemical complexities and similar structures, isolation and quantification of the individual polyphenolic compounds have been challenging. Many traditional techniques (paper chromatography, thin-layer chromatography, column chromatography) have been used. HPLC, with its merits of exacting resolution, ease of use, and short analysis time, has the further advantage that separation and quantification occur simultaneously. A reversed-phase HPLC apparatus equipped with a diode array detector makes possible the easy isolation and separation of many polyphenolics. For enhanced performance of HPLC separation, the polyphenolics should first be isolated into several fractions to effectively separate the individual polyphenolics (Jaworski and Lee, 1987; Oszmianski and Lee, 1990).

Detection of the eluted polyphenolics has been commonly based on the absorptive measurement at characteristic wavelengths. The photodiode array detector has been extensively used for the detection of polyphenolics, mainly because of its collection of online UV-Vis spectra. All polyphenolics absorb in the UV region (Robards and Antolovich, 1997). Examples of UV spectra for polyphenolics are shown in Figure I1.3.11. Two absorption bands are characteristic of flavonoids. Band I, with maximum absorption in the range of 300 to 550 nm, arises from the A aromatic ring (Figure I1.3.4A). Band II, with maximum absorption in the range of 240 to 285 nm, comes from the B ring. Simple phenols, phenolic acids, and hydroxycinnamic acids show their absorption maxima in the range of 230 to 330 nm.

The elution order of polyphenolics may be predicted using a reversed-phase column, which is usually packed with silica-bonded C_8 or C_{18} materials. The more polar polyphenolics are generally eluted first under reversed-phase conditions. Glycosylation in flavonoids increases their polarity and therefore increases their mobility in reversed-phase HPLC. Triglycosides elute before diglycosides, which are followed by monoglycosides and then aglycones. The elution order of benzoic acids, hydroxycinnamic acids, and the aglycones of flavonoids can normally be determined on the basis of the number of polar hydroxyl groups and lipophilic methoxyl groups. The elution order of benzoic acids is as follows: gallic acid, protocatechuic acid, vanillic acid, and syringic acid (Rodríguez-Delgado et al., 2001). The elution order for hydroxycinnamic acids is as follows: caffeic acid, p-coumaric acid, sinapic acid, ferulic acid, and cinnamic acid (Schieber et al., 2001). The elution of flavonoids in order

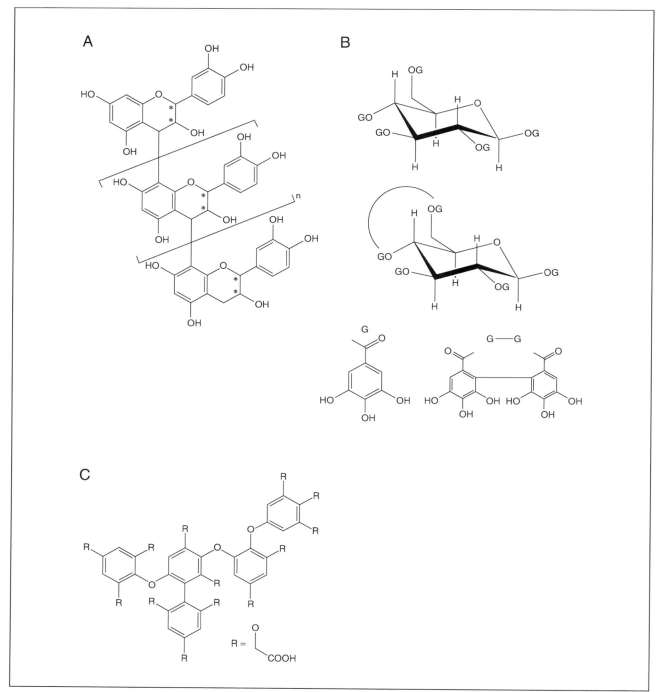

Figure I1.3.10 General structures of (**A**) condensed tannins (n indicates number of subunits), (**B**) hydrolyzable tannins, and (**C**) phlorotannins. Chiral centers indicated by asterisks. G, galloyl.

of decreasing polarity is as follows: catechin, epicatechin, cyanidin, rutin, myricetin, quercetin, and kaempferol. Acylation of polyphenolics reduces their mobility and therefore increases their retention times under the reversed-phase HPLC system.

Mobile-phase elution is frequently a binary system (Merken and Beecher, 2000). Aqueous acetic acid, phosphoric acid, formic acid, per-

chloric acid, or trifluoroacetic acid is used as an aqueous, acidified polar solvent, whereas methanol or acetonitrile is used as a less-polar solvent. Those acids used as modifiers in an aqueous polar solvent are usually employed for the enhancement of the resolution of peaks. Isocratic elution can be used for polyphenolics that are partially isolated from crude extracts. Gradient elution of mobile phases, which gives

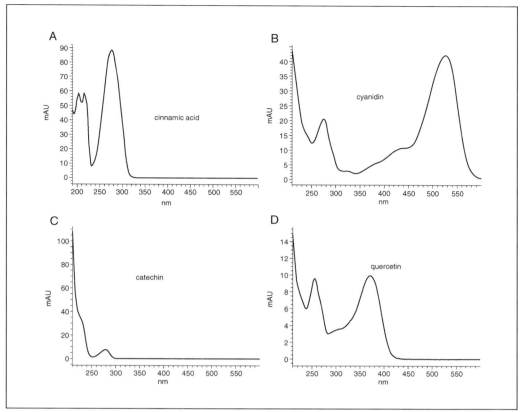

Figure I1.3.11 Sample UV spectra from various polyphenolics classes: (**A**) cinnamic acid (a phenolic acid), (**B**) cyanidin (an anthocyanidin), (**C**) catechin (a flavan-3-ol), and (**D**) quercetin (a flavonol). AU, absorbance units.

enhanced separation efficiency, is employed more commonly in the separation of complex mixtures of polyphenolics from plant materials.

Critical Parameters and Troubleshooting

Reversed-phase HPLC can separate polyphenolics of extracts on the basis of polarity. HPLC easily produces better resolution among chemically similar compounds in extracts than conventional chromatographic methods. The operating temperature of the column during reversed-phase HPLC analysis should be controlled for data reproducibility. A change in temperature produces only a minor effect, however, on band spacing in reversed-phase HPLC and produces essentially no effect in normal-phase HPLC (Lee and Widmer, 1996). A range of ambient temperatures is widely used, and elevated temperatures are often applied. The retention times of the peaks are dependent upon the type of column and the combination of various solvents used in the method.

Samples should be filtered before injection into the HPLC column to prevent it from clogging and thus being damaged. Also, it is recommended that a short guard column be installed before the analytical column. This relatively inexpensive column will capture extract components that may irreversibly adsorb onto the stationary phase of the expensive analytical column. Guard columns need to be replaced before strongly adsorbed contaminants, such as lipids or very hydrophobic molecules, permanently affect the analytical column, causing a loss in column performance. The guard column should be replaced at regular intervals for best results. If column performance deteriorates, the column should be cleaned and regenerated according to the manufacturer's instructions. The reversed-phase column, if not used for a period time, should be saturated and stored with a less-polar solvent such as acetonitrile or aqueous acetonitrile.

Mobile phases containing phosphate are subject to microbial growth, especially molds. These mobile phases should be filter sterilized and stored in autoclaved or sterile containers under refrigeration. Alternatively, a few milligrams of sodium azide per liter of solution may be added to the aqueous mobile phase. An aqueous salt solution used as a mobile phase may cause crystal

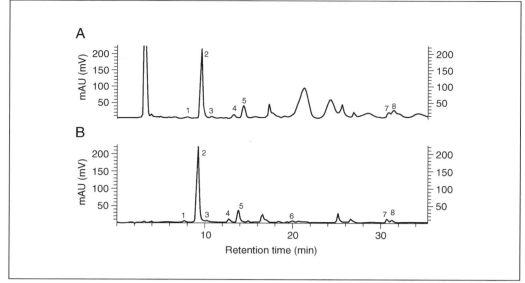

Figure I1.3.12 HPLC chromatograms of polyphenolics in Concord grape extract detected at 280 nm. (**A**) All polyphenolics, including anthocyanins. (**B**) Nonanthocyanin polyphenolics after fractionation. Peak identification: 1, *cis*-caftaric acid; 2, *trans*-caftaric acid; 3, procyanidin B3; 4, *cis*-coutaric acid; 5, *trans*-coutaric acid; 6, epicatechin; 7, quercetin galactoside; 8, quercetin glucoside. Reproduced from Oszmianski and Lee (1990) with permission from the American Society for Enology and Viticulture.

formation or scaling around tubing, which may lead to channeling. The HPLC system should thus be rinsed with water on a regular basis to maintain quality results. The column should always be equilibrated with the initial mobile phase prior to each sample injection.

An accurate sample weight before extraction and the amount of final extract after sample cleanup should be known for an accurate quantification of phenolic compounds extracted from plant materials by HPLC analysis. The characteristic wavelengths for detection of polyphenolics can be selected at the discretion of the experimenter. The solvent gradients described in the Basic and Alternate Protocols can be modified for better resolution.

The determination of polyphenolics may result in interference due to co-elution of phenolic acids and procyanidins. This problem can be eliminated by fractionation of polyphenolics into acidic and neutral polyphenolics prior to sample injection into the HPLC system. Because the fractionation techniques effectively improve the resolution of many polyphenolic peaks in the reversed-phase HPLC system, it is suggested that further characterization and identification of unknown peaks be conducted by additional methods such as mass spectrometry and nuclear magnetic resonance.

Anticipated Results

In reversed-phase HPLC separation of polyphenolics on the basis of polarity, the elution order of polyphenolics may be predicted. The more-polar polyphenolics are generally eluted first under reversed-phase conditions. Glycosylation in flavonoids increases their polarity and therefore their mobility in the reversed-phase system. The elution order of benzoic acids, hydroxycinnamic acids, and aglycones of flavonoids can normally be determined on the basis of the number of polar hydroxyl groups and lipophilic methoxyl groups. For additional information about elution order for various classes of polyphenolics, see Background Information.

After a sample cleanup using a C18 Sep-Pak cartridge, nonanthocyanin polyphenolics from a crude red grape extract were analyzed by reversed-phase HPLC (Oszmianski and Lee, 1990). The removal of 98% to 99% of anthocyanins was achieved. Typical chromatograms of polyphenolics from Concord grapes with and without the isolation of anthocyanins are shown in Figure I1.3.12, which displays the selective fractionation of anthocyanins and nonanthocyanins.

Using the Alternate Protocol, all major polyphenolic compounds are separated with good resolution, which is due to the effective fractionation of acidic and neutral polyphenolics (Jaworski and Lee, 1987; Lee and Jaworski, 1987). The acidic polyphenolics, analyzed by isocratic elution, consisted of *trans*-caftaric acid, *cis*-coutaric acid, and *trans*-coutaric acid, whereas the neutral polyphenolics, analyzed by gradient elution, included catechin, catechingallate, epicatechin, procyanidin B1, procyanidin B2, and procyanidin B3. Some portions of anthocyanin pigments were eliminated during C18 Sep-Pak fractionation. The elution of the residual pigments in the deproteinated juices occurred during the later part of the run and caused no interference with the reversed-phase HPLC analysis.

Time Considerations

The sample preparation techniques described in the Basic and Alternate Protocols require ~1 to 3 hr prior to sample injection into the HPLC. No additional time for sample cleanup is necessary if the sample is directly injected into the HPLC column without solid-phase extraction.

The Basic Protocol requires 60 min of running time for HPLC analysis after each injection. The period between 55 and 60 min allows for column equilibration prior to the next injection. Analysis of the acidic fraction described in the Alternate Protocol is completed within 30 min including column equilibration time. HPLC analysis of the neutral fraction described in the Alternate Protocol needs much more time (60 min) for sample runs. It is desirable to analyze the sample on the same day as the extraction to prevent possible polyphenolic deterioration.

Literature Cited

Dillard, C.J. and German, J.B. 2000. Phytochemicals: Nutraceuticals and human health. *J. Sci. Food Agric.* 80:1744-1756.

Griffith, A.P. and Collison, M.W. 2001. Improved methods for the extraction and analysis of isoflavones from soy-containing foods and nutritional supplements by reversed-phase high-performance liquid chromatography and liquid chromatography-mass spectrometry. *J. Chromatogr. A* 913:397-413.

Gu, L. and Gu, W. 2001. Characterisation of soy isoflavones and screening of novel malonyl glycosides using high-performance liquid chromatography-electrospray ionisation-mass spectrometry. *Phytochem. Anal.* 12:377-382.

Herrmann, K. 1989. Occurrence and content of hydroxycinnamic and hydroxybenzoic acid compounds in foods. *Crit. Rev. Food Sci. Nutr.* 28:315-347.

Hertog, M.G.L., Feskens, E.J.M., Hollman, P.C.H., Katan, M.B., and Kromhout, D. 1993. Dietary antioxidant flavonoids and risk of coronary heart disease: The Zutphen elderly study. *Lancet* 342:1007-1011.

Huang, M.-T. and Ferraro, T. 1992. Phenolic compounds in food and cancer prevention. *In* Phenolic Compounds in Food and Their Effects on Health II: Antioxidants and Cancer Prevention (M.-T. Huang, C.-T. Ho, and C.Y. Lee, eds.), American Chemical Society Symposium Series 507, pp. 8-34. ACS, Washington, D.C.

Jaworski, A.W. and Lee, C.Y. 1987. Fractionation and HPLC determination of grape phenolics. *J. Agric. Food Chem.* 35:257-259.

Lamuela-Raventós, R.M. and Waterhouse, A.L. 1994. A direct HPLC separation of wine phenolics. *Am. J. Enol. Vitic.* 45:1-5.

Lee, C.Y. 2000. Phenolic compounds. *In* Encyclopedia of Food Science and Technology (F.J. Francis, ed.) pp. 1872-1881. John Wiley & Sons, New York.

Lee, C.Y. and Jaworski, A. 1987. Phenolic compounds in white grapes grown in New York. *Am. J. Enol. Vitic.* 38:277-281.

Lee, H.S. and Widmer, B.W. 1996. Phenolic compounds. *In* Handbook of Food Analysis, Vol. 1 (L.M.L. Nollet, ed.) pp. 821-894. Marcel Dekker, New York.

Merken, H.M. and Beecher, G.R. 2000. Measurement of food flavonoids by high-performance liquid chromatography: A review. *J. Agric. Food Chem.* 48:577-599.

Mueller-Harvey, I. 2001. Analysis of hydrolysable tannins. *Anim. Feed Sci. Technol.* 91:3-20.

Oszmianski, J. and Lee, C.Y. 1990. Isolation and HPLC determination of phenolic compounds in red grapes. *Am. J. Enol. Vitic.* 41:204-206.

Robards, K. and Antolovich, M. 1997. Analytical chemistry of fruit bioflavonoids. *Analyst* 122:11R-34R.

Rodríguez-Delgado, M.A., Malovaná, S., Pérez, J.P., Borges, T., and García-Montelongo, F.J. 2001. Separation of phenolic compounds by high-performance liquid chromatography with absorbance and fluorimetric detection. *J. Chromatogr. A* 912:249-257.

Schieber, A., Keller, P., and Carle, R. 2001. Determination of phenolic acids and flavonoids of apple and pear by high-performance liquid chromatography. *J. Chromatogr. A* 910:265-273.

Schofield, P., Mbugua, D.M., and Pell, A.N. 2001. Analysis of condensed tannins: A review. *Anim. Feed Sci. Technol.* 91:21-40.

Spanos, G.A. and Wrolstad, R.E. 1990. Influence of processing and storage on the phenolic composition of Thompson seedless grape juice. *J. Agric. Food Chem.* 38:1565-1571.

Swatsitang, P., Tucker, G., Robards, K., and Jardine, D. 2000. Isolation and identification of phenolic compounds in *Citrus sinensis. Anal. Chim. Acta* 417:231-240.

Tomás-Barberán, F.A. and Clifford, M.N. 2000. Flavanones, chalcones and dihydrochalcones—nature, occurrence and dietary burden. *J. Sci. Food Agric.* 80:1073-1080.

Torres, A.M., Mau-Lastovicka, T., and Rezaaiyan, R. 1987. Total phenolics and high-performance liquid chromatography of phenolic acids of avocado. *J. Agric. Food Chem.* 35:921-925.

Key References

Lamuela-Raventós and Waterhouse, 1994. See above.

Describes solvent gradient conditions for HPLC analysis of wine polyphenolics.

Lee and Jaworski, 1987. See above.

Thoroughly describes sample cleanup of crude extracts and HPLC analysis of neutral and acidic polyphenolics.

Merken and Beecher, 2000. See above.

Extensively reviews the HPLC analysis of flavonoids.

Oszmianski and Lee, 1990. See above.

Thoroughly describes the effective fractionation technique of separating anthocyanins and nonanthocyanins.

Contributed by Dae-Ok Kim and
    Chang Y. Lee
Cornell University
Geneva, New York

Proanthocyanidins: Extraction, Purification, and Determination of Subunit Composition by HPLC

Proanthocyanidins are polymeric flavonoid compounds composed of flavan-3-ol subunits (*UNIT I1.3*), and are responsible for bitterness and astringency in some foods and beverages. This unit describes methods for extracting and purifying proanthocyanidins, and for determining their subunit composition by HPLC. Based upon HPLC results, the average degree of polymerization and the conversion yield for purified proanthocyanidins can be determined.

This unit is composed of three separate procedures describing proanthocyanidin extraction from plant tissue (see Basic Protocol 1), purification (see Basic Protocol 2), and subsequent analysis by reversed-phase HPLC (see Basic Protocol 3). These protocols have been developed and used for analysis of grape skins, grape berries, and grape seeds. Without modification, wine, apples, and pears have also been analyzed using these procedures.

EXTRACTION OF PROANTHOCYANIDINS

This extraction method utilizes an aqueous acetone system to extract proanthocyanidins from whole plant tissue. Acetone is preferred over other extraction solvent systems (most notably methanol) because of its ability to solubilize proanthocyanidin-containing material that is insoluble in methanol.

Materials

Plant material containing proanthocyanidins
66% (v/v) aqueous acetone, HPLC grade
Nitrogen gas

Suitably sized Erlenmeyer flask with stopper or septum
Aluminum foil
Platform shaker
Büchner funnel
Whatman no. 1 filter paper
Suitably sized round-bottom flask
Rotary evaporator with vacuum pump or water aspirator, 40°C

1. Accurately weigh plant material (within ±0.1%) and place into an Erlenmeyer flask. Record the weight of the sample.

   As a general guideline, and based upon grape tissue, 5 to 10 g of plant tissue is placed into a 250-ml Erlenmeyer flask with 100 ml of extraction solvent. To ensure adequate mixing, Erlenmeyer flasks should not be filled to more than 80% of capacity (i.e., 200 ml in a 250-ml Erlenmeyer flask). An approximate yield for this step is 1 to 2 mg of proanthocyanidin extraction per gram of grape berry weight.

2. Add enough 66% (v/v) aqueous acetone to the Erlenmeyer flask to cover the plant tissue and record the volume to within ±0.5%.

3. Equip the Erlenmeyer flask with a stopper or septum and sparge the extraction system well with nitrogen.

4. Cover the flask with aluminum foil and shake 24 hr at room temperature on a platform shaker.

Contributed by James A. Kennedy

Polyphenolics

I1.4.1

5. Remove the flask from the shaker and separate the proanthocyanidin extract from the solid plant material by filtering the mixture through a Büchner funnel equipped with an appropriately sized Whatman no. 1 filter.

   As a general guideline, and based again upon grape tissue, an extraction using 5 to 10 grams of plant tissue can be filtered using a Büchner funnel with a 6-cm-diameter perforated surface and 7-cm-diameter filter paper.

6. Place the filtered proanthocyanidin extract into an appropriately sized round-bottom flask and remove acetone on a rotary evaporator equipped with a water aspirator or vacuum pump, at 40°C.

   Because proanthocyanidins are susceptible to oxidation, the amount of time the flask is left on the rotary evaporator should be limited to the time necessary to remove the acetone. This is most easily determined by observing the rotary evaporator condenser. Because water has a higher surface tension than acetone, condensation of water is observed as fogging of the condenser portion of the rotary evaporator.

7. Store aqueous proanthocyanidin extract at −20°C or cooler.

   The author generally proceeds directly to purification, but has used grape seed extracts stored for up to one week and seen little degradation.

PURIFICATION OF PROANTHOCYANIDINS

This purification step is designed to remove impurities from the proanthocyanidin extract. It utilizes liquid-liquid extraction to remove lipophilic material and monomeric flavan-3-ols, and also adsorption chromatography to remove more hydrophilic material such as organic acids, sugars, and residual flavan-3-ol monomers. Following the steps in this protocol, purified and powdered proanthocyanidins are obtained.

Materials

Aqueous crude proanthocyanidin extract (see Basic Protocol 1)
Appropriate solvents for liquid-liquid extraction (e.g., chloroform and ethyl acetate)
Gel filtration medium (Toyopearl HW-40F; Supelco)
Methanol, HPLC grade
Trifluoroacetic acid (TFA), spectrophotometric grade
66% (v/v) aqueous acetone (HPLC grade) containing 0.1% (v/v) TFA
Dry ice/acetone bath

Separatory funnel
Suitably sized round-bottom flasks
Rotary evaporator with vacuum pump or water aspirator, 40°C
Chromatography column
Freeze drier with freeze-drying flask

Perform liquid extraction

1. Depending on the impurities that are present, perform liquid-liquid extraction of the aqueous proanthocyanidin extract to remove impurities with different solubility properties.

   For example, extract the proanthocyanidin mixture with chloroform to remove chlorophyll, carotenoids, and waxy material. Use ethyl acetate if substantial amounts of flavan-3-ol monomers are present. Tissues that would benefit from a chloroform extraction include leafy tissues that contain chlorophyll (i.e., tea leaves) and seeds that contain oils (i.e., grape seeds). Ethyl acetate would be useful in plants such as apples, berries, grapes, and teas (i.e., tissues known to contain significant amounts of flavan-3-ol monomers).

   In general, the volume of extraction solvent should be 20% (by volume) of the aqueous proanthocyanidin extract, and the extraction should be repeated five times.

2. Briefly remove residual organic solvent on a rotary evaporator equipped with a water aspirator or vacuum pump, at 40°C.

Perform adsorption chromatography

3. Prepare a slurry containing a suitable amount of gel filtration medium in 50% (v/v) aqueous methanol containing 0.1% (v/v) TFA, and pack a suitably sized column with this slurry according to the manufacturer's instructions.

   As a rough guideline, approximately 1 gram of proanthocyanidin can be purified from a column bed volume of 100 cm³ resin.

4. Precondition the column with 50% (v/v) aqueous methanol containing 0.1% (v/v) TFA.

5. Add methanol first and then TFA to the aqueous proanthocyanidin extract to obtain a 50% (v/v) methanol concentration with a 0.1% (v/v) TFA concentration. Apply the extract to the column.

6. Rinse the column with at least 5 column volumes of 50% methanol/0.1% TFA solution.

   The potential impurities will vary according to the plant tissue extracted, and therefore the exact washing volume will vary. It is important to determine the impurities present and their retention properties on the column to minimize impurities in the final proanthocyanidin and maximize proanthocyanidin recovery. For this step, the use of a spectrophotometer is helpful in monitoring the eluate. Some typical impurities and monitoring wavelengths include organic acids (215 nm), flavan-3-ol monomers (280 nm), hydroxycinnamic acids (320 nm), and flavonols (365 nm). Anthocyanins are observable in the visible spectrum.

7. Elute the adsorbed proanthocyanidins with 2 column volumes of 66% (v/v) acetone/0.1% TFA.

   This volume is normally sufficient to elute the adsorbed proanthocyanidins.

8. Transfer the proanthocyanidin-containing eluate into an appropriately sized and preweighed round-bottom flask and remove the acetone on a rotary evaporator.

9. Freeze the aqueous proanthocyanidin-containing solution in a dry ice/acetone bath.

   When freezing, and to minimize the time necessary to freeze dry the sample (next step), maximize the surface area of contact between the flask and the dry ice. This is most easily achieved by rotating the flask within the dry ice/acetone mixture so that the walls of the round-bottom flask are coated with the ice.

10. Transfer to a freeze-drying flask and freeze dry the aqueous proanthocyanidins.

11. Weigh the proanthocyanidins and calculate recovery in mg proanthocyanidins/kg sample (see Basic Protocol 1, step 1).

12. Store proanthocyanidins in a light-protected container at −20°C or cooler.

    Proanthocyanidins will slowly oxidize under freezer conditions. Based upon the storage of grape seed proanthocyanidins at −20°C, ~95% of the initial proanthocyanidins remain after 6 months of storage.

DETERMINATION OF SUBUNIT COMPOSITION BY HPLC

In this protocol, the subunit composition and the conversion yield of purified proanthocyanidins is determined by reversed-phase HPLC. The instrument has been selected because of its ability to run a gradient and because it can acquire UV/Vis spectra, which can be useful for compound identification.

In the initial stages of this analytical method the proanthocyanidins are cleaved by acid catalysis into their constitutive subunits. Proanthocyanidin extension subunits, upon cleavage, form unstable electrophilic intermediates. Phloroglucinol (1,3,5-trihydroxybenzene) is added to the reaction mixture as a nucleophile, where it combines with extension subunit intermediates to form analyzable adducts.

This method analyzes proanthocyanidins after cleavage, as opposed to analyzing intact proanthocyanidins, for the following reasons: cleavage reactions provide compositional information and unambiguous identification of proanthocyanidin-derived material as well as providing information on the proportion of purified material that is unknown. Due to the inherent reactivity and heterogeneity of proanthocyanidins, chromatographic methods that analyze the material while still intact provide only approximate quantitative information and are nebulous in terms of providing compositional information.

Materials

Purified proanthocyanidins (see Basic Protocol 2)
Phloroglucinol solution (see recipe)
40 mM sodium acetate buffer (see recipe)
Mobile phase A: 1% (v/v) acetic acid in HPLC-grade water
Mobile phase B: HPLC-grade methanol
(+)-Catechin hydrate (Sigma)

2-ml borosilicate glass vials with a caps
Water bath at 50°C
High performance liquid chromatograph (HPLC) with:
   HPLC vials
   250 × 4.6–mm Wakosil II 5C18 column (5-μm particle size) and guard column (SGE)
   UV absorption detector set at 280 nm

Table I1.4.1   Retention Properties, Molar Absorptivities, and Response Factors for Common Proanthocyanidin Cleavage Products[a]

Compound	Retention factor (k)	Molar absorptivity (ε_{280})[b]	Relative molar response[c]	Corrected relative mass response[c,d]
(–)-Epigallocatechin-(4β→2)-phloroglucinol[e]	5.8	1344	0.34	0.32
(–)-Epicatechin-(4β→2)-phloroglucinol[e]	9.4	4218	1.06	1.06
(+)-Catechin-(4α→2)-phloroglucinol[e]	9.1	4218	1.06	1.06
(–)-Epicatechin-3-*O*-gallate-(4β→2)-phloroglucinol[e]	13.4	14766	3.70	2.44
(–)-Epicatechin[f]	16.6	3988	1.00	1.00
(+)-Catechin[f]	13.1	3988	1.00	1.00
(–)-Epicatechin-3-*O*-gallate[f]	19.1	12611	3.16	2.07

[a]Reprinted from Kennedy and Jones (2001) with permission from the American Chemical Society.
[b]In methanol.
[c]Relative to catechin.
[d]Not including the phloroglucinol moiety.
[e]Product derived from proanthocyanidin extension subunit.
[f]Product derived from proanthocyanidin terminal subunit.

**Proanthocyanidins:
Extraction,
Purification, and
Composition**

I1.4.4

502

1. Place 5 mg purified proanthocyanidins into a clean 2-ml borosilicate glass vial.

2. Add 1 ml phloroglucinol solution, cap the vial, and dissolve the proanthocyanidins.

3. Place the proanthocyanidin solution into a water bath preheated to 50°C and allow acid catalysis to proceed for 20 min.

4. Remove the vial from the water bath and allow the contents to return to room temperature.

5. Transfer 1 ml of 40 mM sodium acetate buffer to an HPLC vial and add 200 µl proanthocyanidin solution .

6. Inject 20 µl onto an HPLC equipped with a 250 × 4.6–mm, 5-µm Wakosill 5C18 column and a guard column containing the same material. Separate proanthocyanidin subunits according to manufacturer's instructions, with peak detection at 280 nm and the following linear gradients at a flow rate of 1.0 ml/min:

   5% (v/v) mobile phase B for 10 min
   5% to 20% mobile phase B in 20 min
   20% to 40% mobile phase B in 25 min.

7. Before the next injection, wash the column with 90% mobile phase B for 10 min and reequilibrate with 5% mobile phase B for 5 min.

8. Identify proanthocyanidin subunits by comparing their retention factors to those shown in Table I1.4.1. After identification, quantify using a (+)-catechin standard.

   Subunits that react with phloroglucinol are derived from proanthocyanidin extension subunits. Subunits that have not reacted with phloroglucinol are derived from proanthocyanidin terminal subunits or were present as monomeric flavan-3-ols. For quantitation, the sample peak areas of the individual subunits are compared with a (+)-catechin standard, and individual quantities are determined using the relative response factors shown in Table I1.4.1.

9. Determine the conversion yield by summing the mass of the subunits (not including the phloroglucinol moiety) and dividing by the starting mass of proanthocyanidins:

$$\text{conversion yield} = \frac{\left(\dfrac{406\ \text{PAU}}{0.32} + \dfrac{68\ \text{PAU}}{1.06} + \dfrac{1816\ \text{PAU}}{1.06} + \dfrac{264\ \text{PAU}}{1.00} + \dfrac{154\ \text{PAU}}{2.44}\right) \times \left(\dfrac{96\ \text{mg/l}}{460\ \text{PAU}}\right) \times \left(\dfrac{6}{1000}\right)}{5.04\ \text{mg sample}} = 0.85$$

   In this example, individual subunit peak areas (PAU) are divided by their relative mass response (Table I1.4.1); 96 mg/liter and 460 PAU are the concentration and peak area of (+)-catechin, and 6/1000 is the dilution factor.

10. Determine the average degree of polymerization by summing the subunits (in mole equivalents) and dividing by the sum of the terminal subunits (also in mole equivalents):

$$\text{average degree of polymerization} = \frac{\left(\dfrac{406\ \text{PAU}}{0.34} + \dfrac{68\ \text{PAU}}{1.06} + \dfrac{1816\ \text{PAU}}{1.06} + \dfrac{264\ \text{PAU}}{1.00} + \dfrac{154\ \text{PAU}}{3.70}\right)}{\dfrac{264\ \text{PAU}}{1.00}} = 12.41$$

   In this example there is only one terminal subunit.

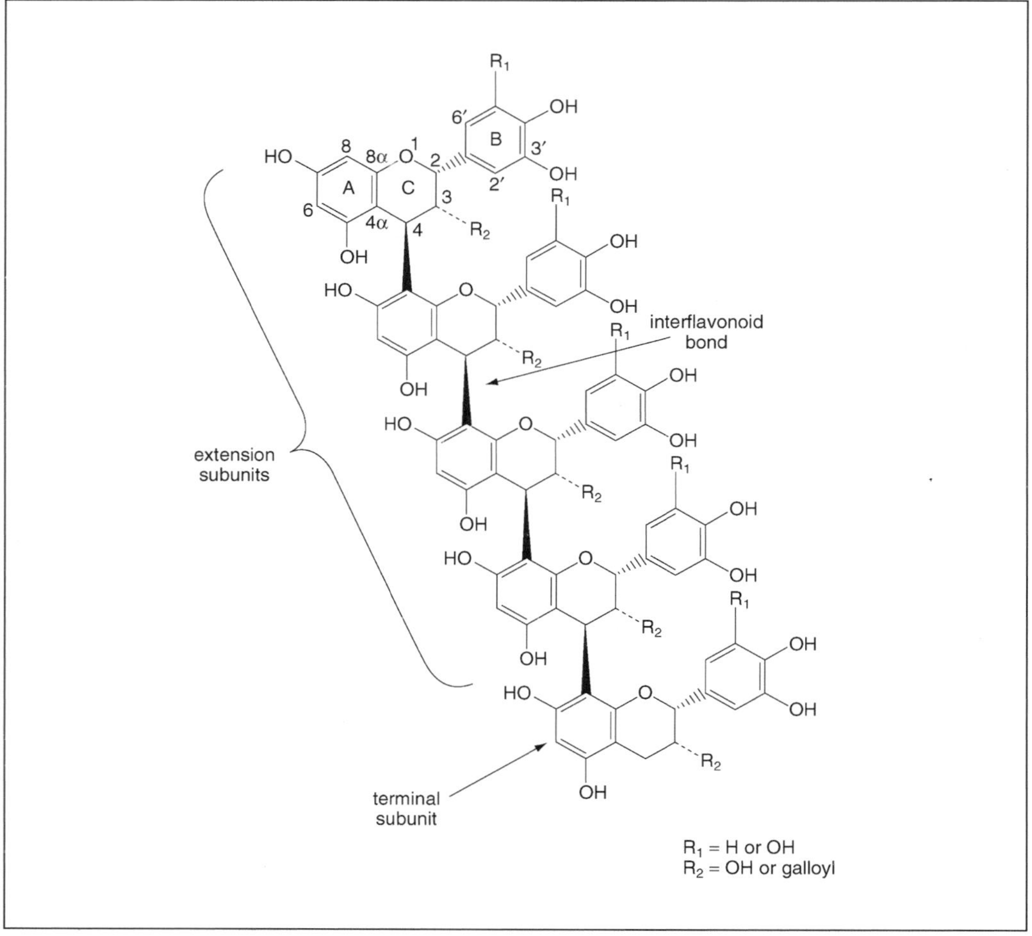

Figure I1.4.1 Generalized proanthocyanidin structure indicating subunit type (extension or terminal) and interflavonoid bond location (4β→8). The most common proanthocyanidin classes in the plant kingdom, as well as in the food and beverage industry, are the procyanidins (3,3′,4′,5,7-pentahydroxyflavans) and prodelphinidins (3,3′,4′,5′,5,7-hexahydroxyflavans). In addition, these proanthocyanidins can be galloylated at C3.

REAGENTS AND SOLUTIONS

Use deionized, distilled water in all recipes and protocol steps. For common stock solutions, see
APPENDIX 2A; for suppliers, see SUPPLIERS APPENDIX.

Phloroglucinol solution

Dissolve 5 g phloroglucinol (1,3,5-trihydroxybenzene) and 1 g ascorbic acid in a minimum amount of methanol (60 ml) and transfer to a 100-ml volumetric flask. Add 0.5 ml of 10 N hydrochloric acid, fill to the mark with methanol, and mix well. Store up to 2 weeks at 4°C. Bring to room temperature before use.

Up to 100 samples can be analyzed with this volume of solution.

CAUTION: *Hydrochloric acid is very corrosive, and 10 N hydrochloric acid releases hydrochloric acid gas when exposed to air. Wear goggles, gloves, and protective clothing, and measure in a well-ventilated area, preferably a fume hood.*

Sodium acetate buffer, 40 mM

Dissolve 0.33 g sodium acetate into a 100-ml volumetric flask filled with approximately 50 ml water. After dissolution, fill to the mark with water and mix well. This solution is intended for immediate use and should be used within 1 day.

Up to 100 samples can be analyzed with this volume of solution.

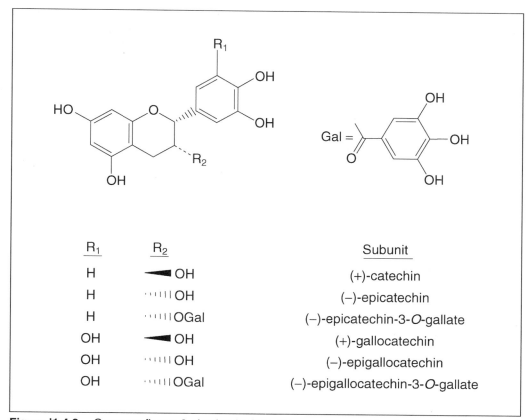

Figure I1.4.2   Common flavan-3-ol subunits found in proanthocyanidins in foods and beverages.

COMMENTARY

Background Information

Proanthocyanidins are polymeric flavonoid compounds composed of flavan-3-ol subunits (Fig. I1.4.1), and are widely distributed in the plant kingdom, including plants that are important as a source of food (Santos-Buelga and Scalbert, 2000). They impart bitter and astringent properties. In addition, these compounds may have potential health effects (Santos-Buelga and Scalbert, 2000).

Proanthocyanidins are so named because under oxidative and acidic conditions they are converted to anthocyanidins, a subclass of flavonoid (for general structure, see Fig. I1.3.4). Historically, they have also been referred to as leucoanthocyanidins, condensed tannins, or simply tannins because of their ability to fix or "tan" leather hides.

Several flavan-3-ol subunits are found in food-based proanthocyanidins, with the most common flavan-3-ol subunit being (–)-epicatechin (Fig. I1.4.2). Flavan-3-ol subunits are linked together by various interflavonoid bonds, with the 4β→8 interflavonoid bond being the most common, followed by the 4β→6 interflavonoid bond. Other types of interfla-

vonoid bonds are also present, although much less common (Hemingway, 1989a). Proanthocyanidins encompass a very large range of molecular weights ranging from the simplest dimer to proanthocyanidins that reportedly exceed 80 subunits in length.

To analyze proanthocyanidins, this unit relies on the susceptibility of the proanthocyanidin interflavonoid bond to acid catalysis. Interflavonoid bonds have different susceptibilities to acid catalysis (Hemingway and McGraw, 1983; Beart et al., 1985). In general, and for procyanidins and prodelphinidins, subunits that are bonded through the benzylic position (C4) are susceptible to acid catalysis. Most natural proanthocyanidins extracted from plant tissue have this type of interflavonoid bond and are therefore suitable candidates for this analytical technique.

Traditional methods for analyzing proanthocyanidins have relied on general colorimetric methods. The main drawback for using these analytical methods is that they are not specific for proanthocyanidins. This can be said for chromatographic methods that analyze intact proanthocyanidins. The selectivity of the

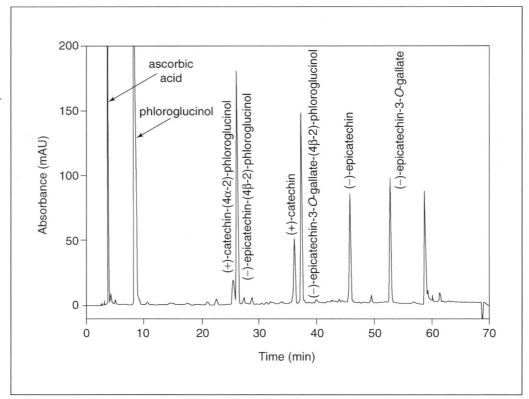

Figure I1.4.3 Sample chromatogram of proanthocyanidin cleavage products from grape seed. Reprinted from Kennedy and Jones (2001) with permission from the American Chemical Society.

method presented here relies on two features of proanthocyanidins: susceptibility to acid-catalyzed cleavage and subsequent attack by strong nucleophiles. Therefore, the products that are formed from acid catalysis in the presence of a nucleophile (phloroglucinol) are clearly derived from proanthocyanidins, and this is the primary advantage that this method has over other analytical methods that are used to measure proanthocyanidins.

Historically, two nucleophiles have been used to trap the cleavage intermediates: benzyl mercaptan and phloroglucinol. Benzyl mercaptan has the distinct disadvantage of having a powerful stench, and therefore it is necessary to conduct reactions involving this reagent in a fume hood. The advantage of using benzyl mercaptan was thought to be the higher conversion yield of proanthocyanidins into their constitutive subunits (Matthews et al., 1997); however, if proper conditions are selected, phloroglucinol is as effective (Kennedy and Jones, 2001).

Critical Parameters and Troubleshooting

Extraction

The extraction of proanthocyanidins is the first step in determining their subunit composition. A number of extraction systems have been investigated in different plant tissues. The most common solvent systems are acetone and methanol with various amounts of water and with or without acid. In general, it has been found that an aqueous acetone system gives the best results in terms of total amount extracted.

Because of the general reactivity of proanthocyanidins (Hemingway, 1989b; McGraw, 1989; Laks, 1989), it is expected that they will quickly become modified once they are extracted into foodstuffs. Specifically, proanthocyanidins are very susceptible to oxidative degradation, and therefore consideration should be given to minimizing oxidation reactions from extraction to analysis. Reduced temperatures and dark conditions are desirable for minimizing oxidation reactions (Cork and Krockenberger, 1991).

In many foods, the oxidation of proanthocyanidins is desirable. Two notable examples are black teas and cocoa. The oxidative modifica-

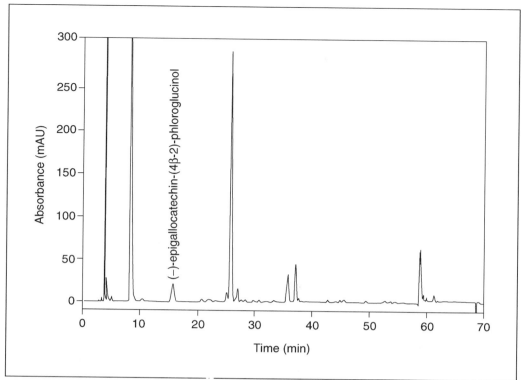

Figure I1.4.4 Sample chromatogram of proanthocyanidin cleavage products from grape skin. Peak identification is identical to Figure I1.4.3, with (–)-epigallocatechin-(4β→2)-phloroglucinol also identified. Reprinted from Kennedy and Jones (2001) with permission from the American Chemical Society.

tion of proanthocyanidins in these foods is considered beneficial because it reduces proanthocyanidin bitterness and astringency. The oxidative products formed from this reaction are resistant to acid catalysis and, as a result, the conversion yield declines with oxidation. For this reason, this analytical method is ideally suited for the analysis of proanthocyanidins in plant tissues, and also in foodstuffs, particularly when oxidation is being monitored.

It is important to determine whether the plant tissues should be extracted whole and for how long. Homogenizing the plant tissue in the extraction solvent will significantly reduce the time needed to complete the extraction; however, this could also lead to a reduction in recovery due to the introduction of impurities and potential loss of proanthocyanidins from complexation and subsequent precipitation with polysaccharides and proteins. In grape seeds, homogenization reduces recovery. This has been attributed to protein precipitation by seed proteins. Cryogenic milling (*UNIT F1.1*) of the plant tissue prior to extraction is an alternative to homogenization but has the same potential pitfalls. Leaving the plant tissue intact can minimize the introduction of impurities and

complexation reactions, but then the extraction time must be increased. The selection of optimal conditions is tissue specific, and should be determined in initial runs.

Purification and subunit composition

It is important to purify proanthocyanidins, particularly for determining their conversion yield. It is also advantageous to do so to eliminate extraneous material that might otherwise react with the proanthocyanidins. A combination of liquid-liquid extraction and adsorption chromatography is effective in removing impurities. The use of chloroform in liquid-liquid extraction is very effective in removing fat-soluble compounds such as carotenoids, chlorophyll, oils, and waxes. These compounds would be expected in leafy plant tissues (carotenoids and chlorophyll) as well as seeds and fruits (oils and waxes). Ethyl acetate is effective in the selective removal of flavan-3-ol monomers, which are also typically present with proanthocyanidins.

It is critically important that the purified proanthocyanidins do not contain water prior to acid-catalyzed cleavage with phloroglucinol. The cleavage of proanthocyanidins into their constitutive subunits generates electrophilic extension subunit intermediates. In this reaction, phloroglucinol, because of its nucleophilicity, is used as a trap to stabilize the extension subunits. Water, if present, competes with phloroglucinol in this reaction, and thus reduces conversion yields, and can also effect the calculated composition of the proanthocyanidin isolate.

When analyzing the cleavage products, it is possible to obtain false positives, specifically with the terminal subunits that will be present as flavan-3-ol monomers. If purification is incomplete, free flavan-3-ol monomer impurities will be present. Although these components are often of interest analytically, they are not, strictly speaking, proanthocyanidins. Analyzing the purified proanthocyanidin before and after acid catalysis in the presence of excess phloroglucinol using the same HPLC method will give some indication of whether or not flavan-3-ol monomers are present.

It is important to monitor the time that the proanthocyanidin is allowed to react with the phloroglucinol solution. The products formed are not stable under acidic conditions, and it is therefore critically important that the reaction not exceed 20 min. Of particular concern are the flavan-3-ol monomers, which degrade more rapidly than the phloroglucinol adducts (Kennedy and Jones, 2001). Excessive degradation of the flavan-3-ol monomers will result in reduced amount of terminal subunits. This in turn will reduce the conversion yield and increase the average degree of polymerization calculated.

Anticipated Results

Extraction

Because of the varied nature of the plant tissues from which the proanthocyanidin extracts are derived, it is difficult to anticipate the expected outcome. As an example of how these procedures can be adapted to specific tissues and analyses, using grape tissues, fruit is harvested and the tissues of interest (e.g., skins and seeds) are removed from the remainder of the berry. They are rinsed well and then extracted as whole tissues using the conditions described in these protocols. For grape skins, a liquid-liquid extraction with chloroform has been successful in the removal of chlorophyll and waxes, yet no extraction with ethyl acetate has been performed because of the small proportion of flavan-3-ol monomers (Kennedy et al., 2001). For grape seeds, these protocols have

been adapted to minimize purification and have been used to monitor proanthocyanidin development (Kennedy et al., 2000).

Purification

Proanthocyanidins following purification are light buff powders, with a varying degree of yellow depending on the level of oxidation that has occurred.

Proanthocyanidins are very susceptible to oxidative degradation. To minimize subsequent oxidation, proanthocyanidins should be stored protected from light and under freezer conditions (−20°C or cooler).

Subunit composition

Example chromatograms are shown in Figures I1.4.3 and I1.4.4. Generally speaking, epicatechin extension subunits are the most prevalent subunits found in plant tissues of interest as foodstuff. Some additional references are given at the end of this unit that provide information on the subunit composition of various plant species.

The expected conversion yield can vary considerably. Proanthocyanidins that are isolated from immature plant tissues during active growth and are judiciously protected from oxidation should have high conversion yields, approaching quantitative conversion. Proanthocyanidins isolated from fully ripened plant tissue or fully senesced plant tissues will have lower conversion yields (roughly 50% to 80% by mass). Conversion yields will also be lower if oxidation has occurred as a result of food processing or prolonged storage.

Time Considerations

Extraction

Extraction time will vary, but as this procedure is written for extraction of whole plant tissue, allow for ~26 hr between the time the extraction is set up and the removal of the acetone following extraction. The extraction time for homogenized or milled tissues is typically less than 15 min per sample.

Purification

Allow 12 to 18 hr to purify proanthocyanidins on the column. Drying time will vary depending on the volume obtained and the freeze-drier that is used.

Subunit composition

The analyst should allow ~5 min per sample for weighing and diluting samples with sodium acetate buffer, followed by 20 min for the catalysis reaction to occur. In addition, each sample run requires 70 min on the HPLC. An experienced analyst should be able to analyze 20 samples and standards in a day.

Literature Cited

Beart, J.E., Lilley, T.H., and Haslam, E. 1985. Polyphenol interactions. Part 2. Covalent binding of procyanidins to proteins during acid-catalysed decomposition; observations on some polymeric proanthocyanidins. *J. Chem. Soc. Perkin Trans. II* 1439-1443.

Cork, S.J. and Krockenberger, A.K. 1991. Methods and pitfalls of extracting condensed tannins and other phenolics from plants: Insights from investigations on *Eucalyptus* leaves. *J. Chem. Ecol.* 17:123-134.

Hemingway, R.W. 1989a. Structural variations in proanthocyanidins and their derivatives. *In* Chemistry and Significance of Condensed Tannins (R.W. Hemingway and J.J. Karchesy, eds.) pp. 83-108. Plenum Press, New York.

Hemingway, R.W. 1989b. Reactions at the interflavonoid bond of proanthocyanidins. *In* Chemistry and Significance of Condensed Tannins (R.W. Hemingway and J.J. Karchesy, eds.) pp. 265-283. Plenum Press, New York.

Hemingway, R.W. and McGraw, G.W. 1983. Kinetics of acid-catalyzed cleavage of procyanidins. *J. Wood Chem. and Tech.* 3:421-425.

Kennedy, J.A. and Jones, G.P. 2001. Analysis of proanthocyanidin cleavage products following acid-catalysis in the presence of excess phloroglucinol. *J. Agric. Food Chem.* 49:1740-1746.

Kennedy, J.A., Troup, G.J., Pilbrow, J.R., Hutton, D.R., Hewitt, D., Hunter, C.R., Ristic, R., Iland, P.G., and Jones, G.P. 2000. Development of seed polyphenols in berries from *Vitis vinifera* L. cv. Shiraz. *Austral. J. Grape Wine Res.* 6:244-254.

Kennedy, J.A., Hayasaka, Y., Vidal, S., Waters, E.J., and Jones, G.P. 2001. Composition of grape skin proanthocyanidins at different stages of berry development. *J. Agric. Food Chem.* 49:5348-5355.

Laks, P.E. 1989. Chemistry of the condensed tannin B-ring. *In* Chemistry and Significance of Condensed Tannins (R.W. Hemingway and J.J. Karchesy, eds.) pp. 249-263. Plenum Press, New York.

Matthews, S., Mila, I., Scalbert, A., Pollet, B., Lapierre, C., Hervé do Penhoat, C.L.M., Rolando, C., and Donnelly, D.M.X. 1997. Method for the estimation of proanthocyanidins based on their acid depolymerization in the presence of nucleophiles. *J. Agric. Food Chem.* 45:1195-1201.

McGraw, G.W. 1989. Reactions at the A-ring of proanthocyanidins. *In* Chemistry and Significance of Condensed Tannins (R.W. Hemingway and J.J. Karchesy, eds.) pp. 227-248. Plenum Press, New York.

Santos-Buelga, C. and Scalbert, A. 2000. Proanthocyanidins and tannin-like compounds—nature, occurrence, dietary intake, and effects on nutrition and health. *J. Sci. Food Agric.* 80:1094-1117.

Key References

Czochanska, Z., Foo, L.Y., Newman, R.H., and Porter, L.J. 1980. Polymeric proanthocyanidins. Stereochemistry, structural units, and molecular weight. *J. Chem. Soc. Perkin I* 2278-2286.

This paper summarizes the various aspects of proanthocyanidin structure.

Foo, L.Y. and Porter, L.J. 1980. The phytochemistry of proanthocyanidin polymers. *Phytochemistry* 19:1747-1754.

This reference summarizes the composition of proanthocyanidins from plants.

Foo, L.Y. and Porter, L.J. 1981. The structure of tannins of some edible fruits. *J. Sci. Food Agric.* 32:711-716.

This reference summarizes the composition of proanthocyanidins from various fruits.

Hemingway, R.W. and Karchesy, J.J. (eds.) 1989. Chemistry and Significance of Condensed Tannins. Plenum Press, New York.

This book in general is a very good reference for these compounds.

Kennedy and Jones, 2001. See above.

Analytical considerations for this unit are detailed in this reference.

Santos-Buelga and Scalbert, 2000. See above.

This review article discusses the presence of proanthocyanidins in foodstuffs and their potential effects on human health.

Contributed by James A. Kennedy
Oregon State University
Corvallis, Oregon

Identification of Flavonol Glycosides Using MALDI-MS

Flavonol glycosides are found ubiquitously throughout higher plants. They are responsible for the yellow pigmentation of many fruits, vegetables, and grains. The flavonol glycoside composition of plants is distinctive, and thus can provide valuable information for taxonomic purposes and authentication of processed food products.

Measurement of specific flavonol glycosides in complex mixtures can sometimes be difficult. A new analytical technique that shows promise in this area is matrix-assisted laser desorption/ionization time-of-flight mass spectrometry (MALDI-MS). This technique has proven successful for identifying flavonol glycosides in a number of food sources including green tea, onions, and almonds (Wang and Sporns, 2000; Frison-Norrie and Sporns, 2002). One major advantage of MALDI-MS is that only rudimentary purification of the sample is required prior to analysis. Also, the analysis is very rapid, generally 1 to 2 min per sample run. Fragmentation is minimal, allowing the molecular weight of the parent ions to be easily determined. Characteristic loss of carbohydrate residues in succession from flavonol glycosides can provide structural insight. Flavonol glycosides are easily ionized and have been shown to out-compete impurities for ionization energy, resulting in the appearance of only matrix peaks and flavonol glycoside peaks in the mass spectrum. However, MALDI-MS is unable to differentiate among structural isomers with identical molecular weights, and can be subject to a high degree of variability among identical runs.

This unit describes procedures for extraction, purification, and identification by MALDI-MS of flavonol glycosides from a plant source. The extraction and purification protocols are not meant to be comprehensive, but rather to offer guidelines for sample preparation prior to a MALDI-MS analysis. The MALDI-MS technique is suggested as a complement to other analytical methods such as HPLC or NMR. Its strength lies in the ability to rapidly screen a number of samples for the presence of flavonol glycosides, which can be identified on the basis of their molecular weights.

EXTRACTION OF FLAVONOL GLYCOSIDES FROM PLANT SOURCES

Methanol is a common solvent for extracting flavonol glycosides from plant material. This procedure involves a simple, rapid, and efficient extraction whereby powdered plant material is stirred with 70% methanol, filtered, and concentrated on a rotary evaporator. As this extraction is not selective, the crude aqueous extract contains contaminants co-extracted with flavonol glycosides. In most cases, however, the impurities do not detract from the quality of the MALDI-MS results.

Materials

Powdered plant material (freeze-dried if fresh sample has high water content, e.g., >30%)
70% (v/v) methanol in water
Whatman no.1 filter paper
Tapered glass funnel
250-ml round-bottom flask
Rotary evaporator with vacuum pump or water aspirator, 35°C

1. Mix ~5 g powdered plant material (accurately weighed and recorded) with 100 ml of 70% methanol using a chemical-resistant stir bar. Stir 30 min.

Half an hour should be sufficient time for extraction of flavonol glycosides because of the high surface area of powdered material and the large volume of solvent. Another method of extraction is to add solvent to the sample in two to three aliquots, decant each through Whatman no. 1 filter paper at the end of a specified time, and combine the filtrates.

Flavonol glycosides are minor constituents in foods, likely constituting less than 1% of the dry weight. In order to achieve sufficient concentration in the extract to detect flavonol glycosides in high-moisture foods (30%) without requiring large volumes of samples and extraction solvents, high-moisture samples such as fruit and leaf material should be freeze-dried first.

2. Separate the flavonol glycoside extract from the insoluble plant material by filtering the slurry by gravity through a Whatman no. 1 filter paper into a 250-ml round-bottom flask.

3. Remove methanol in a rotary evaporator at 35°C under vacuum.

   The presence of flavonol glycosides is indicated by the yellow color of the solution.

   A noticeable reduction in the rate of evaporation and an apparent increase in viscosity indicate that the residual liquid is mainly water. Evaporating to dryness should be avoided, or material may become tightly stuck to the flask.

4. Bring the remaining aqueous extract to a known volume (25 ml) with deionized distilled water. Store for up to 1 to 2 weeks at 4°C.

PURIFICATION OF FLAVONOL GLYCOSIDES FOR MALDI-MS ANALYSIS

Although MALDI-MS can be quite tolerant of impurities, a purification step is often necessary after extraction in order to enhance the response. Considerable amounts of co-extracted material may compete for ionization energy and mitigate the flavonol glycoside signal. Solid-phase extraction using a C18 cartridge is used to enrich the concentration of flavonol glycosides by removing sugars, acids, and other water-soluble impurities. The final methanolic elution may remove other hydrophobic or phenolic compounds along with flavonol glycosides from the column, but the sample will in most cases be sufficiently clean to obtain a significant flavonol glycoside response with MALDI-MS.

Materials

Aqueous flavonol glycoside extract (see Basic Protocol 1)
Methanol
70% (v/v) methanol in water (or 0.01 M NaCl in 70% methanol)
0.22-μm syringe filters
C18 cartridge (e.g., C18 Sep-Pak cartridge, 360 mg sorbent, Waters Chromatography)

1. Filter sample through a 0.22-μm membrane.

2. Condition a C18 cartridge by passing five column volumes of methanol through sorbent bed.

3. Pass five column volumes of deionized distilled water through cartridge to remove remaining methanol.

4. Load 5 ml aqueous flavonol glycoside extract onto the column. Load at a rate of ~1 ml/min to ensure complete adsorption of flavonol glycosides.

   The volume of extract applied to the cartridge depends on the sample concentration, flavonol glycoside concentration, and amount of sorbent packing in the column.

The column packing will become colored. If excessive color leaks through the cartridge, the cartridge has become overloaded and no more sample should be applied.

513

5. Wash cartridge with 30 column volumes deionized distilled water to remove water-soluble impurities (sugars, acids).

6. Elute flavonol glycosides with 2 ml of 70% methanol. Add solvent at a rate of ~1 ml/min to ensure complete elution of flavonol glycosides from the cartridge.

   Sodium chloride may be added to the eluting solvent at a concentration of 0.01 M in order to accentuate sodium adduct ions in the MALDI-TOF mass spectrum and suppress the formation of potassium adduct ions.

7. Store purified flavonol glycoside extract up to 1 to 2 weeks at 4°C.

MALDI-MS IDENTIFICATION OF FLAVONOL GLYCOSIDES

A high-intensity pulse of UV light initiates the matrix-mediated ionization and vaporization of analyte molecules. Analytes undergo only minimal fragmentation by this soft ionization process, resulting in the appearance of abundant molecular ions in the mass spectrum. Only singly charged positive ions are formed, allowing rapid identification of flavonol glycosides based on molecular weight.

Materials

Matrix solution: 2′,4′,6′-trihydroxyacetophenone monohydrate (THAP; Aldrich) dissolved in acetone at 2 mg/100 µl, freshly prepared
Purified aqueous flavonol glycoside extract (see Basic Protocol 2)
0.2 to 2 µl digital micropipettor or microsyringe
Matrix-assisted laser desorption/ionization time-of-flight mass spectrometer (MALDI-TOF-MS) with UV laser (i.e., 337-nm nitrogen laser), MALDI probe, and delayed extraction or reflectron

NOTE: In solution, THAP degrades rapidly at room temperature (within 1 to 2 days). Matrix solution should be made fresh daily.

1. Using a microsyringe or digital micropipettor, load 0.5 µl matrix solution onto each position on the MALDI probe. Allow solvent to evaporate (requires only a few seconds).

   Loading more than 1 µl may cause solution to run outside of the boundaries on the probe.

2. Load 0.5 µl purified aqueous flavonol glycoside extract on top of matrix and allow to air dry (requires ~2 min with a fan).

   Many spectra can be obtained from a single position. However, sometimes non-uniformity in crystallization can prevent a satisfactory spectrum from being generated. For this reason, at least two positions per sample should be spotted to allow a better chance of obtaining an acceptable spectrum.

   To encourage homogeneous crystallization, the sample should be dissolved in an aqueous solution (30% aqueous or more). Samples prepared in high organic solvent concentrations may redissolve the THAP crystal bed, which causes irregular recrystallization. Non-uniform crystal formation is one cause of variability among identical runs.

   Alternatively, the sample and matrix solutions can be mixed 1:1 (v/v) and aliquots of up to 1 µl spotted onto the probe.

3. Insert probe into ion source of the MALDI-MS instrument.

4. Record MALDI mass spectra in positive ion mode. Look for proton, sodium, and potassium adduct ions in the range m/z 250 to 1000. Fragment ions caused by the loss of carbohydrate moieties are typical and can be diagnostic.

For good resolution and signal-to-noise ratios, laser strength should be slightly above the threshold required to generate detectable ions.

Fragmentation patterns of standards can help to determine whether observed peaks are the result of in-source fragmentation or natural occurrence in the sample. Flavonol glycoside standards are available from Extrasynthese and Indo Fine Chemical.

If flavonol glycosides were eluted with excess sodium, potassium adduct ions will likely be suppressed, but there may be the formation of $[M+2Na-H]^+$ ions.

COMMENTARY

Background Information

Flavonol glycosides are water-soluble pigments responsible for the yellow hues of many fruits, vegetables, and cereal grains. It is also thought that consumption of these phytochemicals may confer protective effects against chronic diseases due to their antioxidant properties (Pietta, 2000). Analyzing flavonol glycosides is useful for taxonomic purposes or for authentication of processed products, because the flavonol glycoside composition of individual plants is distinctive. There is considerable structural variability among flavonol glycosides (Fig. I1.5.1). The most common flavonol aglycones are quercetin, kaempferol, and myricetin, with minor variations on these basic structures (i.e., variations in methylation and hydroxylation patterns) encountered more rarely. Sugar residues may occur at any of the hydroxylated positions, although glycosylation

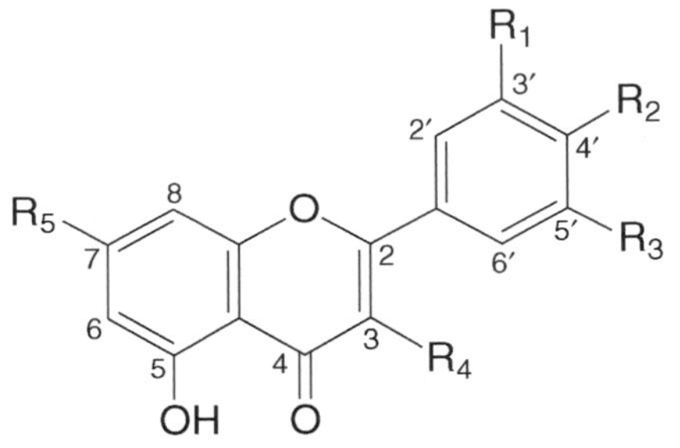

Compound	Mass	R₁	R₂	R₃	R₄	R₅
Astragalin	448.1	H	OH	H	OGlu	OH
Hyperoside	464.1	OH	OH	H	OGal	OH
Isoquercitrin	464.1	OH	OH	H	OGlu	OH
Isorhamnetin	316.1	OMe	OH	H	OH	OH
Kaempferide	300.1	H	OMe	H	OH	OH
Kaempferol	286.1	H	OH	H	OH	OH
Myricetin	318.0	OH	OH	OH	OH	OH
Narcisin	624.2	OMe	OH	H	ORut	OH
Peltatoside	596.1	OH	OH	H	OAraGlu	OH
Quercetin	302.2	OH	OH	H	OH	OH
Quercitrin	448.1	OH	OH	H	ORha	OH
Rhamnetin	316.1	OH	OH	H	OH	OMe
Robinin	740.2	H	OH	H	ORob	ORha
Rutin	610.2	OH	OH	H	ORut	OH

Figure I1.5.1   Structures of common flavonols and flavonol glycosides. Abbreviations: Ara, arabinose; Gal, galactose; Glu, glucose; Me, methyl; Rha, rhamnose; Rob, robinose; Rut, rutinose. Masses given are monoisotopic.

at the 3 position of the flavonol nucleus is the most common (Hollman and Arts, 2000). Glucose is the most prevalent sugar residue, although galactose, arabinose, xylose, and rhamnose are not unusual (Markham, 1982). Other mono-, di-, tri-, and tetrasaccharide derivatives, along with acylated and sulfated derivatives, are less frequently reported (Harborne and Williams, 2001). Altogether, more than 1200 flavonol glycosides have been identified to date (Harborne and Williams, 2001).

Traditionally, paper chromatography was used for the separation and identification of flavonol glycosides. However, in recent years, this methodology has essentially been replaced by high-performance liquid chromatography (HPLC; Merken and Beecher, 2000). Although HPLC is by far the most common method for identifying flavonol glycosides in a mixture, extensive purification prior to analysis may be required, and standards are not always available for definitive identification by comparison of retention times. In addition, separation of chemically similar compounds by chromatography can be laborious. Flavonol glycosides purified by HPLC are often further analyzed by nuclear magnetic resonance (NMR) to provide detailed structural information and confirm the identity. A number of mass spectrometry techniques has also been reported for analyzing flavonol glycosides, although extensive fragmentation, thermal instability, and requirement for derivatization can be prohibitive (Stobiecki, 2000).

MALDI-MS is a new method of flavonol glycoside analysis. The first published account was by Wang and Sporns (2000). Because of the large number of possible flavonol glycosides one could encounter in a sample, the inability of MALDI-MS to differentiate among isomers can be a liability. Definitive identification is best left to NMR analysis. However, MALDI-MS is useful as a rapid screening tool for preliminary identification of flavonol glycosides. Its main advantages include access to molecular weight information of parent ions, minimal sample preparation, and rapid analysis times. If standards are readily available, then HPLC analysis may be used in conjunction to confirm identities. In theory, flavonol glycosides from any type of plant material can be analyzed by MALDI-MS.

It is possible to quantify flavonol glycosides using an internal standard, but the process can

Table I1.5.1 Troubleshooting Guide for MALDI-MS Analysis of Flavonol Glycosides

Problem	Possible cause	Solution
No flavonol glycoside peaks	Co-extracted impurities may be suppressing the signal	Perform additional sample purification step (e.g., column chromatography[a])
	Nonuniform crystallization	Prepare sample in at least 30% aqueous solution
	Sample concentration too high	Dilute the sample
	Sample concentration too low	Concentrate the sample
	Matrix solution concentration too low	Concentrate matrix solution
Poor peak resolution	Laser power too high	Attenuate the laser power
	Detector voltage too high	Attenuate the detector voltage
Signal is very weak	Laser power too low	Increase laser power
	Detector voltage too low	Increase detector voltage
Observed mass does not match theoretical mass	Instrument is not calibrated	Calibrate the instrument using standards (slight variability, ~100 ppm, is normal)
High variability among identical runs	Nonuniform crystallization	Evaporate solvent as quickly as possible; prepare sample in at least 30% aqueous solution
	"Natural" variability	Collect several spectra per sample from different probe positions

[a]One way would be to use HPLC on a reversed-phase column using a gradient of acetonitrile and water with UV detection at 354 nm (Frison-Norrie and Sporns, 2002). Another alternative would be to perform a more selective extraction step using a series of organic solvents followed by column chromatography on silica gel or Sephadex LH 20 (Sang et al., 2002).

be quite complex. One must account for the multiple ions formed for each species, determine response ratios of the standard to all of the analytes, and account for fragmentation patterns, which can be tricky when some fragment ions and some parent ions are identical. For example, the appearance of quercetin glucoside in the mass spectrum may be due to loss of a rhamnose residue from quercetin rutinoside, or to natural occurrence in the sample. This is another reason to verify MALDI-MS results with another method, such as HPLC. For a more detailed discussion of quantification, refer to Frison-Norrie and Sporns (2002).

Critical Parameters

Flavonol glycosides may be non-uniformly distributed throughout the plant. For example, the seedcoats of nuts may be much richer in flavonol glycosides than the flesh. It may be beneficial in such circumstances to extract and analyze the flavonol glycoside-rich components separately to increase the concentration in the extract.

Preparation of the sample is crucial to the success of a MALDI-MS experiment. It is im-portant to achieve a homogenous distribution of the analyte throughout the crystalline matrix structure in order to reduce the sample-to-sample and spot-to-spot variability. Homogeneity may be promoted by techniques such as rapid evaporation of the solvent under vacuum or "sandwich" layering techniques of matrix and analyte solutions, although often air-drying can be sufficient to achieve sufficient uniformity. Also, a partially aqueous sample solution (at least 30%) facilitates "beading" on the sample probe due to surface tension effects, which in turn leads to greater sample concentration and uniformity of the crystal bed. Generally, the matrix should be present in a 10^3 to 10^5 molar excess compared to the analyte.

Troubleshooting

See Table I1.5.1 for troubleshooting guidelines.

Anticipated Results

The highest response is generally observed for protonated ions (Fig. I1.5.2). Alkali cations—[M+Na]$^+$ and [M+K]$^+$—are of lower intensity and appear only for flavonol glycoside

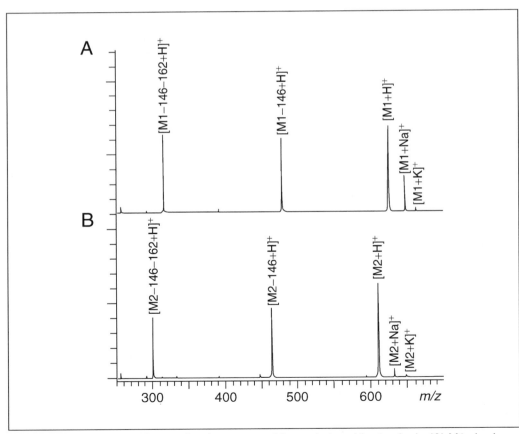

Figure I1.5.2 MALDI-MS natural cation spectra of flavonol glycoside standards. (**A**) M1= isorhamnetin-3-rutinoside (1.5×10^{-3} M in 70% methanol). (**B**) M2 = rutin (1.5×10^{-3} M in 70% methanol). Reprinted from Frison-Norrie and Sporns (2002) with permission from the American Chemical Society.

peaks and not for flavonol aglycone peaks. Larger molecules tend to have a greater preference for alkali adduct incorporation. When excess sodium is added to the extract, formation of $[M+K]^+$ ions will be suppressed, but $[M+2Na-H]^+$ ions may appear (Fig. I1.5.3). Consecutive loss of sugar residues creates characteristic fragment ions of variable intensity. The mass difference between parent ions and fragment ions can offer valuable structural information about the nature of the sugar residue.

Time Considerations

Preparation of the crude aqueous sample extract (Basic Protocol 1) requires 30 min for extraction, 15 min for gravity filtration (~2 min if vacuum filtration is used), and 30 to 40 min for rotary evaporation. This protocol can be completed in less than 2 hr.

Purification of the aqueous sample extract (Basic Protocol 2) requires ~5 min to condition the column, 5 min to load the sample, 5 to 10 min to wash, and 2 min to elute from the column. This protocol can be completed in ~20 min.

The rate-limiting step of MALDI-MS analysis (Basic Protocol 3) is the sample preparation, which involves depositing matrix and sample solutions onto the probe, allowing the solvent to evaporate, and introducing the probe into the ion source. This process may require up to 10 min per sample. From this point, analysis can be carried out in less than 1 min per sample. If multiple spectra are to be collected for each sample, accordingly more time will be required. Manipulation of spectra and peak labeling can be achieved in less than 5 min.

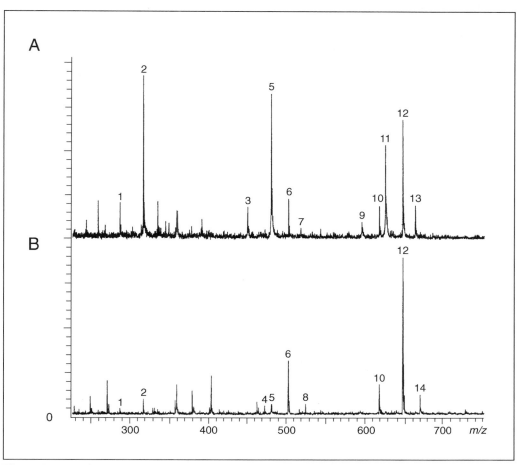

Figure I1.5.3 (**A**) Natural cation MALDI-MS spectrum of almond seedcoat extract. (**B**) MALDI-MS positive ion spectrum of almond seedcoat extract with sodium added at 0.01 M. Peaks: (1) [kaempferol + H]$^+$; (2) [isorhamnetin + H]$^+$; (3) [kaempferol glucoside + H]$^+$; (4) [kaempferol glucoside + Na]$^+$; (5) [isorhamnetin glucoside + H]$^+$; (6) [isorhamnetin glucoside + Na]$^+$; (7) [isorhamnetin glucoside + K]$^+$; (8) [isorhamnetin glucoside + 2Na − H]$^+$; (9) [kaempferol rutinoside + H]$^+$; (10) [kaempferol rutinoside + Na]$^+$; (11) [isorhamnetin rutinoside + H]$^+$; (12) [isorhamnetin rutinoside + Na]$^+$; (13) [isorhamnetin rutinoside + K]$^+$; (14) [isorhamnetin rutinoside + 2Na − H]$^+$.

Literature Cited

Frison-Norrie, S. and Sporns, P. 2002. Identification and quantification of flavonol glycosides in almond seedcoats using MALDI-TOF MS. *J. Agric. Food Chem.* 50:2782-2787.

Harborne, J.B. and Williams, C.A. 2001. Anthocyanins and other flavonoids. *Nat. Prod. Rep.* 18:310-333.

Hollman, P.C.H. and Arts, I.C.W. 2000. Flavonols, flavones and flavanols—nature, occurrence and dietary burden. *J. Sci. Food Agric.* 80:1081-1093.

Markham, K.R. 1982. Techniques of Flavonoid Identification. Academic Press, New York.

Merken, H.M. and Beecher, G.R. 2000. Measurement of food flavonoids by high performance liquid chromatography: A review. *J. Agric. Food Chem.* 48:577-599.

Pietta, P.G. 2000. Flavonoids as antioxidants. *J. Nat. Prod.* 63:1035-1042.

Sang, S.M., Lapsley, K., Jeong, W.S., Lachance, P.A., Ho, C.T., and Rosen, R.T. 2002. Antioxidative phenolic compounds isolated from almond skins (*Prunus amygdalus* Batsch). *J. Agric. Food Chem.* 50:2459-2463.

Stobiecki, M. 2000. Application of mass spectrometry for identification and structural studies of flavonoid glycosides. *Phytochemistry* 54:237-256.

Wang, J. and Sporns, P. 2000. MALDI-TOF MS analysis of food flavonol glycosides. *J. Agric. Food Chem.* 48:1657-1662.

Key References

Frison, S. and Sporns, P. 2002. Variation in the flavonol glycosides composition of almond seedcoats as determined by MALDI-TOF mass spectrometry. *J. Agric. Food Chem.* In press.

Describes use of MALDI-MS as a screening tool for flavonol glycosides in a selection of almond varieties.

Frison-Norrie and Sporns, 2002. See above.

Describes a MALDI-MS methodology for quantifying flavonol glycosides from almond seedcoats.

Wang and Sporns, 2000. See above.

First paper to describe a MALDI-MS methodology for identifying flavonol glycosides in foods.

Contributed by Suzanne Frison and
  Peter Sporns
University of Alberta
Edmonton, Alberta, Canada

Analysis of Isoflavones in Soy Foods

UNIT I1.6

Isoflavones are characteristically found in the family of the Leguminosae, serving such biofunctions as phytoalexins and regulating growth and development of plants (Ingham, 1982). Increasing evidence suggests that consumption of soybean products may significantly impact health (Setchell, 1998, 2001; Birt et al., 2001). The biological activity has been associated, in part, with the presence of isoflavones in soy (Fournier et al., 1998; Setchell and Cassidy, 1999). Analysis of these bioactive compounds in soybean products is an essential part of any research involving soy isoflavones. This unit attempts to provide a reliable method with the most commonly used analytical techniques for this purpose.

Soybeans are the most prominent source of isoflavones in food plants, ranging in concentration from 0.1 to 3.0 mg/g dry weight (Coward et al., 1993). The general structure of isoflavones is shown in Figure I1.6.1, and the common names, chemical names, and chemical structures of the twelve isoflavones in soy are shown in Figure I1.6.2. Because isoflavones are present at small concentrations in soybeans and soy-based food products (see matrices in Table I1.6.1), their efficient extraction prior to analysis is very important. In this unit, extraction of the twelve soy isoflavones is described (see Basic Protocol 1), as is sample preparation before extraction (see Support Protocol 1). This process consists of removing water from the sample and sample grinding or homogenization.

When authentic isoflavone standards are available, the method of choice for analyzing isoflavone extracts from soy foods is generally high-performance liquid chromatography (HPLC) paired with a reversed-phase C18 column and UV spectrophotometer. A mixture of isoflavones in soy food extracts is separated based on the polarity and/or solubility of isoflavones in the column between the stationary (packing material of the column) and mobile (solvent) phases used. All isoflavones exhibit an intense absorption in the UV region of the spectrum, between 240 and 280 nm (Harborne, 1967). Based on the Beer-Lambert law, the concentration of a pure isoflavone compound or an individual isoflavone in a mixture can be calculated using the absorbance measurements obtained and known extinction coefficients (Ollis, 1962). This unit describes techniques for generating calibration curves with reference standards (see Basic Protocol 2), separating and measuring isoflavones using HPLC-UV spectrophotometry (see Basic Protocol 3), preparation of samples for HPLC (see Support Protocol 2), and converting glycosidic isoflavones to their aglycones using enzymatic hydrolysis (see Support Protocol 3).

NOTE: All reagents used in this unit, including water, should be HPLC grade or equivalent.

NOTE: Isoflavones should be protected from light, oxygen, and elevated temperature during and after extraction (see Critical Parameters and Troubleshooting).

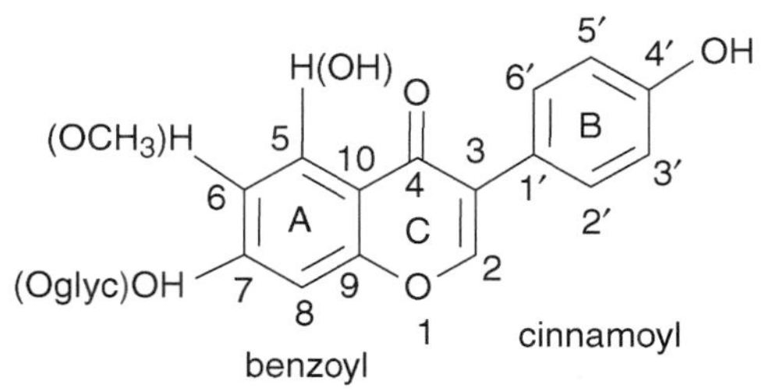

Figure I1.6.1 General structure and ring numbering system applied to known soy isoflavones.

Bioactive Food Components

Contributed by Yu Chu Zhang and Steven J. Schwartz

I1.6.1

519

SOLVENT EXTRACTION OF ISOFLAVONES IN THEIR NATURAL FORMS FROM SOY FOODS

This method focuses on the extraction of soy food samples at room temperature using acidified solvent mixtures. The method is designed to release isoflavones in their natural forms from food matrices based on their polarity and solubility in solvents. Insoluble proteins, carbohydrates, and lipids present in the food are removed from the isoflavone extract using organic solvents, acid, and centrifugation.

Common name	Chemical name	Chemical structure
Daidzein	7,4′-Dihydroxyisoflavone	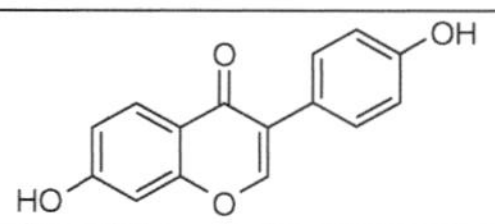
Daidzin	7,4′-Dihydroxyisoflavone 7-glucoside *or* daidzein 7-*O*-glucoside *or* daidzein 7-*O*-β-D-glucopyranoside	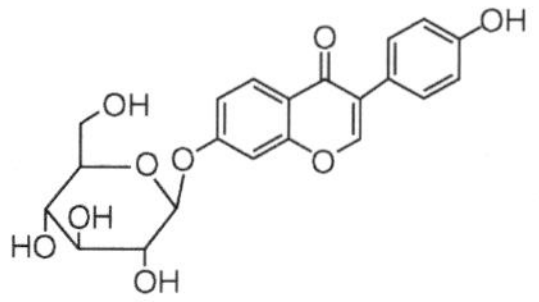
Acetyldaidzin	6″-*O*-Acetyldaidzin	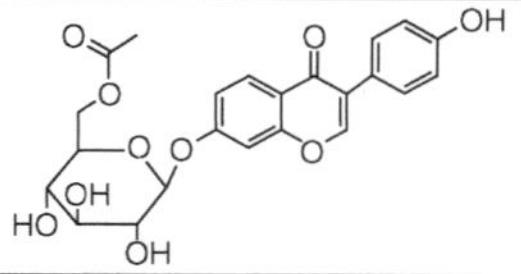
Malonyldaidzin	6″-*O*-Malonyldaidzin	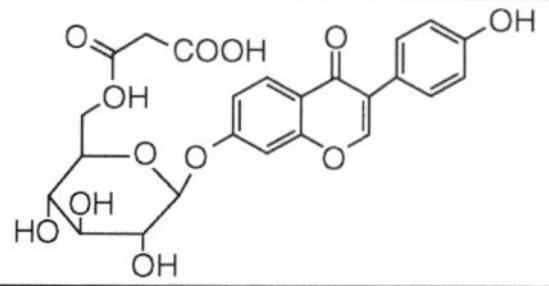
Genistein	5,7,4′-Trihydroxyisoflavone *or* 5,7-dihydroxy-3-(4-hydroxyphenyl)-4*H*-1-benzopyran-4-one	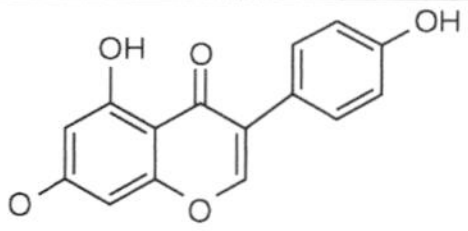
Genistin	5,7,4′-Trihydroxyisoflavone 7-glucoside *or* genistein 7-*O*-glucoside *or* genistein-7-*O*-β-D-glucopyranoside	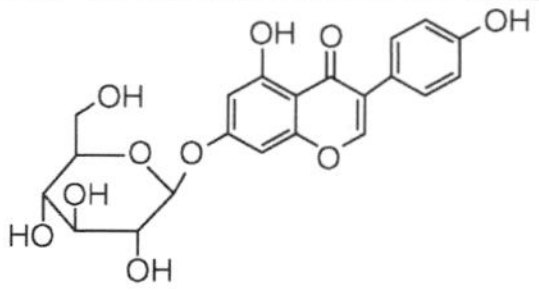
Acetylgenistin	6″-*O*-Acetylgenistin	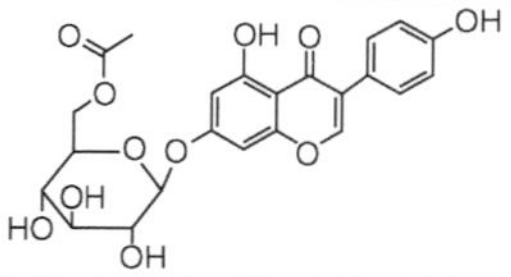
Malonylgenistin	6″-*O*-Malonylgenistein	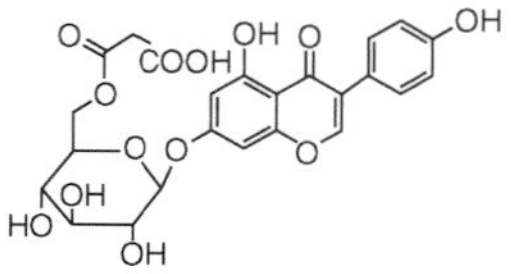

Figure I1.6.2 *(above and at right)* Common name, chemical name, and chemical structure of soy isoflavones.

Prepared food sample (see Support Protocol 1)
Acetonitrile
0.1 M HCl

50-ml centrifuge tube with cap
Ultrasonic bath
Wrist shaker
15-ml centrifuge tube
Analytical pipets

1. In a 50-ml centrifuge tube with cap, carefully measure the prepared food sample and mix it with solvent (i.e., acetonitrile, HCl, and water) as detailed in Table I1.6.2. Break up cellular material using an ultrasonic bath for 10 min.

2. Vortex the mixture for 1 min to homogenize the suspension.

3. Extract on a wrist shaker 2 hr at room temperature (≤25°C).

4. Centrifuge the mixture 30 min at $430 \times g$, room temperature.

5. Transfer an aliquot from the centrifuge tube to a 15-ml test tube. Store the aliquot at 4°C prior to further purification for HPLC analysis.

> *It is recommended that analysis of extracts be conducted within 10 hr of extraction to minimize potential conversion of isoflavones.*

Glycitein	7,4′-Dihydroxy-6-methoxyisoflavone	
Glycitin	7,4′-Dihydroxy-6-methoxyisoflavone-7-D-glucoside	
Acetylglycitin	6″-O-Acetylglycitin	
Malonylglycitin	6″-O-Malonylglycitin	

Table I1.6.1 Matrices of Some Common Soybean Products and Their Isoflavone Content

	Protein ($\%^a$)	Carbohydrate ($\%^a$)	Lipids ($\%^a$)	Isoflavones (mg/g on total basis^b)
Soybeans^c	40	36	19	3.0
Full fat soy flour^d	48	32	11	2.0
Defatted soy flour^e	52	35	1	3.0
Soy milk powder^f	40	32	18	2.8
Soy containing bread^g	40	40	2	0.9

^{a}Percent by weight on total basis.
^{b}Milligrams/gram on total basis from company certificate of analysis unless specified.
^{c}Harvested in Ohio.
^{d}Expeller Soy Flour I.P. from SunRich.
^{e}Defatted soy flour from Cargill Foods.
^{f}Soy Supreme Basic Soy milk powder from SunRich.
^{g}Soy-containing bread from Bavoy.

Table I1.6.2 Solvent Mixture for Extracting Isoflavones from Soy Foods^a

	Sample	Acetonitrile	0.1 M HCl	Water
Dehydrated/solid sample	0.5 g	10 ml	2 ml	4 ml
Liquid sample	2 ml	10 ml	2 ml	2 ml

^{a}Final mixture: ~60% acetonitrile/40% water.

Table I1.6.3 Commercial Sources of Soy Isoflavone Reference Standards^a

Isoflavone	Apin Chemicals	Fisher Scientific	Indo Fine Chemical	LC Laboratories	Sigma Aldrich
Daidzein	X	X	X	X	X
Daidzin	X		X	X	X
Acetyldaidzin				X	
Malonyldaidzin				X	
Genistein	X	X	X	X	X
Genistin	X		X	X	X
Acetylgenistin				X	
Malonylgenistin				X	
Glycitein			X	X	X
Glycitin			X	X	X
Acetylglycitin				X	
Malonylglycitin				X	

^{a}Ordering and product information can be found at the company websites (see *SUPPLIERS APPENDIX*).

SAMPLE PREPARATION FOR EXTRACTION

Extract fresh food samples whenever possible. Grind solid and/or dehydrated soybean products in a grinder or with a mortar and pestle until a fine powder (one that can pass through a 50-mesh screen) is obtained. Mash moist soybean product to a fine paste.

If food samples need to be stored for later extraction, seal in a plastic bag and store up to 10 days at −20°C or lower to prevent isoflavone loss and degradation. Thaw the sample completely at room temperature immediately before extraction. If freeze drying is desired, perform after food samples have been prefrozen in a freeze-dry flask to −20°C or lower.

SUPPORT PROTOCOL 1

PREPARATION OF INDIVIDUAL ISOFLAVONE STANDARDS

BASIC PROTOCOL 2

In this protocol, commercially purchased isoflavone standards (Table I1.6.3) are dissolved in a suitable solvent to prepare solutions in a series of decreasing concentrations. The absorbances are then measured using a UV spectrophotometer set at the isoflavone's maximum wavelength (λ_{max}), and the concentrations of isoflavone standard solutions are calculated using published molar extinction coefficients. The spectrum is also scanned in order to evaluate the fine structure (see Background Information, Spectral fine structure).

Materials

Standard isoflavones (see Support Protocol 3 and Tables I1.6.3 and I1.6.4)
80% (v/v) methanol

5-digit analytical balance
10- and 50-ml volumetric flasks
Ultrasonic bath
0.45-μm filter
Analytical pipets
Amber vials with caps
UV spectrophotometer and quartz/glass cuvette

Table I1.6.4 Isoflavone Stock Standard Solutions and Calibration Ranges

Isoflavones	Weight (mg) in 50 ml stock solution	Stock solution (mg/ml)	Working solutions (μg/ml)
Daidzein	2	0.04	0.4-8
Daidzin	5	0.1	1-20
Acetyldaidzin	2	0.04	0.4-8
Genistein	2	0.04	0.4-8
Genistin	5	0.1	1-20
Acetylgenistin	2	0.04	0.4-8
Glycitein	1	0.02	0.2-4
Glycitin	1	0.02	0.2-4
Acetylglycitin	1	0.02	0.2-4

Bioactive Food Components

I1.6.5

Prepare stock standard solutions

1. Using a 5-digit analytical balance, weigh the amount of each crystalline isoflavone standard given in Table I1.6.4 and transfer to individual 50-ml volumetric flasks.

2. Add 80% methanol to near 50 ml, stopper each flask, and mix well by repeated inversion and by incubating in an ultrasonic bath until complete dissolution is achieved (~10 min).

   Dissolution may be aided by initial addition of a small amount of a more effective solvent (e.g., DMSO) before bringing to volume for spectrophotometric measurement (no more than 1% of the total volume should be adequate). Dissolution can also be aided by using warm solvent ($\leq$50°C).

3. Fill each flask with 80% methanol to the final volume (50 ml). Seal flasks tightly with a cap and store up to 6 months at −20°C.

4. Before use, completely thaw frozen stock solutions at room temperature and then pass through a 0.45-μm filter.

Prepare working standard solutions

5. From each stock solution bottle, accurately transfer 0.1, 0.3, 0.5, 1, and 2 ml solution to 10-ml volumetric flasks using analytical pipets.

6. Dilute to volume using 80% methanol (see Table I1.6.4 for working concentrations). Mix well by repeated inversion of the flasks.

7. Transfer working solutions to tightly capped amber vials and store up to 2 months at −20°C. Completely thaw at room temperature before analysis.

   The working solutions will be used for UV absorbance measurement and for quantification by HPLC. Calibration should be performed at least every 2 months when new working standards are prepared.

Measure absorbance using a UV spectrophotometer

8. Turn on the UV spectrophotometer and allow to warm up at least 30 minutes.

9. Zero the spectrophotometer with an appropriate blank (i.e., 80% methanol).

Table I1.6.5 UV Spectral Pattern of Twelve Soy Isoflavones in 80:20 (v/v) Methanol/Water

Isoflavone	MW	Absorption peaks			λ_{max} (nm)	$\varepsilon_{\lambda max}$ (liter/mol·cm)[a]
Daidzein	254	211.4	249.1	303.6	249	31563
Daidzin	416	216.0	250.3	302.5	250	26830
Acetyldaidzin	458	216.1	249.1	301.5	249	29007
Malonyldaidzin	502	211.4	250.3	302.5	250	26830
Genistein	270	211.4	259.8	327.0	260	35323
Genistin	432	212.6	259.8	327.5	260	30895
Acetylgenistin	474	215.0	260.9	329.8	261	38946
Malonylgenistin	518	215.0	259.8	327.5	260	30895
Glycitein	284	215.0	257.4	321.5	257	25388
Glycitin	446	212.6	258.9	321.5	259	26713
Acetylglycitin	488	212.6	258.6	321.5	259	29595
Malonylglycitin	532	212.6	259.0	321.5	259	26313

[a]Values from Murphy et al. (2002).

10. Fill a quartz/glass cuvette with the standard solution of a specific isoflavone. Measure the absorbance at λ_{max} (see Table I1.6.5). Take reading immediately.

   With the dilution applied here, the absorbance reading should be in the range of 0 to 1.0.

11. Observe the UV spectrum (Table I1.6.5; also see Background Information, Spectral Fine Structure).

12. Repeat steps 9 to 11 for all standards and concentrations.

Calculate isoflavone concentration

13. Calculate the concentration of isoflavone standard solutions based on the Beer-Lambert law.

$$A = c \text{ (mol/liter)} \times l \text{ (cm)} \times \varepsilon \text{ (liter mol}^{-1} \text{ cm}^{-1})$$

where A is absorbance, c is concentration, l is cell pathlength (typically 1 cm), and ε is the molar extinction coefficient (i.e., absorbance of a 1 M solution in a 1-cm light path at a specific wavelength; Table I1.6.5).

GRADIENT SEPARATION AND IDENTIFICATION OF ISOFLAVONES USING A C18 REVERSED-PHASE COLUMN

BASIC PROTOCOL 3

In this protocol, a reversed-phase separation using a C18 HPLC column and gradient mobile phase are used to separate isoflavones in the soy food extract. Reversed-phase HPLC operates on the basis of hydrophilicity and lipophilicity. Reversed-phase C18 columns consist of silica-based packings with covalently bound 18-alkyl chains. The mobile phase carries the sample solution through the column. The chemical and physical interactions of the mobile phase and sample with the column determine the degree of migration and separation of components in the column. In gradient elution, different isoflavone compounds are separated by increasing the strength of the organic solvent. Detection is achieved by monitoring UV absorbance at 260 nm using a UV or photodiode array (PDA) detector. Retention time, UV spectra, and co-elution are used to identify individual isoflavones in food extracts by comparison with pure isoflavone compounds (see Basic Protocol 2).

Materials

Solvent A: 1% (v/v) acetic acid in water, pH >2.0, fresh
Solvent B: 100% acetonitrile
Isoflavone reference standards (see Basic Protocol 2)
Food sample: isoflavone extract from soy foods, after clean-up procedure (see Support Protocol 2)

Sonicator, vacuum filtration device, or in-line vacuum degasser
HPLC system:
   UV/Vis photodiode array (PDA) detector
   Pump: gradient
   Waters 2695 separation module with heating and cooling system
   Waters Nova-Pak C18 reversed-phase column (150 × 3.9 mm, 4-μm i.d., 60-Å pore size)
   Waters Nova-Pak C18 guard column or a guard column with similar packing material
   Injector: automatic
   Computer data system

Additional reagents and equipment for preparing calibration curves (see Basic Protocol 2)

Prepare mobile phase

1. Check pH of solvent A, making sure it is >2.0. Degas both mobile phases via vacuum filtration, ultrasonic agitation, or inline vacuum degasser.

   Acetic acid serves as a modifier to protonate glucosides of isoflavones in the mobile phase.

   While freshly prepared solvent A is preferred, solvent that is up to 2 days old is acceptable.

Set up HPLC apparatus

2. Turn on the UV/Vis PDA detector at least 30 minute before analysis.

3. Prime the gradient pump and flush the lines with mobile phases A and B, following manufacturer's instructions.

4. Program the mobile phase gradient as listed in Table I1.6.6.

5. Set the following HPLC parameters:

   Pump flow rate: 0.6 ml/min
   Column oven temperature: 25°C
   Injection volume: 10 µl
   UV monitor: 260 nm.

   For strong UV absorbers such as isoflavones, total injected amounts of <20 µg and injection volumes of 10 to 30 µl are typical. Larger injection volumes can cause broadening of peaks. The flow rate was developed based on the column inner diameter and mass sensitivity.

6. Equilibrate the system and column with the starting mobile phase composition (i.e., 85% A:15% B) of mobile phase A for 10 column volumes (10 × 1.8 ml/column volume), until the baseline is flat and smooth.

   The small particle size (4 µm) of the bonded phases and small pore size (60 Å) of the C18 column efficiently resolve mixtures of similar isoflavone derivatives in the soy food matrix.

Perform HPLC analysis

7. Inject each working calibration standard and standard mixture (including any optional internal standard) individually into the automatic injector.

   The standard mixture can be prepared by taking a known amount of each standard solution into a sample vial and mixing well. The concentration of each isoflavone in the standard mixture can be calculated with the dilution factor.

   To check the recovery of a chromatographic condition, it is common to spike a known concentration of standard solution into the sample solution. A matrix effect is suspected if the spike recovery is outside the limits of 90% to 110%. Recovery is calculated using the formula $R = [(C_s - C)/S] \times 100$, where R is percent recovery, C_s is spiked sample concentration, C is sample background concentration, and S is concentration equivalent of spike added to sample. If a matrix effect is suspected, adjustment of chromatographic condition must take place.

Table I1.6.6 Mobile Phase Gradients for Isoflavone Analysis Using a Reversed-Phase C18 Column

Time (min)	Solvent A (%)	Solvent B (%)
0	85	15
5	85	15
36	71	29
44	65	35
45	85	15
50	85	15

8. Measure the peak area of each working standard solution at 260 nm. Record the retention time for each peak. Observe the UV spectrum of each peak.

> *Modern HPLC systems commonly include computer software that automatically integrates peak areas. For older systems, manual integration may be necessary.*

> *Besides the retention time, the UV spectra of each peak should be used to confirm the identity of isoflavones (see Background Information, Spectral Fine Structure).*

9. Inject food sample and record data as in step 8.

Analyze results

10. Prepare calibration curves by plotting peak area versus known concentration for each standard solution.

> *The slope of the line will be used to calculate the concentration of the respective isoflavone.*

> *Peak area should be linearly proportional to analyte concentration ($y = ax + b$) with a correlation coefficient >0.98 and an intersect very near the origin. In this equation, y is the peak area, x is the concentration of standard solution, a is the response factor between peak area, and b is the concentration (~0).*

11. Identify isoflavones in the food sample by comparing the retention times of peaks with co-elution of isoflavone standards. Confirm identification by comparing the spectra of isoflavones in the food sample with those of standards (see Background Information).

> *The complete separation of all twelve isoflavones using this chromatographic condition requires ~30 min. The elution order is daidzin, glycitin, genistin, malonyldaidzin, malonylglycitin, acetyldaidzin, acetylglycitin, malonylgenistin, daidzein, glycitein, acetylgenistin, and genistein.*

> *Isoflavone retention and separation are influenced by column temperature. The oven temperature is set at 25°C.*

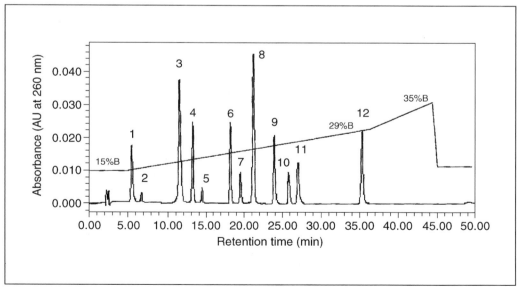

Figure I1.6.3 Gradient HPLC separation of isoflavone standards (see Basic Protocol 3). Peaks: 1, daidzin; 2, glycitin; 3, genistin; 4, malonyldaidzin; 5, malonylglycitin; 6, acetyldaidzin; 7, acetylglycitin; 8, malonylgenistin; 9, daidzein; 10, glycitein; 11, acetylgenistin; 12, genistein. Conditions: Waters Nova-Pak C18 reversed-phase column (150 × 3.9 mm; 4-μm i.d.; 60 Å pore size); mobile phase: 1% acetic acid in water (solvent A) and acetonitrile (solvent B); flow rate: 0.60 ml/min; UV detector: 260 nm; column temperature: 25°C. The dotted line represents the gradient of solvent B.

Bioactive Food Components

12. Use the area of each isoflavone peak and the calibration curves to calculate the isoflavone concentrations in the sample. Calculate the final concentration of isoflavones in soy food by considering the dilution factor during sample preparation.

A sample chromatogram of isoflavone standards is shown in Figure I1.6.3.

SUPPORT
PROTOCOL 2

SAMPLE PREPARATION FOR HPLC ANALYSIS

Soy foods are complex matrices (see Table I1.6.1). After initial aqueous extraction, the crude extract containing isoflavones is concentrated but may contain components that are immiscible with the HPLC mobile phase for direct injection. A clean-up step is necessary to prevent the introduction of particles and precipitates that may block the frit and foul the column.

Materials

Crude isoflavone extract (see Basic Protocol 1)
Methanol

10-ml glass tube with cap
Solvent evaporation apparatus (e.g., nitrogen gas, Speedvac)
Ultrasonic bath
0.20-μm filter

1. Transfer 1 ml supernatant from crude soy extract to a 10-ml glass tube. Dry the aliquot under a nitrogen gas stream or using a Speedvac evaporator.

2. Resuspend the residue in 1 ml of 100% methanol.

3. Ultrasonicate the mixture in a bath for 10 min.

   This step is designed for complete protein precipitation, removal of methanol-insoluble sugars, and isoflavone redissolution in solvent.

4. Store the mixture up to 10 hr at 4°C if injection is not directly conducted.

5. Before injection, vortex the mixture 1 min and filter through a 0.20-μm filter.

SUPPORT
PROTOCOL 3

CONVERTING GLYCOSIDIC ISOFLAVONES TO THEIR AGLYCONES USING ENZYMES

Isoflavone aglycone standards are commercially available. However, the acetyl and malonyl glucosides of isoflavone standards are not readily available due to their lack of stability during transportation and storage. Hydrolysis of isoflavone glycosides removes the sugar moiety from the aglycones, thereby enabling quantification of total isoflavones in soy foods when the authentic standards of glycosidic isoflavones are not available. Hydrolysis is also a useful tool for structural analysis when specificity is achieved.

β-glucosidase (Emulsin) and cellulase (from *Aspergillus niger*) have been used efficiently to convert isoflavone conjugates to their aglycone forms (Franke et al., 1994; Liggins et al., 1998). They remove the β-linked glucose from the 7-hydroxyl group on the isoflavone aglycones with adequate purity. Figure I1.6.4 illustrates the method using enzymatic hydrolysis for isoflavone analysis.

By comparing the chromatogram from enzymatic hydrolysis with that in the absence of enzymatic hydrolysis, the completeness of hydrolysis by enzymes and the total isoflavones in soy food extract can be determined (see Basic Protocol 3).

Crude isoflavone extract (See Basic Protocol 1)
Enzyme: β-glucosidase, almond-isolate *or* cellulase, *Aspergillus niger* extract
0.1 M ammonium acetate buffer, pH 5 (adjust pH with acetic acid)
100% methanol

Solvent evaporation unit (nitrogen gas)
Ultrasonic bath
0.2-μm filter

1. Centrifuge the crude isoflavone extract 30 min at $430 \times g$, 25°C.

2. Take 1 ml supernatant from the crude extract and evaporate the solvent under nitrogen flow.

3. Dissolve 100 Fishman units of enzyme in 5 ml of 0.1 M ammonium acetate buffer, pH 5.

   One Fishman unit is the amount of enzyme necessary to digest 1 μmol substrate in 1 min at pH 5.0, 37°C.

4. Add 1 ml dissolved enzyme solution into the vial containing dried extract. Mix well in an ultrasonic bath for 10 min.

5. Incubate the solution 12 hr in a 37°C water bath.

   Overnight is recommended.

6. Evaporate the solvent under nitrogen flow.

7. Dilute the dried residue with 1 ml of 100% methanol. Vortex and/or place in an ultrasonic bath to ensure solubilization.

8. Pass the sample solution through a 0.2-μm filter immediately before HPLC analysis.

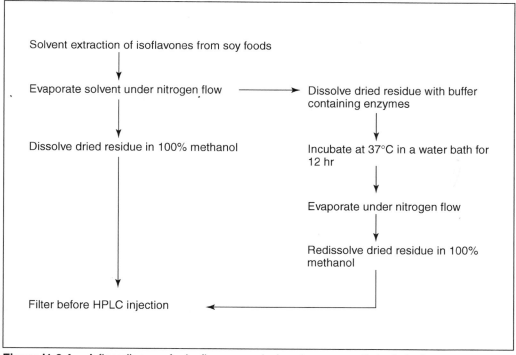

Figure I1.6.4 A flow diagram for isoflavone analysis using enzymatic hydrolysis.

COMMENTARY

Background Information

Isoflavones

Isoflavones comprise two benzene rings (A and B) linked through a heterocyclic pyrane C-ring at the 3 position (Fig. I1.6.1), which distinguishes them from flavones (Ingham, 1982; also see Figures I1.3.7 and I1.3.9 for comparison). The primary isoflavones in soybeans are the genistein, daidzein, and glycitein families (Fig. I1.6.2). Each family consists of its respective aglycone, β-glucoside, malonyl-glucoside, and acetylglucoside. Other isoflavones (i.e., formononetin or biochanin A) are present in only trace amounts (<1.0 µg/g).

Malonylglucosides are the predominant form of isoflavones in unprocessed soybeans. Previous studies have concluded that malonyl-glucosides are heat labile and easily converted to their corresponding acetylglucosides and/or β-glucosides depending on the thermal conditions of processing and preparation (Xu et al., 2002). Enzyme activity may be responsible for the formation of aglycones by hydrolysis of glucosides (Coward et al., 1998). Various processing conditions produce soy products with a wide range of isoflavone content and composition. Recent studies observed that the chemical forms and abundance of isoflavones in soy foods have a significant impact on their bioavailability and biological effects (King, 1998; Izumi et al., 2000; Setchell et al., 2002). It is thus very important to avoid altering the natural forms and abundance of the twelve soy

isoflavones during extraction, identification, and quantification.

Extraction

Soybeans and soybean products contain high levels of protein, carbohydrates, and lipids (Table I1.6.1). As minor components of complex mixtures, isoflavones must first be separated from the bulk of the matrix constituents prior to analysis. Efficient extraction methods for isoflavones should account for their diverse structures, chemical properties, and the food matrix of which they are constituents. This unit describes a practical way of extracting isoflavones from soybean products in their natural forms using readily available solvents and laboratory equipment.

Before extracting a compound of interest from a complex food matrix, it is necessary to obtain knowledge about the physical and chemical nature of the sample—i.e., moisture content, stability in acid and base, and thermal stability. The hydrophobicity of the isoflavone forms is aglycone > acetylglucoside > malonyl-glucoside > β-glucoside, based on their chromatographic behavior on reversed-phase columns in the presence of an acid in the mobile phase to protonate the glycosidic isoflavones. The ester bonds of acetyl- and malonylglucose of isoflavones are labile at elevated temperatures and under acidic or basic conditions. The aqueous solubilities of the isoflavone aglycones are low and are pH dependent due to the acidic nature of their phenolic groups. Conju-

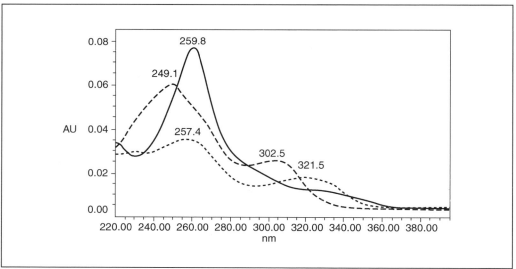

Figure I1.6.5 The UV characteristics of daidzein (dashed line), genistein (solid line), and glycitein (dotted line) in HPLC mobile phases (A: 1% acetic acid in water, B: 100% acetonitrile, A/B = 85:15) at 25°C monitored at 260 nm.

gation to glucose residues increases solubility, while acetylation or malonylation of the glucoses reduces solubility.

Spectral fine structure of soy isoflavones

The differences in the spectral characteristics of individual isoflavones are small but very important in their identification (Harborne, 1967). Isoflavones can be readily distinguished by their UV spectra, which typically exhibit an intense Band II (240 to 280 nm) absorption with only a shoulder or low-intensity peak representing Band I (300 to 330 nm). Band I absorption involves the B ring (cinnamoyl system; Figure I1.6.1), and Band II absorption involves the A ring (benzoyl system). The Band II absorption of isoflavones is relatively unaffected by increased hydroxylation of the B ring. Band II is, however, shifted bathochromically by increased oxygenation in the A ring. In addition to the absorption maxima of the isoflavones, the shape of the spectra provides important information for identification of pure isoflavone standards and isoflavones in soy extract. Fine structures of three soy isoflavone aglycones and their conjugates are shown in Figures I1.6.5 to I1.6.7. Details of the molecular characteristics of isoflavones can be found in *The Chemistry of Flavonoid Compounds* by Ollis (1962).

Isoflavones in solution obey the Beer-Lambert law, where absorbance (*A*) equals concen-

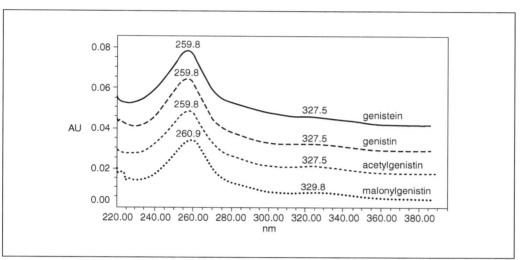

Figure I1.6.6 UV spectra of the genistein family in HPLC mobile phases (A: 1% acetic acid in water, B: 100% acetonitrile, A/B = 85:15) at 25°C, monitored at 220 to 400 nm.

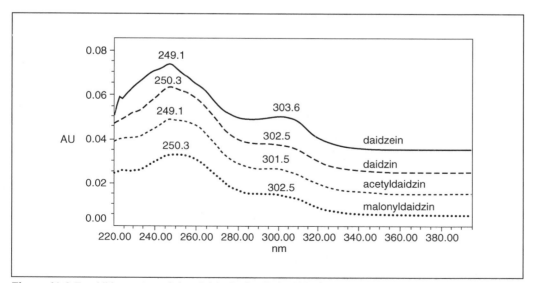

Figure I1.6.7 UV spectra of the daidzein family in HPLC mobile phases (A: 1% acetic acid in water, B: 100% acetonitrile, A/B = 85:15) at 25°C monitored at 220 to 400 nm.

Bioactive Food Components

I1.6.13

tration multiplied by the molar extinction coefficient ($\varepsilon_{\lambda max}$). The molar extinction coefficient is defined as the absorbance of a 1 M solution of isoflavones, in a defined solvent, in a 1-cm-pathlength cuvette, at its maximum wavelength (λ_{max}). This information can be used to quantify the concentration of a certain isoflavone with known absorbance. Table I1.6.5 gives some known values of molar extinction coefficients of soy isoflavones. Variations in laboratory conditions (e.g., different solvents, temperature, absorbance wavelength) may be responsible for differences in reported extinction coefficients. The actual absorption and fine structure will depend on the composition of the mobile phase. A shift of 2 to 3 nm in the maximum absorption wavelength is usual. The spectrum of a specific isoflavone in soy food should be compared with an authentic pure standard. They should be identical for both the λ_{max} and the fine structure.

Isoflavone analysis using HPLC

HPLC is the method of choice for the analysis of isoflavones in soy products. HPLC is fast, reproducible, requires small sample sizes, and can be used for both qualitative and quantitative analysis as well as for separation purposes.

HPLC system. The basic HPLC system comprises a solvent delivery pump, injector, analytical column, and detector. Most isoflavone analyses are performed using a binary gradient, a revered-phase C18 column, and a UV detector. A photodiode array detector is capable of producing a UV/Vis absorption spectrum for each peak and has become the norm in isoflavone analysis. Isocratic elution was reported for isoflavone analysis but is not as effective as a gradient to separate the wide variety of isoflavones within an acceptable elution time. A typical gradient solvent used with reversed-phase C18 columns starts with a high proportion of polar solvents (i.e., water) and gradually increases the proportion of a less polar solvent (i.e., acetonitrile or methanol). The aqueous solvent is usually acidified to prevent ionization of isoflavone glycosides, which can give multiple peaks for some compounds.

Identification. The order of elution of isoflavones is largely independent of minor variations in the solvent system, and thus it is possible to make tentative identifications by comparing relative retention times and co-elution with pure isoflavone standards. The actual retention times will vary between different runs, usually within 1 min. Such variation could be caused by differences in mobile phase prepa-

ration and the HPLC system (i.e., oven temperature, column condition).

Quantification. For accurate quantification, a standard curve of peak area versus concentration should be constructed for each standard using the same chromatographic conditions (e.g., wavelength and solvent) as for the samples under analysis. The concentration range of standard curves should be determined according to both the isoflavone level of soy food samples and dilution factors during sample preparation such that the UV absorbance of the injected sample is within a range of 0 to 1. The appropriate standard curve can then be used to calculate the quantity of isoflavones represented by each HPLC peak in the sample.

Internal standards. The primary function of an internal standard is in the determination of the reliability of extraction, sample preparation, and chromatographic procedure. An internal standard can be added to the original sample. By comparing the peak area of the internal standard in the chromatogram with that of a control, the losses due to sample preparation can be established. Alternatively, an internal standard can be added to the extraction solvent or to the sample prior to HPLC injection to measure the recovery rate of the chromatographic system. A suitable candidate for an internal standard for isoflavone analysis should (1) be similar in structure to isoflavones but not present in soy foods, and (2) be completely separable from isoflavones by the same chromatographic conditions at a specific wavelength. Apigenin, 2,4,4'-trihydroxydeoxybenzoin (THB), flavone, and equilenin have been reportedly used as internal standards.

Hydrolysis

The hydrolytic removal of the carbohydrate component of isoflavone glycosides simplifies the quantitative analysis of the isoflavones. Besides enzymatic hydrolysis (described in Support Protocol 3), acid hydrolysis and alkaline hydrolysis are also reportedly used. Acid hydrolysis is used primarily for cleaving sugars from glycosides, while alkaline hydrolysis finds application in the specific removal of acyl groups from acylated glycosides to produce β-glucosides of isoflavones (Klump et al., 2001). However, these methods are less commonly used due to a variety of reasons, including incomplete hydrolysis of the glycosides and degradation of the unconjugated isoflavones (Liggins et al., 1998).

Extraction

The solvents used for extracting isoflavones from soy foods were chosen according to the solubility of isoflavones and the food matrix involved. The diversity of soy isoflavones in polarity requires the use of a combination of organic solvent and water for extraction. The organic-to-water ratio (10:5) was established based on Murphy's study on solvent selection (Murphy et al., 2002). Water content in the solvent needs to be adjusted according to the moisture content in soy foods. Freeze-drying of samples before extraction simplifies the extraction process, but is not a prerequisite.

Acetonitrile, acetone, ethanol, and methanol have been used to extract isoflavones from soy foods. Among them, acetonitrile proved to be the most efficient (Griffith et al., 2001; Murphy et al., 2002). The solvent is supplemented with 0.1 M HCl to completely un-ionize the isoflavones and to release them from protein complexes by denaturing and precipitating the proteins. Room temperature is recommended for extraction to avoid alteration of the natural forms of the isoflavones. The time for extraction, 2 hr, was chosen for maximum recovery and shortest processing time. Ultrasound is used to aid the extraction process by degrading and weakening the cellular matrix.

Isoflavones are relatively stable compounds, but can degrade under certain conditions. Daidzein and genistein are light sensitive. It is necessary to work in dim light and to avoid exposure of extracts to air. Sample tubes should be capped during extraction. It is recommended to wrap sample flasks with aluminum foil to avoid light. Malonylglycosides of isoflavones are heat labile. Samples should be kept in a refrigerator between preparation and analysis, or kept in a freezer ($-20°C$) for storage.

The crude extract needs to be further purified for HPLC analysis. Direct injection of the crude extract into the HPLC would clog the frit and analytical column with precipitated impurities (i.e., proteins).

Besides solvent extraction, supercritical carbon dioxide has been applied to extract isoflavones (Rostagno et al., 2002). Due to the hydrophobicity of carbon dioxide, supercritical fluid extraction is more suitable for extracting nonpolar aglycones than polar glycosides of isoflavones, and may not be quantitative.

Standard solutions

Follow good laboratory practices for UV spectrophotometry (*UNIT F2.2*). For the overall analysis, it is critical to prepare precise standard curves. Manufacturer's instructions and MSDS's should be considered to minimize hazards and avoid degradation prior to and during sample preparation. Pure isoflavone standards should be completely dissolved in solvent when preparing stock solutions. Ultrasound, addition of a small amount of a more effective solvent, and warming the solvent (up to 50°C) have proved to be efficient and safe methods for aiding dissolution of isoflavone standards without altering their original structure. Stock solutions should be kept at $-20°C$ for storage purposes. The stock solution container needs to be sealed very well to avoid evaporation during storage. Stock solutions that have been stored frozen should be thawed at room temperature before reuse. Ultrasound is necessary to completely dissolve crystals that may develop during storage. All solutions should be filtered before reuse.

Sample preparation

Isoflavone glycosides can be deesterified and decarboxylated to their simpler conjugates under appropriate conditions. The conditions of sample handling and preparation prior to HPLC analysis are critical to ensure that there is no degradation or alteration of analytes. A handling temperature >60°C risks alteration of the original composition of isoflavones in the food matrix and should be avoided. Daylight or white fluorescent light should be avoided during sample preparation. Wrapping glass vials containing isoflavones in aluminum foil is suggested.

When resuspending isoflavones in pure organic solvent, extra care should be taken to completely dissolve extracted isoflavones in the solvent. Solublization can be achieved by ultrasonification and providing sufficient time for the isoflavones to dissolve. Vortexing helps to release and recover isoflavones from precipitated proteins that may stick to the wall of glass vials and containers. All solvents must be of a high degree of purity. All extracted samples should be filtered to remove small particles.

Hydrolysis

Quantitative results for different foods using the hydrolysis method need to be treated with caution. Hydrolysis conditions should be adjusted based on the knowledge of a crude soy food extract—i.e., the range of isoflavone concentration or the aglycone ratios. Partially hydrolyzed products lead to underestimation of the total isoflavones in the extracts.

HPLC analysis

Many parameters are critical to the successful and reproducible separation of isoflavones using HPLC.

Backpressure

Degassing should always be used to avoid outgassing and air bubble formation during HPLC analysis, especially when using a gradient. Backpressure can be used as an indicator for the condition of the column and the system. An abnormally high backpressure can be a sign of a fouled guard column. An abnormally low backpressure can be a sign of a leak between tubing and fittings, which causes poor peak shape. Variations in backpressure during analysis should be <50 psi. Larger variations could be caused by a pump malfunction.

Column care

Column equilibration (~10 column vol. recommended) ensures baseline stability, good peak shape, and reproducible retention times. For the specific column chosen in this unit for isoflavone analysis, at least 18 ml total is needed to equilibrate the system with mobile phase. The common practice is to purge the pump system and connect the inlet end of the column to the injector outlet. The initial pump flow should be set at 0.1 ml/min and increased to 0.6 ml/min in 0.1 ml/min increments. Once a steady backpressure and baseline have been achieved, the column is ready to use. Before injecting samples, it is suggested to run a blank gradient first to clean the column and help check for the possibility of impurity peaks.

For overnight storage, the column should be kept flushing at 0.1 ml/min with a mild solvent (i.e., acetonitrile or methanol). This practice reduces the re-equilibration time the following day to a few minutes rather than an hour or more. For long-term storage, the column should be flushed with its shipping solvent (i.e., acetonitrile for a Waters Nova-Pak RP-C18 column) at 1 ml/min for at least 15 min before being taken off and tightly sealed at both ends.

When contamination collects on a column, poor peak shapes (usually tailing) and extraneous peaks will appear with an increase in backpressure. Therefore, samples should always be filtered before injection. Mobile phases should be filtered a few days after preparation to remove any particles and prevent microbial growth. Nevertheless, after repeated use, the column needs to be cleaned or regenerated, because tightly bound impurities will bind to the stationary phase and reduce the attraction the column packing has for the sample components, thus reducing the retention times for the peaks of interest as the column ages. The manufacturer's instructions should be followed to regenerate the column. To prolong the life of expensive analytical columns, it is recommended that a filter and a guard column be placed before the analytical column to serve as a protective factor.

Solvent preparation

To prepare mobile phases, clean HPLC-grade solvent must be used along with clean solvent bottles. Since most of the organic solvents are volatile, they should be well sealed in solvent bottles after preparation. Acetic acid is subject to evaporation, causing the pH of the aqueous acetic acid solution to shift. Freshly prepared acetic acid in water is recommended for each set of analyses. The calculated pH of 1% acetic acid in water is ~2.7; however, it is suggested to check the pH of the acid solution each time it is prepared to assure its pH is >2.

Anticipated Results

The extraction protocol typically recovers 96% to 108% of isoflavones. The HPLC protocol typically recovers isoflavones in a range between 97% and 105% when tested using spiked reference standards. During reversed-phase HPLC separation, all isoflavones elute from the column in the order of polar to non-polar and are separated within 50 min of run time.

Time Considerations

When freeze-drying is not included, completion of the full extraction process requires ~1/2 day. It is recommended that freeze-drying, which requires at least 12 hr, be performed overnight before the day of extraction.

Calibration prior to HPLC requires 1 or 2 days to perform all necessary steps (i.e., prepare stock solutions, filter, measure absorbance, dilute, and HPLC injection of the working solutions). Once the stock solution and working solutions have been prepared, a calibration check can be performed on a UV spectrophotometer within 1 hr. The sample clean-up procedure for HPLC analysis may require a few hours. Enzymatic hydrolysis requires overnight incubation. Each HPLC analysis requires 50 min. When an autosampler is available, samples can be injected and analyzed overnight. Typically, one technician can extract 12 samples per day to be analyzed automatically overnight.

Literature Cited

Birt, D.F., Hendrich, S., and Wang, W. 2001. Dietary agents in cancer prevention: Flavonoids and isoflavonoids. *Pharmacol. Ther.* 90:157-177.

Coward, L., Barnes, N.C., Setchell, K.D.R., and Barnes, S. 1993. Genistein, daidzein, and their β-glucoside conjugates: Antitumor isoflavones in soybean foods from American and Asian diets. *J. Agric. Food Chem.* 41:1961-1967.

Coward, L., Smith, M., Kirk, M., and Barnes, S. 1998. Chemical modification of isoflavones in soyfoods during cooking and processing. *Am. J. Clin. Nutr.* 68:1486S-1491S.

Fournier, D.B., Erdman, J.W., and Gordon, G.B. 1998. Soy, its components, and cancer prevention: A review of the in vitro, animal and human data. *Cancer Epidemiol. Biomarkers Prev.* 7:1055-1065.

Franke, A., Custer, L.J., Carmencita, M.C., and Narala, K.K. 1994. Quantitation of phytoestrogens in legumes by HPLC. *J. Agric. Food Chem.* 42:1905-1913.

Griffth, A.P. and Collison, M.W. 2001. Improved methods for the extraction and analysis of isoflavones from soy-containing foods and nutritional supplements by reversed-phase high-performance liquid chromatography and liquid chromatography-mass spectrometry. *J. Chromatog. A* 913:397-413.

Harborne, J.B. 1967. Isoflavones. *In* Comparative Biochemistry of the Flavonoids, pp. 91-95. Academic Press, London.

Ingham, J.L. 1982. Phytoalexins from the Leguminosae. *In* Phytoalexins (Bailey and Mansfield, eds.) pp. 21-80. John Wiley & Sons, New York.

Izumi, T., Piskula, M.K., Osawa, S., Obata, A., Tobe, K., Saito, M., Kataoka, S., Kubota, Y., and Kikuchi, M. 2000. Soy isoflavone aglycones are absorbed faster and in higher amounts than their glucosides in human. *J. Nutr.* 130:1695-1699.

King, R.A. 1998. Daidzein conjugates are more bioavailable than genistein conjugates in rats. *Am. J. Clin. Nutr.* 68(Suppl):1496S-1499S

Klump, S., Allred, M., MacDonald, J., and Ballam, J. 2001. Determination of isoflavones in soy and select foods containing soy by extraction, saponification, and LC. Collaborative study. *J. AOAC Int.* 84:1865-1883.

Liggins, J., Bluck, L.J.C., Coward, W.A., and Bingham, S.A. 1998. Extraction and quantification of daidzein and genistein in foods. *Anal. Biochem.* 264:1-7.

Murphy, P.A., Barua, K., and Hauck, C.C. 2002. Solvent extraction selection in the determination of isoflavones in soy foods. *J. Chromatog. B* 777:129-138.

Nguyenle, T., Wang, E., and Cheung, A.P. 1995. An investigation on the extraction and concentration of isoflavones in soy-based products. *J. Pharm. Biomed. Anal.* 14:221-232.

Ollis, W.D. 1962. The isoflavones. *In* The Chemistry of Flavonoid Compounds (T.A. Geissman, ed.) pp. 353-405. MacMillan, New York.

Rostagno, M.A., Arafajo, J.M.A., and Sandi, D. 2002. Supercritical fluid extraction of isoflavones from soybean flour. *Food Chem.* 78:111-117.

Setchell, K.D. 1998. Phytoestrogens: The biochemistry, physiology, and implications for human health of soy isoflavones. *Am. J. Clin. Nutr.* 68(Suppl.):1333S-1346S.

Setchell, K.D. and Cassidy, A. 1999. Dietary isoflavones: Biological effects and relevance to human health. *J. Nutr.* 129:758S-767S.

Setchell, K.D., Brown, N.M., Zimmer-Nechemias, L., Brashear, W.T., Wolfe, B.E., Kirschner, A.S., and Heubi, J.E. 2002. Evidence for lack of absorption of soy isoflavone glycosides in humans, supporting the crucial role of intestinal metabolism for bioavailability. *Am. J. Clin. Nutr.* 76:447-453.

Xu, Z., Wu, Q., and Godber, S. 2002. Stabilities of daidzin, glycitin, genistin, and generation of derivatives during heating. *J. Agric. Food Chem.* 50:7402-7406.

Key References

Griffith and Collison, 2001. See above.

A systematic investigation in extraction and analysis of isoflavones from soy-containing foods and nutritional supplements.

Murphy et al., 2002. See above.

A systematic review and comparison of different solvents and aqueous-to-solvent ratios for isoflavone extraction efficiency.

Contributed by Yu Chu Zhang
   and Steven J. Schwartz
The Ohio State University
Columbus, Ohio

Appendices

1   **Abbreviations and Useful Data**   **A.1A.1**
   1A   Abbreviations Used in This Manual   A.1A.1

2   **Laboratory Stock Solutions, Equipment, and Guidelines**   **A.2A.1**
   2A   Common Buffers and Stock Solutions   A.2A.1
      General Guidelines   A.2A.1
      Storage   A.2A.1
      Selection of Buffers   A.2A.3
      Recipes   A.2A.3
   2B   Laboratory Safety   A.2B.1
      Hazardous Chemicals   A.2B.1
   2C   Standard Laboratory Equipment   A.2C.1

3   **Commonly Used Techniques**   **A.3A.1**
   3A   Introduction to Mass Spectrometry for Food Chemistry   A.3A.1
      Gas Chromatography/Mass Spectrometry (GC/MS)   A.3A.1
      Desorption Ionization Mass Spectrometry   A.3A.2
      Liquid Chromatography/Mass Spectrometry (LC/MS)   A.3A.3
      Tandem Mass Spectrometry (MS/MS) and High Resolution   A.3A.6
      Conclusion   A.3A.7

ABBREVIATIONS AND USEFUL DATA

Abbreviations Used in This Manual

AACC American Association of Cereal Chemists

ACS American Chemical Society

AED atomic emission detection

AEDA aroma extract dilution analysis

AMC 7-amido-4-methylcoumarin

ANS anilinonapththalene sulfonate

AOAC Association of Official Analytical Chemists

AOCS American Oil Chemists' Society

AOM active oxygen method

AOS allene oxide synthase

AP alkaline phosphatase

APCI atmospheric pressure chemical ionization

ASTM American Society for Testing and Materials

ATR attenuated total reflection

AU absorbance units

AV acid value

BCA bicinchoninic acid

BET Brunauer-Emmet-Teller (equation)

BGG bovine gamma globulin

BHA butylated hydroxyanisole

BHC branched hydrocarbons

BHT butylated hydroxytoluene

Brij 35 polyoxyethylene 23-lauren ether

°Brix measure of sugar content as determined by refractometer with a Brix scale

BSA bovine serum albumin

BV biological value

CAPT compensated attached proton test

CD conjugated diene; circular dichroism

CDTA *trans*-1,2-diaminocyclohexane-N,N,N',N'-tetraacetic acid

CETAB cetyltrimethylammonium bromide

Chl *a* and *b* chlorophyll *a* and *b*

CI chemical ionization

CID collision-induced dissociation

CIE Commission Internationale de l'Éclairage (International Commission for Illumination)

CI/MS chemical ionization/mass spectrometry

CLA conjugated linoleic acids

CLSM confocal laser-scanning microscopy

CMC critical micelle concentration

COSY correlation spectroscopy

CP-HPLC chiral-phase high-performance liquid chromatography

CPA *cis*-parinaric acid

CT conjugated triene

CV coefficient of variation

cyd cyanidin

DAD diode array detector

DCI direct exposure chemical ionization

DCM dichloromethane

DEI direct exposure electron impact

Deoxy Mb deoxymyoglobin

DH degree of hydrolysis

DHP dihydroxy pigment

DMF dimethylformamide

DMSO dimethylsulfoxide

DNPH 2,4-dinitrophenylhydrazine

DOPA 3,4-dihydroxyphenylalanine

dpd delphinidin

DPO diphenol oxidase

DQF-COSY double quantum filtered correlation spectroscopy

DSA drop shape analysis

DSC differential scanning calorimetry

DTNB 5,5'-dithiobis(2-nitrobenzoic acid)

DTT dithiothreitol

DVT drop volume tensiometer

EC Enzyme Commission

EDTA ethylenediaminetetraacetic acid

EI electron impact

EI/MS electron impact/mass spectrometry

ELSD evaporative light-scattering detector

EM expressible moisture

EPA (U.S.) Environmental Protection Agency

ERH equilibrium relative humidity

ESI electrospray ionization

FAB/MS fast atom bombardment mass spectrometry

FAME fatty acid methyl ester

FC Folin-Ciocalteau

FDA (U.S.) Food and Drug Administration

FFA free fatty acids

FID free induction decay; flame ionization detection

FOX ferrous oxidation/xylenol orange method

FP fecal protein

FPD flame photometric detection

FPLC fast protein liquid chromatography

FTIR Fourier-transform infrared (spectrometry)

g gravity (in expressions of relative centrifugal force)

GAB Guggenheim-Anderson-DeBoer (equation)

GC gas chromatography

GC/FID gas-liquid chromatography with flame ionization detection

GC/MS gas-liquid chromatography with mass selective detection

GC/O gas chromatography/olfactometry

GLC gas-liquid chromatography

GOPOD glucose oxidase/peroxidase (reagent)

HBS hydroxybenzenesulfonamide

HDPE high-density polyethylene

HEC hydroxyethylcellulose

HEPES *N*-[2-hydroxyethyl]piperazine-*N*′-[2-ethanesulfonic acid]

HIC hydrophobic-interaction chromatography

HMBC heteronuclear multiple bond correlation

HMDS hexamethyldisilazane

HPLC high-performance liquid chromatography

HRGC high-resolution gas chromatography

HRP horseradish peroxidase

HS headspace

HS-SPME headspace solid-phase microextraction

HSQC heteronuclear single quantum coherence

i.d. inner diameter

IDA isotope dilution assay

IEF isoelectric focusing

IgG immunoglobulin G

IPA indole-3-propionic acid

IRMS isotope ratio mass spectrometry

IS internal standard

ISO International Standard Organization

IUB International Union of Biochemistry

IUBMB International Union of Biochemistry and Molecular Biology

IUPAC International Union of Pure and Applied Chemistry

IV iodine value

K_M Michaelis constant

Kat Katal (catalytic unit for enzyme activity)

LC-APCI-MS liquid chromatography/ atmospheric pressure chemical ionization mass spectrometry

LED light-emitting diode

LOX lipoxygenase

LSIMS liquid secondary ion mass spectrometry

LVE linear viscoelastic region

MA malonaldehyde; malondialdehyde

MALDI matrix-assisted laser desorption/ionization

MALDI-TOF MS matrix-assisted laser desorption/ionization time-of-flight mass spectrometer

MCAC metal-chelate affinity chromatography

MCC microcrystalline cellulose

MDGC multidimensional gas chromatography

2-ME 2-mercaptoethanol

Me HODES methyl hydroxyoctadecadienoates

MES 2-(*N*-morpholino)ethanesulfonic acid

MetMb metmyoglobin

MFP metabolic fecal protein

MHP monohydroxy pigment

MOPS 3-(*N*-morpholino)propane sulfonic acid

MS mass spectrometry; mass selective (detection)

MS/MS tandem mass spectrometry

mvd malvidin

MWCO molecular weight cutoff

m/z mass-to-charge ratio

NBS National Bureau of Standards

NIST National Institute of Standards and Technology

NMR nuclear magnetic resonance

NO-heme nitrosylheme

NOESY nuclear Overhäuser enhancement spectroscopy

NPLC normal-phase HPLC

NPR net protein ratio

NPU net protein utilization

OAV odor activity value

o.d. outer diameter

OSI oil stability index

OU odor units

Oxy Mb oxymyoglobin

PBS phosphate-buffered saline

PDA photodiode array

PDCAAS protein digestibility–corrected amino acid score

PDMS polydimethylsiloxane

PE pectinesterase

PEG polyethylene glycol

PER protein efficiency ratio

PGase polygalacturonase

pgd pelargonidin

pI isoelectric point

PIPES piperazine-*N*,*N*′-bis(2-ethanesulfonic acid)

PL pectic lyase

PMSF phenylmethylsulfonyl fluoride

p-NA paranitroanilide
pnd peonidin
psi pounds per square inch
ptd petunidin
PTFE polytetrafluoroethylene
PUFA polyunsaturated fatty acid
PV peroxide value
PVDF polyvinylidene difluoride
PVP polyvinylpyrrolidone
PVPP polyvinylpolypyrolidone
RAS retronasal aroma stimulator
RDA recommended dietary allowance
RF radio frequency
RFI relative fluorescence intensity
RI retention index
RNU relative nitrogen utilization
ROESY rotational nuclear Overhäuser enhancement spectroscopy
RP-HPLC reversed-phase HPLC
RPER relative protein efficiency ratio
RS resistant starch
RT retention time
RVP relative vapor pressure
S sieman (unit of conductance)
SD standard deviation
SDE simultaneous distillation extraction
SDS sodium dodecyl sulfate
SFC solid fat content
SFI solid fat index
SHAM salicylhydroxamic acid
SIM selected ion monitoring
SNIF-NMR site-specific natural isotope fractionation measured by nuclear magnetic resonance spectroscopy
SP-HPLC straight-phase high-performance liquid chromatography

SPME solid-phase microextraction
SV saponification value
TA titratable acidity
TBA thiobarbituric acid
TBARS thiobarbituric acid-reactive substances
TBS Tris-buffered saline
TCA trichloracetic acid
TD true digestibility
TEA triethylamine
TFA trifluoroacetic acid
THF tetrahydrofuran
TLC thin-layer chromatography
TLCK $N\alpha$-p-tosyl-L-lysine chloromethyl ketone
TMCS trimethylchlorosilane imidazole
TMG tetramethylguanidine
TMP 1,1,3,3-tetramethoxypropane
TMS trimethylsilyl
TNBS trinitrobenzenesulfonic acid
TOCSY total correlation spectroscopy
TPA texture profile analysis
TPCK N-tosyl-L-phenylalanine chloromethyl ketone
TRF theoretical relative response factor
Tris tris(hydroxymethyl)aminomethane
Tris·Cl Tris hydrochloride
TTS time-temperature superposition
U unit (of enzyme activity)
UHP ultra high purity
USDA United States Department of Agriculture
UV ultraviolet
WHC water holding capacity
WUA water uptake ability

Abbreviations and Useful Data

A.1A.3

LABORATORY STOCK SOLUTIONS, EQUIPMENT, AND GUIDELINES

Common Buffers and Stock Solutions

This section describes the preparation of buffers and reagents used in the manipulation of nucleic acids.

For preparation of acid and base stock solutions, see Tables A.2A.1 and A.2A.2 as well as individual recipes.

GENERAL GUIDELINES

When preparing solutions, use deionized, distilled water and (for most applications) reagents of the highest grade available. Sterilization is recommended for most applications and is generally accomplished by autoclaving. Materials with components that are volatile, altered or damaged by heat, or whose pH or concentration are critical should be sterilized by filtration through a 0.22-μm filter. In many cases such components are added from concentrated stocks after the solution has been autoclaved. Where specialized sterilization methods are required, this is indicated in the individual recipes.

CAUTION: It is important to follow laboratory safety guidelines and heed manufacturers' precautions when working with hazardous chemicals; consult institutional safety officers and appropriate references for further details.

STORAGE

Most simple stock solutions can be stored indefinitely at room temperature if reasonable care is exercised to keep them sterile; where more rigorous conditions are required, this is indicated in the individual recipes.

Table A.2A.1 Molarities and Specific Gravities of Concentrated Acids and Bases[a]

Acid/base	Molecular weight	% by weight	Molarity (approx.)	1 M solution (ml/liter)	Specific gravity
Acids					
Acetic acid (glacial)	60.05	99.6	17.4	57.5	1.05
Formic acid	46.03	90	23.6	42.4	1.205
		98	25.9	38.5	1.22
Hydrochloric acid	36.46	36	11.6	85.9	1.18
Nitric acid	63.01	70	15.7	63.7	1.42
Perchloric acid	100.46	60	9.2	108.8	1.54
		72	12.2	82.1	1.70
Phosphoric acid	98.00	85	14.7	67.8	1.70
Sulfuric acid	98.07	98	18.3	54.5	1.835
Bases					
Ammonium hydroxide	35.0	28	14.8	67.6	0.90
Potassium hydroxide	56.11	45	11.6	82.2	1.447
Potassium hydroxide	56.11	50	13.4	74.6	1.51
Sodium hydroxide	40.0	50	19.1	52.4	1.53

[a]*CAUTION:* Handle strong acids and bases carefully.

Table A.2A.2 pK_a Values and Molecular Weights for Some Common Biological Buffers[a]

Name	Chemical formula or IUPAC name	pK_a	Useful pH range	Mol. wt. (g/mol)
Phosphoric acid	H_3PO_4	2.12 (pK_{a1})	—	98.00
Citric acid[b]	$C_6H_8O_7$ (H_3Cit)	3.06 (pK_{a1})	—	192.1
Formic acid	HCOOH	3.75	—	46.03
Succinic acid	$C_4H_6O_4$	4.19 (pK_{a1})	—	118.1
Citric acid[b]	$C_6H_7O_7^-$ (H_2Cit^-)	4.74 (pK_{a2})	—	
Acetic acid	CH_3COOH	4.75	—	60.05
Citric acid[b]	$C_6H_6O_7^-$ ($HCit^{2-}$)	5.40 (pK_{a3})	—	
Succinic acid	$C_4H_5O_4^-$	5.57 (pK_{a2})	—	
MES	2-(N-Morpholino]ethanesulfonic acid	6.15	5.5-6.7	195.2
Bis-Tris	bis(2-Hydroxyethyl)iminotris (hydroxymethyl)methane	6.50	5.8-7.2	209.2
ADA	N-(2-Acetamido)-2-iminodiacetic acid	6.60	6.0-7.2	190.2
PIPES	Piperazine-N,N'-bis(2-ethanesulfonic acid)	6.80	6.1-7.5	302.4
ACES	N-(Carbamoylmethyl)-2-amino-ethanesulfonic acid	6.80	6.1-7.5	182.2
Imidazole	1,3-Diaza-2,4-cyclopentadiene	7.00	—	68.08
Diethylmalonic acid	$C_7H_{12}O_4$	7.20	—	160.2
MOPS	3-(N-Morpholino)propanesulfonic acid	7.20	6.5-7.9	209.3
Sodium phosphate, monobasic	NaH_2PO_4	7.21 (pK_{a2})	—	120.0
Potassium phosphate, monobasic	KH_2PO_4	7.21 (pK_{a2})		136.1
TES	N-tris(Hydroxymethyl)methyl-2-aminoethanesulfonic acid	7.40	6.8-8.2	229.3
HEPES	N-(2-Hydroxyethyl)piperazine-N'-(2-ethanesulfonic acid)	7.55	6.8-8.2	238.3
HEPPSO	N-(2-Hydroxyethyl)piperazine-N'-(2-hydroxypropanesulfonic acid)	7.80	7.1-8.5	268.3
Glycinamide HCl	$C_2H_6N_2O \cdot HCl$	8.10	7.4-8.8	110.6
Tricine	N-tris(Hydroxymethyl)methylglycine	8.15	7.4-8.8	179.2
Glycylglycine	$C_4H_8N_2O_3$	8.20	7.5-8.9	132.1
Tris	Tris(hydroxymethyl)aminomethane	8.30	7.0-9.0	121.1
Bicine	N,N-bis(2-Hydroxyethyl)glycine	8.35	7.6-9.0	163.2
Boric acid	H_3BO_3	9.24	—	61.83
CHES	2-(N-Cyclohexylamino)ethane-sulfonic acid	9.50	8.6-10.0	207.3
CAPS	3-(Cyclohexylamino)-1-propane-sulfonic acid	10.40	9.7-11.1	221.3
Sodium phosphate, dibasic	Na_2HPO_4	12.32 (pK_{a3})	—	142.0
Potassium phosphate, dibasic	K_2HPO_4	12.32 (pK_{a3})	—	174.2

[a]Some data reproduced from *Buffers: A Guide for the Preparation and Use of Buffers in Biological Systems* (Mohan, 1997) with permission of Calbiochem.

[b]Available as a variety of salts, e.g., ammonium, lithium, sodium.

A.2A.2

Table A.2A.2 reports pK_a values for some common buffers. Note that polybasic buffers, such as phosphoric acid and citric acid, have more than one useful pK_a value. When choosing a buffer, select a buffer material with a pK_a close to the desired working pH (at the desired concentration and temperature for use). In general, effective buffers have a range of approximately 2 pH units centered about the pK_a value. Ideally the dissociation constant—and therefore the pH—should not shift with a change in concentration or temperature. If the shift is small, as for MES and HEPES, then a concentrated stock solution can be prepared and diluted without adjustment to the pH. Buffers containing phosphate or citrate, however, show a significant shift in pH with concentration change, and Tris buffers show a large change in pH with temperature. For convenience, concentrated stock solutions of these buffers can still be used, provided that a pH adjustment is made *after* any temperature and concentration adjustments. All adjustments to pH should be made using the appropriate base—usually NaOH or KOH, depending on the corresponding free counterion. Tetramethylammonium hydroxide can be used to prepare buffers without a mineral cation. Many common buffers are supplied both as a free acid or base and as the corresponding salt. By mixing precalculated amounts of each, a series of buffers with varying pH values can conveniently be prepared.

RECIPES

Ammonium acetate, 10 M

Dissolve 385.4 g ammonium acetate in 150 ml H_2O
Add H_2O to 500 ml
Sterilize by filtration

Citrate-phosphate buffer (McIlvaine's buffer)

Solution A: 19.21 g/liter citric acid (0.1 M final)
Solution B: 53.65 g/liter $Na_2HPO_4 \cdot 7H_2O$ *or* 71.7 g/liter $Na_2HPO_4 \cdot 12H_2O$

Referring to Table A.2A.3 for desired pH, mix the indicated volumes of solutions A and B, then dilute with water to 100 ml. Filter sterilize, if necessary, using a 0.2 μm filter and store up to 1 month 4°C.

DTT (dithiothreitol), 1 M

Dissolve 1.55 g DTT in 10 ml water and filter sterilize. Store in aliquots at −20°C.

Do not autoclave to sterilize.

EDTA (ethylenediaminetetraacetic acid), 0.5 M (pH 8.0)

Dissolve 186.1 g disodium EDTA dihydrate in 700 ml water. Adjust pH to 8.0 with 10 M NaOH (~50 ml; add slowly). Add water to 1 liter and filter sterilize.

Begin titrating before the sample is completely dissolved. EDTA, even in the disodium salt form, is difficult to dissolve at this concentration unless the pH is increased to between 7 and 8. Heating the solution may also help to dissolve EDTA.

HCl, 1 M

Mix in the following order:
913.8 ml H_2O
86.2 ml concentrated HCl (Table A.2A.1)

KCl, 1 M

74.6 g KCl
H_2O to 1 liter

$MgCl_2$, 1 M

20.3 g $MgCl_2 \cdot 6H_2O$
H_2O to 100 ml

$MgCl_2$ is extremely hygroscopic. Do not store opened bottles for long periods of time.

$MgSO_4$, 1 M

24.6 g $MgSO_4 \cdot 7H_2O$
H_2O to 100 ml

NaCl, 5 M

292 g NaCl
H_2O to 1 liter

NaOH, 10 M

Dissolve 400 g NaOH in 450 ml H_2O
Add H_2O to 1 liter

Potassium acetate buffer, 0.1 M

Solution A: 11.55 ml glacial acetic acid per liter (0.2 M) in water.

Solution B: 19.6 g potassium acetate ($KC_2H_3O_2$) per liter (0.2 M) in water.

Referring to Table A.2A.4 for desired pH, mix the indicated volumes of solutions A and B, then dilute with water to 100 ml. Filter sterilize if necessary. Store up to 3 months at room temperature.

This may be made as a 5- or 10-fold concentrate by scaling up the amount of sodium acetate in the same volume. Acetate buffers show concentration-dependent pH changes, so check the pH by diluting an aliquot of concentrate to the final concentration.

To prepare buffers with pH intermediate between the points listed in Table A.2A.4, prepare closest higher pH, then titrate with solution A.

Table A.2A.3 Preparation of Citrate-Phosphate Buffers

Desired pH	Solution A (ml)	Solution B (ml)
2.6	44.6	5.4
2.8	42.2	7.8
3.0	39.8	10.2
3.2	37.7	12.3
3.4	35.9	14.1
3.6	33.9	16.1
3.8	32.3	17.7
4.0	30.7	19.3
4.2	29.4	20.6
4.4	27.8	22.2
4.6	26.7	23.3
4.8	25.2	24.8
5.0	24.3	25.7
5.2	23.3	26.7
5.4	22.2	27.8
5.6	21.0	29.0
5.8	19.7	30.3
6.0	17.9	32.1
6.2	16.9	33.1
6.4	15.4	34.6
6.6	13.6	36.4
6.8	9.1	40.9
7.0	6.5	43.6

[a]Adapted with permission from Fasman (1989).

Table A.2A.4 Preparation of 0.1 M Sodium and Potassium Acetate Buffers[a]

Desired pH	Solution A (ml)	Solution B (ml)
3.6	46.3	3.7
3.8	44.0	6.0
4.0	41.0	9.0
4.2	36.8	13.2
4.4	30.5	19.5
4.6	25.5	24.5
4.8	20.0	30.0
5.0	14.8	35.2
5.2	10.5	39.5
5.4	8.8	41.2
5.6	4.8	45.2

[a]Adapted by permission from CRC (1975).

Table A.2A.5 Preparation of 0.1 M Sodium and Potassium Phosphate Buffers[a]

Desired pH	Solution A (ml)	Solution B (ml)	Desired pH	Solution A (ml)	Solution B (ml)
5.7	93.5	6.5	6.9	45.0	55.0
5.8	92.0	8.0	7.0	39.0	61.0
5.9	90.0	10.0	7.1	33.0	67.0
6.0	87.7	12.3	7.2	28.0	72.0
6.1	85.0	15.0	7.3	23.0	77.0
6.2	81.5	18.5	7.4	19.0	81.0
6.3	77.5	22.5	7.5	16.0	84.0
6.4	73.5	26.5	7.6	13.0	87.0
6.5	68.5	31.5	7.7	10.5	90.5
6.6	62.5	37.5	7.8	8.5	91.5
6.7	56.5	43.5	7.9	7.0	93.0
6.8	51.0	49.0	8.0	5.3	94.7

[a]Adapted by permission from CRC (1975).

Potassium phosphate buffer, 0.1 M

Solution A: 27.2 g KH_2PO_4 per liter (0.2 M final) in water.

Solution B: 34.8 g K_2HPO_4 per liter (0.2 M final) in water.

Referring to Table A.2A.5 for desired pH, mix the indicated volumes of solutions A and B, then dilute with water to 200 ml. Filter sterilize if necessary. Store up to 3 months at room temperature.

This buffer may be made as a 5- or 10-fold concentrate simply by scaling up the amount of potassium phosphate in the same final volume. Phosphate buffers show concentration-dependent changes in pH, so check the pH of the concentrate by diluting an aliquot to the final concentration.

To prepare buffers with pH intermediate between the points listed in Table A.2A.5, prepare closest higher pH, then titrate with solution A.

Laboratory Stock Solutions, Equipment, and Guidelines

A.2A.5

SDS, 20% (w/v)

Dissolve 20 g SDS (sodium dodecyl sulfate or sodium lauryl sulfate) in water to 100 ml total volume with stirring. Filter sterilize using a 0.45-μm filter.

It may be necessary to heat the solution slightly to fully dissolve the powder.

Sodium acetate, 3 M

Dissolve 408 g sodium acetate trihydrate ($NaC_2H_3O_2\cdot3H_2O$) in 800 ml H_2O
Adjust pH to 4.8, 5.0, or 5.2 (as desired) with 3 M acetic acid (see Table A.2A.1)
Add H_2O to 1 liter
Filter sterilize

Sodium acetate buffer, 0.1 M

Solution A: 11.55 ml glacial acetic acid per liter (0.2 M) in water.

Solution B: 27.2 g sodium acetate ($NaC_2H_3O_2\cdot3H_2O$) per liter (0.2 M) in water.

Referring to Table A.2A.4 for desired pH, mix the indicated volumes of solutions A and B, then dilute with water to 100 ml. Filter sterilize if necessary. Store up to 3 months at room temperature.

This may be made as a 5- or 10-fold concentrate by scaling up the amount of sodium acetate in the same volume. Acetate buffers show concentration-dependent pH changes, so check the pH by diluting an aliquot of concentrate to the final concentration.

To prepare buffers with pH intermediate between the points listed in Table A.2A.4, prepare closest higher pH, then titrate with solution A.

Sodium phosphate buffer, 0.1 M

Solution A: 27.6 g $NaH_2PO_4\cdot H_2O$ per liter (0.2 M final) in water.

Solution B: 53.65 g $Na_2HPO_4\cdot7H_2O$ per liter (0.2 M) in water.

Referring to Table A.2A.5 for desired pH, mix the indicated volumes of solutions A and B, then dilute with water to 200 ml. Filter sterilize if necessary. Store up to 3 months at room temperature.

This buffer may be made as a 5- or 10-fold concentrate by scaling up the amount of sodium phosphate in the same final volume. Phosphate buffers show concentration-dependent changes in pH, so check the pH by diluting an aliquot of the concentrate to the final concentration.

To prepare buffers with pH intermediate between the points listed in Table A.2A.5, prepare closest higher pH, then titrate with solution A.

Tris·Cl, 1 M

Dissolve 121 g Tris base in 800 ml H_2O
Adjust to desired pH with concentrated HCl
Adjust volume to 1 liter with H_2O
Filter sterilize if necessary
Store up to 6 months at 4°C or room temperature

Approximately 70 ml HCl is needed to achieve a pH 7.4 solution, and ~42 ml for a solution that is pH 8.0.

IMPORTANT NOTE: *The pH of Tris buffers changes significantly with temperature, decreasing approximately 0.028 pH units per 1°C. Tris-buffered solutions should be adjusted to the desired pH at the temperature at which they will be used. Because the pK_a of Tris is 8.08, Tris should not be used as a buffer below pH ~7.2 or above pH ~9.0.*

Always use high-quality Tris (lower-quality Tris can be recognized by its yellow appearance when dissolved).

Chemical Rubber Company, 1975. CRC Handbook of Biochemistry and Molecular Biology, Physical and Chemical Data, 3d ed., Vol. 1. CRC Press, Boca Raton, Fla.

Fasman, G.D. (ed.) 1989. Practical Handbook of Biochemistry and Molecular Biology. CRC Press, Boca Raton, Fla.

Mohan, C. (ed.), 1997. Buffers: A Guide for the Preparation and Use of Buffers in Biological Systems, Calbiochem, San Diego, Calif.

Laboratory Stock Solutions, Equipment, and Guidelines

A.2A.7

Laboratory Safety

Persons carrying out the protocols in the laboratory may encounter various hazardous or potentially hazardous materials including: radioactive substances; toxic chemicals and carcinogenic, mutagenic, or teratogenic reagents; and pathogenic and infectious biological agents. Most governments regulate the use of these materials; it is essential that they be used in strict accordance with local and national regulations. Cautionary notes are included in many instances throughout the manual, and some specific guidelines for working safely with chemicals are provided below (and references therein). However, we emphasize that users must proceed with the prudence and precautions associated with good laboratory practice, under the supervision of personnel responsible for implementing laboratory safety programs at their institutions and in compliance with designated guidelines of federal, state, and local officials.

HAZARDOUS CHEMICALS

It is not possible in the space available to list all the precautions to be taken when handling hazardous chemicals. Many texts have been written about laboratory safety; see Literature Cited for a selected list of examples. Obviously, all national and local laws should be obeyed as well as all institutional regulations. Controlled substances are regulated by the Drug Enforcement Administration. By law, Material Safety Data Sheets must be readily available. All laboratories should have a Chemical Hygiene Plan [29CFR Part 1910.1450] and institutional safety officers should be consulted as to its implementation. Help is (or should be) available from your institutional Safety Office. Use it.

Chemicals should be stored properly. For example, flammable chemicals (e.g., ethanol, methanol, acetone, methyl ethyl ketone, petroleum distillates, toluene, benzene, and other materials labeled flammable) should be stored in approved flammable storage cabinets, and flammable chemicals requiring refrigeration should be stored in explosion-proof refrigerators. Oxidizers should be segregated from other chemicals, and corrosive acids (e.g., sulfuric, hydrochloric, nitric, perchloric, and hydrofluoric acids) should also be stored in a separate cabinet, well-removed from the flammable organics.

Facilities should be appropriate for the handling of hazardous chemicals. In particular, hazardous chemicals should only be handled in chemical fume hoods, not in laminar flow cabinets. The functioning of these fume hoods should be periodically checked. Laboratories should also be equipped with safety showers and eye-washing facilities. Again, this equipment should be tested periodically to make sure that it functions correctly. Other safety equipment may be required depending on the nature of the materials being handled. In addition, researchers should be trained in the proper procedures for handling hazardous chemicals as well as other areas of laboratory operations, e.g., handling of compressed gases, use of cryogenic liquids, operation of high voltage power supplies, etc.

Before starting work, have a plan for dealing with spills or accidents; coming up with a good plan on the spur of the moment is difficult. For example, have the appropriate decontaminating or neutralizing agents prepared and close at hand. Small spills can probably be cleaned up by the researcher. In the case of larger spills, the area should be evacuated and help sought from those experienced and equipped for dealing with spills, e.g., your institutional safety department.

Protective equipment should include, at a minimum, eye protection, a lab coat, and gloves. Sandals, open-toed shoes, and shorts should not be worn. In certain circumstances other items of protective equipment may be necessary, e.g., a face shield. Different types of gloves exhibit different chemical resistance properties; listings of these properties are available (Forsberg and Keith, 1989). Gloves should, however, be regarded as the last line of defense and should be changed if they become contaminated, because many types of chemicals pass relatively freely through rubber. If possible, handling procedures should be designed so that gloves do not become contaminated. All common-sense precautions should be observed, e.g., do not pipet by mouth, keep unauthorized persons away from hazardous chemicals, prohibit eating and drinking in the lab, etc.

Order hazardous chemicals only in quantities that are likely to be used in a reasonable time. Buying large quantities at a lower unit cost is no bargain if someone (perhaps you) has to pay to dispose of surplus quantities. Substitute alcohol-filled thermometers for mercury-

---

Contributed by George Lunn

Laboratory Stock Solutions, Equipment, and Guidelines

A.2B.1

filled thermometers. The latter are a hazardous chemical spill waiting to happen.

Although any number of chemicals commonly used in laboratories are toxic if used improperly, the toxic properties of a number of reagents require special attention. Many chemicals are considered carcinogenic, corrosive, flammable, lachrymatory, mutagenic, oxidizing, teratogenic, or toxic. Chemicals labeled carcinogenic range from those accepted by expert review groups as causing cancer in humans to those for which only minimal evidence of carcinogenicity exists. Oxidizers may react violently with oxidizable material, e.g., hydrocarbons, wood, and cellulose. Before using any chemical, thoroughly investigate all of its characteristics. Material Safety Data Sheets are readily available; they list some hazards but vary widely in quality. A number of texts describing hazardous properties are listed in Further Reading. In particular, Sax's Dangerous Properties of Industrial Materials, 8th ed. (Lewis, 1992) and Bretherick's Handbook of Reactive Chemical Hazards, 4th ed. (Bretherick, 1990) give comprehensive listings of known hazardous properties. However, these texts list only the known properties. Many chemicals have been tested only partially or not at all. Prudence dictates, therefore, that unless there is good reason for believing otherwise, all chemicals should be regarded as volatile, highly toxic, flammable human carcinogens and should be handled with care.

Waste should always be disposed of in accordance with all applicable regulations. Waste should be segregated according to institutional requirements, for example, into solid, aqueous, nonchlorinated organic, and chlorinated organic material. A collection (Lunn and Sansone, 1994) of techniques for the disposal of chemicals in laboratories has been published recently. Incorporation of these procedures into laboratory protocols can help to minimize waste disposal problems.

LITERATURE CITED

Bretherick, L. 1990. Bretherick's Handbook of Reactive Chemical Hazards, 4th ed. Butterworths, London.

Forsberg, K. and Keith, L.H. 1989. Chemical Protective Clothing Performance Index Book. John Wiley & Sons, New York.

Lewis, R.J., Sr. 1992. Sax's Dangerous Properties of Industrial Materials, 8th ed. Van Nostrand-Reinhold, New York.

Lunn, G. and Sansone, E.B. 1994. Destruction of Hazardous Chemicals in the Laboratory, 2nd ed. John Wiley & Sons, New York.

KEY REFERENCES

General safety

Freeman, N.T. and Whitehead, J. 1982. Introduction to Safety in the Chemical Laboratory. Academic Press, New York.

Furr, A.K. (ed.) 1990. CRC Handbook of Laboratory Safety, 3rd ed. CRC Press, Boca Raton, Fla.

Fuscaldo, A.A., Erlick, B.J., and Hindman, B. (eds.) 1980. Laboratory Safety, Theory and Practice. Academic Press, New York.

Miller, B.M. (ed.) 1986. Laboratory Safety, Principles and Practices. American Society for Microbiology, Washington, D.C.

Occupational Health and Safety. 1993. National Safety Council, Chicago.

Pal, S.B. (ed.) 1985. Handbook of Laboratory Health and Safety Measures. Kluwer Academic Publishers, Hingham, Mass.

Young, J.A. (ed.) 1987. Improving Safety in the Chemical Laboratory: A Practical Guide. John Wiley & Sons, New York.

Laboratory safety for hazardous chemicals

American Chemical Society, Committee on Chemical Safety. 1990. Safety in Academic Chemistry Laboratories, 5th ed. American Chemical Society, Washington, D.C.

Forsberg and Keith, 1989. See above.

National Research Council, Committee on Hazardous Substances in the Laboratory. 1981. Prudent Practices for Handling Hazardous Chemicals in Laboratories. National Academy Press, Washington, D.C.

Properties and disposal procedures for hazardous chemicals

Aldrich Chemical Co. 2001. Aldrich Catalog Handbook of Fine Chemicals. Aldrich Chemical Co., Milwaukee, Wis.

Bretherick, L. (ed.) 1986. Hazards in the Chemical Laboratory, 4th ed. Royal Society of Chemistry, London.

Bretherick, 1990. See above.

Budavari, S. (ed.) 1996. The Merck Index, 12th ed. Merck & Co., Rahway, N.J.

Lewis, 1992. See above.

Lunn, and Sansone, 1994. See above.

Contributed by George Lunn
Baltimore, Maryland

Standard Laboratory Equipment

Special equipment is itemized in the materials list of each protocol. Listed below are standard pieces of equipment in the modern food science laboratory—i.e., items used extensively in this manual and thus not usually included in the individual materials lists. See *SUPPLIERS APPENDIX* for contact information for commercial vendors of laboratory equipment.

Applicators, cotton-tipped and wooden

Autoclave

Balances, analytical and preparative

Beakers

Biohazard disposal containers and bags

Blender (e.g., Waring Blendor)

Bottles, glass and plastic

Bunsen burners

Centrifuges, low-speed (6,000 rpm) and high-speed (20,000 rpm) refrigerated centrifuges, ultracentrifuge (20,000 to 80,000 rpm), and microcentrifuge that holds standard 0.5- and 1.5-ml microcentrifuge tubes

NOTE: *Centrifuge speeds are provided as g or as rpm (with example rotor models) throughout the manual.*

Cold room (4°C) or cold box

Computer (PC or Macintosh) and printer

Conical centrifuge tubes, 15- and 25-ml plastic

Cuvettes, plastic disposable, glass, and quartz

Darkroom and developing tank, or X-Omat automatic X-ray film developer (Kodak)

Desiccators (including vacuum desiccators) and desiccant

Dry ice

Filtration apparatus, for collecting acid precipitates on nitrocellulose filters or membranes

Flasks, glass (e.g., Erlenmeyer, beveled shaker)

Forceps

Freezers, −20° and −80°C

Gel electrophoresis equipment, horizontal full-size and minigel apparatus, vertical full-size and minigel apparatus for polyacrylamide protein gels, and specialized equipment for two-dimensional protein gels

Grinder (e.g., coffee grinder)

Heat-sealable plastic bags and apparatus

Heating blocks, thermostat-controlled metal heating block that holds test tubes and/or microcentrifuge tubes

Hoods, chemical and microbiological

Hot plates, with or without magnetic stirrer

Gloves, plastic and latex, disposable and asbestos

Graduated cylinders

Ice buckets

Ice maker

Immersion oil for microscopy

Kimwipes, or equivalent lint-free tissues

Lab coats

Laboratory glass ware

Light box, for viewing gels and autoradiograms

Liquid nitrogen and Dewar flask

Magnetic stirrers (with heater is useful)

Markers, including indelible markers and china-marking pencils

Microcentrifuge, Eppendorf-type, maximum speed 12,000 to 14,000 rpm

Microcentrifuge tubes, 1.5-ml and 0.5-ml

Microscope, standard optical model (optionally with epifluorescence or phase-contrast illumination)

Microscope slides and coverslips

Microwave oven, to melt agar and agarose

Mortar and pestle

Muffle furnace

Ovens, drying, vacuum, and microwave

Paper cutter, large size, for 46 × 57-cm Whatman paper sheets

Paper towels

Parafilm

Pasteur pipets and bulbs

pH meter and pH standard solutions

pH paper

Pipet bulbs, or battery-operated pipetting devices—e.g., Pipet-Aid (Drummond Scientific)

Pipets, Pasteur and graduated, glass and plastic, serological (1- to 25-ml)

Pipettors, adjustable delivery, volume ranges 0.5 to 10 µl, 10 to 200 µl, and 200 to 1000 µl

Plastic wrap, UV transparent (e.g., Saran Wrap)

Polaroid camera

Power supplies, 300-V for polyacrylamide gels; 2000- to 3000-V for some applications

Racks, for test tubes and microcentrifuge tubes

Radiation shield, Lucite or Plexiglas

Radioactive waste containers, for liquid and solid waste

Razor blades

Refrigerator, 4°C
Ring stands and rings
Rotator, end-over-end
Rubber bands
Rubber policemen
Rubber stoppers
Safety glasses
Scalpels and blades
Scintillation counter
Scissors
Shakers, orbital and platform
Spectrophotometer, UV and visible
Speedvac evaporator (Savant)
Stir-bars, assorted sizes

Tape, masking and electrician's
Thermometers
Timer
UV transilluminator
Vacuum aspirator
Vacuum line
Volumetric flasks
Vortex mixers
Wash bottles, plastic and glass
Water baths, variable temperature up to 80°C
Water purification equipment, e.g., Milli-Q
  system (Millipore) or equivalent
X-ray film cassettes and intensifying screens

COMMONLY USED TECHNIQUES

APPENDIX 3

Introduction to Mass Spectrometry for Food Chemistry

APPENDIX 3A

Almost a century ago, the first mass spectrometers were used to prove the existence of isotopes of the elements. During the first half of the 20th century, physicists and physical chemists used mass spectrometers to help characterize new elements and the fission products of radioactive elements as they were created or discovered. Other applications included the analysis of isotopic enrichment of elements and their inorganic derivatives. As this era of mass spectrometry reached maturity, by the 1940s, the analysis of organic molecules emerged as a new application of mass spectrometry. Beginning in 1945, organic mass spectrometers using electron impact (EI) ionization became commercially available and were used primarily by the petroleum industry. Toward the late 1950s, organic mass spectrometers began to be used for the analysis of a wider variety of organic molecules, and gradually became a fundamental analytical tool for the characterization of synthetic organic compounds.

During the 1960s, high-resolution, double-focusing magnetic sector instruments became available from multiple manufacturers and were widely used in organic chemistry for exact mass measurements and elemental composition analysis. EI was used for generating structurally significant fragment ions for compound identification, and rules for structure elucidation using mass spectrometry were developed (for a thorough review of EI and ion fragmentation pathways, see McLafferty and Turecek, 1993). Biomedical and food chemistry applications of mass spectrometry were developed during this time. Chemical ionization (CI), which was developed by researchers in the petroleum industry (Field, 1990), was quickly adopted as a softer ionization alternative to EI, useful in reducing fragmentation so that molecular weights could be confirmed more easily. CI became another standard ionization technique for mass spectrometry (see Figure A.3A.1 for a guide to the selection of ionization techniques in mass spectrometry).

GAS CHROMATOGRAPHY/MASS SPECTROMETRY (GC/MS)

With the introduction of computerized data systems for data acquisition, reduction, and storage during the 1960s, the efficiency of mass spectrometric analysis grew rapidly and continues to grow to this day. The use of computers for data reduction and analysis helped gas chromatography/mass spectrometry (GC/MS) become a practical and powerful tool for qualita-

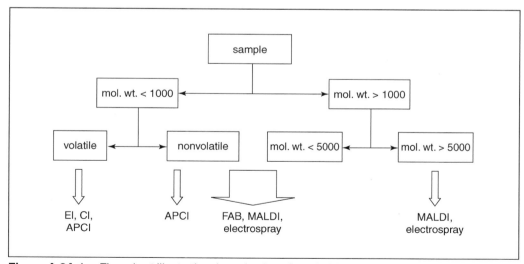

Figure A.3A.1 Flow chart illustrating the selection of a suitable ionization technique for the mass spectrometric analysis of a sample. Abbreviations: APCI, atmospheric pressure chemical ionization; CI, chemical ionization; EI, electron impact; FAB, fast atom bombardment; MALDI, matrix-assisted laser desorption/ionization.

Contributed by Richard B. van Breemen

Commonly Used Techniques

A.3A.1

tive and quantitative analysis of compounds in mixtures. Both EI and CI were immediately useful for GC/MS, since both of these ionization methods require that the analytes be in the gas phase. When capillary GC was incorporated into GC/MS, this technique reached maturity. The advantages of GC/MS include speed, selectivity, and sensitivity. Typically, GC/MS may be used to select, identify, and quantify organic compounds in complex mixtures at the femtomole level. Compounds are selected using a combination of chromatographic separation and mass selection, and when using tandem mass spectrometry (MS/MS; see discussion below), the fragmentation pathway may be used for additional selectivity. The speed of GC/MS is determined by the chromatography step, which typically requires from several minutes to one hour per analysis. Although GC/MS remains important for the analysis of many organic compounds, this technique is limited to volatile and thermally stable compounds (see chromatography/MS selection flow chart in Fig. A.3A.2). Therefore, thermally unstable compounds—including food pigments such as carotenoids and chlorophylls and biomolecules such as proteins, carbohydrates, and nucleic acids—cannot be analyzed in their native forms using GC/MS (for more details regarding GC/MS and its applications, see Watson, 1997).

DESORPTION IONIZATION MASS SPECTROMETRY

During the 1970s and early 1980s, desorption ionization techniques such as field desorption (FD), desorption EI, desorption CI (DCI), and laser desorption were developed to extend the utility of mass spectrometry towards the analysis of more polar and less volatile compounds (see Watson, 1997, for more information regarding desorption ionization techniques including DCI and FD). Although these techniques helped extend the mass range of mass spectrometry beyond a traditional limit of m/z 1000 and toward ions of m/z 5000 (Fig. A.3A.1), the first breakthrough in the analysis of polar, nonvolatile compounds occurred in 1982 with the invention of fast atom bombardment (FAB; Barber et al., 1982). FAB and its counterpart, liquid secondary ion mass spectrometry (LSIMS), facilitate the formation of abundant molecular ions, protonated molecules, and deprotonated molecules of nonvolatile and thermally labile compounds such as peptides, chlorophylls, and complex lipids up to approximately m/z 12,000. FAB and LSIMS use energetic particle bombardment (fast atoms or ions from 3,000 to 20,000 V of energy) to ionize compounds dissolved in nonvolatile matrices such as glycerol or 3-nitrobenzyl alcohol and desorb them from this condensed phase into the gas phase for mass spectrometric analysis. Molecular ions and/or protonated molecules are usually abundant and fragmentation is minimal.

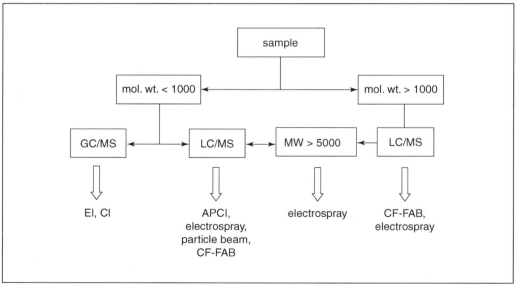

Figure A.3A.2 Selection of chromatography-mass spectrometry system for the analysis of a sample. Abbreviations: APCI, atmospheric pressure chemical ionization; CF, continuous flow; CI, chemical ionization; EI, electron impact; FAB, fast atom bombardment; GC/MS, gas chromatography/mass spectrometry; LC/MS, liquid chromatography/mass spectrometry.

Introduced in the late 1980s, matrix-assisted laser desorption/ionization (MALDI) has helped solve the mass-limit barriers of laser desorption mass spectrometry so that singly charged ions may be obtained up to m/z 500,000 and sometimes higher (Hillenkamp et al., 1991). For most commercially available MALDI mass spectrometers, ions up to m/z 200,000 are readily obtained. Like FAB and LSIMS, MALDI samples are mixed with a matrix to form a solution that is loaded onto the sample stage for analysis. Unlike the other matrix-mediated techniques, the solvent is evaporated prior to MALDI analysis, leaving sample molecules trapped in crystals of solid phase matrix. The MALDI matrix is selected to absorb the pulse of laser light directed at the sample. Most MALDI mass spectrometers are equipped with a pulsed UV laser, although IR lasers are available as an option on some commercial instruments. Therefore, matrices are often substituted benzenes or benzoic acids with strong UV absorption properties. During MALDI, the energy of the short but intense UV laser pulse obliterates the matrix and in the process desorbs and ionizes the sample. Like FAB and LSIMS, MALDI typically produces abundant protonated or deprotonated molecules with little fragmentation.

LIQUID CHROMATOGRAPHY/MASS SPECTROMETRY (LC/MS)

By the time that GC/MS had become a standard technique in the late 1960s, LC/MS was still in the developmental stages. Producing gas-phase sample ions for analysis in a vacuum system while removing the HPLC mobile phase proved to be a challenging task. Early LC/MS techniques included a moving belt interface to desolvate and transport the HPLC eluate into a CI or EI ion source, or a direct inlet system in which the eluate was pumped at a low flow rate of 1 to 3 μl/min into a CI source. However, neither of these systems was robust enough or suitable for a broad enough range of samples to gain widespread acceptance.

Since FAB (or LSIMS) requires that the analyte be dissolved in a liquid matrix, this ionization technique was easily adapted for infusion of solution-phase samples into the FAB ionization source, in an approach known as continuous-flow FAB. Continuous-flow FAB was connected to microbore HPLC columns for LC/MS applications (Ito et al., 1985). Since this method is limited to microbore HPLC applications at flow rates of <10 μl/min

and requires considerable operator intervention, it is not ideal for the analysis of large sample sets. Instead, more robust techniques have been developed to fulfill this requirement. However, continuous-flow FAB is still in use in some laboratories.

Like continuous-flow FAB, the popularity of particle beam interfaces is diminishing, but systems are still available from commercial sources. During particle beam LC/MS, the HPLC eluate is sprayed into a heated chamber connected to a vacuum pump. As the droplets evaporate, aggregates of analyte (particles) form and pass through a momentum separator that removes the lower-molecular-weight solvent molecules. Finally, the particle beam enters the mass spectrometer ion source where the aggregates strike a heated plate from which the analyte molecules evaporate and are ionized using conventional EI or CI ionization. Particle beam LC/MS is limited to the analysis of volatile and thermally stable compounds that are amenable to flash evaporation and EI or CI mass spectrometry. Therefore, this approach is not used for polar compounds in food chemistry such as carbohydrates, sugars, peptides, proteins, or nucleic acids (Fig. A.3A.2).

Since thermospray became the first widely utilized LC/MS technique (during the late 1970s and early 1980s), this technique should be mentioned here. Thermospray facilitates the interfacing of standard analytical HPLC systems at flow rates up to 1 ml/min with mass spectrometers. Although the interface between the HPLC and mass spectrometer is inefficient and exhibits low sensitivity for most analytes, thermospray has been useful for the LC/MS analysis of many types of small molecules. During thermospray, the HPLC eluate is sprayed through a heated capillary into a heated desolvation chamber at reduced pressure. Gas phase ions remaining after desolvation of the droplets are extracted through a skimmer into the mass spectrometer for analysis. The sensitivity of thermospray is poor since there is no mechanism or driving force to enhance the number of sample ions entering the gas phase from the spray during desolvation. Also, thermally labile compounds tend to decompose in the heated source. These problems were solved when thermospray was replaced by electrospray during the late 1980s.

During the 1990s, electrospray ionization (ESI) and atmospheric pressure chemical ionization (APCI) became the standard interfaces for LC/MS. Unlike thermospray, particle beam, or continuous-flow FAB, ESI and APCI interfaces operate at atmospheric pressure and do not depend upon vacuum pumps to remove solvent vapor. As a result, they are compatible with a wide range of HPLC flow rates. Also, no matrix is required. Both APCI and ESI are compatible with a wide range of HPLC columns and solvent systems. Like all LC/MS

systems, the solvent system should contain only volatile solvents, buffers, or ion-pair agents, to reduce fouling of the mass spectrometer ion source. In general, APCI and ESI form abundant molecular ion species (Figures A.3A.1 and A.3A.2). When fragment ions are

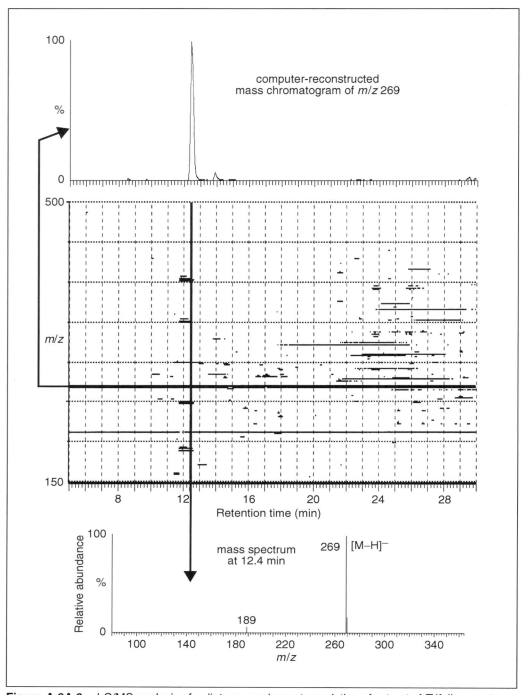

Figure A.3A.3 LC/MS analysis of a dietary supplement consisting of extract of *Trifolium pratense* (red clover). Reversed-phase C18 HPLC and negative ion electrospray ionization mass spectrometry were used with a quadrupole mass spectrometer analyzer (Agilent; also see Table A.3A.1). The map illustrates the abundance of information provided by this hyphenated technique with HPLC mass chromatograms in one dimension and mass spectra in another dimension.

formed, they are usually more abundant in APCI than ESI mass spectra.

The APCI interface uses a heated nebulizer to form a fine spray of the HPLC eluate, which is much finer than the particle beam system but similar to that formed during thermospray. A cross-flow of heated nitrogen gas is used to facilitate the evaporation of solvent from the droplets. The resulting gas-phase sample molecules are ionized by collisions with solvent ions, which are formed by a corona discharge in the atmospheric pressure chamber. Molecular ions, $M^{+\cdot}$ or $M^{-\cdot}$, and/or protonated or deprotonated molecules can be formed. The relative abundance of each type of ion depends upon the sample itself, the HPLC solvent, and the ion source parameters. Next, ions are drawn into the mass spectrometer analyzer for measurement through a narrow opening or skimmer, which helps the vacuum pumps to maintain very low pressure inside the analyzer while the APCI source remains at atmospheric pressure.

During ESI, the HPLC eluate is sprayed through a capillary electrode at high potential (usually 2000 to 7000 V) to form a fine mist of charged droplets at atmospheric pressure. As the charged droplets migrate towards the opening of the mass spectrometer due to electrostatic attraction, they encounter a cross-flow of heated nitrogen that increases solvent evaporation and prevents most of the solvent molecules from entering the mass spectrometer. Molecular ions, protonated or deprotonated molecules, and cationized species such as $[M+Na]^+$ and $[M+K]^+$ can be formed (for additional information on ESI, see Cole, 1997). In addition to singly charged ions, ESI is unique as an ionization technique in that multiply charged species are common and often constitute the majority of the sample ion abundance. The relative abundance of each of these species depends upon the chemistry of the analyte, the pH, the presence of proton-donating or -accepting species, and the levels of trace amounts of sodium or potassium salts in the mobile phase. In contrast, APCI, MALDI, EI, CI, and FAB/LSIMS usually produce singly charged species. A consequence of forming multiply charged ions is that they are detected at lower m/z values (i.e., $|z| > 1$) than the corresponding singly charged species. This has the benefit of allowing mass spectrometers with modest m/z ranges to detect and measure ions of molecules with very high masses. For example, ESI has been used to measure ions with molecular weights of hundreds of thousands or even millions of daltons on mass spectrometers with m/z ranges of only a few thousand (for a review of LC/MS techniques, see Niessen, 1999).

An example of the LC/MS analysis of a plant extract is shown in Figure A.3A.3. In this case, negative ion ESI-MS was used in combination with C18 reversed-phase HPLC separation. Extracts of the botanical *Trifolium pratense* (red clover) are used as dietary supplements by menopausal and post-menopausal women (Liu et al., 2001). The two-dimensional map illustrates the amount of information that may be

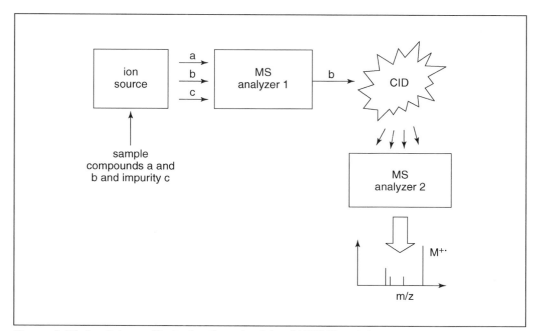

Figure A.3A.4 Scheme illustrating the selectivity of MS/MS and the process by which collision-induced dissociation (CID) facilitates fragmentation of preselected ions.

acquired using hyphenated techniques such as LC/MS. In the time dimension, chromatograms are obtained and a sample computer-reconstructed mass chromatogram is shown for the signal at m/z 269. One intense chromatographic peak was detected in this chromatogram eluting at 12.4 min. In the m/z dimension, the negative ion electrospray mass spectrum recorded at 12.4 min shows a base peak at m/z 269. Based on comparison to authentic standards (data not shown), the ion of m/z 269 was shown to correspond to the deprotonated molecule of genistein, which is an estrogenic isoflavone (Liu et al., 2001). Since almost no fragmentation of the genistein ion was observed, additional characterization would require collision-induced dissociation (CID) and tandem mass spectrometry as discussed in the next section.

TANDEM MASS SPECTROMETRY (MS/MS) AND HIGH RESOLUTION

Desorption ionization techniques like FAB and MALDI and LC/MS ionization techniques like ESI and APCI facilitate the molecular weight determination of a wide range of polar and nonpolar, low- and high-molecular-weight compounds. However, the "soft" ionization character of these techniques means that most of the ion current is concentrated in molecular ions and few structurally significant fragment ions are formed. In order to enhance the amount of structural information in these mass spectra, collision-induced dissociation (CID) may be used to produce abundant fragment ions from molecular ion precursors formed and isolated during the first stage of mass spectrometry. Then, a second mass spectrometry analysis may be used to characterize the resulting product ions. This process is called tandem mass spectrometry or MS/MS and is illustrated in Figure A.3A.4.

Another advantage of the use of tandem mass spectrometry is the ability to isolate a particular ion such as the molecular ion of the analyte of interest during the first mass spectrometry stage. This precursor ion is essentially purified in the gas phase and is free of impurities such as solvent ions, matrix ions, or other analytes. Finally, the selected ion is fragmented using CID and analyzed using a second mass spectrometry stage. In this manner, the resulting tandem mass spectrum contains exclusively analyte ions without impurities that might interfere with the interpretation of the fragmentation patterns. In summary, CID may be used with LC/MS/MS or desorption ionization and MS/MS to obtain structural information such as amino acid sequences of peptides and sites of alkylation of nucleic acids, or to distinguish structural isomers such as β-carotene and lycopene.

The most common types of MS/MS instruments available to researchers in food chemistry include triple quadrupole mass spectrometers and ion traps. Less common but commercially produced tandem mass spectrometers include magnetic sector instruments, Fourier transform ion cyclotron resonance (FTICR) mass spectrometers, and quadrupole time-of-flight (QTOF) hybrid instruments (Table A.3A.1). Beginning in 2001, TOF-TOF tandem mass spectrometers became available from instrument manufacturers. These instruments have the potential to deliver high-resolution tandem mass spectra with high speed and should be compatible with the chip-based chromatography systems now under development.

In addition to MS/MS with CID to obtain structural information, it is also useful to use high-resolution exact mass measurements to confirm the elemental compositions of ions. Essentially, exact mass measurements permit the unambiguous composition analysis of low-molecular-weight compounds (mol. wt. <500) through precise and accurate m/z measurements. The types of mass spectrometers capable of exact mass measurements include magnetic sector mass spectrometers, QTOF hybrid

Table A.3A.1 Types of Mass Spectrometers and Tandem Mass Spectrometers[a]

Instrument	Resolution	m/z Range	Tandem MS
Magnetic sector	100,000	12,000	Low resolution
Quadrupole	<4,000	4,000	None
Triple quadrupole	<4,000	4,000	Low resolution
TOF	15,000	>200,000	None
FTICR	>200,000	<10,000	High resolution
QTOF	12,000	4,000	High resolution
TOF-TOF	15,000	>10,000	High resolution

[a]FTICR, Fourier transform ion cyclotron resonance; QTOF, quadropole time-of-flight; TOF, time-of-flight.

mass spectrometers, reflectron TOF instruments, and FTICR mass spectrometers (Table A.3A.1). Some of these instruments permit the simultaneous use of tandem mass spectrometry and exact mass measurement of fragment ions. These include FTICR instruments, QTOF, and the TOF-TOF.

CONCLUSION

Mass spectrometry has become an essential analytical tool for a wide variety of biomedical applications such as food chemistry and food analysis. Mass spectrometry is highly sensitive, fast, and selective. By combining mass spectrometry with HPLC, GC, or an additional stage of mass spectrometry (MS/MS), the selectivity increases considerably. As a result, mass spectrometry may be used for quantitative as well as qualitative analyses. In this manual, mass spectrometry is mentioned frequently, and extensive discussions of mass spectrometry appear, for example, in units describing the analyses of carotenoids (*UNIT F2.4*) and chlorophylls (*UNIT F4.5*). In particular, these units include examples of LC/MS and MS/MS and the use of various ionization methods.

LITERATURE CITED

Barber, M., Bordoli, R.S., Elliott, G.J., Sedgwick R.D., and Tyler, A.N. 1982. Fast atom bombardment mass spectrometry. *Anal. Chem.* 54:645A-657A.

Cole, R.B. (ed.). 1997. Electrospray Ionization Mass Spectrometry. John Wiley & Sons, New York.

Field, F. 1990. Early days of chemical ionization. *J. Am. Soc. Mass Spectrom.* 1:277-283.

Hillenkamp, F., Karas, M., Beavis, R.C., and Chait, B.T. 1991. Matrix-assisted laser desorption/ionization mass spectrometry of biopolymers. *Anal. Chem.* 63:1193A-1203A.

Ito, Y., Takeuchi, T., Ishii, D., and Goto, M. 1985. Direct coupling of micro high-performance liquid chromatography with fast atom bombardment mass spectrometry. *J. Chromatogr.* 346:161-166.

Liu, J., Burdette, J.E., Xu, H., Gu, C., van Breemen, R.B., Bhat, K.P.L., Booth, N., Constantinou, A.I., Pezzuto, J.M., Fong, H.H.S., Farnsworth, N.R., and Bolton, J.L. 2001. Evaluation of estrogenic activity of plant extracts for the potential treatment of menopausal symptoms. *J. Agric. Food Chem.* 49:2472-2479.

McLafferty, F.W. and Turecek, F. 1993. Interpretation of Mass Spectra, 4th ed. University Science Books, Mill Valley, Calif.

Niessen, W.M. 1999. State-of-the-art in liquid chromatography-mass spectrometry. *J. Chromatogr. A* 856:179-189.

Watson, J.T. 1997. Introduction to Mass Spectrometry, 3rd ed. Lippincott-Raven, Philadelphia, Pa.

KEY REFERENCES

McLafferty and Turecek, 1993. See above.

This classic text describes fragmentation pathways and mechanisms for ions formed using electron impact (EI) ionization. In addition, this edition contains additional information regarding desorption ionization and the corresponding related fragmentation mechanisms.

Watson, 1997. See above.

This textbook provides an overview of biomedical mass spectrometry with particular emphasis on GC/MS and quantitative methods. In addition, descriptions are provided of the various types of mass spectrometers and ionization techniques that are used for biomedical applications.

Contributed by Richard B. van Breemen
University of Illinois at Chicago
Chicago, Illinois

SELECTED SUPPLIERS OF REAGENTS AND EQUIPMENT

Listed below are addresses and phone numbers of commercial suppliers who have been recommended for particular items used in our manuals because: (1) the particular brand has actually been found to be of superior quality, or (2) the item is difficult to find in the marketplace. Consequently, this compilation may not include some important vendors of biological supplies. For comprehensive listings, see *Linscott's Directory of Immunological and Biological Reagents* (Santa Rosa, CA), *The Biotechnology Directory* (Stockton Press, New York), the annual Buyers' Guide supplement to the journal *Bio/Technology*, as well as various sites on the Internet.

A.C. Daniels
72-80 Akeman Street
Tring, Hertfordshire, HP23 6AJ, UK
(44) 1442 826881
FAX: (44) 1442 826880

A.D. Instruments
5111 Nations Crossing Road #8
Suite 2
Charlotte, NC 28217
(704) 522-8415   FAX: (704) 527-5005
http://www.us.endress.com

A.J. Buck
11407 Cronhill Drive
Owings Mill, MD 21117
(800) 638-8673   FAX: (410) 581-1809
(410) 581-1800
http://www.ajbuck.com

A.M. Systems
131 Business Park Loop
P.O. Box 850
Carlsborg, WA 98324
(800) 426-1306   FAX: (360) 683-3525
(360) 683-8300
http://www.a-msystems.com

Aaron Medical Industries
7100 30th Avenue North
St. Petersburg, FL 33710
(727) 384-2323   FAX: (727) 347-9144
http://www.aaronmed.com

Abbott Laboratories
100 Abbott Park Road
Abbott Park, IL 60064
(800) 323-9100   FAX: (847) 938-7424
http://www.abbott.com

ABCO Dealers
55 Church Street Central Plaza
Lowell, MA 01852
(800) 462-3326   (978) 459-6101
http://www.lomedco.com/abco.htm

Aber Instruments
5 Science Park
Aberystwyth, Wales SY23 3AH, UK
(44) 1970 636300
FAX: (44) 1970 615455
http://www.aber-instruments.co.uk

ABI Biotechnologies
See Perkin-Elmer

ABI Biotechnology
See Apotex

Access Technologies
Subsidiary of Norfolk Medical
7350 N. Ridgeway
Skokie, IL 60076
(877) 674-7131   FAX: (847) 674-7066
(847) 674-7131
http://www.norfolkaccess.com

Accurate Chemical and Scientific
300 Shames Drive
Westbury, NY 11590
(800) 645-6264   FAX: (516) 997-4948
(516) 333-2221
http://www.accuratechemical.com

AccuScan Instruments
5090 Trabue Road
Columbus, OH 43228
(800) 822-1344   FAX: (614) 878-3560
(614) 878-6644
http://www.accuscan-usa.com

AccuStandard
125 Market Street
New Haven, CT 06513
(800) 442-5290   FAX: (877) 786-5287
http://www.accustandard.com

Ace Glass
1430 NW Boulevard
Vineland, NJ 08360
(800) 223-4524   FAX: (800) 543-6752
(609) 692-3333

ACO Pacific
2604 Read Avenue
Belmont, CA 94002
(650) 595-8588   FAX: (650) 591-2891
http://www.acopacific.com

Acros Organic
See Fisher Scientific

Action Scientific
P.O. Box 1369
Carolina Beach, NC 28428
(910) 458-0401   FAX: (910) 458-0407

AD Instruments
1949 Landings Drive
Mountain View, CA 94043
(888) 965-6040   FAX: (650) 965-9293
(650) 965-9292
http://www.adinstruments.com

Adaptive Biosystems
15 Ribocon Way
Progress Park
Luton, Bedsfordshire LU4 9UR, UK
(44)1 582-597676
FAX: (44)1 582-581495
http://www.adaptive.co.uk

Adobe Systems
1585 Charleston Road
P.O. Box 7900
Mountain View, CA 94039
(800) 833-6687   FAX: (415) 961-3769
(415) 961-4400
http://www.adobe.com

Advanced Bioscience Resources
1516 Oak Street, Suite 303
Alameda, CA 94501
(510) 865-5872   FAX: (510) 865-4090

Advanced Biotechnologies
9108 Guilford Road
Columbia, MD 21046
(800) 426-0764   FAX: (301) 497-9773
(301) 470-3220
http://www.abionline.com

Advanced ChemTech
5609 Fern Valley Road
Louisville, KY 40228
(502) 969-0000
http://www.peptide.com

Advanced Machining and Tooling
9850 Businesspark Avenue
San Diego, CA 92131
(858) 530-0751   FAX: (858) 530-0611
http://www.amtmfg.com

Advanced Magnetics
See PerSeptive Biosystems

Advanced Process Supply
See Naz-Dar-KC Chicago

Advanced Separation Technologies
37 Leslie Court
P.O. Box 297
Whippany, NJ 07981
(973) 428-9080   FAX: (973) 428-0152
http://www.astecusa.com

Advanced Targeting Systems
11175-A Flintkote Avenue
San Diego, CA 92121
(877) 889-2288   FAX: (858) 642-1989
(858) 642-1988
http://www.ATSbio.com

Advent Research Materials
Eynsham, Oxford OX29 4JA, UK
(44) 1865-884440
FAX: (44) 1865-84460
http://www.advent-rm.com

Advet
Industrivagen 24
S-972 54 Lulea, Sweden
(46) 0920-211887
FAX: (46) 0920-13773

Aesculap
1000 Gateway Boulevard
South San Francisco, CA 94080
(800) 282-9000
http://www.aesculap.com

Affinity Chromatography
307 Huntingdon Road
Girton, Cambridge CB3 OJX, UK
(44) 1223 277192
FAX: (44) 1223 277502
http://www.affinity-chrom.com

Affinity Sensors
See Labsystems Affinity Sensors

Affymetrix
3380 Central Expressway
Santa Clara, CA 95051
(408) 731-5000   FAX: (408) 481-0422
(800) 362-2447
http://www.affymetrix.com

Agar Scientific
66a Cambridge Road
Stansted CM24 8DA, UK
(44) 1279-813-519
FAX: (44) 1279-815-106
http://www.agarscientific.com

A/G Technology
101 Hampton Avenue
Needham, MA 02494
(800) AGT-2535   FAX: (781) 449-5786
(781) 449-5774
http://www.agtech.com

Agen Biomedical Limited
11 Durbell Street
P.O. Box 391
Acacia Ridge 4110
Brisbane, Australia
61-7-3370-6300   FAX: 61-7-3370-6370
http://www.agen.com

Suppliers

1

Agilent Technologies
395 Page Mill Road
P.O. Box 10395
Palo Alto, CA 94306
(650) 752-5000
http://www.agilent.com/chem

Agouron Pharmaceuticals
10350 N. Torrey Pines Road
La Jolla, CA 92037
(858) 622-3000   FAX: (858) 622-3298
http://www.agouron.com

Agracetus
8520 University Green
Middleton, WI 53562
(608) 836-7300   FAX: (608) 836-9710
http://www.monsanto.com

AIDS Research and Reference
Reagent Program
U.S. Department of Health and
   Human Services
625 Lofstrand Lane
Rockville, MD 20850
(301) 340-0245   FAX: (301) 340-9245
http://www.aidsreagent.org

AIN Plastics
249 East Sanford Boulevard
P.O. Box 151
Mt. Vernon, NY 10550
(914) 668-6800   FAX: (914) 668-8820
http://www.tincna.com

Air Products and Chemicals
7201 Hamilton Boulevard
Allentown, PA 18195
(800) 345-3148   FAX: (610) 481-4381
(610) 481-6799
http://www.airproducts.com

ALA Scientific Instruments
1100 Shames Drive
Westbury, NY 11590
(516) 997-5780   FAX: (516) 997-0528
http://www.alascience.com

Aladin Enterprises
1255 23rd Avenue
San Francisco, CA 94122
(415) 468-0433   FAX: (415) 468-5607

Aladdin Systems
165 Westridge Drive
Watsonville, CA 95076
(831) 761-6200   FAX: (831) 761-6206
http://www.aladdinsys.com

Alcide
8561 154th Avenue NE
Redmond, WA 98052
(800) 543-2133   FAX: (425) 861-0173
(425) 882-2555
http://www.alcide.com

Aldrich Chemical
P.O. Box 2060
Milwaukee, WI 53201
(800) 558-9160   FAX: (800) 962-9591
(414) 273-3850   FAX: (414) 273-4979
http://www.aldrich.sial.com

Alexis Biochemicals
6181 Cornerstone Court East, Suite 103
San Diego, CA 92121
(800) 900-0065   FAX: (858) 658-9224
(858) 658-0065
http://www.alexis-corp.com

Alfa Aesar
30 Bond Street
Ward Hill, MA 10835
(800) 343-0660   FAX: (800) 322-4757
(978) 521-6300   FAX: (978) 521-6350
http://www.alfa.com

Alfa Laval
Avenue de Ble 5 - Bazellaan 5
BE-1140 Brussels, Belgium
32(2) 728 3811
FAX: 32(2) 728 3917 or 32(2) 728 3985
http://www.alfalaval.com

Alice King Chatham Medical Arts
11915-17 Inglewood Avenue
Hawthorne, CA 90250
(310) 970-1834   FAX: (310) 970-0121
(310) 970-1063

Allegiance Healthcare
800-964-5227
http://www.allegiance.net

Allelix Biopharmaceuticals
6850 Gorway Drive
Mississauga, Ontario
L4V 1V7 Canada
(905) 677-0831   FAX: (905) 677-9595
http://www.allelix.com

Allentown Caging Equipment
Route 526, P.O. Box 698
Allentown, NJ 08501
(800) 762-CAGE   FAX: (609) 259-0449
(609) 259-7951
http://www.acecaging.com

Alltech Associates
Applied Science Labs
2051 Waukegan Road
P.O. Box 23
Deerfield, IL 60015
(800) 255-8324   FAX: (847) 948-1078
(847) 948-8600
http://www.alltechweb.com

Alomone Labs
HaMarpeh 5
P.O. Box 4287
Jerusalem 91042, Israel
972-2-587-2202   FAX: 972-2-587-1101
US: (800) 791-3904
FAX: (800) 791-3912
http://www.alomone.com

Alpha Innotech
14743 Catalina Street
San Leandro, CA 94577
(800) 795-5556   FAX: (510) 483-3227
(510) 483-9620
http://www.alphainnotech.com

Altec Plastics
116 B Street
Boston, MA 02127
(800) 477-8196   FAX: (617) 269-8484
(617) 269-1400

Alza
1900 Charleston Road
P.O. Box 7210
Mountain View, CA 94043
(800) 692-2990   FAX: (650) 564-7070
(650) 564-5000
http://www.alza.com

Alzet
c/o Durect Corporation
P.O. Box 530
10240 Bubb Road
Cupertino, CA 95015
(800) 692-2990   (408) 367-4036
FAX: (408) 865-1406
http://www.alzet.com

Amac
160B Larrabee Road
Westbrook, ME 04092
(800) 458-5060   FAX: (207) 854-0116
(207) 854-0426

Amaresco
30175 Solon Industrial Parkway
Solon, Ohio 44139
(800) 366-1313   FAX: (440) 349-1182
(440) 349-1313

Ambion
2130 Woodward Street, Suite 200
Austin, TX 78744
(800) 888-8804   FAX: (512) 651-0190
(512) 651-0200
http://www.ambion.com

American Association of
Blood Banks
College of American Pathologists
325 Waukegan Road
Northfield, IL 60093
(800) 323-4040   FAX: (847) 8166
(847) 832-7000
http://www.cap.org

American Bio-Technologies
See Intracel Corporation

American Bioanalytical
15 Erie Drive
Natick, MA 01760
(800) 443-0600   FAX: (508) 655-2754
(508) 655-4336
http://www.americanbio.com

American Cyanamid
P.O. Box 400
Princeton, NJ 08543
(609) 799-0400   FAX: (609) 275-3502
http://www.cyanamid.com

American HistoLabs
7605-F Airpark Road
Gaithersburg, MD 20879
(301) 330-1200   FAX: (301) 330-6059

American International Chemical
17 Strathmore Road
Natick, MA 01760
(800) 238-0001   (508) 655-5805
http://www.aicma.com

American Laboratory Supply
See American Bioanalytical

American Medical Systems
10700 Bren Road West
Minnetonka, MN 55343
(800) 328-3881   FAX: (612) 930-6654
(612) 933-4666
http://www.visitams.com

American Qualex
920-A Calle Negocio
San Clemente, CA 92673
(949) 492-8298   FAX: (949) 492-6790
http://www.americanqualex.com

American Radiolabeled Chemicals
11624 Bowling Green
St. Louis, MO 63146
(800) 331-6661   FAX: (800) 999-9925
(314) 991-4545   FAX: (314) 991-4692
http://www.arc-inc.com

American Scientific Products
See VWR Scientific Products

American Society for
Histocompatibility and
Immunogenetics
P.O. Box 15804
Lenexa, KS 66285
(913) 541-0009   FAX: (913) 541-0156
http://www.swmed.edu/home_pages/ASHI
/ashi.htm

American Type Culture Collection
(ATCC)
10801 University Boulevard
Manassas, VA 20110
(800) 638-6597   FAX: (703) 365-2750
(703) 365-2700
http://www.atcc.org

Amersham
See Amersham Pharmacia Biotech

Amersham International
Amersham Place
Little Chalfont, Buckinghamshire
HP7 9NA, UK
(44) 1494-544100
FAX: (44) 1494-544350
http://www.apbiotech.com

Amersham Medi-Physics
Also see Nycomed Amersham
3350 North Ridge Avenue
Arlington Heights, IL 60004
(800) 292-8514   FAX: (800) 807-2382
http://www.nycomed-amersham.com

Suppliers

2

Amersham Pharmacia Biotech
800 Centennial Avenue
P.O. Box 1327
Piscataway, NJ 08855
(800) 526-3593   FAX: (877) 295-8102
(732) 457-8000
http://www.apbiotech.com

Amgen
1 Amgen Center Drive
Thousand Oaks, CA 91320
(800) 926-4369   FAX: (805) 498-9377
(805) 447-5725
http://www.amgen.com

Amicon
Scientific Systems Division
72 Cherry Hill Drive
Beverly, MA 01915
(800) 426-4266   FAX: (978) 777-6204
(978) 777-3622
http://www.amicon.com

Amika
8980F Route 108
Oakland Center
Columbia, MD 21045
(800) 547-6766   FAX: (410) 997-7104
(410) 997-0100
http://www.amika.com

Amoco Performance Products
See BPAmoco

AMPI
See Pacer Scientific

Amrad
576 Swan Street
Richmond, Victoria 3121, Australia
613-9208-4000
FAX: 613-9208-4350
http://www.amrad.com.au

Amresco
30175 Solon Industrial Parkway
Solon, OH 44139
(800) 829-2805   FAX: (440) 349-1182
(440) 349-1199

Anachemia Chemicals
3 Lincoln Boulevard
Rouses Point, NY 12979
(800) 323-1414   FAX: (518) 462-1952
(518) 462-1066
http://www.anachemia.com

Ana-Gen Technologies
4015 Fabian Way
Palo Alto, CA 94303
(800) 654-4671   FAX: (650) 494-3893
(650) 494-3894
http://www.ana-gen.com

Analox Instruments USA
P.O. Box 208
Lunenburg, MA 01462
(978) 582-9368   FAX: (978) 582-9588
http://www.analox.com

Analytical Biological Services
Cornell Business Park 701-4
Wilmington, DE 19801
(800) 391-2391   FAX: (302) 654-8046
(302) 654-4492
http://www.ABSbioreagents.com

Analytical Genetics Testing Center
7808 Cherry Creek S. Drive, Suite 201
Denver, CO 80231
(800) 204-4721   FAX: (303) 750-2171
(303) 750-2023
http://www.geneticid.com

AnaSpec
2149 O'Toole Avenue, Suite F
San Jose, CA 95131
(800) 452-5530   FAX: (408) 452-5059
(408) 452-5055
http://www.anaspec.com

Ancare
2647 Grand Avenue
P.O. Box 814
Bellmore, NY 11710
(800) 645-6379   FAX: (516) 781-4937
(516) 781-0755
http://www.ancare.com

Ancell
243 Third Street North
P.O. Box 87
Bayport, MN 55033
(800) 374-9523   FAX: (651) 439-1940
(651) 439-0835
http://www.ancell.com

Anderson Instruments
500 Technology Court
Smyrna, GA 30082
(800) 241-6898   FAX: (770) 319-5306
(770) 319-9999
http://www.graseby.com

Andreas Hettich
Gartenstrasse 100
Postfach 260
D-78732 Tuttlingen, Germany
(49) 7461 705 0
FAX: (49) 7461 705-122
http://www.hettich-centrifugen.de

Anesthetic Vaporizer Services
10185 Main Street
Clarence, NY 14031
(719) 759-8490
http://www.avapor.com

Animal Identification and
Marking Systems (AIMS)
13 Winchester Avenue
Budd Lake, NJ 07828
(908) 684-9105   FAX: (908) 684-9106
http://www.animalid.com

Annovis
34 Mount Pleasant Drive
Aston, PA 19014
(800) EASY-DNA   FAX: (610) 361-8255
(610) 361-9224
http://www.annovis.com

Apotex
150 Signet Drive
Weston, Ontario
M9L 1T9 Canada
(416) 749-9300   FAX: (416) 749-2646
http://www.apotex.com

Apple Scientific
11711 Chillicothe Road, Unit 2
P.O. Box 778
Chesterland, OH 44026
(440) 729-3056   FAX: (440) 729-0928
http://www.applesci.com

Applied Biosystems
See PE Biosystems

Applied Imaging
2380 Walsh Avenue, Bldg. B
Santa Clara, CA 95051
(800) 634-3622   FAX: (408) 562-0264
(408) 562-0250
http://www.aicorp.com

Applied Photophysics
203-205 Kingston Road
Leatherhead, Surrey, KT22 7PB
UK
(44) 1372-386537

Applied Precision
1040 12th Avenue Northwest
Issaquah, Washington 98027
(425) 557-1000
FAX: (425) 557-1055
http://www.api.com/index.html

Appligene Oncor
Parc d'Innovation
Rue Geiler de Kaysersberg, BP 72
67402 Illkirch Cedex, France
(33) 88 67 22 67
FAX: (33) 88 67 19 45
http://www.oncor.com/prod-app.htm

Applikon
1165 Chess Drive, Suite G
Foster City, CA 94404
(650) 578-1396   FAX: (650) 578-8836
http://www.applikon.com

Appropriate Technical Resources
9157 Whiskey Bottom Road
Laurel, MD 20723
(800) 827-5931   FAX: (410) 792-2837
http://www.atrbiotech.com

APV Gaulin
100 S. CP Avenue
Lake Mills, WI 53551
(888) 278-4321   FAX: (888) 278-5329
http://www.apv.com

Aqualon
See Hercules Aqualon

Aquarium Systems
8141 Tyler Boulevard
Mentor, OH 44060
(800) 822-1100   FAX: (440) 255-8994
(440) 255-1997
http://www.aquariumsystems.com

Aquebogue Machine and Repair Shop
Box 2055
Main Road
Aquebogue, NY 11931
(631) 722-3635   FAX: (631) 722-3106

Archer Daniels Midland
4666 Faries Parkway
Decatur, IL 62525
(217) 424-5200
http://www.admworld.com

Archimica Florida
P.O. Box 1466
Gainesville, FL 32602
(800) 331-6313   FAX: (352) 371-6246
(352) 376-8246
http://www.archimica.com

Arcor Electronics
1845 Oak Street #15
Northfield, IL 60093
(847) 501-4848

Arcturus Engineering
400 Logue Avenue
Mountain View, CA 94043
(888) 446 7911   FAX: (650) 962 3039
(650) 962 3020
http://www.arctur.com

Argonaut Technologies
887 Industrial Road, Suite G
San Carlos, CA 94070
(650) 998-1350   FAX: (650) 598-1359
http://www.argotech.com

Ariad Pharmaceuticals
26 Landsdowne Street
Cambridge, MA 02139
(617) 494-0400   FAX: (617) 494-8144
http://www.ariad.com

Armour Pharmaceuticals
See Rhone-Poulenc Rorer

Aronex Pharmaceuticals
8707 Technology Forest Place
The Woodlands, TX 77381
(281) 367-1666   FAX: (281) 367-1676
http://www.aronex.com

Artisan Industries
73 Pond Street
Waltham, MA 02254
(617) 893-6800
http://www.artisanind.com

ASI Instruments
12900 Ten Mile Road
Warren, MI 48089
(800) 531-1105   FAX: (810) 756-9737
(810) 756-1222
http://www.asi-instruments.com

Aspen Research Laboratories
1700 Buerkle Road
White Bear Lake, MN 55140
(651) 264-6000   FAX: (651) 264-6270
http://www.aspenresearch.com

Associates of Cape Cod
704 Main Street
Falmouth, MA 02540
(800) LAL-TEST   FAX: (508) 540-8680
(508) 540-3444
http://www.acciusa.com

Astra Pharmaceuticals
See AstraZeneca

AstraZeneca
1800 Concord Pike
Wilmington, DE 19850
(302) 886-3000   FAX: (302) 886-2972
http://www.astrazeneca.com

AT Biochem
30 Spring Mill Drive
Malvern, PA 19355
(610) 889-9300   FAX: (610) 889-9304

ATC Diagnostics
See Vysis

ATCC
See American Type Culture Collection

Athens Research and Technology
P.O. Box 5494
Athens, GA 30604
(706) 546-0207   FAX: (706) 546-7395

Atlanta Biologicals
1425-400 Oakbrook Drive
Norcross, GA 30093
(800) 780-7788 or (770) 446-1404
FAX: (800) 780-7374 or (770) 446-1404
http://www.atlantabio.com

Atomergic Chemical
71 Carolyn Boulevard
Farmingdale, NY 11735
(631) 694-9000   FAX: (631) 694-9177
http://www.atomergic.com

Atomic Energy of Canada
2251 Speakman Drive
Mississauga, Ontario
L5K 1B2 Canada
(905) 823-9040   FAX: (905) 823-1290
http://www.aecl.ca

ATR
P.O. Box 460
Laurel, MD 20725
(800) 827-5931   FAX: (410) 792-2837
(301) 470-2799
http://www.atrbiotech.com

Aurora Biosciences
11010 Torreyana Road
San Diego, CA 92121
(858) 404-6600   FAX: (858) 404-6714
http://www.aurorabio.com

Automatic Switch Company
A Division of Emerson Electric
50 Hanover Road
Florham Park, NJ 07932
(800) 937-2726   FAX: (973) 966-2628
(973) 966-2000
http://www.asco.com

Avanti Polar Lipids
700 Industrial Park Drive
Alabaster, AL 35007
(800) 227-0651   FAX: (800) 229-1004
(205) 663-2494   FAX: (205) 663-0756
http://www.avantilipids.com

Aventis
BP 67917
67917 Strasbourg Cedex 9, France
33 (0) 388 99 11 00
FAX: 33 (0) 388 99 11 01
http://www.aventis.com

Aventis Pasteur
1 Discovery Drive
Swiftwater, PA 18370
(800) 822-2463   FAX: (570) 839-0955
(570) 839-7187
http://www.aventispasteur.com/usa

Avery Dennison
150 North Orange Grove Boulevard
Pasadena, CA 91103
(800) 462-8379   FAX: (626) 792-7312
(626) 304-2000
http://www.averydennison.com

Avestin
2450 Don Reid Drive
Ottawa, Ontario
K1H 1E1 Canada
(888) AVESTIN   FAX: (613) 736-8086
(613) 736-0019
http://www.avestin.com

AVIV Instruments
750 Vassar Avenue
Lakewood, NJ 08701
(732) 367-1663   FAX: (732) 370-0032
http://www.avivinst.com

Axon Instruments
1101 Chess Drive
Foster City, CA 94404
(650) 571-9400   FAX: (650) 571-9500
http://www.axon.com

Azon
720 Azon Road
Johnson City, NY 13790
(800) 847-9374   FAX: (800) 635-6042
(607) 797-2368
http://www.azon.com

BAbCO
1223 South 47th Street
Richmond, CA 94804
(800) 92-BABCO   FAX: (510) 412-8940
(510) 412-8930
http://www.babco.com

Bacharach
625 Alpha Drive
Pittsburgh, PA 15238
(800) 736-4666   FAX: (412) 963-2091
(412) 963-2000
http://www.bacharach-inc.com

Bachem Bioscience
3700 Horizon Drive
King of Prussia, PA 19406
(800) 634-3183   FAX: (610) 239-0800
(610) 239-0300
http://www.bachem.com

Bachem California
3132 Kashiwa Street
P.O. Box 3426
Torrance, CA 90510
(800) 422-2436   FAX: (310) 530-1571
(310) 539-4171
http://www.bachem.com

Baekon
18866 Allendale Avenue
Saratoga, CA 95070
(408) 972-8779   FAX: (408) 741-0944

Baker Chemical
See J.T. Baker

Bangs Laboratories
9025 Technology Drive
Fishers, IN 46038
(317) 570-7020   FAX: (317) 570-7034
http://www.bangslabs.com

Bard Parker
See Becton Dickinson

Barnstead/Thermolyne
P.O. Box 797
2555 Kerper Boulevard
Dubuque, IA 52004
(800) 446-6060   FAX: (319) 589-0516
http://www.barnstead.com

Barrskogen
4612 Laverock Place N
Washington, DC 20007
(800) 237-9192 FAX: (301) 464-7347

BAS
See Bioanalytical Systems

BASF
Specialty Products
3000 Continental Drive North
Mt. Olive, NJ 07828
(800) 669-2273   FAX: (973) 426-2610
http://www.basf.com

Baum, W.A.
620 Oak Street
Copiague, NY 11726
(631) 226-3940   FAX: (631) 226-3969
http://www.wabaum.com

Bausch & Lomb
One Bausch & Lomb Place
Rochester, NY 14604
(800) 344-8815   FAX: (716) 338-6007
(716) 338-6000
http://www.bausch.com

Baxter
Fenwal Division
1627 Lake Cook Road
Deerfield, IL 60015
(800) 766-1077   FAX: (800) 395-3291
(847) 940-6599   FAX: (847) 940-5766
http://www.powerfulmedicine.com

Baxter Healthcare
One Baxter Parkway
Deerfield, IL 60015
(800) 777-2298   FAX: (847) 948-3948
(847) 948-2000
http://www.baxter.com

Baxter Scientific Products
See VWR Scientific

Bayer
Agricultural Division
Animal Health Products
12707 Shawnee Mission Pkwy.
Shawnee Mission, KS 66201
(800) 255-6517   FAX: (913) 268-2803
(913) 268-2000
http://www.bayerus.com

Bayer
Diagnostics Division (Order Services)
P.O. Box 2009
Mishiwaka, IN 46546
(800) 248-2637   FAX: (800) 863-6882
(219) 256-3390
http://www.bayer.com

Bayer Diagnostics
511 Benedict Avenue
Tarrytown, NY 10591
(800) 255-3232   FAX: (914) 524-2132
(914) 631-8000
http://www.bayerdiag.com

Bayer Plc
Diagnostics Division
Bayer House, Strawberry Hill
Newbury, Berkshire RG14 1JA, UK
(44) 1635-563000
FAX: (44) 1635-563393
http://www.bayer.co.uk

BD Immunocytometry Systems
2350 Qume Drive
San Jose, CA 95131
(800) 223-8226   FAX: (408) 954-BDIS
http://www.bdfacs.com

BD Labware
Two Oak Park
Bedford, MA 01730
(800) 343-2035   FAX: (800) 743-6200
http://www.bd.com/labware

BD PharMingen
10975 Torreyana Road
San Diego, CA 92121
(800) 848-6227   FAX: (858) 812-8888
(858) 812-8800
http://www.pharmingen.com

Suppliers

4

BD Transduction Laboratories
133 Venture Court
Lexington, KY 40511
(800) 227-4063   FAX: (606) 259-1413
(606) 259-1550
http://www.translab.com

BDH Chemicals
Broom Road
Poole, Dorset BH12 4NN, UK
(44) 1202-745520
FAX: (44) 1202- 2413720

BDH Chemicals
See Hoefer Scientific Instruments

BDIS
See BD Immunocytometry Systems

Beckman Coulter
4300 North Harbor Boulevard
Fullerton, CA 92834
(800) 233-4685   FAX: (800) 643-4366
(714) 871-4848
http://www.beckman-coulter.com

Beckman Instruments
Spinco Division/Bioproducts Operation
1050 Page Mill Road
Palo Alto, CA 94304
(800) 742-2345   FAX: (415) 859-1550
(415) 857-1150
http://www.beckman-coulter.com

Becton Dickinson Immunocytometry & Cellular Imaging
2350 Qume Drive
San Jose, CA 95131
(800) 223-8226   FAX: (408) 954-2007
(408) 432-9475
http://www.bdfacs.com

Becton Dickinson Labware
1 Becton Drive
Franklin Lakes, NJ 07417
(888) 237-2762   FAX: (800) 847-2220
(201) 847-4222
http://www.bdfacs.com

Becton Dickinson Labware
2 Bridgewater Lane
Lincoln Park, NJ 07035
(800) 235-5953   FAX: (800) 847-2220
(201) 847-4222
http://www.bdfacs.com

Becton Dickinson Primary
Care Diagnostics
7 Loveton Circle
Sparks, MD 21152
(800) 675-0908   FAX: (410) 316-4723
(410) 316-4000
http://www.bdfacs.com

Behringwerke Diagnostika
Hoechster Strasse 70
P-65835 Liederback, Germany
(49) 69-30511   FAX: (49) 69-303-834

Bellco Glass
340 Edrudo Road
Vineland, NJ 08360
(800) 257-7043   FAX: (856) 691-3247
(856) 691-1075
http://www.bellcoglass.com

Bender Biosystems
See Serva

Beral Enterprises
See Garren Scientific

Berkeley Antibody
See BAbCO

Bernsco Surgical Supply
25 Plant Avenue
Hauppague, NY 11788
(800) TIEMANN   FAX: (516) 273-6199
(516) 273-0005
http://www.bernsco.com

Beta Medical and Scientific
(Datesand Ltd.)
2 Ferndale Road
Sale, Manchester M33 3GP, UK
(44) 1612 317676
FAX: (44) 1612 313656

Bethesda Research Laboratories (BRL)
See Life Technologies

Biacore
200 Centennial Avenue, Suite 100
Piscataway, NJ 08854
(800) 242-2599   FAX: (732) 885-5669
(732) 885-5618
http://www.biacore.com

Bilaney Consultants
St. Julian's
Sevenoaks, Kent TN15 0RX, UK
(44) 1732 450002
FAX: (44) 1732 450003
http://www.bilaney.com

Binding Site
5889 Oberlin Drive, Suite 101
San Diego, CA 92121
(800) 633-4484   FAX: (619) 453-9189
(619) 453-9177
http://www.bindingsite.co.uk

BIO 101
See Qbiogene

Bio Image
See Genomic Solutions

Bioanalytical Systems
2701 Kent Avenue
West Lafayette, IN 47906
(800) 845-4246   FAX: (765) 497-1102
(765) 463-4527
http://www.bioanalytical.com

Biocell
2001 University Drive
Rancho Dominguez, CA 90220
(800) 222-8382   FAX: (310) 637-3927
(310) 537-3300
http://www.biocell.com

Biocoat
See BD Labware

BioComp Instruments
650 Churchill Road
Fredericton, New Brunswick
E3B 1P6 Canada
(800) 561-4221   FAX: (506) 453-3583
(506) 453-4812
http://131.202.97.21

BioDesign
P.O. Box 1050
Carmel, NY 10512
(914) 454-6610   FAX: (914) 454-6077
http://www.biodesignofny.com

BioDiscovery
4640 Admiralty Way, Suite 710
Marina Del Rey, CA 90292
(310) 306-9310   FAX: (310) 306-9109
http://www.biodiscovery.com

Bioengineering AG
Sagenrainstrasse 7
CH8636 Wald, Switzerland
(41) 55-256-8-111
FAX: (41) 55-256-8-256

Biofluids
Division of Biosource International
1114 Taft Street
Rockville, MD 20850
(800) 972-5200   FAX: (301) 424-3619
(301) 424-4140
http://www.biosource.com

BioFX Laboratories
9633 Liberty Road, Suite S
Randallstown, MD 21133
(800) 445-6447   FAX: (410) 498-6008
(410) 496-6006
http://www.biofx.com

BioGenex Laboratories
4600 Norris Canyon Road
San Ramon, CA 94583
(800) 421-4149   FAX: (925) 275-0580
(925) 275-0550
http://www.biogenex.com

Bioline
2470 Wrondel Way
Reno, NV 89502
(888) 257-5155   FAX: (775) 828-7676
(775) 828-0202
http://www.bioline.com

Bio-Logic Research & Development
1, rue de l-Europe
A.Z. de Font-Ratel
38640 CLAIX, France
(33) 76-98-68-31
FAX: (33) 76-98-69-09

Biological Detection Systems
See Cellomics or Amersham

Biomeda
1166 Triton Drive, Suite E
P.O. Box 8045
Foster City, CA 94404
(800) 341-8787   FAX: (650) 341-2299
(650) 341-8787
http://www.biomeda.com

BioMedic Data Systems
1 Silas Road
Seaford, DE 19973
(800) 526-2637   FAX: (302) 628-4110
(302) 628-4100
http://www.bmds.com

Biomedical Engineering
P.O. Box 980694
Virginia Commonwealth University
Richmond, VA 23298
(804) 828-9829   FAX: (804) 828-1008

Biomedical Research Instruments
12264 Wilkins Avenue
Rockville, MD 20852
(800) 327-9498
(301) 881-7911
http://www.biomedinstr.com

Bio/medical Specialties
P.O. Box 1687
Santa Monica, CA 90406
(800) 269-1158   FAX: (800) 269-1158
(323) 938-7515

BioMerieux
100 Rodolphe Street
Durham, North Carolina 27712
(919) 620-2000
http://www.biomerieux.com

BioMetallics
P.O. Box 2251
Princeton, NJ 08543
(800) 999-1961   FAX: (609) 275-9485
(609) 275-0133
http://www.microplate.com

Biomol Research Laboratories
5100 Campus Drive
Plymouth Meeting, PA 19462
(800) 942-0430   FAX: (610) 941-9252
(610) 941-0430
http://www.biomol.com

Bionique Testing Labs
Fay Brook Drive
RR 1, Box 196
Saranac Lake, NY 12983
(518) 891-2356   FAX: (518) 891-5753
http://www.bionique.com

Biopac Systems
42 Aero Camino
Santa Barbara, CA 93117
(805) 685-0066   FAX: (805) 685-0067
http://www.biopac.com

Bioproducts for Science
See Harlan Bioproducts for Science

Suppliers

5

Bioptechs
3560 Beck Road
Butler, PA 16002
(877) 548-3235   FAX: (724) 282-0745
(724) 282-7145
http://www.bioptechs.com

BIOQUANT-R&M Biometrics
5611 Ohio Avenue
Nashville, TN 37209
(800) 221-0549 (615) 350-7866
FAX: (615) 350-7282
http://www.bioquant.com

Bio-Rad Laboratories
2000 Alfred Nobel Drive
Hercules, CA 94547
(800) 424-6723   FAX: (800) 879-2289
(510) 741-1000   FAX: (510) 741-5800
http://www.bio-rad.com

Bio-Rad Laboratories
Maylands Avenue
Hemel Hempstead, Herts HP2 7TD, UK
http://www.bio-rad.com

BioRobotics
3-4 Bennell Court
Comberton, Cambridge CB3 7DS, UK
(44) 1223-264345
FAX: (44) 1223-263933
http://www.biorobotics.co.uk

BIOS Laboratories
See Genaissance Pharmaceuticals

Biosearch Technologies
81 Digital Drive
Novato, CA 94949
(800) GENOME1   FAX: (415) 883-8488
(415) 883-8400
http://www.biosearchtech.com

BioSepra
111 Locke Drive
Marlborough, MA 01752
(800) 752-5277   FAX: (508) 357-7595
(508) 357-7500
http://www.biosepra.com

Bio-Serv
1 8th Street, Suite 1
Frenchtown, NJ 08825
(908) 996-2155   FAX: (908) 996-4123
http://www.bio-serv.com

BioSignal
1744 William Street, Suite 600
Montreal, Quebec
H3J 1R4 Canada
(800) 293-4501   FAX: (514) 937-0777
(514) 937-1010
http://www.biosignal.com

Biosoft
P.O. Box 10938
Ferguson, MO 63135
(314) 524-8029 FAX: (314) 524-8129
http://www.biosoft.com

Biosource International
820 Flynn Road
Camarillo, CA 93012
(800) 242-0607 FAX: (805) 987-3385
(805) 987-0086
http://www.biosource.com

BioSpec Products
P.O. Box 788
Bartlesville, OK 74005
(800) 617-3363 FAX: (918) 336-3363
(918) 336-3363
http://www.biospec.com

Biosure
See Riese Enterprises

Biosym Technologies
See Molecular Simulations

Biosys
21 quai du Clos des Roses
602000 Compiegne, France
(33) 03 4486 2275
FAX: (33) 03 4484 2297

Bio-Tech Research Laboratories
NIAID Repository
Rockville, MD 20850
http://www.niaid.nih.gov/ncn/repos.htm

Biotech Instruments
Biotech House
75A High Street
Kimpton, Hertfordshire SG4 8PU, UK
(44) 1438 832555
FAX: (44) 1438 833040
http://www.biotinst.demon.co.uk

Biotech International
11 Durbell Street
Acacia Ridge, Queensland 4110
Australia
61-7-3370-6396
FAX: 61-7-3370-6370
http://www.avianbiotech.com

Biotech Source
Inland Farm Drive
South Windham, ME 04062
(207) 892-3266   FAX: (207) 892-6774

Bio-Tek Instruments
Highland Industrial Park
P.O. Box 998
Winooski, VT 05404
(800) 451-5172   FAX: (802) 655-7941
(802) 655-4040
http://www.biotek.com

Biotecx Laboratories
6023 South Loop East
Houston, TX 77033
(800) 535-6286   FAX: (713) 643-3143
(713) 643-0606
http://www.biotecx.com

BioTherm
3260 Wilson Boulevard
Arlington, VA 22201
(703) 522-1705   FAX: (703) 522-2606

Bioventures
P.O. Box 2561
848 Scott Street
Murfreesboro, TN 37133
(800) 235-8938   FAX: (615) 896-4837
http://www.bioventures.com

BioWhittaker
8830 Biggs Ford Road
P.O. Box 127
Walkersville, MD 21793
(800) 638-8174   FAX: (301) 845-8338
(301) 898-7025
http://www.biowhittaker.com

Biozyme Laboratories
9939 Hibert Street, Suite 101
San Diego, CA 92131
(800) 423-8199   FAX: (858) 549-0138
(858) 549-4484
http://www.biozyme.com

Bird Products
1100 Bird Center Drive
Palm Springs, CA 92262
(800) 328-4139   FAX: (760) 778-7274
(760) 778-7200
http://www.birdprod.com/bird

B & K Universal
2403 Yale Way
Fremont, CA 94538
(800) USA-MICE   FAX: (510) 490-3036

BLS Ltd.
Zselyi Aladar u. 31
1165 Budapest, Hungary
(36) 1-407-2602   FAX: (36) 1-407-2896
http://www.bls-ltd.com

Blue Sky Research
3047 Orchard Parkway
San Jose, CA 95134
(408) 474-0988   FAX: (408) 474-0989
http://www.blueskyresearch.com

Blumenthal Industries
7 West 36th Street, 13th floor
New York, NY 10018
(212) 719-1251   FAX: (212) 594-8828

BOC Edwards
One Edwards Park
301 Ballardvale Street
Wilmington, MA 01887
(800) 848-9800   FAX: (978) 658-7969
(978) 658-5410
http://www.bocedwards.com

Boehringer Ingelheim
900 Ridgebury Road
P.O. Box 368
Ridgefield, CT 06877
(800) 243-0127   FAX: (203) 798-6234
(203) 798-9988
http://www.boehringer-ingelheim.com

Boehringer Mannheim
Biochemicals Division
See Roche Diagnostics

Boekel Scientific
855 Pennsylvania Boulevard
Feasterville, PA 19053
(800) 336-6929   FAX: (215) 396-8264
(215) 396-8200
http://www.boekelsci.com

Bohdan Automation
1500 McCormack Boulevard
Mundelein, IL 60060
(708) 680-3939   FAX: (708) 680-1199

BPAmoco
4500 McGinnis Ferry Road
Alpharetta, GA 30005
(800) 328-4537   FAX: (770) 772-8213
(770) 772-8200
http://www.bpamoco.com

Brain Research Laboratories
Waban P.O. Box 88
Newton, MA 02468
(888) BRL-5544   FAX: (617) 965-6220
(617) 965-5544
http://www.brainresearchlab.com

Braintree Scientific
P.O. Box 850929
Braintree, MA 02185
(781) 843-1644   FAX: (781) 982-3160
http://www.braintreesci.com

Brandel
8561 Atlas Drive
Gaithersburg, MD 20877
(800) 948-6506   FAX: (301) 869-5570
(301) 948-6506
http://www.brandel.com

Branson Ultrasonics
41 Eagle Road
Danbury, CT 06813
(203) 796-0400   FAX: (203) 796-9838
http://www.plasticsnet.net/branson

B. Braun Biotech
999 Postal Road
Allentown, PA 18103
(800) 258-9000   FAX: (610) 266-9319
(610) 266-6262
http://www.bbraunbiotech.com

B. Braun Biotech International
Schwarzenberg Weg 73-79
P.O. Box 1120
D-34209 Melsungen, Germany
(49) 5661-71-3400
FAX: (49) 5661-71-3702
http://www.bbraunbiotech.com

B. Braun-McGaw
2525 McGaw Avenue
Irvine, CA 92614
(800) BBRAUN-2   (800) 624-2963
http://www.bbraunusa.com

B. Braun Medical
Thorncliffe Park
Sheffield S35 2PW, UK
(44) 114-225-9000
FAX: (44) 114-225-9111
http://www.bbmuk.demon.co.uk

Suppliers

6

Brenntag
P.O. Box 13788
Reading, PA 19612-3788
(610) 926-4151   FAX: (610) 926-4160
http://www.brenntagnortheast.com

Bresatec
See GeneWorks

Bright/Hacker Instruments
17 Sherwood Lane
Fairfield, NJ 07004
(973) 226-8450   FAX: (973) 808-8281
http://www.hackerinstruments.com

Brinkmann Instruments
Subsidiary of Sybron
1 Cantiague Road
P.O. Box 1019
Westbury, NY 11590
(800) 645-3050   FAX: (516) 334-7521
(516) 334-7500
http://www.brinkmann.com

Bristol-Meyers Squibb
P.O. Box 4500
Princeton, NJ 08543
(800) 631-5244   FAX: (800) 523-2965
http://www.bms.com

Broadley James
19 Thomas
Irvine, CA 92618
(800) 288-2833   FAX: (949) 829-5560
(949) 829-5555
http://www.broadleyjames.com

Brookhaven Instruments
750 Blue Point Road
Holtsville, NY 11742
(631) 758-3200   FAX: (631) 758-3255
http://www.bic.com

Brownlee Labs
See Applied Biosystems
Distributed by Pacer Scientific

Bruel & Kjaer
Division of Spectris Technologies
2815 Colonnades Court
Norcross, GA 30071
(800) 332-2040   FAX: (770) 847-8440
(770) 209-6907
http://www.bkhome.com

Bruker Analytical X-Ray Systems
5465 East Cheryl Parkway
Madison, WI 53711
(800) 234-XRAY   FAX: (608) 276-3006
(608) 276-3000
http://www.bruker-axs.com

Bruker Instruments
19 Fortune Drive
Billerica, MA 01821
(978) 667-9580   FAX: (978) 667-0985
http://www.bruker.com

BTX
Division of Genetronics
11199 Sorrento Valley Road
San Diego, CA 92121
(800) 289-2465   FAX: (858) 597-9594
(858) 597-6006
http://www.genetronics.com/btx

Buchler Instruments
See Baxter Scientific Products

Buckshire
2025 Ridge Road
Perkasie, PA 18944
(215) 257-0116

Burdick and Jackson
Division of Baxter Scientific Products
1953 S. Harvey Street
Muskegon, MI 49442
(800) 368-0050   FAX: (231) 728-8226
(231) 726-3171
http://www.bandj.com/mainframe.htm

Burleigh Instruments
P.O. Box E
Fishers, NY 14453
(716) 924-9355   FAX: (716) 924-9072
http://www.burleigh.com

Burns Veterinary Supply
1900 Diplomat Drive
Farmer's Branch, TX 75234
(800) 92-BURNS   FAX: (972) 243-6841
http://www.burnsvet.com

Burroughs Wellcome
See Glaxo Wellcome

The Butler Company
5600 Blazer Parkway
Dublin, OH 43017
(800) 551-3861   FAX: (614) 761-9096
(614) 761-9095
http://www.wabutler.com

Butterworth Laboratories
54-56 Waldegrave Road
Teddington, Middlesex
TW11 8LG, UK
(44)(0)20-8977-0750
FAX: (44)(0)28-8943-2624
http://www.butterworth-labs.co.uk

Buxco Electronics
95 West Wood Road #2
Sharon, CT 06069
(860) 364-5558   FAX: (860) 364-5116
http://www.buxco.com

C/D/N Isotopes
88 Leacock Street
Pointe-Claire, Quebec
H9R 1H1 Canada
(800) 697-6254   FAX: (514) 697-6148

C.M.A./Microdialysis AB
73 Princeton Street
North Chelmsford, MA 01863
(800) 440-4980   FAX: (978) 251-1950
(978) 251-1940
http://www.microdialysis.com

Calbiochem-Novabiochem
P.O. Box 12087-2087
La Jolla, CA 92039
(800) 854-3417   FAX: (800) 776-0999
(858) 450-9600
http://www.calbiochem.com

California Fine Wire
338 South Fourth Street
Grover Beach, CA 93433
(805) 489-5144   FAX: (805) 489-5352
http://www.calfinewire.com

Calorimetry Sciences
155 West 2050 North
Spanish Fork, UT 84660
(801) 794-2600   FAX: (801) 794-2700
http://www.calscorp.com

Caltag Laboratories
1849 Bayshore Highway, Suite 200
Burlingame, CA 94010
(800) 874-4007   FAX: (650) 652-9030
(650) 652-0468
http://www.caltag.com

Cambridge Electronic Design
Science Park, Milton Road
Cambridge CB4 0FE, UK
44 (0) 1223-420-186
FAX: 44 (0) 1223-420-488
http://www.ced.co.uk

Cambridge Isotope Laboratories
50 Frontage Road
Andover, MA 01810
(800) 322-1174   FAX: (978) 749-2768
(978) 749-8000
http://www.isotope.com

Cambridge Research Biochemicals
See Zeneca/CRB

Cambridge Technology
109 Smith Place
Cambridge, MA 02138
(617) 441-0600   FAX: (617) 497-8800
http://www.camtech.com

Camlab
Nuffield Road
Cambridge CB4 1TH, UK
(44) 122-3424222
FAX: (44) 122-3420856
http://www.camlab.co.uk/home.htm

Campden Instruments
Park Road
Sileby Loughborough
Leicestershire LE12 7TU, UK
(44) 1509-814790
FAX: (44) 1509-816097
http://www.campden-inst.com/home.htm

Cappel Laboratories
See Organon Teknika Cappel

Carl Roth GmgH & Company
Schoemperlenstrasse 1-5
76185 Karlsrube
Germany
(49) 72-156-06164
FAX: (49) 72-156-06264
http://www.carl-roth.de

Carl Zeiss
One Zeiss Drive
Thornwood, NY 10594
(800) 233-2343   FAX: (914) 681-7446
(914) 747-1800
http://www.zeiss.com

Carlo Erba Reagenti
Via Winckelmann 1
20148 Milano
Lombardia, Italy
(39) 0-29-5231
FAX: (39) 0-29-5235-904
http://www.carloerbareagenti.com

Carolina Biological Supply
2700 York Road
Burlington, NC 27215
(800) 334-5551   FAX: (336) 584-76869
(336) 584-0381
http://www.carolina.com

Carolina Fluid Components
9309 Stockport Place
Charlotte, NC 28273
(704) 588-6101   FAX: (704) 588-6115
http://www.cfcsite.com

Cartesian Technologies
17851 Skypark Circle, Suite C
Irvine, CA 92614
(800) 935-8007
http://cartesiantech.com

Cayman Chemical
1180 East Ellsworth Road
Ann Arbor, MI 48108
(800) 364-9897   FAX: (734) 971-3640
(734) 971-3335
http://www.caymanchem.com

CB Sciences
One Washington Street, Suite 404
Dover, NH 03820
(800) 234-1757   FAX: (603) 742-2455
http://www.cbsci.com

CBS Scientific
P.O. Box 856
Del Mar, CA 92014
(800) 243-4959   FAX: (858) 755-0733
(858) 755-4959
http://www.cbssci.com

CCR (Coriell Cell Repository)
See Coriell Institute for Medical Research

CE Instruments
Grand Avenue Parkway
Austin, TX 78728
(800) 876-6711   FAX: (512) 251-1597
http://www.ceinstruments.com

Suppliers

7

Cedarlane Laboratories
5516 8th Line, R.R. #2
Hornby, Ontario
L0P 1E0 Canada
(905) 878-8891   FAX: (905) 878-7800
http://www.cedarlanelabs.com

CEL Associates
P.O. Box 721854
Houston, TX 77272
(800) 537-9339   FAX: (281) 933-0922
(281) 933-9339
http://www.cel-1.com

Cel-Line Associates
See Erie Scientific

Celite World Minerals
130 Castilian Drive
Santa Barbara, CA 93117
(805) 562-0200   FAX: (805) 562-0299
http://www.worldminerals.com/celite

Cell Genesys
342 Lakeside Drive
Foster City, CA 94404
(650) 425-4400   FAX: (650) 425-4457
http://www.cellgenesys.com

Cell Systems
12815 NE 124th Street, Suite A
Kirkland, WA 98034
(800) 697-1211   FAX: (425) 820-6762
(425) 823-1010

Cellmark Diagnostics
20271 Goldenrod Lane
Germantown, MD 20876
(800) 872-5227   FAX: (301) 428-4877
(301) 428-4980
http://www.cellmark-labs.com

Cellomics
635 William Pitt Way
Pittsburgh, PA 15238
(888) 826-3857   FAX: (412) 826-3850
(412) 826-3600
http://www.cellomics.com

Celltech
216 Bath Road
Slough, Berkshire SL1 4EN, UK
(44) 1753 534655
FAX: (44) 1753 536632
http://www.celltech.co.uk

Cellular Products
872 Main Street
Buffalo, NY 14202
(800) CPI-KITS   FAX: (716) 882-0959
(716) 882-0920
http://www.zeptometrix.com

CEM
P.O. Box 200
Matthews, NC 28106
(800) 726-3331

Centers for Disease Control
1600 Clifton Road NE
Atlanta, GA 30333
(800) 311-3435   FAX: (888) 232-3228
(404) 639-3311
http://www.cdc.gov

CERJ
Centre d'Elevage Roger Janvier
53940 Le Genest Saint Isle
France

Cetus
See Chiron

Chance Propper
Warly, West Midlands B66 1NZ, UK
(44)(0)121-553-5551
FAX: (44)(0)121-525-0139

Charles River Laboratories
251 Ballardvale Street
Wilmington, MA 01887
(800) 522-7287   FAX: (978) 658-7132
(978) 658-6000
http://www.criver.com

Charm Sciences
36 Franklin Street
Malden, MA 02148
(800) 343-2170   FAX: (781) 322-3141
(781) 322-1523
http://www.charm.com

Chase-Walton Elastomers
29 Apsley Street
Hudson, MA 01749
(800) 448-6289   FAX: (978) 562-5178
(978) 568-0202
http://www.chase-walton.com

ChemGenes
Ashland Technology Center
200 Homer Avenue
Ashland, MA 01721
(800) 762-9323   FAX: (508) 881-3443
(508) 881-5200
http://www.chemgenes.com

Chemglass
3861 North Mill Road
Vineland, NJ 08360
(800) 843-1794   FAX: (856) 696-9102
(800) 696-0014
http://www.chemglass.com

Chemicon International
28835 Single Oak Drive
Temecula, CA 92590
(800) 437-7500   FAX: (909) 676-9209
(909) 676-8080
http://www.chemicon.com

Chem-Impex International
935 Dillon Drive
Wood Dale, IL 60191
(800) 869-9290   FAX: (630) 766-2218
(630) 766-2112
http://www.chemimpex.com

Chem Service
P.O. Box 599
West Chester, PA 19381-0599
(610) 692-3026   FAX: (610) 692-8729
http://www.chemservice.com

Chemsyn Laboratories
13605 West 96th Terrace
Lenexa, KS 66215
(913) 541-0525   FAX: (913) 888-3582
http://www.tech.epcorp.com/ChemSyn/
chemsyn.htm

Chemunex USA
1 Deer Park Drive, Suite H-2
Monmouth Junction, NJ 08852
(800) 411-6734
http://www.chemunex.com

Cherwell Scientific Publishing
The Magdalen Centre
Oxford Science Park
Oxford OX44GA, UK
(44)(1) 865-784-800
FAX: (44)(1) 865-784-801
http://www.cherwell.com

ChiRex Cauldron
383 Phoenixville Pike
Malvern, PA 19355
(610) 727-2215   FAX: (610) 727-5762
http://www.chirex.com

Chiron Diagnostics
See Bayer Diagnostics

Chiron Mimotopes Peptide Systems
See Multiple Peptide Systems

Chiron
4560 Horton Street
Emeryville, CA 94608
(800) 244-7668   FAX: (510) 655-9910
(510) 655-8730
http://www.chiron.com

Chrom Tech
P.O. Box 24248
Apple Valley, MN 55124
(800) 822-5242   FAX: (952) 431-6345
http://www.chromtech.com

Chroma Technology
72 Cotton Mill Hill, Unit A-9
Brattleboro, VT 05301
(800) 824-7662   FAX: (802) 257-9400
(802) 257-1800
http://www.chroma.com

Chromatographie
ZAC de Moulin No. 2
91160 Saulx les Chartreux
France
(33) 01-64-54-8969
FAX: (33) 01-69-0988091
http://www.chromatographie.com

Chromogenix
Taljegardsgatan 3
431-53 Mlndal, Sweden
(46) 31-706-20-70
FAX: (46) 31-706-20-80
http://www.chromogenix.com

Chrompack USA
c/o Varian USA
2700 Mitchell Drive
Walnut Creek, CA 94598
(800) 526-3687   FAX: (925) 945-2102
(925) 939-2400
http://www.chrompack.com

Chugai Biopharmaceuticals
6275 Nancy Ridge Drive
San Diego, CA 92121
(858) 535-5900   FAX: (858) 546-5973
http://www.chugaibio.com

Ciba-Corning Diagnostics
See Bayer Diagnostics

Ciba-Geigy
See Ciba Specialty Chemicals or
Novartis Biotechnology

Ciba Specialty Chemicals
540 White Plains Road
Tarrytown, NY 10591
(800) 431-1900   FAX: (914) 785-2183
(914) 785-2000
http://www.cibasc.com

Ciba Vision
Division of Novartis AG
11460 Johns Creek Parkway
Duluth, GA 30097
(770) 476-3937
http://www.cvworld.com

Cidex
Advanced Sterilization Products
33 Technology Drive
Irvine, CA 92618
(800) 595-0200   (949) 581-5799
http://www.cidex.com/ASPnew.htm

Cinna Scientific
Subsidiary of Molecular Research Center
5645 Montgomery Road
Cincinnati, OH 45212
(800) 462-9868   FAX: (513) 841-0080
(513) 841-0900
http://www.mrcgene.com

Cistron Biotechnology
10 Bloomfield Avenue
Pine Brook, NJ 07058
(800) 642-0167   FAX: (973) 575-4854
(973) 575-1700
http://www.cistronbio.com

Clark Electromedical Instruments
See Harvard Apparatus

Clay Adam
See Becton Dickinson Primary Care
Diagnostics

CLB (Central Laboratory
of the Netherlands)
Blood Transfusion Service
P.O. Box 9190
1006 AD Amsterdam, The Netherlands
(31) 20-512-9222
FAX: (31) 20-512-3332

Cleveland Scientific
P.O. Box 300
Bath, OH 44210
(800) 952-7315   FAX: (330) 666-2240
http://www.clevelandscientific.com

Clonetics
Division of BioWhittaker
http://www.clonetics.com
Also see BioWhittaker

Clontech Laboratories
1020 East Meadow Circle
Palo Alto, CA 94303
(800) 662-2566   FAX: (800) 424-1350
(650) 424-8222   FAX: (650) 424-1088
http://www.clontech.com

Closure Medical Corporation
5250 Greens Dairy Road
Raleigh, NC 27616
(919) 876-7800   FAX: (919) 790-1041
http://www.closuremed.com

CMA Microdialysis AB
73 Princeton Street
North Chelmsford, MA 01863
(800) 440-4980   FAX: (978) 251-1950
(978) 251 1940
http://www.microdialysis.com

Cocalico Biologicals
449 Stevens Road
P.O. Box 265
Reamstown, PA 17567
(717) 336-1990   FAX: (717) 336-1993

Coherent Laser
5100 Patrick Henry Drive
Santa Clara, CA 95056
(800) 227-1955   FAX: (408) 764-4800
(408) 764-4000
http://www.cohr.com

Cohu
P.O. Box 85623
San Diego, CA 92186
(858) 277-6700   FAX: (858) 277-0221
http://www.COHU.com/cctv

Cole-Parmer Instrument
625 East Bunker Court
Vernon Hills, IL 60061
(800) 323-4340   FAX: (847) 247-2929
(847) 549-7600
http://www.coleparmer.com

Collaborative Biomedical Products
and **Collaborative Research**
See Becton Dickinson Labware

Collagen Aesthetics
1850 Embarcadero Road
 Palo Alto, CA 94303
(650) 856-0200   FAX: (650) 856-0533
http://www.collagen.com

Collagen Corporation
See Collagen Aesthetics

College of American Pathologists
325 Waukegan Road
Northfield, IL 60093
(800) 323-4040   FAX: (847) 832-8000
(847) 446-8800
http://www.cap.org/index.cfm

Colonial Medical Supply
504 Wells Road
Franconia, NH 03580
(603) 823-9911   FAX: (603) 823-8799
http://www.colmedsupply.com

Colorado Serum
4950 York Street
Denver, CO 80216
(800) 525-2065   FAX: (303) 295-1923
http://www.colorado-serum.com

Columbia Diagnostics
8001 Research Way
Springfield, VA 22153
(800) 336-3081   FAX: (703) 569-2353
(703) 569-7511
http://www.columbiadiagnostics.com

Columbus Instruments
950 North Hague Avenue
Columbus, OH 43204
(800) 669-5011   FAX: (614) 276-0529
(614) 276-0861
http://www.columbusinstruments.com

Computer Associates International
One Computer Associates Plaza
Islandia, NY 11749
(631) 342-6000   FAX: (631) 342-6800
http://www.cai.com

Connaught Laboratories
See Aventis Pasteur

Connectix
2955 Campus Drive, Suite 100
San Mateo, CA 94403
(800) 950-5880   FAX: (650) 571-0850
(650) 571-5100
http://www.connectix.com

Contech
99 Hartford Avenue
Providence, RI 02909
(401) 351-4890   FAX: (401) 421-5072
http://www.iol.ie/~burke/contech.html

Continental Laboratory Products
5648 Copley Drive
San Diego, CA 92111
(800) 456-7741   FAX: (858) 279-5465
(858) 279-5000
http://www.conlab.com

ConvaTec
Professional Services
P.O. Box 5254
Princeton, NJ 08543
(800) 422-8811
http://www.convatec.com

Cooper Instruments & Systems
P.O. Box 3048
Warrenton, VA 20188
(800) 344-3921   FAX: (540) 347-4755
(540) 349-4746
http://www.cooperinstruments.com

Cora Styles Needles 'N Blocks
56 Milton Street
Arlington, MA 02474
(781) 648-6289   FAX: (781) 641-7917

Coriell Cell Repository (CCR)
See Coriell Institute for Medical Research

Coriell Institute for Medical Research
Human Genetic Mutant Repository
401 Haddon Avenue
Camden, NJ 08103
(856) 966-7377   FAX: (856) 964-0254
http://arginine.umdnj.edu

Corion
8 East Forge Parkway
Franklin, MA 02038
(508) 528-4411   FAX: (508) 520-7583
(800) 598-6783
http://www.corion.com

**Corning and
Corning Science Products**
P.O. Box 5000
Corning, NY 14831
(800) 222-7740   FAX: (607) 974-0345
(607) 974-9000
http://www.corning.com

Costar
See Corning

Coulbourn Instruments
7462 Penn Drive
Allentown, PA 18106
(800) 424-3771   FAX: (610) 391-1333
(610) 395-3771
http://www.coulbourninst.com

Coulter Cytometry
See Beckman Coulter

Covance Research Products
465 Swampbridge Road
Denver, PA 17517
(800) 345-4114   FAX: (717) 336-5344
(717) 336-4921
http://www.covance.com

Coy Laboratory Products
14500 Coy Drive
Grass Lake, MI 49240
(734) 475-2200   FAX: (734) 475-1846
http://www.coylab.com

CPG
3 Borinski Road
Lincoln Park, NJ 07035
(800) 362-2740   FAX: (973) 305-0884
(973) 305-8181
http://www.cpg-biotech.com

CPL Scientific
43 Kingfisher Court
Hambridge Road
Newbury RG14 5SJ, UK
(44) 1635-574902
FAX: (44) 1635-529322
http://www.cplscientific.co.uk

CraMar Technologies
8670 Wolff Court, #160
Westminster, CO 80030
(800) 4-TOMTEC
http://www.cramar.com

Crescent Chemical
1324 Motor Parkway
Hauppauge, NY 11788
(800) 877-3225   FAX: (631) 348-0913
(631) 348-0333
http://www.creschem.com

Crist Instrument
P.O. Box 128
10200 Moxley Road
Damascus, MD 20872
(301) 253-2184   FAX: (301) 253-0069
http://www.cristinstrument.com

Cruachem
See Annovis
http://www.cruachem.com

CS Bio
1300 Industrial Road
San Carlos, CA 94070
(800) 627-2461   FAX: (415) 802-0944
(415) 802-0880
http://www.csbio.com

CS-Chromatographie Service
Am Parir 27
D-52379 Langerwehe, Germany
(49) 2423-40493-0
FAX: (49) 2423-40493-49
http://www.cs-chromatographie.de

Cuno
400 Research Parkway
Meriden, CT 06450
(800) 231-2259   FAX: (203) 238-8716
(203) 237-5541
http://www.cuno.com

Curtin Matheson Scientific
9999 Veterans Memorial Drive
Houston, TX 77038
(800) 392-3353   FAX: (713) 878-3598
(713) 878-3500

CWE
124 Sibley Avenue
Ardmore, PA 19003
(610) 642-7719   FAX: (610) 642-1532
http://www.cwe-inc.com

Cybex Computer Products
4991 Corporate Drive
Huntsville, AL 35805
(800) 932-9239   FAX: (800) 462-9239
http://www.cybex.com

Cygnus Technology
P.O. Box 219
Delaware Water Gap, PA 18327
(570) 424-5701   FAX: (570) 424-5630
http://www.cygnustech.com

Cymbus Biotechnology
Eagle Class, Chandler's Ford
Hampshire SO53 4NF, UK
(44) 1-703-267-676
FAX: (44) 1-703-267-677
http://www.biotech@cymbus.com

Cytogen
600 College Road East
Princeton, NJ 08540
(609) 987-8200   FAX: (609) 987-6450
http://www.cytogen.com

Cytogen Research and Development
89 Bellevue Hill Road
Boston, MA 02132
(617) 325-7774   FAX: (617) 327-2405

CytRx
154 Technology Parkway
Norcross, GA 30092
(800) 345-2987   FAX: (770) 368-0622
(770) 368-9500
http://www.cytrx.com

Dade Behring
Corporate Headquarters
1717 Deerfield Road
Deerfield, IL 60015
(847) 267-5300   FAX: (847) 267-1066
http://www.dadebehring.com

Dagan
2855 Park Avenue
Minneapolis, MN 55407
(612) 827-5959   FAX: (612) 827-6535
http://www.dagan.com

Dako
6392 Via Real
Carpinteria, CA 93013
(800) 235-5763   FAX: (805) 566-6688
(805) 566-6655
http://www.dakousa.com

Dako A/S
42 Produktionsvej
P.O. Box 1359
DK-2600 Glostrup, Denmark
(45) 4492-0044   FAX: (45) 4284-1822

Dakopatts
See Dako A/S

Dalton Chemical Laboratoris
349 Wildcat Road
Toronto, Ontario
M3J 253 Canada
(416) 661-2102   FAX: (416) 661-2108
(800) 567-5060 (in Canada only)
http://www.dalton.com

Damon, IEC
See Thermoquest

Dan Kar Scientific
150 West Street
Wilmington, MA 01887
(800) 942-5542   FAX: (978) 658-0380
(978) 988-9696
http://www.dan-kar.com

DataCell
Falcon Business Park
40 Ivanhoe Road
Finchampstead, Berkshire
RG40 4QQ, UK
(44) 1189 324324
FAX: (44) 1189 324325
http://www.datacell.co.uk
In the US:
(408) 446-3575   FAX: (408) 446-3589
http://www.datacell.com

DataWave Technologies
380 Main Street, Suite 209
Longmont, CO 80501
(800) 736-9283   FAX: (303) 776-8531
(303) 776-8214

Datex-Ohmeda
3030 Ohmeda Drive
Madison, WI 53718
(800) 345-2700   FAX: (608) 222-9147
(608) 221-1551
http://www.us.datex-ohmeda.com

DATU
82 State Street
Geneva, NY 14456
(315) 787-2240   FAX: (315) 787-2397
http://www.nysaes.cornell.edu/datu

David Kopf Instruments
7324 Elmo Street
P.O. Box 636
Tujunga, CA 91043
(818) 352-3274   FAX: (818) 352-3139

Decagon Devices
P.O. Box 835
950 NE Nelson Court
Pullman, WA 99163
(800) 755-2751   FAX: (509) 332-5158
(509) 332-2756
http://www.decagon.com

Decon Labs
890 Country Line Road
Bryn Mawr, PA 19010
(800) 332-6647   FAX: (610) 964-0650
(610) 520-0610
http://www.deconlabs.com

Decon Laboratories
Conway Street
Hove, Sussex BN3 3LY, UK
(44) 1273 739241
FAX: (44) 1273 722088

Degussa
Precious Metals Division
3900 South Clinton Avenue
South Plainfield, NJ 07080
(800) DEGUSSA   FAX: (908) 756-7176
(908) 561-1100
http://www.degussa-huls.com

Deneba Software
1150 NW 72nd Avenue
Miami, FL 33126
(305) 596-5644   FAX: (305) 273-9069
http://www.deneba.com

Deseret Medical
524 West 3615 South
Salt Lake City, UT 84115
(801) 270-8440   FAX: (801) 293-9000

Devcon Plexus
30 Endicott Street
Danvers, MA 01923
(800) 626-7226   FAX: (978) 774-0516
(978) 777-1100
http://www.devcon.com

Developmental Studies Hybridoma Bank
University of Iowa
436 Biology Building
Iowa City, IA 52242
(319) 335-3826   FAX: (319) 335-2077
http://www.uiowa.edu/~dshbwww

DeVilbiss
Division of Sunrise Medical Respiratory
100 DeVilbiss Drive
P.O. Box 635
Somerset, PA 15501
(800) 338-1988   FAX: (814) 443-7572
(814) 443-4881
http://www.sunrisemedical.com

Dharmacon Research
1376 Miners Drive #101
Lafayette, CO 80026
(303) 604-9499   FAX: (303) 604-9680
http://www.dharmacom.com

DiaCheM
Triangle Biomedical
Gardiners Place
West Gillibrands, Lancashire
WN8 9SP, UK
(44) 1695-555581
FAX: (44) 1695-555518
http://www.diachem.co.uk

Diagen
Max-Volmer Strasse 4
D-40724 Hilden, Germany
(49) 2103-892-230
FAX: (49) 2103-892-222

Diagnostic Concepts
6104 Madison Court
Morton Grove, IL 60053
(847) 604-0957

Diagnostic Developments
See DiaCheM

Diagnostic Instruments
6540 Burroughs
Sterling Heights, MI 48314
(810) 731-6000   FAX: (810) 731-6469
http://www.diaginc.com

Diamedix
2140 North Miami Avenue
Miami, FL 33127
(800) 327-4565   FAX: (305) 324-2395
(305) 324-2300

DiaSorin
1990 Industrial Boulevard
Stillwater, MN 55082
(800) 328-1482   FAX: (651) 779-7847
(651) 439-9719
http://www.diasorin.com

Diatome US
321 Morris Road
Fort Washington, PA 19034
(800) 523-5874   FAX: (215) 646-8931
(215) 646-1478
http://www.emsdiasum.com

Difco Laboratories
See Becton Dickinson

Digene
1201 Clopper Road
Gaithersburg, MD 20878
(301) 944-7000   (800) 344-3631
FAX: (301) 944-7121
http://www.digene.com

Digi-Key
701 Brooks Avenue South
Thief River Falls, MN 56701
(800) 344-4539   FAX: (218) 681-3380
(218) 681-6674
http://www.digi-key.com

Digitimer
37 Hydeway
Welwyn Garden City, Hertfordshire
AL7 3BE, UK
(44) 1707-328347
FAX: (44) 1707-373153
http://www.digitimer.com

Dimco-Gray
8200 South Suburban Road
Dayton, OH 45458
(800) 876-8353   FAX: (937) 433-0520
(937) 433-7600
http://www.dimco-gray.com

Dionex
1228 Titan Way
P.O. Box 3603
Sunnyvale, CA 94088
(408) 737-0700   FAX: (408) 730-9403
http://dionex2.promptu.com

Display Systems Biotech
1260 Liberty Way, Suite B
Vista, CA 92083
(800) 697-1111   FAX: (760) 599-9930
(760) 599-0598
http://www.displaysystems.com

Diversified Biotech
1208 VFW Parkway
Boston, MA 02132
(617) 965-8557   FAX: (617) 323-5641
(800) 796-9199
http://www.divbio.com

DNA ProScan
P.O. Box 121585
Nashville, TN 37212
(800) 841-4362   FAX: (615) 292-1436
(615) 298-3524
http://www.dnapro.com

DNAStar
1228 South Park Street
Madison, WI 53715
(608) 258-7420   FAX: (608) 258-7439
http://www.dnastar.com

DNAVIEW
Attn: Charles Brenner
http://www.wco.com
~cbrenner/dnaview.htm

Doall NYC
36-06 48th Avenue
Long Island City, NY 11101
(718) 392-4595   FAX: (718) 392-6115
http://www.doall.com

Dojindo Molecular Technologies
211 Perry Street Parkway, Suite 5
Gaithersburg, MD 20877
(877) 987-2667
http://www.dojindo.com

Dolla Eastern
See Doall NYC

Dolan Jenner Industries
678 Andover Street
Lawrence, MA 08143
(978) 681-8000   (978) 682-2500
http://www.dolan-jenner.com

Dow Chemical
Customer Service Center
2040 Willard H. Dow Center
Midland, MI 48674
(800) 232-2436   FAX: (517) 832-1190
(409) 238-9321
http://www.dow.com

Dow Corning
Northern Europe
Meriden Business Park
Copse Drive
Allesley, Coventry CV5 9RG, UK
(44) 1676 528 000
FAX: (44) 1676 528 001

Dow Corning
P.O. Box 994
Midland, MI 48686
(517) 496-4000
http://www.dowcorning.com

Dow Corning (Lubricants)
2200 West Salzburg Road
Auburn, MI 48611
(800) 248-2481   FAX: (517) 496-6974
(517) 496-6000

Dremel
4915 21st Street
Racine, WI 53406
(414) 554-1390
http://www.dremel.com

Drummond Scientific
500 Parkway
P.O. Box 700
Broomall, PA 19008
(800) 523-7480   FAX: (610) 353-6204
(610) 353-0200
http://www.drummondsci.com

Duchefa Biochemie BV
P.O. Box 2281
2002 CG Haarlem, The Netherlands
31-0-23-5319093
FAX: 31-0-23-5318027
http://www.duchefa.com

Duke Scientific
2463 Faber Place
Palo Alto, CA 94303
(800) 334-3883   FAX: (650) 424-1158
(650) 424-1177
http://www.dukescientific.com

Duke University Marine Laboratory
135 Duke Marine Lab Road
Beaufort, NC 28516-9721
(252) 504-7503   FAX: (252) 504-7648
http://www.env.duke.edu/marinelab

DuPont Biotechnology Systems
See NEN Life Science Products

DuPont Medical Products
See NEN Life Science Products

DuPont Merck Pharmaceuticals
331 Treble Cove Road
Billerica, MA 01862
(800) 225-1572   FAX: (508) 436-7501
http://www.dupontmerck.com

DuPont NEN Products
See NEN Life Science Products

Dynal
5 Delaware Drive
Lake Success, NY 11042
(800) 638-9416   FAX: (516) 326-3298
(516) 326-3270
http://www.dynal.net

Dynal AS
Ullernchausen 52,
0379 Oslo, Norway
47-22-06-10-00   FAX: 47-22-50-70-15
http://www.dynal.no

Dynalab
P.O. Box 112
Rochester, NY 14692
(800) 828-6595   FAX: (716) 334-9496
(716) 334-2060
http://www.dynalab.com

Dynarex
1 International Boulevard
Brewster, NY 10509
(888) DYNAREX   FAX: (914) 279-9601
(914) 279-9600
http://www.dynarex.com

Dynatech
See Dynex Technologies

Dynex Technologies
14340 Sullyfield Circle
Chantilly, VA 22021
(800) 336-4543   FAX: (703) 631-7816
(703) 631-7800
http://www.dynextechnologies.com

Dyno Mill
See Willy A. Bachofen

E.S.A.
22 Alpha Road
Chelmsford, MA 01824
(508) 250-7000   FAX: (508) 250-7090

E.W. Wright
760 Durham Road
Guilford, CT 06437
(203) 453-6410   FAX: (203) 458-6901
http://www.ewwright.com

E-Y Laboratories
107 N. Amphlett Boulevard
San Mateo, CA 94401
(800) 821-0044   FAX: (650) 342-2648
(650) 342-3296
http://www.eylabs.com

Eastman Kodak
1001 Lee Road
Rochester, NY 14650
(800) 225-5352   FAX: (800) 879-4979
(716) 722-5780   FAX: (716) 477-8040
http://www.kodak.com

ECACC
See European Collection of Animal Cell
Cultures

EC Apparatus
See Savant/EC Apparatus

Ecogen, SRL
Gensura Laboratories
Ptge. Dos de Maig
9(08041) Barcelona, Spain
(34) 3-450-2601   FAX: (34) 3-456-0607
http://www.ecogen.com

Ecolab
370 North Wabasha Street
St. Paul, MN 55102
(800) 35-CLEAN   FAX: (651) 225-3098
(651) 352-5326
http://www.ecolab.com

ECO PHYSICS
3915 Research Park Drive, Suite A-3
Ann Arbor, MI 48108
(734) 998-1600   FAX: (734) 998-1180
http://www.ecophysics.com

Edge Biosystems
19208 Orbit Drive
Gaithersburg, MD 20879-4149
(800) 326-2685   FAX: (301) 990-0881
(301) 990-2685
http://www.edgebio.com

Edmund Scientific
101 E. Gloucester Pike
Barrington, NJ 08007
(800) 728-6999   FAX: (856) 573-6263
(856) 573-6250
http://www.edsci.com

EG&G
See Perkin-Elmer

Ekagen
969 C Industry Road
San Carlos, CA 94070
(650) 592-4500   FAX: (650) 592-4500

Elcatech
P.O. Box 10935
Winston-Salem, NC 27108
(336) 544-8613   FAX: (336) 777-3623
(910) 777-3624
http://www.elcatech.com

Electron Microscopy Sciences
321 Morris Road
Fort Washington, PA 19034
(800) 523-5874   FAX: (215) 646-8931
(215) 646-1566
http://www.emsdiasum.com

Electron Tubes
100 Forge Way, Unit F
Rockaway, NJ 07866
(800) 521-8382   FAX: (973) 586-9771
(973) 586-9594
http://www.electrontubes.com

Elicay Laboratory Products, (UK) Ltd.
4 Manborough Mews
Crockford Lane
Basingstoke, Hampshire
RG 248NA, England
(256) 811-118   FAX: (256) 811-116
http://www.elkay-uk.co.uk

Eli Lilly
Lilly Corporate Center
Indianapolis, IN 46285
(800) 545-5979   FAX: (317) 276-2095
(317) 276-2000
http://www.lilly.com

ELISA Technologies
See Neogen

Elkins-Sinn
See Wyeth-Ayerst

EMBI
See European Bioinformatics Institute

EM Science
480 Democrat Road
Gibbstown, NJ 08027
(800) 222-0342   FAX: (856) 423-4389
(856) 423-6300
http://www.emscience.com

EM Separations Technology
See R & S Technology

Endogen
30 Commerce Way
Woburn, MA 01801
(800) 487-4885   FAX: (617) 439-0355
(781) 937-0890
http://www.endogen.com

ENGEL-Loter
HSGM Heatcutting Equipment
& Machines
1865 E. Main Street, No. 5
Duncan, SC 29334
(888) 854-HSGM   FAX: (864) 486-8383
(864) 486-8300
http://www.engelgmbh.com

Enzo Diagnostics
60 Executive Boulevard
Farmingdale, NY 11735
(800) 221-7705   FAX: (516) 694-7501
(516) 694-7070
http://www.enzo.com

Enzogenetics
4197 NW Douglas Avenue
Corvallis, OR 97330
(541) 757-0288

The Enzyme Center
See Charm Sciences

Enzyme Systems Products
486 Lindbergh Avenue
Livermore, CA 94550
(888) 449-2664   FAX: (925) 449-1866
(925) 449-2664
http://www.enzymesys.com

Epicentre Technologies
1402 Emil Street
Madison, WI 53713
(800) 284-8474   FAX: (608) 258-3088
(608) 258-3080
http://www.epicentre.com

Erie Scientific
20 Post Road
Portsmouth, NH 03801
(888) ERIE-SCI   FAX: (603) 431-8996
(603) 431-8410
http://www.eriesci.com

ES Industries
701 South Route 73
West Berlin, NJ 08091
(800) 356-6140   FAX: (856) 753-8484
(856) 753-8400
http://www.esind.com

ESA
22 Alpha Road
Chelmsford, MA 01824
(800) 959-5095   FAX: (978) 250-7090
(978) 250-7000
http://www.esainc.com

Ethicon
Route 22, P.O. Box 151
Somerville, NJ 08876
(908) 218-0707
http://www.ethiconinc.com

Ethicon Endo-Surgery
4545 Creek Road
Cincinnati, OH 45242
(800) 766-9534   FAX: (513) 786-7080

Eurogentec
Parc Scientifique du Sart Tilman
4102 Seraing, Belgium
32-4-240-76-76   FAX: 32-4-264-07-88
http://www.eurogentec.com

European Bioinformatics Institute
Wellcome Trust Genomes Campus
Hinxton, Cambridge CB10 1SD, UK
(44) 1223-49444
FAX: (44) 1223-494468

European Collection of Animal
Cell Cultures (ECACC)
Centre for Applied Microbiology &
Research
Salisbury, Wiltshire SP4 0JG, UK
(44) 1980-612 512
FAX: (44) 1980-611 315
http://www.camr.org.uk

Evergreen Scientific
2254 E. 49th Street
P.O. Box 58248
Los Angeles, CA 90058
(800) 421-6261   FAX: (323) 581-2503
(323) 583-1331
http://www.evergreensci.com

Exalpha Biologicals
20 Hampden Street
Boston, MA 02205
(800) 395-1137   FAX: (617) 969-3872
(617) 558-3625
http://www.exalpha.com

Exciton
P.O. Box 31126
Dayton, OH 45437
(937) 252-2989   FAX: (937) 258-3937
http://www.exciton.com

Extrasynthese
ZI Lyon Nord
SA-BP62
69730 Genay, France
(33) 78-98-20-34
FAX: (33) 78-98-19-45

Factor II
1972 Forest Avenue
P.O. Box 1339
Lakeside, AZ 85929
(800) 332-8688   FAX: (520) 537-8066
(520) 537-8387
http://www.factor2.com

Falcon
See Becton Dickinson Labware

Fenwal
See Baxter Healthcare

Filemaker
5201 Patrick Henry Drive
Santa Clara, CA 95054
(408) 987-7000   (800) 325-2747

Fine Science Tools
202-277 Mountain Highway
North Vancouver, British Columbia
V7J 3P2 Canada
(800) 665-5355   FAX: (800) 665 4544
(604) 980-2481   FAX: (604) 987-3299

Fine Science Tools
373-G Vintage Park Drive
Foster City, CA 94404
(800) 521-2109   FAX: (800) 523-2109
(650) 349-1636   FAX: (630) 349-3729

Fine Science Tools
Fahrtgasse 7-13
D-69117 Heidelberg, Germany
(49) 6221 905050
FAX: (49) 6221 600001
http://www.finescience.com

Finn Aqua
AMSCO Finn Aqua Oy
Teollisuustiez, FIN-04300
Tuusula, Finland
358 025851   FAX: 358 0276019

Finnigan
355 River Oaks Parkway
San Jose, CA 95134
(408) 433-4800   FAX: (408) 433-4821
http://www.finnigan.com

Dr. L. Fischer
Lutherstrasse 25A
D-69120 Heidelberg
Germany
(49) 6221-16-0368
http://home.eplus-online.de/
electroporation

Fisher Chemical Company
Fisher Scientific Limited
112 Colonnade Road
Nepean, Ontario K2E 7L6 Canada
(800) 234-7437   FAX: (800) 463-2996
http://www.fisherscientific.com

Fisher Scientific
2000 Park Lane
Pittsburgh, PA 15275
(800) 766-7000   FAX: (800) 926-1166
(412) 562-8300
http://www3.fishersci.com

W.F. Fisher & Son
220 Evans Way, Suite #1
Somerville, NJ 08876
(908) 707-4050   FAX: (908) 707-4099

Fitzco
5600 Pioneer Creek Drive
Maple Plain, MN 55359
(800) 367-8760   FAX: (612) 479-2880
(612) 479-3489
http://www.fitzco.com

5 Prime → 3 Prime
See 2000 Eppendorf-5 Prime
http://www.5prime.com

Flambeau
15981 Valplast Road
Middlefield, Ohio 44062
(800) 232-3474   FAX: (440) 632-1581
(440) 632-1631
http://www.flambeau.com

Fleisch (Rusch)
2450 Meadowbrook Parkway
Duluth, GA 30096
(770) 623-0816   FAX: (770) 623-1829
http://ruschinc.com

Flow Cytometry Standards
P.O. Box 194344
San Juan, PR 00919
(800) 227-8143   FAX: (787) 758-3267
(787) 753-9341
http://www.fcstd.com

Flow Labs
See ICN Biomedicals

Flow-Tech Supply
P.O. Box 1388
Orange, TX 77631
(409) 882-0306   FAX: (409) 882-0254
http://www.flow-tech.com

Fluid Marketing
See Fluid Metering

Fluid Metering
5 Aerial Way, Suite 500
Sayosett, NY 11791
(516) 922-6050   FAX: (516) 624-8261
http://www.fmipump.com

Fluorochrome
1801 Williams, Suite 300
Denver, CO 80264
(303) 394-1000   FAX: (303) 321-1119

Fluka Chemical
See Sigma-Aldrich

FMC BioPolymer
1735 Market Street
Philadelphia, PA 19103
(215) 299-6000   FAX: (215) 299-5809
http://www.fmc.com

FMC BioProducts
191 Thomaston Street
Rockland, ME 04841
(800) 521-0390   FAX: (800) 362-1133
(207) 594-3400   FAX: (207) 594-3426
http://www.bioproducts.com

Forma Scientific
Milcreek Road
P.O. Box 649
Marietta, OH 45750
(800) 848-3080   FAX: (740) 372-6770
(740) 373-4765
http://www.forma.com

Fort Dodge Animal Health
800 5th Street NW
Fort Dodge, IA 50501
(800) 685-5656   FAX: (515) 955-9193
(515) 955-4600
http://www.ahp.com

Fotodyne
950 Walnut Ridge Drive
Hartland, WI 53029
(800) 362-3686   FAX: (800) 362-3642
(262) 369-7000   FAX: (262) 369-7013
http://www.fotodyne.com

Fresenius HemoCare
6675 185th Avenue NE, Suite 100
Redwood, WA 98052
(800) 909-3872
(425) 497-1197
http://www.freseniusht.com

Fresenius Hemotechnology
See Fresenius HemoCare

Fuji Medical Systems
419 West Avenue
P.O. Box 120035
Stamford, CT 06902
(800) 431-1850   FAX: (203) 353-0926
(203) 324-2000
http://www.fujimed.com

Fujisawa USA
Parkway Center North
Deerfield, IL 60015-2548
(847) 317-1088   FAX: (847) 317-7298

Ernest F. Fullam
900 Albany Shaker Road
Latham, NY 12110
(800) 833-4024   FAX: (518) 785-8647
(518) 785-5533
http://www.fullam.com

Gallard-Schlesinger Industries
777 Zechendorf Boulevard
Garden City, NY 11530
(516) 229-4000   FAX: (516) 229-4015
http://www.gallard-schlessinger.com

Gambro
Box 7373
SE 103 91 Stockholm, Sweden
(46) 8 613 65 00
FAX: (46) 8 611 37 31
In the US: **COBE Laboratories**
225 Union Boulevard
Lakewood, CO 80215
(303) 232-6800   FAX: (303) 231-4915
http://www.gambro.com

Garner Glass
177 Indian Hill Boulevard
Claremont, CA 91711
(909) 624-5071   FAX: (909) 625-0173
http://www.garnerglass.com

Garon Plastics
16 Byre Avenue
Somerton Park, South Australia 5044
(08) 8294-5126   FAX: (08) 8376-1487
http://www.apache.airnet.com.au/~garon

Garren Scientific
9400 Lurline Avenue, Unit E
Chatsworth, CA 91311
(800) 342-3725   FAX: (818) 882-3229
(818) 882-6544
http://www.garren-scientific.com

GATC Biotech AG
Jakob-Stadler-Platz 7
D-78467 Constance, Germany
(49) 07531-8160-0
FAX: (49) 07531-8160-81
http://www.gatc-biotech.com

Gaussian
Carnegie Office Park
Building 6, Suite 230
Carnegie, PA 15106
(412) 279-6700   FAX: (412) 279-2118
http://www.gaussian.com

G.C. Electronics/A.R.C. Electronics
431 Second Street
Henderson, KY 42420
(270) 827-8981   FAX: (270) 827-8256
http://www.arcelectronics.com

GDB (Genome Data Base, Curation)
2024 East Monument Street, Suite 1200
Baltimore, MD 21205
(410) 955-9705   FAX: (410) 614-0434
http://www.gdb.org

GDB (Genome Data Base, Home)
Hospital for Sick Children
555 University Avenue
Toronto, Ontario
M5G 1X8 Canada
(416) 813-8744   FAX: (416) 813-8755
http://www.gdb.org

Gelman Sciences
See Pall-Gelman

Gemini BioProducts
5115-M Douglas Fir Road
Calabasas, CA 90403
(818) 591-3530   FAX: (818) 591-7084

Gen Trak
5100 Campus Drive
Plymouth Meeting, PA 19462
(800) 221-7407   FAX: (215) 941-9498
(215) 825-5115
http://www.informagen.com

Genaissance Pharmaceuticals
5 Science Park
New Haven, CT 06511
(800) 678-9487   FAX: (203) 562-9377
(203) 773-1450
http://www.genaissance.com

GENAXIS Biotechnology
Parc Technologique
10 Avenue Ampère
Montigny le Bretoneux
78180 France
(33) 01-30-14-00-20
FAX: (33) 01-30-14-00-15
http://www.genaxis.com

GenBank
National Center for Biotechnology
Information
National Library of Medicine/NIH
Building 38A, Room 8N805
8600 Rockville Pike
Bethesda, MD 20894
(301) 496-2475   FAX: (301) 480-9241
http://www.ncbi.nlm.nih.gov

Gene Codes
640 Avis Drive
Ann Arbor, MI 48108
(800) 497-4939   FAX: (734) 930-0145
(734) 769-7249
http://www.genecodes.com

Genemachines
935 Washington Street
San Carlos, CA 94070
(650) 508-1634   FAX: (650) 508-1644
(877) 855-4363
http://www.genemachines.com

Genentech
1 DNA Way
South San Francisco, CA 94080
(800) 551-2231   FAX: (650) 225-1600
(650) 225-1000
http://www.gene.com

General Scanning/GSI Luminomics
500 Arsenal Street
Watertown, MA 02172
(617) 924-1010   FAX: (617) 924-7327
http://www.genescan.com

General Valve
Division of Parker Hannifin Pneutronics
19 Gloria Lane
Fairfield, NJ 07004
(800) GVC-VALV
FAX: (800) GVC-1-FAX
http://www.pneutronics.com

Genespan
19310 North Creek Parkway, Suite 100
Bothell, WA 98011
(800) 231-2215   FAX: (425) 482-3005
(425) 482-3003
http://www.genespan.com

Gene Therapy Systems
10190 Telesis Court
San Diego, CA 92122
(858) 457-1919   FAX: (858) 623-9494
http://www.genetherapysystems.com

Généthon Human Genome
Research Center
1 bis rue de l'Internationale
91000 Evry, France
(33) 169-472828
FAX: (33) 607-78698
http://www.genethon.fr

Genetic Microsystems
34 Commerce Way
Wobum, MA 01801
(781) 932-9333   FAX: (781) 932-9433
http://www.genticmicro.com

Genetic Mutant Repository
See Coriell Institute for Medical Research

Genetic Research Instrumentation
Gene House
Queenborough Lane
Rayne, Braintree, Essex CM7 8TF, UK
(44) 1376 332900
FAX: (44) 1376 344724
http://www.gri.co.uk

Genetics Computer Group
575 Science Drive
Madison, WI 53711
(608) 231-5200   FAX: (608) 231-5202
http://www.gcg.com

**Genetics Institute/American Home
Products**
87 Cambridge Park Drive
Cambridge, MA 02140
(617) 876-1170   FAX: (617) 876-0388
http://www.genetics.com

Genetix
63-69 Somerford Road
Christchurch, Dorset BH23 3QA, UK
(44) (0) 1202 483900
FAX: (44)(0) 1202 480289
In the US: (877) 436 3849
US FAX: (888) 522 7499
http://www.genetix.co.uk

Gene Tools
One Summerton Way
Philomath, OR 97370
(541) 9292-7840   FAX: (541) 9292-7841
http://www.gene-tools.com

GeneWorks
P.O. Box 11, Rundle Mall
Adelaide, South Australia 5000, Australia
1800 882 555   FAX: (08) 8234 2699
(08) 8234 2644
http://www.geneworks.com

Genome Systems (INCYTE)
4633 World Parkway Circle
St. Louis, MO 63134
(800) 430-0030   FAX: (314) 427-3324
(314) 427-3222
http://www.genomesystems.com

Genomic Solutions
4355 Varsity Drive, Suite E
Ann Arbor, MI 48108
(877) GENOMIC   FAX: (734) 975-4808
(734) 975-4800
http://www.genomicsolutions.com

Genomyx
See Beckman Coulter

Genosys Biotechnologies
1442 Lake Front Circle, Suite 185
The Woodlands, TX 77380
(281) 363-3693   FAX: (281) 363-2212
http://www.genosys.com

Suppliers

13

575

Genotech
92 Weldon Parkway
St. Louis, MO 63043
(800) 628-7730   FAX: (314) 991-1504
(314) 991-6034

GENSET
876 Prospect Street, Suite 206
La Jolla, CA 92037
(800) 551-5291   FAX: (619) 551-2041
(619) 515-3061
http://www.genset.fr

Gensia Laboratories Ltd.
19 Hughes
Irvine, CA 92718
(714) 455-4700   FAX: (714) 855-8210

Genta
99 Hayden Avenue, Suite 200
Lexington, MA 02421
(781) 860-5150   FAX: (781) 860-5137
http://www.genta.com

GENTEST
6 Henshaw Street
Woburn, MA 01801
(800) 334-5229   FAX: (888) 242-2226
(781) 935-5115   FAX: (781) 932-6855
http://www.gentest.com

Gentra Systems
15200 25th Avenue N., Suite 104
Minneapolis, MN 55447
(800) 866-3039   FAX: (612) 476-5850
(612) 476-5858
http://www.gentra.com

Genzyme
1 Kendall Square
Cambridge, MA 02139
(617) 252-7500   FAX: (617) 252-7600
http://www.genzyme.com
See also R&D Systems

Genzyme Genetics
One Mountain Road
Framingham, MA 01701
(800) 255-7357   FAX: (508) 872-9080
(508) 872-8400
http://www.genzyme.com

George Tiemann & Co.
25 Plant Avenue
Hauppauge, NY 11788
(516) 273-0005   FAX: (516) 273-6199

GIBCO/BRL
A Division of Life Technologies
1 Kendall Square
Grand Island, NY 14072
(800) 874-4226   FAX: (800) 352-1968
(716) 774-6700
http://www.lifetech.com

Gilmont Instruments
A Division of Barnant Company
28N092 Commercial Avenue
Barrington, IL 60010
(800) 637-3739   FAX: (708) 381-7053
http://barnant.com

Gilson
3000 West Beltline Highway
P.O. Box 620027
Middletown, WI 53562
(800) 445-7661
(608) 836-1551
http://www.gilson.com

Glas-Col Apparatus
P.O. Box 2128
Terre Haute, IN 47802
(800) Glas-Col   FAX: (812) 234-6975
(812) 235-6167
http://www.glascol.com

Glaxo Wellcome
Five Moore Drive
Research Triangle Park, NC 27709
(800) SGL-AXO5   FAX: (919) 248-2386
(919) 248-2100
http://www.glaxowellcome.com

Glen Mills
395 Allwood Road
Clifton, NJ 07012
(973) 777-0777   FAX: (973) 777-0070
http://www.glenmills.com

Glen Research
22825 Davis Drive
Sterling, VA 20166
(800) 327-4536   FAX: (800) 934-2490
(703) 437-6191   FAX: (703) 435-9774
http://www.glenresearch.com

Glo Germ
P.O. Box 189
Moab, UT 84532
(800) 842-6622   FAX: (435) 259-5930
http://www.glogerm.com

Glyco
11 Pimentel Court
Novato, CA 94949
(800) 722-2597   FAX: (415) 382-3511
(415) 884-6799
http://www.glyco.com

Gould Instrument Systems
8333 Rockside Road
Valley View, OH 44125
(216) 328-7000   FAX: (216) 328-7400
http://www.gould13.com

Gralab Instruments
See Dimco-Gray

GraphPad Software
5755 Oberlin Drive #110
San Diego, CA 92121
(800) 388-4723   FAX: (558) 457-8141
(558) 457-3909
http://www.graphpad.com

Graseby Anderson
See Andersen Instruments
http://www.graseby.com

Grass Instrument
A Division of Astro-Med
600 East Greenwich Avenue
W. Warwick, RI 02893
(800) 225-5167   FAX: (877) 472-7749
http://www.grassinstruments.com

Greenacre and Misac Instruments
Misac Systems
27 Port Wood Road
Ware, Hertfordshire SF12 9NJ, UK
(44) 1920 463017
FAX: (44) 1920 465136

Greer Labs
639 Nuway Circle
Lenois, NC 28645
(704) 754-5237
http://greerlabs.com

Greiner
Maybachestrasse 2
Postfach 1162
D-7443 Frickenhausen, Germany
(49) 0 91 31/80 79 0
FAX: (49) 0 91 31/80 79 30
http://www.erlangen.com/greiner

GSI Lumonics
130 Lombard Street
Oxnard, CA 93030
(805) 485-5559   FAX: (805) 485-3310
http://www.gsilumonics.com

GTE Internetworking
150 Cambridge Park Drive
Cambridge, MA 02140
(800) 472-4565   FAX: (508) 694-4861
http://www.bbn.com

GW Instruments
35 Medford Street
Somerville, MA 02143
(617) 625-4096   FAX: (617) 625-1322
http://www.gwinst.com

H & H Woodworking
1002 Garfield Street
Denver, CO 80206
(303) 394-3764

Hacker Instruments
17 Sherwood Lane
P.O. Box 10033
Fairfield , NJ 07004
800-442-2537   FAX: (973) 808-8281
(973) 226-8450
http://www.hackerinstruments.com

Haemenetics
400 Wood Road
Braintree, MA 02184
(800) 225-5297   FAX: (781) 848-7921
(781) 848-7100
http://www.haemenetics.com

Halocarbon Products
P.O. Box 661
River Edge, NJ 07661
(201) 242-8899   FAX: (201) 262-0019
http://halocarbon.com

Hamamatsu Photonic Systems
A Division of Hamamatsu
360 Foothill Road
P.O. Box 6910
Bridgewater, NJ 08807
(908) 231-1116   FAX: (908) 231-0852
http://www.photonicsonline.com

Hamilton Company
4970 Energy Way
P.O. Box 10030
Reno, NV 89520
(800) 648-5950   FAX: (775) 856-7259
(775) 858-3000
http://www.hamiltoncompany.com

Hamilton Thorne Biosciences
100 Cummings Center, Suite 102C
Beverly, MA 01915
http://www.hamiltonthorne.com

Hampton Research
27631 El Lazo Road
Laguna Niguel, CA 92677
(800) 452-3899   FAX: (949) 425-1611
(949) 425-6321
http://www.hamptonresearch.com

Harlan Bioproducts for Science
P.O. Box 29176
Indianapolis, IN 46229
(317) 894-7521   FAX: (317) 894-1840
http://www.hbps.com

Harlan Sera-Lab
Hillcrest, Dodgeford Lane
Belton, Loughborough
Leicester LE12 9TE, UK
(44) 1530 222123
FAX: (44) 1530 224970
http://www.harlan.com

Harlan Teklad
P.O. Box 44220
Madison, WI 53744
(608) 277-2070   FAX: (608) 277-2066
http://www.harlan.com

Harrick Scientific Corporation
88 Broadway
Ossining, NY 10562
(914) 762-0020   FAX: (914) 762-0914
http://www.harricksci.com

Harrison Research
840 Moana Court
Palo Alto, CA 94306
(650) 949-1565   FAX: (650) 948-0493

Harvard Apparatus
84 October Hill Road
Holliston, MA 01746
(800) 272-2775   FAX: (508) 429-5732
(508) 893-8999
http://harvardapparatus.com

Harvard Bioscience
See Harvard Apparatus

Haselton Biologics
See JRH Biosciences

Hazelton Research Products
See Covance Research Products

Health Products
See Pierce Chemical

Heat Systems-Ultrasonics
1938 New Highway
Farmingdale, NY 11735
(800) 645-9846   FAX: (516) 694-9412
(516) 694-9555

Heidenhain Corp
333 East State Parkway
Schaumberg, IL 60173
(847) 490-1191   FAX: (847) 490-3931
http://www.heidenhain.com

Hellma Cells
11831 Queens Boulevard
Forest Hills, NY 11375
(718) 544-9166   FAX: (718) 263-6910
http://www.helmaUSA.com

Hellma
Postfach 1163
D-79371 Müllheim/Baden, Germany
(49) 7631-1820
FAX: (49) 7631-13546
http://www.hellma-worldwide.de

Henry Schein
135 Duryea Road, Mail Room 150
Melville, NY 11747
(800) 472-4346   FAX: (516) 843-5652
http://www.henryschein.com

Heraeus Kulzer
4315 South Lafayette Boulevard
South Bend, IN 46614
(800) 343-5336
(219) 291-0661
http://www.kulzer.com

Heraeus Sepatech
See Kendro Laboratory Products

Hercules Aqualon
Aqualon Division
Hercules Research Center, Bldg. 8145
500 Hercules Road
Wilmington, DE 19899
(800) 345-0447   FAX: (302) 995-4787
http://www.herc.com/aqualon/pharma

Heto-Holten A/S
Gydevang 17-19
DK-3450 Allerod, Denmark
(45) 48-16-62-00
FAX: (45) 48-16-62-97
Distributed by ATR

Hettich-Zentrifugen
See Andreas Hettich

Hewlett-Packard
3000 Hanover Street
Mailstop 20B3
Palo Alto, CA 94304
(650) 857-1501   FAX: (650) 857-5518
http://www.hp.com

HGS Hinimoto Plastics
1-10-24 Meguro-Honcho
Megurouko
Tokyo 152, Japan
3-3714-7226   FAX: 3-3714-4657

Hitachi Scientific Instruments
Nissei Sangyo America
8100 N. First Street
San Elsa, CA 95314
(800) 548-9001   FAX: (408) 432-0704
(408) 432-0520
http://www.hii.hitachi.com

Hi-Tech Scientific
Brunel Road
Salisbury, Wiltshire, SP2 7PU
UK
(44) 1722-432320
(800) 344-0724 (US only)
http://www.hi-techsci.co.uk

Hoechst AG
See Aventis Pharmaceutical

Hoefer Scientific Instruments
Division of Amersham-Pharmacia Biotech
800 Centennial Avenue
Piscataway, NJ 08855
(800) 227-4750   FAX: (877) 295-8102
http://www.apbiotech.com

Hoffman-LaRoche
340 Kingsland Street
Nutley, NJ 07110
(800) 526-0189   FAX: (973) 235-9605
(973) 235-5000
http://www.rocheUSA.com

Holborn Surgical and Medical
Instruments
Westwood Industrial Estate
Ramsgate Road
Margate, Kent CT9 4JZ UK
(44) 1843 296666
FAX: (44) 1843 295446

Honeywell
101 Columbia Road
Morristown, NJ 07962
(973) 455-2000   FAX: (973) 455-4807
http://www.honeywell.com

Honeywell Specialty Films
P.O. Box 1039
101 Columbia Road
Morristown, NJ 07962
(800) 934-5679   FAX: (973) 455-6045
http://www.honeywell-specialtyfilms.com

Hood Thermo-Pad Canada
Comp. 20, Site 61A, RR2
Summerland, British Columbia
V0H 1Z0 Canada
(800) 665-9555   FAX: (250) 494-5003
(250) 494-5002
http://www.thermopad.com

Horiba Instruments
17671 Armstrong Avenue
Irvine, CA 92714
(949) 250-4811   FAX: (949) 250-0924
http://www.horiba.com

Hoskins Manufacturing
10776 Hall Road
P.O. Box 218
Hamburg, MI 48139
(810) 231-1900   FAX: (810) 231-4311
http://www.hoskinsmfgco.com

Hosokawa Micron Powder Systems
10 Chatham Road
Summit, NJ 07901
(800) 526-4491   FAX: (908) 273-7432
(908) 273-6360
http://www.hosokawamicron.com

HT Biotechnology
Unit 4
61 Ditton Walk
Cambridge CB5 8QD, UK
(44) 1223-412583

Hugo Sachs Electronik
Postfach 138
7806 March-Hugstetten, Germany
D-79229(49) 7665-92000
FAX: (49) 7665-920090

Human Biologics International
7150 East Camelback Road, Suite 245
Scottsdale, AZ 85251
(480) 990-2005   FAX: (480)-990-2155
http://www.humanbiological.com

Human Genetic Mutant Cell
Repository
See Coriell Institute for Medical Research

HVS Image
P.O. Box 100
Hampton, Middlesex TW12 2YD, UK
FAX: (44) 208 783 1223
In the US: (800) 225-9261
FAX: (888) 483-8033
http://www.hvsimage.com

Hybaid
111-113 Waldegrave Road
Teddington, Middlesex TW11 8LL, UK
(44) 0 1784 42500
FAX: (44) 0 1784 248085
http://www.hybaid.co.uk

Hybaid Instruments
8 East Forge Parkway
Franklin, MA 02028
(888)4-HYBAID   FAX: (508) 541-3041
(508) 541-6918
http://www.hybaid.com

Hybridon
155 Fortune Boulevard
Milford, MA 01757
(508) 482-7500   FAX: (508) 482-7510
http://www.hybridon.com

HyClone Laboratories
1725 South HyClone Road
Logan, UT 84321
(800) HYCLONE   FAX: (800) 533-9450
(801) 753-4584   FAX: (801) 750-0809
http://www.hyclone.com

Hyseq
670 Almanor Avenue
Sunnyvale, CA 94086
(408) 524-8100   FAX: (408) 524-8141
http://www.hyseq.com

IBA GmbH
1508 South Grand Blvd.
St. Louis, MO 63104
(877) 422-4624   FAX: (888) 531-6813
http://www.iba-go.com

IBF Biotechnics
See Sepracor

IBI (International Biotechnologies)
See Eastman Kodak
For technical service (800) 243-2555
(203) 786-5600

ICN Biochemicals
See ICN Biomedicals

ICN Biomedicals
3300 Hyland Avenue
Costa Mesa, CA 92626
(800) 854-0530   FAX: (800) 334-6999
(714) 545-0100   FAX: (714) 641-7275
http://www.icnbiomed.com

ICN Flow and Pharmaceuticals
See ICN Biomedicals

ICN Immunobiochemicals
See ICN Biomedicals

ICN Radiochemicals
See ICN Biomedicals

ICONIX
100 King Street West, Suite 3825
Toronto, Ontario
M5X 1E3 Canada
(416) 410-2411   FAX: (416) 368-3089
http://www.iconix.com

ICRT (Imperial Cancer Research
Technology)
Sardinia House
Sardinia Street
London WC2A 3NL, UK
(44) 1712-421136
FAX: (44) 1718-314991

Idea Scientific Company
P.O. Box 13210
Minneapolis, MN 55414
(800) 433-2535   FAX: (612) 331-4217
http://www.ideascientific.com

IEC
See International Equipment Co.

IITC
23924 Victory Boulevard
Woodland Hills, CA 91367
(888) 414-4482   (818) 710-1556
FAX: (818) 992-5185
http://www.iitcinc.com

IKA Works
2635 N. Chase Parkway, SE
Wilmington, NC 28405
(910) 452-7059   FAX: (910) 452-7693
http://www.ika.net

Ikegami Electronics
37 Brook Avenue
Maywood, NJ 07607
(201) 368-9171   FAX: (201) 569-1626

Ikemoto Scientific Technology
25-11 Hongo
3-chome, Bunkyo-ku
Tokyo 101-0025, Japan
(81) 3-3811-4181
FAX: (81) 3-3811-1960

Imagenetics
See ATC Diagnostics

Imaging Research
c/o Brock University
500 Glenridge Avenue
St. Catharines, Ontario
L2S 3A1 Canada
(905) 688-2040   FAX: (905) 685-5861
http://www.imaging.brocku.ca

Imclone Systems
180 Varick Street
New York, NY 10014
(212) 645-1405   FAX: (212) 645-2054
http://www.imclone.com

IMCO Corporation LTD., AB
P.O. Box 21195
SE-100 31
Stockholm, Sweden
46-8-33-53-09   FAX: 46-8-728-47-76
http://www.imcocorp.se

Imgenex Corporation
11175 Flintkote Avenue
Suite E
San Diego, CA 92121
(888) 723-4363   FAX: (858) 642-0937
(858) 642.0978
http://www.imgenex.com

IMICO
Calle Vivero, No. 5-4a Planta
E-28040, Madrid, Spain
(34) 1-535-3960   FAX: (34) 1-535-2780

Immunex
51 University Street
Seattle, WA 98101
(206) 587-0430   FAX: (206) 587-0606
http://www.immunex.com

Immunocorp
1582 W. Deere Avenue
Suite C
Irvine, CA 92606
(800) 446-3063
http://www.immunocorp.com

Immunotech
130, av. Delattre de Tassigny
B.P. 177
13276 Marseilles Cedex 9
France
(33) 491-17-27-00
FAX: (33) 491-41-43-58
http://www.immunotech.fr

Imperial Chemical Industries
Imperial Chemical House
Millbank, London SW1P 3JF, UK
(44) 171-834-4444
FAX: (44)171-834-2042
http://www.ici.com

Inceltech
See New Brunswick Scientific

Incstar
See DiaSorin

Incyte
6519 Dumbarton Circle
Fremont, CA 94555
(510) 739-2100   FAX: (510) 739-2200
http://www.incyte.com

Incyte Pharmaceuticals
3160 Porter Drive
Palo Alto, CA 94304
(877) 746-2983   FAX: (650) 855-0572
(650) 855-0555
http://www.incyte.com

Individual Monitoring Systems
6310 Harford Road
Baltimore, MD 21214

Indo Fine Chemical
P.O. Box 473
Somerville, NJ 08876
(888) 463-6346   FAX: (908) 359-1179
(908) 359-6778
http://www.indofinechemical.com

Industrial Acoustics
1160 Commerce Avenue
Bronx, NY 10462
(718) 931-8000   FAX: (718) 863-1138
http://www.industrialacoustics.com

Inex Pharmaceuticals
100-8900 Glenlyon Parkway
Glenlyon Business Park
Burnaby, British Columbia
V5J 5J8 Canada
(604) 419-3200   FAX: (604) 419-3201
http://www.inexpharm.com

Ingold, Mettler, Toledo
261 Ballardvale Street
Wilmington, MA 01887
(800) 352-8763   FAX: (978) 658-0020
(978) 658-7615
http://www.mt.com

Innogenetics N.V.
Technologie Park 6
B-9052 Zwijnaarde
Belgium
(32) 9-329-1329   FAX: (32) 9-245-7623
http://www.innogenetics.com

Innovative Medical Services
1725 Gillespie Way
El Cajon, CA 92020
(619) 596-8600   FAX: (619) 596-8700
http://www.imspure.com

Innovative Research
3025 Harbor Lane N, Suite 300
Plymouth, MN 55447
(612) 519-0105   FAX: (612) 519-0239
http://www.inres.com

Innovative Research of America
2 N. Tamiami Trail, Suite 404
Sarasota, FL 34236
(800) 421-8171   FAX: (800) 643-4345
(941) 365-1406   FAX: (941) 365-1703
http://www.innovrsrch.com

Inotech Biosystems
15713 Crabbs Branch Way, #110
Rockville, MD 20855
(800) 635-4070   FAX: (301) 670-2859
(301) 670-2850
http://www.inotechintl.com

INOVISION
22699 Old Canal Road
Yorba Linda, CA 92887
(714) 998-9600   FAX: (714) 998-9666
http://www.inovision.com

Instech Laboratories
5209 Militia Hill Road
Plymouth Meeting, PA 19462
(800) 443-4227   FAX: (610) 941-0134
(610) 941-0132
http://www.instechlabs.com

Instron
100 Royall Street
Canton, MA 02021
(800) 564-8378   FAX: (781) 575-5725
(781) 575-5000
http://www.instron.com

Instrumentarium
P.O. Box 300
00031 Instrumentarium
Helsinki, Finland
(10) 394-5566
http://www.instrumentarium.fi

Instruments SA
Division Jobin Yvon
16-18 Rue du Canal
91165 Longjumeau, Cedex, France
(33)1 6454-1300
FAX: (33)1 6909-9319
http://www.isainc.com

Instrutech
20 Vanderventer Avenue, Suite 101E
Port Washington, NY 11050
(516) 883-1300   FAX: (516) 883-1558
http://www.instrutech.com

Integrated DNA Technologies
1710 Commercial Park
Coralville, IA 52241
(800) 328-2661   FAX: (319) 626-8444
http://www.idtdna.com

Integrated Genetics
See Genzyme Genetics

Integrated Scientific Imaging Systems
3463 State Street, Suite 431
Santa Barbara, CA 93105
(805) 692-2390   FAX: (805) 692-2391
http://www.imagingsystems.com

Integrated Separation Systems (ISS)
See OWL Separation Systems

IntelliGenetics
See Oxford Molecular Group

Interactiva BioTechnologie
Sedanstrasse 10
D-89077 Ulm, Germany
(49) 731-93579-290
FAX: (49) 731-93579-291
http://www.interactiva.de

Interchim
213 J.F. Kennedy Avenue
B.P. 1140
Montlucon
03103 France
(33) 04-70-03-83-55
FAX: (33) 04-70-03-93-60

Interfocus
14/15 Spring Rise
Falcover Road
Haverhill, Suffolk CB9 7XU, UK
(44) 1440 703460
FAX: (44) 1440 704397
http://www.interfocus.ltd.uk

Intergen
2 Manhattanville Road
Purchase, NY 10577
(800) 431-4505   FAX: (800) 468-7436
(914) 694-1700   FAX: (914) 694-1429
http://www.intergenco.com

Intermountain Scientific
420 N. Keys Drive
Kaysville, UT 84037
(800) 999-2901   FAX: (800) 574-7892
(801) 547-5047   FAX: (801) 547-5051
http://www.bioexpress.com

International Biotechnologies (IBI)
See Eastman Kodak

International Equipment Co. (IEC)
See Thermoquest

International Institute for the
Advancement of Medicine
1232 Mid-Valley Drive
Jessup, PA 18434
(800) 486-IIAM   FAX: (570) 343-6993
(570) 496-3400
http://www.iiam.org

International Light
17 Graf Road
Newburyport, MA 01950
(978) 465-5923   FAX: (978) 462-0759

International Market Supply (I.M.S.)
Dane Mill
Broadhurst Lane
Congleton, Cheshire CW12 1LA, UK
(44) 1260 275469
FAX: (44) 1260 276007

International Marketing Services
See International Marketing Ventures

International Marketing Ventures
6301 Ivy Lane, Suite 408
Greenbelt, MD 20770
(800) 373-0096   FAX: (301) 345-0631
(301) 345-2866
http://www.imvlimited.com

International Products
201 Connecticut Drive
Burlington, NJ 08016
(609) 386-8770   FAX: (609) 386-8438
http://www.mkt@ipcol.com

Intracel Corporation
Bartels Division
2005 Sammamish Road, Suite 107
Issaquah, WA 98027
(800) 542-2281   FAX: (425) 557-1894
(425) 392-2992
http://www.intracel.com

Invitrogen
1600 Faraday Avenue
Carlsbad, CA 92008
(800) 955-6288   FAX: (760) 603-7201
(760) 603-7200
http://www.invitrogen.com

In Vivo Metric
P.O. Box 249
Healdsburg, CA 95448
(707) 433-4819   FAX: (707) 433-2407

IRORI
9640 Towne Center Drive
San Diego, CA 92121
(858) 546-1300   FAX: (858) 546-3083
http://www.irori.com

Irvine Scientific
2511 Daimler Street
Santa Ana, CA 92705
(800) 577-6097   FAX: (949) 261-6522
(949) 261-7800
http://www.irvinesci.com

ISC BioExpress
420 North Kays Drive
Kaysville, UT 84037
(800) 999-2901   FAX: (800) 574-7892
(801) 547-5047
http://www.bioexpress.com

ISCO
P.O. Box 5347
4700 Superior
Lincoln, NE 68505
(800) 228-4373   FAX: (402) 464-0318
(402) 464-0231
http://www.isco.com

Isis Pharmaceuticals
Carlsbad Research Center
2292 Faraday Avenue
Carlsbad, CA 92008
(760) 931-9200
http://www.isip.com

Isolabs
See Wallac

ISS
See Integrated Separation Systems

J & W Scientific
See Agilent Technologies

J.A. Webster
86 Leominster Road
Sterling , MA 01564
(800) 225-7911   FAX: (978) 422-8959
http://www.jawebster.com

J.T. Baker
See Mallinckrodt Baker
222 Red School Lane
Phillipsburg, NJ 08865
(800) JTBAKER   FAX: (908) 859-6974
http://www.jtbaker.com

Jackson ImmunoResearch
Laboratories
P.O. Box 9
872 W. Baltimore Pike
West Grove, PA 19390
(800) 367-5296   FAX: (610) 869-0171
(610) 869-4024
http://www.jacksonimmuno.com

The Jackson Laboratory
600 Maine Street
Bar Harbor, ME 04059
(800) 422-6423   FAX: (207) 288-5079
(207) 288-6000
http://www.jax.org

Jaece Industries
908 Niagara Falls Boulevard
North Tonawanda, NY 14120
(716) 694-2811   FAX: (716) 694-2811
http://www.jaece.com

Jandel Scientific
See SPSS

Janke & Kunkel
See Ika Works

Janssen Life Sciences Products
See Amersham

Janssen Pharmaceutica
1125 Trenton-Harbourton Road
Titusville, NJ 09560
(609) 730-2577   FAX: (609) 730-2116
http://us.janssen.com

Jasco
8649 Commerce Drive
Easton, MD 21601
(800) 333-5272   FAX: (410) 822-7526
(410) 822-1220
http://www.jascoinc.com

Jena Bioscience
Loebstedter Str. 78
07749 Jena, Germany
(49) 3641-464920
FAX: (49) 3641-464991
http://www.jenabioscience.com

Jencons Scientific
800 Bursca Drive, Suite 801
Bridgeville, PA 15017
(800) 846-9959   FAX: (412) 257-8809
(412) 257-8861
http://www.jencons.co.uk

JEOL Instruments
11 Dearborn Road
Peabody, MA 01960
(978) 535-5900   FAX: (978) 536-2205
http://www.jeol.com/index.html

Jewett
750 Grant Street
Buffalo, NY 14213
(800) 879-7767   FAX: (716) 881-6092
(716) 881-0030
http://www.JewettInc.com

John's Scientific
See VWR Scientific

John Weiss and Sons
95 Alston Drive
Bradwell Abbey
Milton Keynes, Buckinghamshire
MK1 4HF UK
(44) 1908-318017
FAX: (44) 1908-318708

Johnson & Johnson Medical
2500 Arbrook Boulevard East
Arlington, TX 76004
(800) 423-4018
http://www.jnjmedical.com

Johnston Matthey Chemicals
Orchard Road
Royston, Hertfordshire SG8 5HE, UK
(44) 1763-253000
FAX: (44) 1763-253466
http://www.chemicals.matthey.com

Jolley Consulting and Research
683 E. Center Street, Unit H
Grayslake, IL 60030
(847) 548-2330   FAX: (847) 548-2984
http://www.jolley.com

Jordan Scientific
See Shelton Scientific

Jorgensen Laboratories
1450 N. Van Buren Avenue
Loveland, CO 80538
(800) 525-5614   FAX: (970) 663-5042
(970) 669-2500
http://www.jorvet.com

**JRH Biosciences and
JR Scientific**
13804 W. 107th Street
Lenexa, KS 66215
(800) 231-3735   FAX: (913) 469-5584
(913) 469-5580

Jule Bio Technologies
25 Science Park, #14, Suite 695
New Haven, CT 06511
(800) 648-1772   FAX: (203) 786-5489
(203) 786-5490
http://hometown.aol.com/precastgel/index.htm

K.R. Anderson
2800 Bowers Avenue
Santa Clara, CA 95051
(800) 538-8712   FAX: (408) 727-2959
(408) 727-2800
http://www.kranderson.com

Kabi Pharmacia Diagnostics
See Pharmacia Diagnostics

Kanthal H.P. Reid
1 Commerce Boulevard
P.O. Box 352440
Palm Coast, FL 32135
(904) 445-2000   FAX: (904) 446-2244
http://www.kanthal.com

Kapak
5305 Parkdale Drive
St. Louis Park, MN 55416
(800) KAPAK-57   FAX: (612) 541-0735
(612) 541-0730
http://www.kapak.com

Karl Hecht
Stettener Str. 22-24
D-97647 Sondheim
Rhön, Germany
(49) 9779-8080   FAX: (49) 9779-80888

Karl Storz
Köningin-Elisabeth Str. 60
D-14059 Berlin, Germany
(49) 30-30 69 09-0
FAX: (49) 30-30 19 452
http://www.karlstorz.de

KaVo EWL
P.O. Box 1320
D-88293 Leutkirch im Allgäu, Germany
(49) 7561-86-0   FAX: (49) 7561-86-371
http://www.kavo.com/english/startseite.htm

Keithley Instruments
28775 Aurora Road
Cleveland, OH 44139
(800) 552-1115   FAX: (440) 248-6168
(440) 248-0400
http://www.keithley.com

Kemin
2100 Maury Street, Box 70
Des Moines, IA 50301
(515) 266-2111   FAX: (515) 266-8354
http://www.kemin.com

Kemo
3 Brook Court, Blakeney Road
Beckenham, Kent BR3 1HG, UK
(44) 0181 658 3838
FAX: (44) 0181 658 4084
http://www.kemo.com

Kendall
15 Hampshire Street
Mansfield, MA 02048
(800) 962-9888   FAX: (800) 724-1324
http://www.kendallhq.com

Kendro Laboratory Products
31 Pecks Lane
Newtown, CT 06470
(800) 522-SPIN   FAX: (203) 270-2166
(203) 270-2080
http://www.kendro.com

Kendro Laboratory Products
P.O. Box 1220
Am Kalkberg
D-3360 Osterod, Germany
(55) 22-316-213
FAX: (55) 22-316-202
http://www.heraeus-instruments.de

Kent Laboratories
23404 NE 8th Street
Redmond, WA 98053
(425) 868-6200   FAX: (425) 868-6335
http://www.kentlabs.com

Kent Scientific
457 Bantam Road, #16
Litchfield, CT 06759
(888) 572-8887   FAX: (860) 567-4201
(860) 567-5496
http://www.kentscientific.com

Keuffel & Esser
See Azon

Keystone Scientific
Penn Eagle Industrial Park
320 Rolling Ridge Drive
Bellefonte, PA 16823
(800) 437-2999   FAX: (814) 353-2305
(814) 353-2300 Ext 1
http://www.keystonescientific.com

Kimble/Kontes Biotechnology
1022 Spruce Street
P.O. Box 729
Vineland, NJ 08360
(888) 546-2531   FAX: (856) 794-9762
(856) 692-3600
http://www.kimble-kontes.com

Kinematica AG
Luzernerstrasse 147a
CH-6014 Littau-Luzern, Switzerland
(41) 41 2501257   FAX: (41) 41 2501460
http://www.kinematica.ch

Kin-Tek
504 Laurel Street
LaMarque, TX 77568
(800) 326-3627
FAX: (409) 938-3710
http://www.kin-tek.com

Kipp & Zonen
125 Wilbur Place
Bohemia, NY 11716
(800) 645-2065   FAX: (516) 589-2068
(516) 589-2885
http://www.kippzonen.thomasregister.com
/olc/kippzonen

Kirkegaard & Perry Laboratories
2 Cessna Court
Gaithersburg, MD 20879
(800) 638-3167   FAX: (301) 948-0169
(301) 948-7755
http://www.kpl.com

Kodak
See Eastman Kodak

Kontes Glass
See Kimble/Kontes Biotechnology

Kontron Instruments AG
Postfach CH-8010
Zurich, Switzerland
41-1-733-5733   FAX: 41-1-733-5734

David Kopf Instruments
P.O. Box 636
Tujunga, CA 91043
(818) 352-3274   FAX: (818) 352-3139

Kraft Apparatus
See Glas-Col Apparatus

Kramer Scientific Corporation
711 Executive Boulevard
Valley Cottage, NY 10989
(845) 267-5050   FAX: (845) 267-5550

Kulite Semiconductor Products
1 Willow Tree Road
Leonia, NJ 07605
(201) 461-0900   FAX: (201) 461-0990
http://www.kulite.com

Lab-Line Instruments
15th & Bloomingdale Avenues
Melrose Park, IL 60160
(800) LAB-LINE   FAX: (708) 450-5830
FAX: (800) 450-4LAB
http://www.labline.com

Lab Products
742 Sussex Avenue
P.O. Box 639
Seaford, DE 19973
(800) 526-0469   FAX: (302) 628-4309
(302) 628-4300
http://www.labproductsinc.com

LabRepco
101 Witmer Road, Suite 700
Horsham, PA 19044
(800) 521-0754   FAX: (215) 442-9202
http://www.labrepco.com

Lab Safety Supply
P.O. Box 1368
Janesville, WI 53547
(800) 356-0783   FAX: (800) 543-9910
(608) 754-7160   FAX: (608) 754-1806
http://www.labsafety.com

Lab-Tek Products
See Nalge Nunc International

Labconco
8811 Prospect Avenue
Kansas City, MO 64132
(800) 821-5525   FAX: (816) 363-0130
(816) 333-8811
http://www.labconco.com

Labindustries
See Barnstead/Thermolyne

Labnet International
P.O. Box 841
Woodbridge, NJ 07095
(888) LAB-NET1   FAX: (732) 417-1750
(732) 417-0700
http://www.nationallabnet.com

LABO-MODERNE
37 rue Dombasle
Paris
75015 France
(33) 01-45-32-62-54
FAX: (33) 01-45-32-01-09
http://www.labomoderne.com/fr

Laboratory of Immunoregulation
National Institute of Allergy and
Infectious Diseases/NIH
9000 Rockville Pike
Building 10, Room 11B13
Bethesda, MD 20892
(301) 496-1124

Laboratory Supplies
29 Jefry Lane
Hicksville, NY 11801
(516) 681-7711

Labscan Limited
Stillorgan Industrial Park
Stillorgan
Dublin, Ireland
(353) 1-295-2684
FAX: (353) 1-295-2685
http://www.labscan.ie

Labsystems
See Thermo Labsystems

Labsystems Affinity Sensors
Saxon Way, Bar Hill
Cambridge CB3 8SL, UK
44 (0) 1954 789976
FAX: 44 (0) 1954 789417
http://www.affinity-sensors.com

Labtronics
546 Governors Road
Guelph, Ontario
N1K 1E3 Canada
(519) 763-4930   FAX: (519) 836-4431
http://www.labtronics.com

Labtronix Manufacturing
3200 Investment Boulevard
Hayward, CA 94545
(510) 786-3200   FAX: (510) 786-3268
http://www.labtronix.com

Lafayette Instrument
3700 Sagamore Parkway North
P.O. Box 5729
Lafayette, IN 47903
(800) 428-7545   FAX: (765) 423-4111
(765) 423-1505
http://www.lafayetteinstrument.com

Lambert Instruments
Turfweg 4
9313 TH Leutingewolde
The Netherlands
(31) 50-5018461   FAX: (31) 50-5010034
http://www.lambert-instruments.com

Lancaster Synthesis
P.O. Box 1000
Windham, NH 03087
(800) 238-2324   FAX: (603) 889-3326
(603) 889-3306
http://www.lancastersynthesis-us.com

Lancer
140 State Road 419
Winter Springs, FL 32708
(800) 332-1855   FAX: (407) 327-1229
(407) 327-8488
http://www.lancer.com

LaVision GmbH
Gerhard-Gerdes-Str. 3
D-37079
Goettingen, Germany
(49) 551-50549-0
FAX: (49) 551-50549-11
http://www.lavision.de

Lawshe
See Advanced Process Supply

Laxotan
20, rue Leon Blum
26000 Valence, France
(33) 4-75-41-91-91
FAX: (33) 4-75-41-91-99
http://www.latoxan.com

LC Laboratories
165 New Boston Street
Woburn, MA 01801
(781) 937-0777   FAX: (781) 938-5420
http://www.lclaboratories.com

LC Packings
80 Carolina Street
San Francisco, CA 94103
(415) 552-1855   FAX: (415) 552-1859
http://www.lcpackings.com

LC Services
See LC Laboratories

LECO
3000 Lakeview Avenue
St. Joseph, MI 49085
(800) 292-6141   FAX: (616) 982-8977
(616) 985-5496
http://www.leco.com

Lederle Laboratories
See Wyeth-Ayerst

Lee Biomolecular Research
Laboratories
11211 Sorrento Valley Road, Suite M
San Diego, CA 92121
(858) 452-7700

Suppliers

18

The Lee Company
2 Pettipaug Road
P.O. Box 424
Westbrook, CT 06498
(800) LEE-PLUG   FAX: (860) 399-7058
(860) 399-6281
http://www.theleeco.com

Lee Laboratories
1475 Athens Highway
Grayson, GA 30017
(800) 732-9150   FAX: (770) 979-9570
(770) 972-4450
http://www.leelabs.com

Leica
111 Deer Lake Road
Deerfield, IL 60015
(800) 248-0123   FAX: (847) 405-0147
(847) 405-0123
http://www.leica.com

Leica Microsystems
Imneuenheimer Feld 518
D-69120
Heidelberg, Germany
(49) 6221-41480
FAX: (49) 6221-414833
http://www.leica-microsystems.com

Leinco Technologies
359 Consort Drive
St. Louis, MO 63011
(314) 230-9477   FAX: (314) 527-5545
http://www.leinco.com

Leitz U.S.A.
See Leica

LenderKing Metal Products
8370 Jumpers Hole Road
Millersville, MD 21108
(410) 544-8795   FAX: (410) 544-5069
http://www.lenderking.com

Letica Scientific Instruments
Panlab s.i., c/Loreto 50
08029 Barcelona, Spain
(34) 93-419-0709
FAX: (34) 93-419-7145
http://www.panlab-sl.com

Leybold-Heraeus Trivac DZA
5700 Mellon Road
Export, PA 15632
(412) 327-5700

LI-COR
Biotechnology Division
4308 Progressive Avenue
Lincoln, NE 68504
(800) 645-4267   FAX: (402) 467-0819
(402) 467-0700
http://www.licor.com

Life Science Laboratories
See Adaptive Biosystems

Life Science Resources
Two Corporate Center Drive
Melville, NY 11747
(800) 747-9530   FAX: (516) 844-5114
(516) 844-5085
http://www.astrocam.com

Life Sciences
2900 72nd Street North
St. Petersburg, FL 33710
(800) 237-4323   FAX: (727) 347-2957
(727) 345-9371
http://www.lifesci.com

Life Technologies
9800 Medical Center Drive
P.O. Box 6482
Rockville, MD 20849
(800) 828-6686   FAX: (800) 331-2286
http://www.lifetech.com

Lifecodes
550 West Avenue
Stamford, CT 06902
(800) 543-3263   FAX: (203) 328-9599
(203) 328-9500
http://www.lifecodes.com

Lightnin
135 Mt. Read Boulevard
Rochester, NY 14611
(888) MIX-BEST   FAX: (716) 527-1742
(716) 436-5550
http://www.lightnin-mixers.com

Linear Drives
Luckyn Lane, Pipps Hill
Basildon, Essex SS14 3BW, UK
(44) 1268-287070
FAX: (44) 1268-293344
http://www.lineardrives.com

Linscott's Directory
4877 Grange Road
Santa Rosa, CA 95404
(707) 544-9555   FAX: (415) 389-6025
http://www.linscottsdirectory.co.uk

Linton Instrumentation
Unit 11, Forge Business Center
Upper Rose Lane
Palgrave, Diss, Norfolk IP22 1AP, UK
(44) 1-379-651-344
FAX: (44) 1-379-650-970
http://www.lintoninst.co.uk

List Biological Laboratories
501-B Vandell Way
Campbell, CA 95008
(800) 726-3213   FAX: (408) 866-6364
(408) 866-6363
http://www.listlabs.com

LKB Instruments
See Amersham Pharmacia Biotech

Lloyd Laboratories
604 West Thomas Avenue
Shenandoah, IA 51601
(800) 831-0004   FAX: (712) 246-5245
(712) 246-4000
http://www.lloydinc.com

Loctite
1001 Trout Brook Crossing
Rocky Hill, CT 06067
(860) 571-5100   FAX: (860)571-5465
http://www.loctite.com

Lofstrand Labs
7961 Cessna Avenue
Gaithersburg, MD 20879
(800) 541-0362   FAX: (301) 948-9214
(301) 330-0111
http://www.lofstrand.com

Lomir Biochemical
99 East Main Street
Malone, NY 12953
(877) 425-3604   FAX: (518) 483-8195
(518) 483-7697
http://www.lomir.com

LSL Biolafitte
10 rue de Temara
7810C St.-Germain-en-Laye, France
(33) 1-3061-5260
FAX: (33) 1-3061-5234

Ludl Electronic Products
171 Brady Avenue
Hawthorne, NY 10532
(888) 769-6111   FAX: (914) 769-4759
(914) 769-6111
http://www.ludl.com

Lumigen
24485 W. Ten Mile Road
Southfield, MI 48034
(248) 351-5600   FAX: (248) 351-0518
http://www.lumigen.com

Luminex
12212 Technology Boulevard
Austin, TX 78727
(888) 219-8020   FAX: (512) 258-4173
(512) 219-8020
http://www.luminexcorp.com

LYNX Therapeutics
25861 Industrial Boulevard
Hayward, CA 94545
(510) 670-9300   FAX: (510) 670-9302
http://www.lynxgen.com

Lyphomed
3 Parkway North
Deerfield, IL 60015
(847) 317-8100   FAX: (847) 317-8600

M.E.D. Associates
See Med Associates

Macherey-Nagel
6 South Third Street, #402
Easton, PA 18042
(610) 559-9848   FAX: (610) 559-9878
http://www.macherey-nagel.com

Macherey-Nagel
Valencienner Strasse 11
P.O. Box 101352
D-52313 Dueren, Germany
(49) 2421-969141
FAX: (49) 2421-969199
http://www.macherey-nagel.ch

Mac-Mod Analytical
127 Commons Court
Chadds Ford, PA 19317
800-441-7508   FAX: (610) 358-5993
(610) 358-9696
http://www.mac-mod.com

Mallinckrodt Baker
222 Red School Lane
Phillipsburg, NJ 08865
(800) 582-2537   FAX: (908) 859-6974
(908) 859-2151
http://www.mallbaker.com

Mallinckrodt Chemicals
16305 Swingley Ridge Drive
Chesterfield, MD 63017
(314) 530-2172   FAX: (314) 530-2563
http://www.mallchem.com

Malven Instruments
Enigma Business Park
Grovewood Road
Malven, Worchestershire
WR 141 XZ, United Kingdom

Marinus
1500 Pier C Street
Long Beach, CA 90813
(562) 435-6522   FAX: (562) 495-3120

Markson Science
c/o Whatman Labs Sales
P.O. Box 1359
Hillsboro, OR 97123
(800) 942-8626   FAX: (503) 640-9716
(503) 648-0762

Marsh Biomedical Products
565 Blossom Road
Rochester, NY 14610
(800) 445-2812   FAX: (716) 654-4810
(716) 654-4800
http://www.biomar.com

Marshall Farms USA
5800 Lake Bluff Road
North Rose, NY 14516
(315) 587-2295
e-mail: info@marfarms.com

Martek
6480 Dobbin Road
Columbia, MD 21045
(410) 740-0081   FAX: (410) 740-2985
http://www.martekbio.com

Martin Supply
Distributor of Gerber Scientific
2740 Loch Raven Road
Baltimore, MD 21218
(800) 282-5440   FAX: (410) 366-0134
(410) 366-1696

Mast Immunosystems
630 Clyde Court
Mountain View, CA 94043
(800) 233-MAST   FAX: (650) 969-2745
(650) 961-5501
http://www.mastallergy.com

Matheson Gas Products
P.O. Box 624
959 Route 46 East
Parsippany, NJ 07054
(800) 416-2505   FAX: (973) 257-9393
(973) 257-1100
http://www.mathesongas.com

Mathsoft
1700 Westlake Avenue N., Suite 500
Seattle, WA 98109
(800) 569-0123   FAX: (206) 283-8691
(206) 283-8802
http://www.mathsoft.com

Matreya
500 Tressler Street
Pleasant Gap, PA 16823
(814) 359-5060   FAX: (814) 359-5062
http://www.matreya.com

Matrigel
See Becton Dickinson Labware

Matrix Technologies
22 Friars Drive
Hudson, NH 03051
(800) 345-0206   FAX: (603) 595-0106
(603) 595-0505
http://www.matrixtechcorp.com

MatTek Corp.
200 Homer Avenue
Ashland, Massachusetts 01721
(508) 881-6771   FAX: (508) 879-1532
http://www.mattek.com

Maxim Medical
89 Oxford Road
Oxford OX2 9PD
United Kingdom
44 (0)1865-865943
FAX: 44 (0)1865-865291
http://www.maximmed.com

Mayo Clinic
Section on Engineering
Project #ALA-1, 1982
200 1st Street SW
Rochester, MN 55905
(507) 284-2511   FAX: (507) 284-5988

McGaw
See B. Braun-McGaw

McMaster-Carr
600 County Line Road
Elmhurst, IL 60126
(630) 833-0300   FAX: (630) 834-9427
http://www.mcmaster.com

McNeil Pharmaceutical
See Ortho McNeil Pharmaceutical

MCNC
3021 Cornwallis Road
P.O. Box 12889
Research Triangle Park, NC 27709
(919) 248-1800   FAX: (919) 248-1455
http://www.mcnc.org

MD Industries
5 Revere Drive, Suite 415
Northbrook, IL 60062
(800) 421-8370   FAX: (847) 498-2627
(708) 339-6000
http://www.mdindustries.com

MDS Nordion
447 March Road
P.O. Box 13500
Kanata, Ontario
K2K 1X8 Canada
(800) 465-3666   FAX: (613) 592-6937
(613) 592-2790
http://www.mds.nordion.com

MDS Sciex
71 Four Valley Drive
Concord, Ontario
Canada L4K 4V8
(905) 660-9005   FAX: (905) 660-2600
http://www.sciex.com

Mead Johnson
See Bristol-Meyers Squibb

Med Associates
P.O. Box 319
St. Albans, VT 05478
(802) 527-2343   FAX: (802) 527-5095
http://www.med-associates.com

Medecell
239 Liverpool Road
London N1 1LX, UK
(44) 20-7607-2295
FAX: (44) 20-7700-4156
http://www.medicell.co.uk

Media Cybernetics
8484 Georgia Avenue, Suite 200
Silver Spring, MD 20910
(301) 495-3305   FAX: (301) 495-5964
http://www.mediacy.com

Mediatech
13884 Park Center Road
Herndon, VA 20171
(800) cellgro
(703) 471-5955
http://www.cellgro.com

Medical Systems
See Harvard Apparatus

Medifor
647 Washington Street
Port Townsend, WA 98368
(800) 366-3710   FAX: (360) 385-4402
(360) 385-0722
http://www.medifor.com

MedImmune
35 W. Watkins Mill Road
Gaithersburg, MD 20878
(301) 417-0770   FAX: (301) 527-4207
http://www.medimmune.com

MedProbe AS
P.O. Box 2640
St. Hanshaugen
N-0131 Oslo, Norway
(47) 222 00137   FAX: (47) 222 00189
http://www.medprobe.com

Megazyme
Bray Business Park
Bray, County Wicklow
Ireland
(353) 1-286-1220
FAX: (353) 1-286-1264
http://www.megazyme.com

Melles Griot
4601 Nautilus Court South
Boulder, CO 80301
(800) 326-4363   FAX: (303) 581-0960
(303) 581-0337
http://www.mellesgriot.com

Menzel-Glaser
Postfach 3157
D-38021 Braunschweig, Germany
(49) 531 590080
FAX: (49) 531 509799

E. Merck
Frankfurterstrasse 250
D-64293 Darmstadt 1, Germany
(49) 6151-720

Merck
See EM Science

Merck & Company
Merck National Service Center
P.O. Box 4
West Point, PA 19486
(800) NSC-MERCK
(215) 652-5000
http://www.merck.com

Merck Research Laboratories
See Merck & Company

Merck Sharpe Human Health Division
300 Franklin Square Drive
Somerset, NJ 08873
(800) 637-2579   FAX: (732) 805-3960
(732) 805-0300

Merial Limited
115 Transtech Drive
Athens, GA 30601
(800) MERIAL-1   FAX: (706) 548-0608
(706) 548-9292
http://www.merial.com

Meridian Instruments
P.O. Box 1204
Kent, WA 98035
(253) 854-9914   FAX: (253) 854-9902
http://www.minstrument.com

Meta Systems Group
32 Hammond Road
Belmont, MA 02178
(617) 489-9950   FAX: (617) 489-9952

Metachem Technologies
3547 Voyager Street, Bldg. 102
Torrance, CA 90503
(310) 793-2300   FAX: (310) 793-2304
http://www.metachem.com

Metallhantering
Box 47172
100-74 Stockholm, Sweden
(46) 8-726-9696

MethylGene
7220 Frederick-Banting, Suite 200
Montreal, Quebec
H4S 2A1 Canada
http://www.methylgene.com

Metro Scientific
475 Main Street, Suite 2A
Farmingdale, NY 11735
(800) 788-6247   FAX: (516) 293-8549
(516) 293-9656

Metrowerks
980 Metric Boulevard
Austin, TX 78758
(800) 377-5416
(512) 997-4700
http://www.metrowerks.com

Mettler Instruments
Mettler-Toledo
1900 Polaris Parkway
Columbus, OH 43240
(800) METTLER   FAX: (614) 438-4900
http://www.mt.com

Miami Serpentarium Labs
34879 Washington Loop Road
Punta Gorda, FL 33982
(800) 248-5050   FAX: (813) 639-1811
(813) 639-8888
http://www.miamiserpentarium.com

Michrom BioResources
1945 Industrial Drive
Auburn, CA 95603
(530) 888-6498   FAX: (530) 888-8295
http://www.michrom.com

Mickle Laboratory Engineering
Gomshall, Surrey, UK
(44) 1483-202178

Micra Scientific
A division of Eichrom Industries
8205 S. Cass Ave, Suite 111
Darien, IL 60561
(800) 283-4752   FAX: (630) 963-1928
(630) 963-0320
http://www.micrasci.com

MicroBrightField
74 Hegman Avenue
Colchester, VT 05446
(802) 655-9360   FAX: (802) 655-5245
http://www.microbrightfield.com

Micro Essential Laboratory
4224 Avenue H
Brooklyn, NY 11210
(718) 338-3618   FAX: (718) 692-4491

Micro Filtration Systems
7-3-Chome, Honcho
Nihonbashi, Tokyo, Japan
(81) 3-270-3141

Micro-Metrics
P.O. Box 13804
Atlanta, GA 30324
(770) 986-6015   FAX: (770) 986-9510
http://www.micro-metrics.com

Micro-Tech Scientific
140 South Wolfe Road
Sunnyvale, CA 94086
(408) 730-8324   FAX: (408) 730-3566
http://www.microlc.com

Microbix Biosystems
341 Bering Avenue
Toronto, Ontario
M8Z 3A8 Canada
1-800-794-6694   FAX: 416-234-1626
1-416-234-1624
http://www.microbix.com

MicroCal
22 Industrial Drive East
Northampton, MA 01060
(800) 633-3115   FAX: (413) 586-0149
(413) 586-7720
http://www.microcalorimetry.com

Microfluidics
30 Ossipee Road
P.O. Box 9101
Newton, MA 02164
(800) 370-5452   FAX: (617) 965-1213
(617) 969-5452
http://www.microfluidicscorp.com

Microgon
See Spectrum Laboratories

Microlase Optical Systems
West of Scotland Science Park
Kelvin Campus, Maryhill Road
Glasgow G20 0SP, UK
(44) 141-948-1000
FAX: (44) 141-946-6311
http://www.microlase.co.uk

Micron Instruments
4509 Runway Street
Simi Valley, CA 93063
(800) 638-3770   FAX: (805) 522-4982
(805) 552-4676
http://www.microninstruments.com

Micron Separations
See MSI

Micro Photonics
4949 Liberty Lane, Suite 170
P.O. Box 3129
Allentown, PA 18106
(610) 366-7103   FAX: (610) 366-7105
http://www.microphotonics.com

MicroTech
1420 Conchester Highway
Boothwyn, PA 19061
(610) 459-3514

Midland Certified Reagent Company
3112-A West Cuthbert Avenue
Midland, TX 79701
(800) 247-8766   FAX: (800) 359-5789
(915) 694-7950   FAX: (915) 694-2387
http://www.mcrc.com

Midwest Scientific
280 Vance Road
Valley Park, MO 63088
(800) 227-9997   FAX: (636) 225-9998
(636) 225-9997
http://www.midsci.com

Miles
See Bayer

Miles Laboratories
See Serological

Miles Scientific
See Nunc

Millar Instruments
P.O. Box 230227
6001-A Gulf Freeway
Houston, TX 77023
(713) 923-9171   FAX: (713) 923-7757
http://www.millarinstruments.com

MilliGen/Biosearch
See Millipore

Millipore
80 Ashbury Road
P.O. Box 9125
Bedford, MA 01730
(800) 645-5476   FAX: (781) 533-3110
(781) 533-6000
http://www.millipore.com

Miltenyi Biotec
251 Auburn Ravine Road, Suite 208
Auburn, CA 95603
(800) 367-6227   FAX: (530) 888-8925
(530) 888-8871
http://www.miltenyibiotec.com

Miltex
6 Ohio Drive
Lake Success, NY 11042
(800) 645-8000   FAX: (516) 775-7185
(516) 349-0001

Milton Roy
See Spectronic Instruments

Mini-Instruments
15 Burnham Business Park
Springfield Road
Burnham-on-Crouch, Essex CM0 8TE, UK
(44) 1621-783282
FAX: (44) 1621-783132
http://www.mini-instruments.co.uk

Mini Mitter
P.O. Box 3386
Sunriver, OR 97707
(800) 685-2999   FAX: (541) 593-5604
(541) 593-8639
http://www.minimitter.com

Mirus Corporation
505 S. Rosa Road
Suite 104
Madison, WI 53719
(608) 441-2852   FAX: (608) 441-2849
http://www.genetransfer.com

Misonix
1938 New Highway
Farmingdale, NY 11735
(800) 645-9846   FAX: (516) 694-9412
http://www.misonix.com

Mitutoyo (MTI)
See Dolla Eastern

MJ Research
Waltham, MA 02451
(800) PELTIER   FAX: (617) 923-8080
(617) 923-8000
http://www.mjr.com

Modular Instruments
228 West Gay Street
Westchester, PA 19380
(610) 738-1420   FAX: (610) 738-1421
http://www.mi2.com

Molecular Biology Insights
8685 US Highway 24
Cascade, CO 80809-1333
(800) 747-4362   FAX: (719) 684-7989
(719) 684-7988
http://www.oligo.net

Molecular Biosystems
10030 Barnes Canyon Road
San Diego, CA 92121
(858) 452-0681   FAX: (858) 452-6187
http://www.mobi.com

Molecular Devices
1312 Crossman Avenue
Sunnyvale, CA 94089
(800) 635-5577   FAX: (408) 747-3602
(408) 747-1700
http://www.moldev.com

Molecular Designs
1400 Catalina Street
San Leandro, CA 94577
(510) 895-1313   FAX: (510) 614-3608

Molecular Dynamics
928 East Arques Avenue
Sunnyvale, CA 94086
(800) 333-5703   FAX: (408) 773-1493
(408) 773-1222
http://www.apbiotech.com

Molecular Probes
4849 Pitchford Avenue
Eugene, OR 97402
(800) 438-2209   FAX: (800) 438-0228
(541) 465-8300   FAX: (541) 344-6504
http://www.probes.com

Molecular Research Center
5645 Montgomery Road
Cincinnati, OH 45212
(800) 462-9868   FAX: (513) 841-0080
(513) 841-0900
http://www.mrcgene.com

Molecular Simulations
9685 Scranton Road
San Diego, CA 92121
(800) 756-4674   FAX: (858) 458-0136
(858) 458-9990
http://www.msi.com

Monoject Disposable Syringes & Needles/Syrvet
16200 Walnut Street
Waukee, IA 50263
(800) 727-5203   FAX: (515) 987-5553
(515) 987-5554
http://www.syrvet.com

Monsanto Chemical
800 North Lindbergh Boulevard
St. Louis, MO 63167
(314) 694-1000   FAX: (314) 694-7625
http://www.monsanto.com

Moravek Biochemicals
577 Mercury Lane
Brea, CA 92821
(800) 447-0100   FAX: (714) 990-1824
(714) 990-2018
http://www.moravek.com

Moss
P.O. Box 189
Pasadena, MD 21122
(800) 932-6677   FAX: (410) 768-3971
(410) 768-3442
http://www.mosssubstrates.com

Motion Analysis
3617 Westwind Boulevard
Santa Rosa, CA 95403
(707) 579-6500   FAX: (707) 526-0629
http://www.motionanalysis.com

Mott
Farmington Industrial Park
84 Spring Lane
Farmington, CT 06032
(860) 747-6333   FAX: (860) 747-6739
http://www.mottcorp.com

MSI (Micron Separations)
See Osmonics

Multi Channel Systems
Markwiesenstrasse 55
72770 Reutlingen, Germany
(49) 7121-503010
FAX: (49) 7121-503011
http://www.multichannelsystems.com

Multiple Peptide Systems
3550 General Atomics Court
San Diego, CA 92121
(800) 338-4965   FAX: (800) 654-5592
(858) 455-3710   FAX: (858) 455-3713
http://www.mps-sd.com

Murex Diagnostics
3075 Northwoods Circle
Norcross, GA 30071
(707) 662-0660   FAX: (770) 447-4989

MWG-Biotech
Anzinger Str. 7
D-85560 Ebersberg, Germany
(49) 8092-82890 FAX: (49) 8092-21084
http://www.mwg_biotech.com

Myriad Industries
3454 E Street
San Diego, CA 92102
(800) 999-6777 FAX: (619) 232-4819
(619) 232-6700
http://www.myriadindustries.com

Nacalai Tesque
Nijo Karasuma, Nakagyo-ku
Kyoto 604, Japan
81-75-251-1723
FAX: 81-75-251-1762
http://www.nacalai.co.jp

Nalge Nunc International
Subsidiary of Sybron International
75 Panorama Creek Drive
P.O. Box 20365
Rochester, NY 14602
(800) 625-4327 FAX: (716) 586-8987
(716) 264-9346
http://www.nalgenunc.com

Nanogen
10398 Pacific Center Court
San Diego, CA 92121
(858) 410-4600 FAX: (858) 410-4848
http://www.nanogen.com

Nanoprobes
95 Horse Block Road
Yaphank, NY 11980
(877) 447-6266 FAX: (631) 205-9493
(631) 205-9490
http://www.nanoprobes.com

Narishige USA
1710 Hempstead Turnpike
East Meadow, NY 11554
(800) 445-7914 FAX: (516) 794-0066
(516) 794-8000
http://www.narishige.co.jp

National Bag Company
2233 Old Mill Road
Hudson, OH 44236
(800) 247-6000 FAX: (330) 425-9800
(330) 425-2600
http://www.nationalbag.com

National Band and Tag
Department X 35, Box 72430
Newport, KY 41032
(606) 261-2035 FAX: (800) 261-8247
https://www.nationalband.com

National Biosciences
See Molecular Biology Insights

National Diagnostics
305 Patton Drive
Atlanta, GA 30336
(800) 526-3867 FAX: (404) 699-2077
(404) 699-2121
http://www.nationaldiagnostics.com

National Institute of Standards and Technology
100 Bureau Drive
Gaithersburg, MD 20899
(301) 975-NIST FAX: (301) 926-1630
http://www.nist.gov

National Instruments
11500 North Mopac Expressway
Austin, TX 78759
(512) 794-0100 FAX: (512) 683-8411
http://www.ni.com

National Labnet
See Labnet International

National Scientific Instruments
975 Progress Circle
Lawrenceville, GA 300243
(800) 332-3331 FAX: (404) 339-7173
http://www.nationalscientific.com

National Scientific Supply
1111 Francisco Bouldvard East
San Rafael, CA 94901
(800) 525-1779 FAX: (415) 459-2954
(415) 459-6070
http://www.nat-sci.com

Naz-Dar-KC Chicago
Nazdar
1087 N. North Branch Street
Chicago, IL 60622
(800) 736-7636 FAX: (312) 943-8215
(312) 943-8338
http://www.nazdar.com

NB Labs
1918 Avenue A
Denison, TX 75021
(903) 465-2694 FAX: (903) 463-5905
http://www.nblabslarry.com

NEB
See New England Biolabs

NEN Life Science Products
549 Albany Street
Boston, MA 02118
(800) 551-2121 FAX: (617) 451-8185
(617) 350-9075
http://www.nen.com

NEN Research Products, Dupont (UK)
Diagnostics and Biotechnology Systems
Wedgewood Way
Stevenage, Hertfordshire SG1 4QN, UK
44-1438-734831
44-1438-734000
FAX: 44-1438-734836
http://www.dupont.com

Neogen
628 Winchester Road
Lexington, KY 40505
(800) 477-8201 FAX: (606) 255-5532
(606) 254-1221
http://www.neogen.com

Neosystems
380, 11012 Macleod Trail South
Calgary, Alberta
T2J 6A5 Canada
(403) 225-9022 FAX: (403) 225-9025
http://www.neosystems.com

Neuralynx
2434 North Pantano Road
Tucson, AZ 85715
(520) 722-8144 FAX: (520) 722-8163
http://www.neuralynx.com

Neuro Probe
16008 Industrial Drive
Gaithersburg, MD 20877
(301) 417-0014 FAX: (301) 977-5711
http://www.neuroprobe.com

Neurocrine Biosciences
10555 Science Center Drive
San Diego, CA 92121
(619) 658-7600 FAX: (619) 658-7602
http://www.neurocrine.com

Nevtek
HCR03, Box 99
Burnsville, VA 24487
(540) 925-2322 FAX: (540) 925-2323
http://www.nevtek.com

New Brunswick Scientific
44 Talmadge Road
Edison, NJ 08818
(800) 631-5417 FAX: (732) 287-4222
(732) 287-1200
http://www.nbsc.com

New England Biolabs (NEB)
32 Tozer Road
Beverly, MA 01915
(800) 632-5227 FAX: (800) 632-7440
http://www.neb.com

New England Nuclear (NEN)
See NEN Life Science Products

New MBR
Gubelstrasse 48
CH8050 Zurich, Switzerland
(41) 1-313-0703

Newark Electronics
4801 N. Ravenswood Avenue
Chicago, IL 60640
(800) 4-NEWARK FAX: (773) 907-5339
(773) 784-5100
http://www.newark.com

Newell Rubbermaid
29 E. Stephenson Street
Freeport, IL 61032
(815) 235-4171 FAX: (815) 233-8060
http://www.newellco.com

Newport Biosystems
1860 Trainor Street
Red Bluff, CA 96080
(530) 529-2448 FAX: (530) 529-2648

Newport
1791 Deere Avenue
Irvine, CA 92606
(800) 222-6440 FAX: (949) 253-1800
(949) 253-1462
http://www.newport.com

Nexin Research B.V.
P.O. Box 16
4740 AA Hoeven, The Netherlands
(31) 165-503172
FAX: (31) 165-502291

NIAID
See Bio-Tech Research Laboratories

Nichiryo
230 Route 206
Building 2-2C
Flanders, NJ 07836
(877) 548-6667 FAX: (973) 927-0099
(973) 927-4001
http://www.nichiryo.com

Nichols Institute Diagnostics
33051 Calle Aviador
San Juan Capistrano, CA 92675
(800) 286-4NID FAX: (949) 240-5273
(949) 728-4610
http://www.nicholsdiag.com

Nichols Scientific Instruments
3334 Brown Station Road
Columbia, MO 65202
(573) 474-5522 FAX: (603) 215-7274
http://home.beseen.com
technology/nsi_technology

Nicolet Biomedical Instruments
5225 Verona Road, Building 2
Madison, WI 53711
(800) 356-0007 FAX: (608) 441-2002
(608) 273-5000
http://nicoletbiomedical.com

N.I.G.M.S. (National Institute of
General Medical Sciences)
See Coriell Institute for Medical Research

Nikon
Science and Technologies Group
1300 Walt Whitman Road
Melville, NY 11747
(516) 547-8500 FAX: (516) 547-4045
http://www.nikonusa.com

Nippon Gene
1-29, Ton-ya-machi
Toyama 930, Japan
(81) 764-51-6548
FAX: (81) 764-51-6547

Noldus Information Technology
751 Miller Drive
Suite E-5
Leesburg, VA 20175
(800) 355-9541 FAX: (703) 771-0441
(703) 771-0440
http://www.noldus.com

Nordion International
See MDS Nordion

Suppliers

22

North American Biologicals (NABI)
16500 NW 15th Avenue
Miami, FL 33169
(800) 327-7106   (305) 625-5305
http://www.nabi.com

North American Reiss
See Reiss

Northwestern Bottle
24 Walpole Park South
Walpole, MA 02081
(508) 668-8600   FAX: (508) 668-7790

NOVA Biomedical
Nova Biomedical 200
Prospect Street Waltham, MA 02454
(800) 822-0911   FAX: (781) 894-5915
http://www.novabiomedical.com

Novagen
601 Science Drive
Madison, WI 53711
(800) 526-7319   FAX: (608) 238-1388
(608) 238-6110
http://www.novagen.com

Novartis
59 Route 10
East Hanover, NJ 07936
(800)526-0175   FAX: (973) 781-6356
http://www.novartis.com

Novartis Biotechnology
3054 Cornwallis Road
Research Triangle Park, NC 27709
(888) 462-7288   FAX: (919) 541-8585
http://www.novartis.com

Nova Sina AG
Subsidiary of Airflow Lufttechnik GmbH
Kleine Heeg 21
52259 Rheinbach, Germany
(49) 02226 920-0
FAX: (49) 02226 9205-11

Novex/Invitrogen
1600 Faraday
Carlsbad, CA 92008
(800) 955-6288   FAX: (760) 603-7201
http://www.novex.com

Novo Nordisk Biochem
77 Perry Chapel Church Road
Franklington, NC 27525
(800) 879-6686   FAX: (919) 494-3450
(919) 494-3000
http://www.novo.dk

Novo Nordisk BioLabs
See Novo Nordisk Biochem

Novocastra Labs
Balliol Business Park West
Benton Lane
Newcastle-upon-Tyne
Tyne and Wear NE12 8EW, UK
(44) 191-215-0567
FAX: (44) 191-215-1152
http://www.novocastra.co.uk

Novus Biologicals
P.O. Box 802
Littleton, CO 80160
(888) 506-6887   FAX: (303) 730-1966
http://www.novus-biologicals.com/
main.html

NPI Electronic
Hauptstrasse 96
D-71732 Tamm, Germany
(49) 7141-601534
FAX: (49) 7141-601266
http://www.npielectronic.com

NSG Precision Cells
195G Central Avenue
Farmingdale, NY 11735
(516) 249-7474   FAX: (516) 249-8575
http://www.nsgpci.com

Nu Chek Prep
109 West Main
P.O. Box 295
Elysian, MN 56028
(800) 521-7728   FAX: (507) 267-4790
(507) 267-4689

Nuclepore
See Costar

Numonics
101 Commerce Drive
Montgomeryville, PA 18936
(800) 523-6716   FAX: (215) 361-0167
(215) 362-2766
http://www.interactivewhiteboards.com

NYCOMED AS Pharma
c/o Accurate Chemical & Scientific
300 Shames Drive
Westbury, NY 11590
(800) 645-6524   FAX: (516) 997-4948
(516) 333-2221
http://www.accuratechemical.com

Nycomed Amersham
Health Care Division
101 Carnegie Center
Princeton, NJ 08540
(800) 832-4633   FAX: (800) 807-2382
(609) 514-6000
http://www.nycomed-amersham.com

Nyegaard
Herserudsvagen 5254
S-122 06 Lidingo, Sweden
(46) 8-765-2930

Ohmeda Catheter Products
See Datex-Ohmeda

Ohwa Tsusbo
Hiby Dai Building
1-2-2 Uchi Saiwai-cho
Chiyoda-ku
Tokyo 100, Japan
03-3591-7348   FAX: 03-3501-9001

Oligos Etc.
9775 S.W. Commerce Circle, C-6
Wilsonville, OR 97070
(800) 888-2358   FAX: (503) 6822D1635
(503) 6822D1814
http://www.oligoetc.com

Olis Instruments
130 Conway Drive
Bogart, GA 30622
(706) 353-6547   (800) 852-3504
http://www.olisweb.com

Olympus America
2 Corporate Center Drive
Melville, NY 11747
(800) 645-8160   FAX: (516) 844-5959
(516) 844-5000
http://www.olympusamerica.com

Omega Engineering
One Omega Drive
P.O. Box 4047
Stamford, CT 06907
(800) 848-4286   FAX: (203) 359-7700
(203) 359-1660
http://www.omega.com

Omega Optical
3 Grove Street
P.O. Box 573
Brattleboro, VT 05302
(802) 254-2690   FAX: (802) 254-3937
http://www.omegafilters.com

Omnetics Connector Corporation
7260 Commerce Circle
East Minneapolis, MN 55432
(800) 343-0025   (763) 572-0656
Fax: (763) 572-3925
http://www.omnetics.com/main.htm

Omni International
6530 Commerce Court
Warrenton, VA 20187
(800) 776-4431   FAX: (540) 347-5352
(540) 347-5331
http://www.omni-inc.com

Omnion
2010 Energy Drive
P.O. Box 879
East Troy, WI 53120
(262) 642-7200   FAX: (262) 642-7760
http://www.omnion.com

Omnitech Electronics
See AccuScan Instruments

Oncogene Research Products
P.O. Box Box 12087
La Jolla, CA 92039-2087
(800) 662-2616   FAX: (800) 766-0999
http://www.apoptosis.com

Oncogene Science
See OSI Pharmaceuticals

Oncor
See Intergen

Online Instruments
130 Conway Drive, Suites A & B
Bogart, GA 30622
(800) 852-3504   (706) 353-1972
(706) 353-6547
http://www.olisweb.com

Operon Technologies
1000 Atlantic Avenue
Alameda, CA 94501
(800) 688-2248   FAX: (510) 865-5225
(510) 865-8644
http://www.operon.com

Optiscan
P.O. Box 1066
Mount Waverly MDC, Victoria
Australia 3149
61-3-9538 3333   FAX: 61-3-9562 7742
http://www.optiscan.com.au

Optomax
9 Ash Street
P.O. Box 840
Hollis, NH 03049
(603) 465-3385   FAX: (603) 465-2291

Opto-Line Associates
265 Ballardvale Street
Wilmington, MA 01887
(978) 658-7255   FAX: (978) 658-7299
http://www.optoline.com

Orbigen
6827 Nancy Ridge Drive
San Diego, CA 92121
(866) 672-4436   (858) 362-2030
(858) 362-2026
http://www.orbigen.com

Oread BioSaftey
1501 Wakarusa Drive
Lawrence, KS 66047
(800) 447-6501   FAX: (785) 749-1882
(785) 749-0034
http://www.oread.com

Organomation Associates
266 River Road West
Berlin, MA 01503
(888) 978-7300   FAX: (978)838-2786
(978) 838-7300
http://www.organomation.com

Organon
375 Mount Pleasant Avenue
West Orange, NJ 07052
(800) 241-8812   FAX: (973) 325-4589
(973) 325-4500
http://www.organon.com

Organon Teknika (Canada)
30 North Wind Place
Scarborough, Ontario
M1S 3R5 Canada
(416) 754-4344   FAX: (416) 754-4488
http://www.organonteknika.com

Organon Teknika Cappel
100 Akzo Avenue
Durham, NC 27712
(800) 682-2666 FAX: (800) 432-9682
(919) 620-2000 FAX: (919) 620-2107
http://www.organonteknika.com

Oriel Corporation of America
150 Long Beach Boulevard
Stratford, CT 06615
(203) 377-8282 FAX: (203) 378-2457
http://www.oriel.com

OriGene Technologies
6 Taft Court, Suite 300
Rockville, MD 20850
(888) 267-4436 FAX: (301) 340-9254
(301) 340-3188
http://www.origene.com

OriginLab
One Roundhouse Plaza
Northhampton, MA 01060
(800) 969-7720 FAX: (413) 585-0126
http://www.originlab.com

Orion Research
500 Cummings Center
Beverly, MA 01915
(800) 225-1480 FAX: (978) 232-6015
(978) 232-6000
http://www.orionres.com

Ortho Diagnostic Systems
Subsidiary of Johnson & Johnson
1001 U.S. Highway 202
P.O. Box 350
Raritan, NJ 08869
(800) 322-6374 FAX: (908) 218-8582
(908) 218-1300

Ortho McNeil Pharmaceutical
Welsh & McKean Road
Spring House, PA 19477
(800) 682-6532
(215) 628-5000
http://www.orthomcneil.com

Oryza
200 Turnpike Road, Unit 5
Chelmsford, MA 01824
(978) 256-8183 FAX: (978) 256-7434
http://www.oryzalabs.com

OSI Pharmaceuticals
106 Charles Lindbergh Boulevard
Uniondale, NY 11553
(800) 662-2616 FAX: (516) 222-0114
(516) 222-0023
http://www.osip.com

Osmonics
135 Flanders Road
P.O. Box 1046
Westborough, MA 01581
(800) 444-8212 FAX: (508) 366-5840
(508) 366-8210
http://www.osmolabstore.com

Oster Professional Products
150 Cadillac Lane
McMinnville, TN 37110
(931) 668-4121 FAX: (931) 668-4125
http://www.sunbeam.com

Out Patient Services
1260 Holm Road
Petaluma, CA 94954
(800) 648-1666 FAX: (707) 762-7198
(707) 763-1581

OWL Scientific Plastics
See OWL Separation Systems

OWL Separation Systems
55 Heritage Avenue
Portsmouth, NH 03801
(800) 242-5560 FAX: (603) 559-9258
(603) 559-9297
http://www.owlsci.com

Oxford Biochemical Research
P.O. Box 522
Oxford, MI 48371
(800) 692-4633 FAX: (248) 852-4466
http://www.oxfordbiomed.com

Oxford GlycoSystems
See Glyco

Oxford Instruments
Old Station Way
Eynsham
Witney, Oxfordshire OX8 1TL, UK
(44) 1865-881437
FAX: (44) 1865-881944
http://www.oxinst.com

Oxford Labware
See Kendall

Oxford Molecular Group
Oxford Science Park
The Medawar Centre
Oxford OX4 4GA, UK
(44) 1865-784600
FAX: (44) 1865-784601
http://www.oxmol.co.uk

Oxford Molecular Group
2105 South Bascom Avenue, Suite 200
Campbell, CA 95008
(800) 876-9994 FAX: (408) 879-6302
(408) 879-6300
http://www.oxmol.com

OXIS International
6040 North Cutter Circle
Suite 317
Portland, OR 97217
(800) 547-3686 FAX: (503) 283-4058
(503) 283-3911
http://www.oxis.com

Oxoid
800 Proctor Avenue
Ogdensburg, NY 13669
(800) 567-8378 FAX: (613) 226-3728
http://www.oxoid.ca

Oxoid
Wade Road
Basingstoke, Hampshire RG24 8PW, UK
(44) 1256-841144
FAX: (4) 1256-814626
http://www.oxoid.ca

Oxyrase
P.O. Box 1345
Mansfield, OH 44901
(419) 589-8800 FAX: (419) 589-9919
http://www.oxyrase.com

Ozyme
10 Avenue Ampère
Montigny de Bretoneux
78180 France
(33) 13-46-02-424
FAX: (33) 13-46-09-212
http://www.ozyme.fr

PAA Laboratories
2570 Route 724
P.O. Box 435
Parker Ford, PA 19457
(610) 495-9400 FAX: (610) 495-9410
http://www.paa-labs.com

Pacer Scientific
5649 Valley Oak Drive
Los Angeles, CA 90068
(323) 462-0636 FAX: (323) 462-1430
http://www.pacersci.com

Pacific Bio-Marine Labs
P.O. Box 1348
Venice, CA 90294
(310) 677-1056 FAX: (310) 677-1207

Packard Instrument
800 Research Parkway
Meriden, CT 06450
(800) 323-1891 FAX: (203) 639-2172
(203) 238-2351
http://www.packardinst.com

Padgett Instrument
1730 Walnut Street
Kansas City, MO 64108
(816) 842-1029

Pall Filtron
50 Bearfoot Road
Northborough, MA 01532
(800) FILTRON FAX: (508) 393-1874
(508) 393-1800

Pall-Gelman
25 Harbor Park Drive
Port Washington, NY 11050
(800) 289-6255 FAX: (516) 484-2651
(516) 484-3600
http://www.pall.com

PanVera
545 Science Drive
Madison, WI 53711
(800) 791-1400 FAX: (608) 233-3007
(608) 233-9450
http://www.panvera.com

Parke-Davis
See Warner-Lambert

Parr Instrument
211 53rd Street
Moline, IL 61265
(800) 872-7720 FAX: (309) 762-9453
(309) 762-7716
http://www.parrinst.com

Partec
Otto Hahn Strasse 32
D-48161 Munster, Germany
(49) 2534-8008-0
FAX: (49) 2535-8008-90

PCR
See Archimica Florida

PE Biosystems
850 Lincoln Centre Drive
Foster City, CA 94404
(800) 345-5224 FAX: (650) 638-5884
(650) 638-5800
http://www.pebio.com

Pel-Freez Biologicals
219 N. Arkansas
P.O. Box 68
Rogers, AR 72757
(800) 643-3426 FAX: (501) 636-3562
(501) 636-4361
http://www.pelfreez-bio.com

Pel-Freez Clinical Systems
Subsidiary of Pel-Freez Biologicals
9099 N. Deerbrook Trail
Brown Deer, WI 53223
(800) 558-4511 FAX: (414) 357-4518
(414) 357-4500
http://www.pelfreez-bio.com

Peninsula Laboratories
601 Taylor Way
San Carlos, CA 94070
(800) 650-4442 FAX: (650) 595-4071
(650) 592-5392
http://www.penlabs.com

Pentex
24562 Mando Drive
Laguna Niguel, CA 92677
(800) 382-4667 FAX: (714) 643-2363
http://www.pentex.com

PeproTech
5 Crescent Avenue
P.O. Box 275
Rocky Hill, NJ 08553
(800) 436-9910 FAX: (609) 497-0321
(609) 497-0253
http://www.peprotech.com

Peptide Institute
4-1-2 Ina, Minoh-shi
Osaka 562-8686, Japan
81-727-29-4121 FAX: 81-727-29-4124
http://www.peptide.co.jp

Peptide Laboratory
4175 Lakeside Drive
Richmond, CA 94806
(800) 858-7322 FAX: (510) 262-9127
(510) 262-0800
http://www.peptidelab.com

Suppliers

24

Peptides International
11621 Electron Drive
Louisville, KY 40299
(800) 777-4779   FAX: (502) 267-1329
(502) 266-8787
http://www.pepnet.com

Perceptive Science Instruments
2525 South Shore Boulevard, Suite 100
League City, TX 77573
(281) 334-3027   FAX: (281) 538-2222
http://www.persci.com

Perimed
4873 Princeton Drive
North Royalton, OH 44133
(440) 877-0537   FAX: (440) 877-0534
http://www.perimed.se

Perkin-Elmer
761 Main Avenue
Norwalk, CT 06859
(800) 762-4002   FAX: (203) 762-6000
(203) 762-1000
http://www.perkin-elmer.com
See also PE Biosystems

PerSeptive Bioresearch Products
See PerSeptive BioSystems

PerSeptive BioSystems
500 Old Connecticut Path
Framingham, MA 01701
(800) 899-5858   FAX: (508) 383-7885
(508) 383-7700
http://www.pbio.com

PerSeptive Diagnostic
See PE Biosystems
(800) 343-1346

Pettersson Elektronik AB
Tallbacksvagen 51
S-756 45 Uppsala, Sweden
(46) 1830-3880   FAX: (46) 1830-3840
http://www.bahnhof.se/~pettersson

Pfanstiehl Laboratories, Inc.
1219 Glen Rock Avenue
Waukegan, IL 60085
(800) 383-0126   FAX: (847) 623-9173
http://www.pfanstiehl.com

PGC Scientifics
7311 Governors Way
Frederick, MD 21704
(800) 424-3300   FAX: (800) 662-1112
(301) 620-7777   FAX: (301) 620-7497
http://www.pgcscientifics.com

Pharmacia Biotech
See Amersham Pharmacia Biotech

Pharmacia Diagnostics
See Wallac

Pharmacia LKB Biotech
See Amersham Pharmacia Biotech

Pharmacia LKB Biotechnology
See Amersham Pharmacia Biotech

Pharmacia LKB Nuclear
See Wallac

Pharmaderm Veterinary Products
60 Baylis Road
Melville, NY 11747
(800) 432-6673
http://www.pharmaderm.com

Pharmed (Norton)
Norton Performance Plastics
See Saint-Gobain Performance Plastics

PharMingen
See BD PharMingen

Phenomex
2320 W. 205th Street
Torrance, CA 90501
(310) 212-0555   FAX: (310) 328-7768
http://www.phenomex.com

PHLS Centre for Applied
Microbiology and Research
See European Collection of Animal
Cell Cultures (ECACC)

Phoenix Flow Systems
11575 Sorrento Valley Road, Suite 208
San Diego, CA 92121
(800) 886-3569   FAX: (619) 259-5268
(619) 453-5095
http://www.phnxflow.com

Phoenix Pharmaceutical
4261 Easton Road, P.O. Box 6457
St. Joseph, MO 64506
(800) 759-3644   FAX: (816) 364-4969
(816) 364-5777
http://www.phoenixpharmaceutical.com

Photometrics
See Roper Scientific

Photon Technology International
1 Deerpark Drive, Suite F
Monmouth Junction, NJ 08852
(732) 329-0910   FAX: (732) 329-9069
http://www.pti-nj.com

Physik Instrumente
Polytec PI
23 Midstate Drive, Suite 212
Auburn, MA 01501
(508) 832-3456   FAX: (508) 832-0506
http://www.polytecpi.com

Physitemp Instruments
154 Huron Avenue
Clifton, NJ 07013
(800) 452-8510   FAX: (973) 779-5954
(973) 779-5577
http://www.physitemp.com

Pico Technology
The Mill House, Cambridge Street
St. Neots, Cambridgeshire
PE19 1QB, UK
(44) 1480-396-395
FAX: (44) 1480-396-296
http://www.picotech.com

Pierce Chemical
P.O. Box 117
3747 Meridian Road
Rockford, IL 61105
(800) 874-3723   FAX: (800) 842-5007
FAX: (815) 968-7316
http://www.piercenet.com

Pierce & Warriner
44, Upper Northgate Street
Chester, Cheshire CH1 4EF, UK
(44) 1244 382 525
FAX: (44) 1244 373 212
http://www.piercenet.com

Pilling Weck Surgical
420 Delaware Drive
Fort Washington, PA 19034
(800) 523-2579   FAX: (800) 332-2308
http://www.pilling-weck.com

PixelVision
A division of Cybex Computer Products
14964 NW Greenbrier Parkway
Beaverton, OR 97006
(503) 629-3210   FAX: (503) 629-3211
http://www.pixelvision.com

P.J. Noyes
P.O. Box 381
89 Bridge Street
Lancaster, NH 03584
(800) 522-2469   FAX: (603) 788-3873
(603) 788-4952
http://www.pjnoyes.com

Plas-Labs
917 E. Chilson Street
Lansing, MI 48906
(800) 866-7527   FAX: (517) 372-2857
(517) 372-7177
http://www.plas-labs.com

Plastics One
6591 Merriman Road, Southwest
P.O. Box 12004
Roanoke, VA 24018
(540) 772-7950   FAX: (540) 989-7519
http://www.plastics1.com

Platt Electric Supply
2757 6th Avenue South
Seattle, WA 98134
(206) 624-4083   FAX: (206) 343-6342
http://www.platt.com

Plexon
6500 Greenville Avenue
Suite 730
Dallas,TX 75206
(214) 369-4957   FAX: (214) 369-1775
http://www.plexoninc.com

Polaroid
784 Memorial Drive
Cambridge, MA 01239
(800) 225-1618   FAX: (800) 832-9003
(781) 386-2000
http://www.polaroid.com

Polyfiltronics
136 Weymouth St.
Rockland, MA 02370
(800) 434-7659   FAX: (781) 878-0822
(781) 878-1133
http://www.polyfiltronics.com

Polylabo Paul Block
Parc Tertiare de la Meinau
10, rue de la Durance
B.P. 36
67023 Strasbourg Cedex 1
Strasbourg, France
33-3-8865-8020
FAX: 33-3-8865-8039

PolyLC
9151 Rumsey Road, Suite 180
Columbia, MD 21045
(410) 992-5400   FAX: (410) 730-8340

Polymer Laboratories
Amherst Research Park
160 Old Farm Road
Amherst, MA 01002
(800) 767-3963   FAX: (413) 253-2476
http://www.polymerlabs.com

Polymicro Technologies
18019 North 25th Avenue
Phoenix, AZ 85023
(602) 375-4100   FAX: (602) 375-4110
http://www.polymicro.com

Polyphenols AS
Hanabryggene Technology Centre
Hanaveien 4-6
4327 Sandnes, Norway
(47) 51-62-0990
FAX: (47) 51-62-51-82
http://www.polyphenols.com

Polysciences
400 Valley Road
Warrington, PA 18976
(800) 523-2575   FAX: (800) 343-3291
http://www.polysciences.com

Polyscientific
70 Cleveland Avenue
Bayshore, NY 11706
(516) 586-0400   FAX: (516) 254-0618

Polytech Products
285 Washington Street
Somerville, MA 02143
(617) 666-5064   FAX: (617) 625-0975

Polytron
8585 Grovemont Circle
Gaithersburg, MD 20877
(301) 208-6597   FAX: (301) 208-8691
http://www.polytron.com

Popper and Sons
300 Denton Avenue
P.O. Box 128
New Hyde Park, NY 11040
(888) 717-7677   FAX: (800) 557-6773
(516) 248-0300   FAX: (516) 747-1188
http://www.popperandsons.com

Porphyrin Products
P.O. Box 31
Logan, UT 84323
(435) 753-1901   FAX: (435) 753-6731
http://www.porphyrin.com

Portex
See SIMS Portex Limited

Powderject Vaccines
585 Science Drive
Madison, WI 53711
(608) 231-3150   FAX: (608) 231-6990
http://www.powderject.com

Praxair
810 Jorie Boulevard
Oak Brook, IL 60521
(800) 621-7100
http://www.praxair.com

Precision Dynamics
13880 Del Sur Street
San Fernando, CA 91340
(800) 847-0670   FAX: (818) 899-4-45
http://www.pdcorp.com

Precision Scientific Laboratory
Equipment
Division of Jouan
170 Marcel Drive
Winchester, VA 22602
(800) 621-8820   FAX: (540) 869-0130
(540) 869-9892
http://www.precisionsci.com

Primary Care Diagnostics
See Becton Dickinson Primary
Care Diagnostics

Primate Products
1755 East Bayshore Road, Suite 28A
Redwood City, CA 94063
(650) 368-0663   FAX: (650) 368-0665
http://www.primateproducts.com

5 Prime → 3 Prime
See 2000 Eppendorf-5 Prime
http://www.5prime.com

Princeton Applied Research
PerkinElmer Instr.: Electrochemistry
801 S. Illinois
Oak Ridge, TN 37830
(800) 366-2741   FAX: (423) 425-1334
(423) 481-2442
http://www.eggpar.com

Princeton Instruments
A division of Roper Scientific
3660 Quakerbridge Road
Trenton, NJ 08619
(609) 587-9797   FAX: (609) 587-1970
http://www.prinst.com

Princeton Separations
P.O. Box 300
Aldephia, NJ 07710
(800) 223-0902   FAX: (732) 431-3768
(732) 431-3338

Prior Scientific
80 Reservoir Park Drive
Rockland, MA 02370
(781) 878-8442   FAX: (781) 878-8736
http://www.prior.com

PRO Scientific
P.O. Box 448
Monroe, CT 06468
(203) 452-9431   FAX: (203) 452-9753
http://www.proscientific.com

Professional Compounding Centers of America
9901 South Wilcrest Drive
Houston, TX 77099
(800) 331-2498   FAX: (281) 933-6227
(281) 933-6948
http://www.pccarx.com

Progen Biotechnik
Maass-Str. 30
69123 Heidelberg, Germany
(49) 6221-8278-0
FAX: (49) 6221-8278-23
http://www.progen.de

Prolabo
A division of Merck Eurolab
54 rue Roger Salengro
94126 Fontenay Sous Bois Cedex
France
33-1-4514-8500
FAX: 33-1-4514-8616
http://www.prolabo.fr

Proligo
2995 Wilderness Place
Boulder, CO 80301
(888) 80-OLIGO   FAX: (303) 801-1134
http://www.proligo.com

Promega
2800 Woods Hollow Road
Madison, WI 53711
(800) 356-9526   FAX: (800) 356-1970
(608) 274-4330   FAX: (608) 277-2516
http://www.promega.com

Protein Databases (PDI)
405 Oakwood Road
Huntington Station, NY 11746
(800) 777-6834   FAX: (516) 673-4502
(516) 673-3939

Protein Polymer Technologies
10655 Sorrento Valley Road
San Diego, CA 92121
(619) 558-6064   FAX: (619) 558-6477
http://www.ppti.com

Protein Solutions
391 G Chipeta Way
Salt Lake City, UT 84108
(801) 583-9301   FAX: (801) 583-4463
http://www.proteinsolutions.com

Prozyme
1933 Davis Street, Suite 207
San Leandro, CA 94577
(800) 457-9444   FAX: (510) 638-6919
(510) 638-6900
http://www.prozyme.com

PSI
See Perceptive Science Instruments

Pulmetrics Group
82 Beacon Street
Chestnut Hill, MA 02167
(617) 353-3833   FAX: (617) 353-6766

Purdue Frederick
100 Connecticut Avenue
Norwalk, CT 06850
(800) 633-4741   FAX: (203) 838-1576
(203) 853-0123
http://www.pharma.com

Purina Mills
LabDiet
P. O. Box 66812
St. Louis, MO 63166
(800) 227-8941   FAX: (314) 768-4894
http://www.purina-mills.com

Qbiogene
2251 Rutherford Road
Carlsbad, CA 92008
(800) 424-6101   FAX: (760) 918-9313
http://www.qbiogene.com

Qiagen
28159 Avenue Stanford
Valencia, CA 91355
(800) 426-8157   FAX: (800) 718-2056
http://www.qiagen.com

Quality Biological
7581 Lindbergh Drive
Gaithersburg, MD 20879
(800) 443-9331   FAX: (301) 840-5450
(301) 840-9331
http://www.qualitybiological.com

Quantitative Technologies
P.O. Box 470
Salem Industrial Park, Bldg. 5
Whitehouse, NJ 08888
(908) 534-4445   FAX: 534-1054
http://www.qtionline.com

Quantum Appligene
Parc d'Innovation
Rue Geller de Kayserberg
67402 Illkirch, Cedex, France
(33) 3-8867-5425
FAX: (33) 3-8867-1945
http://www.quantum-appligene.com

Quantum Biotechnologies
See Qbiogene

Quantum Soft
Postfach 6613
CH-8023
Zürich, Switzerland
FAX: 41-1-481-69-51
profit@quansoft.com

Questcor Pharmaceuticals
26118 Research Road
Hayward, CA 94545
(510) 732-5551   FAX: (510) 732-7741
http://www.questcor.com

Quidel
10165 McKellar Court
San Diego, CA 92121
(800) 874-1517   FAX: (858) 546-8955
(858) 552-1100
http://www.quidel.com

R-Biopharm
7950 Old US 27 South
Marshall, MI 49068
(616) 789-3033   FAX: (616) 789-3070
http://www.r-biopharm.com

R. C. Electronics
6464 Hollister Avenue
Santa Barbara, CA 93117
(805) 685-7770   FAX: (805) 685-5853
http://www.rcelectronics.com

R & D Systems
614 McKinley Place NE
Minneapolis, MN 55413
(800) 343-7475   FAX: (612) 379-6580
(612) 379-2956
http://www.rndsystems.com

R & S Technology
350 Columbia Street
Peacedale, RI 02880
(401) 789-5660   FAX: (401) 792-3890
http://www.septech.com

RACAL Health and Safety
See 3M
7305 Executive Way
Frederick, MD 21704
(800) 692-9500   FAX: (301) 695-8200

Radiometer America
811 Sharon Drive
Westlake, OH 44145
(800) 736-0600   FAX: (440) 871-2633
(440) 871-8900
http://www.rameusa.com

Radiometer A/S
The Chemical Reference Laboratory
kandevej 21
DK-2700 Brnshj, Denmark
45-3827-3827   FAX: 45-3827-2727

Radionics
22 Terry Avenue
Burlington, MA 01803
(781) 272-1233   FAX: (781) 272-2428
http://www.radionics.com

Radnoti Glass Technology
227 W. Maple Avenue
Monrovia, CA 91016
(800) 428-l4l6   FAX: (626) 303-2998
(626) 357-8827
http://www.radnoti.com

Rainin Instrument
Rainin Road
P.O. Box 4026
Woburn, MA 01888
(800)-4-RAININ   FAX: (781) 938-1152
(781) 935-3050
http://www.rainin.com

Suppliers

26

Rank Brothers
56 High Street
Bottisham, Cambridge
CB5 9DA UK
(44) 1223 811369
FAX: (44) 1223 811441
http://www.rankbrothers.com

Rapp Polymere
Ernst-Simon Strasse 9
D 72072 Tübingen, Germany
(49) 7071-763157
FAX: (49) 7071-763158
http://www.rapp-polymere.com

Raven Biological Laboratories
8607 Park Drive
P.O. Box 27261
Omaha, NE 68127
(800) 728-5702   FAX: (402) 593-0995
(402) 593-0781
http://www.ravenlabs.com

Razel Scientific Instruments
100 Research Drive
Stamford, CT 06906
(203) 324-9914   FAX: (203) 324-5568

Reagents International
See Biotech Source

Receptor Biology
10000 Virginia Manor Road, Suite 360
Beltsville, MD 20705
(888) 707-4200   FAX: (301) 210-6266
(301) 210-4700
http://www.receptorbiology.com

Regis Technologies
8210 N. Austin Avenue
Morton Grove, IL 60053
(800) 323-8144   FAX: (847) 967-1214
(847) 967-6000
http://www.registech.com

Reichert Ophthalmic Instruments
P.O. Box 123
Buffalo, NY 14240
(716) 686-4500   FAX: (716) 686-4545
http://www.reichert.com

Reiss
1 Polymer Place
P.O. Box 60
Blackstone, VA 23824
(800) 356-2829   FAX: (804) 292-1757
(804) 292-1600
http://www.reissmfg.com

Remel
12076 Santa Fe Trail Drive
P.O. Box 14428
Shawnee Mission, KS 66215
(800) 255-6730   FAX: (800) 621-8251
(913) 888-0939   FAX: (913) 888-5884
http://www.remelinc.com

Reming Bioinstruments
6680 County Route 17
Redfield, NY 13437
(315) 387-3414   FAX: (315) 387-3415

RepliGen
117 Fourth Avenue
Needham, MA 02494
(800) 622-2259   FAX: (781) 453-0048
(781) 449-9560
http://www.repligen.com

Research Biochemicals
1 Strathmore Road
Natick, MA 01760
(800) 736-3690   FAX: (800) 736-2480
(508) 651-8151   FAX: (508) 655-1359
http://www.resbio.com

Research Corporation Technologies
101 N. Wilmot Road, Suite 600
Tucson, AZ 85711
(520) 748-4400   FAX: (520) 748-0025
http://www.rctech.com

Research Diagnostics
Pleasant Hill Road
Flanders, NJ 07836
(800) 631-9384   FAX: (973) 584-0210
(973) 584-7093
http://www.researchd.com

Research Diets
121 Jersey Avenue
New Brunswick, NJ 08901
(877) 486-2486   FAX: (732) 247-2340
(732) 247-2390
http://www.researchdiets.com

Research Genetics
2130 South Memorial Parkway
Huntsville, AL 35801
(800) 533-4363   FAX: (256) 536-9016
(256) 533-4363
http://www.resgen.com

Research Instruments
Kernick Road Pernryn
Cornwall TR10 9DQ, UK
(44) 1326-372-753
FAX: (44) 1326-378-783
http://www.research-instruments.com

Research Organics
4353 E. 49th Street
Cleveland, OH 44125
(800) 321-0570   FAX: (216) 883-1576
(216) 883-8025
http://www.resorg.com

Research Plus
P.O. Box 324
Bayonne, NJ 07002
(800) 341-2296   FAX: (201) 823-9590
(201) 823-3592
http://www.researchplus.com

Research Products International
410 N. Business Center Drive
Mount Prospect, IL 60056
(800) 323-9814   FAX: (847) 635-1177
(847) 635-7330
http://www.rpicorp.com

Research Triangle Institute
P.O. Box 12194
Research Triangle Park, NC 27709
(919) 541-6000   FAX: (919) 541-6515
http://www.rti.org

Restek
110 Benner Circle
Bellefonte, PA 16823
(800) 356-1688   FAX: (814) 353-1309
(814) 353-1300
http://www.restekcorp.com

Rheodyne
P.O. Box 1909
Rohnert Park, CA 94927
(707) 588-2000   FAX: (707) 588-2020
http://www.rheodyne.com

Rhone Merieux
See Merial Limited

Rhone-Poulenc
2 T W Alexander Drive
P.O. Box 12014
Research Triangle Park, NC 08512
(919) 549-2000   FAX: (919) 549-2839
http://www.Rhone-Poulenc.com
Also see Aventis

Rhone-Poulenc Rorer
500 Arcola Road
Collegeville, PA 19426
(800) 727-6737   FAX: (610) 454-8940
(610) 454-8975
http://www.rp-rorer.com

Rhone-Poulenc Rorer
Centre de Recherche de Vitry-Alfortville
13 Quai Jules Guesde, BP14 94403
Vitry Sur Seine, Cedex, France
(33) 145-73-85-11
FAX: (33) 145-73-81-29
http://www.rp-rorer.com

Ribi ImmunoChem Research
563 Old Corvallis Road
Hamilton, MT 59840
(800) 548-7424   FAX: (406) 363-6129
(406) 363-3131
http://www.ribi.com

RiboGene
See Questcor Pharmaceuticals

Ricca Chemical
448 West Fork Drive
Arlington, TX 76012
(888) GO-RICCA   FAX: (800) RICCA-93
(817) 461-5601
http://www.riccachemical.com

Richard-Allan Scientific
225 Parsons Street
Kalamazoo, MI 49007
(800) 522-7270   FAX: (616) 345-3577
(616) 344-2400
http://www.rallansci.com

Richelieu Biotechnologies
11 177 Hamon
Montral, Quebec
H3M 3E4 Canada
(802) 863-2567   FAX: (802) 862-2909
http://www.richelieubio.com

Richter Enterprises
20 Lake Shore Drive
Wayland, MA 01778
(508) 655-7632   FAX: (508) 652-7264
http://www.richter-enterprises.com

Riese Enterprises
BioSure Division
12301 G Loma Rica Drive
Grass Valley, CA 95945
(800) 345-2267   FAX: (916) 273-5097
(916) 273-5095
http://www.biosure.com

Robbins Scientific
1250 Elko Drive
Sunnyvale, CA 94086
(800) 752-8585   FAX: (408) 734-0300
(408) 734-8500
http://www.robsci.com

Roboz Surgical Instruments
9210 Corporate Boulevard, Suite 220
Rockville, MD 20850
(800) 424-2984   FAX: (301) 590-1290
(301) 590-0055

Roche Diagnostics
9115 Hague Road
P.O. Box 50457
Indianapolis, IN 46256
(800) 262-1640   FAX: (317) 845-7120
(317) 845-2000
http://www.roche.com

Roche Molecular Systems
See Roche Diagnostics

Rocklabs
P.O. Box 18-142
Auckland 6, New Zealand
(64) 9-634-7696
FAX: (64) 9-634-7696
http://www.rocklabs.com

Rockland
P.O. Box 316
Gilbertsville, PA 19525
(800) 656-ROCK   FAX: (610) 367-7825
(610) 369-1008
http://www.rockland-inc.com

Rohm
Chemische Fabrik
Kirschenallee
D-64293 Darmstadt, Germany
(49) 6151-1801   FAX: (49) 6151-1802
http://www.roehm.com

Roper Scientific
3440 East Brittania Drive, Suite 100
Tucson, AZ 85706
(520) 889-9933   FAX: (520) 573-1944
http://www.roperscientific.com

Rosetta Inpharmatics
12040 115th Avenue NE
Kirkland, WA 98034
(425) 820-8900   FAX: (425) 820-5757
http://www.rii.com

ROTH-SOCHIEL
3 rue de la Chapelle
Lauterbourg
67630 France
(33) 03-88-94-82-42
FAX: (33) 03-88-54-63-93

Rotronic Instrument
160 E. Main Street
Huntington, NY 11743
(631) 427-3898   FAX: (631) 427-3902
http://www.rotronic-usa.com

Roundy's
23000 Roundy Drive
Pewaukee, WI 53072
(262) 953-7999   FAX: (262) 953-7989
http://www.roundys.com

RS Components
Birchington Road
Weldon Industrial Estate
Corby, Northants NN17 9RS, UK
(44) 1536 201234
FAX: (44) 1536 405678
http://www.rs-components.com

Rubbermaid
See Newell Rubbermaid

SA Instrumentation
1437 Tzena Way
Encinitas, CA 92024
(858) 453-1776   FAX: (800) 266-1776
http://www.sainst.com

Safe Cells
See Bionique Testing Labs

Sage Instruments
240 Airport Boulevard
Freedom, CA 95076
831-761-1000   FAX: 831-761-1008
http://www.sageinst.com

Sage Laboratories
11 Huron Drive
Natick, MA 01760
(508) 653-0844   FAX: 508-653-5671
http://www.sagelabs.com

Saint-Gobain Performance Plastics
P.O. Box 3660
Akron, OH 44309
(330) 798-9240   FAX: (330) 798-6968
http://www.nortonplastics.com

San Diego Instruments
7758 Arjons Drive
San Diego, CA 92126
(858) 530-2600   FAX: (858) 530-2646
http://www.sd-inst.com

Sandown Scientific
Beards Lodge
25 Oldfield Road
Hampden, Middlesex TW12 2AJ, UK
(44) 2089 793300
FAX: (44) 2089 793311
http://www.sandownsci.com

Sandoz Pharmaceuticals
See Novartis

Sanofi Recherche
Centre de Montpellier
371 Rue du Professur Blayac
34184 Montpellier, Cedex 04
France
(33) 67-10-67-10
FAX: (33) 67-10-67-67

Sanofi Winthrop Pharmaceuticals
90 Park Avenue
New York, NY 10016
(800) 223-5511   FAX: (800) 933-3243
(212) 551-4000
http://www.sanofi-synthelabo.com/us

Santa Cruz Biotechnology
2161 Delaware Avenue
Santa Cruz, CA 95060
(800) 457-3801   FAX: (831) 457-3801
(831) 457-3800
http://www.scbt.com

Sarasep
(800) 605-0267   FAX: (408) 432-3231
(408) 432-3230
http://www.transgenomic.com

Sarstedt
P.O. Box 468
Newton, NC 28658
(800) 257-5101   FAX: (828) 465-4003
(828) 465-4000
http://www.sarstedt.com

Sartorius
131 Heartsland Boulevard
Edgewood, NY 11717
(800) 368-7178   FAX: (516) 254-4253
http://www.sartorius.com

SAS Institute
Pacific Telesis Center
One Montgomery Street
San Francisco, CA 94104
(415) 421-2227   FAX: (415) 421-1213
http://www.sas.com

Savant/EC Apparatus
A ThermoQuest company
100 Colin Drive
Holbrook, NY 11741
(800) 634-8886   FAX: (516) 244-0606
(516) 244-2929
http://www.savec.com

Savillex
6133 Baker Road
Minnetonka, MN 55345
(612) 935-5427

Scanalytics
Division of CSP
8550 Lee Highway, Suite 400
Fairfax, VA 22031
(800) 325-3110   FAX: (703) 208-1960
(703) 208-2230
http://www.scanalytics.com

Schering Laboratories
See Schering-Plough

Schering-Plough
1 Giralda Farms
Madison, NJ 07940
(800) 222-7579   FAX: (973) 822-7048
(973) 822-7000
http://www.schering-plough.com

Schleicher & Schuell
10 Optical Avenue
Keene, NH 03431
(800) 245-4024   FAX: (603) 357-3627
(603) 352-3810
http://www.s-und-s.de/english-index.html

Science Technology Centre
1250 Herzberg Laboratories
Carleton University
1125 Colonel Bay Drive
Ottawa, Ontario
K1S 5B6 Canada
(613) 520-4442   FAX: (613) 520-4445
http://www.carleton.ca/universities/stc

Scientific Instruments
200 Saw Mill River Road
Hawthorne, NY 10532
(800) 431-1956   FAX: (914) 769-5473
(914) 769-5700
http://www.scientificinstruments.com

Scientific Solutions
9323 Hamilton
Mentor, OH 44060
(440) 357-1400   FAX: (440) 357-1416
http://www.labmaster.com

Scion
82 Worman's Mill Court, Suite H
Frederick, MD 21701
(301) 695-7870   FAX: (301) 695-0035
http://www.scioncorp.com

Scott Specialty Gases
6141 Easton Road
P.O. Box 310
Plumsteadville, PA 18949
(800) 21-SCOTT   FAX: (215) 766-2476
(215) 766-8861
http://www.scottgas.com

Scripps Clinic and Research
Foundation
Instrumentation and Design Lab
10666 N. Torrey Pines Road
La Jolla, CA 92037
(800) 992-9962   FAX: (858) 554-8986
(858) 455-9100
http://www.scrippsclinic.com

SDI Sensor Devices
407 Pilot Court, 400A
Waukesha, WI 53188
(414) 524-1000   FAX: (414) 524-1009

Sefar America
111 Calumet Street
Depew, NY 14043
(716) 683-4050   FAX: (716) 683-4053
http://www.sefaramerica.com

Seikagaku America
Division of Associates of Cape Cod
704 Main Street
Falmouth, MA 02540
(800) 237-4512   FAX: (508) 540-8680
(508) 540-3444
http://www.seikagaku.com

Sellas Medizinische Gerate
Hagener Str. 393
Gevelsberg-Vogelsang, 58285
Germany
(49) 23-326-1225

Sensor Medics
22705 Savi Ranch Parkway
Yorba Linda, CA 92887
(800) 231-2466   FAX: (714) 283-8439
(714) 283-2228
http://www.sensormedics.com

Sensor Systems LLC
2800 Anvil Street, North
Saint Petersburg, FL 33710
(800) 688-2181   FAX: (727) 347-3881
(727) 347-2181
http://www.vsensors.com

SenSym/Foxboro ICT
1804 McCarthy Boulevard
Milpitas, CA 95035
(800) 392-9934   FAX: (408) 954-9458
(408) 954-6700
http://www.sensym.com

Separations Group
See Vydac

Sepracor
111 Locke Drive
Marlboro, MA 01752
(877)-SEPRACOR   (508) 357-7300
http://www.sepracor.com

Sera-Lab
See Harlan Sera-Lab

Sermeter
925 Seton Court, #7
Wheeling, IL 60090
(847) 537-4747

Serological
195 W. Birch Street
Kankakee, IL 60901
(800) 227-9412   FAX: (815) 937-8285
(815) 937-8270

Seromed Biochrom
Leonorenstrasse 2-6
D-12247 Berlin, Germany
(49) 030-779-9060

Suppliers

28

Serotec
22 Bankside
Station Approach
Kidlington, Oxford OX5 1JE, UK
(44) 1865-852722
FAX: (44) 1865-373899
In the US: (800) 265-7376
http://www.serotec.co.uk

Serva Biochemicals
Distributed by Crescent Chemical

S.F. Medical Pharmlast
See Chase-Walton Elastomers

SGE
2007 Kramer Lane
Austin, TX 78758
(800) 945-6154   FAX: (512) 836-9159
(512) 837-7190
http://www.sge.com

Shandon/Lipshaw
171 Industry Drive
Pittsburgh, PA 15275
(800) 245-6212   FAX: (412) 788-1138
(412) 788-1133
http://www.shandon.com

Sharpoint
P.O. Box 2212
Taichung, Taiwan
Republic of China
(886) 4-3206320
FAX: (886) 4-3289879
http://www.sharpoint.com.tw

Shelton Scientific
230 Longhill Crossroads
Shelton, CT 06484
(800) 222-2092   FAX: (203) 929-2175
(203) 929-8999
http://www.sheltonscientific.com

Sherwood-Davis & Geck
See Kendall

Sherwood Medical
See Kendall

Shimadzu Scientific Instruments
7102 Riverwood Drive
Columbia, MD 21046
(800) 477-1227   FAX: (410) 381-1222
(410) 381-1227
http://www.ssi.shimadzu.com

Sialomed
See Amika

Siemens Analytical X-Ray Systems
See Bruker Analytical X-Ray Systems

Sievers Instruments
Subsidiary of Ionics
6060 Spine Road
Boulder, CO 80301
(800) 255-6964   FAX: (303) 444-6272
(303) 444-2009
http://www.sieversinst.com

SIFCO
970 East 46th Street
Cleveland, OH 44103
(216) 881-8600   FAX: (216) 432-6281
http://www.sifco.com

Sigma-Aldrich
3050 Spruce Street
St. Louis, MO 63103
(800) 358-5287   FAX: (800) 962-9591
(800) 325-3101   FAX: (800) 325-5052
http://www.sigma-aldrich.com

Sigma-Aldrich Canada
2149 Winston Park Drive
Oakville, Ontario
L6H 6J8 Canada
(800) 5652D1400   FAX: (800)
2652D3858
http://www.sigma-aldrich.com

Silenus/Amrad
34 Wadhurst Drive
Boronia, Victoria 3155 Australia
(613)9887-3909   FAX: (613)9887-3912
http://www.amrad.com.au

Silicon Genetics
2601 Spring Street
Redwood City, CA 94063
(866) SIG SOFT   FAX: (650) 365 1735
(650) 367 9600
http://www.sigenetics.com

SIMS Deltec
1265 Grey Fox Road
St. Paul, Minnesota 55112
(800) 426-2448   FAX: (615) 628-7459
http://www.deltec.com

SIMS Portex
10 Bowman Drive
Keene, NH 03431
(800) 258-5361   FAX: (603) 352-3703
(603) 352-3812
http://www.simsmed.com

SIMS Portex Limited
Hythe, Kent CT21 6JL, UK
(44)1303-260551
FAX: (44)1303-266761
http://www.portex.com

Siris Laboratories
See Biosearch Technologies

Skatron Instruments
See Molecular Devices

SLM Instruments
See Spectronic Instruments

SLM-AMINCO Instruments
See Spectronic Instruments

Small Parts
13980 NW 58th Court
P.O. Box 4650
Miami Lakes, FL 33014
(800) 220-4242   FAX: (800) 423-9009
(305) 558-1038   FAX: (305) 558-0509
http://www.smallparts.com

Smith & Nephew
11775 Starkey Road
P.O. Box 1970
Largo, FL 33779
(800) 876-1261
http://www.smith-nephew.com

SmithKline Beecham
1 Franklin Plaza, #1800
Philadelphia, PA 19102
(215) 751-4000   FAX: (215) 751-4992
http://www.sb.com

Solid Phase Sciences
See Biosearch Technologies

SOMA Scientific Instruments
5319 University Drive, PMB #366
Irvine, CA 92612
(949) 854-0220   FAX: (949) 854-0223
http://somascientific.com

Somatix Therapy
See Cell Genesys

Sonics & Materials
53 Church Hill Road
Newtown, CT 06470
(800) 745-1105   FAX: (203) 270-4610
(203) 270-4600
http://www.sonicsandmaterials.com

Sonosep Biotech
See Triton Environmental Consultants

Sorvall
See Kendro Laboratory Products

Southern Biotechnology Associates
P.O. Box 26221
Birmingham, AL 35260
(800) 722-2255   FAX: (205) 945-8768
(205) 945-1774
http://SouthernBiotech.com

SPAFAS
190 Route 165
Preston, CT 06365
(800) SPAFAS-1   FAX: (860) 889-1991
(860) 889-1389
http://www.spafas.com

Specialty Media
Division of Cell & Molecular Technologies
580 Marshall Street
Phillipsburg, NJ 08865
(800) 543-6029   FAX: (908) 387-1670
(908) 454-7774
http://www.specialtymedia.com

Spectra Physics
See Thermo Separation Products

Spectramed
See BOC Edwards

SpectraSource Instruments
31324 Via Colinas, Suite 114
Westlake Village, CA 91362
(818) 707-2655   FAX: (818) 707-9035
http://www.spectrasource.com

Spectronic Instruments
820 Linden Avenue
Rochester, NY 14625
(800) 654-9955   FAX: (716) 248-4014
(716) 248-4000
http://www.spectronic.com

Spectrum Medical Industries
See Spectrum Laboratories

Spectrum Laboratories
18617 Broadwick Street
Rancho Dominguez, CA 90220
(800) 634-3300   FAX: (800) 445-7330
(310) 885-4601   FAX: (310) 885-4666
http://www.spectrumlabs.com

Spherotech
1840 Industrial Drive, Suite 270
Libertyville, IL 60048
(800) 368-0822   FAX: (847) 680-8927
(847) 680-8922
http://www.spherotech.com

SPSS
233 S. Wacker Drive, 11th floor
Chicago, IL 60606
(800) 521-1337   FAX: (800) 841-0064
http://www.spss.com

SS White Burs
1145 Towbin Avenue
Lakewood, NJ 08701
(732) 905-1100   FAX: (732) 905-0987
http://www.sswhiteburs.com

Stag Instruments
16 Monument Industrial Park
Chalgrove, Oxon OX44 7RW, UK
(44) 1865-891116
FAX: (44) 1865-890562

Standard Reference Materials
Program
National Institute of Standards and
Technology
Building 202, Room 204
Gaithersburg, MD 20899
(301) 975-6776   FAX: (301) 948-3730

Starna Cells
P.O. Box 1919
Atascandero, CA 93423
(805) 466-8855   FAX: (805) 461-1575
(800) 228-4482
http://www.starnacells.com

Starplex Scientific
50 Steinway
Etobieoke, Ontario
M9W 6Y3 Canada
(800) 665-0954   FAX: (416) 674-6067
(416) 674-7474
http://www.starplexscientific.com

State Laboratory Institute of
Massachusetts
305 South Street
Jamaica Plain, MA 02130
(617) 522-3700   FAX: (617) 522-8735
http://www.state.ma.us/dph

Suppliers

29

Stedim Labs
1910 Mark Court, Suite 110
Concord, CA 94520
(800) 914-6644   FAX: (925) 689-6988
(925) 689-6650
http://www.stedim.com

Steinel America
9051 Lyndale Avenue
Bloomington, MN 55420
(800) 852 4343   FAX: (952) 888-5132
http://www.steinelamerica.com

Stem Cell Technologies
777 West Broadway, Suite 808
Vancouver, British Columbia
V5Z 4J7 Canada
(800) 667-0322   FAX: (800) 567-2899
(604) 877-0713   FAX: (604) 877-0704
http://www.stemcell.com

Stephens Scientific
107 Riverdale Road
Riverdale, NJ 07457
(800) 831-8099   FAX: (201) 831-8009
(201) 831-9800

Steraloids
P.O. Box 689
Newport, RI 02840
(401) 848-5422   FAX: (401) 848-5638
http://www.steraloids.com

Sterling Medical
2091 Springdale Road, Ste. 2
Cherry Hill, NJ 08003
(800) 229-0900   FAX: (800) 229-7854
http://www.sterlingmedical.com

Sterling Winthrop
90 Park Avenue
New York, NY 10016
(212) 907-2000   FAX: (212) 907-3626

Sternberger Monoclonals
10 Burwood Court
Lutherville, MD 21093
(410) 821-8505   FAX: (410) 821-8506
http://www.sternbergermonoclonals.com

Stoelting
502 Highway 67
Kiel, WI 53042
(920) 894-2293   FAX: (920) 894-7029
http://www.stoelting.com

Stovall Lifescience
206-G South Westgate Drive
Greensboro, NC 27407
(800) 852-0102   FAX: (336) 852-3507
http://www.slscience.com

Stratagene
11011 N. Torrey Pines Road
La Jolla, CA 92037
(800) 424-5444   FAX: (888) 267-4010
(858) 535-5400
http://www.stratagene.com

Strategic Applications
530A N. Milwaukee Avenue
Libertyville, IL 60048
(847) 680-9385   FAX: (847) 680-9837

Strem Chemicals
7 Mulliken Way
Newburyport, MA 01950
(800) 647-8736   FAX: (800) 517-8736
(978) 462-3191   FAX: (978) 465-3104
http://www.strem.com

StressGen Biotechnologies
Biochemicals Division
120-4243 Glanford Avenue
Victoria, British Columbia
V8Z 4B9 Canada
(800) 661-4978   FAX: (250) 744-2877
(250) 744-2811
http://www.stressgen.com

Structure Probe/SPI Supplies
(Epon-Araldite)
P.O. Box 656
West Chester, PA 19381
(800) 242-4774   FAX: (610) 436-5755
http://www.2spi.com

Süd-Chemie Performance Packaging
101 Christine Drive
Belen, NM 87002
(800) 989-3374   FAX: (505) 864-9296
http://www.uniteddesiccants.com

Sumitomo Chemical
Sumitomo Building
5-33, Kitahama 4-chome
Chuo-ku, Osaka 541-8550, Japan
(81) 6-6220-3891
FAX: (81)-6-6220-3345
http://www.sumitomo-chem.co.jp

Sun Box
19217 Orbit Drive
Gaithersburg, MD 20879
(800) 548-3968   FAX: (301) 977-2281
(301) 869-5980
http://www.sunboxco.com

Sunbrokers
See Sun International

Sun International
3700 Highway 421 North
Wilmington, NC 28401
(800) LAB-VIAL   FAX: (800) 231-7861
http://www.autosamplervial.com

Sunox
1111 Franklin Boulevard, Unit 6
Cambridge, Ontario
N1R 8B5 Canada
(519) 624-4413   FAX: (519) 624-8378
http://www.sunox.ca

Supelco
See Sigma-Aldrich

SuperArray
P.O. Box 34494
Bethesda, MD 20827
(888) 503-3187   FAX: (301) 765-9859
(301) 765-9888
http://www.superarray.com

Surface Measurement Systems
3 Warple Mews, Warple Way
London W3 ORF, UK
(44) 20-8749-4900
FAX: (44) 20-8749-6749
http://www.smsuk.co.uk/index.htm

SurgiVet
N7 W22025 Johnson Road, Suite A
Waukesha, WI 53186
(262) 513-8500   (888) 745-6562
FAX: (262) 513-9069
http://www.surgivet.com

Sutter Instruments
51 Digital Drive
Novato, CA 94949
(415) 883-0128   FAX: (415) 883-0572
http://www.sutter.com

Swiss Precision Instruments
1555 Mittel Boulevard, Suite F
Wooddale, IL 60191
(800) 221-0198   FAX: (800) 842-5164

Synaptosoft
3098 Anderson Place
Decatur, GA 30033
(770) 939-4366   FAX: 770-939-9478
http://www.synaptosoft.com

SynChrom
See Micra Scientific

Synergy Software
2457 Perkiomen Avenue
Reading, PA 19606
(800) 876-8376   FAX: (610) 370-0548
(610) 779-0522
http://www.synergy.com

Synteni
See Incyte

Synthetics Industry
Lumite Division
2100A Atlantic Highway
Gainesville, GA 30501
(404) 532-9756   FAX: (404) 531-1347

Systat
See SPSS

Systems Planning and Analysis (SPA)
2000 N. Beauregard Street
Suite 400
Alexandria, VA 22311
(703) 931-3500
http://www.spa-inc.net

3M Bioapplications
3M Center
Building 270-15-01
St. Paul, MN 55144
(800) 257-7459   FAX: (651) 737-5645
(651) 736-4946

**T Cell Diagnostics and
T Cell Sciences**
38 Sidney Street
Cambridge, MA 02139
(617) 621-1400

TAAB Laboratory Equipment
3 Minerva House
Calleva Park
Aldermaston, Berkshire RG7 8NA, UK
(44) 118 9817775
FAX: (44) 118 9817881

Taconic
273 Hover Avenue
Germantown, NY 12526
(800) TAC-ONIC   FAX: (518) 537-7287
(518) 537-6208
http://www.taconic.com

Tago
See Biosource International

TaKaRa Biochemical
719 Alliston Way
Berkeley, CA 94710
(800) 544-9899   FAX: (510) 649-8933
(510) 649-9895
http://www.takara.co.jp/english

Takara Shuzo
Biomedical Group Division
Seta 3-4-1
Otsu Shiga 520-21, Japan
(81) 75-241-5100
FAX: (81) 77-543-9254
http://www.Takara.co.jp/english

Takeda Chemical Products
101 Takeda Drive
Wilmington, NC 28401
(800) 825-3328   FAX: (800) 825-0333
(910) 762-8666   FAX: (910) 762-6846
http://takeda-usa.com

TAO Biomedical
73 Manassas Court
Laurel Springs, NJ 08021
(609) 782-8622   FAX: (609) 782-8622

Tecan US
P.O. Box 13953
Research Triangle Park, NC 27709
(800) 33-TECAN   FAX: (919) 361-5201
(919) 361-5208
http://www.tecan-us.com

Techne
University Park Plaza
743 Alexander Road
Princeton, NJ 08540
(800) 225-9243   FAX: (609) 987-8177
(609) 452-9275
http://www.techneusa.com

Technical Manufacturing
15 Centennial Drive
Peabody, MA 01960
(978) 532-6330   FAX: (978) 531-8682
http://www.techmfg.com

Technical Products International
5918 Evergreen
St. Louis, MO 63134
(800) 729-4451   FAX: (314) 522-6360
(314) 522-8671
http://www.vibratome.com

Suppliers

30

Technicon
See Organon Teknika Cappel

Techno-Aide
P.O. Box 90763
Nashville, TN 37209
(800) 251-2629   FAX: (800) 554-6275
(615) 350-7030
http://www.techno-aid.com

Ted Pella
4595 Mountain Lakes Boulevard
P.O. Box 492477
Redding, CA 96049
(800) 237-3526   FAX: (530) 243-3761
(530) 243-2200
http://www.tedpella.com

Tekmar-Dohrmann
P.O. Box 429576
Cincinnati, OH 45242
(800) 543-4461   FAX: (800) 841-5262
(513) 247-7000   FAX: (513) 247-7050

Tektronix
142000 S.W. Karl Braun Drive
Beaverton, OR 97077
(800) 621-1966   FAX: (503) 627-7995
(503) 627-7999
http://www.tek.com

Tel-Test
P.O. Box 1421
Friendswood, TX 77546
(800) 631-0600   FAX: (281)482-1070
(281)482-2672
http://www.isotex-diag.com

TeleChem International
524 East Weddell Drive, Suite 3
Sunnyvale, CA 94089
(408) 744-1331   FAX: (408) 744-1711
http://www.gst.net/~telechem

Terrachem
Mallaustrasse 57
D-68219 Mannheim, Germany
0621-876797-0   FAX: 0621-876797-19
http://www.terrachem.de

Terumo Medical
2101 Cottontail Lane
Somerset, NJ 08873
(800) 283-7866   FAX: (732) 302-3083
(732) 302-4900
http://www.terumomedical.com

Tetko
333 South Highland Manor
Briarcliff, NY 10510
(800) 289-8385   FAX: (914) 941-1017
(914) 941-7767
http://www.tetko.com

TetraLink
4240 Ridge Lea Road
Suite 29
Amherst, NY 14226
(800) 747-5170   FAX: (800) 747-5171
http://www.tetra-link.com

TEVA Pharmaceuticals USA
1090 Horsham Road
P.O. Box 1090
North Wales, PA 19454
(215) 591-3000   FAX: (215) 721-9669
http://www.tevapharmusa.com

Texas Fluorescence Labs
9503 Capitol View Drive
Austin, TX 78747
(512) 280-5223   FAX: (512) 280-4997
http://www.teflabs.com

The Nest Group
45 Valley Road
Southborough, MA 01772
(800) 347-6378   FAX: (508) 485-5736
(508) 481-6223
http://world.std.com/~nestgrp

ThermoCare
P.O. Box 6069
Incline Village, NV 89450
(800) 262-4020
(775) 831-1201

Thermo Labsystems
8 East Forge Parkway
Franklin, MA 02038
(800) 522-7763   FAX: (508) 520-2229
(508) 520-0009
http://www.finnpipette.com

Thermometric
Spjutvagen 5A
S-175 61 Jarfalla, Sweden
(46) 8-564-72-200

Thermoquest
IEC Division
300 Second Avenue
Needham Heights, MA 02194
(800) 843-1113   FAX: (781) 444-6743
(781) 449-0800
http://www.thermoquest.com

Thermo Separation Products
Thermoquest
355 River Oaks Parkway
San Jose, CA 95134
(800) 538-7067   FAX: (408) 526-9810
(408) 526-1100
http://www.thermoquest.com

Thermo Shandon
171 Industry Drive
Pittsburgh, PA 15275
(800) 547-7429   FAX: (412) 899-4045
http://www.thermoshandon.com

Thermo Spectronic
820 Linden Avenue
Rochester, NY 14625
(585) 248-4000   FAX: (585) 248-4200
http://www.thermo.com

Thomas Scientific
99 High Hill Road at I-295
Swedesboro, NJ 08085
(800) 345-2100   FAX: (800) 345-5232
(856) 467-2000   FAX: (856) 467-3087
http://www.wheatonsci.com/html/nt/
Thomas.html

Thomson Instrument
354 Tyler Road
Clearbrook, VA 22624
(800) 842-4752   FAX: (540) 667-6878
(800) 541-4792   FAX: (760) 757-9367
http://www.hplc.com

Thorn EMI
See Electron Tubes

Thorlabs
435 Route 206
Newton, NJ 07860
(973) 579-7227   FAX: (973) 383-8406
http://www.thorlabs.com

Tiemann
See Bernsco Surgical Supply

Timberline Instruments
1880 South Flatiron Court, H-2
P.O. Box 20356
Boulder, CO 80308
(800) 777-5996   FAX: (303) 440-8786
(303) 440-8779
http://www.timberlineinstruments.com

Tissue-Tek
A Division of Sakura Finetek USA
1750 West 214th Street
Torrance, CA 90501
(800) 725-8723   FAX: (310) 972-7888
(310) 972-7800
http://www.sakuraus.com

Tocris Cookson
114 Holloway Road, Suite 200
Ballwin, MO 63011
(800) 421-3701   FAX: (800) 483-1993
(636) 207-7651   FAX: (636) 207-7683
http://www.tocris.com

Tocris Cookson
Northpoint, Fourth Way
Avonmouth, Bristol BS11 8TA, UK
(44) 117-982-6551
FAX: (44) 117-982-6552
http://www.tocris.com

Tomtec
See CraMar Technologies

TopoGen
P.O. Box 20607
Columbus, OH 43220
(800) TOPOGEN
FAX: (800) ADD-TOPO
(614) 451-5810   FAX: (614) 451-5811
http://www.topogen.com

Toray Industries, Japan
Toray Building 2-1
Nihonbash-Muromach
2-Chome, Chuo-Ku
Tokyo, Japan 103-8666
(03) 3245-5115   FAX: (03) 3245-5555
http://www.toray.co.jp

Toray Industries, U.S.A.
600 Third Avenue
New York, NY 10016
(212) 697-8150   FAX: (212) 972-4279
http://www.toray.com

Toronto Research Chemicals
2 Brisbane Road
North York, Ontario
M3J 2J8 Canada
(416) 665-9696   FAX: (416) 665-4439
http://www.trc-canada.com

TosoHaas
156 Keystone Drive
Montgomeryville, PA 18036
(800) 366-4875   FAX: (215) 283-5035
(215) 283-5000
http://www.tosohaas.com

Towhill
647 Summer Street
Boston, MA 02210
(617) 542-6636   FAX: (617) 464-0804

Toxin Technology
7165 Curtiss Avenue
Sarasota, FL 34231
(941) 925-2032   FAX: (9413) 925-2130
http://www.toxintechnology.com

Toyo Soda
See TosoHaas

Trace Analytical
3517-A Edison Way
Menlo Park, CA 94025
(650) 364-6895   FAX: (650) 364-6897
http://www.traceanalytical.com

Transduction Laboratories
See BD Transduction Laboratories

Transgenomic
2032 Concourse Drive
San Jose, CA 95131
(408) 432-3230   FAX: (408) 432-3231
http://www.transgenomic.com

Transonic Systems
34 Dutch Mill Road
Ithaca, NY 14850
(800) 353-3569   FAX: (607) 257-7256
http://www.transonic.com

Travenol Lab
See Baxter Healthcare

Tree Star Software
20 Winding Way
San Carlos, CA 94070
800-366-6045
http://www.treestar.com

Trevigen
8405 Helgerman Court
Gaithersburg, MD 20877
(800) TREVIGEN   FAX: (301) 216-2801
(301) 216-2800
http://www.trevigen.com

Trilink Biotechnologies
6310 Nancy Ridge Drive
San Diego, CA 92121
(800) 863-6801   FAX: (858) 546-0020
http://www.trilink.biotech.com

Tripos Associates
1699 South Hanley Road, Suite 303
St. Louis, MO 63144
(800) 323-2960 FAX: (314) 647-9241
(314) 647-1099
http://www.tripos.com

Triton Environmental Consultants
120-13511 Commerce Parkway
Richmond, British Columbia
V6V 2L1 Canada
(604) 279-2093 FAX: (604) 279-2047
http://www.triton-env.com

Tropix
47 Wiggins Avenue
Bedford, MA 01730
(800) 542-2369 FAX: (617) 275-8581
(617) 271-0045
http://www.tropix.com

TSI Center for Diagnostic Products
See Intergen

2000 Eppendorf-5 Prime
5603 Arapahoe Avenue
Boulder, CO 80303
(800) 533-5703 FAX: (303) 440-0835
(303) 440-3705

Tyler Research
10328 73rd Avenue
Edmonton, Alberta
T6E 6N5 Canada
(403) 448-1249 FAX: (403) 433-0479

UBI
See Upstate Biotechnology

Ugo Basile Biological Research Apparatus
Via G. Borghi 43
21025 Comerio, Varese, Italy
(39) 332 744 574
FAX: (39) 332 745 488
http://www.ugobasile.com

UltraPIX
See Life Science Resources

Ultrasonic Power
239 East Stephenson Street
Freeport, IL 61032
(815) 235-6020 FAX: (815) 232-2150
http://www.upcorp.com

Ultrasound Advice
23 Aberdeen Road
London N52UG, UK
(44) 020-7359-1718
FAX: (44) 020-7359-3650
http://www.ultrasoundadvice.co.uk

UNELKO
14641 N. 74th Street
Scottsdale, AZ 85260
(480) 991-7272 FAX: (480)483-7674
http://www.unelko.com

Unifab Corp.
5260 Lovers Lane
Kalamazoo, MI 49002
(800) 648-9569 FAX: (616) 382-2825
(616) 382-2803

Union Carbide
10235 West Little York Road, Suite 300
Houston, TX 77040
(800) 568-4000 FAX: (713) 849-7021
(713) 849-7000
http://www.unioncarbide.com

United Desiccants
See Süd-Chemie Performance Packaging

United States Biochemical
See USB

United States Biological (US Biological)
P.O. Box 261
Swampscott, MA 01907
(800) 520-3011 FAX: (781) 639-1768
http://www.usbio.net

Universal Imaging
502 Brandywine Parkway
West Chester, PA 19380
(610) 344-9410 FAX: (610) 344-6515
http://www.image1.com

Upchurch Scientific
619 West Oak Street
P.O. Box 1529
Oak Harbor, WA 98277
(800) 426-0191 FAX: (800) 359-3460
(360) 679-2528 FAX: (360) 679-3830
http://www.upchurch.com

Upjohn
Pharmacia & Upjohn
http://www.pnu.com

Upstate Biotechnology (UBI)
1100 Winter Street, Suite 2300
Waltham, MA 02451
(800) 233-3991 FAX: (781) 890-7738
(781) 890-8845
http://www.upstatebiotech.com

USA/Scientific
346 SW 57th Avenue
P.O. Box 3565
Ocala, FL 34478
(800) LAB-TIPS FAX: (352) 351-2057
(3524) 237-6288
http://www.usascientific.com

USB
26111 Miles Road
P.O. Box 22400
Cleveland, OH 44122
(800) 321-9322 FAX: (800) 535-0898
FAX: (216) 464-5075
http://www.usbweb.com

USCI Bard
Bard Interventional Products
129 Concord Road
Billerica, MA 01821
(800) 225-1332 FAX: (978) 262-4805
http://www.bardinterventional.com

UVP (Ultraviolet Products)
2066 W. 11th Street
Upland, CA 91786
(800) 452-6788 FAX: (909) 946-3597
(909) 946-3197
http://www.uvp.com

V & P Scientific
9823 Pacific Heights Boulevard, Suite T
San Diego, CA 92121
(800) 455-0644 FAX: (858) 455-0703
(858) 455-0643
http://www.vp-scientific.com

Valco Instruments
P.O. Box 55603
Houston, TX 77255
(800) FOR-VICI FAX: (713) 688-8106
(713) 688-9345
http://www.vici.com

Valpey Fisher
75 South Street
Hopkin, MA 01748
(508) 435-6831 FAX: (508) 435-5289
http://www.valpeyfisher.com

Value Plastics
3325 Timberline Road
Fort Collins, CO 80525
(800) 404-LUER FAX: (970) 223-0953
(970) 223-8306
http://www.valueplastics.com

Vangard International
P.O. Box 308
3535 Rt. 66, Bldg. #4
Neptune, NJ 07754
(800) 922-0784 FAX: (732) 922-0557
(732) 922-4900
http://www.vangard1.com

Varian Analytical Instruments
2700 Mitchell Drive
Walnut Creek, CA 94598
(800) 926-3000 FAX: (925) 945-2102
(925) 939-2400
http://www.varianinc.com

Varian Associates
3050 Hansen Way
Palo Alto, CA 94304
(800) 544-4636 FAX: (650) 424-5358
(650) 493-4000
http://www.varian.com

Vector Core Laboratory/
National Gene Vector Labs
University of Michigan
3560 E MSRB II
1150 West Medical Center Drive
Ann Arbor, MI 48109
(734) 936-5843 FAX: (734) 764-3596

Vector Laboratories
30 Ingold Road
Burlingame, CA 94010
(800) 227-6666 FAX: (650) 697-0339
(650) 697-3600
http://www.vectorlabs.com

Vedco
2121 S.E. Bush Road
St. Joseph, MO 64504
(888) 708-3326 FAX: (816) 238-1837
(816) 238-8840
http://database.vedco.com

Ventana Medical Systems
3865 North Business Center Drive
Tucson, AZ 85705
(800) 227-2155 FAX: (520) 887-2558
(520) 887-2155
http://www.ventanamed.com

Verity Software House
P.O. Box 247
45A Augusta Road
Topsham, ME 04086
(207) 729-6767 FAX: (207) 729-5443
http://www.vsh.com

Vernitron
See Sensor Systems LLC

Vertex Pharmaceuticals
130 Waverly Street
Cambridge, MA 02139
(617) 577-6000 FAX: (617) 577-6680
http://www.vpharm.com

Vetamac
Route 7, Box 208
Frankfort, IN 46041
(317) 379-3621

Vet Drug
Unit 8
Lakeside Industrial Estate
Colnbrook, Slough SL3 0ED, UK

Vetus Animal Health
See Burns Veterinary Supply

Viamed
15 Station Road
Cross Hills, Keighley
W. Yorkshire BD20 7DT, UK
(44) 1-535-634-542
FAX: (44) 1-535-635-582
http://www.viamed.co.uk

Vical
9373 Town Center Drive, Suite 100
San Diego, CA 92121
(858) 646-1100 FAX: (858) 646-1150
http://www.vical.com

Victor Medical
2349 North Watney Way, Suite D
Fairfield, CA 94533
(800) 888-8908 FAX: (707) 425-6459
(707) 425-0294

Virion Systems
9610 Medical Center Drive, Suite 100
Rockville, MD 20850
(301) 309-1844 FAX: (301) 309-0471
http://www.radix.net/~virion

VirTis Company
815 Route 208
Gardiner, NY 12525
(800) 765-6198 FAX: (914) 255-5338
(914) 255-5000
http://www.virtis.com

Visible Genetics
700 Bay Street, Suite 1000
Toronto, Ontario
M5G 1Z6 Canada
(888) 463-6844 (416) 813-3272
http://www.visgen.com

Vitrocom
8 Morris Avenue
Mountain Lakes, NJ 07046
(973) 402-1443 FAX: (973) 402-1445

VTI
7650 W. 26th Avenue
Hialeah, FL 33106
(305) 828-4700 FAX: (305) 828-0299
http://www.vticorp.com

VWR Scientific Products
200 Center Square Road
Bridgeport, NJ 08014
(800) 932-5000 FAX: (609) 467-5499
(609) 467-2600
http://www.vwrsp.com

Vydac
17434 Mojave Street
P.O. Box 867
Hesperia, CA 92345
(800) 247-0924 FAX: (760) 244-1984
(760) 244-6107
http://www.vydac.com

Vysis
3100 Woodcreek Drive
Downers Grove, IL 60515
(800) 553-7042 FAX: (630) 271-7138
(630) 271-7000
http://www.vysis.com

W&H Dentalwerk Bürmoos
P.O. Box 1
A-5111 Bürmoos, Austria
(43) 6274-6236-0
FAX: (43) 6274-6236-55
http://www.wnhdent.com

Wako BioProducts
See Wako Chemicals USA

Wako Chemicals USA
1600 Bellwood Road
Richmond, VA 23237
(800) 992-9256 FAX: (804) 271-7791
(804) 271-7677
http://www.wakousa.com

Wako Pure Chemicals
1-2, Doshomachi 3-chome
Chuo-ku, Osaka 540-8605, Japan
81-6-6203-3741 FAX: 81-6-6222-1203
http://www.wako-chem.co.jp/egaiyo/
index.htm

Wallac
See Perkin-Elmer

Wallac
A Division of Perkin-Elmer
3985 Eastern Road
Norton, OH 44203
(800) 321-9632 FAX: (330) 825-8520
(330) 825-4525
http://www.wallac.com

Waring Products
283 Main Street
New Hartford, CT 06057
(800) 348-7195 FAX: (860) 738-9203
(860) 379-0731
http://www.waringproducts.com

Warner Instrument
1141 Dixwell Avenue
Hamden, CT 06514
(800) 599-4203 FAX: (203) 776-1278
(203) 776-0664
http://www.warnerinstrument.com

Warner-Lambert
Parke-Davis
201 Tabor Road
Morris Plains, NJ 07950
(973) 540-2000 FAX: (973) 540-3761
http://www.warner-lambert.com

Washington University Machine Shop
615 South Taylor
St. Louis, MO 63310
(314) 362-6186 FAX: (314) 362-6184

Waters Chromatography
34 Maple Street
Milford, MA 01757
(800) 252-HPLC FAX: (508) 478-1990
(508) 478-2000
http://www.waters.com

Watlow
12001 Lackland Road
St. Louis, MO 63146
(314) 426-7431 FAX: (314) 447-8770
http://www.watlow.com

Watson-Marlow
220 Ballardvale Street
Wilmington, MA 01887
(978) 658-6168 FAX: (978) 988 0828
http://www.watson-marlow.co.uk

Waukesha Fluid Handling
611 Sugar Creek Road
Delavan, WI 53115
(800) 252-5200 FAX: (800) 252-5012
(414) 728-1900 FAX: (414) 728-4608
http://www.waukesha-cb.com

WaveMetrics
P.O. Box 2088
Lake Oswego, OR 97035
(503) 620-3001 FAX: (503) 620-6754
http://www.wavemetrics.com

Weather Measure
P.O. Box 41257
Sacramento, CA 95641
(916) 481-7565

Weber Scientific
2732 Kuser Road
Hamilton, NJ 08691
(800) FAT-TEST FAX: (609) 584-8388
(609) 584-7677
http://www.weberscientific.com

Weck, Edward & Company
1 Weck Drive
Research Triangle Park, NC 27709
(919) 544-8000

Wellcome Diagnostics
See Burroughs Wellcome

Wellington Laboratories
398 Laird Road
Guelph, Ontario
N1G 3X7 Canada
(800) 578-6985 FAX: (519) 822-2849
http://www.well-labs.com

Wesbart Engineering
Daux Road
Billingshurst, West Sussex
RH14 9EZ, UK
(44) 1-403-782738
FAX: (44) 1-403-784180
http://www.wesbart.co.uk

Whatman
9 Bridewell Place
Clifton, NJ 07014
(800) 631-7290 FAX: (973) 773-3991
(973) 773-5800
http://www.whatman.com

Wheaton Science Products
1501 North 10th Street
Millville, NJ 08332
(800) 225-1437 FAX: (800) 368-3108
(856) 825-1100 FAX: (856) 825-1368
http://www.algroupwheaton.com

Whittaker Bioproducts
See BioWhittaker

Wild Heerbrugg
Juerg Dedual Gaebrisstrasse 8 CH
9056 Gais, Switzerland
(41) 71-793-2723
FAX: (41) 71-726-5957
http://www.homepage.swissonline.net/
dedual/wild_heerbrugg

Willy A. Bachofen
AG Maschinenfabrik
Utengasse 15/17
CH4005 Basel, Switzerland
(41) 61-681-5151
FAX: (41) 61-681-5058
http://www.wab.ch

Winthrop
See Sterling Winthrop

Wolfram Research
100 Trade Center Drive
Champaign, IL 61820
(800) 965-3726 FAX: (217) 398-0747
(217) 398-0700
http://www.wolfram.com

World Health Organization
Microbiology and Immunology Support
20 Avenue Appia
1211 Geneva 27, Switzerland
(41-22) 791-2602
FAX: (41-22) 791-0746
http://www.who.org

World Precision Instruments
175 Sarasota Center Boulevard
International Trade Center
Sarasota, FL 34240
(941) 371-1003 FAX: (941) 377-5428
http://www.wpiinc.com

Worthington Biochemical
Halls Mill Road
Freehold, NJ 07728
(800) 445-9603 FAX: (800) 368-3108
(732) 462-3838 FAX: (732) 308-4453
http://www.worthington-biochem.com

WPI
See World Precision Instruments

Wyeth-Ayerst
2 Esterbrook Lane
Cherry Hill, NJ 08003
(800) 568-9938 FAX: (858) 424-8747
(858) 424-3700

Wyeth-Ayerst Laboratories
P.O. Box 1773
Paoli, PA 19301
(800) 666-7248 FAX: (610) 889-9669
(610) 644-8000
http://www.ahp.com

Xenotech
3800 Cambridge Street
Kansas City, KS 66103
(913) 588-7930 FAX: (913) 588-7572
http://www.xenotechllc.com

Xeragon
19300 Germantown Road
Germantown, MD 20874
(240) 686-7860 FAX: (240)686-7861
http://www.xeragon.com

Xillix Technologies
300-13775 Commerce Parkway
Richmond, British Columbia
V6V 2V4 Canada
(800) 665-2236 FAX: (604) 278-3356
(604) 278-5000
http://www.xillix.com

Xomed Surgical Products
6743 Southpoint Drive N
Jacksonville, FL 32216
(800) 874-5797   FAX: (800) 678-3995
(904) 296-9600   FAX: (904) 296-9666
http://www.xomed.com

Yakult Honsha
1-19, Higashi-Shinbashi 1-chome
Minato-ku Tokyo 105-8660, Japan
81-3-3574-8960

Yamasa Shoyu
23-8 Nihonbashi Kakigaracho
1-chome, Chuoku
Tokyo, 103 Japan
(81) 3-479 22 0095
FAX: (81) 3-479 22 3435

Yeast Genetic Stock Center
See ATCC

Yellow Spring Instruments
See YSI

YMC
YMC Karasuma-Gojo Building
284 Daigo-Cho, Karasuma Nisihiirr
Gojo-dori Shimogyo-ku
Kyoto, 600-8106, Japan
(81) 75-342-4567
FAX: (81) 75-342-4568
http://www.ymc.co.jp

YSI
1725-1700 Brannum Lane
Yellow Springs, OH 45387
(800) 765-9744   FAX: (937) 767-9353
(937) 767-7241
http://www.ysi.com

Zeneca/CRB
See AstraZeneca
(800) 327-0125   FAX: (800) 321-4745

Zivic-Miller Laboratories
178 Toll Gate Road
Zelienople, PA 16063
(800) 422-LABS   FAX: (724) 452-4506
(800) MBM-RATS   FAX: (724) 452-5200
http://zivicmiller.com

Zymark
Zymark Center
Hopkinton, MA 01748
(508) 435-9500   FAX: (508) 435-3439
http://www.zymark.com

Zymed Laboratories
458 Carlton Court
South San Francisco, CA 94080
(800) 874-4494   FAX: (650) 871-4499
(650) 871-4494
http://www.zymed.com

Zymo Research
625 W. Katella Avenue, Suite 30
Orange, CA 92867
(888) 882-9682   FAX: (714) 288-9643
(714) 288-9682
http://www.zymor.com

Zynaxis Cell Science
See ChiRex Cauldron

Suppliers

34

INDEX

A

Abbreviations, commonly used, 539–541
Absorbance. see Absorption; Absorptivity
Absorption, of light, color analysis, 203–215
Absorption, of volatile odorants, SPME, 301–312
Absorption spectrophotometry
  anthocyanins, 19–30
  betalains, 123–124
  carotenoids, 19–30, 81–90
  cooked and cured meats, 131–138
  fresh meats, 139–150
  isoflavones, 523–527, 530–532
  meat pigments, cooked and cured meats, 131–138
Absorptivity
  anthocyanins, 21–23 (table)
  definition, 124
  values for
    betanin and vulgaxanthin-I, 124
    proanthocyanidin cleavage products, 502 (table)
Acetic acid concentrations, organic standards from juices, 358 (table)
Acetone extraction
  of anthocyanins, 7–9, 12–16
  of polyphenolics, 478
  of proanthocyanidins, 499–500, 506–509
Acid/base titration, quantifying aldehydes in lemon oil, 287–288
Acidic fractions, isolation of phenolics
  with C18 cartridges, 475–476, 478, 481
  by RP-HPLC, 486–488
Acids. see also Amino acids; specific listings
  gallic, in colorimetry for total phenolics, 468
  nonvolatile, HPLC analysis, 351–362
  organic
    concentrations in selected juices, 360 (table)
    preparation for HPLC, 356–357
    standards, recipes and concentrations, 357–358
  sorbic, total phenolics analysis, 469
Acid tastants
  HPLC of nonvolatile acids, 351–362
  pH measurement, 343, 346–349
  titratable activity
    characteristics of, 348
    colorimetric, 343, 345–346
    potentiometric, 343–345
Acrylamide gels. see Polyacrylamide gel electrophoresis
Acylation, anthocyanins in HPLC analysis, 41–44
Adsorption
  aroma compounds, volatile traps, 237–242
  proanthocyanidins, purification of, 500–501, 507–508
  solid-phase, 7, 10, 11 (fig.)
  SPME methods, 301–312
Alcohol, carotenoid extraction, 77

Aldehydes. see also Carbonyl compounds
  citrus oils, composition and properties, 291–299
  citrus oils, quantification
    by acid/base titration, 287–288
    by hydroxylamine titration, 287
    with N-hydroxybenzenesulfonamide, 288–289
Alkaline hydrolysis. see Saponification
Almond seedcoat extracts, cation spectra in MALDI-MS, 517–518 (figs.)
ANS. see 1-Anilinonaphthalene-8-sulfonic acid
Anthocyanidins
  common, 24 (table), 41
  definition, 41, 491
  HPLC analysis, 38–39, 42 (table)
  structure, 490 (fig.)
Anthocyanins
  acetone, 7–9, 12–16
  chloroform partitioning, 7–9, 13–15
  common, 24 (table)
  content in some foods, 29 (table)
  decomposition, 15
  in fruits, 29 (table)
  HPLC, 33–44, 483–498
  methanol, 9, 13, 16
  molar absorptivity, 21–23 (table), 28
  molecular weights, 24 (table)
  NMR
    analysis by, 47–67
    purification, 10–12, 14–15, 473–475, 478–479, 481
    spectral data, 50–54
    structure, 20 (fig.), 33 (fig.), 490 (fig.)
  UV/Vis spectrophotometry, 19–30
Antioxidants, carotenoid extraction, 79
APCI. see Atmospheric pressure chemical ionization
Apple juice, acid concentrations
  in juice, 358 (table)
  in standards for HPLC, 358 (table)
Apples
  polyphenolics extraction, 478
  texture profile analysis, 418, 421 (table), 422
Apricot juice, acid concentrations in, 360 (table)
Aroma compounds. see also Citrus oils; Flavor analysis
  common compounds and labeled analogs, 251–252 (table)
  extraction and sampling
    concentration, volatile traps for, 237–240
    distillation/extraction, 235–237, 240, 242
    headspace sampling, 225–226
  GC analysis
    identification and quantification, 245–254, 329–340
    isotope dilution assay, 248–250
    with olfactometry (GC/O), 329–340
    retention indices, 229–230, 247 (fig.)
    stereodifferentiation, 257–276

  mouth simulators
    model mouth, 317–318
    retronasal aroma simulator (RAS), 314–316
  odor activity value (OAV), 266, 271
  odor spectrum value (OSV), 271, 274
  solid-phase microextraction, 301–312
  solvent extraction, 227–229
Atmospheric pressure chemical ionization (APCI)
  characterized, 555–561
  used with LC/MS
    carotenoid analysis, 111, 113–114, 117–118
    chlorophyll analysis, 193–198
Atomic emission detector, for volatile compounds, 250
ATR-FTIR. see Attenutated total reflection
Autoxidation. see also Oxidation

B

Back extrusion, compressive measurement, 412–413, 415
Banana, textural analysis, 400
Barreling, in compressive measurement, 402
BCA. see Bicinchoninic acid
Beer-Lambert law, 85
Beer's Law, 136
Beets, betalains extracted from, 126–128
Benzoic acids
  HPLC separation, 483–498
  structure, 489 (fig.)
Betacyanins
  HPLC quantification, 124–128
  spectrophotometry, 123–124
Betalains
  beets, extraction from, 126–128
  HPLC quantification, 124–128
  spectrophotometry, 123–124
Betaxanthins, spectrophotometry, 123–124
BHA. see Butylated hydroxyanisole
BHT. see Butylated hydroxytoluene
Binding, water. see Water retention
Bingham viscosity model, 373
Blackberries, polyphenolics extraction, 478
Black current juice, acid concentrations in standards for HPLC, 358 (table)
Bleaching agents
  anthocyanins, 27–28
  carotenoid detection, 89
Blotting. see Electroblotting; Immunoblotting
Blueberries, HPLC analysis, 42
Bostwick consistometer, 391–392
Bradford protein assay. see Coomassie dye binding assays
Brix
  in consistency measurement, 391–392
  for dilutions, serial, 28–29
  juices, common, standard values for, 354 (table)

Browning
  anthocyanins, 25–27
  meats, 139–150
Buckwheat, rutin/polyphenolics
  extraction, 478

C

Cabbage, HPLC analysis, 42
Cannon-Fenske viscometer, 385–389
Capsanthin, 78
Capsorubin, 78
Carbonated beverages, 348–349
α-Carotenes
  extraction of, 78
  separation of, 104
β-Carotenes
  extinction coefficients, 89
  extraction of, 78
  spectral characteristics, 85–87 (fig.)
Carotenoids. see also α-Carotene;
      β-Carotene
  characteristics of
    elemental composition and masses,
      118 (table)
    molecular weights, 82 (table)
    spectral characteristics, 82–83
      (tables), 84–87 (figs.)
  commercial sources of, 81 (table)
  degradation of, 89
  extraction of, 19–30, 73–80, 165–170
  HPLC of
    calibration, 93–95, 105
    sample preparation, 95–97, 103–104
    separation, 91–92, 96–102, 105
    standards preparation, 93–95
  mass spectrometry
    APCI LC/MS, 111, 113–114, 117–118
    EI/MS and CI/MS, 107–108
    ESI LC/MS, 110–111, 113–114, 116
    FAB-MS and LSIMS, 108–109, 112,
      116 (fig.)
    MALDI-TOF-MS, 109–110, 113–116
  prepurification by crystallization, 75–77
  reference systems, fresh vs. dry weights,
      176–178
  UV/Vis spectrophotometry of, 81–90,
      91–105, 171–178
Carreau-Yasuda viscosity model, 375, 382
Carrots, texture profile analysis, 417, 421
      (table)
Casson viscosity model, 374
Catalysis in pectic enzyme assays. see also
      Enzyme activity assays
Catechins (flavan-3-ols)
  HPLC separation, 483–498
  structure, 490 (fig.), 491
Cations, spectra of flavonol glycosides in
      MALDIMS
  of almond seedcoat extracts, 517–518
      (figs.)
  of standards, 516 (fig.)
C18 cartridges
  flavonol glycosides fractionation with,
      512–513
  polyphenolics fractionation with,
      473–476
  solid-phase purification of anthocyanins,
      10–11, 14–15
C18 columns for HPLC
  anthrocyanin analysis

polymeric, sample preparation and
      separation, 35–37
  silica, sample preparation and
      separation, 34–35
  chlorophyll analysis, 34–37
    column cleaning, 184 (table), 185
    nonpolar derivatives, 180–182
    polar derivatives, 182–185 (table)
    system setup, 180
  of nonvolatile acids, 351–353, 360 (fig.)
CD. see Circular dichroism
Centrifugation, in anthocyanin
      extraction, 16
CF-FAB. see Continuous-flow fast atom
      bombardment mass spectrometry
Cheese
  compressive rate, force/deformation
      curve
    cheddar, 400 (fig.)
    mozzarella, 403 (fig.)
  viscosity measurement, 376
Chemical ionization (CI), mass
      spectrometry. see also Desorption
      chemical ionization
  of carotenoids, 107–108
  description, 555–561
  of smell chemicals, 232
Chemically induced gels, dynamic tests
      measuring rheology, 439
Chemorheology, for complex fluids,
      433–435
Cherry juice (sour), acid concentrations in,
      360 (table)
Chiral odorants, stereodifferentiation using
      high-resolution gas chromatography
      (HRGC)
  chiral columns, selection and care of,
      272–273
  enantiomer composition, determination
      of, 258–264
  instrumentation and conditions, 257, 272
  sensory discrimination using
    gas chromatography/olfactory (GC-O),
      264–266, 272, 274
    multidimensional gas chromatography
      (MDGC), 267–268, 269–274
CHIRBASE, 269
Chlorophyllides, 155–163
Chlorophylls
  chromatography of, general, 160–161
  classes and derivatives of, 155–160
  content
    interpretation of, 176–178
    values, 156–157 (table), 177 (table)
  conversion to pheophytins, 158–161,
      166, 169–170
  extraction, 160, 165–170
  mass spectrometry, 191–199
  pheophytin standards for, 186
  post-chromatography detection
      methods, 161
  reference systems, fresh vs. dry weights,
      176–178
  separation by C18 RP-HPLC, 179–189
  structure of, 155 (fig.), 159 (fig.)
Chocolate, viscosity measurement, 376
Chroma, in color analysis, 203–204,
      211–214. see also Color
Chromatography. see also Gas
      chromatography; Gas
      chromatography/olfactory;

High-performance liquid
      chromatography; Thinlayer
      chromatography
  of aroma compounds, GC/O, 329–340
  of chlorophylls, 160–161, 179–189
  of citrus oils, 278–280, 296 (figs.)
  of flavonol glycosides, 515
  hydrosulfite, removal of, 145
  of polyphenolics, 471–481, 483–498
  of proanthocyanidins, 499–509
  of soy isoflavones, 525–528, 532–535
Chromophores, characterized, 85
CI. see Chemical ionization
CID. see Collision-induced dissociation
CIE. see Commission Internationale de
      l'Eclairage
CIE illuminants, 204–205
CIELAB values for color analysis, 209–215
CIELCH values for color analysis, 211–215
CIE X,Y,Z values for color analysis,
      207–211
Cinnamic acids
  HPLC separation, 483–498
  structure, 489 (fig.)
Citric acid concentrations
  in organic standards for selected juices,
      358 (table)
  in selected juices, 360 (table)
Citrus oils
  analysis by
    gas chromatography, 278–280,
      296–297 (figs.)
    polarimetry, 282
    pycnometry, 281
    refractive index, 281–282, 295 (table)
  components of, 292–294 (tables)
  determination of
    aldehyde content, 287–289, 299
      (table)
    limonene purity, 280
    market value, 291 (table)
    moisture content, 286
    physical and chemical properties, 295
      (table)
  quantification of
    oil from dry peel or pellet, 285, 299
      (table)
    oil from press liquor and molasses, 286
    press cake oil, 285
    total oil, from whole fruit or wet peel,
      282–284
    volatile esters, 289–290
CMC ellipse for color analysis, 214–215
Cold-pressed citrus oils
  aldehyde content, 292 (table), 296 (fig.)
  ester content, 292 (table), 295 (table)
  quantitation of, 285–286
Collision-induced dissociation (CID)
  carotenoid analysis by MS, 107
  chlorophyll analysis with tandem MS,
      191–192, 194 (fig.), 196–197
Color analysis
  CIELAB, 209–215
  CIELCH, 211–215
  CIE X,Y,Z, 207–211
  CMC, 214–215
Colorants and pigments
  anthocyanins, 19–30
  betalains, 123–129
  carotenoids, 83–84, 165–178
  color analysis, 203–215

cooked and cured, 131–138
fresh, 139–150
measurement strategies, 139–150
in meats, cooked and cured, 131–138
Color density, 25. *see also* Colorimetry
color difference equations, 211–214
definition, 203–204
flavonol glycoside analysis, 511–517
human observer for, 206–207
Hunter L,a,b, 209–213
light source, 205
measurement strategies, 139–150
object, 205–206
observer situation, 204–207
standard color observers, 207
Colorimetric titration, acid tastants, 343, 345–346
Colorimetry. *see also* Color analysis
Folin-Ciocalteau (FC), for total phenolics, 463–469
meat discoloration, 143–144, 150
Column chromatography
anthocyanin extraction, 14
carotenoid extraction, 73, 75 (fig.), 78–79
Commission Internationale de l'Eclairage (CIE). *see also* Color
CIELAB values, 209–215
CIELCH, 211–215
CIE X,Y,Z values, 207–211
CMC ellipse, 214–215
color difference equations, 211–214
illuminant standards, 204–205
standard color observers, 207
Complex fluids. *see* Fluids, complex
Compounds, volatile. *see* Volatile compounds
Compressive measurement
barreling, 402
engineering strain, 401
food rheology, 401
force/deformation curves, 405–411 (figs.)
Hooke's Law, 401
rupture point, 408 (fig.)
sampling, 401–402
strength of specimen, 408 (fig.)
stress, 401–402
textural with special fixtures
sample preparation, 415
using back extrusion, 412–413, 415
using cone penetrometer, 405–409, 413–414
using puncture probe, 405–409, 414
using Warner-Bratzler, Kramer, or wirecutting fixtures, 409–415
texture profile analysis (TPA)
apples, 418, 421, 422 (table)
carrots, 417, 421 (table)
compression cycle, 421
gelatin gel, 419, 422
ground beef, cooked, 418, 422 (table)
instrumentation, 419–420 (table), 422–423
potatoes, 417, 420 (table)
sample preparation, 421
uniaxial compression, 402–403
Concord grapes, chromatogram of polyphenolics, 496 (fig.)
Cone-and-plate geometry, 376
Cone pentrometer, compressive measurement, 405–409, 413–414

Consent form, for gas chromatography/ olfactrometry (GC/O) study, 338–339 (fig.)
Continuous-flow fast atom bombardment (CF-FAB)
carotenoid analysis by MS, 108–109, 112
MS method, 556 (fig.), 557
Cookie dough
complex viscosity as function of temperature, 434 (fig.)
LVE range of shortening, 431 (fig.)
Copper, synthesis of Cu2+ pheophytin standards, 186
CPA. *see* cis-parinaric acid
Cranberry juice
acid concentrations in, 360 (table)
HPLC analysis, 35 (fig.), 41–45
Creep tests, 449–456
Cross viscosity model, 375, 382
α-Cryptoxanthin, separation of, 104
β-Cryptoxanthin
extraction of, 78
separation of, 104
Crystallization, prepurification of carotenoids
basic protocol, 75–76, 79
removal of water using alcohol, 77
removal of water using vacuum oven, 76–77
Cyanidin, HPLC analysis, 41–45
Cyclodextrin
chiral odorant analysis, 269
columns, 270 (table), 273
Cyd-3-glu, 42

D

β-Damascenone, 233
DCI. *see* Desorption chemical ionization
Delphinidin
HPLC analysis, 41
spectral characteristics, 28
Desorption, aroma compounds, volatile traps, 239–240, 242
Desorption mass spectrometry
of chlorophylls, 193–198
theory and applications, 555–561
Detection frequency method, for GC/O analysis, 334–335
Deuterium lamps, 102
DH. *see* Degree of hydrolysis; Protein hydrolysis
Dichloromethane (DCM), 241, 731
Diethyl ether, 241
Diffuse reflection, in color analysis, 205 (fig.), 206
Diffusion of emulsions. *see also* Ostwald ripening
Dilution analysis
for GC flavor study, 303, 306–307
in GC/O analysis, 332–333
Diode-array detectors (DAD), carotenoid separation, 103
Discoloration in fresh meat
colorimetry of meat surfaces, 143
myoglobin, isolation of, 144–145
oxymyoglobin, preparation of, 145–146
sensory assessment of appearance, 143
spectrophotometry of metmyoglobin in meat extracts, 139–140

on meat surfaces, 141–142
Dispersion, viscosity
Casson model, 374
non-Newtonian fluids, 376
Dissociation, collision-induced (CID), with tandem MS of chlorophylls, 191–192, 194 (fig.), 196–197
DNPH. *see* 2,4-Dinitrophenylhydrazine
Dough
cookie dough
LVE range of shortening, 431 (fig.)
viscosity as function of temperature, 434 (fig.)
viscosity measurement, 376
DPH. *see* 1,6-Diphenyl-1,3,5-hexatriene
Drinking water, geosmin contamination, 231
Drip loss in solid foods. *see also* Water retention
DSA. *see* Drop shape analysis
DVT. *see* Drop volume tensiometer
Dynamic testing of complex fluids, 427–437

E

EAI. *see* Emulsifying activity index
EI. *see* Electron impact ionization
Elasticity, gel rheology tests, 427–437, 439–446. *see also* Viscoelasticity
Elastic modulus, in compressive measurement, 397, 403
Electronic transitions of amino acids, CD analysis. *see also* Circular dichroism
Electron impact, in volatile compound quantitation, 253
Electron impact (EI) ionization
for carotenoids, 107–108
characterized, 555–557, 559
for smell chemicals, 232
Electrophoresis. *see* Polyacrylamide gel electrophoresis
Electrospray ionization, interface for LC/MS
of carotenoids, 110–111, 113–114, 116–118
characterized, 557–559
of chlorophylls, 193–198
Emulsion
stability/instability of, 434 (fig.)
temperature stability of, 434 (fig.)
viscosity
Casson model, 374
generally, 369
non-Newtonian fluids, 376
Enantiomers
purity, 270
ratio, 274
sensory evaluation, 274
stereodifferentiation, 257–276
Endopeptidases. *see* Proteinases
Enzymatic hydrolysis, of isoflavones, 528–529
Equilibrium flow tests, 380–382
Equilibrium relative humidity (ERH). *see also* Water activity
Equipment, standard laboratory, 553–554
ERH. *see* Equilibrium relative humidity
ESI. *see* Electrospray ionization
Essence oils. *see* specific listings
Esters, in citrus oils, volatile

content in specific oils, 292 (table),
    294–295 (tables)
determination of, 289–290
Ethanol
    anthocyanin extraction, 13
    carotenoid extraction, 78
    polyphenolic extraction, 478
EVI. *see* Emulsion volume index
Expressible moisture. *see also* Water
    retention
Extinction coefficient. *see* Molar absorptivity
Extraction
    anthocyanins
        acetone, 7–9, 12–16
        methanol, 9, 13, 16
        solid phase, 38–41
    betalains, 126–128
    carotenoids
        discussion, 73–74
        using diethyl ether, 165–170
    chlorophylls, 160, 165–170
    pigments of cooked and cured
        meats, 137
    polyphenolics
        homogenizer-assisted, 472–473,
            477–481
        solvents, choice of, 478
        ultrasound-assisted, 471–472,
            477–481
    simultaneous distillation/extraction,
        aroma compounds, 235–237
    solid-phase, 104
    solvent, in smell chemical sampling,
        227–229, 231–233
    soy isoflavones, 520–521, 522 (table)
    SPME methods for flavor analysis,
        301–312
Extracts, ground meat, 139–140, 146

F

FAB-MS. *see* Fast atom bombardment
    mass spectrometry
FAMES. *see* Fatty acid methyl esters
Far-UV CD spectra. *see also* Circular
    dichroism
Fast atom bombardment mass
    spectrometry (FAB-MS). *see also*
    Continuous-flow fast atom
    bombardment
    of carotenoids, 108–109, 116 (fig)
    characterized, 556–557, 559
    of chlorophylls, 161, 191–192
Fats. *see also* Fatty acids; Lipid Oxidation;
    Lipids
    oscillatory response, 429–432
Fatty acids. *see also* Fats
FD. *see* Field desorption
FD&C Red No. 3 and No. 40, 29
Ferroprotoporphyrin, 131
Fiber coatings, for SPME, 302 (table),
    309–310, 312
FID. *see* Flame ionization detection
Field desorption (FD) in mass
    spectrometry, 556
Flame ionization detection (FID)
    aroma compounds, 245–254
    smell chemicals, 232
Flame photometric detector, volatile
    compounds, 250
Flavan-3-ols (catechins)

HPLC separation, 483–498
removal from proanthocyanidins,
    500–501
structure, 490 (fig.), 491
Flavanones
    HPLC separation, 483–498
    structure of, 491 (fig.)
Flavones
    HPLC separation, 483–498
    structure of, 491, 492 (fig.)
Flavonoids. *see also* Polyphenolics
    anthocyanins, 7–17, 19–30
    classification and structures of, 13 (fig.),
        477 (fig.), 490–493 (figs.),
        519–521 (figs.)
    extraction
        all polyphenolics, 471–481
        flavonol glycosides, 511–517
        isoflavones, 520–523
        proanthocyanidins, 499–509
    fractionation, 471–481
    HPLC separation
        all polyphenolics, 483–498
        anthocyanins, 483–498
        isoflavones, 525–528
        proanthocyanidins, subunit
            composition, 499–509
    MALDI-MS, flavonol glycosides,
        511–517
    NMR, anthocyanins, 47–67
    UV/Vis spectrophotometry, isoflavones,
        523–525
Flavonol glycosides
    chromatographic analysis, 515
    extraction, 511–512
    MALDI-MS analysis, 511–517
    purification with C18 cartridge, 512–513
    structures of, 491, 492 (fig.), 514 (fig.)
Flavonols. *see also* Flavonol glycosides
    HPLC separation, 483–498
    structures, 491, 492 (fig.), 514 (table)
Flavor analysis. *see also* Aroma
    compounds
    of citrus oils, 277–299
    isotope dilution assay, 245–248,
        250, 253
    model mouth, 317–318
    mouth simulators, 313–325
    retronasal aroma simulator (RAS),
        314–316
    solid-phase microextraction, 301–312
Fluids, complex
    discussion of, 427–429, 435–437
    dynamic or oscillatory tests, 427–437,
        439–446
    LVE range determination, 430–432,
        435–437
    mechanical spectrum or fingerprint,
        432–433
    oscillatory responses, specific, 427–429
        (figs.)
    steady state determination, 429–430
    stress/strain responses, specific,
        430–432 (figs.)
    structural changes and chemorheology,
        433–435
    time-temperature superposition, 433, 434
        (fig.)
Fluorescence, in chlorophyll detection, 161
Folin-Ciocalteu (FC) colorimetry for total
    phenolics, 463–464, 466–469

Food deterioration, chlorophylls, 157–159
Fourier transform ion cyclotron resonance
    (FTICR) mass spectrometers, 560
Fractionation
    anthocyanin extraction, 14
    chemical ionization mass
        spectroscopy, 108
    polyphenolics, 473–476
Fruit juices
    acid concentrations in, 360 (table)
    acid concentrations in standards for
        HPLC, 358 (table)
    anthocyanins, HPLC analysis, 33–44
    Brix, concentration, 28–29
    Brix, standards for, 354 (table)
    consistency measurement, 391–392
    recipes for HPLC standards, 357–358
    viscosity measurement, 376, 387–388
Fruits
    anthocyanins in, 29 (table)
    chlorophylls, extraction, 160, 165–170
    chromatographic separation of, 179–189
    citrus oils, analysis of, 277–299
    extraction, 160, 165–170
    flavonol glycosides, extraction and
        MALDIMS, 511–517
    formation, detection and analysis,
        155–163
    mass spectrometry, 191–199
    polyphenolics
        extraction, 471–481
        RP-HPLC, 483–498
    sample preparation, 88
    total phenolics in, 463–469
    UV/Vis spectrophotometry, 171–178
FTICR. *see* Fourier transform ion cyclotron
    resonance
FTIR. *see* Fourier transform infrared
    spectroscopy
Furaneol, 240
Furfural, 240

G

Galacturonic acid, apple juice, HPLC
    standard concentrations,
    358 (table)
Gallic acid equivalents (GAE) in FC
    colorimetry for total phenolics, 468
Gas chromatography (GC)
    aroma compounds, GC-FID and GC-MS,
        245–254
    of citrus oils
        for limonene purity, rapid, 280
        qualitative, 278–279, 296 (figs.)
        quantitative, 279–280
    flame ionization detection (FID)
        aroma compounds, 245–254
        smell chemicals, 232
    with olfactometry (GC/O), 329–340
    olfactory, smell chemicals
        characteristics of, 230–232
        direct headspace sampling, 226
        retention indexing, 229–230, 232
    SPME for, 301–312
Gas chromatography-mass spectrometry
    (GC-MS)
    aroma compounds, volatile compounds,
        245–254
    general, 555–556
Gas chromatography/olfactory (GC/O),

chiral odorant analysis, 264–266, 269–274

Gas chromatography/olfactrometry (GC/O)
  consent form, example, 338–339 (fig.)
  detection frequency method, 334–335
  dilution analysis method, 332–333
  direct sniffing, basic method, 329–331
  posterior intensity method, 335–336
  sample data
    grapes, Niagara, 331 (table), 336 (fig.)
    volatile compounds, 335 (fig.)
  sniff port, diagram, 329 (fig.)
  time intensity method, 333–334
Gas-phase SPME analysis, 301–312. see also Solid-phase microextraction
GC. see Gas chromatography
Gelatin gel, compressive measurement, texture profile analysis, 419, 422
Gelation process in protein dispersions
  heat-induced, 439, 445
  monitored by gel rheology, 439–447
Gel point, 445. see also Gelation
Gel rheology
  concentration dependence, 445 (fig.), 446
  oscillatory and dynamic tests, 427–437, 439–446
Gels. see Polyacrylamide gel electrophoresis
Geometry, in non-Newtonian fluid viscosity measurement
  material selection, 377
  setting up, 377–378
  size, selection factors, 376
  type, selection factors, 376–377
Globin hemochrome, detection in meats, 134–135
Glycosides. see Flavonol glycosides
GOPOD. see Glucose oxidase/peroxidase/4-aminoantipyrine
Gradient separation, of carotenoids using C30 carotenoid column, 99–101
Grains, flavonol glycosides, extraction and MALDIMS, 511–517
Grapefruit juice, acid concentrations
  in juice, 360 (table)
  in standards for HPLC, 358 (table)
Grapefruit oil. see also Citrus oils
  aldehyde quantification, 288–289, 299 (table)
  market value and limonene content, 291 (table)
  physical and chemical properties, 293–295, 299 (table)
Grape/grapes. see also Grape juice; Wine
  Concord
    extraction of, 231, 233
    HPLC, 39 (fig.)
    polyphenolics chromatogram, 496
  Niagara
    GC/O sample data, 331 (table), 336 (fig.)
    polyphenolics chromatogram, 487–488 (figs.)
    polyphenolics extraction, 478
  red, HPLC, 42
  skin, proanthocyanidin cleavage products from, 507 (fig.)
  total phenolics determination, 463–469
Grape juice

acid concentrations
  in juice, 360 (table)
  in standards for HPLC, 358 (table)
  Concord, HPLC chromatogram, 37 (fig.)
Grease, oscillatory response
  LVE range of shortening, 430–432
  structure recovery, 429 (fig.)
Green beans, zinc-pheophytins in, 159
Green peas, compressive measurement, 410
Green vegetables
  carotenoids, extracts/extraction, 89, 165–170
  chlorophylls, extraction, 160, 165–170
  chromatographic separation of, 179–189
  extraction, 160, 165–170
  extracts/extraction, 89, 165–170
  formation, detection, and analysis, 155–163
  mass spectrometry, 191–199
  UV/Vis spectrophotometry, 171–178
  UV-vis spectroscopic analysis, 171–178
Ground beef, cooked, texture profile analysis, 418, 422 (table)
Ground meat, metmyoglobin analysis, 139–140, 146, 148

H

Hahn Echo. See also Spin echo
Hamlin oil. see Citrus oil; Orange
Headspace extraction. see also Dilution analysis; Solid-phase microextraction
  comparison of methods, 307 (table)
  quantification of, 303–306
  SPME, 302–303
Heat-induced gels, dynamic tests measuring rheology, 439
Heme, measurement of
  globin hemochrome, 134–135
  nitrosylheme, 131–133
  total heme, 133–134
Hering theory of color vision, 209
High-performance liquid chromatography (HPLC). see also Reversed-phase HPLC
  of acid tastants, 351–362
  of anthocyanidins, 38–39
  of anthocyanins, 33–44
  of betalains, 124–128
  of carotenoids, 91–105
  C18 columns, 179–189
  of chlorophylls
    general characterization, 160–161
    LC/MS, 191–198
  of isoflavones, 525–528, 532–535
  in LC/MS, 191–198, 557–559, 561
  of nonvolatile acids, 351–362
  of polyphenolics, 483–498
  of proanthocyanidins, subunit composition, 499–509
High-resolution gas chromatography (HRGC), chiral odorant analysis, 258–259, 269
High-resolution tandem mass spectrometry (HRMS), 560
Hooke's law, 401
HPLC. see High-performance liquid chromatography

HPX-87H column, for HPLC of nonvolatile acids, 351, 354–356, 361 (fig.)
Hue, in color analysis, 203–204, 211–214. see also Color
Human subjects/studies. see also specific studies
  consent form, 338–339 (fig.)
  eye, spectral response to color, 206–207 (fig.)
  for GC/O using direct sniffing, 329, 338–339
Hunter L,a,b values in color analysis, 209–213
Hydrochloric acid, 12–13, 15
Hydrolysis. see also Hydrophobicity; Protein hydrolysis; Water activity; Water content; Water retention
  acid, anthocyanin analysis, 41–42
  enzymatic, isoflavones, 528–529
Hydrometry. see also Hydration; Hydrolysis; Hydrophobicity; Water content; Water Retention
Hydrophobicity. see also Hydration; Hydrometry; Water activity; Water content; Water retention
Hydroxybenzoic acid, structure of, 476 (fig.)
Hydroxycinnamic acid, structure of, 476 (fig.)
Hydroxylamine titration, quantifying aldehydes in citrus oils, 287
Hydroxyphenylacetic acid, structure of, 476 (fig.)

I

IAEDANS. see N-Iodoacetyl- N′-(5-sulfo-1-naphthyl)ethylenediamine
1,5-I-AEDANS. see N-Iodoacetyl- N′-(5-sulfo-1-naphthyl)ethylenediamine
Illumination. see also Commission Internationale de l'Eclairage (CIE)
  light source for color analysis, 205
Infrared spectroscopy. see also Fourier transform infrared spectroscopy
International Commission on Illumination. see Commission Internationale de l'Eclairage (CIE)
Ionization techniques
  chlorophylls, 191–199
  in mass spectrometry of carotenoids
    CI, 107–110, 112–116
    ESI, 110–111, 113–114, 116–118
    FAB, 108–109, 112–113, 116 (fig.), 117
    MALDI, 107, 113–115
Isobetanin, 128
Isocitric acid concentrations
  in organic standards for selected juices, 358 (table)
  in selected juices, 360 (table)
Isocratic carotenoid separation
  using spherisorb ODS2, 98–99
  using wide-pore polymeric C18, 91–92
Isocratic gradient liquid chromatography, betalains, 124–125, 128 (fig.)
Isoflavones
  commercially available, 522 (table)
  content in common soy products, 522 (table)
  conversion to aglycones, 528–529
  HPLC of, 483–498, 525–528, 532–535

preparation of, 523–525
solvent extraction and preparation, 520–523, 530–531
spectral characteristics, 523–525, 530–531
structure of, 492, 493 (fig.), 519–521 (figs.), 530
Isolation
  of anthocyanins, 7
  of aroma compounds, 225–243
  of betalains, 126–127
  of carotenoids, 72–80, 165–170
  of chlorophylls, 165–170
  of flavonol glycosides, 512–513
  of isoflavones, 520–523
  of polyphenolics, 471–481
  of proanthocyanidins, 499–501
  of total myoglobin, 144–146
Isotope dilution assays (IDA), aroma compounds
  identification, 245–248
  quantitation, 245, 250, 253–254
Isotope ratio mass spectrometry (IRMS), food odorant analysis, 269
Italian lemons, 295 (table)

J

Juices, selected
  _Brix standards for, 354 (table)
  organic acid concentrations in, 360 (table)
Juice standards
  organic, acid concentrations for HPLC, 358 (table)
  recipes, 357–358

K

Ketchup, viscosity measurement, 376
Kramer shear press, compressive measurement, 409–412, 414–415
Kubelka–Munk equation, 141–142

L

Labile nature of chlorophylls, 159–160
Laboratory guidelines
  equipment list, general, 553–554
  laboratory safety, hazardous chemicals, 551–552
  mass spectrometry, 555–561
Laboratory stock solutions and equipment
  acids and bases, molarities and specific gravities, 543 (table)
  buffers
    molecular weights, 544 (table)
    pKa values, 544 (table)
    selection of, 545
  general guidelines, 543
  recipes
    ammonium acetate, 10 M, 545
    citrate-phosphate buffer (McIlvaine's buffer), 545, 546 (table)
    DTT (dithiothreitol), 1 M, 545
    EDTA (ethylenediaminetetraacetic acid), 0.5 M (pH 8.0), 545
    HCl, 1 M, 545
    KCl, 1 M, 545
    MgCl2, 1 M, 546

MgSO4, 1 M, 546
NaCl, 5 M, 546
NaOH, 10 M, 546
potassium acetate buffer, 0.1 M, 546, 547 (table)
potassium phosphate buffer, 0.1 M, 547
SDS, 20% (w/v), 548
sodium acetate, 3 M, 548
sodium acetate buffer, 0.1 M, 547 (table), 548
sodium phosphate buffer, 0.1 M, 548
Tris-Cl, 1 M, 548
storage, 543
Lactic acid concentrations, in organic standards for selected juices, 358 (table)
Lambert–Beer law, used to determine total carotenoid content, 175–176
Laminar flow, viscosity and, 369
LC/MS. see Liquid chromatography/mass spectrometry
Lemon juice, acid concentrations
  in juice, 360 (table)
  in standards for HPLC, 358 (table)
Lemon oil. see also Citrus oils
  aldehyde quantification, 287–288
  market value and limonene content, 291 (table)
  physical and chemical properties, 295 (table), 296 (fig.), 299 (table)
  volatile compounds, for flavor, 294 (table)
Lightness difference in color analysis, 212
Light source. see Color; Illumination
Lime oil. see also Citrus oils
  aldehyde content, 294 (table), 297 (fig.), 298 (table)
  market value and limonene content, 291 (table)
  physical and chemical properties, 295 (table), 299 (table)
Limonene in citrus oils
  content in specific oils, 291 (table)
  purity by rapid chromatography, 280
Linear viscoelastic (LVE) region, determination
  in complex fluids, 430–435, 439–447
  mathematics of, 435–437
Lipid composition. see also Lipids
Lipids. See also Lipid composition; Lipid oxidation/stability
Liquid chromatography. see High-performance liquid chromatography; Liquid chromotography/mass spectrometry
Liquid chromatography/mass spectrometry (LC/MS)
  of carotenoids, 110–111
  characterized, 557–560
  of chlorophylls, 191–194, 194
Liquids. see also Fluids
  complex, dynamic or oscillatory testing of, 427–437, 439–446
  pure, viscosity measurement, 386–387
Liquid secondary ion mass spectrometry (LSIMS)
  of carotenoids, 108–109
  characterized, 556–557, 559
  of chlorophylls, 191–192
LSIMS. see Liquid secondary ion mass spectrometry

Lutein
  extraction of, 78
  separation of, 104
  spectral characteristics, 87 (fig.)
LVE region. see Linear viscoelastic region
Lycopene
  extraction of, 78
  separation of, 104
  spectral characteristics, 87 (fig.)

M

MA. see Malonaldehyde
MALDI-MS. see Matrix-assisted laser desorption/ionization mass spectrometry
Malic acid concentrations
  in organic standards for selected juices, 358 (table)
  in selected juices, 360 (table)
Malonaldehyde (MA). see also 2-Thiobarbituric acid (TBA) assays
Malondialdehyde. see Malonaldehyde
Maltol, 233
Malvidin, HPLC analysis, 41
Mandarin orange oil. see Citrus oils; Orange
Market value, of citrus oils, 287–289
Mass fragmentography, 232
Mass spectrometry (MS). see also specific MS methods
  APCI
    of aroma compounds, 245–254
    of carotenoids, 111
    of chlorophylls, 193–194
  of carotenoids, 107–118, 107–119
  chemical ionization of carotenoids, 107–108
  of chlorophylls, 161, 191–199
  continuous-flow fast atom bombardment (CF-FAB)
    of carotenoids, 108–109, 116 (fig.)
    of chlorophylls, 191–192
  desorption ionization methods, 555 (fig.), 556–557
  electron impact of carotenoids, 107–108
  ESI
    of carotenoids, 110–111
    of chlorophylls, 193–194
  FAB
    of carotenoids, 108–109, 116 (fig.)
    of chlorophylls, 161, 191–192
    of flavonol glycosides, 511–517
  GC/MS
    of aroma chemicals, 245–254
    characterized, 555–556
  LC/MS
    of carotenoids, 108–111, 110–111
    characterized, 557–560
  LSIMS, of carotenoids, 108–109
  MALDI-TOF
    of carotenoids, 109–110
    of flavonol glycosides, 513–514
  tandem (MS/MS)
    of carotenoids, 107, 112–113
    characterized, 559 (fig.), 560
    of chlorophylls, 161
Mastication. see also Compressive measurement
Matrix, for FAB, 112, 115, 117

Matrix-assisted laser desorption/ionization mass spectrometry (MALDI-MS)
carotenoid analysis, 107, 113–116
characterized, 556–557, 559
flavonol glycosides, 511–517
TOF, 116
Meat
cooked or cured, discoloration of, 131–138
fresh, discoloration of, 139–150
globin hemochrome, detection of, 134–135
metmyoglobin, measurement of, 139–126
myoglobin, isolation, 144–145
nitrosylheme concentration, 131–133
oxymyoglobin, preparation, 145–146
pigments
absorbance spectrophotometry, 131–134, 139–140
classified, 132 (table)
colorimetry, 143
reflectance spectrophotometry, 134–135, 141–142
total heme concentration, 133–134
Metallochlorophylls
formation and detection of, 158–161
synthesis of pheophytin standards, 186
Methanol
of anthocyanins, 9, 13, 16
extraction
of chlorophylls, 160
of flavonol glycosides, 511–512
of polyphenolics, 471–473, 477–481
Methyl anthranilate, 231
Metmyoglobin
measurement, 139–142
total myoglobin, isolation, 144–145
Metric hue difference in color analysis, 212
Mexican lime oil. see Citrus oils; Lime oil
Milk, milk fat, elastic modulus, 397, 403
Moisture analysis. see also Expressible moisture; Water content; Water retention
content determination in citrus oils, 286
Molar absorptivity. see Absorptivity
Molarities, acids and bases, 543
Molasses, quantification, 286
Molecular weights
anthocyanins, 24 (table)
buffers, 544 (table)
carotenoids, 82
Mononitrosylhemochrome, 131
Mouth simulators
model mouth, 317–318, 320–321, 323–325
RAS, 313–316, 321–324
MS. see Mass spectrometry
MS/MS. see Mass spectrometry
Multidimensional gas chromatography (MDGC), of chiral odorants, 260–264, 267–268, 269–274
Munsell system describing color, 203 (fig.), 204
Mustard, viscosity measurement, 376
Myoglobin, isolation from fresh meat, 144–145

N

Near-UV CD spectra. see also Circular dichroism

Neutral phenolics, fractionation
fractionation, 475–476, 478, 481
RP-HPLC, 486–488
Newtonian viscosity model, 374
N-Hydroxybenzenesulfonamide (HBS) assay, 288–289
Niagara grapes. see Grapes, Niagara
Ninhydrin reaction, with amino acid in SPME/GC, 301 (fig.)
Nitrogen
carotenoid analysis, 88
nitrogen phosphorus detector, volatile compound identification, 250
Nitrosylheme pigments, detection in meats, 131–133
NMR. see Nuclear magnetic resonance
Nonequilibrium ramped tests, 378–380, 382–383
Non-Newtonian fluids
complete flow curve, 373 (fig.)
measurement of, 370–371
rheogram data, 372
shear rate, 382–383
time-dependent behavior, 378, 383
viscosity, generally, 369, 371
viscosity measurement
complete flow curve, 379
equilibrium flow tests, 380–382
geometry set-up and selection factors, 376–378
nonequilibrium ramped or stepped flow tests, 375, 378–380, 382–383
sample loading, 378
yield stress, 375
Nonpolar derivatives of chlorophylls, 180–182, C18 RPHPLC of
C18 RPHPLC of, 180–182
Nuclear magnetic resonance, of anthocyanins, 47–67
Nuts. see also specific listings
almond seedcoat extracts, cation spectra in MALDI-MS, 517–518 (figs.)

O

Octadecyl-bonded stationary phase. see C18 columns
Odor activity value (OAV), 250, 266, 271
Odorants and Odors. see Aroma compounds
Odor spectrum value (OSV), 271, 274
ODS2, carotenoid separation, 98–99
Oils
citrus oil analysis, 277–299
olive oil, polyphenolics extraction, 478
Oil stability index (OSI). see also Lipid oxidation/stability
Olfactometry with GC (GC/O). see Gas chromatography/ olfactometry
Olive oil, polyphenolics extraction, 478
Onions, polyphenolics extraction, 478
Optical activity of citrus oils, 295 (table)
Optical asymmetry in CD analysis of proteins. see also Circular dichroism; Ellipticity
Orange juice, acid concentrations
in juice, 360 (table)
in standards for HPLC, 358 (table)
Orange oils. see also Citrus oils

aldehyde quantification, 288–289, 299 (tables)
early-mid and Valencia, 291, 292 (table)
flavor components, 291–299
Hamlin, 299 (tables)
mandarin and tangerine, 291–293, 295 (table)
market value and limonene content, 291 (table)
physical and chemical properties, 295 (table), 299 (table)
Organic acids
in juices, preparation for HPLC, 356–357
standards for HPLC, 357, 358 (table), 360 (table)
Oscillatory testing of complex fluids, 427–437
Oxidation, myoglobin, 147
Oxygen activation. see Oxidation
Oxymyoglobin, preparation, 145–146

P

PAGE. see Polyacrylamide gel electrophoresis
Paraffins, chromatographic separation, 229–230
Passion fruit juices, acid concentrations in, 360 (table)
Pastes, viscosity/consistency measurement, 376, 387–388, 391–392
Pastry shortening, LVE range of, 431 (fig.)
PDA. see Photodiode array
PE. see Pectinesterase
Peel-oil content, citrus, 282, 285, 299 (table)
Pelargonidin
HPLC analysis, 39, 41
spectral characteristics, 28
Peonidin, HPLC analysis, 41
Peptidases. see also Proteinases
Peptide hydrolase. see Peptidases
Petunidin, HPLC analysis, 41
PGase. see Polygalacturonase
pH
differential, anthocyanins, 19–20, 26–27
measurements, acid tastants, 343, 346–349
myoglobin, 147
Phenolic acids. see also Polyphenolics
HPLC analysis of, 483–488 (fig.)
isolation using C18 cartridges, 475–476, 478, 481
structure of, 476 (fig.), 477, 489 (fig.)
Phenolics. see Polyphenolics
Phenylacetic acids
HPLC separation, 483–498
structure, 489 (fig.)
Phenylethylalcohol, 240
4-Phenylspiro[furan-2(3 H),1-phthalan]-3, 3-dione. see Fluorescamine
Pheophorbides, 155–163. see also Chlorophylls
Pheophytins. see also Chlorophylls
absorption characteristics, 169
chlorophyll conversion to, 166, 169–170
formation and detection of, 155–163
synthesis of standards, 186
Photodiode array (PDA), chlorophyll detection, 161

Photoelastic modulation in CD analysis of. *see also* Circular dichroism

Photosynthetic pigment. *see* Carotenoids; Chlorophylls

Phytochemicals, RP-HPLC of polyphenolics, 483–498

Pigments and colorants
anthocyanins, 7–68, 471–481–I1.3
betalains, 123–129
carotenoids, 72–118, 165–178
color analysis, 203–215
in cooked and cured meats, 131–138
in fresh meats, 139–150
polyphenolics, 463–517
procyanodins, 499–509

Pineapple juice, acid concentrations
in juice, 360 (table)
in standards for HPLC, 358 (table)

Pineapple oil, peel-oil content and aldehyde composition, 299 (tables). *see also* Citrus oils

pKa values, buffers, 544 (table)

PL. *see* Pectic lyase

Plane-polarized radiation, CD analysis of proteins. *see also* Circular dichroism

Plants, isoflavone analysis, 519–535

Poisson's ratio, 403

Polar derivatives of chlorophylls, C18 RPHPLC of, 182–185 (table)

Polarimetry, analysis of citrus oils, 282

Polyacrylamide gel electrophoresis (PAGE). *see also* Electroblotting; Immunoblotting

Polydisperse system for emulsion diffusion. *see also* Ostwald ripening

Polymer coatings, for SPME, 302 (table), 309–310, 312

Polymeric color indices, anthocyanins, 25–26, 29

Polyphenolics. *see also* Flavonoids; Phenolic acids; Tannins
anthocyanins, 7–17, 19–30
classification and structures of, 13 (fig.), 476–477 (fig.), 489–494 (figs.), 519–521 (figs.)
colorimetry, total polyphenolic, 463–465
definition, I.0.1, 476
extraction
all polyphenolics, 471–481
flavonol glycosides, 511–517
isoflavones, 520–523
proanthocyanidins, 499–509
fractionation, 471–481
HPLC separation
all polyphenolics, 483–498
isoflavones, 525–528
proanthocyanidins, subunit composition, 499–509
MALDI-MS, flavonol glycosides, 511–517
NMR, anthocyanins, 47–67
total, determination of, 463–469
UV/Vis spectrophotometry
isoflavones, 523–525
total polyphenolics, 465

Polyphenol oxidase. *see* Diphenol oxidases

Polyvinylpyrrolidone (PVP), 14

Posterior intensity method, for GC/O analysis, 335–336

Potatoes, compressive measurement,

texture profile analysis, 417, 420 (table)

Potentiometric titration, acid tastants, 343–345

Power Law viscosity model, 374–375

Prebetanin, 128

Precipitation, myoglobins, 144–145

Preformed gels, dynamic tests measuring rheology, 439

Press cake oil, quantification, 285

Press liquor, quantification, 286

Primary antibodies. *see* Antibodies

Proanthocyanidins. *see also* Polyphenolics
acetone extraction, 499–500, 506–507
cleavage products characterized, 502 (table), 507 (fig.)
definition, 499, 505
purification, 500–501, 507–508
subunit composition by HPLC, 502–509

Processed solid foods, drip loss measurement. *see also* Frozen foods

6-Propionyl-2-dimethylaminonapthalene. *see* Prodan

Proteinases, in electrophoresis. *See also* Peptidases

Protein efficiency ratio (PER). *see* Protein quality analysis

Proteins, gelation, monitored by gel rheology, 439–447

Proteolyic enzymes. *see* Peptidases; Proteinases

Prunes, polyphenolics extraction, 478

Puncture probe, compressive measurement, 405–409, 414

Purge and trap sampling, SPME, 308–309

Purification
anthocyanins, 10–12, 14–16
flavonol glycosides with C18 cartridges, 512–513
myoglobins, 144–145
polyphenolics
with C18 cartridges, 473–476, 478–479, 481
by RP-HPLC, 483–488
of proanthocyanidins, 500–501, 507–508

Pycnometry analysis of citrus oils, 281

Pyropheophorbides, 155–163. *see also* Chlorophylls

Pyropheophytins, 155–163. *see also* Chlorophylls

Q

Quadrupole time-of-flight (QTOF) hybrid mass spectrometers, 560

Quinic acid concentrations
in organic acid standards for apple juice, 358 (table)
in selected juices, 360 (table)

R

Radishes, HPLC analysis, 42, 44

Raspberry juice, acid concentrations in, 360 (table)

Reagents and Solutions
acidifed acetone, 135
adsorbent, 78

ammonium acetate, 10 M, 545
aqueous acetone, 135
arsenious oxide (As2O3) solution, 290
artificial saliva, 318–319
bisulfite solution, 26
bromphenol blue indicator, 290
citrate-phosphate buffer (McIlvaine's buffer), 545, 546 (table)
DTT (dithiothreitol), 1 M, 545
EDTA (ethylenediaminetetraacetic acid), 0.5 M (pH 8.0), 545
gallic acid calibration standards, 466
HCl, 1 M, 545
H2SO4, 0.005 M, 357
hydrocarbon standard, 268–269
hydroxylamine solution, 0.5 N, 290
indexing standards solution, 50 ng/μl, 231
KCl, 1 M, 545
KH2PO4, 0.05 M, pH 2.40, 357
0.4 M, pH 4.5, 26
MgCl2, 1 M, 546
MgSO4, 1 M, 546
NaCl, 5 M, 546
NaOH, 10 M, 546
NaOH solution, 0.1N, 348
organic acid standard solutions, 357–358
paraffin stock solution, 1000 ppm, 231
phloroglucinol solution, 504
potassium acetate buffer, 0.1 M, 546, 547 (table)
potassium bromide/bromate (KBr/KBRO3) solution, 0.1 N, 290
potassium chloride buffer, 0.025 M, pH 1.0, 26
potassium phosphate buffer, 0.1 M, 547
saliva, artificial, 318–319
SDS, 20% (w/v), 548
sodium acetate, 3 M, 548
sodium acetate buffer, 40 mM, 504
sodium carbonate solution, 466
sodium phosphate buffer, 0.1 M, 548
standardized NaOH solution, 0.1N, 348
Tris-Cl, 1 M, 548

Red wine, total phenolics, 463–469

Reflectance spectrophotometry, meat pigments
cooked meats, 134–135
fresh meats, 141–142

Reflection/reflectance, in color analysis, 205 (fig.), 206

Refractive index, citrus oil analysis, 281–282, 295 (table)

Response factors of proanthocyanidin cleavage products, 502 (table)

Retention index
aroma compounds, volatiles, 247 (fig.)
enantiomers, 258
flavor of RAS sample, 313 (fig.)
smell chemicals, 229–230, 232

Retention of water. *see* Water retention

Retention properties of proanthocyanidin cleavage products, 502 (table)

Retention time, smell chemicals, 229

Retinol, separation using spherisorb ODS2, 91, 98–99

Retronasal aroma simulator (RAS), 313–316, 321–324

Reversed-phase high-performance liquid chromatography (RP-HPLC)
carotenoid separation, 102–103

chlorophyll analysis, 160–161
chlorophyll separation by C18 columns, 180–185 (table)
C18 method in LC/MS, 558 (fig.), 559–560
isoflavone analysis, gradient C18 RPHPLC, 525–528, 532–535
of polyphenolics, 483–498
Rheogram, 372, 374
Rheology
  capillary viscometry, 391–393
  compressive measurements, 397–415
  creep tests, 449–456
  oscillatory/dynamic tests of complex fluids, 427–437, 439–446
Rheometers
  data, complete flow curve, 374
  function of, 371–372, 377, 382
  rotational, for complex fluid tests, 427–437, 439–447
Ring numbering of isoflavones, 519 (fig.), 530
Rotational rheometer, complex fluids tested, using, 427–437, 439–447
RP-HPLC. see High-performance liquid chromatography; Reversed-phase high performance liquid chromatography
Rutin, extraction from buckwheat, 478

S
Saccharides. see Disaccharides; Monosaccharides; Oligosaccharides; other specific listings; Polysaccharides
Safety guidelines, lipid analysis, 446, 449–450 See also Laboratory guidelines
Sampling, smell chemicals, 225–229, 231–233
Saponification
  of acylated anthocyanins, 40, 44
  carotenoid analysis, 89
  for carotenoid extraction, 19–20, 24
  in carotenoid sample preparation, 95, 97, 104
SDS-PAGE. see Polyacrylamide gel electrophoresis
Sectilometers, 414–415
Selected ion monitoring (SIM), 232
Serums, viscosity measurement, 387–388
Shear
  gel dynamic response to, 440 (fig.)
  stress, 401
Shear-thickening materials, 370
Shear-thinning materials, 370, 383
Shear viscosity
  complex, 370–371
  defined, 369
  range of viscosity and, 371–372
  simple, 370–371
  thinning vs. thickening, 369–370
Shortening, oscillatory response and LVE, range, 430–432
Simultaneous distillation extraction (SDE), aroma compounds, 235–237, 240, 242
Sisko viscosity model, 375
Site-specific natural isotope fractionation by nuclear magnetic resonance

spectroscopy (SNIF-NMR), food odorant analysis, 269
Smell chemicals. see Aroma compounds
Sniff port, diagram of, 329 (fig.)
Sodium dodecyl sulfate, in SDS-PAGE. see Polyacrylamide gel electrophoresis
Soft drinks, mechanical spectrum in viscosity tests, 432 (fig.)
Solid-phase extraction (SPE)
  anthocyanins, 38–41
  carotenoids, 104
Solid-phase microextraction (SPME)
  basics of, 301, 307–308
  enantiomers, 258
  fiber coatings and maintenance, 302 (table), 309–310, 312
  GC desorption depth, 310
  headspace extraction, 302–306, 307 (table), 308
  from liquid samples, 303
  purge and trap sampling, 308–309
  RAS sample vs. one from sealed container, 313 (fig.)
  solvent extraction, 308
  static headspace sampling, 308
  stir-bar sorptive extraction, 309
  submersion method, 303
Solubility, of chlorophylls, 159–160
Soluble solids (Brix), standard for common juices, 354 (table)
Solutions, pure, viscosity measurement, 386–387
Sorbic acid, spectral analysis for total phenolics, 469
Soret band, 147
Sour cherry juice, acid concentrations in, 360 (table)
Soybeans, isoflavones
  commonly found, 520–521, 522 (table)
  conversion to aglycones, 528–529
  extraction and preparation, 521–523
  HPLC analysis, 525–528
Specific gravities
  acids and bases, 543 (table)
  citrus oils, 295 (table)
Spectral power distribution curves of CIE illuminants, 204 (fig.)
Spectral response of the human eye to color, 206–207 (fig.)
Spectrometry, mass spectrometry
  instrumentation, 560 (table)
  overview, 555–561
Spectrophotometry. see also Absorption spectrophotometry; Reflectance spectrophotometry; Spectrometry; Spectroscopy
  of betalains, 123–124, 128
  of carotenoids, 81–90, 165–178
  of chlorophylls, 160, 165–178
  in color analysis, 207–209
  of isoflavones, 523–527, 530–532
  of meat pigments, 131–151
Spectroscopy. see also Nuclear magnetic resonance
Specular reflection (gloss), in color analysis, 205 (fig.), 206
Spin Echo. See also Hahn Echo
SPME. see Solid-phase microextraction
Standard laboratory equipment, 553–554
Standards
  carotenoid, 81–83

pheophytins, synthesis of Cu2+ and Zn2+, 186
Static headspace, SPME, 308
Stepped flow tests, 378–380, 382–383
Stir-bar sorptive extraction, SPME, 309
Storage
  for buffers and stock solutions, 543
  hazardous chemicals, 551
Strain response. see Stress/strain responses
Strawberry, HPLC chromatogram, 39 (fig.)
Strawberry juice, acid concentrations in, 360 (table)
Stress/strain responses
  relaxation tests, 449–456
  viscoelasticity of complex fluids, 430–435
Structure
  of chlorophylls, 155–159
  of flavonol glycosides, 514 (fig.)
  of fluids, complex, 433–435
  of isoflavones, 519–521 (figs.), 530
  of polyphenolics, 488–493
  of proanthocyanidins, subunit composition, 502–509
Submersion SPME, 303, 306–307. see also Solid-phase microextraction
Succinic acid concentrations in apple juice, standards, 358 (table)
Sucrose, viscosity of sucrose solutions, 388 (table)
Sugar correction in FC colorimetry of wines, 467–468 (table)
Sugars. see also Disaccharides; Monosaccharides; Oligosaccharides; Polysaccharides; specific listings
Surface tension and interfacial properties. see also Interfaces
Surfactants. see also Interfacial tension
Suspensions, rheometry, oscillatory and dynamic tests, 427–437, 439–446

T
Tandem mass spectrometry. see Mass spectrometry
Tangerine oil. see Citrus; Orange
Tannins, structure of, 477, 492–494 (fig.). see also Polyphenolics
Tartaric acid concentrations
  in grape juice, 360 (table)
  in grapes, pineapple juice, and wine, 359
  in organic standards for selected juices, 358 (table)
Taste. see Flavor analysis
TBA. see 2-Thiobarbituric acid
TBARS. see 2-Thiobarbituric acid reactive substances
TBHQ. see t-butylhydroquinone
TCA. see Trichloroacetic acid
Tea, polyphenolics extraction, 478
Temperature, complex viscosity as function of, 433, 434 (fig.)
Tensiometers. see also Interfacial tension
Tension. see Interfacial tension
Textile profile analysis (TFA)
  apples, 418, 421 (table), 422
  carrots, 417, 421 (table)
  compression cycle, 421
  gelatin gel, 419, 422

ground beef, cooked, 418, 422 (table)
  instrumentation, 419–421 (table),
    422–423
  potatoes, 417, 421 (table)
  sample preparation, 421
Thaw drip loss in solid foods. *see also*
    Frozen foods; Water retention
Thin-layer chromatography (TLC)
  antocyanins analysis, 41
  chlorophyll analysis, 160
Thixotropy, 370, 383
Time intensity method, for GC/O analysis,
    333–334
Time-of-flight tandem mass spectrometers
    (TOF/TOF), 560
Time-temperature superposition (TTS) in
    gel rheology, 433, 434 (fig.), 446
Titration
  acid/base aldehyde assay, 287–288
  hydroxylamine aldehyde assay, 287
TMP. *see* 1,1,3,3-Tetramethoxypropane
TNBS. *see* Trinitrobenzenesulfonic acid
    hydrate
Tocopherols and tocotrienols, separation
    using spherisorb ODS2, 91, 98–99
Tocotrienols. *see* Tocopherols and
    tocotrienols
TOF. *see* Time-of-flight
Total heme pigments, in meats, 133–134,
    136–138
Triacylglycerol acylhydrolase. *See also*
    Lipases
Trifolium pratense (red clover), 558 (fig.),
    559–560
Tristimulus values, in color analysis
  CIELAB, 209–215
  CIELCH, 211–215
  CIE X,Y,Z, 207–211
  Hunter L,a,b, 209–213
TTAB. *see* Tetradecyltrimethylammonium
    bromide
TTS. *see* Time-temperature superposition

U

Ultraviolet (UV). *see also*
    Spectrophotometry
  detectors, for HPLC, 486–487, 525–527
  MALDI, absorption properties, 557
  polyphenolics classed by UV spectra,
    495 (fig.)
  spectral patterns of isoflavones, 524
    (table), 530–532 (figs.)
Uniaxial compression, 402–403
Uptake ability. *see* Water retention; Water
    uptake ability
UV. *see* Ultraviolet
UV/Vis spectrophotometry, 160, 171–178.
    see Absorption spectrophotometry;
    Reflectance spectrophotometry;
    Spectrometry; Spectroscopy

V

Vacuum oven, carotenoid extraction, 76–77
Valencia oil. *see* Citrus; Orange
Vanillin, 240
Vapor pressure. *see* Water activity
Vegetable oils, carotenoid extraction, 78
Vegetables
  anthocyanins in, 29 (table)
  carotenoid analysis, sample
    preparation, 88
  chlorophyll analysis, extraction, 160,
    165–170
  chromatographic separation, 179–189
  extraction, 160, 165–170
  flavonol glycosides, extraction and
    MALDIMS, 511–517
  formation and detection, 155–163
  mass spectrometry, 191–199
  polyphenolics, RP-HPLC, 483–498
  total phenolics, FC colorimetry and
    spectral analysis, 463–469
  UV/Vis spectrophotometry, 171–178
VERI-GREEN beans, zinc-pheophytins
    in, 159
Viscoelasticity of complex fluids, 427–437,
    439–446
Viscometers, 370–371
Viscosity
  apparent, 369–370
  capillary viscometry, 386–388
  consistency measurement, juices and
    pastes, 391–392
  defined, 369
  gel rheology, 427–437, 439–447
  mechanical spectrum for soft drinks,
    432 (fig.)
  non-Newtonian fluids
    complete flow curve, 373 (fig.), 379
    equilibrium flow tests, 380–382
    geometry of, 376–378
    measurement of, 370–371
    nonequilibrium ramped or stepped flow
      tests, 378–380, 382–383
    shear rate, 382–383
    time-dependent behavior, 378, 383
  oscillatory response in viscous foods,
    427 (fig.)
  range, 371–372
  rheogram, 372, 374
  rheology, 427–437
  rheometers, 371–372, 374
  shear, 369–372
  time-dependent behavior, 370
  yield stress, 370
  zero-shear, 374
Volatile compounds
  GC/O sample data, 335 (fig.)
  identification of, 245–248, 250, 253
  quantitation, 245–251, 253
Volatile traps, aroma compounds, 237–242

VPM. *see* Vapor pressure manometer
Vulgaxantin-I, spectrophotometric
    determination, 123–124, 128

W

Warner-Bratzler cutting fixture,
    compressive measurement,
    409–412, 414–415
Water activity. *see also* Hydration;
    Hydrolysis; Hydrometry;
    Hydrophobicity; Water content;
    Water retention
Water content. *see also* Hydration;
    Hydrolysis; Hydrometry;
    Hydrophobicity; Water activity;
    Water retention
  in chlorophyll/carotenoid, 159–160,
    176–178
Water holding capacity (WHC). *see* Water
    retention
Water retention in solid foods. *see also*
    Hydration; Hydrolysis; Hydrometry;
    Hydrophobicity; Water activity;
    Water content
Water uptake ability (WUA)
    see also Water retention; WHC (Water
    holding capacity) (see Water
    retention)
Water vapor pressure. *see* Water activity
Western blot. *see* Immunoblot
White wine, total phenolics analysis,
    463–469
Williamson viscosity model, 375
Wine. *see also* Grapes
  tartaric acid concentrations in, 359
  total phenolic determination, 463–469
Wire cutting fixtures, compressive
    measurement, 409–412, 414–415
WUA. *see* Water uptake ability

X

Xanthophylls, normal-phase separation,
    101–102
Xenon lamps, 102

Y

Yield stress, 370, 375
Young's modulus of elasticity, 397, 403

Z

Zeaxanthin
  HPLC analysis, 78
  spectral fine structure, 85–86
Zero-shear viscosity, 374, 382
Zinc-pheophytins. *see also* Chlorophylls
  synthesis of Zn2+ pheophytin
    standards, 186

3